Encyclopedia of
Textiles, Fibers, and
Nonwoven Fabrics

Encyclopedia of Textiles, Fibers, and Nonwoven Fabrics

ENCYCLOPEDIA REPRINT SERIES

Editor: Martin Grayson

A WILEY-INTERSCIENCE PUBLICATION

JOHN WILEY & SONS

NEW YORK • CHICHESTER • BRISBANE • TORONTO • SINGAPORE

Copyright © 1984 by John Wiley & Sons, Inc.

Library of Congress Cataloging in Publication Data

Main entry under title:

Encyclopedia of textiles, fibers, and nonwoven fabrics.
  (Encyclopedia reprint series)
  Articles reprinted from the Kirk-Othmer Encyclopedia of
chemical technology.
  "A Wiley-Interscience publication."
  Bibliography: p.
  Includes index.
  1. Textile fibers—Dictionaries. 2. Textile fabrics—
Dictionaries. 3. Nonwoven fabrics—Dictionaries.
I. Grayson, Martin. II. Encyclopedia of
chemical technology.

TS1309.E53   1984   677′.003′21   84-13213
ISBN 0-471-81461-X

Printed in the United States of America

10 9 8 7 6 5 4 3

CONTENTS

Acrylic and Modacrylic Fibers, 1
Aramid Fibers, 32
Cellulose Acetate and Triacetate
    Fibers, 62
Coated Fabrics, 90
Cotton, 99
Fibers, Chemical, 118
Fibers, Elastomeric, 136
Fibers, Multicomponent, 152
Fibers, Vegetable, 172
Flame Retardants for Textiles, 188
Leatherlike Materials, 212
Metal Fibers, 231
Nonwoven Textile Fabrics
    Spunbonded, 252
    Staple fibers, 284

Novoloid Fibers, 305
Olefin Fibers, 318
Polyamides, Fibers, 347
Polyester Fibers, 381
Rayon, 399
Recreational Surfaces, 424
Refractory Fibers, 438
Silk, 451
Textiles
    Survey, 459
    Finishing, 466
Tire Cords, 499
Wool, 519

EDITORIAL STAFF

CONTRIBUTORS

B. A. Kottes Andrews, *United States Department of Agriculture, New Orleans, Louisiana,*
     Cotton
W. S. Boston, *CSIRO, Belmont, Victoria, Australia,* Wool
D. R. Buchanan, *North Carolina State University, Raleigh, North Carolina,* Olefin fibers
F. P. Civardi, *Inmont Corporation, Clifton, New Jersey,* Leatherlike materials
Gerald W. Davis, *Fiber Industries, Inc., Charlotte, North Carolina,* Polyester fibers
Ines V. de Gruy, *United States Department of Agriculture, New Orleans, Louisiana, isiana,*
     Cotton

George L. Drake, Jr., *United States Department of Agriculture, New Orleans, Louisiana,* Flame retardants for textiles

Arthur Drelich, *Chicopee Division, Johnson and Johnson, Milltown, New Jersey,* Nonwoven textile fabrics, staple fibers

W. F. Hamner, *Monsanto Company, Research Triangle Park, North Carolina,* Recreational surfaces

Joseph S. Hayes, Jr., *American Kynol, Inc., New York, New York,* Novoloid fibers

Eric S. Hill, *Fiber Industries, Inc., Charlotte, North Carolina,* Polyester fibers

P. H. Hobson, *Monsanto Triangle Park Development Center, Inc., Research Triangle Park, North Carolina,* Acrylic and modacrylic fibers

G. Frederick Hutter, *Inmont Corporation, Clifton, New Jersey,* Leatherlike materials

Charles D. Livengood, *North Carolina State University, Raleigh, North Carolina,* Silk

John Lundberg, *Georgia Institute of Technology, Atlanta, Georgia,* Rayon

John N. McGovern, *The University of Wisconsin, Madison, Wisconsin,* Fibers, vegetable

A. L. McPeters, *Monsanto Triangle Park Development Center, Inc., Research Triangle Park, North Carolina,* Acrylic and modacrylic fibers

W. C. Miller, *Manville Corporation, Denver, Colorado,* Refractory fibers

T. A. Orofino, *Monsanto Company, Research Triangle Park, North Carolina,* Recreational surfaces

Timothy V. Peters, *Consultant, Princeton, New Jersey,* Fibers, elastomeric

K. Porter, *ICI Fibres, Harrogate, North Yorkshire, UK,* Fibers, multicomponent; Nonwoven textile fabrics, spunbonded

J. Preston, *Monsanto Triangle Park Development Center, Inc., Research Triangle Park, North Carolina,* Aramid fibers

Ludwig Rebenfeld, *Textile Research Institute, Princeton, New Jersey,* Textiles, survey

John A. Roberts, *Arco Ventures Co., Troy, Michigan,* Metal fibers

William J. Roberts, *Consultant, Bernardsville, New Jersey,* Fibers, chemical

J. R. Sanders, *Celanese Fibers Company, Charlotte, North Carolina,* Cellulose acetate and triacetate fibers

J. H. Saunders, *Monsanto Company, St. Louis, Missouri,* Polyamides, fibers

George A. Serad, *Celanese Fibers Company, Charlotte, North Carolina,* Cellulose acetate and triacetate fibers

Leonard Skolnik, *BF Goodrich Co., University Heights, Ohio,* Tire cords

Fred N. Teumac, *Uniroyal, Inc., Mishaka, Indiana,* Coated fabrics

Albin Turbak, *Georgia Institute of Technology, Atlanta, Georgia,* Rayon

Sidney L. Vail, *Southern Regional Research Center, USDA, New Orleans, Louisiana,* Textiles, finishing

PREFACE

This volume is another in the series of carefully chosen selections from the world-renowned Kirk Othmer *Encyclopedia of Chemical Technology* designed to provide specific audiences with articles grouped by a central theme. Although the 25-volume Kirk-Othmer Encyclopedia is widely available, many readers and users of this key reference tool have expressed interest in having selected articles in their specialty collected for handy desk reference or teaching purposes. In response to this need, we have chosen all of the original, complete articles related to textiles, fibers, and non-woven fabrics to make up this new volume. The full texts, tables, figures, and reference materials from the original work have been reproduced here unchanged. All articles are by industrial or academic experts in their field and the final work represents the result of careful review by competent specialists and the thorough editorial processing of the professional Wiley staff. Introductory information from the Encyclopedia concerning nomenclature, SI units and conversion factors, and related information has been provided as a further guide to the contents.

Alphabetical organization, extensive cross references, and a complete index further enhance the utility of this book. The articles have been prepared by leading authorities from industries, universities, and research institutes. The contents should be of interest to all those engaged in the manufacture and use of natural and synthetic fibers, textiles, and fabrics as well as those specifically engaged in the application of this technology. The contents range over such diverse fields as acrylic and modacrylic fibers, aramid fibers, cellulose acetate and triacetate fibers, coated fabrics, cotton, elastomeric, synthetic and vegetable fibers, flame retardants for textiles, synthetic leather, metal fibers, nonwoven textile fabrics, olefin fibers, refractory fibers, silk, tire cords, textile finishing, wool, etc. This book should be an important research tool, desk-top information resource, and supplementary reading asset for teaching professionals and their students.

M. GRAYSON

NOTE ON CHEMICAL ABSTRACTS SERVICE REGISTRY NUMBERS AND NOMENCLATURE

Chemical Abstracts Service (CAS) Registry Numbers are unique numerical identifiers assigned to substances recorded in the CAS Registry System. They appear in brackets in the *Chemical Abstracts* (CA) substance and formula indexes following the names of compounds. A single compound may have many synonyms in the chemical literature. A simple compound like phenethylamine can be named β-phenylethylamine or, as in *Chemical Abstracts*, benzeneethanamine. The usefulness of the *Encyclopedia* depends on accessibility through the most common correct name of a substance. Because of this diversity in nomenclature careful attention has been given the problem in order to assist the reader as much as possible, especially in locating the systematic CA index name by means of the Registry Number. For this purpose, the reader may refer to the CAS Registry Handbook-Number Section which lists in numerical order the Registry Number with the *Chemical Abstracts* index name and the molecular formula; eg, **458-88-8**, Piperidine, 2-propyl-, (*S*)-, $C_8H_{17}N$; in the *Encyclopedia* this compound would be found under its common name, coniine [*458-88-8*]. The Registry Number is a valuable link for the reader in retrieving additional published information on substances and also as a point of access for such on-line data bases as Chemline, Medline, and Toxline.

In all cases, the CAS Registry Numbers have been given for title compounds in articles and for all compounds in the index. All specific substances indexed in *Chemical Abstracts* since 1965 are included in the CAS Registry System as are a large number of substances derived from a variety of reference works. The CAS Registry System identifies a substance on the basis of an unambiguous computer-language description of its molecular structure including stereochemical detail. The Registry Number is a machine-checkable number (like a Social Security number) assigned in sequential order to each substance as it enters the registry system. The value of the number lies in the fact that it is a concise and unique means of substance identification, which is independent of, and therefore bridges, many systems of chemical nomenclature. For polymers, one Registry Number is used for the entire family; eg, polyoxyethylene (20) sorbitan monolaurate has the same number as all of its polyoxyethylene homologues.

Registry numbers for each substance are provided in the third edition cumulative index and appear as well in the annual indexes (eg, Alkaloids shows the Registry Number of all alkaloids (title compounds) in a table in the article as well, but the intermediates have their Registry Numbers shown only in the index). Articles such as Analytical methods, Batteries and electric cells, Chemurgy, Distillation, Economic evaluation, and Fluid mechanics have no Registry Numbers in the text.

Cross-references are inserted in the index for many common names and for some systematic names. Trademark names appear in the index. Names that are incorrect, misleading or ambiguous are avoided. Formulas are given very frequently in the text to help in identifying compounds. The spelling and form used, even for industrial names, follow American chemical usage, but not always the usage of *Chemical Abstracts* (eg, *coniine* is used instead of *(S)-2-propylpiperidine*, *aniline* instead of *benzenamine*, and *acrylic acid* instead of *2-propenoic acid*).

There are variations in representation of rings in different disciplines. The dye industry does not designate aromaticity or double bonds in rings. All double bonds and aromaticity are shown in the *Encyclopedia* as a matter of course. For example, tetralin has an aromatic ring and a saturated ring and its structure appears in the

Encyclopedia with its common name, Registry Number enclosed in brackets, and parenthetical CA index name, ie, tetralin, [*119-64-2*] (1,2,3,4-tetrahydronaphthalene). With names and structural formulas, and especially with CAS Registry Numbers the aim is to help the reader have a concise means of substance identification.

CONVERSION FACTORS, ABBREVIATIONS, AND UNIT SYMBOLS

SI Units (Adopted 1960)

A new system of measurement, the International System of Units (abbreviated SI), is being implemented throughout the world. This system is a modernized version of the MKSA (meter, kilogram, second, ampere) system, and its details are published and controlled by an international treaty organization (The International Bureau of Weights and Measures) (1).

SI units are divided into three classes:

BASE UNITS

length	meter[†] (m)
mass[‡]	kilogram (kg)
time	second (s)
electric current	ampere (A)
thermodynamic temperature[§]	kelvin (K)
amount of substance	mole (mol)
luminous intensity	candela (cd)

[†] The spellings "metre" and "litre" are preferred by ASTM; however "-er" are used in the Encyclopedia.

[‡] "Weight" is the commonly used term for "mass."

[§] Wide use is made of "Celsius temperature" (t) defined by

$$t = T - T_0$$

where T is the thermodynamic temperature, expressed in kelvins, and $T_0 = 273.15$ K by definition. A temperature interval may be expressed in degrees Celsius as well as in kelvins.

SUPPLEMENTARY UNITS

plane angle	radian (rad)
solid angle	steradian (sr)

DERIVED UNITS AND OTHER ACCEPTABLE UNITS

These units are formed by combining base units, supplementary units, and other derived units (2–4). Those derived units having special names and symbols are marked with an asterisk in the list below:

Quantity	Unit	Symbol	Acceptable equivalent
*absorbed dose	gray	Gy	J/kg
acceleration	meter per second squared	m/s²	
*activity (of ionizing radiation source)	becquerel	Bq	1/s
area	square kilometer	km²	
	square hectometer	hm²	ha (hectare)
	square meter	m²	
*capacitance	farad	F	C/V
concentration (of amount of substance)	mole per cubic meter	mol/m³	
*conductance	siemens	S	A/V
current density	ampere per square meter	A/m²	
density, mass density	kilogram per cubic meter	kg/m³	g/L; mg/cm³
dipole moment (quantity)	coulomb meter	C·m	
*electric charge, quantity of electricity	coulomb	C	A·s
electric charge density	coulomb per cubic meter	C/m³	
electric field strength	volt per meter	V/m	
electric flux density	coulomb per square meter	C/m²	
*electric potential, potential difference, electromotive force	volt	V	W/A
*electric resistance	ohm	Ω	V/A
*energy, work, quantity of heat	megajoule	MJ	
	kilojoule	kJ	
	joule	J	N·m
	electron volt[†]	eV[†]	
	kilowatt hour[†]	kW·h[†]	
energy density	joule per cubic meter	J/m³	

[†] This non-SI unit is recognized by the CIPM as having to be retained because of practical importance or use in specialized fields (1).

Quantity	Unit	Symbol	Acceptable equivalent
*force	kilonewton	kN	
	newton	N	kg·m/s²
*frequency	megahertz	MHz	
	hertz	Hz	1/s
heat capacity, entropy	joule per kelvin	J/K	
heat capacity (specific), specific entropy	joule per kilogram kelvin	J/(kg·K)	
heat transfer coefficient	watt per square meter kelvin	W/(m²·K)	
*illuminance	lux	lx	lm/m²
*inductance	henry	H	Wb/A
linear density	kilogram per meter	kg/m	
luminance	candela per square meter	cd/m²	
*luminous flux	lumen	lm	cd·sr
magnetic field strength	ampere per meter	A/m	
*magnetic flux	weber	Wb	V·s
*magnetic flux density	tesla	T	Wb/m²
molar energy	joule per mole	J/mol	
molar entropy, molar heat capacity	joule per mole kelvin	J/(mol·K)	
moment of force, torque	newton meter	N·m	
momentum	kilogram meter per second	kg·m/s	
permeability	henry per meter	H/m	
permittivity	farad per meter	F/m	
*power, heat flow rate, radiant flux	kilowatt	kW	
	watt	W	J/s
power density, heat flux density, irradiance	watt per square meter	W/m²	
*pressure, stress	megapascal	MPa	
	kilopascal	kPa	
	pascal	Pa	N/m²
sound level	decibel	dB	
specific energy	joule per kilogram	J/kg	
specific volume	cubic meter per kilogram	m³/kg	
surface tension	newton per meter	N/m	
thermal conductivity	watt per meter kelvin	W/(m·K)	
velocity	meter per second	m/s	
	kilometer per hour	km/h	
viscosity, dynamic	pascal second	Pa·s	
	millipascal second	mPa·s	
viscosity, kinematic	square meter per second	m²/s	
	square millimeter per second	mm²/s	

Quantity	Unit	Symbol	Acceptable equivalent
volume	cubic meter	m^3	
	cubic decimeter	dm^3	L(liter) (5)
	cubic centimeter	cm^3	mL
wave number	1 per meter	m^{-1}	
	1 per centimeter	cm^{-1}	

In addition, there are 16 prefixes used to indicate order of magnitude, as follows:

Multiplication factor	Prefix	Symbol	Note
10^{18}	exa	E	
10^{15}	peta	P	
10^{12}	tera	T	
10^9	giga	G	
10^6	mega	M	
10^3	kilo	k	
10^2	hecto	h[a]	[a] Although hecto, deka, deci, and centi
10	deka	da[a]	are SI prefixes, their use should be
10^{-1}	deci	d[a]	avoided except for SI unit-mul-
10^{-2}	centi	c[a]	tiples for area and volume and
10^{-3}	milli	m	nontechnical use of centimeter,
10^{-6}	micro	μ	as for body and clothing
10^{-9}	nano	n	measurement.
10^{-12}	pico	p	
10^{-15}	femto	f	
10^{-18}	atto	a	

For a complete description of SI and its use the reader is referred to ASTM E 380 (4) and the article Units and Conversion Factors in Vol. 23.

A representative list of conversion factors from non-SI to SI units is presented herewith. Factors are given to four significant figures. Exact relationships are followed by a dagger. A more complete list is given in ASTM E 380-79(4) and ANSI Z210.1-1976 (6).

Conversion Factors to SI Units

To convert from	To	Multiply by
acre	square meter (m^2)	4.047×10^3
angstrom	meter (m)	$1.0 \times 10^{-10\dagger}$
are	square meter (m^2)	$1.0 \times 10^{2\dagger}$
astronomical unit	meter (m)	1.496×10^{11}
atmosphere	pascal (Pa)	1.013×10^5
bar	pascal (Pa)	$1.0 \times 10^{5\dagger}$
barn	square meter (m^2)	$1.0 \times 10^{-28\dagger}$
barrel (42 U.S. liquid gallons)	cubic meter (m^3)	0.1590

† Exact.

To convert from	*To*	*Multiply by*
Bohr magneton (μ_β)	J/T	9.274×10^{-24}
Btu (International Table)	joule (J)	1.055×10^3
Btu (mean)	joule (J)	1.056×10^3
Btu (thermochemical)	joule (J)	1.054×10^3
bushel	cubic meter (m³)	3.524×10^{-2}
calorie (International Table)	joule (J)	4.187
calorie (mean)	joule (J)	4.190
calorie (thermochemical)	joule (J)	4.184†
centipoise	pascal second (Pa·s)	1.0×10^{-3}†
centistokes	square millimeter per second (mm²/s)	1.0†
cfm (cubic foot per minute)	cubic meter per second (m³/s)	4.72×10^{-4}
cubic inch	cubic meter (m³)	1.639×10^{-5}
cubic foot	cubic meter (m³)	2.832×10^{-2}
cubic yard	cubic meter (m³)	0.7646
curie	becquerel (Bq)	3.70×10^{10}†
debye	coulomb·meter (C·m)	3.336×10^{-30}
degree (angle)	radian (rad)	1.745×10^{-2}
denier (international)	kilogram per meter (kg/m)	1.111×10^{-7}
	tex‡	0.1111
dram (apothecaries')	kilogram (kg)	3.888×10^{-3}
dram (avoirdupois)	kilogram (kg)	1.772×10^{-3}
dram (U.S. fluid)	cubic meter (m³)	3.697×10^{-6}
dyne	newton (N)	1.0×10^{-5}†
dyne/cm	newton per meter (N/m)	1.0×10^{-3}†
electron volt	joule (J)	1.602×10^{-19}
erg	joule (J)	1.0×10^{-7}†
fathom	meter (m)	1.829
fluid ounce (U.S.)	cubic meter (m³)	2.957×10^{-5}
foot	meter (m)	0.3048†
footcandle	lux (lx)	10.76
furlong	meter (m)	2.012×10^{-2}
gal	meter per second squared (m/s²)	1.0×10^{-2}†
gallon (U.S. dry)	cubic meter (m³)	4.405×10^{-3}
gallon (U.S. liquid)	cubic meter (m³)	3.785×10^{-3}
gallon per minute (gpm)	cubic meter per second (m³/s)	6.308×10^{-5}
	cubic meter per hour (m³/h)	0.2271
gauss	tesla (T)	1.0×10^{-4}
gilbert	ampere (A)	0.7958
gill (U.S.)	cubic meter (m³)	1.183×10^{-4}
grad	radian	1.571×10^{-2}
grain	kilogram (kg)	6.480×10^{-5}
gram-force per denier	newton per tex (N/tex)	8.826×10^{-2}
hectare	square meter (m²)	1.0×10^4†

† Exact.

‡ See footnote on p. xiv.

To convert from	To	Multiply by
horsepower (550 ft·lbf/s)	watt (W)	7.457×10^2
horsepower (boiler)	watt (W)	9.810×10^3
horsepower (electric)	watt (W)	$7.46 \times 10^{2\dagger}$
hundredweight (long)	kilogram (kg)	50.80
hundredweight (short)	kilogram (kg)	45.36
inch	meter (m)	$2.54 \times 10^{-2\dagger}$
inch of mercury (32°F)	pascal (Pa)	3.386×10^3
inch of water (39.2°F)	pascal (Pa)	2.491×10^2
kilogram-force	newton (N)	9.807
kilowatt hour	megajoule (MJ)	$3.6^\dagger$
kip	newton (N)	4.48×10^3
knot (international)	meter per second (m/s)	0.5144
lambert	candela per square meter (cd/m²)	3.183×10^3
league (British nautical)	meter (m)	5.559×10^3
league (statute)	meter (m)	4.828×10^3
light year	meter (m)	9.461×10^{15}
liter (for fluids only)	cubic meter (m³)	$1.0 \times 10^{-3\dagger}$
maxwell	weber (Wb)	$1.0 \times 10^{-8\dagger}$
micron	meter (m)	$1.0 \times 10^{-6\dagger}$
mil	meter (m)	$2.54 \times 10^{-5\dagger}$
mile (statute)	meter (m)	1.609×10^3
mile (U.S. nautical)	meter (m)	$1.852 \times 10^{3\dagger}$
mile per hour	meter per second (m/s)	0.4470
millibar	pascal (Pa)	1.0×10^2
millimeter of mercury (0°C)	pascal (Pa)	$1.333 \times 10^{2\dagger}$
minute (angular)	radian	2.909×10^{-4}
myriagram	kilogram (kg)	10
myriameter	kilometer (km)	10
oersted	ampere per meter (A/m)	79.58
ounce (avoirdupois)	kilogram (kg)	2.835×10^{-2}
ounce (troy)	kilogram (kg)	3.110×10^{-2}
ounce (U.S. fluid)	cubic meter (m³)	2.957×10^{-5}
ounce-force	newton (N)	0.2780
peck (U.S.)	cubic meter (m³)	8.810×10^{-3}
pennyweight	kilogram (kg)	1.555×10^{-3}
pint (U.S. dry)	cubic meter (m³)	5.506×10^{-4}
pint (U.S. liquid)	cubic meter (m³)	4.732×10^{-4}
poise (absolute viscosity)	pascal second (Pa·s)	$0.10^\dagger$
pound (avoirdupois)	kilogram (kg)	0.4536
pound (troy)	kilogram (kg)	0.3732
poundal	newton (N)	0.1383
pound-force	newton (N)	4.448
pound force per square inch (psi)	pascal (Pa)	6.895×10^3
quart (U.S. dry)	cubic meter (m³)	1.101×10^{-3}
quart (U.S. liquid)	cubic meter (m³)	9.464×10^{-4}
quintal	kilogram (kg)	$1.0 \times 10^{2\dagger}$

† Exact.

To convert from	To	Multiply by
rad	gray (Gy)	$1.0 \times 10^{-2\dagger}$
rod	meter (m)	5.029
roentgen	coulomb per kilogram (C/kg)	2.58×10^{-4}
second (angle)	radian (rad)	4.848×10^{-6}
section	square meter (m^2)	2.590×10^{6}
slug	kilogram (kg)	14.59
spherical candle power	lumen (lm)	12.57
square inch	square meter (m^2)	6.452×10^{-4}
square foot	square meter (m^2)	9.290×10^{-2}
square mile	square meter (m^2)	2.590×10^{6}
square yard	square meter (m^2)	0.8361
stere	cubic meter (m^3)	$1.0^{\dagger}$
stokes (kinematic viscosity)	square meter per second (m^2/s)	$1.0 \times 10^{-4\dagger}$
tex	kilogram per meter (kg/m)	$1.0 \times 10^{-6\dagger}$
ton (long, 2240 pounds)	kilogram (kg)	1.016×10^{3}
ton (metric)	kilogram (kg)	$1.0 \times 10^{3\dagger}$
ton (short, 2000 pounds)	kilogram (kg)	9.072×10^{2}
torr	pascal (Pa)	1.333×10^{2}
unit pole	weber (Wb)	1.257×10^{-7}
yard	meter (m)	$0.9144^{\dagger}$

Abbreviations and Unit Symbols

Following is a list of commonly used abbreviations and unit symbols appropriate for use in the *Encyclopedia*. In general they agree with those listed in *American National Standard Abbreviations for Use on Drawings and in Text (ANSI Y1.1) (6)* and *American National Standard Letter Symbols for Units in Science and Technology (ANSI Y10) (6)*. Also included is a list of acronyms for a number of private and government organizations as well as common industrial solvents, polymers, and other chemicals.

Rules for Writing Unit Symbols (4):

1. Unit symbols should be printed in upright letters (roman) regardless of the type style used in the surrounding text

2. Unit symbols are unaltered in the plural.

3. Unit symbols are not followed by a period except when used as the end of a sentence.

4. Letter unit symbols are generally written in lower-case (eg, cd for candela) unless the unit name has been derived from a proper name, in which case the first letter of the symbol is capitalized (W, Pa). Prefix and unit symbols retain their prescribed form regardless of the surrounding typography.

5. In the complete expression for a quantity, a space should be left between the numerical value and the unit symbol. For example, write 2.37 lm, *not* 2.37lm, and 35 mm, *not* 35mm. When the quantity is used in an adjectival sense, a hyphen is often used, for example, 35-mm film. *Exception:* No space is left between the numerical value and the symbols for degree, minute, and second of plane angle, and degree Celsius.

6. No space is used between the prefix and unit symbols (eg, kg).

7. Symbols, not abbreviations, should be used for units. For example, use "A," not "amp," for ampere.

8. When multiplying unit symbols, use a raised dot:

N·m for newton meter

In the case of W·h, the dot may be omitted, thus:

Wh

An exception to this practice is made for computer printouts, automatic typewriter work, etc, where the raised dot is not possible, and a dot on the line may be used.

9. When dividing unit symbols use one of the following forms:

$$\text{m/s } or \text{ m·s}^{-1} or \frac{\text{m}}{\text{s}}$$

In no case should more than one slash be used in the same expression unless parentheses are inserted to avoid ambiguity. For example, write:

$$\text{J/(mol·K) } or \text{ J·mol}^{-1} \cdot \text{K}^{-1} or \text{ (J/mol)/K}$$

but *not*

J/mol/K

10. Do not mix symbols and unit names in the same expression. Write:

joules per kilogram *or* J/kg *or* J·kg^{-1}

but *not*

joules/kilogram *nor* joules/kg *nor* joules·kg^{-1}

ABBREVIATIONS AND UNITS

A	ampere	AIP	American Institute of Physics
A	anion (eg, H*A*); mass number		
a	atto (prefix for 10^{-18})	AISI	American Iron and Steel Institute
AATCC	American Association of Textile Chemists and Colorists	alc	alcohol(ic)
		Alk	alkyl
ABS	acrylonitrile–butadiene–styrene	alk	alkaline (not alkali)
		amt	amount
abs	absolute	amu	atomic mass unit
ac	alternating current, *n.*	ANSI	American National Standards Institute
a-c	alternating current, *adj.*		
ac-	alicyclic	AO	atomic orbital
acac	acetylacetonate	AOAC	Association of Official Analytical Chemists
ACGIH	American Conference of Governmental Industrial Hygienists	AOCS	American Oil Chemists' Society
ACS	American Chemical Society	APHA	American Public Health Association
AGA	American Gas Association		
Ah	ampere hour	API	American Petroleum Institute
AIChE	American Institute of Chemical Engineers	aq	aqueous
AIME	American Institute of Mining, Metallurgical, and Petroleum Engineers	Ar	aryl
		ar-	aromatic
		as-	asymmetric(a

ASH-RAE	American Society of Heating, Refrigerating, and Air Conditioning Engineers		coml	commercial(ly)
			cp	chemically pure
			cph	close-packed hexagonal
ASM	American Society for Metals		CPSC	Consumer Product Safety Commission
ASME	American Society of Mechanical Engineers			
			cryst	crystalline
ASTM	American Society for Testing and Materials		cub	cubic
			D	debye
at no.	atomic number		D-	denoting configurational relationship
at wt	atomic weight			
av(g)	average		**d**	differential operator
AWS	American Welding Society		*d-*	*dextro-*, dextrorotatory
b	bonding orbital		da	deka (prefix for 10^1)
bbl	barrel		dB	decibel
bcc	body-centered cubic		dc	direct current, *n*.
BCT	body-centered tetragonal		d-c	direct current, *adj*.
Bé	Baumé		dec	decompose
BET	Brunauer-Emmett-Teller (adsorption equation)		detd	determined
			detn	determination
bid	twice daily		Di	didymium, a mixture of all lanthanons
Boc	*t*-butyloxycarbonyl			
BOD	biochemical (biological) oxygen demand		dia	diameter
			dil	dilute
bp	boiling point		DIN	Deutsche Industrie Normen
Bq	becquerel		*dl-*; DL-	racemic
C	coulomb		DMA	dimethylacetamide
°C	degree Celsius		DMF	dimethylformamide
C-	denoting attachment to carbon		DMG	dimethyl glyoxime
			DMSO	dimethyl sulfoxide
c	centi (prefix for 10^{-2})		DOD	Department of Defense
c	critical		DOE	Department of Energy
ca	circa (approximately)		DOT	Department of Transportation
cd	candela;current density; circular dichroism			
			DP	degree of polymerization
CFR	Code of Federal Regulations		dp	dew point
			DPH	diamond pyramid hardness
cgs	centimeter–gram–second		dstl(d)	distill(ed)
CI	Color Index		dta	differential thermal analysis
cis-	isomer in which substituted groups are on same side of double bond between C atoms			
			(*E*)-	entgegen; opposed
			ϵ	dielectric constant (unitless number)
cl	carload		*e*	electron
cm	centimeter		ECU	electrochemical unit
cmil	circular mil		ed.	edited, edition, editor
cmpd	compound		ED	effective dose
CNS	central nervous system		EDTA	ethylenediaminetetraacetic acid
CoA	coenzyme A			
COD	chemical oxygen demand		emf	electromotive force

emu	electromagnetic unit		grd	ground
en	ethylene diamine		Gy	gray
eng	engineering		H	henry
EPA	Environmental Protection Agency		h	hour; hecto (prefix for 10^2)
epr	electron paramagnetic resonance		ha	hectare
			HB	Brinell hardness number
eq.	equation		Hb	hemoglobin
esca	electron-spectroscopy for chemical analysis		hcp	hexagonal close-packed
			hex	hexagonal
esp	especially		HK	Knoop hardness number
esr	electron-spin resonance		hplc	high pressure liquid chromatography
est(d)	estimate(d)		HRC	Rockwell hardness (C scale)
estn	estimation		HV	Vickers hardness number
esu	electrostatic unit		hyd	hydrated, hydrous
exp	experiment, experimental		hyg	hygroscopic
ext(d)	extract(ed)		Hz	hertz
F	farad (capacitance)		^{i}(eg, Pr^i)	iso (eg, isopropyl)
F	faraday (96,487 C)		i-	inactive (eg, i-methionine)
f	femto (prefix for 10^{-15})		IACS	International Annealed Copper Standard
FAO	Food and Agriculture Organization (United Nations)		ibp	initial boiling point
			IC	inhibitory concentration
fcc	face-centered cubic		ICC	Interstate Commerce Commission
FDA	Food and Drug Administration			
FEA	Federal Energy Administration		ICT	International Critical Table
			ID	inside diameter; infective dose
FHSA	Federal Hazardous Substances Act		ip	intraperitoneal
			IPS	iron pipe size
fob	free on board		IPTS	International Practical Temperature Scale (NBS)
fp	freezing point			
FPC	Federal Power Commission		ir	infrared
FRB	Federal Reserve Board		IRLG	Interagency Regulatory Liaison Group
frz	freezing			
G	giga (prefix for 10^9)		ISO	International Organization for Standardization
G	gravitational constant = 6.67×10^{11} N·m^2/kg^2			
			IU	International Unit
g	gram		IUPAC	International Union of Pure and Applied Chemistry
(g)	gas, only as in H$_2$O(g)			
g	gravitational acceleration		IV	iodine value
gc	gas chromatography		iv	intravenous
gem-	geminal		J	joule
glc	gas-liquid chromatography		K	kelvin
g-mol wt; gmw	gram-molecular weight		k	kilo (prefix for 10^3)
			kg	kilogram
GNP	gross national product		L	denoting configurational relationship
gpc	gel-permeation chromatography			
			L	liter (for fluids only)(5)
GRAS	Generally Recognized as Safe		l-	$levo$-, levorotatory

(l)	liquid, only as in $NH_3(l)$	ms	mass spectrum
LC_{50}	conc lethal to 50% of the animals tested	mxt	mixture
		μ	micro (prefix for 10^{-6})
LCAO	linear combination of atomic orbitals	N	newton (force)
		N	normal (concentration); neutron number
LCD	liquid crystal display		
lcl	less than carload lots	N-	denoting attachment to nitrogen
LD_{50}	dose lethal to 50% of the animals tested		
		n (as n_D^{20})	index of refraction (for 20°C and sodium light)
LED	light-emitting diode		
liq	liquid	^{n} (as Bu^n), n-	normal (straight-chain structure)
lm	lumen		
ln	logarithm (natural)	n	neutron
LNG	liquefied natural gas	n	nano (prefix for 10^9)
log	logarithm (common)	na	not available
LPG	liquefied petroleum gas	NAS	National Academy of Sciences
ltl	less than truckload lots		
lx	lux	NASA	National Aeronautics and Space Administration
M	mega (prefix for 10^6); metal (as in MA)		
		nat	natural
M	molar; actual mass	NBS	National Bureau of Standards
$\overline{M}_w$	weight-average mol wt		
$\overline{M}_n$	number-average mol wt	neg	negative
m	meter; milli (prefix for 10^{-3})	NF	*National Formulary*
m	molal	NIH	National Institutes of Health
m-	meta		
max	maximum	NIOSH	National Institute of Occupational Safety and Health
MCA	Chemical Manufacturers' Association (was Manufacturing Chemists Association)		
		nmr	nuclear magnetic resonance
		NND	New and Nonofficial Drugs (AMA)
MEK	methyl ethyl ketone		
meq	milliequivalent	no.	number
mfd	manufactured	NOI-(BN)	not otherwise indexed (by name)
mfg	manufacturing		
mfr	manufacturer	NOS	not otherwise specified
MIBC	methyl isobutyl carbinol	nqr	nuclear quadruple resonance
MIBK	methyl isobutyl ketone	NRC	Nuclear Regulatory Commission; National Research Council
MIC	minimum inhibiting concentration		
min	minute; minimum	NRI	New Ring Index
mL	milliliter	NSF	National Science Foundation
MLD	minimum lethal dose	NTA	nitrilotriacetic acid
MO	molecular orbital	NTP	normal temperature and pressure (25°C and 101.3 kPa or 1 atm)
mo	month		
mol	mole		
mol wt	molecular weight	NTSB	National Transportation Safety Board
mp	melting point		
MR	molar refraction	O-	denoting attachment to

	oxygen	rds	rate determining step
o-	ortho	ref.	reference
OD	outside diameter	rf	radio frequency, *n.*
OPEC	Organization of	r-f	radio frequency, *adj.*
	Petroleum Exporting	rh	relative humidity
	Countries	RI	Ring Index
o-phen	*o*-phenanthridine	rms	root-mean square
OSHA	Occupational Safety and	rpm	rotations per minute
	Health Administration	rps	revolutions per second
owf	on weight of fiber	RT	room temperature
Ω	ohm	ˢ (eg,	secondary (eg, secondary
P	peta (prefix for 10^{15})	Buˢ);	butyl)
p	pico (prefix for 10^{-12})	*sec-*	
p-	para	S	siemens
p	proton	(*S*)-	sinister (counterclockwise
p.	page		configuration)
Pa	pascal (pressure)	*S*-	denoting attachment to
pd	potential difference		sulfur
pH	negative logarithm of the	*s*-	symmetric(al)
	effective hydrogen ion	s	second
	concentration	(s)	solid, only as in $H_2O(s)$
phr	parts per hundred of resin	SAE	Society of Automotive
	(rubber)		Engineers
p-i-n	positive-intrinsic-negative	SAN	styrene–acrylonitrile
pmr	proton magnetic resonance	sat(d)	saturate(d)
p-n	positive-negative	satn	saturation
po	per os (oral)	SBS	styrene–butadiene–styrene
POP	polyoxypropylene	sc	subcutaneous
pos	positive	SCF	self-consistent field;
pp.	pages		standard cubic feet
ppb	parts per billion (10^9)	Sch	Schultz number
ppm	parts per million (10^6)	SFs	Saybolt Furol seconds
ppmv	parts per million by volume	SI	Le Système International
ppmwt	parts per million by weight		d'Unités (International
PPO	poly(phenyl oxide)		System of Units)
ppt(d)	precipitate(d)	sl sol	slightly soluble
pptn	precipitation	sol	soluble
Pr (no.)	foreign prototype (number)	soln	solution
pt	point; part	soly	solubility
PVC	poly(vinyl chloride)	sp	specific; species
pwd	powder	sp gr	specific gravity
py	pyridine	sr	steradian
qv	quod vide (which see)	std	standard
R	univalent hydrocarbon	STP	standard temperature and
	radical		pressure (0°C and 101.3
(*R*)-	rectus (clockwise		kPa)
	configuration)	sub	sublime(s)
r	precision of data	SUs	Saybolt Universal
rad	radian; radius		seconds

syn	synthetic	Twad	Twaddell
^{t} (eg, Bu^t), t-, tert-	tertiary (eg, tertiary butyl)	UL	Underwriters' Laboratory
		USDA	United States Department of Agriculture
T	tera (prefix for 10^{12}); tesla (magnetic flux density)	USP	*United States Pharmacopeia*
		uv	ultraviolet
t	metric ton (tonne); temperature	V	volt (emf)
		var	variable
TAPPI	Technical Association of the Pulp and Paper Industry	*vic-*	vicinal
		vol	volume (not volatile)
tex	tex (linear density)	vs	versus
T_g	glass-transition temperature	v sol	very soluble
tga	thermogravimetric analysis	W	watt
THF	tetrahydrofuran	Wb	weber
tlc	thin layer chromatography	Wh	watt hour
TLV	threshold limit value	WHO	World Health Organization (United Nations)
trans-	isomer in which substituted groups are on opposite sides of double bond between C atoms		
		wk	week
		yr	year
TSCA	Toxic Substance Control Act	(*Z*)-	zusammen; together; atomic number
TWA	time-weighted average		

Non-SI (Unacceptable and Obsolete) Units		*Use*
Å	angstrom	nm
at	atmosphere, technical	Pa
atm	atmosphere, standard	Pa
b	barn	cm^2
bar†	bar	Pa
bbl	barrel	m^3
bhp	brake horsepower	W
Btu	British thermal unit	J
bu	bushel	m^3; L
cal	calorie	J
cfm	cubic foot per minute	m^3/s
Ci	curie	Bq
cSt	centistokes	mm^2/s
c/s	cycle per second	Hz
cu	cubic	exponential form
D	debye	C·m
den	denier	tex
dr	dram	kg
dyn	dyne	N
dyn/cm	dyne per centimeter	mN/m
erg	erg	J
eu	entropy unit	J/K
°F	degree Fahrenheit	°C; K
fc	footcandle	lx
fl	footlambert	lx
fl oz	fluid ounce	m^3; L
ft	foot	

† Do not use bar (10^5Pa) or millibar (10^2Pa) because they are not SI units, and are accepted internationally only for a limited time in special fields because of existing usage.

Non-SI (Unacceptable and Obsolete) Units   *Use*

ft·lbf	foot pound-force	J
gf/den	gram-force per denier	N/tex
G	gauss	T
Gal	gal	m/s^2
gal	gallon	m^3; L
Gb	gilbert	A
gpm	gallon per minute	(m^3/s); (m^3/h)
gr	grain	kg
hp	horsepower	W
ihp	indicated horsepower	W
in.	inch	m
in. Hg	inch of mercury	Pa
in. H$_2$O	inch of water	Pa
in.-lbf	inch pound-force	J
kcal	kilogram-calorie	J
kgf	kilogram-force	N
kilo	for kilogram	kg
L	lambert	lx
lb	pound	kg
lbf	pound-force	N
mho	mho	S
mi	mile	m
MM	million	M
mm Hg	millimeter of mercury	Pa
mμ	millimicron	nm
mph	mile per hour	km/h
μ	micron	μm
Oe	oersted	A/m
oz	ounce	kg
ozf	ounce-force	N
η	poise	Pa·s
P	poise	Pa·s
ph	phot	lx
psi	pound-force per square inch	Pa
psia	pound-force per square inch absolute	Pa
psig	pound-force per square inch gauge	Pa
qt	quart	m^3; L
°R	degree Rankine	K
rd	rad	Gy
sb	stilb	lx
SCF	standard cubic foot	m^3
sq	square	exponential form
thm	therm	J
yd	yard	m

BIBLIOGRAPHY

1. The International Bureau of Weights and Measures, BIPM (Parc de Saint-Cloud, France) is described on page 22 of Ref. 4. This bureau operates under the exclusive supervision of the International Committee of Weights and Measures (CIPM).
2. *Metric Editorial Guide (ANMC-78-1)* 3rd ed., American National Metric Council, 5410 Grosvenor Lane, Bethesda, Md. 20814, 1981.
3. *SI Units and Recommendations for the Use of Their Multiples and of Certain Other Units (ISO 1000-1981)*, American National Standards Institute, 1430 Broadway, New York, N. Y. 10018, 1981.
4. Based on *ASTM E 380-82 (Standard for Metric Practice)*, American Society for Testing and Materials, 1916 Race Street, Philadelphia, Pa. 19103, 1982.
5. *Fed. Regist.*, Dec. 10, 1976 (41 FR 36414).
6. For ANSI address, see Ref. 3.

R. P. LUKENS
American Society for Testing and Materials

ACRYLIC AND MODACRYLIC FIBERS

An acrylic fiber is defined (1) as a manufactured fiber in which the fiber-forming substance is any long-chain synthetic polymer composed of at least 85 wt % acrylonitrile units. A modacrylic fiber is defined as one composed of less than 85 but at least 35 wt % acrylonitrile units. In Europe, modacrylic fibers were originally defined as those containing 50–85 mass % acrylonitrile units, but in 1977 the International Organization for Standardization (2) adopted the broader 35–85 mass % acrylonitrile units definition also. The nature of the other components is not specified for either fiber. Commercially, however, the only modacrylics marketed have 25–60% of monomers, such as vinyl chloride or vinylidene chloride, and consequently possess a high degree of flame resistance.

ACRYLIC FIBERS

Research on synthetic fiber production from polyacrylonitrile (PAN) started in Germany, in the early 1930s and independently later in the United States. Polymer degradation prevented conventional melt spinning and no solvents suitable for solution spinning had been known. In 1938, H. Rein of I. G. Farbenindustrie described fibers obtained from the polymer dissolved in aqueous salt solutions (3), including those of sodium thiocyanate. The Du Pont Company studied the effectiveness of many solvents (4) and produced experimental fibers for military applications in 1942.

The first commercial acrylic fiber was announced by Du Pont in 1949, under the trade name of Orlon and production of a continuous filament yarn started in 1950. Orlon staple production started in 1952 as did production of Acrilan staple by the Chemstrand Corporation (now Monsanto Textiles Company). Acrylic fiber production has grown rapidly. World production of acrylic and modacrylic fibers reached 130,000 metric tons in 1960 (5), 400,000 t in 1965 (6), 1.0 Mt in 1970, and 1.4 Mt in 1975 (7). Acrylic fibers account for a predominant percentage of these totals. Growth occurred primarily in the United States during the 1950s, but subsequent expansion has included Western Europe and Japan. For a tabulation of acrylic fiber producers by country see Table 1 (7). World production capacity for the two fiber types in 1976 exceeded 2.0 Mt (7).

Structure

Since acrylic fibers contain at least 85 wt % polyacrylonitrile, many of their properties are determined by the inherent chemical nature or physical behavior of long chain PAN molecules in oriented structures (see Acrylonitrile polymers).

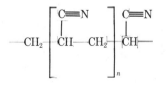

Hearle and Greer (8) summarized the literature through 1970 on the molecular structure of PAN and Meredith reviewed the gross morphology of acrylic fibers in 1975

Table 1. Acrylic Fiber Producers, 1976

Country and producer	Fiber trademarks	Fiber[a] types
Argentina		
Hisisa Argentina S.A.I.C.I.F	Cashmilon	S
Austria		
Chemiefaser Lenzing A.G.	Redolen	S
Belgium		
Akso Belge S.A. Fabelta	Acribel	S
Brazil		
Fisiba	Triana	S
Noracryl S.A.	Cashmilon	S
Rhodia Indústrias	Crylor, Tercryl	S
Bulgaria		
Kombinat "Dimitar Domov"	Bulana	S
Canada		
Du Pont of Canada	Orlon	S
China, Peoples Republic of		
Synthetic Fiber Plant (2)		S
China, Taiwan		
Formosa Plastics (2)	Tairylon, Taylilan	S
Tung HWA Synthetic Fiber	Town Flower	S
France		
Courtaulds, S.A. (2)	Courtelle	S
Rhone-Poulenc-Textile	Crylor	S
German Democratic Republic		
Chemiefaserwerk "Friedrich Engels"	Wolpryla	S
VEB Filmfabrik	Wolpryla	S
Germany, Federal Republic of		
Bayer, A.G.	Dralon, Ultraspan	S
Hoechst, A.G.	Dolan	S
Monsanto (Deutschland) G.m.b.H.	Acrilan	S, tow
Greece		
Vetlans Naoussa S.A.		S
Vomvicryl S.A.		S, tow, tops
Hungary		
Magyar Viscosagyár	Crumeron	S
India		
Indian Petrochemicals		S
J. K. Synthetics	Jekrilon	S
Iran		
Polyacryl Iran		S
Ireland, Republic of		
Asahi Synthetics	Cashmilon	S
Israel		
Israel Chemical Fibers	Acrilan	S, tow
Italy		
ANIC Societá per Azioni	Euroacril	S, tow
Fibra Del Tirso		S, tow
Montefibre, S.p.A.	Leacril	S, tow
SNIA Viscosa (2)	Velicren	S
Societá Italina Resine	Siracril	S
Japan		
Asahi Chemical Industry (2)	Cashmilon, Pewlon, Cerister	S, CF

Table 1 (*continued*)

Country and producer	Fiber trademarks	Fiber[a] types
Japan Exlan	Exlan	S, CF
Kanebo Acrylic Fibers	Kanebo Acryl	S, tow
Mitsubishi Rayon	Vonnel, Silpalon, Finel	S, CF, tow
Toho Beslon	Beslon	S, tow
Toray Industries	Toraylon	S
Korea, Republic of		
Hanil Synthetic Fiber	Cashmilon	S
Tae Kwang Industrial	Exlan, Acelan	S
Mexico		
Celanese Mexicana		S, tow, tops
Celulosa y Derivados	Crysel	S, tow, tops
Fibras Sinteticas	Fisisa	S, tow, tops
Netherlands		
Du Pont de Nemours N.V.	Orlon	S
Peru		
Bayer Industrial	Dralon	S
Poland		
Chemitex-Anilana	Anilana	S
Portugal		
Fibras Sintéticas de Portugal		S
Romania		
Uzina de Fibre	Melana	S, tops
Spain		
Cyanenka	Crilenka	S, tow
Montefibre Hispania	Leacril	S
Turkey		
Akrilik Kimya Sanayll	Aksa	S
Union of Soviet Socialist Republics		
state owned (5)	Nitron	S, tow
United Kingdom		
Courtaulds	Courtelle, Neochrome	S
Du Pont Limited	Orlon	S, tow
Monsanto Limited	Acrilan	S, tow, tops
United States of America		
American Cyanamid	Creslan	S, tow
Dow Badische	Zefran, Zefkrome	S, tow
E. I. du Pont de Nemours & Co., Inc. (2)	Orlon	S, tow
Monsanto Company	Acrilan, Fina	S, tow
Yugoslavia		
Organsko Hemijska	Makrolan, Malon	S

[a] S = staple; CF = continuous filament.

(9). The PAN molecule is usually described as a stiff rod with a diameter of 0.6 nm, although it can bend (8). Bohn and co-workers (10) picture the polymer molecule as randomly twisted or kinked due to the strong steric and dipolar repulsive forces between adjacent nitrile groups (11). The primary valence bonds in the polymer restrain the effect of these repulsive forces and lead to a helical chain conformation and a stiff rodlike structure (10). The nitrile groups are randomly distributed around the "rod surface" at different angles to its axis. These rodlike molecules attract each other strongly because of electrical dipolar forces, but the random geometry of the nitrile groups prevents a regularly spaced pattern of attractive forces between molecules. Consequently, the molecules pack together with varying degrees of lateral order but with little longitudinal order along the fiber axis. The x-ray diffraction patterns of solution-spun PAN fibers reveal only a moderate degree of crystalline order, but an orthorhombic unit cell has been proposed (12) containing 24 monomer units (6 in the base and 4 along the fiber axis) and a length along the fiber axis of 0.7096 nm. The calculated density of PAN in this cell would be 1199 g/L. In fibers, low angle x-ray scattering indicates the density is reduced by the presence of microvoids (13). Most commercial acrylic fibers contain 5–10% of other monomers with bulky side groups that are added to further distort the individual chains and disrupt interchain bonding. The presence of these comonomers reduces transition temperatures of the polymers (14). As a result, the fibers shrink at lower temperatures and exhibit a lower resistance to extension in hot water.

Physical Properties

Acrylic fibers are sold in the form of staple, tow, and a small amount of continuous filament. Some producers also sell "tops" in which the tow has been stretch-broken or cut to form a sliver. Staple lengths may vary from 25 to 150 mm depending on the spinning system used to convert the fiber into spun yarn. Fiber size is expressed in weight per unit length or linear density. The traditional unit, denier, is the weight in grams of 9000 meters. The new unit, tex, is the weight in grams of 1000 meters. Specific stress on the fiber during extension or at break has been written as g/den or gf/tex. The unit for specific stress (modulus and tenacity) in the International System (SI) is newtons per tex (N/tex); a conversion formula for the various units is found in reference 15.

$$1 \text{ N/tex} = 102 \text{ gf/tex} = 11.33 \text{ g/den} \tag{1}$$

The tensile properties of acrylic fibers will vary according to end use. Morton and Hearle (16) have published a compilation of the properties of natural and synthetic fibers which includes a standard acrylic and is a valuable source for comparing different fiber types. Tensile properties of seven acrylic fibers are listed in Table 2. These were selected to include different production processes, varied end uses, and the commercial tex range. The values shown are considered typical of commercial products, but do not represent producer specifications. Tenacity varies from 0.19 to 0.32 N/tex and elongation to break from 33 to 64%. The only apparent pattern is the lower tenacity and higher elongation of the larger tex fibers. Boiling water shrinkage of standard acrylic fibers is low, but may be much higher for special types. High bulk fibers, which have been stretched and cooled to retain a frozen-in strain, are designed to shrink 15–30% in boiling water. Some bicomponent fibers shrink as they develop their helical crimp during hot-wet treatment after manufacture (17). The density of most fibers is about 1170 g/L and increases with the incorporation of halogens. The tensile properties of acrylic fibers are little affected by water at normal temperatures. Like other synthetic fibers, their resistance to extension (modulus) decreases at higher temperatures and the elongation increases. The effect of dry heat on the stress–strain properties of a typical acrylic is shown in Figure 1. The effects of moisture and heat are shown in Figure 2 (18).

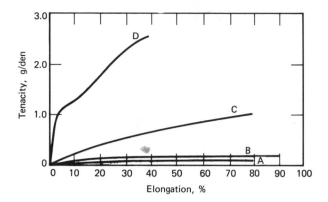

Figure 1. Effect of temperature on stress–strain curves of a three denier acrylic fiber. A, in air at 204°C; B, in air at 149°C; C, in air at 93°C; D, in air at 21°C, 65% rh. To convert g/den to N/tex multiply by 0.0882.

Cross-sections of six acrylic fibers are shown in Figure 3. A dogbone shape is usually indicative of a dry spinning process. Fibers with round or bean-like shapes are normally wet spun. Some bicomponent fibers have unusual lobed or mushroom shapes.

Chemical Properties

A number of countries now have legislation and tests to specify the flame retardance of textiles for critical end uses. Standard acrylics are acceptable for many applications. The general flame resistance test for carpets in the United States, DOC FF1-70, specifies a maximum burn distance after ignition from a methenamine pill (19). Most acrylic fibers produced specifically for carpet use contain sufficient amounts of a halogen to pass this test and also the Steiner Tunnel Test (20) in many carpet constructions. Recently, the flooring Radiant Panel Test has been accepted by many regulatory agencies to assess the flammability of carpets in corridors and exitways of public buildings. One producer (20) reports that his fiber in properly constructed carpets will meet the proposed criterion of this test in a glue-down (with no pad) installation, which is desirable for these areas. Where higher degrees of flame retardancy are required, blends of acrylic and modacrylic fibers are used (see Flame-retardant textiles; Flame retardants).

Acrylic fibers have outstanding resistance to sunlight, compared to other fibers (Fig. 4), and to attack by insects or microorganisms in the soil (21). These properties make them particularly valuable for outdoor uses. Pigmented fibers with excellent color stability to light are also available. As shown below, acrylics are resistant to most ordinary chemicals.

Agent	*Effect*
acids, inorganic	
weak	little effect
strong	may swell; will dissolve in concentrated HNO_3 or H_2SO_4

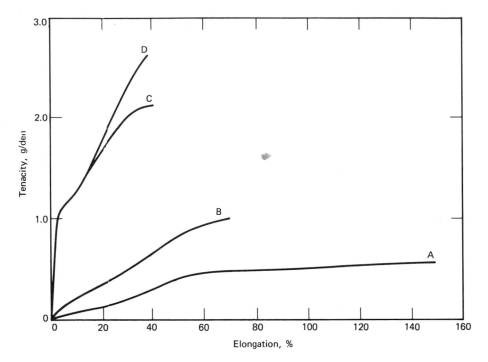

Figure 2.   Effect of water and temperature on stress–strain curves of a three denier acrylic fiber. A, in water at 93°C; B, in water at 71°C; C, in water at 21°C; D, in air at 21°C, 65% rh. To convert g/den to N/tex, multiply by 8.826 × 10⁻².

alkalies
   weak   little effect
   strong   degraded by hot, concentrated alkalies
organic solvents   negligible for common solvents
oxidizing agents   little effect
dry cleaning solvents no effect

Fiber Analysis and Identification

    A comprehensive article on the analysis of acrylic and modacrylic fibers has been published (23). A general review of the chemical testing and analysis of synthetic fibers was issued in 1971 (24). Composition of an acrylic or modacrylic fiber can be determined by a combination of physical and chemical methods. Elemental analyses are used to determine the percentages of C, N, O, H, S, Br, Cl, Na, and K, which are the common elements found. The presence of most comonomers can be established by ir, uv, or nmr techniques if known polymers or absorption spectra are available for comparison. Acidic groups can be measured by potentiometric titration in dimethylformamide (25). Organic additives are often extracted by solvents which do not dissolve the polymer.

    Schemes for the identification of textile fibers have been issued by the Textile Institute (26) and the American Association of Textile Chemists and Colorists (27). These attempt to cover all natural and synthetic fibers, although the complexity of

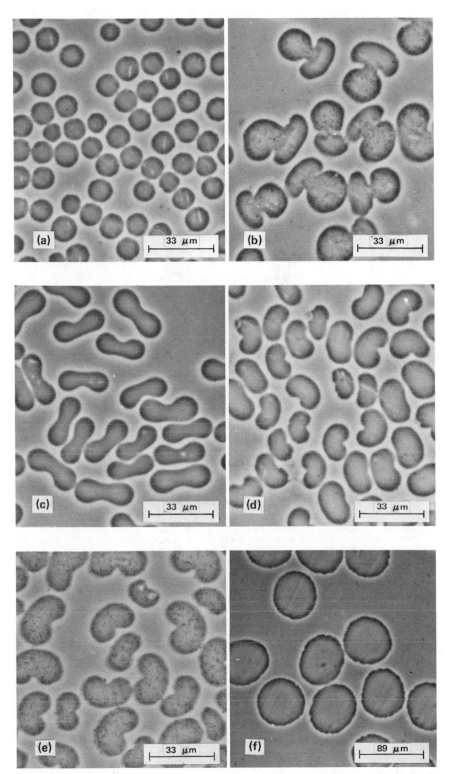

Figure 3. Cross-sections of acrylic fibers, 600X phase contrast. **(a)**, Creslan 61, 0.17 tex; **(b)**, Orlon 21, 0.33 tex; **(c)**, Dralon, 0.33 tex; **(d)**, Crylor, 0.33 tex; **(e)**, Leacril, 0.33 tex; **(f)**, Acrilan B-96, 1.7 tex (225X).

Table 2. Physical Properties of Selected Acrylic Fibers [a]

Fiber	Acrilan B-16	Orlon 42	Euro-acril	Cour-telle	Dow type 500[b]	Acrilan B-96	Zefran 253A
producer	Mon-santo	Du Pont	ANIC	Court-aulds	Dow Badische	Mon-santo	Dow Badische
end use	textile	textile	textile	textile	industrial	carpet	carpet
tex	0.13	0.38	0.34	0.50	0.67	1.67	1.67
denier	(1.2)	(3.4)	(3.1)	(4.5)	(6.0)	(15.0)	(15.0)
tenacity, N/tex	0.32	0.26	0.24	0.22	0.31	0.19	0.20
g/den	(3.6)	(3.0)	(2.7)	(2.5)	(3.5)	(2.2)	(2.3)
elongation, %	42	33	35	64	33	56	62
relative knot tenacity, %	90	81	91	93		80	82
initial modulus, N/tex	3.9	4.0	3.6	3.3		1.9	1.8
g/den	(44)	(45)	(41)	(37)		(21)	(20)
boiling water shrinkage, %	1.0	0.7	0.7	1.2	<1.0	1.0	1.0

[a] Tensile properties measured at 65% rh and 21°C. Extension rate, 100% per minute.
[b] Source: *Dow Badische Technical Bulletin A-10.*

new fiber types makes the task increasingly difficult. Moncrieff has a chapter (28) on the identification of man-made fibers and Stratmann (29) has proposed a classification scheme for fibers according to their solubility in five solvents. Acrylics can be distinguished from other commercial fibers by their solubility in hot dimethylformamide and their density of 1120–1200 g/L. Spandex fibers may fall in this group but are easily recognized by their very high reversible elongations (see Fibers, elastomeric). Modacrylic fibers are also soluble (or form a plastic mass) in hot dimethylformamide, have a density of 1240–1400 g/L, and contain 10–35% total halogen. Vinyon fibers behave similarly but contain over 48% chlorine. The characteristic nitrile band at about 2240 cm^{-1} in the ir also clearly separates acrylic or modacrylic fibers from other types.

Manufacture

Composition and Polymerization.   As already stated, most polymers for acrylic fibers contain minor amounts of various monomers, in addition to acrylonitrile. Neutral monomers, such as methyl acrylate (the most widely used), methyl methacrylate, and vinyl acetate, are copolymerized with acrylonitrile to increase solubility of the polymer in various solvents and the diffusion rate of dyes into the fiber. Small amounts of ionic monomers are often included, usually along with a neutral monomer, to enhance dyeability. Sodium styrene sulfonate and sodium methallyl sulfonate are commonly used to provide dye sites for basic dyes, and vinylpyridines and other basic monomers confer dyeability with acid dyes. Halogen-containing monomers are added to improve flame resistance.

These copolymers are generally made by either heterogeneous or solution polymerization. Acrylonitrile is moderately water soluble, but the polymer is insoluble in both acrylonitrile and water. This makes aqueous heterogeneous polymerization a natural choice. Both batch and continuous processes are used. In a typical continuous process, monomer, water, and initiator are fed to a continuously-stirred, overflow reactor at atmospheric pressure and a temperature of 30–70°C. The slurry of polymer,

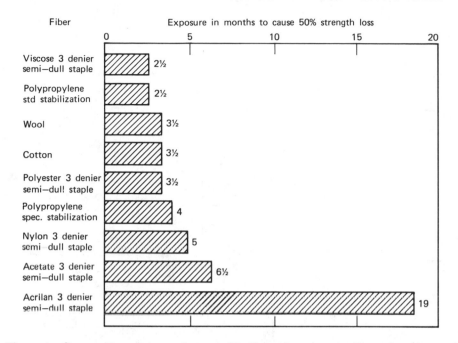

Figure 4. Comparative resistance of yarns to Florida outdoor exposure. Except for polypropylene, all fibers were unstabilized and the results compare stability of the polymers. Resistance of the other fibers can be increased by the use of stabilizers but acrylics are still superior. For viscose, polyester, nylon, acetate, and Acrilan, 6.35 cm staple, 30/2 yarn; for wool, 64s cut 6.35 cm, 16/2 yarn; for cotton, combed, 30/2 yarn. Data for polypropylene from Ref. 22. To convert den to kg/m multiply by 1.111×10^{-7}

water, and unreacted monomer is filtered; polymer is washed and dried; monomer is recovered from the filtrate and returned to the reactor (Fig. 5). Advantages of this process are simplicity, control, and applicability to a variety of polymer compositions. Initiators are usually water soluble redox combinations such as persulfate–bisulfite, activated by a small amount of iron, or chlorate–sulfite (see Initiators).

Solution polymerization is used to prepare acrylic polymers directly in a form suitable for wet or dry spinning. Solvents include dimethyl sulfoxide, dimethylform-amide, and aqueous solutions of zinc chloride or various thiocyanates. Preferred initiators are various organic peroxides and azo compounds. Ultraviolet light may also be used as an initiator (30). The principal advantage of solution polymerization is reduced manufacturing cost through elimination of process steps (31).

Mass or bulk polymerization can also be used to prepare acrylic polymers when precautions are taken to prevent auto-catalytic polymerization which can become explosive. A key requirement described in recent patents (32–34) is an initiator system that produces an easily stirrable monomer–polymer slurry from which heat can be readily removed. Typical conditions listed for continuous polymerization using a cumene hydroperoxide–sodium methyl sulfite redox initiator, are a reaction temperature of 30–60°C and a one hour residence time.

Drawing on the results of several investigators, Dainton and colleagues suggested a mechanism for the heterogeneous aqueous polymerization of acrylonitrile (35). This treatment reconciles seemingly contradictory observations of a decrease in monomer reaction order as the monomer concentration increases. The presence of two or three separate phases in heterogeneous polymerization makes a rigorous kinetic treatment

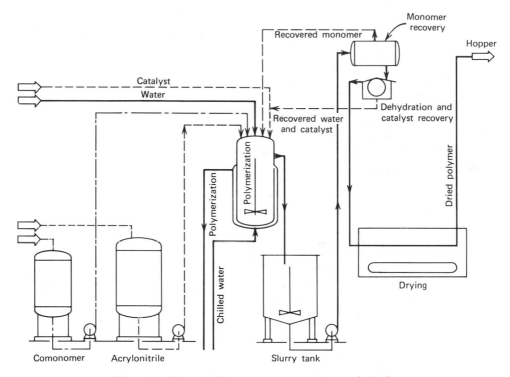

Figure 5.   Flow sheet for aqueous-suspension acrylonitrile.

difficult indeed. Using the simplifying assumption of homogeneous polymerization, Peebles (36) derived a kinetic model for persulfate–bisulfite initiated polymerization which correctly predicts the relationship between initiator concentration and polymer end groups (sulfonate and sulfate). Because of the assumption of homogeneity, it does not accurately predict molecular weight distribution.

Acrylonitrile copolymers used for acrylic fibers are white, easy flowing powders. The viscosity average molecular weight calculated by the Cleland and Stockmayer relationship (37) as shown below

$$[\eta] = 2.33 \times 10^{-4} \, M_v{}^{0.75} \tag{2}$$

is commonly in the range of 100,000–150,000. The polymers decompose before melting, but indirect methods indicate a melting point of 340°C for PAN and 260–280°C for copolymers containing 90–95% acrylonitrile. Above 180°C exothermic decomposition becomes rapid. It results from intrapolymerization of nitrile groups along the chain to form cyclic structures (38). This characteristic, undesirable for acrylic fiber because it causes discoloration, is used advantageously for the manufacture of graphite fibers (39) (see Carbon).

Solvents and Solution Preparation.   Of fourteen solvents given for polyacrylonitrile (40), seven are used commercially. They are listed in Table 3 with the approximate percentage of world fiber production based on each solvent. Dimethylformamide is the principal solvent since it is used in both dry and wet spinning processes (see Formic acid). Second is dimethylacetamide (see Acetic acid). Aqueous solutions of sodium thiocyanate and nitric acid are the most important inorganic solvents. Fiber production

Table 3.   Commercial Processes for Acrylic Fiber Production

Spinning process	Solvent	World capacity, %
dry	dimethylformamide	22
wet	dimethylformamide	11
wet	dimethylacetamide	23
wet	aqueous NaSCN	23
wet	aqueous HNO_3	12
wet	aqueous $ZnCl_2$	4
wet	dimethyl sulfoxide	3
wet	ethylene carbonate	2

from sulfuric acid solution has also been reported (41) but no commercial processes are known. Effective solvents for polyacrylonitrile must break the intermolecular nitrile–nitrile attraction; they are all highly polar, aprotic compounds. Solubility and solution properties are affected by the polymer composition and molecular weight. The presence of 3–10% neutral comonomers increases acrylic polymer solubility and decreases solution viscosity at a given concentration. Molecular weight, composition, spinning concentration, and coagulation conditions are adjusted to give acceptable fiber formation and good tensile properties. Typical solvent and solution compositions are:

Solvent	Polymer in solution, %
organic, dry spun	25–32
organic, wet spun	20–28
NaSCN, 45–55% in water	10–15
$ZnCl_2$, 55–65% in water	8–12
HNO_3, 65–75% in water	8–12

Processes for preparation of the spinning solution vary. Where solution polymerization is employed, no washing or dissolving steps are needed, although unreacted monomers are generally removed. Polymer made by heterogeneous aqueous polymerization may be washed and used without drying. With inorganic solvent systems, the wet polymer cake is dissolved directly (42). For organic solvent systems, two vaporization methods are described for solution preparation. The wet filter cake is first slurried with solvent. The slurry is then either passed through a wiped-film evaporator at low pressures to remove water and form a solution (43) or into a tubular heat exchanger to dissolve the polymer and then through a vaporization zone to remove water (44). In those processes where polymer is dried prior to use, solutions are prepared continuously from dry polymer and solvent in a mixing device and the slurry is passed through a heat exchanger (45).

Numerous soluble and insoluble solution additives are used for a variety of purposes in both acrylic and modacrylic fibers. Dull or semi-dull fibers are delustered with TiO_2. Colored fibers may be produced by the incorporation of either insoluble pigments or soluble cationic dyestuffs which react with dye sites in the fiber during coagulation. Flame resistance of fibers is increased by additives containing chlorine, bromine, or phosphorus. Antimony compounds (46–47) are used for their synergistic effect on flame resistance in fibers containing a halogen.

Heat stabilizers (qv) are added to minimize discoloration of the solution prior to spinning and the fiber during production. Typical stabilizers include organic acids and sequestering agents (48), organophosphorus compounds (49–50), tin compounds (51), zinc compounds (52), and zirconium salts (53). Other additives may be used to improve fiber light stability. In general, heat and light stabilizers are particularly useful in modacrylic fiber production.

Fiber Production

Acrylic fibers are produced either by wet spinning (78%) or dry spinning (22%). Each major spinning system has been developed over 20–30 years to produce competitive and commercially useful products. Development includes (1) optimizing fiber structure and properties, (2) increasing production rate as determined by linear speed, filament tex and tow size, and (3) improving process stability and fiber uniformity. The combination of process conditions to achieve all these features simultaneously is strongly influenced by the desired filament size or tex. For example, spinnerette-hole size is varied with fiber tex, and larger filaments require longer times in process steps where diffusion is important. Fiber tex in all spinning systems is controlled by the material balance

$$\text{tex/filament} = 1.075 \times 10^4 (p W_2 Q / S V_1) \qquad (3)$$

where tex = g/km; p = solution density, g/L; W_2 = weight fraction of polymer in the spinning solution; Q = flow rate per spinnerette hole in L/s; V_1 = first godet (take-out roll) velocity (after coagulation bath or below dry tower) in m/s; S = overall stretch ratio after first godet including relaxation shrinkage. These parameters also affect fiber structure and properties in a complex interaction with other conditions that will be discussed later. Optimizing fiber structure is a primary objective in many process studies. Acrylic fiber structure is best characterized by a combination of microscopic observation, determination of gel-fiber density, and measurement of internal surface area by gas adsorption (54). The average size of fibrils and pores in the fiber at each process stage can be calculated from the density and surface area results.

The major difference between wet and dry spinning occurs in the initial fiber formation stage, therefore, this step will be discussed separately. Subsequent process steps, such as washing, orientation, drying, relaxation, and finish application, are common to both wet and dry spinning, and are combined for discussion under Tow Processing.

Wet Spun Fiber Formation. In wet spinning, fine streams of viscous polymer solution are extruded through spinnerette holes into a liquid bath containing a coagulant that is miscible with the solvent. Fiber formation occurs rapidly as the polymer comes out of solution and is drawn onto the first godet or take-out roll. Each of the seven processes shown in Table 3 employs a different solvent, nevertheless, some features are common to all. In each case, a basic coagulation bath consists of the solvent diluted with water. Water as a coagulant has advantages in cost, process simplicity, and minimum environmental concerns. Another common feature is the spinning of many filaments from each spinnerette. Spinnerettes may have 10,000–60,000 holes with hole sizes from about 0.05–0.38 mm (2–15 mil) in diameter. The holes are often divided into segments to improve coagulant diffusion across the spinnerette face.

Fiber structure and ultimate fiber properties are largely determined by the choice of coagulation conditions, consequently, much research effort has been devoted to understanding and optimizing this critical step. Craig and co-workers (54) used the techniques mentioned earlier to investigate the structure of fibers wet-spun from solutions of dimethylacetamide, aqueous zinc chloride, and aqueous sodium thiocyanate into dilute coagulation baths of the respective solvent. The fibers, prepared from the same polymer, had similar structures. The freshly coagulated fibers were described as an unoriented, fibrillar network consisting of voids and fibrils with diameters on the order of 20 nm (Fig. 6). Superimposed on this network in some fibers were large radial macrovoids. This fibrillar network was oriented by subsequent stretching and the fibrils were also elongated. The oriented structure collapsed during drying to produce a compact fibrillar network without residual pores. Knudsen (55) found in the dimethylacetamide process that the best tensile properties, abrasion resistance, and fatigue life were obtained when the fiber leaving the coagulation bath had a high gel density, a homogeneous internal structure with small fibrils, and no macrovoids. These three characteristics are now generally recognized as guidelines for optimizing fiber structure.

Wet spinning is a three-component system (polymer–solvent–nonsolvent) in

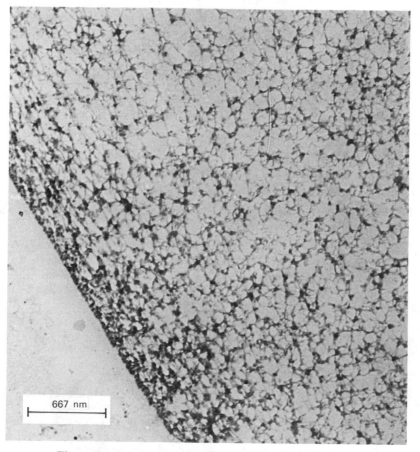

667 nm

Figure 6.   Gel structure of unoriented acrylic fiber, 31,500X.

which two transitions may occur, gelation and phase separation (56). Gelation is the gradual and continuous transition of the fluid solution into a homogeneous elastic gel. To achieve the desired dense, homogeneous fiber structure, gelation should precede phase separation or solidification. In studies of the reversible gelation of concentrated solutions of acrylic polymers, Paul (57–58) found that gelation tendency increases as polymer concentration increases and as the percent of neutral comonomer in the polymer decreases. Gelation also occurs more rapidly as temperature is decreased. These observations relate directly to effects observed in fiber spinning. Macrovoids were eliminated in one system either by reducing spin bath temperature or by increasing solution solids from 20 to 27.5% (55). Reducing spin-bath temperature is reported to increase gel density and decrease fibrillar size in the dimethylacetamide (55), nitric acid (59), sodium thiocyanate (60), and dimethyl sulfoxide (61) processes. A large effect of neutral comonomer on gel density and average pore size was found by Terada (62) in the dimethyl sulfoxide process. Fibers spun from polyacrylonitrile had a mercury porosimeter density of 450 g/L and an average pore size of 44.4 nm. The gel density decreased and the average pore size increased in fibers spun from copolymers containing 5 and 10% methyl acrylate. The same study found that incorporation of acidic comonomers, such as styrene sulfonate or allyl sulfonate, in the polymer increased gel density and reduced average pore size in the fibers.

Spin-bath composition and diffusion rates of the solvent and coagulant also have an important influence on fiber structure. Diffusion coefficients have been reported for the dimethylacetamide (63), dimethylformamide (64), and dimethyl sulfoxide (61) processes. In the latter system, diffusion rates of dimethyl sulfoxide were higher than those of water. The diffusion rates of both increased as spin-bath temperature increased and as linear velocity of the filaments in the bath increased. As the diffusion rates increased, fiber structure became poorer as evidenced by lower gel density and larger fibrillar size. This agrees with observed adverse effects of spin-bath temperature and spinning speed on fiber properties. Optimum fiber properties were obtained with 55% dimethyl sulfoxide in the coagulation bath, which is in the range often reported for the other organic solvent systems. For inorganic solvent systems, aqueous spin baths containing 30–35% $ZnCl_2$ (65), 28–36% HNO_3 (66), and 11% NaSCN (67) are reported.

Rheological properties of the spinning solution influence fiber structure and also have an important effect on spinning speeds and process stability (see Rheological measurements). A filament freely extruded into the coagulation bath (with no extension imposed by the first godet) swells considerably as it exits from the spinnerette hole. This die-swell results primarily from the elastic properties of the spinning solution rather than its viscosity (68). The extent of die-swell increases with the apparent shear rate, $\dot{\gamma}_{wa}$, at the capillary wall (69–70) as calculated from the equation below where R is hole radius.

$$\dot{\gamma}_{wa} = \frac{4Q}{\pi R^3} \tag{4}$$

The freely extruded filament after maximum die-swell should be almost strain-free (68). Based on this concept, the velocity of the freely extruded fiber, V_f, has proved a very useful parameter in explaining phenomena observed in wet spinning. For example, the maximum first godet speed that can be taken without the filaments breaking increases steadily as V_f increases (69). Since spinning speed is partly determined by the first godet velocity and spinning stability is a function of the ratio (maximum first godet speed)/(actual first godet speed) it can be seen that V_f may limit production rates and affect spinning stability. Although early workers described the fiber leaving the coagulation bath as unoriented, acrylic fibers do acquire orientation as they form and are drawn through the coagulation bath (71). The apparent draw ratio can be determined (72) and often correlates with the ratio $(V_l$ actual$)/V_f$.

Cross-sections of acrylic fibers wet-spun from organic solvent systems are often bean-shaped as a result of initial skin formation around the periphery of the swollen filament. As the fiber interior deflates, the skin must fold. Fibers from the inorganic solvent systems are normally round. Shaped fibers can be produced from nonround holes by control of coagulation conditions (72–73). To maintain a nonround shape, the tension applied to the filament must be great enough to eliminate die-swell (a high V_l/V_f ratio). High V_l/V_f ratios may also increase macrovoid formation. A variety of fiber cross-sections have been described in the patent literature (74–76).

Special Wet Spinning Processes. Three modified processes have been described for the production of acrylic fibers. The use of nonaqueous coagulants, such as glycerol, was recognized as early as 1943 (77). Coagulation baths containing a polyalkylene glycol and solvent produced fiber with very high gel densities and small fibrils, which led to superior tensile properties (78–79). The high molecular weight glycols diffuse very slowly and a fine gel structure is formed as the solvent diffuses out of the fiber. The use of hexanetriol (80), low molecular weight alcohols (81), and cumene–paraffin mixtures (82) have also been reported. In a second process variant known as "gel state" spinning, a hot, concentrated polymer solution is extruded downward through air and the filaments allowed to gel by cooling before they pass into an aqueous bath for solvent extraction (83). Homogeneous, macrovoid-free fibers are claimed. This system is different from the dry jet-wet spinning process, in which the spinnerette is positioned only a few centimeters above the coagulation bath surface (84–85), and no gelation occurs in the air. Improved fiber properties and high spinning speeds are claimed (86–87) for dry jet-wet spinning. The high spinning speeds achieved with nonaqueous coagulants or by dry jet-wet spinning could be particularly useful for making continuous filament acrylic products.

Dry Spun Fiber Formation. In dry spinning, fiber formation occurs as solvent is evaporated from streams of polymer solution drawn downward through heated gas. The spinning cell consists of a vertical jacketed cylinder with the spinnerette positioned in the center of the top and the fiber exit at the bottom. The cell may be 5–10 m long and 13–23 cm in diameter. In one mode of operation, the heated gas is drawn into the top of the column, it flows downward with the filaments, and is withdrawn near the bottom of the cell (88). In another process, the column is divided into three heated zones with the inlet gas flowing upward countercurrent to the filaments and being withdrawn above each zone (89). The solvent-rich gas is cooled after removal from the tower to recover solvent from the recycled inert gas. Below the tower exit, finish may be applied to the vertical threadline by contact with a rotating roll wetted with a finish solution. The filament bundle is next wrapped around a godet for several turns and then directed to subsequent process steps. The velocity of the filaments is controlled by the speed of the godet and the undrawn filament tex is established at this point. The material balance, equation 3, shown previously is applicable to dry spinning; therefore, this group of factors must be adjusted to produce a final fiber of the desired tex. As in wet spinning, process optimization and production capacity are greatly influenced by the final fiber tex. The spinnerette may have 300–900 holes arranged in concentric circles. The number of holes is limited by the need to maintain filament separation. Turbulence in the tower is also minimized so that the filaments will not touch each other prematurely and fuse. Other important conditions that must be considered in dry spinning process development include gas-flow rate, gas temperature, solvent-gas ratio, and the explosive limits, heat input from the walls, tower length, and the solvent boiling point. The principal solvent used for dry spinning acrylics, dimethylformamide, has a high boiling point of 153°C and complete solvent removal is difficult at desirable spinning rates (88). Depending on filament tex, exit speed, and tower length, residual solvent in the fiber may be as high as 50%.

Fiber formation in the polymer–solvent systems used for dry spinning results only from gelation (56). Other transitions do not occur because the systems are miscible, temperatures are high, and the solvent in the fibers lowers their transition temperatures. Gelation occurs gradually as the solvent evaporates and the concentration of polymer in the fiber increases. Homogeneous gelation is desirable for optimum fiber structure. The structure of dry-spun fibers was compared to those prepared from the same polymer by three wet spinning processes (54). Tower conditions were varied to produce fibers with 1, 10, and 20% residual solvent. The gel densities of the three dry-spun samples were much higher than those of the wet-spun samples and approached collapsed-fiber density as the residual solvent decreased. Very little fibrillar structure was apparent in electron micrographs of the unoriented, dry-spun fibers. During stretching, the dry-spun samples developed an oriented fibrillar structure similar to that of the oriented wet-spun samples.

Monocomponent dry-spun fibers often have dogbone cross-sections. This results from rapid solvent removal and skin formation around the periphery of the filament. Round cross-sections can be produced by use of a static zone with a high solvent–gas ratio in the top of the tower (90). Acrylic fibers with a trilobal shape are also produced.

Processing of dry-spun fibers after the godet varies. When residual solvent and exit speeds are low, the bundle of filaments can be drawn from heated rolls (89), since no washing is required. With exit speeds of 200–400 m/min, direct introduction into an aqueous tow treatment process is not practical. In one process, the solvent laden filaments from multiple towers are combined into a rope, which is piddled (laid in a transverse pattern) into a container (91). Multiple ropes are subsequently combined prior to tow processing.

Tow Processing. Wet-spun filaments from several spinnerettes may be combined after leaving the coagulation bath and drawn directly into the tow line. The weight of water and solvent in the tow will be 100–300% of the fiber weight. Ropes of dry-spun fibers, which were previously collected, will be combined to form a tow of the desired size. The dry-spun tow may contain 10–50% solvent based on fiber weight. Tows from the different wet and dry spinning systems all undergo the essential steps of washing, orientation or stretching, finish application, and drying accompanied by collapse, relaxation, and crimping, but the sequence of these steps varies. Figure 7 is a simplified flow diagram which illustrates the features of several tow processes described in the patent literature. After the common initial steps of washing, stretching, and finish application at least three different procedures have been described. In one, the fibers are collapsed and dried with no shrinkage permitted. In another, the fibers are dried and collapsed under conditions where controlled shrinkage occurs. In the third, the fibers are collapsed and dried without restraint, so that free relaxation occurs. In some cases, the dried fiber will undergo additional stretch and relaxation steps. Finally, the fibers are either packaged as tow or cut into staple and baled.

In the washing step, virtually all the solvent and minor impurities are removed. The important factors in washing are: fiber size; fiber porosity and diffusion rates within the fiber; amount and temperature of the wash liquor; and penetration of the wash liquid through the tow bundle. In one process (92), the tow is led in and out of a series of extraction baths while being drawn. Bath liquor is also sprayed with high impact on the fibrous tow at points above the surface of the bath. In another method, a flat ribbon of tow is passed through a chamber which is designed to force the water to pass back and forth through the tow at high velocities (93–94). Still another washing technique involves porous cylinders and squeeze rolls (95). Generally, fresh water enters at the last stage and flows countercurrent to the tow. To reduce pollution and

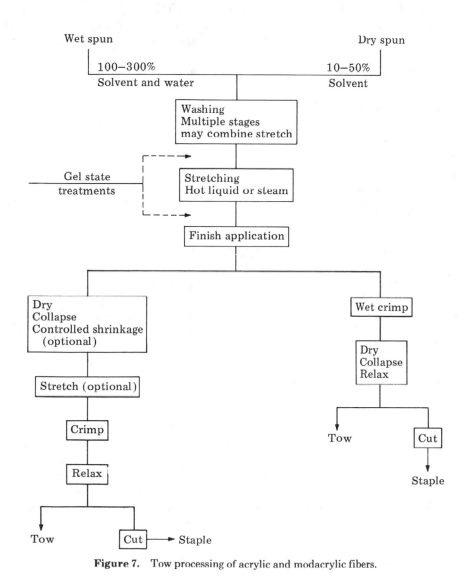

Figure 7.   Tow processing of acrylic and modacrylic fibers.

manufacturing costs, the wash water is usually recovered; therefore, low water usage is economically important.

Operating performance in the wet-tow stretching step is critically affected by temperature. One study (96), found a rapid increase in maximum draw ratio as the water bath temperature increased above 65°C, the fiber's approximate glass-transition temperature. Higher bath temperatures also reduced stress on the fiber at a normal draw ratio; this is desirable for process stability.

The unoriented fiber after washing and, to a lesser extent, the wet stretched fiber have a porous structure in which diffusion can occur rapidly. Consequently, a variety of in-line treatments have been developed either before or after stretching. In-line dyeing processes with cationic dyes have been described for dry-spun fibers (97–99) and wet-spun fibers from three different solvent systems (100–102). In some cases, dye pickup is increased by drawing the fibers in the dye bath. In-line treatments are also used to add flame retardants (67,103), apply crosslinking agents (104), incorporate dye receptor polymers (105), and add ultraviolet light stabilizers (106). Finish is often applied to the fibers before drying. Standard finishes contain a lubricant to prevent filament sticking during drying, an antistatic agent and perhaps a softener to improve fiber hand. Producer finishes are partially removed in later scouring or dyeing operations.

Fiber tows may be dried and collapsed by contact with a heated roll surface, by hot air flow through the tow on a perforated cylinder, or by hot air flow as the tow on a belt moves through an enclosed oven. When drying is accompanied by relaxation, special conditions of temperature and humidity may be necessary to obtain complete collapse (107).

The fibrillar network observed in wet-spun fibers leaving the coagulation bath undergoes major changes as the fibers are oriented, dried, and collapsed. One study (108) concludes that the fibrillar elements remain intact during stretching and collapse, although network junctures are destroyed and new junctures form. Another (109) found that structural changes in both steps are affected by the polymer composition and the gel density of the coagulated fiber. Cross-sections show that collapse during drying starts around the periphery of the fiber and moves inward (110). The collapse phenomena is probably best explained by a capillary attraction mechanism (110). As mentioned earlier, the structure of dry-spun fibers also changes during orientation and collapse. When dry-spun fibers were stretched in boiling water (54), their gel density decreased and a measurably increased surface area was found. The oriented fibrillar structure was similar to that observed in wet-spun fibers. Drying and collapse then produced a structure very much like that of wet-spun fibers.

Relaxation of acrylic fibers after orientation is an essential part of most processes. Relaxation increases fiber elongation, decreases the tendency to fibrillate, and increases dye diffusion rates into the fiber. These changes result primarily from a deorientation of the amorphous regions in the fiber structure (111). Relaxation may be taken at various steps in the tow process. The drawn, uncollapsed fiber will shrink in boiling water. The unrestrained fiber will also shrink during fiber drying and collapse. After collapse, boiling water shrinkage is reduced and higher temperatures are required for relaxation. Acrylic fibers are usually relaxed either wet or in steam since moisture reduces the temperature required. Relaxation of the collapsed fiber may be in-line or by batches in an autoclave (112). Unrestrained relaxation is important for the full development of helical crimp in bicomponent fibers. The helical crimp develops because of differential shrinkages of the two polymers in the fiber (113). Fiber shrinkage during relaxation may range from 10–40% depending on the amount of prior orientation, the fiber structure, and the intended use.

Bicomponent Fibers.   Bicomponent acrylic fibers have become a major product type since their introduction in the early 1960s (114). These fibers are composed of two polymers in separate areas of the cross-section. Each polymer is continuous along the fiber length and the two are permanently joined at an interface. The polymers are selected to shrink or swell differently in response to heat or moisture. As a result, when the fibers are properly treated, differential shrinkage will occur and a spiral or helical crimp will form. The extent of crimp development will depend on (1) the shrinkage differential between the components, (2) the distribution of the components in the fiber, and (3) translational restraints which may inhibit crimp development (113). The theoretical mechanics of crimp development in bicomponent filaments has been analyzed (115).

Bicomponent fibers may be produced in any of the spinning processes described above. Special equipment is required prior to the spinnerette to channel two separate solution streams into each hole. One spinnerette assembly, described for dry spinning, keeps the solutions separate until they enter the spinnerette hole (116). This requires annular sources of solution supply which are separated by a wall that extends to a concentric knife-edge just above each spinnerette hole. Different techniques may be used in wet spinning, where each spinnerette contains a much larger number of holes. One assembly brings the two solution streams together in channels prior to the spinnerette holes (117). Another method makes use of the principle that viscous polymer solutions can be merged into a common flow stream without significant mixing between the two components (118). This method permits the use of simpler mixing units some distance prior to the spinnerette (119). Furthermore, the pattern of the mixed stream flow can be varied to produce a range of bicomponent types.

Two different types of bicomponent acrylic fibers are produced. One contains different amounts of water-ionizable groups in the two polymers; consequently the hydrophilic side swells more in water. Crimp develops when the fiber is dried after wet-heat treatment. This spiral crimp is water-reversible; ie, it decreases on wetting out and reforms on drying. In the second bicomponent type, the polymer pairs contain different amounts of nonionic comonomers. Helical crimp develops when the oriented fiber is heated to a temperature where adequate differential shrinkage between the two polymers occurs. Once formed, the crimp in this type is permanent. Each bicomponent type has certain advantages in processing, esthetics, and product care. The properties of each can be modified by process changes (120–122).

Spinning processes have also been described for monocomponent fibers which develop a permanent helical crimp (123–124). These usually involve two-stage coagulation or other techniques to produce an asymmetric fiber structure with pronounced skin-core differences. The helical crimp develops when the fiber is relaxed after orientation.

Moisture Absorbent Acrylic Fibers.   A novel acrylic fiber has been announced (125) which absorbs 30–50% of water compared to 5% for standard acrylics. This increase in absorption results from fiber porosity rather than a chemical change. Each fiber contains a core with a large number of tiny capillaries that run the length of the fiber. Around the core there is a dense skin, to protect the core in textile processing, but with channels to allow moisture passage into the core. The fiber has a lower density than normal acrylics and is reported to increase wearer comfort.

Melt Extrusion. As pointed our earlier, the polymers used for acrylic and modacrylic fibers cannot be melt spun because they degrade too rapidly at temperatures below their melting point. Still, the economic desirability of a process without solvent, fiber washing, or solvent recovery has led to numerous studies of melt extrusion. Polymer melting points can be reduced by adding 10–60 wt % of certain plasticizers (126–128), or solvent (129) and the resulting melts can be extruded to form fibers. Part of the economic advantage is lost, however, since the plasticizer or solvent must be extracted and recovered. The melting point of acrylic polymers can also be reduced by the addition of water under pressure. Early attempts to spin fibers from polymer– water melts under pressure led to fibrillar material suitable for making paper but not for textile purposes. Recently, it was found that when the polymer–water mixture contains only enough water to hydrate a certain percentage of the nitrile groups, the melt can be extruded to produce textile filaments (130). Loop properties of these fibers were improved by the addition to the melt of 3–7% of a compatible solvent, based on polymer. Another approach to the production of fibers from polymer–water melts is to extrude the melt into a pressurized solidification zone (131). Conditions in this zone are controlled to minimize water evaporation from the filaments as they pass through and to permit very high draw ratios (exit velocity/velocity in spinnerette hole). No subsequent drawing step is necessary for fiber orientation and the usual relaxation conditions can be employed.

Dyeing. Acrylic fibers are dyed mainly with cationic dyes. Bright, fast colors are achieved that have high esthetic appeal. Disperse dyes can also be used to get a good range of colors, not as bright. A small quantity of acrylic spun from special polymers containing basic groups is dyed with anionic dyes. Dye sites for cationic dyes are acidic polymer end groups from polymerization initiators (132) and others added by copolymerization with acidic monomers. Diffusion of dye to these sites is negligible below the glass-transition temperature (about 65°C in water) but is rapid at 96–100°C, the normal dyeing temperature. No carriers are required to accelerate dyeing (see Dye carriers); instead, retarders are commonly added to reduce dyeing rate and improve uniformity of coloring. Typical dyeing time is 1–2 h, but can be shortened by dyeing under pressure.

An oversimplified but useful view of the mechanism of dyeing with cationic dyes has three steps: surface adsorption, diffusion into the fiber, and dye-fiber bonding (133). Proper attention to the implications of each step is necessary for satisfactory dyeing. In hot water, the tensile modulus of acrylic fibers is low (about 5×10^{-2} N/tex at 80°C in water and 4 N/tex when dry at 20°C). Because of this, special care must be exercised in dyeing to prevent fiber, yarn, or fabric distortion.

The fiber is readily dyed as staple (stock dyed), in yarn on cones (package dyed), or as skeins. As previously mentioned, some is dyed during fiber spinning (producer dyed) or as tow, using special tow dyeing equipment. Fabric or garment dyeing (piece dyeing) is used on acrylic knitted goods. To a lesser extent it is also used on woven materials, particularly where acrylics are blended with other fibers such as polyesters. Modern continuous dyeing equipment for carpets is designed so that minimal distortion of the carpet fiber occurs. Acrylic carpets can be dyed or printed on such equipment, especially those made from fiber modified to dye rapidly. Final appearance is improved by including a pile erection step after drying.

Transfer printing is a recently developed process in which sublimable dyes are transferred to the fabric by heat from a printed design on paper. Though acrylics in general are not well suited for this process, current research offers promise for the future (134). Dyeing in nonaqueous liquids (solvent dyeing) is also a useful process for coloring acrylics; fiber or fabric distortion is reduced, as is loss of fiber luster. Both processes require less energy than aqueous dyeing and greatly reduce water pollution (see Dyes, application).

Uses

Acrylic fibers are used extensively in the apparel and home furnishings markets; a relatively small amount goes into industrial uses. Within the apparel market major uses are for sweaters, single and double knit jersey fabrics, hosiery and craft yarns, and pile and fleece goods. In home furnishings the major uses are for carpets, blankets, curtains and drapes. A breakdown of United States shipments for 1972–1975 is shown in Table 4.

Apparel. The sweater market has been a major one for acrylic fibers since early in their development. Sweaters knitted of acrylics are warm, resilient, have a desirable feel or hand, which can be varied to match different types of wool. They have bright, fast colors; are bulky and lightweight; and are easily washed. Good bulk and resilience are achieved by using either high bulk or bicomponent fibers. High bulk fibers, which shrink on heating, are blended with regular, nonshrinking fibers and made into yarn. When the yarn is heated, the high bulk fiber shrinks, buckling the nonshrinking fiber and producing a lofty yarn. Bicomponent fiber makes a bulky, resilient yarn and sweater because it has a three-dimensional spiral fiber crimp.

The same features of warmth, bulk, resilience, and bright colors are important for use of acrylics in socks and craft yarns. For socks, nylon or polyester is usually blended with the acrylic to increase abrasion resistance. The large amount of craft or hand knitting yarn goes mainly into sweaters, afghans, and rugs.

Major quantities of acrylics are used for single and double knit jersey fabrics. The single knit fabrics are mostly used in women's dresses and blouses, and men's shirts. Women's dresses, pantsuits, skirts, and slacks, and men's slacks are the markets for double knits. Blending with polyester fiber is frequently done to enhance wrinkle resistance and easy care of these garments.

High luster and resilience make acrylic fibers especially desirable for pile and fleece fabrics. These fabrics simulate at a lower cost the appearance and properties of natural furs (see Furs, synthetic). They are used for men's and women's coats, both

Table 4.   United States Shipments of Acrylic and Modacrylic Fibers [a]

Material	1972	1973	1974	1975
hosiery, sweater, and craft yarns	62.6	83.5	77.1	86.2
pile and fleece	26.8	32.6	32.2	29.9
other circular and flat knit	59.4	68.9	47.6	31.3
blankets	10.9	17.7	14.1	15.9
other broadwoven	14.5	17.2	8.2	5.9
carpet face yarns	63.5	76.7	51.7	37.6
other	6.4	3.6	2.7	5.0
Total	*244.1*	*300.2*	*233.6*	*211.8*

[a] Shipments in thousands of metric tons (kt). Source: *Textile Organon.*

as the outer shell and warm inner-lining. Acrylic fiber dyed during manufacturing is particularly well-suited for this market because of brightness of color and high luster.

Home Furnishings.   In the United States, a major use of acrylic fiber has been for carpets and rugs. The fiber was first used as a replacement for wool because of bulk, resilience, woollike feel, and resistance to staining. Modified acrylics have since been developed to increase flame resistance, and also to make carpets dyeable on continuous equipment. Colored acrylic carpets usually are made from dyed fiber or yarn. Some are made from fiber dyed or pigmented during fiber spinning. Ease of fiber distortion under hot-wet conditions and a relatively long time required for dyeing (as compared with nylon) make dyeing in carpet form difficult. Special acrylic carpet fibers are available which dye faster, distort less, and can be readily dyed or printed in carpet form on continuous equipment.

Acrylic fiber makes excellent blankets. Warmth, lightness, ease of cleaning, coloration, and resistance to attack by moths and mildew are the important fiber characteristics. In the United States, it is the dominant fiber for this large home furnishing use.

Draperies, curtains, and awnings are together a significant outlet for acrylic fibers. They are used because of ease of coloration, colorfastness, dimensional stability, and exceptional resistance to loss of strength when exposed to sunlight. Acrylic fibers used for awnings are usually colored with pigments during fiber manufacture because these have excellent dyed lightfastness.

MODACRYLIC FIBERS

The first modacrylic fiber resulted from research at Union Carbide on copolymers of vinyl chloride (135). A continuous filament product was introduced in 1948 and Dynel staple fiber was marketed in 1950 (136). Tennessee Eastman introduced Verel in 1956 (137) and Kanegafuchi produced Kanekalon in Japan the same year (138). Courtaulds announced its modacrylic (later named Teklan) in the United Kingdom in 1962 (139). During this period, modacrylic fiber production had grown slowly for specialty applications or for use in blends with other fibers. This trend continued through 1968 (140). Since that time many countries have passed legislation requiring the use of flame-retardant textiles in end uses such as children's sleepwear and furnishings in public buildings (see Flame-retardant textiles). These laws spurred the development of new modacrylics with properties suitable for a range of textile applications. In 1976 there were nine known producers in five countries (see Table 5).

Structure

Since modacrylic fibers usually contain more than 50 mol % acrylonitrile, their properties are related to those of acrylic fibers. One important difference, resulting from the presence of the halogen-containing monomers, is the lower maximum temperatures at which the modacrylics are dimensionally stable. Fiber heat stability will depend on the type and amount of halogen-containing monomer present and can be influenced by process conditions. Halogen also affects other fiber properties, such as moisture regain, light stability, and resistance to degradation on heating.

Table 5.   Modacrylic Fiber Producers, 1976

Country and producer	Trademarks and names
Germany, Federal Republic of	
Bayer, A.G.	Dralon MA
Japan	
Kanegafuchi	Kanekalon, Kanecaron
Mitsubishi Rayon	Vonnel H-704
Toray Industries	Toraylon Unfla
Union of Soviet Socialist Republics	
Moscow Textile Institute	Saniv
United Kingdom	
Courtaulds	Teklan
United States of America	
E. I. du Pont de Nemours & Co., Inc.	Orlon 775F
Monsanto Company	SEF; Elura wigs
Tennessee Eastman	Verel

Physical Properties

Modacrylic fibers are sold as staple and tow. Textile fibers range from 0.13 to 1.67 tex (1.2–15 den) and wig fibers from 3.33 to 5.55 tex (30–50 den). Cross-sections of five textile fibers and one wig fiber are shown in Figure 8 (see Wigs). The pigment-like particles in Orlon 775F, Verel, and S-06 may be antimony oxide. Some producers offer modacrylics with and without antimony.

Physical properties of seven modacrylic fibers are shown in Table 6. These data were obtained from measurements on commercial products. Although the tenacities and elongations vary, all are quite satisfactory for textile (or wig) purposes. There is a considerable temperature range (173–206°C) in which dry heat will shrink different

Table 6.   Physical Properties of Modacrylic Fibers [a]

Fiber	Verel	Kanekalon SE	SEF S-06	Teklan	Orlon 775F	Dralon MA	S-32
Producer	Tennessee Eastman	Kanega-fuchi	Mon-santo	Court-aulds	Du Pont	Bayer	Mon-santo
denier	3.0	3.2	2.9	3.8	3.2	2.9	50.0
tex	0.33	0.36	0.32	0.42	0.36	0.32	5.6
tenacity, N/tex	0.24	0.24	0.19	0.22	0.19	0.22	0.22
g/den	(2.7)	(2.7)	(2.2)	(2.5)	(2.2)	(2.5)	(2.5)
elongation, %	35	32	52	40	47	35	28
relative knot tenacity, %	54	58	87	77	90	65	
boiling water shrinkage, %	4.3	2.0	1.4	5.2	1.4		2.8
dry heat shrinkage temperature for 10%, °C	187	173	206	175	184	181	
moisture regain, %	2.7	1.2	2.5	0.9	1.2		
density, g/L	1360	1250	1350	1360	1240		

[a] Tensile properties measured at 65% rh and 21°C. Extension rate, 100% per minute.

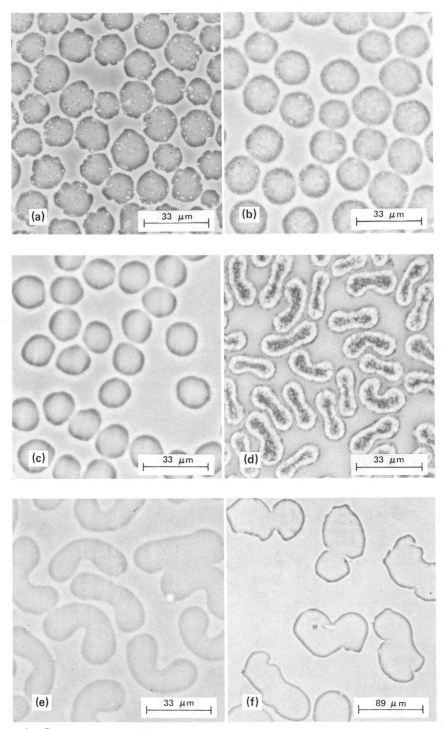

Figure 8. Cross-sections of modacrylic fibers, 600X phase contrast. (a), SEF S-06, 0.33 tex; (b), Orlon 775F, 0.33 tex; (c), Teklan, 0.40 tex; (d), Verel, 0.33 tex; (e), Kanekalon S, 0.33 tex; (f), S-32 wig fiber, 5.5 tex (225X).

24

fibers 10%; Dynel, however, will shrink 10% at 141°C. This reflects efforts to overcome a problem once considered typical of modacrylics. Dynel is no longer in production (141). The wide range in density (1240–1360 g/L) may result from the use of different halogen-containing monomers as well as the amount present. The tensile properties of modacrylic fibers are little affected by water at ambient temperatures. Like acrylic fibers, their resistance to extension will decrease at high temperatures under wet or dry conditions.

Most modacrylic fibers are quite lustrous. Semi-dull fiber types are permanently delustered by the presence of TiO_2 or antimony oxide pigments. In both types, micropores may open in the fiber structure during boiling water treatment and cause additional delustering. The fibers are usually restored to their original luster after dyeing by drying at 110–130°C. The tendency to deluster during dyeing has been considerably reduced in some of the newer modacrylics.

Chemical Properties

The most important property of modacrylic fibers is their flame retardance. The commercial significance of this property has increased rapidly with consumer concern and government regulations. Flammability tests are usually made on fabrics, garments, carpets, drapes, etc; therefore, results will depend on several factors, in addition to fiber composition. Conclusions about the flame retardancy of the modacrylic fibers can be drawn only in relation to specific products and tests. One producer states (142) that using their fiber, "the standards FF3-71 and FF5-74 for children's sleepwear (including sleepers, robes, nightgowns, and pajamas), FF1-70 and FF2-70 for scatter rugs, FF4-74 for mattresses (tapes only), and the California Fire Marshal Test for draperies, can be met by many properly constructed fabrics." Fabrics containing the other modacrylics are also reported to pass these or similar stringent tests in other countries. Since the official definition of modacrylic fibers neither requires the presence of flame retardant materials nor defines a degree of flame retardancy, any new fiber should be evaluated in desired fabrics by the appropriate tests.

Like the acrylics, modacrylics are highly resistant to biological agents such as insects and microorganisms. They are also stable to aqueous solutions of most salts, acids, and bases. They are generally resistant to dry cleaning solvents but some will be dissolved by acetone and other ketones. Their lightfastness is good but not equal to that of the acrylics. On prolonged heating, most modacrylics will yellow due to the loss of hydrogen halides, but short exposures have little effect on tensile properties.

Manufacture

Composition and Polymerization.   All modacrylic fibers of commercial importance are made from copolymers of acrylonitrile and a halogen-containing monomer, with or without additional monomers to enhance dyeability and other physical properties. The two most commonly used are vinyl chloride and vinylidene chloride. Others, such as vinyl bromide (143), vinylidene bromide (144), and 2,3-dibromopropyl acrylate (145), can be used but are more expensive. As with acrylic fibers, monomers incorporated to increase ease of dyeing with basic dyes usually contain sulfonic acid groups. Examples are sodium salts of these acids: styrene sulfonic acid (145), methallyl sulfonic acid (146), and p-methallyloxybenzene sulfonic acid (147). Dye sites can also be provided by monomers, such as itaconic acid, which contain carboxyl groups, or by olefinically unsaturated sulfonamides (148). Various other monomers are copolymerized to increase fiber-dye diffusion rate.

Polymers for modacrylic fibers can be made by emulsion, heterogeneous, and solution polymerization, both batch and continuous. Initiation is usually with the persulfate–bisulfite–iron redox system. Polymer separation, washing, and drying follow standard procedures. In solution polymerization, with dimethylformamide or dimethyl sulfoxide as solvents, azo and peroxy initiators can be used, as well as light (149). After removing unreacted monomers, these solutions are spun directly into fibers.

Using halogen-containing monomers introduces certain polymerization complexities not generally encountered in preparing acrylic polymers. Their low boiling point (−12°C for vinyl chloride and 38°C for vinylidene chloride) may require pressurized polymerization vessels. The monomers act as chain-transfer agents by loss of halogen and limit polymer molecular weight. Their reactivity ratios are usually much lower than acrylonitrile, requiring that for a given amount in the polymer a much higher percentage be fed to the reactor, and the unreacted fraction recovered for reuse. Because of its toxicity, special precautions must be taken in handling vinyl chloride (see Industrial hygiene and toxicology).

Solvents and Solution Preparation.   Based on the literature (150) and published patents, acetone, dimethylacetamide, dimethylformamide, and dimethyl sulfoxide are used for the production of modacrylic fibers. Solubility characteristics of acrylonitrile–vinyl halide copolymers vary with the molar composition and to a lesser extent with the specific halogen. The composition of Dynel (60 wt % vinyl chloride) was partly determined by its solubility in acetone (135). This solubility was considered surprising since neither polyacrylonitrile nor poly(vinyl chloride) of a desirable molecular weight is soluble in acetone. Similarly, copolymers of vinylidene chloride containing less than about 50% acrylonitrile (150) are soluble in acetone. Dimensional stability of the fibers at elevated temperatures increases, however, as the acrylonitrile content increases above 50%. Copolymers containing more than 50–55% acrylonitrile are insoluble in acetone but are soluble in other organic solvents. The use of dimethylacetamide (151), dimethylformamide (145,152), dimethyl sulfoxide (153), and acetonitrile (154) is described. Aqueous thiocyanate (155) and nitric acid (156) are also reported. Inspection of the patent literature indicates that producers often employ a common solvent for both acrylic and modacrylic fibers. Spinning solution preparation procedures are similar to those discussed earlier for acrylic fibers. In most cases, solution additives are similar to those described for acrylic fibers.

Fiber Production.   Modacrylic fibers are spun by conventional wet and dry spinning techniques. The equipment, the factors involved in process development and product optimization, and the sequence of process steps are generally the same as those discussed under spinning of acrylic fibers. Two of the solvents mentioned above for modacrylic polymers, acetone and acetonitrile, are both low boiling and can be used for either wet spinning or dry spinning. Acetone's low boiling point and low flash temperature may cause difficulty in complying with federal regulations for handling flammable solvents (157). In dry spinning, inert gases can be used to keep the composition of the tower gas outside the explosive limits. Control of the conditions above wet spinning baths and process equipment is more difficult.

Very little has been published on unique features of spinning processes for modacrylic fibers. A review article (140) discussed the importance of the orientation and relaxation steps in processing of a Dynel composition. Generally, the lower glass-transition temperature of the modacrylics allows the use of lower temperatures for orientation and relaxation. The more thermoplastic modacrylics may be partially relaxed by heat treatment at constant length, but some shrinkage is usually employed. A number of patents have described process improvements to minimize delustering during dyeing. These often involve changes in polymer composition, plus carefully defined spinning conditions. In one example (158), a copolymer containing an N-alkyl acrylamide was added to the spinning solution of the modacrylic polymer and the fiber was coagulated in a normal solvent water bath. The coagulated fiber was not homogeneous, but by drawing 1.2 to 2 times in the presence of solvent, heating with solvent present during drying, redrawing, and relaxing in dry heat, a lustrous fiber was produced which delustered very little in boiling water. Another example (154) describes the addition of a copolymer containing glycidyl methacrylate to the modacrylic spinning solution. The fiber was coagulated in multiple baths where the solvent concentration was much higher in the second bath than in the first. The use of nonaqueous coagulants and the production of bicomponent (159) modacrylic fibers have also been revealed.

Uses

As a result of flammability legislation and the improvement in their textile performance, modacrylics are now used in a wide range of fabrics. The largest volume sales are in the apparel and home furnishings markets. Wig and industrial applications are also important.

Children's sleepwear is a major application for modacrylics in the apparel market. The warmth, softness, and resiliency of modacrylics make them especially desirable for garments of this type. They are also used in adult robes and sleepwear where flame retardancy is desired. Blends with other fibers, especially polyester, offer additional advantages. The blend fabrics have superior abrasion resistance and wash and wear performance, as well as adequate flame resistance. One producer sells preblended mixtures of modacrylic and polyester fibers (142).

Pile fabrics are the second major apparel use for modacrylic fibers. These sliver-knit products are used as linings for coats and jackets or in outerwear for coats, parkas, collars, and other accessories. Modacrylics have long held a share of this market because of their flame resistance, luster, and controllable response to polishing, the final step in fabric finishing. Pile fabrics are polished by contact with a heated roll which removes crimp in the fiber tips and may also cause fiber shrinkage. Modacrylic fibers are produced with varied degrees of crimp stability for this end use. The standard fibers will not shrink at polishing temperatures but high shrinkage variants are available. As in the case of acrylic fibers, producer-dyed modacrylics are especially desirable for pile fabrics. Blends of acrylic and modacrylic fibers are often used.

In the home furnishings market, modacrylic fibers are used for draperies and curtains which must pass flame-retardant standards. Their superior drape, good dimensional stability, easy soil and stain release, and colorfastness make modacrylics highly desirable for these applications. Modacrylics are blended with other fibers in scatter rugs and carpets to provide increased flame resistance and desirable esthetics. Proposed standards of flame resistance for upholstery may also increase the use of modacrylics for this purpose.

Modacrylic fibers are used in a number of industrial applications where their flame retardancy, chemical resistance, and other properties are desirable. Paint rollers, battery separators, and protective uniforms for industrial workers are specific examples. Modacrylic fabrics may also be molded to form shaped articles such as speaker grilles.

Wigs are an important specialty end use for modacrylic fibers. Their luster, hand, heat-settability, and curl retention make the modacrylics an excellent substitute for human hair. Wig fibers are generally producer-colored and sold in a small tow form (see Wigs).

Fiber Prices

List prices of typical acrylic and modacrylic fibers from 1970 through 1976 are shown in Table 7. Each column is for a specific fiber type from one producer. The relative stability of fiber prices is noteworthy because raw material, labor, and utility costs have increased substantially since 1970. Producers have offset these increases to some extent by improved process technology and higher productivity.

Table 7.   Acrylic and Modacrylic Fiber Prices ($/kg) [a]

Year	Acrylic		Modacrylic	
	Apparel[b]	Carpet[c]	Apparel[b]	Carpet[d]
1970	1.96	1.41	1.72	1.43
1971	1.72	1.41	1.65	1.43
1972	1.72	1.41	1.65	1.43
1973	1.72	1.41	1.65	1.43
1974	1.76	1.37	1.65	1.52
1975	1.76	1.42	1.76	1.63
1976	1.90	1.42		

[a] Source for the years 1970–1975: is reference 160.
[b] Three-denier staple.
[c] Fifteen-denier staple.
[d] Sixteen-denier staple.

BIBLIOGRAPHY

"Textile Fibers" in *ECT* 1st ed., Vol. 13: "Acrilan, Orlon, X-51," pp. 824–830, by P. M. Levin, E. I. du Pont de Nemours & Co., Inc.; "Dynel and Vinyon," pp. 831–836, by H. L. Carolan, Union Carbide and Carbon Corporation. "Acrylic and Modacrylic Fibers" in *ECT* 2nd ed., Vol. 1, pp. 313–338, by D. W. Chaney, Chemstrand Research Center, Inc.

1. *Rules and Regulations Under the Textile Fiber Products Identification Act, effective March 3, 1960,* U.S. Federal Trade Commission, Washington, D. C., 1960, p. 4.
2. *Man-made fibres-Generic names, ISO-2076, 1977 (E),* International Organization for Standardization, Switzerland.
3. U.S. Pat. 2,140,921 (Dec. 20, 1938), H. Rein (to I. G. Farbenindustrie).
4. R. C. Houtz, *Text. Res. J.* **20,** 786 (1950).
5. *Textile Organon* **32,** 88 (1961).
6. *Textile Organon* **37,** 98 (1966).

7. *Textile Organon* **47,** 69 (1976).
8. J. W. S. Hearle and R. Greer, *Text. Prog.* **2**(4), 88 (1970).
9. R. Meredith, *Text. Prog.* **7**(4), 21 (1975).
10. C. R. Bohn, J. R. Schaefgen, and W. O. Statton, *J. Polym. Sci.* **55,** 540 (1961).
11. W. R. Krigbaum and N. Tokita, *J. Polym. Sci.* **43,** 467 (1960).
12. B. G. Colvin and P. Storr, *Eur. Polym. J.* **10,** 337 (1974).
13. M. Takahashi and Y. Nukushina, *J. Polym. Sci.* **56,** 519 (1962).
14. S. Minami, *Appl. Polym. Symp.* **25,** 145 (1974).
15. W. E. Morton and J. W. S. Hearle, *Physical Properties of Textile Fibers,* 2nd ed., John Wiley & Sons, Inc., New York, 1975, p. 643.
16. Ref. 15, pp. 265–305.
17. *Bulletin OR-185, Shrinkage of Orlon Acrylic Fiber,* E. I. du Pont de Nemours & Co., Inc., Mar. 1975.
18. *Bulletin AP-14, The Properties of Acrilan-16,* Monsanto, Aug. 1965.
19. *ASTM D-2859 70T, Annual Book of ASTM Standards,* Pt 32, American Society of Testing and Materials, Easton, Md., 1975.
20. *Bulletin TT-54, Acrilan Acrylic Carpet Staple,* Monsanto, Aug. 1976.
21. *Acrylics Outdoors,* bulletin, Monsanto, Oct. 1967.
22. *Proceedings, Symposium on Polypropylene Fibers, Sept. 17–18, 1964,* sponsored by Southern Research Institute.
23. R. H. Heidner and M. E. Gibson, "Acrylic and Modacrylic Fibers," in F. S. Snell and C. L. Hilton, eds., *Encyclopedia of Industrial Chemical Analysis,* Vol. 4, Interscience Publishers, a division of John Wiley & Sons, Inc., New York, 1967, pp. 219–360.
24. G. C. East, *Text. Prog.* **3**(1), 67 (1971).
25. J. R. Kirby and A. J. Baldwin, *Anal. Chem.* **40**(4), 689 (1968).
26. J. E. Ford, G. Pearson, and R. M. Smith, eds., *Identification of Textile Materials,* 6th ed., Textile Institute, Manchester, Eng., 1970.
27. "AATCC Test Method 20-1973, Fibers in Textiles: Identification," *Technical Manual* **50,** 50 (1974).
28. R. W. Moncrieff, *Man-Made Fibers,* 6th ed., John Wiley & Sons, Inc., New York, 1975.
29. M. Stratmann, *Z. Gesamte Textilind.* **72,** 13 (1970).
30. P. Ellwood, *Chem. Eng.* **76,** 90 (1969).
31. H. Burger, V. Grobe, E. Peter, and H. Schönherr, *Faserforsch. Textiletech.* **18,** 503 (1967).
32. U.S. Pat. 3,787,365 (Jan. 22, 1974), P. Melacini, L. Patron, A. Moretti, and R. Tedesco (to Montefibre).
33. U.S. Pat. 3,821,178 (June 28, 1974), P. Melacini, L. Patron, A. Moretti, and R. Tedesco (to Montefibre).
34. U.S. Pat. 3,879,360 (Apr. 22, 1975), L. Patron, A. Moretti, R. Tedesco, and R. Pasqualetto (to Montefibre).
35. F. S. Dainton and co-workers, *J. Polym. Sci.* **34,** 209 (1959).
36. L. H. Peebles, Jr., *J. Appl. Polym. Sci.* **17,** 113 (1973).
37. R. L. Cleland and W. H. Stockmayer, *J. Polym. Sci.* **18,** 473 (1955).
38. H. N. Friedlander, L. H. Peebles, Jr., J. Brandrup, and J. R. Kirby, *Macromolecules* **1,** 79 (1968).
39. P. E. Morgan, *Text. Prog.* **8,** 72 (1976).
40. W. Fester, "Physical Constants of Polyacrylonitrile," in J. Brandrup and E. H. Immergut, eds., *Polymer Handbook,* 2nd ed., John Wiley & Sons, Inc., New York, 1975, p. V38.
41. U.S. Pat. 3,251,796 (May 17, 1966), W. Saar, E. N. Petersen, and W. Irion (to Phrix-Werke).
42. U.S. Pat. 2,605,246 (1952), A. Cresswell and P. W. Cummings (to American Cyanamid).
43. U.S. Pat. 3,630,986 (Dec. 28, 1971), A. Mison and P. Tarbouriech (to Rhone-Poulenc).
44. U.S. Pat. 3,969,305 (July 13, 1976), A. A. Armstrong (to Monsanto).
45. U.S. Pat. 3,010,932 (Nov. 28, 1961), R. S. Stoveken (to E. I. du Pont de Nemours & Co., Inc.).
46. U.S. Pat. 3,847,864 (Nov. 12, 1974), J. D. Chase and R. L. Potter (to American Cyanamid).
47. U.S. Pat. 3,657,179 (Apr. 18, 1972), P. C. Yates (to E. I. du Pont de Nemours & Co., Inc.).
48. U.S. Pat. 3,383,350 (May 14, 1968), B. M. Pettyjohn (to E. I. du Pont de Nemours & Co., Inc.).
49. U.S. Pat. 3,784,511 (Jan. 8, 1974), J. R. Kirby (to Monsanto).
50. U.S. Pat. 3,784,512 (Jan. 8, 1974), J. C. Masson (to Monsanto).
51. Neth. Pat. Appl. 65-1113 (Feb. 28, 1966), (to Farbenfabriken Bayer).
52. U.S. Pat. 3,436,364 (Apr. 1, 1969), H. Logemann, E. Roos, and C. Suling (to Farbenfabriken bayer).

53. U.S. Pat. 3,296,171 (Jan. 3, 1967), J. H. Hennes and C. R. Pfeifer (to The Dow Chemical Co.).
54. J. P. Craig, J. P. Knudsen, and V. F. Holland, *Text. Res. J.* **32,** 435 (1962).
55. J. P. Knudsen, *Text. Res. J.* **33,** 13 (1963).
56. A. Ziabicki, *Sen'i Gakkaishi* **26,** 158 (1970).
57. D. R. Paul, *J. Appl. Polym. Sci.* **11,** 439 (1967).
58. D. R. Paul, *J. Appl. Polym. Sci.* **11,** 1719 (1967).
59. U.S. Pat. 2,907,096 (Oct. 6, 1959), P. Halbig.
60. U.S. Pat. 3,491,179 (Jan. 20, 1970), S. N. Chinai and L. H. Schwind (to American Cyanamid).
61. K. Terada, *Sen'i Gakkaishi* **29,** T-345 (1973).
62. K. Terada, *Sen'i Gakkaishi* **29,** T-451 (1973).
63. D. R. Paul, *J. Appl. Polym. Sci.* **12,** 383 (1968).
64. V. Groebe and H. Heyer, *Faserforsch. Textiletech.* **19,** 313 (1968).
65. U.S. Pat. 3,346,685 (Oct. 10, 1967), R. D. Crozier, R. E. Harder, J. H. Hood, and R. B. Hurley (to The Dow Chemical Co.).
66. U.S. Pat. 3,147,322 (Sept. 1, 1964), Y. Fujisaki, C. Nakayama, and H. Kobayashi (to Asahi Kasei Kogyo).
67. U.S. Pat. 3,944,384 (Mar. 16, 1976), D. J. Poynton (to Courtaulds Limited).
68. W. E. Fitzgerald and J. P. Craig, *Appl. Polym. Symp.* **6,** 67 (1967).
69. D. R. Paul, *J. Appl. Sci.* **12,** 2273 (1968).
70. C. D. Han and L. Segal, *J. Appl. Polym. Sci.* **14,** 2999 (1970).
71. D. R. Paul, *J. Appl. Polym. Sci.* **13,** 817 (1969).
72. D. R. Paul and A. L. McPeters, *J. Appl. Polym. Sci.,* to be published.
73. C. D. Han and J. Y. Park, *J. Appl. Polym. Sci.* **17,** 187 (1973).
74. U.S. Pat. 3,621,087 (Nov. 16, 1971), M. Shimamura and co-workers, (to Toyo Rayon).
75. U.S. Pat. 3,801,691 (Apr. 2, 1974), E. Brigmanis (to American Cyanamid).
76. U.S. Pat. 3,786,125 (Jan. 15, 1974), K. Shimoda and T. Kusumose (to Japan Exlan).
77. U.S. Pat. 2,426,719 (Sept. 2, 1947), W. W. Watkins (to E. I. du Pont de Nemours & Co., Inc.).
78. U.S. Pat. 3,124,629 (Mar. 10, 1964), J. P. Knudsen (to Monsanto).
79. U.S. Pat. 3,088,188 (May 7, 1963), J. P. Knudsen (to Monsanto).
80. W. Dohrn, *Faserforsch. Textiletech.* **25,** 28 (1974).
81. U.S. Pat. 3,758,659 (Sept. 11, 1973), H. Takeda (to Toray).
82. U.S. Pat. 3,449,485 (June 10, 1969), R. M. Costa and F. Codignola (to Societa Italiana Resine).
83. M. Zwick, *Appl. Polym. Sym.* **6,** 109 (1967).
84. U.S. Pat. 2,957,748 (Oct. 25, 1960), F. Lieseberg (to Badische-Aniline).
85. U.S. Pat. 3,080,210 (Mar. 5, 1963), P. A. Ucci (to Monsanto).
86. U.S. Pat. 3,523,150 (Aug. 4, 1970), R. E. Vigneault (to Monsanto).
87. U.S. Pat. 3,701,820 (Oct. 31, 1972), K. Kuratani and K. Fukushima (to Japan Exlan).
88. U.S. Pat. 2,954,271 (Sept. 27, 1960), L. Cenzato (to E. I. du Pont de Nemours & Co., Inc.).
89. U.S. Pat. 2,811,409 (Oct. 29, 1957), J. W. Clapp and C. B. Mather (to Eastman Kodak).
90. U.S. Pat. 2,975,022 (Mar. 14, 1961), R. D. Euler (to E. I. du Pont de Nemours & Co., Inc.).
91. U.S. Pat. 3,767,360 (Oct. 23, 1973), P. P. Singh (to E. I. du Pont de Nemours & Co., Inc.).
92. U.S. Pat. 3,725,523 (Apr. 3, 1973), B. C. Bowen (to E. I. du Pont de Nemours & Co., Inc.).
93. U.S. Pat. 3,353,381 (Nov. 21, 1967), E. A. Taylor, Jr. (to Monsanto).
94. U.S. Pat. 3,791,788 (Feb. 12, 1974), E. A. Taylor (to Monsanto).
95. Jpn. Pat. JA 7328969 (Sept. 6, 1973), (to Toray).
96. A. L. McPeters and D. R. Paul, *Appl. Polym. Symp.* **25,** 159 (1974).
97. U.S. Pat. 3,932,571 (Jan. 13, 1976), L. O. Dworjanyn (to E. I. du Pont de Nemours & Co., Inc.).
98. U.S. Pat. 3,944,386 (Mar. 16, 1976), W. K. Wilkinson (to E. I. du Pont de Nemours & Co., Inc.).
99. U.S. Published Pat. Appl. B 458,060 (Apr. 4, 1974), U. Reinehi, A. Nogaj, and G. Holzing (to Bayer).
100. U.S. Pat. 3,483,576 (Dec. 16, 1969), K. Nakagawa and K. Kuratani (to American Cyanamid).
101. U.S. Pat. 3,111,357 (Feb. 19, 1963), A. R. Wirth, S. A. Murdock, and G. B. Berry (to The Dow Chemical Co.).
102. U.S. Pat. 3,907,498 (Sept. 23, 1975), D. F. Bittle and A. L. McPeters (to Monsanto).
103. Z. A. Zgibneva and co-workers, *Fibre Chem.* **5**(6), 659 (1973).
104. U.S. Pat. 3,544,262 (Dec. 1, 1970), S. O. Harris and G. K. Miller (to American Cyanamid).
105. U.S. Pat. 3,296,341 (Jan. 3, 1967), H. P. Briar and R. P. Hurley (to The Dow Chemical Co.).
106. U.S. Pat. 3,281,260 (Oct. 25, 1966), P. A. Ucci (to Monsanto).
107. U.S. Pat. 2,984,912 (May 23, 1961), T. H. Robertson and G. K. Klausner (to American Cyanamid).
108. J. P. Bell and J. H. Dumbleton, *Text. Res. J.* **41,** 196 (1971).

109. K. Terada, *Sen'i Gakkaishi* **28,** 504 (1972).
110. K. Terada, *Sen'i Gakkaishi* **29,** T-120 (1973).
111. P. H. Hobson, *Faserforsch Textiltech.* **22,** 80 (1971).
112. U.S. Pat. 3,895,908 (July 22, 1975), R. E. Harder and L. B. Ticknor (to Dow Badische).
113. W. E. Fitzgerald and J. P. Knudsen, *Papers of the 51st Annual Conference of the Textile Institute 1966,* the Textile Institute, Manchester, Eng., 1966, p. 134.
114. E. M. Hicks, Jr., J. F. Ryan, Jr., R. B. Taylor, Jr., and R. L. Tichenor, *Text. Res. J.* **30,** 675 (1960).
115. B. S. Gupta and W. George, *Text. Res. J.* **45,** 338 (1975).
116. U.S. Pat. 3,039,525 (June 19, 1962), L. H. Belck and K. G. Siedschlag, Jr. (to E. I. du Pont de Nemours & Co., Inc.).
117. U.S. Pat. 3,182,106 (May 4, 1965), Y. Fujita, K. Shimoda, and K. Zoda (to American Cyanamid).
118. W. E. Fitzgerald and J. P. Knudsen, *Text. Res. J.* **37,** 447 (1967).
119. U.S. Pat. 3,295,552 (Jan. 3, 1967), F. B. Powell and B. K. Polk (to Monsanto).
120. U.S. Pat. 3,038,236 (June 12, 1962), A. L. Breen (to E. I. du Pont de Nemours & Co., Inc.).
121. U.S. Pat. 3,547,763 (Dec. 15, 1970), H. A. Hoffman, Jr. (to E. I. du Pont de Nemours & Co., Inc.).
122. U.S. Pat. 3,864,447 (Feb. 4, 1975), H. Sekiguchi, N. Tsutsui, and T. Sumi (to Japan Exlan).
123. U.S. Pat. 3,737,504 (June 5, 1973), P. Herrbach and A. Breton (to CTA-Compagnie).
124. Neth. Pat. 1,369,951 (Oct. 9, 1974), H. Monmaerts and A. E. Dubois (to AKZO).
125. *Chem. Eng. News* **54**(50), 22 (1976).
126. Jpn. Pat. Appl. J5 0029-822 (Mar. 25, 1975) (to Toho Beslon).
127. U.S. Pat. 3,082,056 (Mar. 19, 1963), J. R. Caldwell (to Eastman Kodak).
128. Can. Pat. 907,771 (Aug. 15, 1972), G. A. Serad (to Celanese).
129. U.S. Pat. 3,655,857 (Apr. 11, 1972), T. C. Bohrer and A. E. Champ (to Celanese).
130. U.S. Pat. 3,896,204 (July 22, 1975), A. Goodman and M. A. Suwyn (to E. I. du Pont de Nemours & Co., Inc.).
131. Fr. Pat. 2,216,372 (Aug. 30, 1974), H. Porosoff (to American Cyanamid).
132. W. Beckman, *Z. Gesamte Textil Ind.* **71,** 603 (1969); see also R. H. Peters, *Textile Chemistry,* Vol. III, Elsevier Scientific Publishing Co. Inc., New York, 1975, pp. 549–579.
133. Y. Tsuda, *J. Appl. Polym. Sci.* **5,** 104 (1961).
134. F. Schlaippi, *Present and Future Developments in Transfer Printing, paper presented at 46th TRI Annual Research and Technology Conference, 1976.*
135. E. W Rugeley, T. A. Feild, Jr., and G. H. Fremon, *Ind. Eng. Chem.* **40,** 1724 (1948).
136. F. Adams, *Rayon Synth. Text.* **31**(1), 63 (1950).
137. H. W. Coover, Jr., W. R. Ivey, and W. C. Wooten, *Text. Res. J.* **27,** 745 (1957).
138. *Textile Organon* **27,** 93 (1956).
139. *Daily News Record* (31), 1, 31 (1962).
140. R. K. Kennedy, "Modacrylic Fibers," in H. F. Mark, N. G. Gaylord, and N. M. Bikales, eds., *Encyclopedia of Polymer Science and Technology,* Vol. 8, Interscience Publishers, a division of John Wiley & Sons, Inc., New York, 1968, p. 812.
141. D. Machalaba, *Daily News Record* (4), 1,15 (Dec. 12, 1974).
142. *Bulletin M-8, Monsanto Modacrylic Fibers,* Monsanto, 1975.
143. Brit. Pat. 1,090,580 (Dec. 30, 1965) (to Monsanto).
144. Jpn. Pat. Appl. 7,131,807 (May 9, 1968) (to Mitsubishi).
145. U.S. Pat. 3,748,302 (Nov. 17, 1971), E. B. Jones (to E. I. du Pont de Nemours & Co., Inc.).
146. West Ger. Pat. Appl. 2,128,004 (June 5, 1971) (to Bayer).
147. Brit. Pat. 1,130,000 (Dec. 30, 1965) (to Monsanto).
148. West Ger. Pat. Appl. 2,043,402 (Sept. 2, 1970) (to Bayer).
149. Brit. Pat. 1,271,334 (Apr. 19, 1972) (to Toray).
150. J. Atkinson, *Text. Prog.* **8,** 65 (1976).
151. Brit. Pat. 1,099,845 (Jan. 17, 1968) (to Monsanto).
152. West Ger. Pat. Appl. 2,043,402 (Sept. 2, 1970) (to Bayer).
153. Jpn. Pat. Appl. 72,08771 (Oct. 11, 1968) (to Toray).
154. U.S. Pat. 3,943,223 (Mar. 9, 1976), K. Kozuka and co-workers, (to Kanegafuchi).
155. West Ger. Pat. Appl. 2,424,599 (Japan, May 29, 1973) (to Exlan).
156. Jpn. Pat. Appl. 49,007524 (May 31, 1972) (to Asahi).
157. "Standard for Dip Tanks Containing Flammable or Combustible Liquids," in *National Fire Codes,* Vol. 1, National Fire Protection Association, Boston, Mass., 1968–1969, Section 34.
158. Fr. Pat. 2,067,642 (Aug. 20, 1971) (to Kanegafuchi).
159. U.S. Pat. 3,751,332 (Aug. 7, 1973) F. F. Chen (to Eastman Kodak).
160. "Acrylic and Modacrylic Fibers," *Chemical Economics Handbook,* Stanford Research Institute, Menlo Park, Calif., 1975, p. 543.3524R.

General References

P. H. Hobson, "Acrylic and Modacrylic Fibers: Development, Use, and Potential," *Text. Chem. Color.* **4,** 232 (1972).

D. J. Poynton, "Acrylic Fibers" and J. Atkinson, "Modacrylic Fibers," *Text. Prog.* 8(1), 51 (1976).

F. McNeirney, "1976 Fiber Supply Sources Directory," a list of United States acrylic and modacrylic fibers by type number, denier, length, luster, etc, *Mod. Text.* 24 (Sept. 1976).

E. Cernia, "Acrylic Fibers" and R. K. Kennedy, "Modacrylic Fibers," in H. F. Mark, S. M. Atlas, and E. Cernia, eds., *Man-Made Fibers Science and Technology,* Vol. 3, John Wiley & Sons, Inc., New York, pp. 135–243.

P. H. HOBSON
A. L. MCPETERS
Monsanto Triangle Park Development Center, Inc.

ARAMID FIBERS

Aromatic polyamides are formed by reactions that lead to the formation of amide linkages between aromatic rings. In practice this generally means the reaction of aromatic diamines and aromatic diacid chlorides in an amide solvent. From solutions of these polymers it is possible to produce fibers of exceptional heat and flame resistance and fibers of good to quite remarkable tensile strength and modulus. Because the physical property differences between fibers of aromatic and aliphatic polyamides are greater than those between other existing generic classes of fibers, a new generic term for fibers from aromatic polyamides was requested by the DuPont Company in 1971. Subsequently, the generic term aramid was adopted (1974) by the United States Federal Trade Commission for designating fibers of the aromatic polyamide type: aramid—a manufactured fiber in which the fiber-forming substance is a long-chain synthetic polyamide in which at least 85% of the amide (—CO—NH—) linkages are attached directly to two aromatic rings.

At the same time that the new generic term became effective, the generic term for nylon was amended as follows: nylon—a manufactured fiber in which the fiber-forming substance is a long-chain synthetic polyamide in which less than 85% of the amide (—CO—NH—) linkages are attached directly to two aromatic rings.

Aromatic polyamide–imides have also been considered as aramids by the ISO which has proposed the definition: synthetic linear macromolecules made from aromatic groups joined by amide linkages, in which at least 85% of the amide linkages are joined directly to two aromatic rings and in which imide groups may be substituted for up to 50% of the amide groups. In this article, amide–imides are not discussed as aramid fibers and only those polymers having amide condensation linkages are reviewed (see also Polyamides; Heat-resistant polymers; Flame-retardant textiles).

There is no recognized systematic nomenclature for aromatic polyamides as exists for aliphatic polyamides. However, acronyms or the initial letters of the various monomers are frequently combined (see equations 1 and 3), with the initials of the diamine moieties preceding those of the diacid moieties. The letters T and I have the commonly-accepted significance of standing for, respectively, terephthalamide and isophthalamide.

The first aramid fiber to be developed, known experimentally as HT-1 (1) and almost certainly based on poly(*m*-phenyleneisophthalamide), was commercialized under the trademark Nomex by the DuPont Company in 1967. This fiber was introduced for applications requiring heat resistance far in excess of that possessed by conventional synthetic fibers. In Japan, Teijin introduced Conex in the early 1970s and has disclosed (2) the composition to be poly(*m*-phenyleneisophthalamide), ie, apparently the same as that used for Nomex. Conex production, however, appears to be quite low at present, and is probably only a fraction of that of Nomex. Fenilon (or Phenylone), also based on MPD-I is produced in the Soviet Union for both civilian and military as well as space exploration needs. Although the quantity of Fenilon produced has not been announced, it is almost certain that production is less than 4500 metric tons per year.

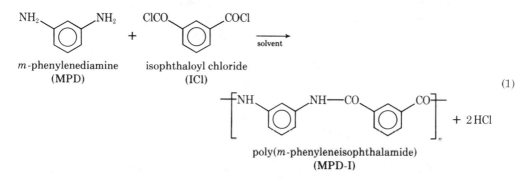

m-phenylenediamine   isophthaloyl chloride
(MPD)   (ICl)

(1)

poly(*m*-phenyleneisophthalamide)
(MPD-I)

An aramid fiber of high strength and exceptionally high modulus was introduced by the DuPont Company in 1970 for use in tires under the experimental name of Fiber B (3). This early high modulus aramid tire cord (qv) appears to have been based on poly(*p*-benzamide) (PPB) spun from an organic solvent. Later, another version of

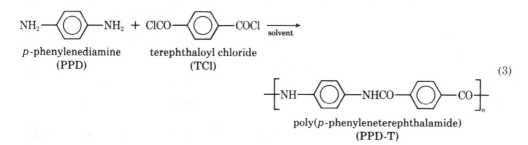

(2)

poly(*p*-benzamide)
(PPB)

Fiber B, apparently based on poly(*p*-phenyleneterephthalamide) (PPD-T) and probably spun from sulfuric acid, was introduced (4). This second generation fiber was of considerably higher strength (by nearly two-fold) than that of Fiber B.

p-phenylenediamine   terephthaloyl chloride
(PPD)   (TCl)

(3)

poly(*p*-phenyleneterephthalamide)
(PPD-T)

An even higher modulus fiber intended for use in rigid composites was introduced under the experimental name of PRD-49; it too is believed to be composed essentially of PPD-T, the higher modulus of PRD-49 being achieved through hot-drawing of this fiber. Upon announcement of construction of commercial facilities for these fibers, the trademarks Kevlar-29 and Kevlar-49, respectively, were introduced. In early 1975, the Akzo Company of the Netherlands announced their intention to commercialize an aramid fiber under the name Arenka; this fiber too is probably based on PPD-T.

Properties

Aramid fibers do not melt in the conventional sense because decomposition generally occurs simultaneously. An endothermic peak in the differential thermal analysis (DTA) thermogram for aramid fibers can be obtained and the values are generally >400°C and can be as high as 550°C. Glass transitions range from ca 250 to >400°C. Weight loss, as determined by thermogravimetric analysis (tga) in an inert gas, begins at about 425°C for most aramids although some of the rod-like polymers do not lose substantial weight to 550°C.

Aramid fibers are characterized by medium to ultra-high tensile strength, medium to low elongation, and moderately high modulus to ultra-high modulus. Most of these fibers have been reported to be either highly crystalline (5–6) or crystallizable (7) and densities for crystalline fibers range from about 1.35 to 1.45 g/cm^3. Poly(m-phenyleneisophthalamide) fiber of low orientation has a density of ca 1.35 g/cm^3 and hot-drawn fiber has a density of ca 1.38 g/cm^3; Kevlar, poly(p-phenyleneterephthalamide), has a density of ca 1.45 g/cm^3.

Since most of the work on aramid fibers has been directed toward the development of heat and flame-resistant fibers, and ultra high-strength–high-modulus fibers, it is possible to categorize them in these terms. In general, polymers useful for the former contain a high portion of meta-oriented phenylene rings and polymers useful for the latter fibers contain principally para-oriented phenylene rings. The meta-oriented polymers are generally considered to be chain-folding polymers. On the other hand, small angle x-ray diffraction studies of the rod-like para-oriented polymers show no evidence of chain-folding for this class of polymers. However, for purposes of discussion, the two classes of polymers will be referred to here as heat and flame-resistant fibers and high-strength–high-modulus fibers, respectively.

Heat and Flame-Resistant Aramid Fibers. The fiber properties of some aramid fibers having medium strength and elongation-to-break (expressed in %) are given in Table 1. No attempt has been made to present data for all of the fibers reported in the literature. Only the commercially important or potentially important compositions and fibers of those compositions that show significantly different properties from the simple aramids are considered. Although random copolyamides (eg, mixed isophthalamides and terephthalamides) have been reported, no clearcut advantage has been demonstrated for fibers from such polymers. On the contrary, lower softening points and lower crystallinity are usually observed.

Tensile Properties at Elevated Temperatures. The tensile properties at elevated temperatures of selected aramid fibers are given in Table 2. These fibers have tensile strengths at 250°C that are characteristic of conventional textile fibers at room temperature, and have useful tenacities to >300°C. In contrast, nylon-6,6 fiber loses almost all of its strength at about 205°C. The tensile moduli of the aramid fibers fall off considerably at 300–350°C; the modulus values are still substantial, however, even

Table 1. Tensile Properties of Selected Heat and Flame Resistant Aramid Fibers

Structure	CAS Registry No.	Mol formula	$\eta_{inh}{}^a$	Method of prep^b	Spun from^c	Spinning method^d	T/E/M$_i{}^e$	Ref.
AB Type								
(1)	[25735-77-7]	$(C_7H_5NO)_n$	1.25	S	DMAC–LiCl	D	0.51/10/10.4	8
				S	DMAC–LiCl	W	0.35/14/	9
AA–BB Type								
(2)	[24938-60-1]	$(C_{14}H_{10}N_2O_2)_n$	1.86	S	DMAC–CaCl$_2$	D	0.49/28/10.0	10
			2.11	S	DMAC–CaCl$_2$	W	0.69/24/10.8	11
(3)	[31808-02-3]	$(C_{15}H_{10}N_2O_4)_n$	2.01	S	DMAC	W	0.35/9.0/7.1	12
(4)	[25670-09-1]	$(C_{22}H_{18}N_2O_2)_n$		S	DMSO	D	0.81/6.8/16.1	13
(5)	[26026-92-6]	$(C_{20}H_{14}N_2O_3)_n$	2.32^f				0.37/8.5/7.9	14
(6)	[26854-93-3]	$(C_{30}H_{14}N_2O_3)_n$	1.54^f				0.62/6.0/10.6	14
(7)	[25667-73-6]	$(C_{21}H_{16}N_2O_2)_n$	2.51	S	DMF	D	0.36/20/	15
(8)	[32027-57-9]	$(C_{28}H_{20}N_2O_3)_n$	2.84	S	DMAC–LiCl	D	0.32/22/2.6	16

35

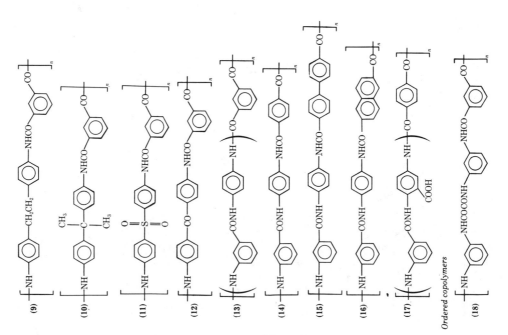

Ordered copolymers

	CAS No.	Formula			Solvent			
(9)	[31986-64-8]	(C$_{22}$H$_{18}$N$_2$O$_2$)$_n$		S	DMAC–LiCl	W	0.33/30/4.6	15
(10)	[27880-27-9]	(C$_{23}$H$_{20}$N$_2$O$_2$)$_n$	1.80	I	DMF	D	0.32/28/	
(11)	[26100-95-8]	(C$_{20}$H$_{14}$N$_2$O$_4$S)$_n$					0.35/17/4.1	17
(12)	[26026-95-9]	(C$_{21}$H$_{14}$N$_2$O$_3$)$_n$	1.26	S		D	0.27/23/2.5	16
(13)	[29153-49-9]	(C$_{21}$H$_{15}$N$_3$O$_3$)$_n$	2.1	S		W	0.50/23/6.8	18
(14)	[52303-33-0]	(C$_{21}$H$_{15}$N$_3$O$_3$)$_n$	2.7	S		W	0.47/20/7.4	18
(15)	[52303-35-2]	(C$_{27}$H$_{19}$N$_3$O$_3$)$_n$	2.0	S		W	0.24/10/7.5	18
(16)	[52303-34-1]	(C$_{25}$H$_{17}$N$_3$O$_3$)$_n$	2.4	S		W	0.38/41/8.6	18
(17)	[65749-41-9]	(C$_{22}$H$_{15}$N$_3$O$_5$)$_n$	1.97	S	DMAC	W	0.46/8.8/10.5	19
(18)	[65749-42-0]	(C$_{22}$H$_{16}$N$_4$O)$_n$	2.48	S	DMAC–LiCl	W	0.26/24/4.7	

Table 1 (*continued*)

Structure	CAS Registry No.	Mol formula	$\eta_{inh}{}^a$	Method of prep ^{b}	Spun from ^{c}	Spinning method ^{d}	$T/E/M_i{}^e$	Ref.
(19) [structure]	[65749-43-1]	$(C_{28}H_{20}N_4O_4)_n$		S	DMAC–CaCl$_2$	D	0.36/33/7.9	10
(20) [structure]	[27290-59-1] and [27307-23-9]	$(C_{28}H_{20}N_4O_4)_n$		S	DMAC–LiCl	W	0.45/37/7.9	20
(21) [structure]	[25035-07-8]	$(C_{28}H_{20}N_4O_4)_n$		S		W	0.53/23/8.9	21
(22) [structure]	[65749-44-2]	$(C_{34}H_{24}N_4O_4)_n$		S		W	0.52/16/8.2	21
(23) [structure]	[32195-67-8]	$(C_{22}H_{22}N_4O_4)_n$		S		W	0.56/19/8.3	21
(24) [structure]	[32195-68-9]	$(C_{28}H_{20}N_4O_4)_n$		S		W	0.55/20/6.2	21
Copolymers of limited order								
(25) [structure]	[29153-28-6]	$(C_{15}H_{12}N_2O_2)_n$	1.37	I	DMAC–LiCl		0.12/96/2.8	15
(26) [structure]	[59789-54-7]	$(C_{16}H_{13}N_3O_3)_n$	1.33	S	DMAC–LiCl	W	0.26/26/4.7	

^{a} Determined in sulfuric acid at 30°C on a 0.5 g/100 mL solution. $\eta_{inh} = \ln \eta_{rel}/c$; $\eta_{rel} = \eta_{soln}/\eta_{solvent}$; c = conc of solvent.

^{b} S = solution polymerization; I = interfacial polymerization.

^{c} DMAC = dimethylacetamide; DMSO = dimethyl sulfoxide; DMF = dimethylformamide.

^{d} D = dry spinning; W = wet spinning.

^{e} T = tenacity, N/tex (= 11.33 gf/den); E = elongation-to-break, %; M_i = initial modulus, N/tex.

^{f} Specific viscosity on 0.5% solution.

Table 2. Tensile Properties of Aramid Fibers at Elevated Temperatures

Structure no.	T/E/M$_i$[a] at room temp	T/E/M$_i$[a] at 250°C	T/E/M$_i$[a] at 300°C	Zero strength temp[b], °C	Ref.
(1)	0.51/10/10.4		0.26/7.5/7.9	480–550	8
(2)	0.49/28/10.0	0.27/22/5.6	0.11/23/6.6	440	10
(4)	0.81/6.8/16.1		0.23/12/2.9		13
(19)	0.36/33/7.9	0.24/48/3.4	0.14/51/2.3	490	10
(20)	0.45/37/7.9		0.13/22/0.97	>400	20
(21)	0.53/23/8.9		0.18/11/4.4	485	21
(22)	0.52/16/8.2	0.14/15/5.1		>450	21
(23)	0.56/19/8.3		0.18/19/	455	21
(24)	0.55/20/6.2	0.29/21/4.1	0.20/19/3.2	470	21
(13)	0.42/27/6.4[c]	0.21/41/2.5			22
	0.76/28/[d]				

[a] T/E/M$_i$ = tenacity, N/tex; elongation-to-break, %; initial modulus, N/tex (= 11.33 gf/den).
[b] Temperature at which the fiber breaks under a load of 8.8 mN/tex (0.1 gf/den).
[c] Tensile properties of polymer of η_{inh} 1.1; fibers prepared by dry spinning.
[d] Tensile properties of polymer of η_{inh} 2.8; fibers prepared by wet spinning.

at these temperatures (Table 2). Surprisingly, in general the elongations of these fibers tend to be less at 300°C than at room temperature.

Tensile Properties after Prolonged Aging at Elevated Temperatures. It is important to be able to predict the long term performance of aramid fibers at elevated temperatures. For this purpose, evaluation of the tensile properties after heat-aging at intermediate temperatures is more meaningful than determining the tensile properties at greatly elevated temperatures. Heat-aging performance in air (thermo-oxidation) is a particularly severe test of stability because of the high surface-to-volume ratio of the fibers.

The effect of heat-aging at 300°C in air on the fibers is given in Table 3. In general, aramid fibers retain useful tensile properties for 1–2 weeks under the severe conditions indicated. Nomex in the form of paper has been reported to have a useful lifetime of only around 40 h at 300°C for a certain electrical application, but a useful lifetime of about 1400 h at 250°C. At 177°C, this fiber retained 80% of its strength after exposure to air for several thousand hours (23).

Table 3. Physical Properties of Aramid Fibers after Prolonged Exposure in Air at 300°C

Structure no.	T/E/M$_i$[a], d 0	1	7	14	35	Ref.
(1)	0.51/10/10.4				0.31/4/11.5	8
(2)	0.49/28/10.0	0.26/9.9/8.6	0.15/5.7/6.8	0.063/1.6/4.1		10
(19)	0.36/33/7.9	0.30/32/6.3	0.088/3.1/4.4	0.063/1.4/5.1		10
(20)	0.45/37/7.9	0.29/25/8.3	0.23/13/7.9			20
(21)	0.53/23/8.9	0.31/9.0/6.5	0.18/3.0/6.6			20
	0.48/15/11.5	0.29/14/8.8		0.16/5/8.8		24
(23)	0.49/25/7.9		0.11/2.5/4.7[b]			25
(24)	0.57/20/6.9	0.40/17/5.9	0.12/16/			21

[a] T/E/M$_i$ = tenacity, N/tex; elongation-to-break, %; initial modulus, N/tex (= 11.33 gf/den).
[b] Determined after nine days.

Heat-aging almost always results in a decrease in the elongation of the fibers, the loss being roughly proportional to the decrease in tenacity. Frequently, the limit of usefulness of a fiber, such as in flexible products, is even more dependent upon elongation than tenacity, so that loss of elongation is of greater concern.

Flame Resistance. Aramid fibers characteristically burn only with difficulty and they do not melt like nylon-6,6 or polyester fibers. They are useful in a number of applications requiring high flame resistance. Upon burning, the aramid fibers produce a thick char which acts as a thermal barrier and prevents serious burns to the skin. Limiting oxygen index (LOI) values, generally accepted as a measure of flame resistance for polymers, are quite high for aramid fibers (Table 4) (see Flame retardants). For purposes of comparison, Table 4 contains some LOI values for several fibers containing either alkylene chains or heterocycles in combination with arylene rings and amide linkages. From Table 4, it is interesting to compare the LOI values of poly[(5-carboxy-*m*-phenylene)isophthalamide] (3) and poly(*m*-phenyleneisophthalamide) (2), which have the same structure except for a pendant COOH group in the former. The LOI values given in Table 4 and those usually reported are for ignition from the top of the sample. Bottom ignition of fabric samples generally gives considerably lower LOI values. This fact has been dramatically illustrated by Sprague (27) who reported LOI values for Kevlar-49, Nomex, and Durette (X-400) fabrics (28) of 24.5, 26.0, and 36.0, respectively, when lighted from the top with a wick. LOI values for the same samples ignited from the bottom with a wick were only 16.5, 17.0, and 18.0, respectively.

Some aramid fibers, such as Nomex, shrink away from a flame or a high heat source. Durette fabrics, based on Nomex and treated with hot chlorine gas or other chemical reagents to promote surface cross-linking which stabilizes the fibers, were developed for greater dimensional stability (28). HT-4, an experimental aramid fiber from DuPont, has high LOI values and high dimensional stability in a flame (29). Although the structure of the HT-4 polymer has not been disclosed, patents assigned to the DuPont Company have described the spinning of PPD-T from sulfuric acid followed by incorporation of tetrakis(hydroxymethyl)phosphonium chloride (THPC) and subsequently cross-linking the THPC by means of melamine. PPD-T alone has high dimensional stability but an LOI value comparable to Nomex (ie, 28–30); addition of 1% phosphorus raises the LOI value to 40–42. A process which diffuses THPC and a melamine–formaldehyde resin (Aerotex UM) into never-dried Nomex followed by a heat treatment, has been described in a recent patent (30).

Electrical Properties. Wholly aromatic polyamides have high volume resistivities and high dielectric strengths and, significantly, they retain these properties at elevated temperatures to a high degree. Accordingly, they have considerable potential as high temperature dielectrics particularly for use in motors and transformers. The high temperature electrical properties of aramid fibers are exemplified by breakdown voltages of 76 V/mm at temperatures up to 180°C. By comparison, the breakdown voltage for poly(tetrafluoroethylene) is only 57 V/mm at 150°C and for nylon-6,6 is merely 3 V/mm at 150°C (23) (see Insulation, electric).

Paper of Nomex has almost twice the dielectric strength of high quality rag paper, and retains its useful electrical properties at higher operating temperatures than the upper limit for the rag paper (about 105°C) (see Synthetic paper). In the early years of the commercialization of Nomex it is almost certain that the production of Nomex in the form of paper equaled or outstripped the production of Nomex in the form of fiber.

Table 4. LOI[a] Values for Selected Aramid Fibers[b,c]

	Structure	CAS Registry No	Mol formula	LOI
(18)				22.0–23.0
(9)				22.5–23.0
(27)		[65749-40-8]	$(C_{24}H_{20}N_4O_4)_n$	23.0–23.5
(11)				25.5–26.0
(28)		[59789-57-0]	$(C_{23}H_{17}N_3O_5)_n$	26.0–26.5
(2)				28.5–29.0
(29)		[24938-64-5]	$(C_{14}H_{10}N_2O_2)_n$	28.5–29.0
(26)				29.0–30.0
(30)		[59789-55-8]	$(C_{22}H_{15}N_3O_5)_n$	31.0–31.5
(31)		[25670-03-5]	$(C_{34}H_{20}N_4O_5)_n$	32.0
(3)				32.0–32.5

[a] Limiting oxygen index.
[b] Determined on thin knits.
[c] Ref. 26.

Resistance to Chemicals, Uv, and Ionizing Radiation. The chemical resistance of aramid fibers is, in general, very good (Tables 5 and 6). Although these fibers are much more resistant to acid than nylon-6,6 fibers, they are not as acid resistant as polyester fibers, except at elevated temperatures, particularly in the range of pH 4–8. Their resistance to strong base is comparable to that of nylon-6,6 fibers and hydrolytic stability of the aramids is superior to that of polyester and comparable to nylon-6,6.

Aramid fibers, like their aliphatic counterparts, are susceptible to degradation by ultraviolet light. Flanking the amide linkage on each side by arylene rings confers no special immunity to degradation by uv irradiation. The mechanism of photodegradation of the aramids is quite different from that of the aliphatic polyamides and presents greater problems as regards the development of color. To date, photo quenchers, which have been highly successful in the stabilization of aliphatic polyamides, have not been found for aromatic polyamides. Nor have uv screening agents been found which are effective and efficient at reasonable levels (= 1%) (see Uv absorbers).

The resistance of aramids to ionizing radiation is greatly superior to that of nylon-6,6 (Table 7). Nomex fiber retained 76% of its original strength after exposure to 155 C/g in a Van de Graff generator, whereas nylon-6,6 fiber was destroyed by the same exposure (23). The greater resistance of Nomex to degradation in comparison with the nylon-6,6 fiber is just as marked upon exposure to gamma radiation and x-rays. That excellent radiation resistance is a characteristic property of the aramid fibers is evidenced by fibers of two ordered aromatic copolyamides that actually gained about 10% strength after exposure to a dose of 30 kGy of gamma radiation.

Dyeability. Aramid fibers are exceedingly difficult to dye, probably as a consequence of their very high glass transition temperatures. A procedure for dyeing Nomex has been published (31) that makes use of a dye carrier (qv) at an elevated temperature (ca 120°C) in pressurized beam and jet-dyeing machines. Cationic dyes, which are used exclusively, must be selected carefully in order to achieve reasonable light fastness; a list of recommended dyes for Nomex has been given in a DuPont technical information bulletin (31).

Table 5. Chemical Resistance of Fibers of Aromatic Ordered Copolyamides[a]

Chemical	Strength	Temp, °C	Time, h	Strength retained, % Polymer (21)[b]	Strength retained, % Polymer (23)[b]
dimethylacetamide	100%	21	264	81	99.5
sodium hypochlorite	0.5%	21	8	96	89
			264	55	44
sodium hydroxide	16N	21	8	100	84
			264	71	74
	4N	100	8	33	38
			22	c	19
sulfuric acid	20N	21	8	95	70
			264	76	70
	4N	100	8	51	51
			96	d	e

[a] Ref. 25.
[b] From Table 1.
[c] Too weak to test.
[d] Disintegrated.
[e] Weak, but still in fiber form.

Table 6.   Chemical Resistance of Nomex High Temperature Resistant Aramid Fiber[a]

Chemical	Concentration, %	Temp, °C	Time, h	Effect on breaking strength[b]
Acids				
formic	90	21	10	none
hydrochloric	10	95	8	none
	35	21	10	appreciable
	35	21	100	appreciable
sulfuric	10	21	100	none
	10	60	1000	moderate
	70	21	100	none
	70	95	8	appreciable
Alkalies				
ammonium hydroxide	28	21	100	none
sodium hydroxide	10	21	100	none
	10	60	100	slight
	10	95	8	appreciable
	40	21	10	none
	50	60	100	degraded
Miscellaneous chemicals				
dimethylformamide	100	70	168	none
perchloroethylene	100	70	168	none
phenol	100	21	10	none
sodium chlorite	0.5	21	10	none
	0.5	60	100	moderate

[a] Ref. 23.

[b] None, 0–9% strength loss; slight, 10–24% strength loss; moderate, 25–44% strength loss; appreciable, 45–79% strength loss; degraded, 80–100% strength loss.

Table 7.   Resistance of Fibers to Ionizing Radiation Degradation[a]

Radiation	Tenacity retained, %	
	Nomex	Nylon-6,6
beta radiation (Van de Graff)		
52 C/g	81	29
155 C/g	76	0
gamma radiation (Brookhaven Pile)		
52 C/g	70	32
520 C/g	45	0
x-ray		
50 kV	85	22
100 kV	73	0

[a] Ref. 23.

Ultra High-Strength–High-Modulus Aramid Fibers. Although the rod-like aramids have very high melting points (generally above 500°C with decomposition) and are heat-resistant polymers, they are not particularly useful as heat-resistant fibers. The reason is that fibers from the rod-like polymers have rather low elongations, and they lose elongation rapidly after being heated at an elevated temperature, thereby becoming too brittle to be especially useful. For very short-term exposure, retention of strength at elevated temperatures is outstanding, eg, in the case of Kevlar, tenacity is greater than 0.9 N/tex (10 gf/den) at about 300°C (32). The chemical resistance (32) of Kevlar is similar to that of Nomex (Table 6).

The flame resistance of the aramid fibers from rod-like polymers is high, but a flame tends to propagate along a vertically-held fiber of this class because of the grid formed by the residual char. However, incorporation of phosphorus, as pointed out earlier, has been used to raise the flame resistance to very high levels for fiber from poly(p-phenyleneterephthalamide) (PPD-T) which has not been highly drawn during spinning. The high flame resistance of PPD-T fabrics containing phosphorus coupled with the inherent high resistance to shrinkage of PPD-T fabrics can contribute significantly to providing insulation and protection from burns.

The fiber properties of some aramid fibers having ultra-high strength and high modulus are given in Table 8. The rod-like polymers (Table 8) used to prepare these fibers are of the AB, AA–BB, ordered copolymer and limited copolymer types. Terlon, Vnivlon, and SVM, high-strength–high-modulus fibers developed in the Soviet Union, have properties which would suggest that they are poly(p-benzamide) (PPB) spun from an organic solvent and poly(p-phenyleneterephthalamide) (PPD-T) and a heterocycle–amide copolymer of PPD-T spun from sulfuric acid (33–35).

Apparently the great regularity necessary for the attainment of good tensile properties in homopolymers that chain-fold is not necessary for the rod-like polymers which probably exist in the chain-extended state. In Tables 9 and 10, the tensile properties of selected aramid fibers based on random copolymers are given. These properties show that random copolymers yield fibers having tensile properties comparable to homopolymers (see Copolymers).

As shown for fibers from polyamide–hydrazides (36), the introduction of a few mol % of rings that are not rigid chain-extending types does not necessarily detract from the tensile properties of aramid fibers. In fact, on balance, fiber tensile properties may be enhanced considerably; note, eg, the excellent balance of properties for fiber (52), Table 9, and fiber (61), Table 10.

Two important features of the ultra high-strength–high-modulus fibers should be noted here: (1) the fibers are of fine diameter, usually less than 7×10^{-7} kg/m (6 den) per filament and generally more nearly about 1.7–2.2×10^{-7} kg/m (1.5–2 den) per filament, and (2) the fibers are produced from rod-like polymers exhibiting extremely high intrinsic and inherent viscosity values although molecular weights of these polymers may be moderate. The situation regarding fine linear densities is not unlike that for glass fibers which must be of very small cross-sectional area in order to exhibit good mechanical properties in flexure.

The stress-strain curves for the ultra high-strength–high-modulus fibers show a strong similarity to those for glass and steel. On a specific basis (ie, taking into account the lower specific gravity of the aramid fibers in comparison to glass and steel) the aramid fibers are stronger and stiffer than glass and steel. These properties suggest that the organic fibers should be quite useful in the reinforcement of rigid and flexible composites. This is indeed the case. Thus Kevlar fiber may be used as a tire cord as a replacement for glass and steel and Kevlar-49 may be used competitively with the lower modulus types of graphite fibers (see Ablative materials; Carbon; Composite materials).

Table 8. Tensile Properties of the Rod-like Aramid Fibers

Structure	CAS Registry No.	Mol formula	η_{inh}[a]	Spun from[b]	Solids, %	Spinning method[c]	Tex[d]	T/E/M$_i$[e]	W-T-B[f]	Remarks[g]	Ref.
AB Type											
(32)	[24991-08-0]	$(C_7H_5NO)_n$	1.67	O (—)		D		0.72/3.1/44.9		spun from anisotropic layer of dope	37
			2.36	O (A)		W	0.54	0.64/8.1/25.0			5
			same dope	O (I)		W	2.53	0.11/9.0/5.6		spun from isotropic layer of dope	5
AA–BB Type											
(33)	[24938-64-5]	$(C_{14}H_{10}N_2O_2)_n$	3.7	A (A)	18.0	DJ-W	0.11	1.7/4.0/50.3	36.5		6
(34)	[27307-20-6]	$(C_{18}H_{12}N_2O_2)_n$	4.8	A (A)		DJ-W	0.15	2.1/5.0/42.7			6
			3.6	A (A)		DJ-W	0.20	2.8/3.7/81.2	55.1	hot-drawn	38
(35)	[26402-76-6]	$(C_{14}H_8Cl_2N_2O_2)_n$	3.6	A (A)	3.9	DJ-W	0.22	15.0/4.9/41.5	35.6	hot-drawn	6
			1.59	A (—)		W	0.10	0.65/2.4/34.1			5
(36)	[28779-61-5]	$(C_{26}H_{18}N_2O_2)_n$	2.0	O (—)		D		0.24/3.6/8.6		hot-drawn	39
(37)	[27252-16-0]	$(C_{20}H_{14}N_4O_2)_n$	3.24	A (—)		DJ-W	0.17	0.62/4.9/29.2		hot-drawn	40
(38)	[65749-45-3]	$(C_8H_6N_2O_2)_n$	0.9	O (—)		W		0.35/3.0/39.7		hot-drawn	41
(39)	[37357-28-1]	$(C_{15}H_{12}N_2O_2)_n$	4.46	O (—)		W	0.24	0.86/10.2/25.3			5
			3.1	A (A)		DJ-W		1.4/2.5/68.8	19.4		6
(40)	[26402-76-6]	$(C_{14}H_8Cl_2N_2O_2)_n$	3.77	O (—)			0.36	0.77/5.6/21.3			5
			same dope				0.30	1.3/3.5/43.2		hot-drawn	5

44

Table 8 (continued)

No.	Structure	CAS Registry No.	Mol formula	η_{inh}[a]	Spun from[b]	Solids %	Spinning method[c]	Tex[d]	T/E/M_i[e]	W-T-B[f]	Remarks[g]	Ref.
(41)	(structure)	[51257-61-7]	$(C_{14}H_9ClN_2O_2)_n$	3.1	A (A)	20.0	DJ-W	3.9	1.6/6.5/32.7	49.4		6
(42)	(structure)	[31801-22-6]	$(C_{14}H_9ClN_2O_2)_n$	3.7	A (A)	20.0	DJ-W	1.9	1.9/4.8/56.5	43.8		6
(43)	(structure)	[65749-46-4]	$(C_{13}H_9N_3O_2)_n$	5.3	A (A)	20.0	DJ-W	3.5	1.6/5.8/41.5	46.2		6
(44)	(structure)	[65761-30-0]	$(C_{28}H_{20}N_4O_4)_n$	2.4 / same dope	A (A)	3.9	W / W	6.1 / 6.3	0.42/12.4/14.7 / 0.48/0.9/51.5		hot-drawn	5 / 5
(45)	(structure)	[65749-48-6]	$(C_{28}H_{13}Cl_2N_4O_4)_n$	3.25 / same dope	O (—)		W / W	1.49 / 1.36	0.94/6.6/30.3 / 1.2/1.9/67.3		hot-drawn	5 / 5
(46)	(structure)	[65749-49-7]	$(C_{36}H_{29}N_6O_4)_n$	7.3[h] / same dope	O (—)				0.58/18.6/17.4 / 0.86/2.4/51.8		hot-drawn	42 / 42
(47)	(structure)	[52270-04-9]	$(C_{22}H_{13}N_4O_4)_n$	3.1 / same dope	O (—)		DJ-W / DJ-W		0.27/7/11.7 / 0.89/ /22.1		hot-drawn	43 / 43

Copolymers of limited order

No.	Structure	CAS Registry No.	Mol formula	η_{inh}[a]	Spun from[b]	Solids %	Spinning method[c]	Tex[d]	T/E/M_i[e]	W-T-B[f]	Remarks[g]	Ref.
(48)	(structure)	[29153-47-7]	$(C_{21}H_{15}N_3O_3)_n$	6.6 / 3.6 same dope / 4.3	O (—) / A (A) / A (A)	10.0 / / 20.0	DJ-W / W / W / DJ-W	5.6 / 6.7 / 4.2 / 1.6	0.65/8.3/25.2 / 0.57/11/15.3 / 0.89/1.2/73.0 / 1.5/3.9/54.7	43.8 / / / 31.6	hot-drawn / hot-drawn	36 / 5 / 5 / 6
(49)	(structure)	[65749-50-0]	$(C_{27}H_{13}N_3O_3)_n$	2.4 same dope	A (A)	11.0	DJ-W / DJ-W	7.2	0.49/9/19.1 / 1.2/2.4/62.4		hot-drawn	5 / 5

[a] Determined in sulfuric acid, 0.5 g/100 mL, at 30°C. η_{inh} = 1a η_{rel}/c; η_{rel} = η_{soln}/$\eta_{solvent}$; c = conc of solvent.
[b] O = organic solvent; A = acid (H_2SO_4); (I) = isotropic dope; (A) = anisotropic dope.
[c] D = dry spun; W = wet spun; DJ-W = dry-jet–wet spun.
[d] Tex (= 9 den) per filament.
[e] T = tenacity, N/tex; E = elongation-to-break, %; M_i = initial modulus, N/tex (= 11.33 gf/den).
[f] W-T-B = work-to-break, J/g (= 11.33 × 10^{-3} gf·cm/(den·cm)).
[g] Fiber properties are for as-spun fiber except as noted.
[h] Determined in DMAC–LiCl.

45

Table 9. Fibers from Random Copolyterephthalamides of p-Phenylenediamine (PPD) and Various Diamines[a]

Diamine	Mol formula	Mol %	η_{inh}[b]	Tex[c]	T/E/M_i[d]	W-T-B[e]	Ref.
(50) NH_2—⬡—CH_2—⬡—NH_2	$C_{13}H_{14}N_2$	5	3.8	0.27	1.3/4.1/51.2	30.0	6
(51) NH_2—⬡—CH_2CH_2—⬡—NH_2	$C_{14}H_{16}N_2$	7.5	3.4	0.23	1.9/3.8/64.4	38.1	6
(52) NH_2—⬡—O—⬡—NH_2	$C_{12}H_{12}N_2O$	5	4.3	0.29	2.1/6.2/45.9	68.9	6
(53) ⬡ NH_2/NH_2	$C_6H_8N_2$	5	4.2	0.39	1.5/5.4/40.6	42.1	6
(54) ⬡ NH_2/Cl	$C_6H_7ClN_2$	25	5.7	0.48	1.9/6.9/30.9	65.6	6
(55) NH_2—⬡—⬡—NH_2 (CH₃/CH₃)	$C_{14}H_{16}N_2$	5	3.4	0.36	1.5/4.9/48.5	39.7	6
(56) NH_2—⬡—NHCOCONH—⬡—NH_2	$C_{14}H_{14}N_4O_2$	50	6.6[f]		0.69/2.4/33.8[g]		44
		60	4.3[f]		1.0/1.8/52.8[h]		44
					1.2/1.4/75.2[g]		44

[a] Spun from sulfuric acid except as noted.
[b] Determined at 30°C in 0.5 g fiber dissolved in 100 mL sulfuric acid. $\eta_{inh} = \ln \eta_{rel}/c$; $\eta_{rel} = \eta_{soln}/\eta_{solvent}$; c = conc of solvent.
[c] Tex (= 9 den) per filament.
[d] T = tenacity, N/tex; E = elongation-to-break, %; M_i = initial modulus, N/tex (= 11.33 gf/den).
[e] W-T-B = work-to-break, J/g (= 11.33 × 10⁻³ gf·cm/(den·cm))
[f] Determined on polymer in DMAC–LiCl.
[g] Spun from organic solvent and hot drawn.
[h] Double hot-draw.

46

Table 10. Fibers from Random Copolyterephthalamides of p-Phenylenediamine (PPD) and Various Diacid Chlorides[a,b]

Diacid chloride	Mol formula	Mol %	η_{inh}[c]	Tex[d]	T/E/M$_i$[e]	W-T-B[f]
(57) ClCO—⟨⟩—⟨⟩—COCl	$C_{14}H_8Cl_2O_2$	55	5.3	0.38	1.9/5.5/60.9	56.5
(58) ClCO—CH=CH—COCl	$C_4H_2Cl_2O_2$	40	3.3	0.18	1.6/5.7/49.4	48.5
(59) ClCO—⟨⟩—N=N—⟨⟩—COCl	$C_{14}H_8Cl_2N_2O_2$	5	3.4	0.34	1.7/4.6/44.1	39.7
(60) ClCO—⟨⟩(Cl)—COCl	$C_8H_3Cl_3O_2$	5	5.5	0.47	2.0/4.2/51.2	40.6
(61) ClCO—⟨⟩—COCl	$C_8H_4Cl_2O_2$	5	3.9	0.17	1.8/4.6/51.2	43.2
(62) ClCO—⟨⟩—COCl (cyclohexane)	$C_8H_{10}Cl_2O_2$	25	4.3	0.16	1.9/4.5/47.7	45.0
		50	3.4	0.23	1.6/4.9/43.2	41.5

[a] Fibers spun from sulfuric acid.
[b] Ref. 6.
[c] Determined at 30°C on 0.5 g of fiber dissolved in 100 mL of sulfuric acid. $\eta_{inh} = \ln \eta_{rel}/c$; $\eta_{rel} = \eta_{soln}/\eta_{solvent}$; c = conc of solvent.
[d] Tex (= 9 den) per filament.
[e] T = tenacity, N/tex; E = elongation-to-break, %; M$_i$ = initial modulus, N/tex (= 11.33 gf/den).
[f] W-T-B = work-to-break, J/g (= 11.33×10^{-3} gf·cm/(den·cm)).

47

Synthesis

Methods of Preparation. The usual methods for the preparation of aliphatic polyamides are unsuited to the preparation of high molecular weight, wholly aromatic polyamides. However, two general synthetic methods are presently available for the preparation of medium to high molecular weight polymer: low temperature polycondensation and direct polycondensation in solution using phosphites, especially in the presence of metal salts (45).

Low temperature (ie, <100°C) polycondensation procedures developed over the past two decades håve led to the preparation of a very large number of high molecular weight polyamides otherwise unobtainable. Morgan, in an excellent text, has reviewed the literature on low temperature polycondensation methods up to 1965 (46). The principal low temperature methods are interfacial polycondensation and solution polycondensation.

Interfacial Polymerization. The interfacial method for the preparation of polyamides is an adaptation of the well-known Schotten-Baumann reaction: the diacid chloride is dissolved or dispersed in an inert, water-immiscible organic solvent which is preferably a swelling agent for the polymer, and the diamine is dissolved or dispersed along with a proton acceptor in the aqueous phase. The two-phase system may be stirred or unstirred, but higher molecular weight polymers are generally obtained when the system is stirred rapidly. In the latter process, the use of an emulsifying agent is usually helpful. The polymer is collected and dried; it can then be dissolved in a suitable solvent and fabricated.

Solution Polymerization. The most significant polymerization process for aromatic polymers is the low temperature polycondensation of diacid chlorides and diamines in amide solvents. This type of polymerization is often more convenient to carry out than interfacial polycondensation and has the further advantage of usually providing a solution of the polymer amenable to direct fabrication of fibers. Nomex is prepared in solution and spun from the solution after neutralization (Fig. 1). Another distinct advantage of solution polycondensation over the interfacial method results from the peculiar solubility characteristics of certain wholly aromatic polyamides: although such polymers are soluble as made in solution by low temperature polycondensation, they often cannot be redissolved in solvents (other than strong acids) once precipitated. Interfacial polymerization gives a precipitated product which cannot be readily fabricated.

In solution polycondensation the polymerization medium is a solvent for at least one of the reactants and is a solvent or swelling agent for the polymer, preferably a solvent. The solvent must be inert or relatively so to the reactants. To ensure completeness of the reaction, an acid acceptor, usually a tertiary amine, is used unless the solvent itself is such an acceptor. The better solvents are amides such as dimethylacetamide (DMAC), N-methylpyrrolidinone (NMP), hexamethylphosphoric triamide (HPT), and tetramethylurea (TMU). Dimethylformamide (DMF) is not useful because of its rapid and irreversible reaction with acid chlorides, nor is dimethyl sulfoxide (DMSO) useful because of its sometimes violent reaction with acid chlorides.

For the solution polymerization of many aromatic polyamides, there is no known organic solvent sufficiently powerful to keep the polymer in solution as its molecular weight builds up. The solvating power of many organic solvents is greatly increased, however, by the addition of inorganic salts such as lithium chloride and calcium chloride. Use of these salt-containing solvents has permitted preparation of many high

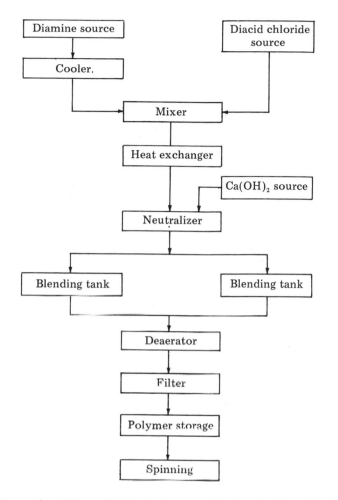

Figure 1.   Preparation of Nomex (7). Reactants were *m*-phenylenediamine and isophthaloyl chloride; solvent, dimethyacetamide; polymer(MPD-I) of η_{inh} = 1.65 produced.

molecular weight polymers not obtainable using organic solvents alone. Also, mixtures of the basic amide solvents, such as DMAC with a minor amount of the corresponding hydrochloride, are much better solvents for aromatic polyamides than are the amide solvents alone. This is advantageous since hydrogen chloride is produced in the condensation of aromatic diamines with diacid chlorides and, when amide solvents are used, the desired salt–solvent combination is produced. Moreover, the quantity of the hydrochloride salt increases as the polycondensation proceeds, thereby increasing the solvating power of the solvent at the same time that the molecular weight of the polymer is building up.

Certain of the rod-like aromatic polyamides apparently are best prepared in mixed solvents, especially in specific ratios of these component solvents. Thus it was reported (47) that the highest molecular weight for poly(*p*-phenyleneterephthalamide) had been obtained in a 1:2 weight ratio of HPT to NMP; the optimum ratio of DMAC to HPT was obtained (47) at about 1:1.4. These weight ratios correspond to molar ratios of, respectively, 1:3 and 2:3 (Fig. 2). Preparation in DMAC and NMP or mixtures of

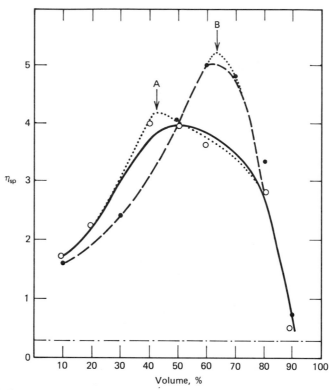

Figure 2. Relationship of molecular weight, as indicated by η_{sp}, for PPD-T prepared in the synergistic solvent mixtures of DMAC–HPT, —; and NMP–HPT, — —. Neither DMAC nor NMP alone or in any combination, · — ·, produce high molecular weight PPD-T. (The original drawing (47) has been modified by the · · · · lines to show the maxima in the curves.) A, the molar ratio of DMAC:HPT, 3:2; B, the molar ratio of NMP:HPT, 3:1.

these solvents at all ratios give low molecular weight polymer, thus showing no synergistic solvent effect (Fig. 2). Addition of salts to the mixtures of DMAC–HPT and NMP–HPT useful for preparation of PPD-T results in lower molecular weight polymer, presumably because the salts complex with the individual solvents, thereby breaking up the complex between the solvent pairs.

For aromatic polyamides that are not of the rod-like type, a solution of 15–25% polymer solids is typically prepared. This level of polymer solids, ie, a concentration of about 0.5 mol/L, yields polymer of optimum molecular weight (Fig. 3). However, for the synthesis of rod-like polymers, eg, PPD-T, it has been pointed out that molecular weight increases (Fig. 3) with concentration only up to about 0.25 mol/L or about 6–7% solids and decreases rapidly above this concentration (48–49). It may be significant that above this concentration rod-like polymers can yield anisotropic solutions in organic solvents but below this level isotropic solutions are obtained (50). In the case of the preparation of polymer at high solids levels, it is possible that the growing closely-packed polymer chains, as in the anisotropic state, may become inaccessible to the monomers, thereby limiting molecular weight.

The preparation of PPD-T, used to spin Kevlar fiber, is shown schematically in Figure 4.

Recently, the direct polycondensation of aromatic dicarboxylic acids and aromatic

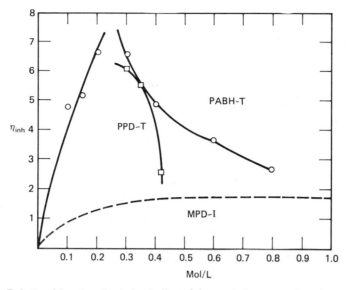

Figure 3. Relationship of mol wt (as indicated by η_{inh}) for aromatic polymers (PABH-T =

$$\left[\!\!\left(NH\!-\!\!\langle\bigcirc\rangle\!-\!CONHNH\right)\!\!-\!CO\!-\!\langle\bigcirc\rangle\!-\!CO\right]_n ; \text{PPD-T data from Ref. 6).}$$

diamines in amide solvents using aryl phosphites in the presence of pyridine has been demonstrated (45). Addition of metal salts, eg, LiCl, to the amide solvents has made it possible to keep the polymer in solution formed during the reaction and thereby allow for the build-up of moderately high molecular weight polymer as indicated by the inherent viscosity values obtained ($\eta_{inh} = 1.0 \pm 0.1$). High molecular weight, wholly aromatic polyamides have not been made using the phosphorylation reaction, presumably because of the reaction of by-product phenol to yield phenyl ester.

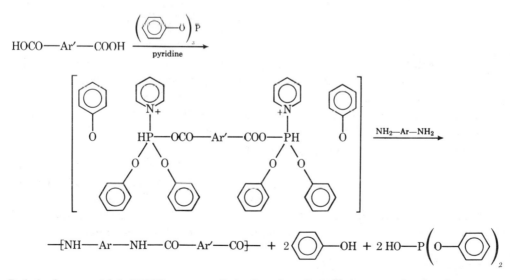

Poly(p-benzamide) (PPB) prepared via the phosphorylation reaction has been prepared having an inherent viscosity value of 1.7, but this value is high because of the rod-like character of PPB; the actual molecular weight of PPB is probably no higher than that of poly(m-benzamide) (PMB) having an inherent viscosity of 0.4 (45).

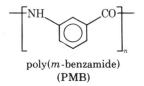

poly(m-benzamide)
(PMB)

Preparation of Selected Classes of Polymers. Several classes of fiber-forming wholly aromatic polyamides have been prepared. In addition to the conventional AB and AA–BB classes, some classes of AA–BB polymers of more complex structure have been made. Synthesis of the latter involves the use of diamines or diacids containing preformed amide or heterocyclic units, or pendant functional groups which do not participate in the polycondensation. Only polymers from which fibers have been prepared will be considered here.

A–B Polyamides. Low temperature polycondensation of reactive A–B monomers has been used to obtain high molecular weight wholly aromatic A–B polyamides. Polymers derived from m- or p-aminobenzoic acid have been prepared from the m- or p-aminobenzoyl chloride hydrochloride by the addition of a base under special conditions (8,37,52–53). The m- and p-aminobenzoyl chloride hydrochlorides are synthesized via reaction of the aminobenzoic acids with thionyl chloride to yield the thionylaminobenzoyl chlorides which react with HCl under anhydrous conditions.

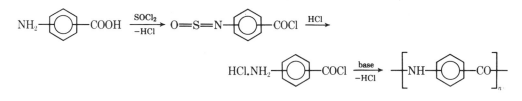

Apparently considerable attention to technique is important in the A–B polyamide polycondensation reaction because one worker reported (8) that interfacial polymerization of m-aminobenzoyl chloride hydrochloride yielded a polymer of η_{inh} 0.7 and another reported (52) a polymer of η_{inh} 2.36. The use of solution techniques by the same workers gave polymers of η_{inh} 1.37 and η_{inh} 0.3, respectively.

Use of the interfacial polycondensation method employing p-aminobenzoyl chloride hydrochloride yields only p-aminobenzoic acid and its dimer (53). PPB of η_{inh} 0.7 is obtained upon the addition of a tertiary organic base, eg, pyridine, to a slurry of monomer in dry dioxane (53), and high molecular weight PPB is obtained by polymerization in an amide solvent at low temperature using either p-aminobenzoyl chloride hydrochloride (37) or 4(4′-aminobenzamido) benzoyl chloride hydrochloride (54).

$$\text{HCl.NH}_2 \text{—} \bigcirc \text{—CONH—} \bigcirc \text{—COCl} \longrightarrow \text{PPB}$$

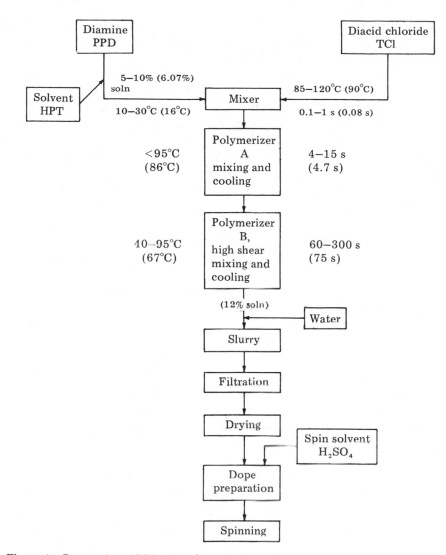

Figure 4. Preparation of PPD-T, used to spin Kevlar fiber (6,51). Values in parentheses are from an actual example in Ref. 51; polymer of η_{inh} = 5.3 was produced.

AA–BB Polyamides. AA–BB polyamides are readily prepared from aromatic diamines and diacid chlorides.

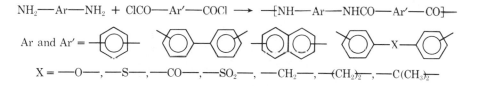

Polyamides having phenylene rings bridged by short alkylene chains are included because from a property point of view they closely approach wholly aromatic polymers and because most of the literature concerning wholly aromatic polyamides is interwoven with these polymers. Polymers containing pendant methyl or chloro groups on the phenylene units have also been included.

The interfacial polycondensation technique was the first low temperature process to yield high molecular weight wholly aromatic polyamides. The preparation of poly(benzidine terephthalamide) [25667-70-3] prepared via this route, although of relatively low molecular weight, demonstrated that this low temperature route to wholly aromatic polyamides was feasible (55). Although high molecular weight polymer can usually be prepared by the interfacial polycondensation method, doing so is frequently not without difficulty.

Polar cycloaliphatic and heterocyclic solvents that are either miscible or partially miscible with the aqueous phase are most generally useful; consequently weak acid acceptors are preferred to minimize hydrolysis of the acid chloride. The rod-like polymers, because of their low solubility in and resistance to swelling by organic solvents, apparently are not prepared in high molecular weight by the interfacial technique.

In general, high molecular weight wholly aromatic polyamides can be obtained more consistently by low temperature solution polycondensation than by the interfacial method. The great value of the low temperature solution polymerization method has been shown amply by the numerous reports of high molecular weight, fiber-forming polymers of widely varying structure. In general, however, wholly aromatic polyamides will not remain in solution when prepared by solution polymerization unless either inorganic salts, such as lithium chloride or calcium chloride, are present or a basic solvent, such as dimethylacetamide or N-methylpyrrolidinone, which forms a salt with hydrogen chloride, is used (56). The products of low temperature solution polycondensation may be used in a convenient, economical, and direct process for the spinning of fibers.

The monomers for rod-like polymers are mixed in solution, but they frequently form insoluble gel structures at high molecular weight (5–6). In order to obtain high molecular weight polymer, shearing action, such as may be obtained with a high speed food blender, is necessary. A process confining mixing of the monomers and the necessary shearing action has been described (51). The polymer then may be washed free of solvent, dried, and redissolved in acids, such as sulfuric, chlorosulfuric, fluorosulfuric or hydrofluoric, for the preparation of fibers by wet-spinning methods (5–6).

AA–BB polyamides having reactive substituents on the rings can be prepared provided that the reactive substituent is either masked during the polycondensation or is unreactive under the conditions of the polycondensation. The polymer from isophthaloyl chloride and 3,5-diaminobenzoic acid, DAB-I, is an example (12).

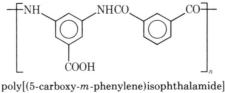

poly[(5-carboxy-m-phenylene)isophthalamide]
(DAB-I)

AA–BB Polyamides from Intermediates Containing Preformed Amide Linkages.
Ordered copolyamides (10,20) are readily prepared by low temperature polycondensation from diamines or diacid chlorides containing preformed amide linkages because
no reorganization of structural units is possible under the conditions of polymerization.
Polymers of this type may be prepared by both the interfacial and the solution
methods, but more consistent results and polymers of higher inherent viscosities are
generally obtained by the solution technique. The preparation of ordered copolyamides
may be illustrated by the following examples:

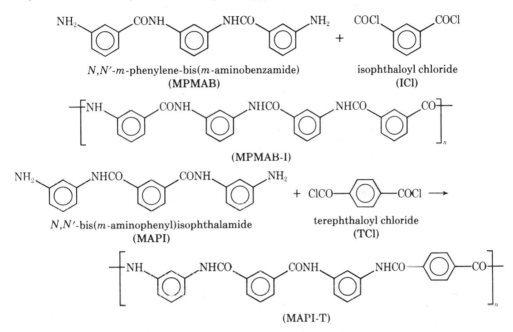

N,N′-m-phenylene-bis(*m*-aminobenzamide)
(MPMAB)

isophthaloyl chloride
(ICl)

(MPMAB-I)

N,N′-bis(*m*-aminophenyl)isophthalamide
(MAPI)

terephthaloyl chloride
(TCl)

(MAPI-T)

Polycondensation of MPMAB with terephthaloyl chloride yields MPMAB-T and
polycondensation of MAPI with isophthaloyl chloride yields MPD-I (Nomex).

Diamines, such as MPMAB, were prepared by reaction of aromatic diamines with
nitroaroyl chlorides followed by reduction of the dinitro intermediate. Diamines, such
as MAPI, may be prepared by reaction of nitroaniline with a diacid chloride followed
by reduction.

A series of copolyamides having somewhat less order (the so-called copolymers
of limited order) may be prepared (22) by the low temperature polycondensation of
diaminobenzanilides with diacid chlorides; both interfacial and solution methods have
been investigated (22). Because the diamine moiety can enter the polymer chain
head-to-head or head-to-tail fashion, the diamine moiety is placed in parentheses
within the polymer repeat unit.

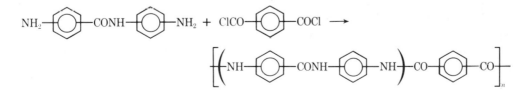

AA–BB Polyamides from Intermediates Containing Preformed Heterocyclic Units.
Polymers characterized structurally by an ordered sequence of amide and heterocyclic
units in the polymer chain may be prepared by low temperature solution polycondensation of an aromatic diacid chloride and an aromatic diamine, one or both of which
contain a heterocyclic unit (57). The preparation of these polymers may be illustrated
by the following examples employing heterocyclic diamines:

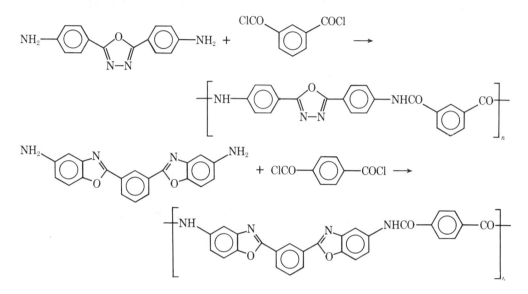

Diacid chlorides containing preformed heterocycles, eg, oxadiazole groups or imide
groups, can also be used to prepare amide–heterocyclic copolymers.

A series of copolyamides that probably have some degree of disorder along the
chain (amide-heterocyclic polymers of limited order) can be prepared by the low
temperature polycondensation of unsymmetrical heterocyclic diamines. The preparation of such polymers may be illustrated by the following example:

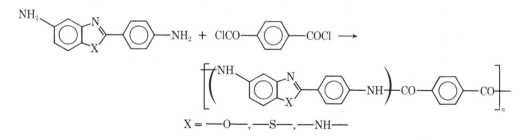

$$X = \text{---O---}, \text{---S---}, \text{---NH---}$$

Polymers of this type and copolymers with *p*-phenylenediamine (PPD) based
on a benzimidazolediamine appear to have been employed by Soviet workers to make
a high-strength–high-modulus fiber.

The ordered amide–heterocyclic copolymers and the amide–heterocyclic copolymers of limited order appear to fall into the definition of aramids, but they are
not discussed further.

Production

All wholly aromatic polyamide fibers reported here have been spun from solutions of the polymers. Both dry-spinning and wet-spinning techniques appear to be equally satisfactory for fiber formation from polymers that are soluble in organic solvents although properties may vary with the method of spinning. The mol wt of a given polymer generally should be higher when spun by the wet method and should be 60,000 or greater for the best balance of initial tensile properties and retention of tensile properties after the fiber is placed in service. Polyamides prepared by solution poly-condensation can be conveniently spun from the resulting solution (provided that it is stable), circumventing the need for isolating and redissolving the polymer. The ability to spin the polymer solution as prepared directly is crucial for those polymers that do not redissolve (or redissolve only with difficulty) in organic solvents after isolation in the dry state and that are not stable in strong acid solvents. Polymers prepared by interfacial polymerization must necessarily be dissolved prior to spinning since solutions are not produced directly by this method.

Almost all wholly aromatic polyamides require the presence of a salt, either organic or inorganic, in the amide solvents to dissolve the polymers or prevent them from precipitating if prepared by solution polycondensation. If organic salts such as DMAC hydrochloride are the solvating salts during the course of polymerization, it is common practice to neutralize the hydrogen chloride produced during polycondensation with an inorganic base, eg, lithium hydroxide or calcium oxide, prior to spinning to prevent corrosion of spinning equipment. In the case of dry spinning (58), the salt remains in the spun fibers and must be leached subsequently in order to obtain fiber with optimum mechanical properties. In wet spinning most of the salt diffuses from the fiber during spinning and thus a separate step for leaching the salt is not required. Fiber containing any residual salt has poorer thermal stability and poorer electrical properties than salt-free fiber.

Wet spinning (11,54) in the conventional manner, immersing the jet in the coagulation bath, and the dry jet or so-called air-gap method (6,59–60), leaving the dry jet above the coagulation bath, have both been described in the patent literature for the spinning of aromatic polyamides. Wet spinning permits the use of both organic and inorganic solvents. With some inorganic solvents, dry spinning would be impossible.

The yarn must be dried and it is usually hot drawn no matter if spun by a wet or dry process. An important exception to the need for a hot-drawing step is described below. In general, the wet-spinning processes lend themselves better to continuous in-line processing than dry spinning when salts are used.

Certain rod-like polymers, such as poly(p-phenyleneterephthalamide) (PPD-T) and poly(p-benzamide) (PPB), yield anisotropic solutions when the concentration reaches a critical level (5–6,50). The basis for the anisotropy is the formation of liquid crystals (qv), ie, close-packed aggregates of the rod-like molecules mutually aligned within a given packet. Hot drawing of fibers spun from such solutions is not necessary for the attainment of high strength (6); however, hot drawing of these fibers can result in more than a two-fold increase in initial modulus but at a sacrifice in elongation-to-break without an attendant appreciable increase in tenacity.

Because the rod-like aramid polymer molecules have a very high effective volume in solution, it is not surprising that at relatively low concentrations in polar organic solvents and in inorganic solvents (such as sulfuric and hydrofluoric acids), viscous solutions are obtained. For certain of these polymers it has been demonstrated (61) that the viscosity of spinning solutions (dopes) may be *decreased* by *increasing* the polymer solids concentration above a certain critical value by changing the solution from the isotropic to the anisotropic state.

The importance of using anisotropic dopes instead of isotropic ones for the spinning of certain types of polymers has been recognized for some time (62–63). More recently, this principle has been applied to the spinning of aramids.

Economic Aspects

The two types of aramid fibers—heat-resistant fibers, eg, Nomex and Conex, and ultra high-strength–high-modulus fibers, eg, Kevlar and Arenka—are relatively expensive compared to other synthetic fibers, such as nylon and polyester, which are produced in very large volume. The reasons for this are the relatively low volume production of both types of aramid fibers, the high cost of the monomers employed, and high processing costs. The announced annual DuPont capacity for Nomex was about 4500 metric tons; of the actual production, more than one-half went to the manufacture of electrical insulating paper. Expansion of Nomex production to more than 9100 t/yr was completed in 1975. The announced capacity of the Kevlar facilities is 23,000 t/yr and semiworks production was reported to be 27,000 t/yr in 1975.

The price of Nomex bright, 2.2×10^{-7} kg/m (2 den) continuous filament yarns, ie, 2.2–13.3×10^{-5} kg/m (200–1200 den), in early 1975 was in the range of \$16.28–19.58/kg, and that of 1.7–2.2×10^{-7} kg/m (1.5–2.0 den) staple was in the range of \$9.24–9.39/kg. The price of 1.1×10^{-6} kg/m (10 den) Nomex staple for use in carpets was \$10.45/kg. The price of Kevlar tire cord currently is about \$8.69/kg and the prices of Kevlar-29 and Kevlar-49 vary widely according to yarn denier. The price of Kevlar-29 varies from a low of \$17.75/kg for 1.6×10^{-4} kg/m (1420 den) yarn to \$49.61/kg for 2.1×10^{-5} kg/m (190 den) yarn. For Kevlar-49, the price range is about \$19–60/kg for yarns ranging from 1.6×10^{-4} kg/m (1420 den) to 2.1×10^{-5} kg/m (190 den).

Although the annual production of aramid fibers is not great, the total dollar volume is nevertheless substantial and rising. Because of rather high monomer and processing costs coupled with low volume production, aramid fibers are relatively expensive and probably will remain so compared to other synthetic fibers which are produced in very large volume.

The introductory price of Kevlar tire cord was \$6.27/kg but was increased in 1975 to \$6.93/kg and in early 1977 to \$8.69/kg. PRD-49 (now Kevlar-49) was introduced at a price of about \$220/kg, but the price was soon cut to \$40–60/kg; current (1977) prices for Kevlar-29 and Kevlar-49 have ranged from about \$18–60/kg. Further price reductions might be expected with high volume production.

Health and Safety Factors

Toxicology. In recent inhalation studies at the Haskell Laboratory for Toxicology and Industrial Medicine of the DuPont Company, hexamethylphosphoric triamide (HPT) used in the preparation of rod-like polyamide, eg, poly(*p*-phenyleneterephthalamide) (PPD-T), has been found to be carcinogenic in rats (64). Accordingly, any work with HPT should be carried out in an efficient hood to avoid the inhalation of vapors. Since HPT is absorbed by the skin, care should be taken to avoid contact with HPT. Despite the fact that the HPT solvent is washed from the PPD-T prior to drying the polymer and despite the fact that the fibers spun from sulfuric acid are rigorously washed with water, some HPT solvent apparently persists in the fiber. However, because HPT is so tightly bound (most likely by hydrogen bonding to the amide linkage), it seems unlikely that it would be readily lost from the fiber and present an inhalation problem.

Kevlar presents no evidence of skin sensitization when tested on guinea pig skin under occluded conditions or in a standard 200-subject prophetic patch test on human skin, also under occluded conditions (32). In the human patch test, no skin irritation was observed after 48 h of continuous contact. Some irritation, probably from mechanical causes, was observed after 144 h of continuous occluded contact (32). No adverse mechanical or chemical skin effects are to be expected from the usual industrial handling of either Kevlar or Nomex.

For Nomex, and especially for Kevlar, as with any fine, fibrous material, inhalation of the "fly" generated in certain textile handling operations should be avoided.

Safety. Because of its unusual high strength and cutting resistance, caution should be exercised in splicing, handling, or cutting Kevlar (32). Manual cutting and splicing should be attempted only with stationary yarns to avoid possible injury from entanglement in moving yarn or fabric (32).

Applications

Typical applications for aramid fibers are listed according to the distinctive properties of these fibers. A particularly useful discussion of the relationship of Kevlar properties to applications is in a paper by Wilfong and Zimmerman (65).

Heat resistance. Filter bags for hot stack gases (see Air pollution control methods); press cloths for industrial presses (eg, application of permanent press finishes to cotton and cotton–polyester garments); home ironing board covers; sewing thread for high-speed sewing; insulation paper for electrical motors and transformers (see Insulation, electric; Synthetic paper); braided tubing for insulation of wires; paper-makers' dryer belts.

Flame resistance. Industrial protective clothing (eg, pants, shirts, coats, and smocks for workers in laboratories, foundries, chemical plants, and petroleum refineries); welder's clothing and protective shields; fire department turnout coats, pants, and shirts; jump suits for forest fire fighters; flight suits for pilots of the Armed Services (specified by the United States Army, Navy, and Air Force); auto racing drivers' suits; pajamas and robes, particularly for persons who are nonambulatory; mailbags; carpets, upholstery, and drapes (specified on some aircraft and on all ships of the United States Navy); cargo covers, boat covers, and tents (see Flame-retardant textiles).

Dimensional stability. Reinforcement of fire hose and V belts by aramid fibers of moderately high modulus, eg, Nomex; conveyor belts.

Ultra high-strength–high-modulus. Tire cord for use in tire carcasses and as the belt in bias-belted and radial-belted tires (see Tire cords); V belts; cables; parachutes; body armor; and in rigid reinforced plastics in general and more specifically in high performance boats and aircraft (interior trim, exterior fairings, control surfaces, structure) and other vehicles of transportation; in radomes and antenna components; in circuit boards; in filament-wound vessels; in fan blades; in sporting goods (skis, golf clubs, surf boards).

Electrical resistivity.   Electrical insulation (paper and fiber).

Chemical inertness.   Use in filtration (qv).

Permselective properties.   Hollow-fiber permeation separation membranes to purify sea water, brackish water, or to make separation of numerous types of salts and water (see Hollow-fiber membranes) (65–67).

BIBLIOGRAPHY

1. L. K. McCune, *Text. Res. J.* **32**(9), 762 (1962).
2. T. Ono, *Jpn. Text. News* **243,** 71 (Feb. 1975).
3. J. W. Hannell, *Polym. News* **1**(1), 8 (1970).
4. R. E. Wilfong and J. Zimmerman, *J. Appl. Polym. Sci.* **17,** 2039 (1973).
5. U.S. Pat. 3,671,542 (June 20, 1972), S. L. Kwolek (to E. I. du Pont de Nemours & Co., Inc.).
6. U.S. Pat. 3,767,756 (Oct. 23, 1973), H. Blades (to E. I. du Pont de Nemours & Co., Inc.).
7. U.S. Pat. 3,287,324 (Nov. 22, 1966), W. Sweeny (to E. I. du Pont de Nemours & Co., Inc.).
8. U.S. Pat. 3,472,819 (Oct. 14, 1969), C. W. Stephens (to E. I. du Pont de Nemours & Co., Inc.).
9. A. S. Semenova and E. A. Vasil'eva-Sokolova, *Fibre Chem. (USSR)* **5,** 470 (1969).
10. U.S. Pat. 3,049,518 (Aug. 14, 1962), C. W. Stephens (to E. I. du Pont de Nemours & Co., Inc.).
11. U.S. Pat. 3,079,219 (Feb. 26, 1963), F. W. King (to E. I. du Pont de Nemours & Co., Inc.).
12. H. E. Hinderer, R. W. Smith, and J. Preston, *Appl. Polym. Symp.* **21,** 1 (1973).
13. Belg. Pat. 569,760 (1958), (to E. I. du Pont de Nemours & Co., Inc.).
14. E. P. Krasnov and co-workers, *Fibre Chem. (USSR)* **5,** 28 (1972).
15. U.S. Pat. 3,094,511 (June 18, 1963), H. W. Hill, Jr., S. L. Kwolek, and W. Sweeny (to E. I. du Pont de Nemours & Co., Inc.).
16. U.S. Pat. 3,354,123 (Nov. 21, 1967), P. W. Morgan (to E. I. du Pont de Nemours & Co., Inc.).
17. *Khim. Volokna* (4), 78 (1969).
18. J. Preston, R. W. Smith, and S. M. Sun, *J. Appl. Polym. Sci.* **16,** 3237 (1972).
19. H. E. Hinderer and J. Preston, *Appl. Polym. Symp.* **21,** 11 (1973).
20. J. Preston, *J. Polym. Sci. A-1* **4,** 529 (1966).
21. J. Preston, R. W. Smith, and C. J. Stehman, *J. Polym. Sci. C.* **19,** 29 (1967).
22. J. Preston and R. W. Smith, *J. Polym. Sci. B* **4,** 1033 (1966).
23. *Properties of Nomex High Temperature Resistant Nylon Fiber, NP-33 Bulletin,* E. I. du Pont de Nemours & Co., Inc., Wilmington, Del., Oct. 1969.
24. J. O. Weiss, H. S. Morgan, and M. R. Lilyquist, *J. Polym. Sci. C* **19,** 29 (1967).
25. F. Dobinson and J. Preston, *J. Polym. Sci. A-1* **4,** 2093 (1966).
26. J. Preston, *Polym. Eng. Sci.* **16**(5), 298 (1976).
27. B. S. Sprague, *paper presented at the 163rd National ACS Meeting, Boston, Mass., Apr. 1972.*
28. C. E. Hathaway and C. L. Early, *Appl. Polym. Symp.* **21,** 101 (1973).
29. J. C. Shivers and R. A. A. Hentschel, *Text. Res. J.* **44**(9), 665 (1974).
30. U.S. Pat. 3,519,355 (Apr. 13, 1976), B. R. Baird (to E. I. du Pont de Nemours & Co., Inc.).
31. *Dyeing and Finishing Nomex Type 450 Aramid, Bulletin NX-1,* E. I. du Pont de Nemours & Co., Inc., Wilmington, Del., May 1976.
32. *Properties of Industrial Filament Yarns of Kevlar Aramid Fiber for Tires and Mechanical Rubber Goods, Bulletin K-1,* E. I. du Pont de Nemours & Co., Inc., Wilmington, Del., Dec. 1974.
33. *Khim. Volokna* (6), 20 (1972).
34. T. S. Sokolova and co-workers, *Khim. Volokna* (3), 25 (1974).
35. G. I. Kudryavtsev and co-workers, *Khim. Volokna* (6), 70 (1974).
36. J. Preston, H. S. Morgan and W. B. Black, *J. Macromol. Sci. Chem.* **A7**(1), 325 (1973).
37. U.S. Pat. 3,600,350 (Aug. 17, 1971), S. L. Kwolek (to E. I. du Pont de Nemours & Co., Inc.).
38. U.S. Pat. 3,869,429 (Mar. 4, 1975), H. Blades (to E. I. du Pont de Nemours & Co., Inc.).
39. U.S. Pat. 3,296,201 (Jan. 3, 1967), C. W. Stephens (to E. I. du Pont de Nemours & Co., Inc.).
40. U.S. Pat. 3,804,791 (Apr. 16, 1974), P. W. Morgan (to E. I. du Pont de Nemours & Co., Inc.).
41. U.S. Pat. 3,932,365 (Jan. 13, 1976), R. Penisson (to Rhône-Poulenc-Textile).
42. H. C. Bach and H. E. Hinderer, *Polym. Prepr. Am. Chem. Soc. Div. Polym. Chem.* **11**(1), 334 (1970).
43. U.S. Pat. 3,770,704 (Nov. 6, 1973), F. Dobinson (to Monsanto Company).
44. U.S. Pat. 3,738,964 (June 12, 1973), F. Dobinson and F. M. Silver (to Monsanto Company).

45. N. Yamazaki, M. Matsumoto, and F. Higashi, *J. Polym. Sci. Polym. Chem. Ed.* **13**, 1373 (1975).
46. P. W. Morgan, *Condensation Polymers: By Interfacial and Solution Methods,* Interscience Publishers, a division of John Wiley & Sons, Inc., New York, 1965.
47. A. A. Federov and co-workers, *Viskomol Soyedin Ser. B* **15**(1), 74 (1973).
48. J. Preston, *Polym. Eng. Sci.* **15**(3), 199 (1975).
49. T. I. Bair, P. W. Morgan, and F. L. Killian, *Polym. Prepr. Am. Chem. Soc. Div. Polym. Chem.* **17**(1), 59 (1976).
50. S. P. Papkov and co-workers, *J. Polym. Sci. Polym. Phys. Ed.* **12**, 1753 (1974).
51. U.S. Pat. 3,850,888 (Nov. 26, 1974), J. A. Fitzgerald and K. K. Likhyani (to E. I. du Pont de Nemours & Co., Inc.).
52. U.S. Pat. 3,203,933 (Aug. 31, 1965), W. A. Huffman, R. W. Smith, and W. T. Dye, Jr., (to Monsanto Company).
53. U.S. Pat. 3,225,011 (Dec. 21, 1965), J. Preston and R. W. Smith (to Monsanto Company).
54. U.S. Pat. 3,541,056 (Nov. 17, 1970), J. Pikl (to E. I. du Pont de Nemours & Co., Inc.).
55. U.S. Pat. 2,831,834 (Apr. 22, 1958), E. E. Magat (to E. I. du Pont de Nemours & Co., Inc.).
56. U.S. Pat. 3,063,966 (Nov. 13, 1962), S. L. Kwolek, P. W. Morgan, and W. R. Sorensen (to E. I. du Pont de Nemours & Co., Inc.).
57. J. Preston and W. B. Black, *J. Polym. Sci. B* **4**, 267 (1966).
58. U.S. Pat. 3,360,598 (Dec. 26, 1967), C. R. Earnhart (to E. I. du Pont de Nemours & Co., Inc.).
59. U.S. Pat. 3,414,645 (Dec. 3, 1968), H.S. Morgan (to Monsanto Company).
60. U.S. Pat. 3,642,706 (Feb. 15, 1972), H. S. Morgan (to Monsanto Company).
61. S. W. Kwolek and co-workers, *Polym. Prepr. Am. Chem. Soc. Div. Polym. Chem.* **17**(1), 53 (1976).
62. U.S. Pat. 3,089,749 (May 14, 1963), D. G. H. Ballard (to Courtaulds, Ltd.).
63. U.S. Pat. 3,121,766 (Feb. 18, 1964), D. G. H. Ballard and J. D. Griffths (to Courtaulds, Ltd.).
64. J. A. Zapp, Jr., *Science* **190**, 422 (1975).
65. R. E. Wilfong and J. Zimmerman, *Appl. Polym. Symp.* **31**, 1 (1977).
66. U.S. Pat. 3,567,632 (Mar. 2, 1971), J. W. Richter and H. H. Hoehn.
67. U.S. Pat. 3,775,361 (Nov. 27, 1973), J. H. Jensen (to E. I. du Pont de Nemours & Co., Inc.).

General References

O. E. Snider and R. J. Richardson, "Polyamide Fibers" in N. Bikales, ed., *Encyclopedia of Polymer Science and Technology,* Vol. 10, Interscience Publishers, a division of John Wiley & Sons, Inc., New York, 1969, pp. 347–460.

W. Sweeny and J. Zimmerman, "Polyamides" in N. Bikales, ed., *Encyclopedia of Polymer Science and Technology,* Vol. 10, Interscience Publishers, a division of John Wlley & Sons, Inc., New York, 1969, pp. 483–597.

W. B. Black and J. Preston, "Fiber-Forming Aromatic Polyamides" in H. F. Mark, S. M. Atlas, and F. Cernia, eds., *Man-Made Fibers: Science and Technology,* Vol. 2, Interscience Publishers, a division of John Wiley & Sons, Inc., New York, 1968, pp. 297–364.

W. B. Black, "Wholly Aromatic High-Modulus Fibers" in C. E. H. Bawn, ed., *MTP Int. Rev. Sci., Phys. Chem. Series 2, Macromol. Sci. Vol.,* Chapt. 2, Butterworths, London, 1975, pp. 34–122.

W. B. Black and J. Preston, eds., *High-Modulus Wholly Aromatic Fibers,* Marcel Dekker, Inc., New York, 1973.

G. B. Carter and V. T. J. Schenk, "Ultra-High Modulus Organic Fibres" in I. M. Ward, ed., *Structure and Properties of Oriented Polymers,* Halsted Press, a division of John Wiley & Sons, Inc., New York, 1975, pp. 454–491.

J. Preston, "Synthesis and Properties of Rod-Like Condensation Polymers" in A. Blumstein, ed., *Liquid Crystalline Order in Polymers,* Academic Press, Inc., New York, 1978.

J. PRESTON
Monsanto Triangle Park Development Center, Inc.

CELLULOSE ACETATE AND TRIACETATE FIBERS

Acetate fiber is the generic name for a cellulose acetate [9004-35-7] fiber which is a partially acetylated cellulose, also known as secondary acetate. The desirability and wide textile use of acetate lies in its uniformly high quality, color, and styling versatility, drape, hand, and other favorable esthetic properties (see Economic Aspects). Triacetate is the generic name for cellulose triacetate [9012-09-3] fiber, also known as primary acetate. It is an almost completely acetylated cellulose. Acetate and triacetate differ only moderately in the degree of acetyl substitution on cellulose. Yet they have different chemical and physical properties. Triacetate fiber is hydrophobic; heat treatment develops a high degree of crystallinity which can be used to impart desirable fabric performance characteristics.

Cellulose acetate is the reaction product of cellulose and acetic anhydride (see Cellulose derivatives, esters). The degree of polymerization (DP) of the cellulose used for esterification is generally around 1000–1500, whereas the esterification product DP is around 300. There is evidence that the DP of cellulose acetate is relatively independent of the DP of the cellulose (1). Commercial triacetate is not a precise chemical entity because acetylation is not quite complete.

Secondary cellulose acetate is obtained by acid catalyzed hydrolysis of the triacetate to an average degree of substitution of 2.4 acetyl groups per glucose unit. The primary acetyl groups hydrolyze more readily than the secondary but the exact ratio of primary and secondary substitution depends upon the hydrolysis conditions.

Two separate terms, acetyl value (%) and combined acetic acid (%), are used to specify the degree of acetylation. Since the formula weights are 43 for the acetyl groups (CH_3CO) and 60 for the acid (CH_3COOH), the two terms are always in the ratio of 43:60. The relation of acetyl value and combined acetic acid to the number of hydroxyls acetylated per glucose residue ($C_6H_{10}O_5$) is:

$$\text{combined acetic acid, \%} = \frac{60}{43} \times (\text{acetyl value, \%})$$

$$= \frac{60 \,(\text{acetyls per glucose unit}) \times 100}{162 + 42 \,(\text{acetyls per glucose unit})}$$

Acetyls per glucose	Acetyl value, %	Combined acetic acid, %
2	35.0	48.8
2.5	40.3	56.2
3	44.8	62.5

Commercial cellulose triacetate has a combined acetic acid content of 61.5%, corresponding to 2.92 acetyls per glucose unit. Cellulose acetate with 2.4 acetyl groups per glucose unit has a combined acetic acid content of approximately 55%.

Commercial cellulose acetates contain small amounts of residual free carboxyl and sulfate groups as well as acetyl and hydroxyl groups. The thermal stability of the cellulose acetate depends upon the level and form (free acid or salt) of the sulfate groups. The salt form is ordinarily more stable and commercial acetates are carefully neutralized in the final process steps. Multivalent cations may form cross-links between residual sulfate and carboxyl groups, causing an artificially high solution viscosity. This limits the concentration of cellulose acetate used in solutions for yarn manufacture. The purity of the cellulose used to prepare the cellulose acetate also has an influence on solution rheology (2).

Properties

Structure. Cellulose acetate fibers have a low degree of crystallinity and orientation even after heat treatment. Cellulose triacetate, however, develops considerable crystallinity after heat treatment. This is apparent from x-ray diffraction diagrams which show increased crystalline order in cellulose triacetate that has been heat treated at 240°C for 1 min. Many improved fiber performance characteristics of triacetate over acetate are a result of the increased internal structural order.

Appearance and Color. Fibers of both cellulose acetate and triacetate have a bright, lustrous appearance. A duller, whiter yarn is produced by the addition of pigments (qv) such as titanium dioxide at 1–2%. The bright, nonpigmented fiber reflects light specularly whereas the dull, pigmented fiber reflects diffusely. The application of twist to a fiber bundle reduces the specular reflection and makes it more diffuse. Acetate and triacetate fibers have essentially the same light absorption characteristics in the visible spectrum. Absorption increases slightly in the ultraviolet region.

The exceptionally high level of whiteness of both types of fibers (3) is obtained by careful selection of high purity wood pulp, manufacturing conditions, and pigment inclusion and permits bright, clean, pure colors to be obtained when the fibers are dyed.

Dyeing Characteristics. Disperse dyes (qv) are most frequently used for cellulose acetate and triacetate fibers. They are high melting, crystalline compounds that have low solubility in the dye bath even at high temperatures. They are milled to very small particle size to permit effective dispersion without agglomeration in the dye bath. A small quantity of disperse dye is dissolved in the aqueous medium and then diffuses into the fiber to give a uniform color. The rate-determining step in the dyeing operation is the slow diffusion of the large dye molecules into the fibers. Properties such as dye bath temperature and fiber composition modify the dyeing rate. Triacetate fibers are dyed more slowly than acetate fibers but dye carriers (qv) can be used to accelerate the rate of dyeing. Some typical carriers are based on butyl benzoate, methyl salicylate, diallyl phthalate, and diethyl phthalate (4–14).

Selection of the appropriate azo, anthraquinone, or diphenylamine disperse dye ensures good colorfastness. Fading inhibitors are used to resist the effects of nitrogen oxides and ozone. Triacetate fabrics are heat-treated to raise the safe ironing temperature, drive the dye further into the fiber to gain further gas-fading resistance, and improve colorfastness.

Inherently colored acetate and triacetate yarns are produced by incorporating colored pigments (inorganic or organic) or soluble dyes in the acetate solution prior to spinning. Solution-dyed acetate and triacetate yarns are extremely colorfast to washing, dry cleaning, sunlight, perspiration, sea water, and crocking, and usually surpass the performance of vat dyed yarns. In addition, acetate and triacetate are susceptible to gas- or fume-fading conventionally; such fading is absent with solution dying. Unfortunately, the additional cost of producing solution-dyed acetate fiber has restrained its market acceptance.

Specific Gravity. Fiber cross sections are often irregular and an immersion technique is commonly used to measure specific gravity of fibers. The immersion fluid may interact with the fiber by absorption or surface wetting, and the selection of a suitable fluid is critical for accurate measurements. Values for acetate have been reported (15–17) for different solvents over a range of 1.306 with carbon tetrachloride to 1.415 with n-heptaldehyde. The values of 1.32 for acetate and 1.30 for triacetate are accepted for fibers of combined acetic acid contents of 55 and 61.5%, respectively.

Refractive Index.   The refractive index parallel to the fiber axis (ϵ) is 1.478 for acetate and 1.472 for triacetate. The index perpendicular to the axis (ω) is 1.473 for acetate and 1.471 for triacetate. The difference between ϵ and ω (birefringence) is very low for acetate fiber and practically undetectable for triacetate.

Absorption and Swelling Behavior.   The absorption of moisture by acetate and triacetate fibers generally depends upon the relative humidity to which the fibers are exposed. However, it varies according to whether equilibrium is approached from the dry or the wet side. This hysteresis effect is noted over the entire range of relative humidities as shown in Figure 1 (18).

Additional moisture regain isotherms for acetate fiber have been reported (19). Heat-treated triacetate fiber has a lower regain than nonheat-treated fiber, and values in the range of 2.5–3.2% have been observed (4–5,20–21).

In the United States, commercial regain is used to calculate commercial weights of yarns or fibers. In effect, commercial moisture regain is added to the weight of bone-dry fiber to account for the moisture normally found in textile fibers. The percentage of commercial regain, taken from ASTM D 1909-68, is 6.5 for acetate fiber and 3.5 for triacetate. In Europe and elsewhere, the Bureau International Pour La Standardisation De La Rayonne Et Des Fibres Synthetiques (BISFA) rules (21) are used for determining the commercial weight of acetate and other man-made fibers. The basic premise differs from that of the ASTM in that the BISFA include a percentage for the normal finish present on the fiber as shipped as well as a figure to account for the moisture normally found in fibers. Thus the commercial weight of acetate according to the BISFA rules includes a figure of 9% in determining the commercial weight of acetate fiber, based on the bone-dry, finish-free weight of the fiber.

Percentage of water imbibition is an important property in ease-of-care and quick-drying fabrics. This value is determined by measuring the moisture remaining in the fiber when equilibrium is established between a fiber and air at 100% rh while the fiber is being centrifuged at forces up to 1000 g. Average recorded values are: acetate, 24%; nonheat-treated triacetate, 16%; heat-treated triacetate, 10%.

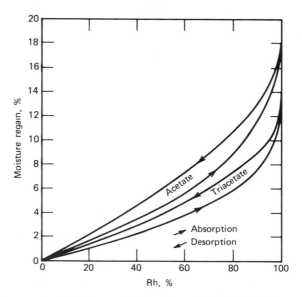

Figure 1.   Moisture regains (moisture content on a bone-dry basis) of cellulose acetate and triacetate fibers on absorption and desorption at 22°C (12,19).

Absorption of water by fibers causes swelling roughly proportional to moisture content. Although there is considerable disagreement among investigators in the field, the average value for the increase in length of acetate fibers due to water absorption is about 1%, and the average increase in diameter is approximately 10%. The corresponding values for the swelling of triacetate fiber are lower than those for acetate. Tesi (22) reported a 1.5% cross-sectional area increase for heat-treated triacetate fiber and 4.0% for nonheat-treated fiber.

Thermal Behavior. Acetate, like other thermoplastic fibers, sticks, softens, and even melts when ironed at high temperatures. Sticking and softening temperatures of all fibers depends on such factors as yarn diameter, fabric construction, and general fabric geometry, and varies with different test procedures. Sticking and softening temperatures are not necessarily directly related to the fiber melting point; acetate softens and sticks in the range of 190–205°C but fuses at approximately 260°C. The apparent shining or glazing temperature is usually lower than the sticking temperature and is also influenced by moisture content, fabric construction, and color. (When ironing acetate fabrics the sole-plate temperature of hand irons should not exceed 170–180°C.) The sticking and glazing temperatures of nonheat-treated triacetate fiber are in the same range as those of acetate; but heat-treated triacetate fibers have considerably higher sticking and glazing temperatures. The latter fabrics can be ironed at temperatures as high as 240°C. The melting point of triacetate is approximately 300°C.

Acetate and triacetate, because they are thermoplastic fibers, exhibit changes in mechanical properties as a function of temperature. However, within the range of normal climatic temperatures, the mechanical properties are not altered significantly. As temperature is raised, the modulus of acetate and triacetate fibers is reduced, and the fibers extend more readily under stress (Fig. 2).

Acetate and triacetate are weakened by prolonged exposure to elevated temperatures in air (Fig. 3).

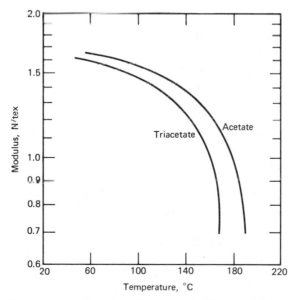

Figure 2. Tensile modulus as a function of temperature. To convert N/tex to gf/den, multiply by 11.33.

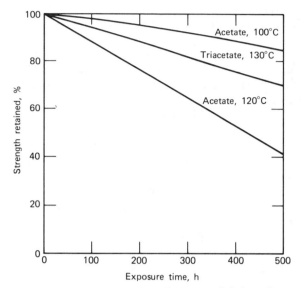

Figure 3.   Effect of dry heat exposure on acetate and triacetate (lubricant-free yarns tested at standard conditions of 65% rh and 21°C).

Light Stability.   Resistance of textile fibers to sunlight degradation depends upon the wavelength of the incident light, relative humidity, and atmospheric fumes. Although fibers cannot be accurately rated for their resistance to sunlight degradation in terms of Langley units or hours of exposure, certain generalizations appear justified. Acetate and triacetate fibers, when exposed under glass, behave similarly to cotton and rayon, ie, they are somewhat more resistant than unstabilized pigmented nylon and silk and appreciably less resistant than acrylic and polyester fibers. When acetate and triacetate are exposed to direct weathering, their resistance is lowered as compared with exposure under glass. Additional information on actinic degradation is described in refs. 23–29.

Certain pigments, particularly carbon black, offer protection from Fade-Ometer sunlight exposure, as indicated for acetate in Figure 4. Special yarn additives may be used to improve light stability for a specific use such as drapery fabrics (29).

Electrical Behavior.   Because acetate and triacetate yarns readily develop static charges, it is sometimes desirable to apply an antistatic finish to aid in textile processing. Both yarns have been used for electrical insulation after lubricants and other finishing agents have been removed by solvent extraction, followed by washing with water. Table 1 contains data from a report of the British Cotton Industry Association (30) and lists the resistivity in $M\Omega \cdot cm$ of various fibers over a range of relative humidities (45–95%) (see Antistatic agents).

Cellulose acetate has a high electrical resistivity. Table 2 shows comparative resistivity of scoured acetate and triacetate taffeta fabrics.

Mechanical Properties.   The mechanical and esthetic properties of textile fabrics prepared from fibers and yarns depends greatly on the physical form of the fibers themselves and the geometrical construction of the fabric. Some important performance criteria include: hand, drape, wrinkle resistance and recovery, strength, and flexibility. The interactions among fibers and yarns in a fabric array are complex and the most straightforward way of describing the mechanical properties of the fabric is to refer to the inherent mechanical properties of the acetate. Fiber mechanical properties are described by the stress–strain and recovery behavior under conditions of tensile, torsional, bending, and shear loading.

Tensional Properties.   Stress–strain curves typical of commercial acetate and triacetate yarns are shown in Figures 5 and 6. The curve is indicative of most com-

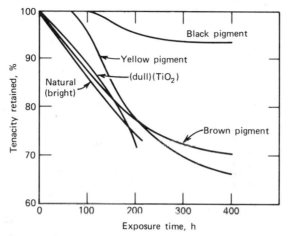

Figure 4.   Effect of colored pigments on tenacity retained by cellulose acetate fiber after carbon arc Fade-Ometer exposure. Tenacity measured at standard conditions of 65% rh and 21°C.

Table 1.   Resistivity (MΩ·cm) of Various Fibers in Commercial and Purified States[a]

Rh, %	Acetate Commercial	Acetate Purified	Nylon Commercial	Nylon Purified	Cotton Commercial	Cotton Washed[b]	Rayon Commercial	Rayon Purified
45	967,000	81,500,000	813,000	6,430,000	149		543	1,720
50	662,000	21,600,000	585,000	3,200,000	64		235	680
55	424,000	6,040,000	387,000	1,430,000	30		95	266
60	256,000	1,650,000	208,000	525,000	14	530	36	93
65	150,000	448,000	104,000	193,000	6.0	150	13	34
70	74,000	126,000	43,000	70,000	2.4	38	4.6	12
75	28,900	33,200	14,500	20,000	1.02	11.2	1.8	4.4
80	7,200	9,000	4,000	6,000	0.33	3.1	0.63	1.5
85	1,610	2,460	863	1,290	0.106	0.91	0.23	0.52
90	160	370	120	180	0.024	0.23	0.06	0.13
95	11	39	8.5	18.5			0.012	0.02

[a] Ref. 30.
[b] Water-washed only.

Table 2.   Resistivity of Acetate and Triacetate Fabrics[a,b]

Fabric	Resistivity, Ω·cm
cellulose acetate	1.27
cellulose triacetate, nonheat-treated	3.81
heat-treated	15.2
specially scoured and heat-treated	1016

[a] Ref. 5.
[b] ASTM Method D 1000-55T.

mercial acetate and triacetate yarns although some high tenacity triacetate and acetate fibers are being produced. Table 3 shows some mechanical properties characteristic of both acetate and triacetate commercial yarn and fibers.

The ability of a material to resist deformation under an applied tensile stress is

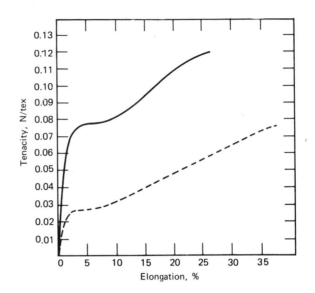

Figure 5. Typical stress–strain curves for bright acetate yarn under standard and wet conditions. Instron tensile tester rate at 60%/min rate of extension, 3.9 cm gage length. ——, 65% rh, 21°C; - - - - -, wet, 21°C. To convert N/tex to gf/den, multiply by 11.33.

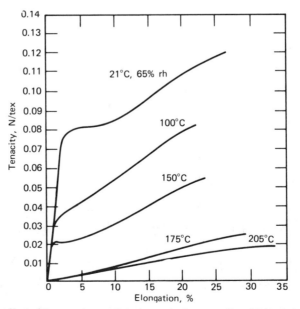

Figure 6. The effect of temperature on tne stress–strain properties of triacetate yarn. Instron tensile tester rate at 60%/min rate of extension, 3.9 cm gage length (6). To convert N/tex to gf/den, multiply by 11.33.

Table 3.   Tenacity and Elongation of Commercial Acetate and Triacetate Yarns and Fibers[a]

Properties	Acetate and triacetate
tenacity, N/tex[b]	
standard conditions[c]	0.10–0.12
wet	0.07–0.09
bone-dry	0.12–0.14
knot, standard conditions[c]	0.09–0.10
loop, standard conditions[c]	0.09–0.10
elongation at break, %	
standard conditions[c]	25–45
wet	35–50

[a] Ref. 5.
[b] To convert N/tex to gf/den, multiply by 11.33.
[c] 65% rh, 21°C.

measured by the modulus of elasticity (Young's modulus). In viscoelastic materials, the apparent modulus of a textile fiber is defined as the ratio of stress to strain in the initial, linear portion of the stress–strain curve. The apparent modulus at low strain levels is directly related to many of the mechanical performance characteristics of textile products. The modulus of elasticity can be affected by drawing (elongating) the fiber and by other manufacturing procedures. Values for commercial acetate and triacetate fibers are generally 2.2–4.0 N/tex (25–45 gf/den).

The wet modulus of fibers at various temperatures is significant for textile applications because it influences the degree of creasing and mussiness caused by laundering. Figure 7 shows the change with temperature of the wet modulus of acetate, triacetate, and a number of other fibers (31).

The ability of a fiber to absorb energy during straining is measured by the area under the stress–strain curve. This property is also known as toughness or work of rupture. Table 4, according to Meredith (32), lists the work of rupture of acetate in comparison with other textile fibers. Work of rupture of triacetate fiber is essentially the same as that of acetate.

A fiber which is strained and allowed to recover will give up a portion of the work absorbed during straining. The ratio of the work recovered to the total work absorbed (measured by the respective areas under the stress–strain and stress–recovery curves) is designated as resilience.

The elongation of a stretched fiber is best described as a combination of instantaneous extension and a time dependent extension or creep. This is called viscoelastic behavior and is common to many textile fibers, including acetate. Conversely, recovery of viscoelastic fibers is typically described as a combination of immediate elastic recovery, delayed recovery, and permanent set or secondary creep. The permanent set is residual extension which is not recoverable. Susich and Backer (33) have described these three components of recovery for acetate (Table 5).

Temperature and moisture content of the fiber affect viscous behavior and hence modify the stress–strain relationship. Most stress–strain data are reported under standard conditions of 21°C and 65% rh.

Strains of more than 10% are avoided in textile processing so that the dimensional or shape stability of the resultant fabric will be acceptable.

Bending.   The bending properties of a fiber generally depend on the viscoelastic behavior of the material. However, in most textile applications the radius of curvature

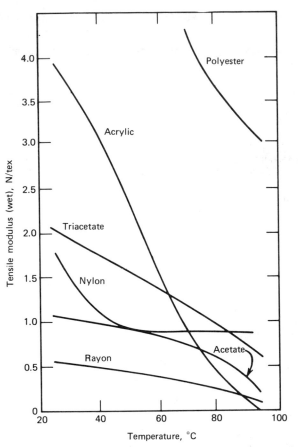

Figure 7. Effect of water temperature on wet modulus of fibers. To convert N/tex to gf/den, multiply by 11.33.

Table 4. Work of Rupture of Acetate and of Some Other Common Fibers[a]

Fibers	Work of rupture (toughness), N/tex[b]
acetate	0.022
cotton	0.010
nylon	0.076
rayon (viscose regular)	0.023
silk	0.072
wool	0.032

[a] Ref. 32.
[b] To convert N/tex to gf/den, multiply by 11.33.

Table 5. Elongation Recovery of Acetate Fibers[a]

Fiber	Immediate elastic recovery, %	Delayed recovery, %	Permanent set, %
acetate multifilament			
at 50% of breaking tenacity	74	26	0
at breaking point	14	16	70
acetate staple yarn			
at 50% of breaking tenacity	58	42	0
at breaking point	12	18	70

[a] Ref. 33.

Table 6.   Flexural Rigidity of Cellulose Acetate Fibers as a Function of Tex

Tex[a]	Flex rigidity, g·m^2
6.7	83
3.3	27
2.2	8.9
1.7	7.4
1.1	2.5
0.56	0.56
0.42	0.35

[a] To convert tex to den, multiply by 9.

of bending is relatively great, and the imposed strains are of a low order of magnitude. As a first approximation, it is possible to examine the bending properties by the classical methods. The bending stiffness or flexural rigidity of a fiber is the product of the bending modulus and the moment of inertia of the cross section. Thus for fibers of round cross section and constant modulus, the flexural rigidity varies directly with the square of the tex. Table 6 gives data on the flexural rigidity of acetate fibers as a function of fiber tex.

Torsion and Shear.   For a discussion of torsional and shear properties of acetate and other fibers, see Kaswell (34).

Chemical Properties.   Under slightly acidic or basic conditions at room temperature, acetate and triacetate fibers are very resistant to chlorine bleach at the concentrations normally encountered in laundering.

Triacetate fiber is significantly more resistant than acetate to alkalies encountered in normal textile operations. It is recommended that temperatures no higher than 85°C or a pH above 9.5 be used when dyeing acetate. Under normal scouring and dyeing

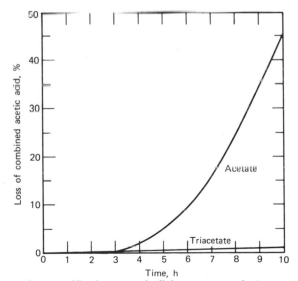

Figure 8.   Comparative saponification rates of cellulose acetate and triacetate at scouring conditions of pH 9.5–9.8 and 95°C (23).

conditions, alkalies up to pH 9.5 and temperatures up to 96°C may be used with tri-acetate with little saponification or delustering. Heat-treated triacetate fiber has even greater alkali resistance, as shown in Figure 8. Strong alkalies and boiling temperatures saponify triacetate as well as acetate fiber.

Acetate and triacetate are essentially unaffected by dilute solutions of weak acids. Strong mineral acids cause serious degradation of both fibers. The results of exposure of nonheat-treated and heat-treated triacetate taffeta fabrics to various chemical reagents have been reported (5).

Acetate and triacetate fibers are not affected by the perchloroethylene dry cleaning solutions normally used in the United States and Canada. Trichloroethylene, employed to a limited extent in Great Britain and Europe, softens triacetate.

Resistance to Microorganisms and Insects. The resistance (based on soil-burial tests) of triacetate to microorganisms is very high, approaching that of polyester, acrylic, and nylon fibers. Figure 9 shows comparative soil-burial test results on acetate, triacetate, and cotton. Neither acetate nor triacetate fiber is readily attacked by moths or carpet beetles; however, there have been a few instances in which test larvae have cut through acetate to get at wool fibers, or have damaged acetate contaminated with foreign substances containing starches.

Manufacture of Cellulose Acetate and Triacetate Flakes

Acetate and triacetate flakes are prepared by the esterification of high purity chemical cellulose with acetic anhydride (35–37). High purity cellulose is required because the solution properties of the resultant acetate or triacetate flake may be adversely affected by impurities even at low concentrations; extrusion processes require the polymer to flow freely through small diameter capillaries (eg, 30–80 μm). Wood pulp has replaced cotton linters as a source of chemical cellulose except for special plastic-grade acetates where color and clarity are very important.

Wood contains 40–50% cellulose, hemicellulose, lignin, and extractives. The noncellulose impurities must be removed to produce an acetylation-grade wood pulp with a purity of 95–98% α-cellulose. A strong economic incentive to cope with the higher

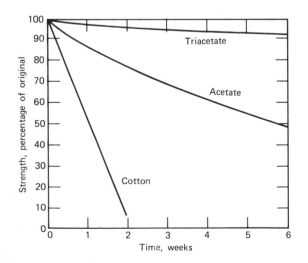

Figure 9.   Resistance to biological attack as measured by residual tenacity after various periods of soil burial (12).

level of hemicellulose impurities in lower purity pulp has led to continued technical interest (38–39). Although several processes have been developed (40–42), none is commercial.

The other raw materials for the manufacture of cellulose acetate, acetic acid, acetic anhydride, and sulfuric acid, are items of commerce and can be obtained in high purity and uniformity.

Secondary Acetate Processes.   Three major processes are used to produce cellulose acetate. A solution process is most common (Fig. 10); acetylation is obtained with acetic anhydride using glacial acetic acid as the solvent. The sulfuric acid catalyst increases the solubility of the partially esterified cellulose in acetic acid until sufficient acetyl groups are added to achieve solubility of the cellulose acetate in acetic acid. Two major variants of this process are high catalyst (10–15 wt % sulfuric acid based on cellulose) and low catalyst (<7 wt % sulfuric acid) procedures. The second most common process is the solvent process. Methylene chloride is substituted for all or part of the acetic acid and acts as a solvent for the triacetate as it is formed. Perchloric acid is frequently used as a catalyst as sulfuric acid is not required to form a soluble intermediate. A third method for producing secondary acetate is a heterogeneous process (also called nonsolvent or Schering process) in which an inert organic liquid, such as benzene or ligroin, is used as a nonsolvent to prevent the acetylated cellulose from dissolving as it is formed. The cellulose ester produced is never solubilized, hence its physical form is similar to the original cellulose fiber. More detailed information on each of the above three processes has been presented by Malm and Hiatt (43).

To obtain a secondary acetate which is soluble in acetone, it is necessary to completely acetylate the cellulose during the dissolution step and then hydrolyze it, while still dissolved, to the required acetyl value. The overall process may be separated into four main steps: (1) preparation of cellulose for acetylation, (2) acetylation, (3) hydrolysis, and (4) recovery of cellulose acetate and solvents.

Preparation of Cellulose for Acetylation.   Wood pulp is customarily supplied in rolls weighing up to 300 kg but can be obtained in bales of individual sheets. The pulp sheet must be fluffed with a disk refiner to generate a large surface area for acetylation because the rate and completeness of acetylation depends on the accessibility of the

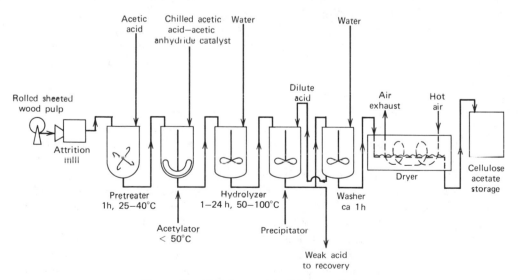

Figure 10.   Cellulose acetate flake manufacture.

cellulose (44–45). The fluffed pulp is treated to further increase its accessibility to the acetylation solution. Several agents can be used for this step (called pretreatment) but an acetic acid–water mixture is most common. Sufficient water for pretreatment is usually present when the moisture content of the pulp is ca 6% of the weight of cellulose. Pretreatment generally involves agitating the pulp–acetic acid mixture for about 1 h at 25–40°C.

The ratio of pulp to pretreat acid depends upon the catalyst level used. An activation stage is added to the low catalyst acetylation procedure. An acetic acid–sulfuric acid mixture is introduced until the sulfuric acid concentration is 1–2% of the pulp weight. The activation period may last for 1–2 h during which the degree of polymerization of the cellulose is reduced. Activation time and temperature are selected so as to achieve the desired degree of polymerization in the acetate flake product. The high catalyst procedure does not usually involve an activation step. The degree of polymerization of the acetate flake is controlled by the conditions selected during acetylation and hydrolysis. The pulp is charged to the acetylation reactor after pretreatment and activation.

Acetylation. The acetylation mixture, consisting of acetic anhydride (esterifying agent), acetic acid (solvent), and sulfuric acid (catalyst), is precooled prior to its addition to the pretreated cellulose. Precooling provides a heat sink for two exothermic reactions that occur in the acetylation process, the esterification of cellulose with acetic anhydride liberates 1.03 kJ/g (246 cal/g) of cellulose, and the reaction of acetic anhydride with water from the pretreat mix generates 3.30 kJ/g (789 cal/g) of H_2O. Because heat is generated quickly and the mixture is viscous, a jacketed vessel does not provide the necessary cooling capacity. A separate vessel called a crystallizer is used to prechill the acetylation mixture and freeze some of the acetic acid. The heat of fusion of the uniformly dispersed acetic acid crystals and the low temperature of the mixture provide readily accessible heat sinks; this is particularly important in the early stages of acetylation where a rapid temperature rise would reduce the degree of polymerization of the cellulose. A cold brine circulates through the jacket of the acetylizer to chill the equipment before esterification. In the methylene chloride process, the required heat transfer is provided by refluxing the solvent. For either process, a 5–15 wt % excess of acetic anhydride assures complete reaction. A series of simultaneous, complex reactions occur during acetylation and a large body of literature is available describing both the solution chemistry and acetylation reaction (35–37, 46–49).

The cellulose chain is partially depolymerized during acetylation by the action of the sulfuric acid catalyst. High temperature and a large amount of catalyst accelerate depolymerization.

Temperatures in excess of 50°C are avoided. At the end point, microscopic examination of the solution should reveal no undissolved residues. The acetylation reaction is terminated by adding water to destroy the excess anhydride and provide a level of 5–10% total water for hydrolysis. A 15–25% cellulose acetate concentration is typical.

Hydrolysis. The number of acetyl groups present in each anhydroglucose unit at the end of acetylation is slightly less than 3.0 and must be reduced to around 2.4 to prepare secondary cellulose acetate. The number of acetyl groups is reduced and the combined sulfate groups minimized by acid hydrolysis under controlled conditions of time, temperature, and acidity. The sulfate groups hydrolyze more easily than the acetyl groups. The sulfuric acid formed increases the acidity of the reaction solution. In the high catalyst acetylation case, a portion of the sulfuric acid is neutralized, eg, by the addition of sodium acetate or magnesium acetate to reduce the free sulfuric acid content and prevent excessive depolymerization of the cellulose acetate. Hydrolysis temperature, normally 50–100°C, is obtained by direct steam injection (50) and reaction time varies from 1 to 24 h.

Flake Recovery. Precipitation, washing, and drying are the final steps in flake preparation. The precipitation procedure varies according to the product desired. In flake preparation, the hydrolyzed cellulose acetate solution is brought to incipient precipitation by mixing it with a stream of dilute (10–15%) acetic acid. Additional dilute acetic acid is rapidly introduced and the solution vigorously agitated. To obtain a powder precipitate, the agitated solution is slowly diluted until precipitation occurs. Another process involves extrusion of the hydrolyzed solution through small holes into a precipitating acid bath; this produces fine strands which are then cut into pellets.

The precipitated cellulose acetate is physically separated from the dilute (25–35%) acetic acid. The flake is sent to a washer which removes the acetic acid and salts remaining from sulfuric acid neutralization. The wet flake is dried to a moisture level of 1–5% in a suitable commercial dryer. The dilute acetic acid that results from the washing and precipitation steps can be used directly in other stages of the process or recovered for reuse. Efficient recovery and recycle of the dilute acetic acid is an economic necessity.

A pressure stabilization step may be required if thermal stability and nonyellowing on heating are critical considerations (eg, in thermoplastic molding application). The acetate flake is heated in deionized water at ≤ 1.4 MPa (200 psi) for ca 1 h. This treatment removes residual sulfate groups by hydrolysis.

Acetate and triacetate flakes are white amorphous solids produced in granular, flake, powder, or fiber form. They are used as raw materials in the preparation of fibers, films, and plastics. Flake density varies with physical form and ranges from 100–500 kg/m^3 in loose bulk. Acetate flake is shipped by trailer truck or railroad freight car. Smaller shipping quantities are packaged in multiwall paper bags. The bag may include a vapor-barrier layer to protect against a change in moisture content due to atmospheric humidity.

Acid Recovery. Approximately 4.0–4.5 kg of acetic acid per kg of cellulose acetate is used in the homogeneous process. Approximately 0.5 kg is consumed in the product and the remaining 3.5–4.0 kg is recovered. Acetic acid for recovery leaves the main process sequence as an aqueous solution of 25–35% acetic acid. It may also contain dissolved salts from sulfuric acid neutralization, and dissolved and suspended low molecular weight cellulose and hemicellulose acetates. Suspended material is removed in a settling tank or filter. Acetic acid is recovered from the clarified weak acid stream by solvent extraction. Several frequently used organic solvents are benzene, ethyl acetate, and methyl ethyl ketone. The organic extract upper layer from the extractor is sent to a distillation column. The aqueous raffinate phase, containing most of the inorganic salts, is discarded. In the distillation column, the extraction solvent is taken overhead and glacial acetic acid is the bottom product. The energy requirements for acid recovery depend upon the specific solvent used and may be in the range of 4.2–10.5 kJ/g (1–2.5 kcal/g) of acid recovered. A portion of the acetic acid may be subsequently converted to acetic anhydride required for acetylation. This is achieved by catalytic pyrolysis and produces acetic anhydride in good yield and at a low cost (see Acetic acid).

Batch Preparation of Triacetate. The batch triacetate process differs from the preparation of secondary acetate in that there is little hydrolysis and the temperature is in the range of 50–100°C. The triacetate hydrolysis step (or desulfation) removes only the sulfate groups from the polymer by slow addition of a dilute acetic acid solution containing sodium or magnesium acetate (51–52) or triethanolamine (53) to neutralize the liberated sulfuric acid. Meanwhile, the temperature is kept above the peak reaction temperature. The cellulose triacetate product has a combined acetic acid content of 61.5%. Flake recovery is similar to the secondary acetate process.

Sulfuric acid levels as low as 1% can be used for acetylation in the methylene chloride process. Only a minor amount of desulfation is required to reach a combined acetic acid level of 62.0%. If perchloric acid is used as the catalyst, the nearly theoretical value of 62.5% combined acetic acid is obtained. Standard flake recovery methods are used.

The nonsolvent process can also be used to prepare triacetates of nearly theoretically combined acetic acid. The catalyst is normally perchloric acid. However, because the original cellulose was never dissolved, the triacetate retains a fibrous appearance.

Continuous Flake Manufacturing Processes.   All processes described thus far have incorporated batch pretreatment, acetylation, and hydrolysis stages. A continuous process offers potential economic advantages owing to reduced labor requirements, lower energy consumption, a more uniform, higher quality product, lower raw materials consumption, and complete automation. Major disadvantages include: more highly specialized equipment, increased maintenance problems with the more sophisticated equipment, and the entire production capability is lost if there are malfunctions in the separate process stages of pretreatment, acetylation, or hydrolysis.

Several continuous processes have been described in the patent literature (54–71). The Societé Rhodiaceta continuous solution process is used commercially to produce both secondary acetate and triacetate. A continuous triacetate process is operated in the United States by Celanese Corporation.

The basic processing stages are identical to those of the batch process but some equipment and materials handling techniques are different. The initial pretreatment segment is actually a batch operation. The cellulose is weighed and combined with a measured amount of 80–90% aq acetic acid. This mixture is slurried 10–15 min to produce a 2–5% suspension of cellulose in the liquid. The slurry is sent to a holding tank which continuously feeds the subsequent processing steps. The water in the pretreatment mixture produces a completely swollen, highly reactive cellulose. It is much more reactive toward acetylation than batch-pretreated cellulose. However, the water must be removed from the pretreated pulp slurry prior to acetylation to reduce the consumption of acetic anhydride. This is accomplished by countercurrent leaching with glacial acetic acid in a multistage, belt-type extractor. In each successive stage the cellulose slurry is contacted with a more concentrated acetic acid solution. The number of stages is selected to reduce the water content to a specified low level at the extractor exit. A reduced pressure is applied at each stage to reduce liquor carryover between stages. The pretreated cellulose mat leaving the extractor is further deliquored prior to entering the acetylizer.

The acetylizer is a specialized, high energy input reactor with a range of materials-handling capability. It receives matted pulp, acetic acid, catalyst, and acetic anhydride, and produces a highly viscous, homogenous solution of cellulose triacetate. The acetylizer must provide: thorough mixing and kneading action, and positive transport of material through the reactor; a plug flow residence time distribution to ensure complete reaction and uniform flake properties; and good heat transfer capability to control the exothermic acetylation reaction. The viscous triacetate solution from the acetylizer is pumped into a high speed blender where water or a dilute acetic acid solution stops the reaction.

The hydrolysis step consists of a series of retention tanks which provide a controlled time–temperature exposure to reach the desired combined acetic acid and residual sulfate level. The remainder of the process (precipitation, washing, and drying) is continuous and similar to the batch processes previously described.

Manufacture of Cellulose Acetate and Triacetate Fibers

Extrusion Processes.   Polymer solutions are converted into fiber form by spinning or extrusion. The dry extrusion process is used primarily for acetate and triacetate. A solution of cellulose acetate or triacetate polymer in a volatile solvent is forced through a multihole spinneret into a cabinet of warm air; the fibers are formed by evaporation of the solvent. In wet extrusion, a polymer solution is forced through a spinneret into a nonsolvent liquid which coagulates the filaments and removes the solvent. In melt extrusion, molten polymer is forced into air which cools the strands into filaments.

The dry-extrusion process consists of four main operations: (1) dissolution of the cellulose acetate in a volatile solvent, (2) filtration of the solution to remove insoluble matter, (3) extrusion of the solution to form fibers, and (4) lubrication and take-up of the yarn on a suitable package.

Acetone is the universal solvent for secondary acetate in dry extrusion. The optimum concentration for an acetate spinning solution falls between the highest possible solids concentration and the resulting high solution viscosity. Though higher concentrations of solids produce fibers with better properties, practical limits of viscosity are quickly reached. If water is added to the acetone solvent, the solution viscosity exhibits a minimum at approximately 9:1 acetone to water. Typical solvent composition is about 95% acetone and 5% water, typical solids concentration in the extrusion solution is 20–30% depending on the polymer molecular weight. The viscosity of the solution at room temperature is about 100–300 Pa·s (1000–3000 P).

The solubility of acetates of different combined acetic acid content in an acetone–water and acetic acid–water solution is shown in Figures 11 and 12. Cellulose triacetate is insoluble in acetone and another solvent system must be used for dry extrusion. Triacetate solvents include chlorinated hydrocarbons such as methylene chloride, methyl acetate, acetic acid, dimethylformamide, and dimethyl sulfoxide. Methylene chloride containing 5–15% methanol is often employed.

Acetate and triacetate dissolution and handling operations are conducted in fully enclosed systems for several reasons. Both acetone and methylene chloride are expensive, acetone is highly flammable in air, and both solvents have strict worker exposure levels administered under the Occupational Safety and Health Act. Ultimately, all solvent vapor must be collected and recycled through the process. This recovery operation represents a significant operating cost.

Acetate or triacetate flake is charged to large, heavy-duty mixers along with solvent and a filter aid such as wood pulp fibers. Concentration, temperature, and mixing uniformity are closely controlled through several mixing stages. If a delustered or dull fiber is desired, 1–2% of a finely ground titanium dioxide pigment may be added in the mixing process or by injection of a pigment slurry after filtration. In the latter method, particular care must be exercised to achieve thorough mixing and maintain strict control of the pigment slurry. The polymer concentration and the composition of the extrusion solvent strongly affect the uniformity and tensile properties of the acetate fiber and must be closely controlled.

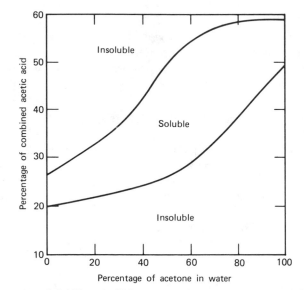

Figure 11.   Solubility of cellulose acetate in acetone–water at 25°C (72).

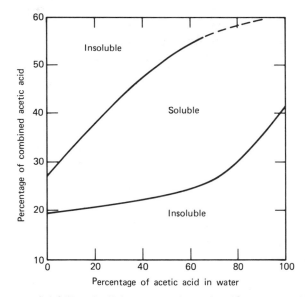

Figure 12.   Solubility of cellulose acetate in acetic acid–water at 25°C (72).

For successful extrusion of acetate filaments, the polymer solution should pass through spinnerets whose holes range from 30–80 μm dia. Consequently, the solution must be free of unacetylated cellulose, undissolved gels, and dirt. Multistage filtration, usually consisting of plate-and-frame filter presses with fabric and paper filter media, removes extraneous matter before extrusion. The acetate solution is heated during filtration to reduce viscosity and increase flow rate. The solution may be allowed to degas in holding tanks between each stage of filtration.

Extrusion of the acetate solution is a precisely controlled process. The filtered, preheated solution is delivered to the spinneret at constant volume by very accurate metering pumps. A separate pump for each spinneret position ensures uniform fiber formation. The spinnerets are made of stainless steel or another suitable metal and may contain from thirteen to several hundred holes which are precision-made to close size and shape tolerances. Auxiliary filters are used both prior to the fixture which holds the spinneret and in the spinneret itself as a final precaution against particulate matter in the extrusion solution. A schematic diagram of the extrusion process is shown in Figure 13.

Prior to entering the spinneret the extrusion dope solution is heated to lower the solution viscosity to an appropriate value and to provide some of the heat necessary to flash the solvent from the extruded filament. A thermostatically-controlled heat exchanger may be used to heat the dope or the filter–spinneret assembly may be located inside the heated extrusion cabinet.

The heated polymer solution emerges as a plurality of filaments from the spinneret into a column of warm air. Instantaneous loss of solvent from the surface of the incipient filament causes a solid skin to form over the still liquid interior of the filament. As the filament is heated by the warm air, more solvent evaporates and is carried away. More than 80% of the solvent can be removed during the brief residence time of less than one second in the hot air column. The air column or cabinet height is 2–8 m de-

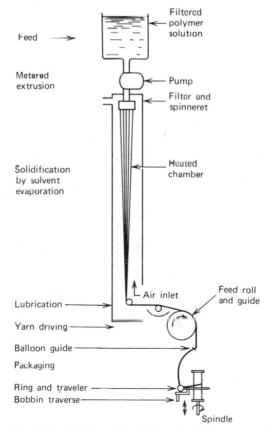

Figure 13.   Dry extrusion—dry spinning of cellulose acetate fibers.

pending on the extent of drying required and the spinning speed. The air flow may be concurrent or countercurrent to the direction of fiber movement. The fiber properties are contingent upon the solvent removal rate and precise air flow and temperature control is necessary.

A feedroll applies tension to the bundle of acetate fibers, or yarn, to withdraw them from the extrusion cabinet. The product of a spinning position is called a continuous filament yarn (as distinguished from staple). Cellulose acetate yarns are produced from 4 to >100 tex (36 to >900 den). Feedroll speed, metering pump output, and cabinet conditions must be carefully balanced to produce a yarn of specified and uniform denier.

A finish or lubricant gives the extruded yarn the frictional and antistatic properties required for further processing. The finish is applied at levels of 1–5% as the yarn exits from the cabinet. The formulation of a lubricant depends upon the intended use of the yarn; many proprietary types are used. The lubricated yarn, containing only a small amount of residual solvent, is taken up on a ring twister which inserts just enough twist to avoid the handling difficulties of untwisted yarn on a bobbin. Yarn with no twist may be wound on a cylindrical tube. Instead of applying twist to the yarn to aid in handling, the yarn filaments may be compacted or entangled by passing the yarn through a device which intermixes the filaments with air jets (73–75).

The solvent used to form the extrusion dope, which is evaporated during the extrusion process, must be recovered. Adsorption on activated carbon or refrigeration to condense the solvent are the usual processes. Final purification is by distillation. Recovery is about 99% efficient. Approximately 3 kg of acetone must be recovered per kg of acetate yarn produced. Recovery of solvent from triacetate extrusion is similar but about 4 kg of methylene chloride solvent is used per kg of triacetate yarn extruded.

Only a small quantity of triacetate yarn is made by wet extrusion (66) since extrusion speeds are much lower than for dry extrusion and the process is not attractive for producing filament yarns. A solution of the triacetate or a stabilized reaction product in acetic acid is extruded into a nonsolvent bath, such as water, and the filaments form as the triacetate precipitates from solution. A large number of holes in the spinneret (eg, >1000) improves processing economics.

A melt extrusion process is not used to any great extent for fiber formation although a small quantity of triacetate yarn has been produced by this procedure. The residence time of the polymer at the melting point must be minimized to prevent degradation of the acetate.

Types of Yarns and Fibers.   Many different acetate and triacetate continuous filament yarns, staples, and tows are manufactured. The variant properties are tex (wt (g) of a 1000 m filament) or denier (wt (g) of a 9000 m filament), cross-sectional shape, and number of filaments. Individual filament deniers (denier per filament or dpf) are usually in the 2–4 dpf range (0.2–0.4 tex per filament). Common continuous filament yarns have total denier of 55, 60, 75, and 150 (6.1, 6.7, 8.3, and 16.7 tex, respectively). However, different fabric properties can be obtained by varying the filament count (dpf) to reach the total denier.

Though the cross-sectional shape of the spinneret hole directly influences the cross-sectional shape of the fiber, the shapes are not identical. Round holes produce filaments with an approximately round cross section, but with crenulated edges; triangular holes produce filaments in the form of a Y. Other cross sections can be achieved by adjusting the extrusion temperatures. Different cross sections are responsible for a variety of esthetics in the finished fabric such as hand, luster, or cover. Some yarn types may also include chemical additives to provide resistance to sunlight degradation and to give fire retardant properties. These additives are usually added to the acetate solution before spinning.

A metier is an array of individual extrusion positions on one common machine. There are usually 100–200 such positions. The yarn is collected at each position on a package (eg, bobbin, tube, pirn) and removed from the machines at regular intervals to maintain a constant amount of yarn on each package. The package may contain 0.5–7 kg of yarn. As previously indicated, bobbin yarn may contain a low level of twist (about 0.08 turns per centimeter) whereas yarn taken up on tubes may have zero twist. The yarn is transferred from bobbins to different packages for sale. The product may contain twist levels of 0.3–8 turns per centimeter. Compacted yarn is presently more popular than low twist yarn.

Yarn Packages. The principal package types used by the textile industry are tubes, cones, and beams, although specialized packages are available for specific products.

Tubes are wrapped with 1.0–4.0 kg of yarn. The package is built on winders to provide package integrity and easy removal. Some packages are provided with a magazine wrap at the start of winding so that customers can automatically change packages. Zero-twist entangled yarn is packaged on tubes for use in circular knits and on tricot and section beams for warp knits.

Cones contain 0.5–4.0 kg of yarn. The tip of the cone tube must have a smooth finish to prevent damage to the yarn which is drawn over the top. Again, a magazine wrap may be provided for automatic package transfer. Both compacted and twisted yarns are packaged on cones.

Beams (large spools) are usually constructed of an aluminum alloy and vary from 50 to 170 cm in length and from 50 to 90 cm in flange diameter. A beam holds 100–700 kg of yarn. The beam most commonly used by the warp-knitting trade is 107 cm in length and 53–76 cm in dia. Section beams for weaving are usually 137 cm in length and 76 cm in dia. Both types of beams are parallel wound with a large number (up to 2400) of individual yarn ends. The length of yarn on beams varies with yarn denier, beam capacity, and intended use. Lengths are ordinarily in the range of 11,000–78,000 m. When beam winding is complete, the ends are taped in position and the beam is wrapped with a protective cover.

Staple and Tow. The same basic extrusion technology that produces continuous filament yarn also produces staple and tow. The principal difference is that spinnerets with more holes are used, and instead of winding the output of each spinneret on an individual package, the filaments from a number of spinnerets are gathered together into a ribbonlike strand, or tow. A mechanical crimping device uniformly plaits the tow into a carton from which it can be continuously withdrawn without tangling.

Staple is produced by cutting the tow (which may be crimped) into short lengths (usually 4–5 cm) resembling short, natural fibers. Acetate and triacetate staple is shipped in 180–365-kg bales. Conventional staple processing technology used with natural fibers is used to process acetate and triacetate staple into spun yarn.

Economic Aspects

Although cellulose acetate is the second oldest man-made fiber, it continues to be an important factor in the textile industry, with 317,000 metric tons produced worldwide in 1976 (Table 7) (76). Prior to the mid 1950s it had the second largest volume of the man-made fibers but it has subsequently been surpassed by polyester, nylon, and acrylic fibers (see Fibers, man-made and synthetic). It is sold at a relatively

Table 7.   World Production of Cellulose Acetate Textile Fibers (1968–1976)[a,b]

Year	Production, thousands of metric tons		
	Filament	Staple	Total
1968	388	35	423
1969	399	32	431
1970	402	25	427
1971	404	20	424
1972	377	21	398
1973	399	19	418
1974	359	14	373
1975	315	9	324
1976	310	7	317

[a] Textile Economics Bureau.
[b] Ref. 76.

low price. Triacetate was introduced to the American market in 1954. The major textile applications of both acetate and triacetate fibers are in women's apparel and home furnishing fabrics (see Textiles). A list of trade names and manufacturers is shown in Table 8.

Although the use of acetate fiber for textile applications has generally declined, the production of cellulose acetate tow for cigarette filters rose from 136,000 to 257,000 metric tons in the 1969–1976 period (Table 9) (76). Because of its superior filtration, impact on cigarette taste, and cost, acetate is projected to supply up to 90% of the growing filter cigarette market.

A list of major acetate and triacetate producers, primary trade names, and production levels is given in ref. 76. The combined annual world acetate production (filament, staple, and tow) has been approximately 575,000 metric tons in the 1968–1976 period. Production in the United States accounts for approximately 51% of the total over that period. Other major acetate producing countries are the United Kingdom, Japan, and the Union of Soviet Socialist Republics.

Health Effects

Both acetate and triacetate remain undigested and cause no harmful reactions when ingested. Toxic effects, skin irritations, or allergic reactions attributable to acetate or triacetate fibers have never been reported.

Uses

The two major markets for cellulose acetate are textiles and cigarette filters. A unique combination of desirable esthetics and low cost accounts for the demand in textiles.

Textiles. Approximately 50% of the acetate and triacetate filament in the United States has been used for tricot knitting, 40% for woven fabrics, 5% for circular knits, and the remaining 5% for other applications (77). This distribution changes according to textile market trends. The principal markets are women's apparel (eg, dresses, blouses, lingerie, robes, housecoats, ribbons) and decorative household applications (eg, draperies, bedspreads, and ensembles). Acetate has been replacing rayon filament in liner fabrics for men's suits and has been evaluated for nonwovens (78–80) (see Nonwoven textiles; Textiles).

Table 8.   Trade Names and Manufacturers of Acetate and Triacetate

Trade name[a]	Manufacturer and country
Acele	E. I. du Pont de Nemours & Co., Inc., U.S. (closed in 1977)
Acesil	Montedison Fibre S.p.A., Italy
Albene	Rhodiaseta, Argentina S.A., Argentina
Amcel	Amcel Europe S.A., Belgium
Arnel (ta)	Amcel Europe S.A., Belgium
Arnel (ta)	Celanese Canada Limited, Canada
Arnel (ta)	Celanese Fibers Company, U.S.
Ashi	Asahi Chisso Acetate Co., Ltd., Japan
Avisco	Avtex Fibers Inc., U.S.
Carolan (ta)	Mitsubishi Acetate Co., Ltd., Japan
Celanese	Celanese Fibers Company, U.S.
Celanese	Celanese Venezolana S.A., Venzuela
Celaren	Rayon y Celanese Peruana, S.A., Peru
Celcorta	Celanese Mexicana, S.A., Mexico
Chrysella	Courtaulds Ltd., Australia
Dicel	British Celanese Ltd. (subsidiary of Courtalds), U.K.
Eslon	Diacel Co., Ltd., Japan
Estron	Tennessee Eastman Co., (Division of Eastman Kodak Company), U.S.
Krasil	Ravi Rayon, Ltd., Pakistan
Lanalbene	Montedison Fibre S.p.A., Italy
Lonzona	Lonzona, Gesellschaft für Acetaprodukte m.b.H., FRG
Loteyarn	Teijin, Ltd., Japan
Novaceta	Montedison Fibre S.p.A., Italy
Rhodia	Deutsche Rhodiaceta A.G., FRG
Rhodia	Rhodia Industrias Quimical e Texteis, S.A., Brazil
Rhodia	Société Rhodiaceta, France
Rhodia	Deutsche Rhodiaceta A.G., FRG
Tri-A-Faser (ta)	Deutsche Rhodiaceta A.G., FRG
Tricel (ta)	British Celanese Ltd. (subsidiary of Courtalds), U.K.
Trilan (ta)	Celanese Canada Limited, Canada
Trilbene (ta)	Société Rhodiaceta, France
Velion	Industrias Del Acetato De Celulosa, S.A., Spain

[a] ta, triacetate

Acetate and triacetate fibers have lower strength and abrasion resistance than most other man-made fibers and are frequently used with nylon or polyester in combination yarns. The latter can be used in markets formerly unavailable to 100% acetate fabrics (eg, men's shirts). Combination yarns can be prepared by twisting or by air entanglement and bulking. Yarns prepared by air-entanglement and bulking have unique characteristics and esthetics that permit their use in casement and upholstery fabric markets. With chemical additives, both acetate and triacetate fibers can pass current United States Government flame retardant fabric legislation (eg, DOC FF 3-71) (see Flame-retardant textiles).

Triacetate can be used in place of secondary acetate in many apparel applications but has the added advantage of ease-of-care properties. A particularly important application of triacetate is in velour and suede-like fabrics for robes and dresses. These fabrics offer superb esthetic qualities at reasonable cost. Triacetate is also desirable for print fabrics as it produces bright, sharp colors.

Table 9.   World Production of Cellulose Acetate Cigarette Filter Tow (1968–1976)[a], Thousands of Metric Tons

	1968	1970	1972	1974	1976
Europe[b]	25	35	43	48	59
United States	73	84	102	124	134
other American countries[c]	11	15	19	25	27
Asia[d]	19	20	26	30	37
Total	128	154	190	227	257

[a] Textile Economics Bureau.
[b] Belgium, France, FRG, U.K.
[c] Brazil, Canada, Colombia, Mexico, Venezuela.
[d] Japan, South Korea.

Cigarette Tow.   Acetate fiber used in the production of cigarette filters is supplied in the form of tow (81). Tow is a continuous band composed of several thousand filaments held loosely together by crimp, a wave configuration set into the band during manufacture (Fig. 14). A tow is formed by combining the output of a large number of spinnerets and crimping the collection of filaments to create an integrated band of continuous fibers. The tow is then dried and baled. The wide range of available acetate filter tow products makes it possible to control selected properties in the finished cigarette filter rod.

Tow Properties.   An individual tow item is characterized and identified by the following parameters, which are determined by controlling certain variables in the manufacturing process.

Cross Section.   The shape of the filament cross section is related to the shape of the minute orifices in the spinneret used to form the filament. Filament cross sections currently used are shown in Figure 15.

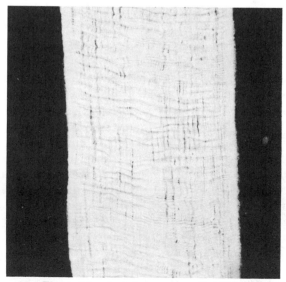

Figure 14.   Acetate tow as supplied (81).

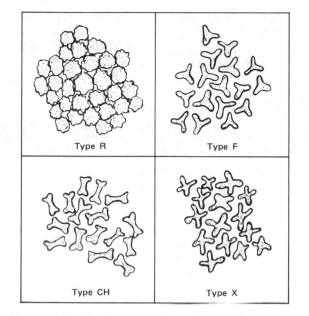

Figure 15.   Filament cross sections of acetate filter tows (81).

Denier.   Denier (1.111×10^{-7} kg/m or tex = 1.0×10^{-6} kg/m) is a measure of linear density, and is defined as the weight in grams of a 9000 m length of yarn. In filter-tow processing, there are several denier terms to consider: denier per filament (dpf) or 0.1111 tex per filament; total denier (TD) of the uncrimped tow (the product of the dpf multiplied by the number of filaments in the tow band), and crimped total denier (somewhat higher than the total denier).

A tow item described as 8.0 dpf (0.89 tex per filament), 50,000 TD, may therefore be interpreted as: an uncrimped tow band that weighs 50,000 g for each 9000 m of length and which is composed of 6250 individual filaments ($50,000 \div 8.0$) each weighing 8.0 g for each 9000 m in length.

Crimp.   The crimp imparted to the tow is normally manifested by a sawtooth or sinusoidal wave shape. Because the filaments are usually crimped as a group, the crimp in parallel fibers is in lateral registry, ie, with the ridges and troughs of the waves aligned, as shown in Figure 16. The presence of crimp in the tow is necessary for two reasons: (1) to ensure that the tow can be packaged, processed, and handled easily, and (2) to impart bulk to the finished filter. To achieve the latter, it is necessary in production of the filters to open the tow band (Fig. 17) to the desired bulk so that the fibers completely fill the paper wrap without voids and soft spots. Several tow-opening systems are used to reposition the crimp out of lateral registry to create bulk while retaining the crimp in the individual filaments.

Tow Characterization.   A linear relationship, constant for any given tow item, exists between the weight of tow in a cigarette filter rod and certain of its properties such as pressure drop and smoke removal efficiency. By preparing filter rods over a range of rod weights, eg, testing the rods for pressure drop, then plotting the results on rectilinear graph paper, a linear curve characteristic of the specific tow item is generated. This characterization provides a means of determining whether the required filter characteristics can be achieved, and if so, the weight of a specific tow item necessary

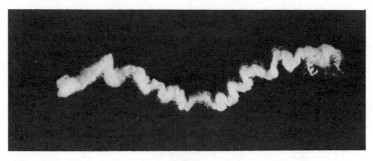

Figure 16.   Section of tow showing crimp configuration (81).

Figure 17.   Opened acetate tow (81).

to achieve them. By this kind of application, tows can be used to design cigarette filter rods with the desired performance properties.

Effect of Denier and Cross Section.   Physical parameters of the tow (eg, denier per filament, total denier, crimp, and cross section) have a marked effect on the physical properties and performance of the finished filter rod.

Other Potential Applications.   Additional applications for acetate and triacetate fibers, based on use of their unique properties as well as mastery of their inherent deficiencies, are being explored. Cellulose acetate has been a beneficial membrane material in reverse osmosis (qv) and hollow acetate fibers are now being explored for this application (82–85) (see Hollow-fiber membranes). Water soluble acetate fibers can be produced by selecting the appropriate acetyl value (86). Techniques of improving the antistatic characteristics of acetate fiber (87–93) which might make it competitive with cotton in critical antistatic applications (such as hospital operating rooms) are being developed (see Antistatic agents). Other studies aimed at improved water and solvent resistance (94–97), abrasion resistance (98–101), grafting (102), selective adsorption (103), and timed release of additives (104) offer the potential for developing future applications.

BIBLIOGRAPHY

"Acetate Fibers" treated in *ECT* 1st ed., under "Rayon and Acetate Fibers," Vol. 11, pp. 552–569, by G. W. Seymour and B. S. Sprague, Celanese Corporation of America; "Acetate and Triacetate Fibers" in *ECT* 2nd ed., Vol. 1, pp. 109–138, by L. I. Horner and A. F. Tesi, Celanese Fibers Company, Division of Celanese Corporation of America.

1. *J. Polym. Sci. C* 11, 161 (1965).
2. C. J. Malm and L. J. Tanghe, *Tappi* 46(10), 629 (1963).
3. N. F. Getchell and co-workers, *Am. Dyest. Rep.* 45, 845 (1956).
4. A. Mellor and H. C. Olpin, *J. Soc. Dyers Colour.* 71, 817 (1955).
5. *Triacetate-General Information and Physical and Chemical Properties, Technical Bulletin TBT30*, Celanese Fibers Marketing Co., Charlotte, N.C., 1974.
6. T. Vickerstaff and E. Water, Jr., *J. Soc. Dyers Colour.* 58, 116 (1942).
7. C. L. Bird and co-workers, *J. Soc. Dyers Colour.* 70, 68 (1954).
8. R. K. Fourness, *J. Soc. Dyers Colour.* 72, 513 (1956).
9. F. Fortess and V. S. Salvin, *Tex. Res. J.* 28, 1009 (1958).
10. F. Fortess, *Am. Dyest. Rep.* 44, 524 (1955).
11. *The Physical and Chemical Properties and Dyeability Characteristics of Acetate Filament Yarns and Staple Fiber, Technical Bulletin TBA 8*, Celanese Fibers Marketing Co., Charlotte, N.C., 1974.
12. *Dyeing Triacetate Fabrics, Technical Bulletin TBT 24*, Celanese Fibers Marketing Co., Charlotte, N.C., 1972.
13. R. J. Mann, *J. Soc. Dyers Colour.* 76, 665 (1960).
14. J. Boulton, *J. Soc. Dyers Colour.* 71, 451 (1955).
15. F. Fortess, *Text. Res. J.* 19, 23 (1949).
16. P. M. Heertjes, W. Colthof, and H. I. Waterman, *Rec. Trav. Chim.* 52, 305 (1933).
17. C. J. Malm, C. R. Fordyce, and H. A. Tanner, *Ind. Eng. Chem.* 34, 430 (1942).
18. D. K. Beever and L. Valentine, *J. Text. Inst.* 49, T95 (1958).
19. R. K. Toner, C. F. Bowen, and J. C. Whitwell, *Text. Res. J.* 17, 14 (1947).
20. *Tricel Technical Service Manual*, British Celanese, Ltd., Coventry, Eng., 1958.
21. *BISFA, Internationally Agreed Methods of Testing Regenerated Cellulose and Acetate Continuous Filament Yarn*, 1970 ed., Bureau International Pour La Standardisation De La Rayonne Et Des Fibres Synthetiques, Basel, Switz., 1971.
22. A. F. Tesi, *Am. Dyest. Rep.* 45, 512 (1956).
23. L. G. Ray, Jr., *Text. Res. J.* 22, 144 (1952).
24. H. M. Fletcher, *Am. Dyest. Rep.* 38, 603 (1949).
25. M. L. Staples and C. J. Brown, paper presented at the *Fourth Canadian Textile Seminar*, Queens College, Kingston, Ontario, 1954, p. 132.
26. G. S. Egerton, *J. Soc. Dyers Colour.* 65, 765 (1949).
27. L. Hochstaedter, *Text. Res. J.* 28, 78 (1958).
28. M. Fels, *J. Text. Inst.* 51, 648 (1960).
29. R. C. Harrington, Jr., and C. A. Jarrett, *Mod. Text.* (4), 67 (1963).
30. *Effect of Conditioning Humidity on the Electrical Resistance of Rayon Yarns*, British Cotton Industry Research Association, London, Eng., 1945.
31. J. C. Guthrie, *J. Text. Inst.* 48(6), T193 (1957).
32. R. Meredith, *J. Text. Inst.* 37, T107 (1945).
33. G. Susich and S. Backer, *Text. Res. J.* 21, 482 (1951).
34. E. R. Kaswell, *Textile Fibers, Yarns, and Fabrics*, Reinhold Publishing Corp., New York, 1953, p. 57.
35. L. Segal, *Cellulose and Cellulose Derivatives* in N. M. Bikales and L. Segal, eds., *High Polymers Series*, Vol. V, Wiley-Interscience, New York, 1971, Chapt. XVII-A.
36. G. D. Hiatt and W. J. Rebel in ref 35, Chapt. VII-B.
37. C. L. Smart and C. N. Zellner in ref. 35, Chapt. XIX-C.
38. P. E. Gardner and M. Y. Chang, *Tappi* 57(8), 71 (1974).
39. J. D. Wilson and R. S. Tabke, *Tappi* 57(8), 77 (1974).
40. U.S. Pat. 3,846,403 (Nov. 5, 1974), K. B. Gibney and R. S. Evans (to Canadian Cellulose Company, Ltd.).

41. Can. Pat. 973,174 (Aug. 19, 1975), K. B. Gibney, B. E. Grisack, and R. S. Evans (to Canadian Cellulose Company, Ltd.).
42. Can. Pat. 975,764 (Oct. 7, 1975), K. B. Gibney and R. S. Evans (to Canadian Cellulose Company, Ltd.).
43. C. J. Malm and G. D. Hiatt, *Cellulose and Cellulose Derivatives* in E. Ott, H. M. Spurlin, and M. W. Graffin, eds., *High Polymers Series,* 2nd ed., Vol. V, Pt. II, Wiley-Interscience, New York, 1954.
44. E. Barabash, A. J. Rosenthal, and B. B. White, *Tappi* **38**(12), 745 (1955).
45. E. Dyer and H. D. Williams, *Tappi* **40**(1), 14 (1957).
46. A. Casadevall and co-workers, *Bull. Soc. Chim. Fr. Mem.,* 187 (1974); 196 (1964); 204 (1960); 719 (1970); 1850 (1970); 1856 (1970).
47. S. A. Kadyroba, *Dokl. Akad. Uzb. SSR* **26**(10), 29 (1969).
48. C. J. Clemett, *J. Chem. Soc. B,* 2202 (1971).
49. I. B. Gorovaya, N. A. Kozlov, and O. G. Tarakanov, *Zh. Prikl. Khim.* (*Leningrad*) **46**(4), 870 (1973).
50. U.S. Pat. 2,539,586 (Jan. 30, 1951), M. E. Martin, T. M. Andrews, and A. R. Franck (to Celanese Corporation).
51. U.S. Pat. 2,259,462 (Oct. 21, 1941), C. L. Fletcher (to Eastman Kodak Co.).
52. Brit. Pat. 566,863 (Feb. 26, 1945), H. Dreyfus.
53. U.S. Pat. 3,525,734 (Aug. 25, 1970), A. Rajon (to Société Rhodiaceta).
54. U.S. Pat. 2,603,634 (July 15, 1952), G. W. Seymour, B. B. White, and M. Plunguian (to Celanese Corporation).
55. U.S. Pat. 2,603,638 (July 15, 1952), G. W. Seymour, B. B. White, and M. Plunguian (to Celanese Corporation).
56. U.S. Pat. 2,731,247 (Jan. 17, 1956), C. Hudry (to Société Rhodiaceta).
57. U.S. Pat. 2,778,820 (Jan. 22, 1957), R. Clevy and J. Robin (to Société Rhodiaceta).
58. U.S. Pat. 2,790,796 (Apr. 30, 1957), R. Clevy and J. Robin (to Société Rhodiaceta).
59. U.S. Pat. 2,801,237 (July 30, 1957), R. Clevy and J. Robin (to Société Rhodiaceta).
60. U.S. Pat. 2,854,445 (Sept. 30, 1957), R. Clevy and J. Robin (to Société Rhodiaceta).
61. U.S. Pat. 2,854,446 (Sept. 30, 1957), R. Clevy and J. Robin (to Société Rhodiaceta).
62. U.S. Pat. 2,966,485 (Dec. 27, 1960), K. C. Laughlin, R. J. Osborne, and J. G. Santangelo (to Celanese Corporation).
63. Can. Pat. 609,900 (Dec. 6, 1960), K. C. Laughlin, R. J. Osborne, and J. G. Santangelo (to Celanese Corporation).
64. U.S. Pat. 3,040,027 (June 19, 1962), H. Bates, F. Hindley, and W. Popiolek (to British Celanese, Ltd.).
65. H. Genevray and J. Robin, *Pure Appl. Chem.* **14,** 489 (1967).
66. J. Corviere, *Faserforsch. Textiltech.* **22,** 71 (1971).
67. U.S. Pat. 3,631,023 (Dec. 28, 1971), C. Horne, Jr., and C. J. Howell, Jr., (to Celanese Corporation).
68. U.S. Pat. 3,755,297 (Aug. 28, 1973), K. C. Campbell and co-workers (to Celanese Corporation).
69. S. A. Kadyroba, *Prom. Arm.* **11**(A), 34 (1969).
70. Brit. Pat. 1,323,200 (July 11, 1973), F. M. Mikhalsky and co-workers.
71. U.S.S.R. Pat. 319,227 (C1.C08b) (Dec. 5, 1975), F. M. Mikhalsky and co-workers; *Otkrytiya Izobret. Prom. Obraztsy Tovarnye Znaki,* **52**(45), 185 (1975).
72. C. J. Malm and co-workers, *Ind. Eng. Chem.* 49, 79 (1957).
73. U.S. Pat. 2,985,995 (May 30, 1961), W. W. Bunting, Jr., and T. L. Nelson (to E. I. du Pont de Nemours & Co., Inc.).
74. U.S. Pat. 3,110,151 (Nov. 12, 1963), W. W. Bunting, Jr., and T. L. Nelson (to E. I. du Pont de Nemours & Co., Inc.).
75. U.S. Pat. 3,364,537 (Jan. 23, 1968), W. W. Bunting, Jr., and T. L. Nelson (to E. I. du Pont de Nemours & Co., Inc.).
76. *Text. Organon* **48,** 82 (1977).
77. *Knitting Times,* (Mar. 8, 1976).
78. A. A. Lukoshaitis and co-workers, *Fibre Chem.* **6,** 441 (1974); Y. A. Matskevichene and co-workers, *Fibre Chem.* **6,** 446 (1974).
79. R. R. Rhinehart, *International Nonwoven and Disposable Association, Technical Symposium Paper,* 25 (Mar. 1975).
80. Ger. Pat. 2,502,519 (July 31, 1975), C. K. Arisaka and co-workers (to Diacel, Ltd.).
81. *World Smoking Products Technical Bulletin WSP 2.1, Acetate Tow Production and Characterization,* Celanese Fibers Marketing Co., Charlotte, N.C., 1974.

82. G. Rakhmanberdiev and co-workers, *Zh. Prikl. Khim.* (*Leningrad*) **46**(2), 416 (1973).

83. U.S. Pat. 3,763,299 (Oct. 2, 1973), W. S. Stephen (to FMC Company).

84. G. Rakhmanberdiev and co-workers, *Fibre Chem.* **6**, 219 (1974).

85. Brit. Pat. 1,418,115 (Dec. 17, 1975), R. L. Leonard (to Monsanto).

86. U.S. Pat. 3,482,011 (Dec. 2, 1969), T. C. Bohrer (to Celanese Corporation).

87. O. G. Pikovskaya and Z. G. Serebryakova, *Fibre Chem.* **2**, 378 (1970).

88. F. A. Ismailov and co-workers, *Fibre Chem.* **4**, 584 (1972).

89. P. A. Chakhoya and co-workers, *Fibre Chem.* **5**, 184 (1973).

90. A. V. Kuchmenko, *Fibre Chem.* **5**, 210 (1973).

91. Brit. Pat. 1,381,334 (May 31, 1973), W. Ueno, H. Kawaguchi, and N. Minagawa (to Fuji Photo Film Co., Ltd.).

92. A. P. Fedotov and co-workers, *Fibre Chem.* **6**, 87 (1974).

93. M. N. Skarnlite and Y. Y. Shlyazhas, *Fibre Chem.* **6**, 557 (1974).

94. *Text. World* **123**(9), 37 (1973).

95. U.S. Pat. 3,816,150 (June 11, 1974), K. Ishii and co-workers (to Daicel, Ltd.).

96. U.S. Pat. 3,839,517 (Oct. 1, 1974), A. F. Turbak and J. R. Thelman (to International Telephone and Telegraph Corp.—Rayonier).

97. U.S. Pat. 3,839,528 (Oct. 1, 1974), A. F. Turbak and J. R. Thelman (to International Telephone and Telegraph Corp.—Rayonier).

98. G. F. Kiseleva and co-workers, *Fibre Chem.* **4**, 257 (1972).

99. M. Papikyan and co-workers, *Fibre Chem.* **4**, 441 (1972).

100. M. Mirzaev and co-workers, *Khim. Volokna* (3), 21 (1975).

101. Brit. Pat. 1,414,395 (Nov. 19, 1975), N. V. Mikhailov and co-workers.

102. M. A. Siahkolah and W. K. Walsh, *Text. Res. J.* **44**(11), 895 (1974).

103. Ger. Pat. 2,507,551 (Feb. 28, 1975), B. V. Chandler and R. L. Johnson (to Commonwealth Scientific Org.).

104. U.S. Pat. 3,846,404 (Nov. 5, 1974), L. D. Nichols (to Moleculon Research Corp.).

GEORGE A. SERAD
J. R. SANDERS
Celanese Fibers Company

COATED FABRICS

A coated fabric is a construction that combines the beneficial properties of a textile and a polymer. The textile (fabric) provides tensile strength, tear strength, and elongation control. The coating is chosen to provide protection against the environment in the intended use. A polyurethane might be chosen to protect against abrasion or a polychloroprene (Neoprene) to protect against oil (see Urethane polymers; Elastomers, synthetic).

Textile Component

The vast majority of textiles that are used for coating are purchased by the coating company. Several large textile manufacturers have divisions that specialize in industrial fabrics. These companies perform extensive development on substrates and can provide advice on choosing the correct substrate, hand samples, pilot yardage, and ultimately production requirements. A listing of suppliers of industrial fabrics can be found in *Davison's Textile Blue Book* (1).

Fiber.   For many years cotton (qv) and wool (qv) were used as primary textile components, contributing the properties of strength, elongation control, and esthetics. Although the modern coated fabrics industry began by coating wool to make boots, cotton has been used more extensively. Cotton constructions, including sheetings, drills, sateens and knits, command a major share of the market. Cotton is easily dyed, absorbs moisture, withstands high temperature without damage, and is stronger wet than dry. This latter property renders cotton washable; it can also be drycleaned because of its resistance to solvents (see Drycleaning).

Polyester, by itself and in combination with cotton, is used extensively in coated fabrics. Polyester produces fibers that are smooth, crisp, and resilient. Since moisture does not penetrate polyester, it does not affect the size or shape of the fiber. Polyester resists chemical and biological attack. Because of its thermoplastic nature, the heat required for adhesion to this smooth fiber can also create shrinkage during coating (see Polyester fibers; Fibers, man-made and synthetic).

Nylon is the strongest of the commonly-used fibers. Since it is both elastic and resilient, articles made with nylon will return to their original shape. There is a degree of thermoplasticity so that articles can be shaped and then heated to retain that shape. Nylon fibers are smooth, very nonabsorbent, and will not soil easily. Nylon resists chemical and biological action. Nylon substrates are used in places where very high strength is required. Lightweight knits and taffetas are thinly coated with polyurethane or poly(vinyl chloride) and used extensively in apparel. In coating, PVC does not adhere well to nylon (see Polyamides).

Rayon and glass fibers are the least used because of their poor qualities. Rayon's strength approaches that of cotton but its smoother fibers make adhesion more difficult. Rayon has a tendency to shrink more than cotton, which makes processing more difficult. Glass fibers offer very low elongation, very high strength, and have a tendency to break under compression. Therefore, glass fabric is only used where support with low stretch is required and where the object is not likely to be flexed. For instance, glass might be used to support a lead-filled vinyl compound for sound dampening (see Glass; Insulation, acoustic).

Textile Construction. There are many choices in textile construction. The original, and still the most commonly used, is the woven fabric. Woven fabrics have three basic constructions: the plain weave, the satin weave, and the twill weave. The plain weave is by far the strongest because it has the tightest interlacing of fibers; it is used most often (2). Twill weaving produces distinct surface appearances and is used for styling effects. Because it is the weakest of the wovens, satin weave is used principally for styling. Woven nylon or heavy cotton are used for tarpaulin substrates. For shoe uppers, and other applications where strength is important, woven cotton fabrics are used.

Knitted fabrics are used where moderate strength and considerable elongation are required. Where cotton yarn formerly dominated the knit market, it has recently been replaced by polyester–cotton yarn, and polyester yarn and filament. Where high elongation is required, nylon is used. Knits are predominantly circular jersey; however, patterned knits are becoming more and more prevalent. When a polymeric coating is put on a knit fabric, the stretch properties are somewhat less than that of the fabric. Stretch and set properties are important for upholstering and forming. The main use of knit fabrics is in apparel, automotive and furniture upholstery, shoe liners, boot shanks—any place elongation is required.

Many types of nonwoven fabrics are used as substrates (3) (see Nonwoven textiles). The wet web process gives a nonwoven fabric with paperlike properties; low elongation, low strength, and poor drape. When these substrates are coated, the papery characteristics show through the coating, and fabric esthetics are not satisfactory. The nonwovens prepared by laying dry webs, compressing by needle punching, and then impregnating from 50–100% with a rubbery material, resemble in many respects split leather. These materials are used for shoe liners (see Leather-like materials). It is difficult to achieve uniformity of stretch and strength in two directions as well as a smooth surface; therefore, a high quality nonwoven of this type is very expensive. Spunbonded nonwovens are available in both polyester and nylon in a range of weights. The strength qualities are very high and elongation is low. Since they are quite stiff, these materials are used where strength and price are the major considerations. A lightly needled, low density nonwoven was marketed in 1970 and in the last three years has gained prominence in the coated fabrics industry. It is used in weights from 60–180 g and can be prepared from either polypropylene or polyester fibers. The light needling combined with careful orientation of the fibers and selection of the fiber length gives very good strength and more balanced stretch. Optionally, a thin layer of polyester-based polyurethane foam can be needled into the nonwoven to improve the surface coating properties. The furniture upholstery market was the first to accept this product. A 0.4 mm poly(vinyl chloride) skin on this nonwoven replaced expanded PVC on knit fabric at approximately half the cost. The finished product is softer, plumper, and may have better wear characteristics. A major automobile company has introduced this type of seating in one line of cars, and apparel manufacturers have exhibited interest in these types of construction.

Post Finishes of the Textile Component. The construction that results from either weaving or knitting is called a greige good. Other steps are required before the fabric can be coated: scouring to remove surface impurities; and heat setting to correct width and minimize shrinkage during coating.

Optional treatments include: dyeing if a colored substrate is required; napping of cotton and polyester–cotton blends for polyurethane coated fabrics; flame resistance treatments; bacteriostatic finishes for hygienic applications; and mildew treatments for applications in high humidity (see Textiles).

Polymer Component

Rubber and Synthetic Elastomers.   For many years coated fabrics consisted of natural rubber (qv) on cotton cloth. Natural rubber is possibly the best all-purpose rubber but some characteristics such as poor resistance to oxygen and ozone attack, reversion and poor weathering, and low oil and heat resistance, limit its use in special application areas (see also Elastomers, synthetic).

Polychloroprene (Neoprene) introduced in 1933 rapidly gained prominence as a general purpose synthetic elastomer having oil, weather and flame resistance. The introduction of new elastomers in solid or latex form was accelerated by World War II. Currently, in addition to natural rubber and polychloroprene, other polymers in use include: styrene–butadiene (SBR), polyisoprene, polyisobutylene (Vistanex), isobutylene–isoprene copolymer (Butyl), polysulfides (Thiokol), polyacrylonitrile (Paracril), silicones, chlorosulfonated polyethylene (Hypalon), poly(vinyl butyral), acrylic polymers, polyurethanes, ethylene–propylene copolymer (Royalene), fluorocarbons (Viton), polybutadiene, polyolefins, and many more. Copolymerizations and physical blends make the number available staggering (see Copolymers; Olefin polymers; Polymers containing sulfur; Acrylic ester polymers; Vinyl polymers; Fluorine compounds; Acrylonitrile polymers; Silicon compounds; Urethane polymers).

In fact, the number of commercially available polymers in use is well over 1000. In each class there are several variations manifesting a wide range of properties. DuPont supplies about 24 types of polychloroprene. B. F. Goodrich supplies about 140 types of acrylonitrile elastomers (Hycar) and the same holds true for all the other types of coating polymers.

Most elastomers are vulcanizable; they are processed in the plastic state and cross-linked to provide elasticity after being put into final form. With the number of elastomer coatings available today almost any use requirement can be met. If there are limitations, they lie in the areas of processability and cost.

Elastomers are applied to the textile by either calendering or solution coating. Thin coatings are applied from solution and thicker coating by direct calendering.

A natural rubber-based formulation is shown in Table 1.

SBR (styrene–butadiene rubber) has replaced natural rubber in many applications because of price and availability. It has good aging properties, abrasion resistance and flexibility at low temperatures. A typical SBR-based formula is shown in Table 2.

Table 1.   A Typical Natural Rubber Compound

Component	Parts
smoked sheet	100.00
stearic acid	1.00
ZnO	3.00
agerite white antioxidant	0.50
P-33 black	10.00
CaCO$_3$	75.00
clay	50.00
sulfur	0.75
methyl zimate } accelerators	0.25
telloy }	0.50
Total	*200.00*

Table 2.   A Typical SBR Compound

Component	Parts
SBR	100.0
processing aid	5.0
stearic acid	2.5
ZnO	3.0
agerite white antioxidant	0.5
tackifier	20.0
CaCO$_3$	75.0
P-33 black	10.0
clay	75.0
sulfur	2.5
methyl zimate } accelerators	0.5
tuex	
Total	294.0

Neoprene offers resistance to oil, weathering, is inherently nonburning and is processable on either calenders or coaters. The cost of Neoprene and its reduced availability in recent years have led to the development of substitutes. Nitrile rubber–PVC blends and nitrile rubber–EPDM (ethylene–propylene–diene monomer) perform on an equivalent basis. The blends do not discolor like Neoprene, and light-colored decorative fabrics can be made.

A typical Neoprene-based formulation is shown in Table 3.

This mixture can be calendered or dissolved in toluene to 25–60% solids for coating.

Isobutylene–isoprene elastomer (Butyl) has high resistance to oxidation, resists chemical attack and is the elastomer most impervious to air. These properties suggest its use for protective garments, inflatables, and roofing.

Chlorosulfonated polyethylene (Hypalon) resists ozone, oxygen, and oxidizing agents. In addition it has nonchalking weathering properties and does not discolor, permitting pigmentation for decorative effects.

Nitrile elastomers (acrylonitrile–butadiene copolymers) have high resistance to oils at up to 120°C. If higher temperature protection is required, a polyacrylate elastomer can be employed up to 200°C.

Polyurethane. Polyurethanes have a number of important applications in coated fabrics. The most striking is footwear uppers because polyurethanes are lighter weight than vinyl polymers and have better abrasion resistance and strength. Polyure-

Table 3.   A Typical Neoprene Formulation

Component	Parts
polychloroprene	100.0
stearic acid	1.5
antioxidant	2.0
MgO	4.0
clay	66.0
SRF black	22.0
circo oil	10.0
petrolatum	1.0
ZnO	5.0
ethylenethiourea	0.5
Total	212.0

thane-coated fabrics can be decorated to look like leather (qv). Earlier attempts to produce poromerics (coatings that transmit moisture much like leather) were not commercially successful because, although they approach leather in cost, they did not match it in comfort. However, poromerics are still available. Most of the urethane-coated fabrics are used in women's footwear where styling is important and lightweight is desirable. These products usually consist of 0.05 mm of polyurethane on a napped woven cotton fabric. The result is a lightweight product 0.88 mm thick that has good abrasion and scuff resistance (4). Urethane-coated fabrics have not been successful in either men's or children's shoes because greater toughness is required.

Low-weight coatings of polyurethane on very low-weight nylon fabric produce products suitable for apparel. This lightweight product, used for windbreakers and industrial clothing, resists water, provides thermal insulation, and has good drape. Coatings of urethane on heavier nylon structures are used for industrial tarpaulins to provide protection from the elements and extreme toughness. Polyurethane coatings have had limited application to furniture upholstery and practically none on automobile seating.

Poly(vinyl Chloride). By far the most important polymer used in coated fabrics is poly(vinyl chloride). This relatively inexpensive polymer resists aging processes readily, resists burning, and is very durable. It can be compounded readily to improve processing, aging, burning properties, softness, etc. In addition, it can be decorated to fit the required use. PVC-coated fabrics are used for window shades, book covers, furniture upholstery, automotive upholstery and trim, wall covering, apparel, conveyor belts, shoe liners, and shoe uppers. These few uses require millions of meters of coated fabrics each year and demonstrate the diverse properties of PVC coatings. Tables 4 and 5 show typical PVC formulations.

Table 4.   A Typical Compound for Calendering PVC

Component	Parts
poly(vinyl chloride) resin (calender grade)	100.00
epoxy plasticizer	5.00
dioctyl phthalate	35.00
polymeric plasticizer	35.00
BaCdZn stabilizer	3.00
TiO_2 (pigment)	15.00
calcium carbonate (filler)	20.00
stearic acid (lubricant)	0.25
Total	*213.25*

Table 5.   A Typical Plastisol PVC Formulation

Component	Parts
poly(vinyl chloride) resin (dispersion grade)	100.00
epoxy plasticizer	4.00
dioctyl phthalate	70.00
BaCdZn stabilizer	2.50
lampblack (pigment)	2.00
calcium carbonate (filler)	25.00
lecithin (wetting agent)	1.00
Total	*204.50*

Processing

Coated fabrics can be prepared by lamination, direct calendering, direct coating or transfer coating (see Coating processes). The basic problem in coating is to bring the polymer and the textile together without altering undesirably the properties of the textile. Almost any technique in applying polymers to a textile requires having the polymer in a fluid condition, which requires heat. Therefore, damage to the synthetic or thermoplastic fabric may occur.

Calendering. The polymer is combined in a Banbury mill with a filler, stabilizing agents, pigments, and plasticizers and brought to 150–170°C. The mixture ("compound") temperature is adjusted on warming mills and calendered directly onto a preheated fabric. The object is to get the required amount of adhesion without driving the compound into the fabric excessively, which would cause a clothy appearance and lower the stretch and tear properties of the coated fabric.

Coating. Coating operations require a much more fluid compound. Rubbers are dissolved in solvents. In the case of PVC, fluidity is achieved by adding plasticizers and making a plastisol. If lower viscosity is required, an organosol is made by adding solvent to the plasticized PVC. After the ingredients are mixed and brought to a coating head, the mixture is applied by either knife, knife-over-roll, or reverse roll coaters. Unless the fabric is very dense, or a high degree of penetration is desired, the coating cannot be placed directly on the textile. Transfer coating limits the penetration into the fabric. The mixture is coated directly on the release paper and penetration is limited either by the viscosity of the coating or partial solidification (gelling) of the coating prior to application of the textile. Most polyurethane coated fabrics are transfer coated. Expanded vinyl-coated fabrics consist of a wear layer, an expanded layer, and the textile substrate. The wear layer is coated on release paper (see Abherents) and gelled. A layer of vinyl-based compound containing a chemical blowing agent such as azodicarbonamide is applied. The fabric is placed on top of the second layer and sufficient heat is applied to decompose the blowing agent causing the expansion (5).

Lamination. In lamination a film is prepared by calendering or extrusion. It is adhered to the textile at a laminator either with an adhesive or by sufficient heat to melt the film (see Laminated and reinforced plastics).

Post Treatment. Coated fabrics can be decorated by printing with an ink. Usually the appearance of a textile or leather is the goal. The inks are applied as low-solid solutions by metal rotogravure rolls. Warm air drying is carried out in an oven. Because the ink dries rapidly, multiple print heads can be used (see Ink).

If a textured surface is desired, the coated fabric is heated to soften it and pressure is applied by an engraved embossing roll. Printing usually precedes embossing so that a flat surface is presented for printing. Special effects are obtained by embossing first and then printing or wiping the high points (see Printing processes).

The final layer is called the slip. Most coatings are tacky enough to stick to themselves (block) during stacking or rolling. The main purpose of the slip is to prevent blocking (see Abherents). Slips can be formulated as shown in Table 6 to improve abrasion resistance, seal the surface, adjust color and adjust gloss. Slips are low-solid solutions that are applied by metal rotogravure rolls. Air drying leaves about 200 g of solids per 100 m^2 of coated fabric.

Table 6.   A Typical Slip Formulation for PVC-Coated Fabric

Component	Parts
vinyl chloride–vinyl acetate copolymer	100.00
polymethacrylate resin(s)	96.00
vinyl stabilizer	2.00
silica gel	18.00
methyl ethyl ketone	620.00
xylol	350.00
Total	*1186.00*

Economic Aspects

Poly(vinyl chloride) is the principal polymer employed (see Vinyl polymers). In the United States alone the consumption of PVC for calendered-coated fabrics was 28,000 metric tons in 1975 and 45,000 t in 1976. Coating of paper and textiles consumed 75,000 t in 1976.

The United States consumption of polyurethane for fabric coatings was 4500 t in 1975 and 5000 t in 1976 (6).

Health and Safety Factors

Some materials used in coating operations have been identified by the United States government as being hazardous to the workers' health.

Even when coatings are applied by extrusion and calendering, consideration should be given to handling the materials, evolution of gases during heating and post-finishes. For instance, there are strict regulations on exposure to vinyl chloride monomer. Emptying bags or bulk transfer must be monitored. The regulations do not apply to the handling or use of fabricated products made from poly(vinyl chloride).

When a coating machine is employed, attention must be given to exposure of the operator to the solvents.

In addition, particulate irritants such as asbestos (qv), pigments, and reactive chemicals are often involved.

Coating operation should not be initiated without consulting the *Federal Regulations on Occupational Safety and Health Standards, Subpart Z, Toxic and Hazardous Substances* (7).

Uses

Table 7 lists uses of coated fabrics and demonstrates how combinations of textiles and polymers can give significantly different products.

Table 7. Uses of Coated Fabrics [a]

Substrate	Coating	Use
nylon tricot nylon sheeting cotton sheeting	PVC, PU, SBR, Neoprene	clothing
polyester–cotton sheeting	PVC	
napped cotton drill	PU	shoe uppers
nonwovens (high density)	PU	
nonwovens (medium high density)	PVC	shoe liner-insoles
cotton knits	PVC	
polyester nonwovens (light density)	PVC	
polypropylene nonwovens (light density)	PVC	
cotton knits	PVC	furniture upholstery
polyester knits	PVC	
polyester–cotton knits	PVC	
napped cotton drills	PU	
nylon Helanca knits	PVC	
cotton single knits	PVC	auto upholstery
polyester–cotton pattern knits	PVC	
polyester nonwoven (light density on PU foam)	PVC	
polyester nonwoven (light density)	PVC	
polyester stitched nonwoven	PVC	landau tops
polyester knit	PVC	
asbestos nonwoven	PVC	floor covering
polyester spunbonded	PVC PE	wallcovering
cotton sheeting glass scrim rayon scrim	PVC, SBR, Neoprene, silicone rubber, etc	tapes
polyester drill	PVC, SBR, Neoprene, natural rubber	hospital sheeting
absorbent cotton	PVC	Band-Aids
nylon scrim	PVC	window shades
cotton scrim	PVC	wallpaper
cotton sheeting polyester sheeting polyester–cotton sheeting	acrylic	lined drapes
glass scrim polyester scrim	lead-filled PVC, barytes-filled PVC, barytes-filled SBR	acoustical barriers
paperlike nonwovens	PVC	air and oil filters
dyed rayon drill	expanded PVC	soft-side luggage
rayon drill cotton drill polyester–cotton drill polyester drill nonwovens	PVC	luggage
nylon woven polyester woven	PVC, Neoprene, Hypalon, PU	tarpaulins
nylon woven cotton woven	PVC	awning
nylon scrim polyester scrim	PVC, EPDM, Hypalon, Butyl	pond and ditch liner
glass woven polyester woven	Neoprene, PVC, Hypalon	air supported structures

[a] PVC = poly(vinyl chloride); PU = polyurethane; EPDM = ethylene–propylene–diene-modified rubber; SBR = styrene–butadiene–rubber; PE = polyethylene.

BIBLIOGRAPHY

"Coated Fabrics" in *ECT* 1st ed., Vol. 4, pp. 134–144, H. B. Gausebeck, Armour Research Foundation of Illinois Institute of Technology; "Coated Fabrics" in *ECT* 2nd ed., pp. 679–690, by D. G. Higgins, Waldron-Hartig Division of Midland-Ross Corporation.

1. *Davison's Textile Blue Book,* Davison Publishing Co., Ridgewood, N.J., 1977.
2. N. J. Abbott, T. E. Lannefeld, and R. J. Brysson, *J. Coated Fibrous Mater.* **1,** 4 (July 1971).
3. S. P. Suskind, *J. Coated Fibrous Mater.* **2,** 187 (Apr. 1973).
4. H. L. Gee, *J. Coated Fabr.* **4,** 205 (Apr. 1975).
5. W. G. Joslyn, *Rubber Age* **106,** 49 (Feb. 1974).
6. *Mod. Plast.* **54,** 49 (Jan. 1977).
7. *Code of Federal Regulations, Title 29,* Chapter XVII, Section 1910.93 of Subpart G redesignated as 1910.1000 at 40 FR23072, U.S. Government Printing Office, Washington, D.C., May 28, 1975.

General References

F. J. Beaulieu and M. D. Troxler, "Substrates for Coated Apparel Applications," *J. Coated Fibrous Mater.* **2,** 214 (Apr. 1973).
"1977 Manmade Fiber Deskbook," *Mod. Text.* **2,** 16 (Mar. 1977).
"Generic Description of Major U.S. Manmade Fibers," *Mod. Text.* **58,** 17 (Mar. 1977).
"Names and Addresses of U.S. Manmade Fiber Producers," *Mod. Text.* **58,** 30 (Mar. 1977).
"77–78 Buyers Guide," *Text. World* **127,** (July 1977).
R. M. Murray and D. C. Thompson, *The Neoprenes,* E. I. du Pont de Nemours & Co., Inc., Wilmington, Del., 1963.
M. Morten, *Rubber Technology,* Van Nostrand Reinhold Co., New York, 1973.
J. Bunten, "Performance Requirements of Urethane Coated Fabrics," *J. Coated Fabr.* **5,** 35 (July 1975).
D. Popplewell and L. G. Hole, "Urethane Coated Fabrics," *J. of Coated Fabrics* **3,** 55 (July 1973).
H. A. Sarvetnick, *Polyvinyl Chloride,* Van Nostrand Reinhold Co., New York, 1969.
"Manufacturing Handbook and Buyer's Guide 1977/78," *Plast. Technol.* **23,** (Mid-May 1977).
Davison's Textile Blue Book, Davison Publishing Co., Ridgewood, N.J., 1977.
Rubber Red Book, Palmerton Publishing Co., New York, 1977.

FRED N. TEUMAC
Uniroyal, Inc.

COTTON

The story of cotton predates recorded history, and although the actual origin of cotton is still unknown, there is evidence that it existed in Egypt as early as 12,000 BC. Its use in cloth in 3000 BC was indicated by archeological findings and recorded evidence exists of its cultivation in India as far back as 700 BC. In the fifth century BC, Herodotus wrote of trees growing wild in India bearing wool of a softness and beauty equivalent to that of the sheep; clothes made from this tree wool were described as garments of extraordinary perfection.

It is said that Alexander the Great introduced Indian cotton into Egypt in the 4th century BC, and from there it spread to Greece, Italy, and Spain. During the year 700 AD, China began growing cotton as a decorative plant, and 798 AD saw its introduction into Japan. Early explorers in Peru found cotton cloth on exhumed mummies that dated to 200 BC. Cotton was found in North America by Columbus in 1492. About three hundred years later the first cotton mill was built at Beverly, Mass., and in 1794 Eli Whitney was granted a patent for the invention of the cotton gin.

Cotton is the most important vegetable fiber used in spinning (see Fibers, vegetable). Its origin, breeding, morphology, and chemistry have been described in innumerable publications (1–4). It is a member of the Malvaceae or mallow family, a plant of the genus *Gossypium*, and is widely grown in warm climates the world over.

The average cotton plant is a herbaceous shrub having a normal height of 1.2–1.8 m, although some tree varieties reach a maximum height of 4.5–6.0 m. The most important species included in the genus *Gossypium* are *hirsutum, barbadense, arboreum,* and *herbaceum.*

G. hirsutum originated in Central America and is found in the Mayan culture of Mexico. It is the species that comprises all of the many varieties of American Upland cotton of commerce. *G. hirsutum,* a shrubby plant that reaches a maximum height of 1.8 m, is probably the origin of the green-seeded (ie, bearing green fuzz fibers) cotton of former times grown so extensively in the southern United States.

G. barbadense, originally from the Incan civilization, grows from black seeds and ranges in height from 1.8–4.5 m. It is the species with longest staple and includes Sea Island, Egyptian Giza strains, American–Egyptian, and Tanguis cottons; of these, Sea Island is the longest and silkiest of the commercial cottons.

G. arboreum, the tree wool of India, grows as tall as 4.5–6.0 m and includes both Indian and Asiatic varieties. Its seeds are covered with greenish-gray fuzz fibers below the white lint fibers.

G. herbaceum, the original cotton of India, averages 1.2–1.8 m. The fiber is grayish-white and grows from a seed encased in gray fuzz fibers. This species includes the short staple cottons of Asia and some of China, as well as most native Indian varieties. It is also grown commercially in Iran, Iraq, Turkey, and the USSR.

The most favorable growing conditions for cotton include a warm climate (17–27°C, mean temperature), where fairly moist and loamy, rather than rich, soil is an important factor; seed planted in dry soil produces fine and strong, but short, fibers of irregular lengths and shapes.

Under normal climatic conditions, cotton seeds germinate in 7–10 d. Flower buds (known as squares) appear in 35–45 d followed by open flowers 21–25 d later. After one day the cotton boll begins to grow rapidly if the flower has been fertilized. The mature boll opens 45–90 d after flowering, depending on variety and environmental conditions. Within the boll are 3–5 divisions called locks, each of which normally has 7–9 seeds that are covered both with lint and with fuzz fibers (Fig. 1). The fuzz fibers form a short, shrubby undergrowth beneath the lint hairs on the seed. Each seed contains at least 10,000 fibers and there are close to 500,000 fibers in each boll. The usual range of planting time in the United States extends from the beginning of March to the end of May; harvest time is in the late summer or early fall.

The cotton fiber is a single cell that originates in the epidermis of the seed coat at about the time the flower opens; it first emerges on the broad, or chalazal end of the seed and progresses by degrees to the sharp, or micropylar end. As the boll matures, the fiber grows until it attains its maximum length, which averages about two thousand five hundred times its width (Fig. 2). During this time, the cell is composed of only a thin wall that is covered with a waxy, pectinaceous material and that encloses the

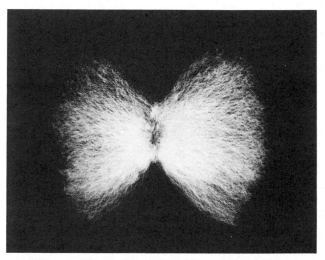

Figure 1.   Cotton butterfly with lint and fuzzy fibers.

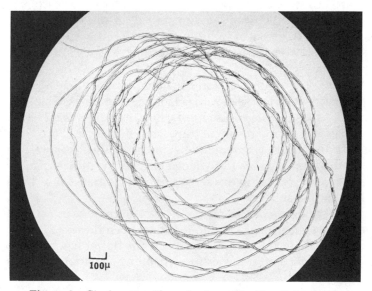

Figure 2.   Single cotton fibers showing ratio of length to width.

protoplasm or plant juices. In ca 17–25 d after the flower opens, when the boll is half mature, the fiber virtually attains its full length, at which time it begins to deposit layers of cellulose on the inside of the thin casing, or primary wall. The pattern of deposition is such that one layer of cellulose is formed each day in a centripetal manner until the mature fiber has developed a thick secondary wall of cellulose, from the primary wall to the lumen, or central canal. The fiber now consists of three main parts: primary wall, secondary wall, and lumen. At the end of the growing period when the boll bursts open, the fibers dry out and collapse, forming shrivelled, twisted, flattened tubes.

The seed hairs of cultivated cottons are divided into two groups—fuzz and lint—that may be distinguished on the basis of such characteristics as length, width, pigmentation, and strength of adherence to the seed. The growth of fuzz fibers is much the same as that of lint, but they are usually about 0.33 cm long compared with the 2.5 cm average length of lint fibers, and are twice as thick, or about 32 μm (Fig. 3). Their

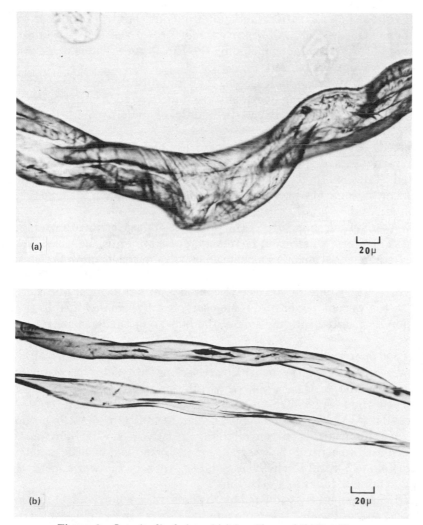

Figure 3.   Longitudinal view of (**a**) fuzz fiber and (**b**) lint fibers.

color range is from greenish-brown to gray. After lint fibers have been ginned off the seed, the fuzz fibers or linters are left and must be removed by a machine similar to the regular cotton gin. Linters are an important source of cellulose for chemical purposes and are used in upholstery and batting.

Cultivation and Production

At present, the chief cotton-growing countries of the world are the Union of Soviet Socialist Republics (22%), the People's Republic of China (19%), the United States (11%), India (9%), Pakistan (4%), Brazil (4%), Turkey (4%), and Egypt (3%) (5). In the Western hemisphere, cotton is cultivated from the equator to about 37° N latitude and 32° S latitude; in the Eastern hemisphere, the limits extend to 47° N latitude and 30° S latitude.

The United States produces approximately 11% of the world's cotton. Although the actual size of the crop has changed very little, improved fertilizers, insecticides, and harvesting practices have increased the yield per hectare so much that only about half as much area of land is needed today as in 1930.

Cultivation of cotton differs markedly from one country to another, depending on methods of mechanization. Approximately 70% of the United States cotton is rain-grown, but western states, including the Rio Grande Valley of Texas, have increased their yield by irrigation.

Through cotton-breeding research, the United States has developed varieties that today remain the leading fine quality cottons. Two such varieties are Stoneville and Deltapine. Lankart, Acala, Paymastir, and Coker follow to complete the six leading cottons in the United States (6).

Fertilizers. Perhaps the greatest progress in fertilization of cotton fields is reflected in techniques of application. Throughout all of the cotton growing regions of the United States, methods of applying nutrients must be tailored to the needs of designated areas, some of which also require special attention in the types of fertilizers (qv) applied.

Characteristics of soils differ, as do practices of past fertilization, cropping, and irrigation. Therefore, an efficient fertilization program must be based on results of soil tests. Requirements include adequate amounts of nitrogen, phosphorus, potassium, and boron.

However, regardless of soil needs, methods of fertilizer application must be adapted to the various regions. In some cases, fertilizer is included in the seedbed preparation or inserted into the side of the beds, and the beds are rerun immediately.

Pests and Insecticides. Perhaps the most destructive pests of the cotton plant are the boll weevil and the bollworm, which are serious threats to the cotton industry in countries around the world. The boll weevil, after migrating from Mexico around 1892, within 30 years had spread over the entire cotton belt. The domestic cotton crop lost to the weevil is worth 200 million dollars a year; in addition, about 75 million dollars a year is spent for pesticides to control this destructive pest (7). Unfortunately, some insecticides used to control the weevil kill many beneficial insects, among them insects that help to control the bollworm and the tobacco budworm, pests that cause another 200 million dollar loss in cotton.

In 1916 calcium arsenate dusted by airplane was used to control the boll weevil; however, throughout many developments in effective insecticides, such as organophosphates, the boll weevil began to build up resistance to poisons that were formerly effective (see Insect control technology).

Several years ago, a pilot program to eradicate the boll weevil, carried out on an 8100-ha region of Mississippi, Alabama, and Louisiana, led to plans for beltwide eradication of this devastating pest (7). The plan of attack on the weevil has been mapped out, and it is believed that success can be realized by 1980.

One of the most destructive insects is the pink bollworm, which overwinters as diapausing (hibernating) larvae in the soil. After feeding on the late-blooming bolls, the larvae drop to the ground and hibernate for the winter, emerging as adults in the spring to lay eggs on the early cotton blooms. The eggs hatch and the new larvae bore into the fresh cotton bolls, go through molting stages, bore their way out, and drop to the ground. Throughout the growing season the cycle repeats itself, to the destruction of vast numbers of cotton plants in a single field.

A technique to combat this pest has been developed by agricultural research scientists in cooperation with the Arizona Agricultural Experiment Station in Phoenix. The number of overwintering pink bollworms is limited by reduction of their food supply late in the year. Growth regulators are used to prevent late boll formation; this has no effect on the existing bolls and causes no loss in harvesting, as most of the bolls fail to mature anyway. Termination of late-fruiting cotton is carried out by applying chemical formulations. Along with the chemical treatments, other eradication practices include early stalk shredding, early and deep tillage, and winter irrigation that drowns diapausing larvae (8).

To the problems involving the boll weevil and the pink bollworm must be added the efforts expended on research to eliminate other insects injurious to the cotton plant. Included in the list are aphids, leafhoppers, lygus bugs, mites, white flies, black fleahoppers, thrips, cutworms, and leafminers (9).

Harvesting. When the ripened boll bursts and the cotton dries and fluffs, it is ready for harvesting. Prior to the actual operation, however, the plants may be prepared by treatment with harvest-aid chemicals, classified as defoliants or desiccants. Defoliants usually cause an abcission or shedding of leaves earlier than normal, preventing further plant development. Desiccants, on the other hand, rapidly kill the plant by causing an immediate loss of water from the tissue. The dried leaves remain on the plant. Conditions of both plants and weather are of major importance in the success of these harvest-aid programs (see Herbicides; Plant-growth substances).

Until about the middle of this century, cotton was picked entirely by hand, and in many foreign countries this practice still exists. However, in the United States, mechanical harvesting essentially has replaced handpicking; in fact, over 99% of the cotton in the United States is harvested by machine.

The first type of mechanical harvester was the stripper, a machine that extracts from each plant all of the cotton in the field, regardless of maturity. As a result, large quantities of unopened bolls, stalks, and leaves are caught up by the stripper.

The spindle harvester straddles the cotton and extracts it from the bolls in one or two rows simultaneously. The lint from the open bolls is wound onto the spindle, from which it is removed and carried to a large container.

As a general rule, harvesting should take place when the relative humidity is 60% or less and cotton is not wet with dew. Storage of damp cotton on trailers affects the quality of the cotton and germination of the seed. Since the gin cannot process seed cotton at the same rate at which it is harvested, storage should be under optimum conditions.

Ginning. The greatest number of cotton gins in existence in the United States was in 1902; 30,948 gins, the majority on plantations, processed 10.6 million bales (of ca 217 kg each) of cotton (10). Since 1902 the number of gins has declined and the average number of bales handled has correspondingly increased. In 1975, 2856 active gins handled a crop of 8,151,223 bales for an average of 2854 bales per gin plant (11).

Mechanical harvesting systems have reduced the harvesting and ginning period from 4–5 mo to ca 6 wk of intensive operation (12). In 1957, the typical United States cotton gin had an average production rate of about 6 bales per hour; the present design capacity of new gins is in excess of 20 bales per hour.

Most of the United States gins are now operated as corporations or as cooperatives serving many cotton producers, in contrast to the plantation ownership pattern. Automatic devices do the work faster, more efficiently and more economically than hand labor. Examples of increased efficiency are (*1*) high-volume bulk seed cotton handling systems to get cotton into the gin, (*2*) reduced processing time for drying, ginning, and cleaning cotton, and (*3*) automated bale packaging devices.

Commercial Classification

Classing of cotton involves describing the quality of cotton in terms of grade and staple length in accordance with the official cotton standards of the United States. Classing in this way indicates the cotton's spinning utility and value to the buyers and sellers and provides information for evaluating efficiency of production, harvesting, ginning, and processing practices. About 95% of the cotton produced in the United States is classed by grade and staple length. This classification is often supplemented by inclusion of "Micronaire reading" to indicate the fineness of the cotton.

There are 37 grades for American Upland cotton within seven general color classifications in addition to below-grade cotton. Color (qv) is defined in terms of hue, lightness, and chroma. Color classifications are from white, actually a pale cream color, to light-spotted, spotted, tinged, yellow-stained, to light gray or gray, the last two indicating cotton that was left too long unpicked, with weather and possible fungal damage. There are ten grades for American–Egyptian cotton. In addition to color, leaf and ginning preparation are also used to establish grade. Leaf includes dried pieces of leaf, stem, stalk, seed coat, and other contaminants. Leaf is divided into two general groups based on size large leaf, and pin or pepper leaf. The penalty is greater for a large amount of fine trash because it is so difficult to remove. Ginning preparation describes the degree of smoothness and relative nappiness or neppiness of the ginned lint.

Classification of cotton in terms of staple length is in accordance with official United States standards and regulations. Current official standards for staple length include 14 standards for the range 2.1–3.2 cm (including most American Upland cotton), four standards covering 3.3–3.8 cm (including most American–Egyptian cotton), and four covering 3.8–4.4 cm (including most American Sea Island cotton). Cotton below 2.1 cm is classed as "below 13/16 in." (2.1 cm).

Physical Properties

Length is the most important dimension of the cotton fiber. The length of staple of any cotton is the length by measurement of a representative sample of fiber, without regard to quality or value, at 65% rh and 21°C (13). An experienced classifier determines the length by parallelizing a typical portion of fibers from pulls drawn from a sample.

Variations of length are peculiar to classes of cotton and range from less than 2.5 cm for short staple Upland varieties, and 2.6–2.8 cm for medium staple Uplands, to >2.85 cm for long staple (14).

Shorter staple varieties tend to be coarser, whereas the longer varieties such as Pima and Sea Island, which can measure up to 5.7 cm in length, are fine, silky, and soft.

Diameter of the fiber is an inherited characteristic that can be greatly influenced by soil and weather. This fiber dimension may best be observed in cross sections of bundles of fine, medium, or coarse fibers. Whereas the typical shape has been described as resembling that of the kidney bean, the shapes range from circular to elliptical to linear in most varieties (15).

Uniformity within a sample is largely dependent on variety and degree of maturity; hence diameter measurements must be calculated. Use of a bidiameter scale gives values for both the major and minor axes of the fiber cross sections. Other measurements from which diameter may be derived are wall thickness, lumen axes, or combinations of these.

Fineness is defined as a relative measure of size, diameter, linear density, or weight per unit length expressed as micrograms per inch (μg/in.) [0.394 μg/cm] (16), and is sometimes referred to as linear density. Today, however, the term Micronaire reading is accepted in the United States as an added criterion for marketing. It is a measure of the resistance of a plug of cotton to air flow and indicates the fineness of the cotton (16).

Strength of the cotton fiber has been attributed mostly, if not entirely, to the cellulose (qv) it contains and, in particular, to the molecular chain length and orientation of the cellulose (15) (see also Biopolymers). Fiber bundle testing and single fiber testing (17) have contributed much basic knowledge to this important property.

Directly associated with the strength of the fiber is the degree of maturity it has attained. Average maturity acceptable for mill usage is from 75–80%. When this percentage drops, an excess of immature fibers causes higher picker and card waste, formation of neps, and production of yarns that grade low in appearance.

Morphology

The cotton fiber is tapered for a short length at the tip, and along its entire length is twisted frequently, with the direction of twist reversing occasionally. These twists are referred to as convolutions and it is believed that they are important in spinning because they contribute to the natural interlocking of fibers in a yarn.

Cotton is essentially 95% cellulose (Table 1). The noncellulosic materials, consisting mostly of waxes, pectinaceous substances, and nitrogenous matter, are located to a large extent in the primary wall, with small amounts in the lumen.

Of the noncellulosic substances in cotton, protein normally occurs in the largest amounts. This protein is apparently the protoplasmic residue left behind on the gradual drying up of the living cell. It occurs almost entirely in the lumen. On the assumption that the cell content is approximately the same for each fiber, it has been suggested that the nitrogen content might be used as an indirect measure of wall thickening. However, no practical test of sufficient accuracy has been developed.

Table 1.   Composition of Typical Cotton Fibers

Constituent	Composition, % of dry weight	
	Typical	Range
cellulose	94.0	88.0–96.0
protein (% N × 6.25)[a]	1.3	1.1–1.9
pectic substances	1.2	0.7–1.2
ash	1.2	0.7–1.6
wax	0.6	0.4–1.0
total sugars	0.3	
pigment	trace	
others	1.4	

[a] Standard method of estimating percent protein from nitrogen content (% N).

Most of the pectin in the cotton fiber is in the primary wall. Total removal of the pectin substances is accomplished readily by scouring, which does not change greatly the properties of the cotton.

The wax of most cottons is a soft, low melting, complex mixture having some differences among varieties. The noncommercial, strongly pigmented green lint cotton may contain up to about 17% wax of high melting point. Because the wax becomes established in fibers, largely if not wholly, during the first phase of development, the wax content expressed as a percentage of the whole fiber mass decreases as the fiber maturity or degree of wall thickening increases. Removal of cotton wax by chemical treatment increases the friction between fibers and between fiber and metal.

Although the mature cotton fiber is considered as having a primary wall, a secondary wall, and a lumen, further examination of its structure shows a cuticle and a winding layer (Fig. 4). The cellulose of the primary wall exists as a woven network of microfibrils in and on which are deposited noncellulosic materials that form the cuticle. Just beneath the primary wall is the winding layer, which is also the first layer of the secondary wall. The winding layer appears to comprise a single layer of fibrillar bundles composed of microfibrils and oriented at an angle to the fiber axis. The main body of the fiber consists of cellulose fibrils packed tightly in a solid cylinder, which, under certain conditions of chemical swelling, can be induced to separate into more or less concentric layers. These layers seem to have a finer and more regular structure than does the winding layer, since the 20–50 secondary wall layers have cellulose microfibrils compactly parallelized along the axis of the fiber (see Cellulose).

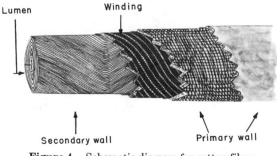

Figure 4.   Schematic diagram for cotton fiber.

The gross morphology of cotton, which refers to the relatively large structural elements above, is visible in the electron microscope. The microfibrillate structure includes pores, channels, and cavities that play an important role in the chemical modification of cotton. The arrangement of fibrils follows a spiral pattern and at times reverses itself; it is believed that regions of low strength along the length of the fiber occur within the reversal zones.

Microfibrils of the secondary wall are 10–40 nm in width, and these in turn are composed of elementary fibrils (crystallites) 3–6 nm wide.

Chemical modification of the cotton fiber must be achieved within the physical framework of this rather complicated architecture. Uniformity of reaction and distribution of reaction products are inevitably influenced by rates of diffusion, swelling and shrinking of the whole fiber, and by distension or contraction of the fiber's individual structural elements during finishing processes.

Chemical Properties

Although the cotton fiber, on drying, collapses from its round, never-dried shape and much of its water is removed, moisture is retained tenaciously in cotton. This moisture is expressed either as moisture content (amount of moisture as a percentage of original sample mass) or more commonly as moisture regain (amount of moisture as a percentage of oven-dry sample). Under ordinary atmospheric conditions, moisture regain is 7–11%. Because the cotton fiber has such a high cellulose content, the chemical properties are essentially those of the cellulose polymer. Indeed, the standard cellulose adopted by a committee of the ACS is simply a purified cotton. Like other forms of cellulose, the molecular chains of cotton cellulose consist of anhydroglucose units joined by 1–4 linkages (see Carbohydrates). These glycosidic linkages characterize the cotton cellulose as a polysaccharide, and are cleaved during hydrolysis, acetolysis, or oxidation resulting in shorter chains. If degradation is extensive enough, cellobiose or glucose derivatives are produced.

The cotton cellulose is a naturally occurring polymer; therefore, the molecular cellulose chains are of varying length. A measure of the average chain length can be obtained by determining the fluidity (or its reciprocal, the viscosity) of a solution of cotton. Solvents for this purpose include cuprammonium hydroxide solutions, phosphoric acid, nitric acid, quaternary ammonium bases, cadmium ethylenediamine hydroxide, and the one most used currently, cupriethylenediamine hydroxide (18). Low fluidities indicate cotton cellulose of high molecular weight; an increase in fluidity from that of untreated cotton usually signifies hydrolytic or oxidative damage to the cellulose. Oxidative damage also may be detected by increased adsorption of such dyes as methylene blue, by determination of copper number, or by titration of the resultant carboxyl group. If oxidation has proceeded to a great extent, marked alkali solubility results.

In addition to the length of cotton cellulose chains and to the chain length distribution, there is another factor that characterizes cotton cellulose. This is the degree of accessibility of the long, threadlike, polymeric molecules of cellulose within the elementary fibril of the cotton. Until the 1960s the fibrils were thought to divide into well-ordered crystalline and less organized amorphous regions. These dense crystalline regions, composed of highly hydrogen-bonded elementary fibrils, were believed to be the basis for the characteristic x-ray crystal diagram of the cellulose. The amorphous regions, thought to be composed of random, nonparallel polymeric aggregates, were responsible for the diffuse background in the x-ray crystal diagram. Thus relative amounts of crystalline and amorphous cellulose have been estimated from the x-ray diagram. Recent interpretations suggest that differences observed in the x-ray diagram of cellulose arise from structural defects that provide surfaces of varying accessibility, implying that cellulose is paracrystalline material (19–20). Also, whereas estimates formerly were made of crystalline and amorphous regions from rates and extent of reactions with hydrolytic and oxidative chemicals, it is now believed that these data merely indicate differences in accessibility (21).

These molecular chains that are associated into the elementary fibril are not parallel to the fiber axis; the spiral angle is approximately the same as that of the microfibrillar units of which the molecular chains form the ultimate unit. Furthermore, the helical configuration is irregular, due to frequent reversals in spiral angle. Chemical reactivity of the cellulose molecular chains within this complex physical structure is influenced by the location of the molecules on surfaces of the elementary fibril. Highly bonded surfaces result in inaccessible regions; surfaces under strain from spiral or reversal demands provide the most accessible regions.

Reactions.   One of the earliest known modifications of cotton was mercerization. Traditionally, the process employed a cold concentrated sodium hydroxide treatment of yarn or woven fabric, followed by washing and a mild acetic acid neutralization. Maintaining the fabric under tension during the entire procedure was integral to the expected properties. The resultant mercerized cotton of commerce has improved luster and dyeability, and to a lesser extent, improved strength. A recent variation of this procedure substitutes hot sodium hydroxide that is allowed to cool while the cotton remains immersed in the caustic solution. More thorough initial penetration increases the efficiency of the mercerizing process. If the cotton is allowed to shrink freely during contact with mercerizing caustic, slack mercerization takes place; this technique produces a product with greatly increased stretch (stretch cotton) that has found application in both medical and apparel fields.

Effects similar to those from sodium hydroxide mercerization have been produced by exposure of the cotton to volatile primary amines or to ammonia. A procedure that utilizes liquid ammonia has found commercial adaptation (22). Improvements in luster and strength are similar to those from sodium hydroxide mercerization, but dyeability is not enhanced. A distinct difference exists chemically between cotton mercerized in sodium hydroxide and that mercerized in liquid ammonia. Under laboratory conditions of sodium hydroxide mercerization, Cellulose I of native cotton is converted to Cellulose II; under conditions of mercerization with liquid ammonia followed by nonaqueous quenching, Cellulose III is the product (23). Figure 5 shows the characteristic diffractograms (CuK$_2$ radiation) of native cellulose and cellulose mercerized with sodium hydroxide and liquid ammonia. With both treatments, there is increased accessibility, but differences in dye receptivity presumably result from differences in swelling loci between sodium hydroxide and liquid ammonia treatments.

The hydroxyl groups on the 2, 3, and 6 positions of the anhydroglucose residue are quite reactive (24), and by virtue of their functionality provide sites for much of the current modification of cotton cellulose to impart special properties. The two most common classes into which modifications fall include esterification and etherification of the cotton cellulose hydroxyls, as well as addition reactions with certain unsaturated compounds to produce a cellulose ether (see Cellulose derivatives, ethers).

Cellulose esters can be subdivided further into inorganic and organic esters (25). Of the three most common inorganic esters, cellulose nitrate, phosphate, and sulfate,

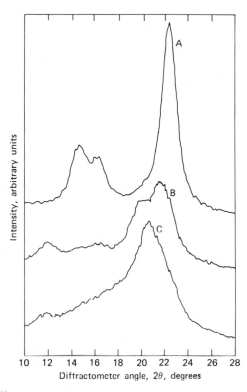

Figure 5.   X-ray diffractograms: A, native; B, NaOH mercerized; and C, NH₃ mercerized cellulose (15).

only the cellulose sulfate is soluble in water. Cellulose sulfate attains water solubility at a degree of substitution (DS) of 3, indicating esterification of all three hydroxyls, whereas the sodium salt of cellulose sulfate is soluble in hot and cold water with a DS of only 0.33. Sodium cellulose sulfate is used in applications requiring suspension, thickening, stabilizing, and film-forming properties. A recent addition to cellulose-reactive dyestuffs is the class of phosphonic acid and phosphoric acid dyestuffs; these attach to cotton through esterification by the phosphonic acid or phosphoric acid group of the dyestuff (26). Organic esters of cotton cellulose, with two notable exceptions, are only of academic interest, although partial esterification of cotton by fatty acids has been reported to increase resiliency (27). A large class of cellulose-reactive dyestuffs in commercial use attach to the cellulose through an alkali-catalyzed esterification by the chlorotriazine moiety of the dyestuff:

$$\text{cellulose—OH} + \text{Cl—dyestuff} \rightarrow \text{cellulose—O—dyestuff} + \text{HCl}$$

Acetylation of cotton to an acetyl content slightly greater than 21% produces a material with greatly increased resistance to fungal and microbiological degradation, in addition to a tolerance of high temperatures not exhibited by native cotton; fibrous appearance and physical properties are unchanged by the acetylation. X-ray diffractograms indicate that, at this extent of substitution, only accessible regions of the cotton are involved in the acetylation (28). Differences in reaction rates between formation of inorganic and organic esters of cellulose depend on the availability to the reagents of the cellulose hydroxyls within the microstructure of cotton. Because only the most accessible regions can be reached by the organic reagents, organic esterification proceeds layer by layer. Spreading apart of the microstructure by swelling during inorganic esterification creates new accessible regions and allows essentially simultaneous reaction throughout the layers of the fibrillar structure (see Cellulose derivatives, esters).

By far the most important commercial modifications of cotton cellulose occur through etherification. For example, commercial modification of cotton to impart durable-press, smooth drying, or shrinkage resistance properties involves cross-linking adjacent cellulose chains through amidomethyl ether linkages. This cross-linking is commonly achieved via a pad–bake process. Although methylene, or oligomeric, cross-links from a pad–bake formaldehyde treatment have been produced, the resultant fabric exhibits severe strength loss. Most reagents for cross-linking cotton cellulose are di- or polyfunctional amidomethylol compounds (see Amino resins). The general formula for these compounds is shown with equations for synthesis of a methylol agent and its reaction with cellulose in Figure 6 (29). Commercially available cross-linking agents are dimethylolurea, dimethylolethyleneurea, dimethyloldihydroxyethyleneurea, dimethylolpropyleneurea, dimethylol alkylcarbamate, tetramethylolacetylenediurea, and methylolated melamine. The cross-linking proceeds via either Lewis or Bronsted catalysis (30) by a carbonium ion mechanism. Gross effects of the cross-linking are increased resiliency, manifested in wrinkle resistance, smooth drying properties, and greater shape holding properties, all important to cotton textiles, and conversely, reduced extensibility, strength, and moisture regain. These effects are observed at substitutions of 0.04–0.05 cross-links per anhydroglucose residue (31). The use of liquid ammonia treatment of cotton fabric followed by cross-linking attenuates the strength loss as well as an accompanying loss in abrasion resistance; a

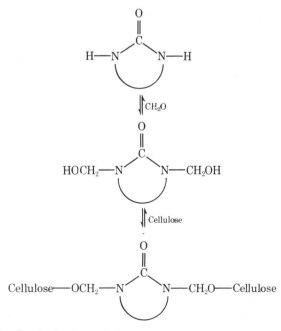

Figure 6.   Synthesis of a methylol agent and its reaction with cellulose (19).

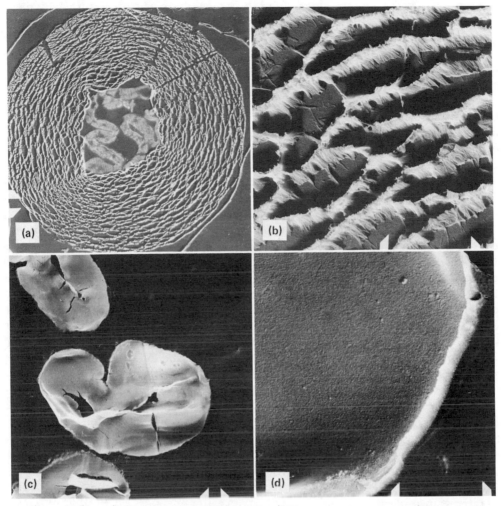

Figure 7. Methacrylate expansion patterns: (a) and (b), cross sections of unmodified cotton fiber at two levels of magnification; (c) and (d), expansion-resistant cross sections of cross-linked fiber (23). The gage marks on all four photographs indicate 1 μm.

result of this unique combination of existing technologies is a reappearance of all-cotton fabrics in the woven shirting and high-fashion knitted fabric markets (32).

The microstructural effects of cross-linking via a pad–bake process can be seen in the response of the microfibrils and elementary fibrils to applied loads and to swelling agents. Resistance of cross-linked fibers to a methacrylate layer-expansion treatment that separates lamellae and reveals pore structure in untreated cotton is shown in Figure 7 (33). Although the ultimate chemical reaction—etherification of cellulose—is the same, the amount of moisture present in the fiber at the time of etherification influences both the response of the microstructure of the etherified fiber to swelling and solvation, and the physical properties of the cotton product (34). Thus resistance to layer-expansion treatment of the cotton fiber cross-linked with formaldehyde in a water medium that allows free fiber swelling lies between that of untreated fibers and pad–bake cross-linked fibers (35). Response to layer expansion by the cotton fiber cross-linked with formaldehyde in the vapor phase, with moisture equivalent to normal regain, is similar to that by a conventionally cross-linked fiber (36). Cross-linking cotton fibers in a swollen state increases resiliency under wet but not under ambient conditions. As the amount of water present in the fiber at the time of reaction is decreased, wet resiliency decreases, and resiliency under ambient conditions increases.

Base-catalyzed reactions of cotton cellulose with either mono- or diepoxides to form cellulose ethers also result in fabrics with increased resiliency. Monoepoxides, believed to result only in cellulose hydroxyalkyl ethers or linear graft polymers (37), produce marked improvement in resiliency under wet conditions, but little improvement under ambient conditions. Difunctional epoxides, capable of cross-linking the cellulose, can produce increases in resiliency under both wet and ambient conditions (38). Microscopical examination of ultrathin cross sections from fabrics finished with both mono- and diepoxides indicates much less cross-linking in the former. Not only resiliency can be imparted through epoxide etherification; oil and water repellency can be imparted by reactions of monomeric perfluoro epoxides with cotton. Etherification of cotton with ethyleneimine has also provided the basis for imparting special properties to cotton, with the end product dependent on the attached group. Another base-catalyzed etherification of cotton is the cross-linking reaction between bis(hydroxyethyl) sulfone and cellulose; fabrics possessing increased resiliency under both wet and ambient conditions are obtained. The earliest application of sulfone cross-links to cotton textiles was the reaction of divinyl sulfone under alkaline conditions (39). However, the hazard of working with the vinyl compound led to modifications of the sulfone agent to replace the vinyl groups with more stable precursors such as the β-thiosulfatoethyl or β-sulfatoethyl (40) and β-hydroxyethyl groups (41). Etherification of cotton by divinyl sulfone and its precursors also forms the basis for another large class of fiber-reactive dyestuffs with the general formula: dyestuff—$SO_2CH{=}CH_2$ (42).

Other cotton cellulose ethers that have been prepared include carboxymethyl, carboxyethyl, hydroxyethyl, cyanoethyl, sulfoethyl, and aminoethyl (aminized cotton). Most of these ethers, with the exception of cyanoethylated and aminized cotton, are of interest in applications requiring solubility in water or alkali (26). In addition, ethers with pendant acid or basic groups have ion-exchange properties (43). Aminized cotton is of interest because it introduces into the cotton basic groups that provide sites for attachment of acid dyes. Simultaneous aminization of cotton and dyeing with an acid dyestuff marked the first successful attempt at dye attachment to cellulose through an ether linkage (44) (see Cellulose derivatives).

The flameproofing of cotton will not be discussed extensively here (see Flame-retardants; Flame-retardant textiles). Although certain cellulose esters such as the ammonium salt of phosphorylated cotton and cellulose phosphate are flame resistant, the attachment of most currently used durable polymeric flame-retardants for cotton is through an ether linkage to the cellulose at a relatively low degree of substitution (DS).

The development of water-repellent cellulose ethers has been reviewed by Marsh (45) (see Waterproofing). A typical example of a commercial etherification for waterproofing cotton is with stearamidomethylpyridinium chloride:

$$\underset{\underset{\text{Cl}^-}{|}}{C_{17}H_{35}\overset{\overset{\displaystyle O}{\|}}{C}\!-\!\underset{H}{N}CH_2\!-\!\overset{+}{N}\!\!\bigcirc} + \text{cell}\!-\!\text{OH} \longrightarrow C_{17}H_{35}\overset{\overset{\displaystyle O}{\|}}{C}\!-\!\underset{H}{N}CH_2O\!-\!\text{cell} + HCl + N\!\!\bigcirc$$

N-Substituted, long chain alkyl monomethylol cyclic ureas have also been used to waterproof cotton through etherification. Other water repellent finishes for cotton are produced by cross-linked silicone films (46). In addition to the polymerization of the phosphorus-containing polymers on cotton to impart flame retardancy, and of silicone to impart water repellency, polyfluorinated polymers have been successfully applied to cotton to impart oil repellency. Chemical attachment to the cotton is not necessary for durability; oil repellency occurs because of the low surface energy of a fluorinated surface (47).

One of the earliest examples of etherification of cellulose by an unsaturated compound through vinyl addition is the cyanoethylation (qv) of cotton (48). This base-catalyzed reaction with acrylonitrile, a Michael addition, proceeds as follows:

$$CH_2\!=\!CHCN + \text{cell}\!-\!OH \rightarrow \text{cell}\!-\!OCH_2CH_2CN$$

For most textile uses, a DS less than 1 is desirable. Cyanoethylation can be used to impart a wide variety of properties to the cotton fabric such as rot resistance, heat and acid resistance, and receptivity to acid and acetate dyes. Acrylonitrile (qv) has also been radiation-polymerized onto cotton with a ^{60}Co source. Microscopical examination of ultrathin sections of the product shows the location of the polymer is within the fiber (49). Examination of the ir spectrum of cotton containing polymerized acrylonitrile indicates grafting does occur at the hydroxyl site of the cellulose (50). Another monomer grafted onto cellulose by irradiation is styrene (qv). Chemical properties, mechanisms, and textile properties of these graft polymers of cellulose are summarized in ref. 51. Graft polymerization onto cotton has also been induced by both chemical (52) and photochemical (53) initiation (see Radiation curing).

The effects of high energy radiation on cotton properties have also been investigated, eg, the effects of gamma radiation on cotton are described in refs. 54–56. Depolymerization of the cellulose occurs with increasing energy absorption; carbonyl formation, carboxyl formation, and chain cleavage occur in a ratio of 20:1:1. With these chemical changes, there is a corresponding increase in solubility in water and alkali and a decrease in fiber strength. In addition, the gamma-irradiated cotton was found to possess base ion-exchange properties. Irradiation of cotton with near ultraviolet light (325–400 nm) causes formation of cellulose free radicals and mild oxidative degradation of the cotton (57). Carbonyl and carboxyl contents of the cotton cellulose increase, and DP and tensile strength decrease, with increasing time of irradiation (58) (see Photochemical technology). The induction of cellulose free radicals by near uv irradiation forms the basis for photofinishing with vinyl monomers to produce graft polymers on the cotton:

$$\text{cell}\!-\!H \xrightarrow{h\nu} \text{cell} + H\cdot$$

$$\text{cell}\cdot + n\,M \rightarrow \text{cell}\!-\!(M)_{n-1}M\cdot$$

Another interesting reaction of cotton cellulose occurs with an ionized atmosphere. This is essentially a surface reaction. Glow discharge treatment of cotton yarn in air increase water absorbancy and strength (59), and surface-dependent properties of cotton fabric are drastically changed by exposure to low temperature–low pressure argon plasma generated by radio-frequency radiation (60). Because only a few extremely high energy electrons (10–15 eV) are generated, ambient temperature is maintained in the chamber. Light microscopy indicates a smoother surface, although scanning electron microscopy shows no change from native cotton. Spectral changes show some oxidation of the cotton, a decreased carbon to oxygen ratio. Free radicals similar to those from ^{60}Co radiation are formed. In addition, highly charged species are also formed, allowing such usually inert monomers as benzene to be polymerized onto the cotton; there is great capacity for bond cleavage. An increased rate of wetting and drying indicates potential end uses requiring absorbancy. The cohesiveness and fiber friction of cotton sliver was increased temporarily through air–trace chlorine corona treatments at 95°C and atmospheric pressure (61–62). With a 15 kV electrode voltage at a frequency of 2070 Hz, no chemical effects on the cotton could be noted. Dyeability, hand, and wettability were unaffected. The increase in cohesiveness was used to produce yarns with increased strength, abrasion resistance, and greater spinnability (63). Thus yarns of significantly lower twist can be produced with strength equal to, or higher than, untreated cotton yarns of higher twist.

Insolubilization of compounds within textiles parallels the history of man; the direct dyeing technique for cotton was highly advanced in the Bronze Age (see Dyes, natural). With the exception of fiber-reactive dyestuffs discussed earlier, all other cotton dyes—substantive, vat and sulfur—are insolubilized within the fiber, the latter two after an oxidizing step (see Dyes and dye intermediates). Insoluble metal oxides have been used to flameproof cotton. Certain zirconium compounds have been insolubilized on cotton to render the fabric microbial resistant (64) or mildew resistant (65) via a mineral dyeing process (see Textiles). Insolubilization, along with five other methods for imparting antimicrobial properties to cotton, are described in ref. 66. These methods can all be classified under one or more of the chemical reactions of cotton cited earlier; they include fiber reactions to form metastable bonds, grafting through thermosetting agents, formation of coordination compounds, ion-exchange methods, polymer formation with possible grafting, and a regeneration process. In the regeneration process, a compound is chemically bound to the cotton cellulose. This compound is capable of reversible reaction with antimicrobial species present in end-use conditions such as ozone, light, peroxide from water and ozone, or chemicals in detergents. As the antimicrobial species are consumed, newly present species recombine to regenerate the finish. Promising antimicrobial cotton products include halodeoxy cellulose fabrics (67) and cotton fabrics containing peroxide complexes of zirconyl acetate (68).

Economic Aspects

Marketing.   There are several routes by which cotton fiber in the United States changes ownership from the grower to its final destination as a fiber, at the cotton mill. The grower may sell cotton directly to a spinning mill under a grower contract, or the grower may sell the cotton to a gin, or broker, or commission firm. Some growers, after ginning their cotton, may sell through a cooperative organization or may place the cotton in a depository as collateral under the Commodity Credit Corporation Loan Program, to be either withdrawn on repayment of the loan plus interest, or forfeited for sale by the government. These intermediate buyers then sell the cotton to shippers,

Table 2.   World Production of Cotton, in 1000 Bales [a,b]

Area	Year[c]		
	1968	1973	1978
North America	10,272	16,938	17,618
South America	4,091	4,810	4,806
Western Europe	769	916	926
Eastern Europe and the U.S.S.R	9,500	11,505	12,670
Asia, Oceania, and Australia	18,355	23,365	24,049
Africa	4,759	5,879	5,155
Total	*47,746*	*63,413*	*65,224*

[a] Ref. 70.

[b] 1 bale = 216.8 kg.

[c] A year begins Aug. 1st of previous year to July 31st of year given.

Table 3.   World Consumption and United States Prices of Cotton [a]

Factor	Year		
	1968	1973	1978
consumption, t	11,471	13,105	11,452
price, $/kg	0.58–0.73	0.86–1.01	1.39–1.72

[a] Ref. 71.

who in turn sell to foreign mills or to domestic mills that have not purchased the cotton directly from the grower (69).

World Production and Prices.   World production, consumption, and prices are shown in Tables 2 and 3.

Through the *Universal Cotton Standards of Agreement* (72), the official cotton standards of the United States for the grade of Upland cotton are recognized by 14 cotton associations and exchanges in 10 major consuming countries of Europe and Asia. The major differences between the United States system and those of other countries is the smaller number of grades established and the factors that influence quality, such as variety, moisture, and geographic background. In some instances, staple length is included as an attribute of grade, whereas in the United States, grade and staple length are separate criteria of the classification system.

Health and Safety Factors

Byssinosis.   Byssinosis, also called mill fever or brown lung disease, is a pulmonary ailment, similar to bagassosis (see Bagasse) or silicosis, that is alleged to develop on repeated inhalation of cotton dust. A small percentage of textile workers exposed to cotton dust contract byssinosis. Other natural fibers whose dust afflicts workers with byssinosis are flax, sisal, soft hemp, and perhaps jute. Incidence of the disease can be controlled with adequate regulation of dust levels in work areas (see Air pollution control methods). OSHA has issued standards for control of cotton dust (73).

BIBLIOGRAPHY

"Cotton" in *ECT* 1st ed., Vol. 4, pp. 563–578, by Kyle Ward, Jr., R. B. Evans, Mary L. Rollins, Barkley Meadows, and Ines de Gruy, Southern Regional Research Laboratory, U.S. Department of Agriculture; "Cotton" in *ECT* 2nd ed., Vol. 6, pp. 376–412, by E. Lord, The Cotton Silk & Man-Made Fibres Research Association, Shirley Institute, Manchester.

1. W. L. Balls, *The Development and Properties of Raw Cotton,* A. & C. Black, Ltd., London, Eng., 1915.
2. C. B. Purves in E. Ott, H. M. Spurlin, and M. W. Graefflin, eds., *Cellulose and Cellulose Derivatives,* 2nd ed., Pt. 1, Interscience Publishers, Inc., New York, 1954, pp. 29–53.
3. R. D. Preston, *The Molecular Architecture of Plant Cell Walls,* John Wiley & Sons, Inc., New York, 1952.
4. H. B. Brown and J. O. Ware, *Cotton,* 3rd ed., McGraw-Hill Book Co., New York, 1958.
5. International Cotton Advisory Committee, *Cotton World Statistics* **30**(6, Pt. 2), 8 (Jan. 1977).
6. *Cotton Varieties Planted 1972–1976,* U.S. Agricultural Marketing Service, Memphis, Tenn., Sept. 1976.
7. G. A. Slater, *Cotton Int. (Memphis)* **43,** 90, 130, 138 (1976).
8. *Agric. Res.* **24**(12), 8 (June 1976).
9. *Cotton Int. (Memphis)* **43,** 55, 58 (1976).
10. U.S. Bur. of the Census, *Cotton Production in the United States; Crop of 1970,* U.S. Government Printing Office, Washington, D.C., 1971.
11. U.S. Bur. of the Census, *Cotton Ginnings in the United States; Crop of 1975,* U.S. Government Printing Office, Washington, D.C., 1976.
12. A. C. Griffin, private communication, Mar. 1977.
13. W. H. Fortenberry in D. S. Hamby, ed., *The American Cotton Handbook,* 3rd ed., Vol. 1, Interscience Publishers, Inc., New York, 1965, pp. 110–131.
14. *U.S. Cotton Handbook,* Cotton Council International and National Cotton Council of America, Washington, D.C., 1976.
15. M. L. Rollins in ref. 13, pp. 44–81.
16. L. A. Fiori and J. Compton in ref. 13, pp. 132–205.
17. J. N. Grant, *Text. Res. J.* **26,** 74 (1956).
18. E. Heuser, *The Chemistry of Cellulose,* John Wiley & Sons, Inc., New York, 1944, pp. 575–607.
19. S. Haworth and co-workers, *Carbohydr. Res.* **10,** 1 (1969).
20. R. Jeffries and co-workers, *Cellul. Chem. Technol.* **3,** 255 (1969).
21. R. Jeffries, J. G. Roberts, and R. N. Robinson, *Text. Res. J.* **38,** 234 (1968).
22. Brit. Pat. 1,084,612 (Sept. 27, 1967) (to Sentralinstitutt for Industriell Forskning and Norsk Tekstil-forskningsinstitutt); Brit. Pat. 1,136,417 (Dec. 11, 1968), R. M. Gailey, (to J. and P. Coats, Ltd.).
23. T. A. Calamari, Jr., and co-workers, *Text. Chem. Color.* **3,** 235 (1971).
24. E. Heuser, *Text. Res. J.* **20,** 828 (1950); T. E. Timell, *Sven. Papperstidn.* **56,** 483 (1953).
25. R. M. Reinhardt and J. D. Reid, *Text. Res. J.* **27,** 59 (1957).
26. L. A. Graham and C. A. Suratt, *Am. Dyest. Rep.* **67**(7), 36 (1978).
27. J. B. McKelvey, R. J. Berni, and R. R. Benerito, *Text. Res. J.* **34,** 1102 (1964).
28. C. F. Goldthwait, E. M. Buras, Jr., and A. S. Cooper, Jr., *Text. Res. J.* **21,** 831 (1951).
29. J. G. Frick, Jr., *Chem. Technol.* **1,** 100 (1971).
30. A. G. Pierce, R. M. Reinhardt, and R. M. H. Kullman, *Text. Res. J.* **46,** 420 (1976).
31. J. G. Frick, Jr., B. A. Kottes Andrews, and J. D. Reid, *Text. Res. J.* **30,** 495 (1960).
32. S. A. Heap, *Colourage* **24**(7), 15 (1977).
33. M. L. Rollins and co-workers, *Norelco Rep.* **13,** 119, 132 (1966).
34. W. A. Reeves, R. M. Perkins, and L. H. Chance, *Text. Res. J.* **30,** 179 (1960).
35. S. P. Rowland, M. L. Rollins, and I. V. de Gruy, *J. Appl. Polym. Sci.* **10,** 1763 (1966).
36. A. M. Cannizzaro and co-workers, *Text. Res. J.* **40,** 1087 (1970).
37. J. B. McKelvey, B. G. Webre, and E. Klein, *Text. Res. J.* **29,** 918 (1959).
38. R. R. Benerito and co-workers, *J. Polym. Sci. Part A* **1,** 3407 (1963).
39. U.S. Pat. 2,524,399 (Oct. 3, 1950), D. L. Schoene and V. S. Chambers (to U.S. Rubber Co.).
40. G. C. Tesoro, P. Linden, and S. B. Sello, *Text. Res. J.* **31,** 283 (1961).
41. Can. Pat. 625,790 (Aug. 15, 1961), R. O. Steele (to Rohm and Haas Co.).

42. R. H. Peters, *Textile Chemistry III, The Physical Chemistry of Dyeing,* Elsevier Scientific Publishing Co., New York, 1975, pp. 624–629.

43. J. D. Guthrie, *Ind. Eng. Chem.* **44,** 2187 (1952).

44. J. D. Guthrie, *Am. Dyest. Rep.* **41**(1), P13, 30 (1952).

45. J. T. Marsh, *An Introduction to Textile Finishing,* 2nd ed., Chapman and Hall, London, Eng., 1966, pp. 458–494.

46. C. M. Welch, J. B. Bullock, and M. F. Margavio, *Text. Res. J.* **37,** 324 (1967).

47. E. J. Grajeck and W. H. Petersen, *Text. Res. J.* **32,** 320 (1962).

48. G. C. Daul, R. M. Reinhardt, and J. D. Reid, *Text. Res. J.* **25,** 246 (1955).

49. J. C. Arthur, Jr., and R. J. Demint, *Text. Res. J.* **30,** 505 (1960).

50. J. C. Arthur, Jr., and R. J. Demint, *Text. Res. J.* **31,** 988 (1961).

51. J. C. Arthur, Jr., and F. A. Blouin, *Am. Dyest. Rep.* **51**(26), 1024 (1962).

52. H. H. St. Mard, C. Hamalainen, and A. S. Cooper, Jr., *Am. Dyest. Rep.* **56**(5), 24 (1967).

53. A. H. Reine, N. A. Portnoy, and J. C. Arthur, Jr., *Text. Res. J.* **43,** 638 (1973).

54. F. A. Blouin and J. C. Arthur, Jr., *Text. Res. J.* **28,** 198 (1958).

55. J. C. Arthur, Jr., *Text. Res. J.* **28,** 204 (1958).

56. R. J. Demint and J. C. Arthur, Jr., *Text. Res. J.* **29,** 276 (1959).

57. J. C. Arthur, Jr., *U.S. Agric. Res. Serv. South. Reg. Rep.* **ARS-S-64,** 6 (1975).

58. J. C. Arthur, Jr., and O. Hinojosa, *Appl. Polym. Symp.* **26,** 147 (1975).

59. R. B. Stone and J. R. Barrett, Jr., *Text. Bull.* **88**(1), 65 (1962).

60. H. Z. Jung, T. L. Ward, and R. R. Benerito, *Text. Res. J.* **47,** 217 (1977).

61. W. J. Thorsen, *Text. Res. J.* **41,** 331 (1971).

62. *Ibid.,* p. 455.

63. D. P. Thibodeaux and H. R. Copeland, *Text. Chem. Process. Conf., 15th,* U.S.D.A. South. Reg. Res. Center, New Orleans, Louisiana, 1975, pp. 36–40.

64. C. J. Conner and co-workers, *Text. Res. J.* **34,** 347 (1954).

65. C. J. Conner and co-workers, *Text. Res. J.* **31,** 94 (1967).

66. D. D. Gagliardi, *Am. Dyest. Rep.* **51,** P49 (1962).

67. T. L. Vigo, D. J. Daigle, and C. M. Welch, *Text. Res. J.* **43,** 715 (1973).

68. T. L. Vigo, G. F. Danna, and C. M. Welch, *Text. Chem. Color.* **9,** 77 (1977).

69. R. S. Corkern, *Cotton Wool Situation (U.S.D.A. Economic Research Service)* **CWS-3,** 34 (Dec. 1975).

70. International Cotton Advisory Committee, *Cotton-World Statistics* **25**(6), 8 (1972); **31**(6), 8 (1978).

71. *Ibid.,* **22**(7), 14 (1968); **24**(6), 12 (1971); **28**(6), 34 (1975); **32**(6), 34 (1979).

72. *The Classification of Cotton Miscellaneous Publication No. 310,* U.S. Department of Agriculture, Washington, D.C., Jan. 1965, pp. 11–15.

73. *Fed. Reg.* **43**(122), 27350–27463 (June 23, 1978).

B. A. KOTTES ANDREWS
INES V. DE GRUY
United States Department of Agriculture

FIBERS, CHEMICAL

The chemical fiber industry, a lineal descendant of prehistoric use of natural fibers in textiles, is now an important industry in its own right, a large segment of the chemical process industry, and heavily dependent on petrochemicals. Its history begins with the patent of an artificial silk by Count Hilaire de Chardonnet in France in 1885. Chardonnet silk was made by two of the three major spinning processes in use today. It was a regenerated cellulose fiber (now called rayon) made by converting a nitrated cellulose into fiber form and then chemically regenerating the cellulose to avoid the unacceptable flammability of cellulose nitrate. Initially, it was produced by extruding an alcohol solution of cellulose nitrate through small holes into water to effect fiber coagulation. In a later process it was generated by similarly extruding an alcohol—ether polymer solution into air, forming filaments that hardened as a result of solvent evaporation.

It is appropriate to introduce some definitions here. The word spinning as applied to chemical fibers refers to the process by which they are formed from polymeric substances. Chardonnet's first process is representative of wet spinning, in which a solution of a polymer is forced through one or more holes in a device called a jet or spinneret into a suitable coagulating bath. The individual thread-like strands are called filaments. Chardonnet's second process is called dry spinning, wherein a polymer solution is forced through the spinneret holes into a chamber where the solvent is evaporated from the filaments.

Chardonnet's fiber was derived from a naturally occurring long-chain polymeric material, cellulose. As the chemical fiber industry grew steadily from the 1880s to the 1930s, it continued to be based on use of naturally occurring polymers, primarily cellulose. Research in the 1920s and 1930s, particularly that of Staudinger and Carothers, laid the foundation for further advances, especially the development of the synthetic fibers. Staudinger, a 1953 Nobel prize winner, proved polymers were molecules of very high molecular weight. Carothers clearly defined the two major types of polymerization, condensation and addition; showed how condensation polymers of high molecular weight could be synthesized; and demonstrated that certain of these could be formed into filaments that exhibited the property of cold drawing and crystallization, resulting in oriented, strong fibers (1). Carothers' elegant research in DuPont's laboratories led in 1939 to two major innovations in the chemical fiber field. One was the commercialization of nylon fiber, the first truly synthetic fiber based on a polymer itself. The second was the technique of melt spinning. Instead of being dissolved as for wet or dry spinning, the nylon polymer was melted by heating, and the viscous liquid was forced through the spinneret holes to form filaments that hardened as they cooled.

The generally used term man-made, referred to here as chemical fibers, as applied to fibers differentiates between the natural fibers and those produced by chemical and physical means in manufacturing processes. The advent of nylon, and the subsequent proliferation of fibers from synthetic polymers, has brought about the subdivision of chemical fibers into two broad classes: those made from natural polymers such as the cellulosics (see Cellulose acetate and triacetate fibers; Rayon; Fibers, vegetable), and those derived from synthesized polymers. The latter, which are called synthetics (or sometimes noncellulosics), now represent the majority of chemical fibers. In this category fall the nylons, the polyesters, the acrylics, the polyolefins, and many others (see Acrylic and modacrylic fibers; Olefin fibers: Polyamides; Polyester fibers).

By 1977 world production of chemical fibers closely approached world production of natural fibers, and the production of synthetic fibers had grown to well over twice that of the cellulosics, with polyester taking the lead. Table 1 shows the changes in volume of various fibers between 1967 and 1977.

For polymers, whether naturally occurring or synthesized, to be suitable for fiber production they must meet several conditions. They must be capable of conversion into fibrous form and must be of sufficiently high molecular weight to permit development of adequate fiber properties. In order to develop such adequate fiber properties they must be capable of being oriented so that the long polymer molecules tend to lie parallel to the fiber axis.

The fiber is normally drawn (stretched) either during or subsequent to spinning to effect orientation. Drawing frequently imparts or enhances crystallinity resulting in enhanced fiber tensile strength. The choice of spinning process is primarily dictated by the nature of the polymer used and secondarily, in some cases, by the nature of the products desired. If the spinneret has a single hole, a monofilament is produced. Such monofilaments are desirable for certain uses, eg, fishing lines, sheer hosiery. Usually, however, the spinneret has many holes. For textile use multifilament yarns are produced, each thread line or "end" containing from fewer than 10 to 100 or more individual filaments depending on end-use, eg, apparel, upholstery, carpeting. Yarns of this type are usually referred to simply as filament yarns. Except for silk, however, natural fibers occur in relatively short lengths, usually measured in centimeters. These are called staple fibers, and yarns spun from them tend to have different aesthetic

Table 1. World Textile Fiber Production, 1967 and 1977[a], Thousands of Metric Tons

	1967	1977
Natural fibers		
raw cotton	10,180	14,138
raw wool	1,562	1,390
raw silk	32	49
Total	*11,774*	*15,577*
Chemical fibers		
cellulosics[b]	3,428	3,535
synthetics		
polyester	754	4,233
nylon	1,317	2,939
acrylic	542	1,765
olefin	150	949
other[c]	110	110
Sub total	*2,873*	*9,996*
Total	*6,301*	*13,531*

[a] Adapted from Textile Economics Bureau Data (2–3).

[b] Rayon and acetate, including acetate cigarette filter tow.

[c] Predominantly synthetic.

qualities from continuous-filament yarns. Chemical fibers are, in many cases, desired in staple form, either for use by themselves or for blending with other fibers. In the production of staple chemical fibers, polyester being preeminent, the spinneret contains hundreds or thousands of holes, and a very large bundle of tens of thousands of filaments, called a tow, is produced by combining output from several spinnerets. Such tows, after stretching, are crimped, ie, mechanically given a wavy form, and then cut into staple fiber of any length desired.

The physical, mechanical, and chemical properties vary widely among the various chemical fibers. They can also be varied to a degree in a given type of fiber by the method of spinning used and by the post-treatment, particularly extent of drawing. Key physical properties include the strength of the fiber, the degree of elongation or extensibility at the breaking point, and the modulus, or degree of stiffness. These are generally expressed relative to the fiber tex (denier times 0.1111), a measure of size defined as the weight in grams of 1000 (9000 for denier) meters of yarn or filament. The tex and the decitex, a unit ca 11% greater than denier, have come into use in Europe. Fibers of different densities thus have different cross-sectional areas for a given tex. Most of the chemical fibers in use today, except for glass and certain specialty products, have densities varying from slightly below 1.0 to ca 1.5 g/mL. The tenacity of a fiber or yarn is its breaking strength in grams divided by the dtex (or denier). Elongation at break is measured as a percentage of original length extended, and the modulus (or initial modulus), expressed in grams per dtex (or denier), is the force required to extend the fiber initially, divided by the degree of extension. As the properties of most fibers are affected by absorbed moisture and also by temperature, physical testing is done under controlled and defined conditions; ie, 21°C and 65% relative humidity (rh). Chemical fibers range widely in moisture regain from the dry state, from near zero for polypropylene and below 1% for polyester to ca 13% for rayon. Other properties significant to various uses include abrasion resistance, elasticity, resistance to drycleaning solvents and other chemicals, ease of dyeing and types of dyestuffs accepted, and dimensional stability, or resistance to shrinkage and wrinkling. Strength is important in industrial applications such as tire cord (qv). Abrasion resistance is significant in carpet-face yarns, but not particularly so in women's apparel. Though some of these properties influence what is called hand, the latter is a somewhat elusive, yet very important, property of a fabric. It denotes the tactile and aesthetic sensations experienced in feeling a fabric.

In addition to the several broad-based synthetic fibers introduced in the 1940s and the 1950s, there are now many high-performance specialty fibers introduced as a result of continuing research and development. For reference, Table 2 summarizes generic names and definitions established by the U.S. Federal Trade Commission for a number of chemical fibers. Several of them are discussed in more detail below.

Chemical Fibers Based on Natural Polymers

Rayon. Two other processes have long since superseded Chardonnet's method of regenerating cellulose, the cuprammonium process and the viscose process. They are based on two different methods of solubilizing the relatively intractable cellulose molecule. The cuprammonium, or Bemberg process, is currently of minor present significance. It is based on dissolving relatively purified cellulose, such as that from cotton linters or woodpulp, in an aqueous solution of ammonia, caustic soda, and basic

Table 2.   Relevant F.T.C. Generic Names for Manufactured Textile Fibers

Generic name	Definition of fiber-forming substance[a,b]
acetate	cellulose acetate; triacetate where not less than 92% of the cellulose is acetylated
acrylic	at least 85% acrylonitrile units
aramid	polyamide in which at least 85% of the amide linkages are directly attached to two aromatic rings
azlon	regenerated naturally occurring proteins
glass	glass
modacrylic	less than 85% but at least 35% acrylonitrile units
novoloid	at least 85% cross-linked novolac
nylon	polyamide in which less than 85% of the amide linkages are directly attached to two aromatic rings
nytril	at least 85% long chain polymer of vinylidene dinitrile where the latter represents not less than every other unit in the chain
olefin	at least 85% ethylene, propylene, or other olefin units
polyester	at least 85% ester of a substituted aromatic carboxylic acid, including but not restricted to substituted terephthalate units and para-substituted hydroxybenzoate units
rayon	regenerated cellulose with less than 15% chemically combined substituents
saran	at least 80% vinylidene chloride
spandex	elastomer of at least 85% of a segmented polyurethane
vinal	at least 50% vinyl alcohol units and at least 85% total vinyl alcohol and acetal units
vinyon	at least 85% vinyl chloride units

[a] All percentages are by weight.

[b] Except for acetate, azlon, glass, novoloid, and rayon, the fiber-forming substance is described as a long chain synthetic polymer of the composition noted.

copper sulfate, followed by wet spinning into water. The regenerated cellulose fibers are stretched, extracted with sulfuric acid solution, washed, and dried. The process, first commercialized in Germany in 1899, was based on work by a Swiss chemist, Schweizer, in 1857 (see Cellulose; Rayon).

In 1891 the British workers Cross, Bevan, and Beadle found that cellulose is soluble in a mixture of caustic soda and carbon disulfide. The viscose process based on this finding was commercialized in 1903 in the United Kingdom (1910 in the United States) and is today predominant in rayon manufacture worldwide. It is a very complex process and its commercial utilization represented a triumph of ingenuity and technology. Basically, relatively purified wood-pulp cellulose, called dissolving pulp or chemical cellulose, is treated first with caustic soda, then with carbon disulfide, to form a xanthate which is, in turn, soluble in aqueous alkali solution. Control of conditions in the spinning solution, called viscose, and of the bath into which the viscose is extruded, importantly affect final fiber properties. The concentrations of sulfuric acid, sodium sulfate, zinc sulfate and other components of the spinning bath are especially important. The coagulated fiber formed in the spin bath is stretched, washed, and dried. For staple fiber production, the process sequence includes crimping and cutting. Hydraulic and coagulation circumstances limit the linear speed at which such wet spinning can be effected to not much over 100 m/min. Because of this, the process is best suited to the production of large total tex (denier) continuous filament yarns or of very large tows for production of staple fibers.

Rayon fibers, particularly staple fibers, serve in a wide variety of uses, alone and in combination with other fibers. High-tenacity, high total tex (denier) filament is used in tire cord and other industrial uses, although rayon's once-dominant tire cord position has long since yielded to nylon and polyester fibers.

Rayon staple is produced in a variety of types, including polynosic and high wet-modulus types, these retaining greater strength and dimensional stability when wet. Rayon staple is used in apparel, household goods, and various nonwoven fabrics (see Nonwoven textiles).

Despite the maturity of rayon, new product variants continue to be developed. Among the most recent developments are a highly crimped high wet-modulus item (4) closer to cotton in fabric hand and cover (opacity) either by itself or in blends with polyester staple, a higher absorptive capacity variant (5) for use in certain nonwoven products, and a hollow viscose rayon variant (6) that also provides more cotton-like hand and cover.

Cellulose Acetate. Unlike rayon, cellulose acetate [9004-35-7] is dry-spun, predominantly in continuous filament form. The fiber-forming base polymer is made by acetylating dissolving pulp with acetic anhydride usually in acetic acid solvent with sulfuric acid as catalyst. The fully acetylated cellulose initially produced is partially hydrolyzed to produce secondary acetate. Following catalyst removal, precipitation of the acetate flake, washing, and drying, the flake is dissolved in acetone containing a small amount of water. This spinning "dope" is subjected to extensive filtration and then dry-spun to evaporate the acetone solvent. The filament yarn so produced is wound at the bottom of the chamber on a bobbin or other suitable package for such further handling as end-uses dictate. Linear spinning speeds up to or exceeding 1000 m/min can be achieved.

Acetate fabrics have attractive appearance, pleasant hand, and excellent whiteness before dyeing. A wide variety of textile fabrics, both knit and woven, are produced. Women's apparel in which the relatively poor abrasion resistance of acetate fibers is not a particular problem, draperies, and upholstery are major use areas. Although acetate textile filament yarn is still an important chemical fiber, its production appears to have peaked ca 1970–1971. Since then acetate production has declined accompanied by closing of some plants and abandonment of acetate production by some manufacturers (7). A relatively small amount of acetate staple is also produced.

Cigarette filters represent a major and still growing use for acetate tow. The tow is a continuous-filament product containing several thousand filaments and made by combining the requisite number of multifilament yarns from several spinning chambers into a tow that is then crimped and baled for use in cigarette-making. Acetate filters predominate worldwide because of their combination of appearance, taste, and economics (see Cellulose acetate and triacetate fibers).

Cellulose Triacetate. Cellulose triacetate [9012-09-3] filament became a useful specialty fiber in the 1950s, initially in the United States and the United Kingdom. Its production is similar to that of acetate, differing in that the initial reaction product is not hydrolyzed to a product with less than 92% of the hydroxyl groups esterified, and the spinning solution for the triacetate flake is made with a mixture of methylene chloride and a small amount of methanol or ethanol. Its importance is that, unlike regular cellulose acetate, it can be heat-treated to induce a degree of crystallization, which produces dimensional stability and related "ease-of-care" characteristics permitting, for example, water washing. As a result heat-treated triacetate resembles such fully synthetic fibers as nylon or polyester.

Other Fibers Based on Natural Polymers. Natural polymers other than cellulose that are capable of producing fibrous structures in nature suggested themselves as bases for chemical fibers. Two types of such noncellulosic chemical fibers reached commercialization in the 1930s and 1940s. Although neither is any longer produced in the United States or of much commercial significance elsewhere, these fibers deserve mention.

Alginate Fibers. Both monofilament and multifilament fibers can be produced by wet spinning an aqueous sodium alginate solution based on alginic acid from seaweed (see Gums). The spin bath is acid to produce alginic acid fibers or a bath containing calcium ion is used to produce calcium alginate [9005-35-0] fibers. The latter was commercialized in the United Kingdom during World War II. These fibers are unsuitable for ordinary textile use, as they are very sensitive to weak soap or alkali solutions. Quite small amounts have been made for special uses in which the fiber is dissolved out of a textile structure. Production is understood never to have exceeded a hundred metric tons annually.

Protein Fibers. Just as cotton (qv) and flax (see Fibers, vegetable) are natural cellulose fibers so wool (qv) and silk (qv) are protein fibers in which polymeric amino acids provide the fibrous structure. These polypeptides are natural polyamides, analogous to the synthetic nylons, particularly nylon-6, though differing in that they are polymers of alpha amino acids (see Biopolymers).

Regenerated protein fibers have been made commercially since the 1930s. Several have been based on casein from milk, the fibers being produced by wet spinning (see Milk products). Most of these fibers are treated with formaldehyde to improve wet strength and reduce alkali sensitivity. The casein fibers have relatively high moisture absorption, similar to that of wool, and have been primarily used in staple form for blending with wool. A small amount is still made outside the United States.

The other regenerated protein fibers, based on wet-spinning proteins from peanuts, corn meal, and soybeans are chiefly of historical interest.

Synthetic Fibers

Polyamide Fibers. The importance of the development of nylon can hardly be overstated. It opened the way to a great number of synthetic fibers having a wide variety of chemical and physical properties. It freed the chemical fiber industry from the restrictions inherent in the use of natural polymers, permitting controlled tailoring of polymers.

Nylon-6,6. Nylon-6,6 [9011-55-6] is so designated because both the adipic acid and the diamine from which the polymer is synthesized contain 6 carbon atoms. Although coal tar intermediates were used, these starting materials are now derived from other intermediates based ultimately on petroleum hydrocarbons. The fiber-forming polymer is formed by heating nylon-6,6 salt under pressure to remove water as the adipic acid and diamine condense to form a long-chain, linear polymer in what is called a melt polymerization. When the desired average molecular weight is reached, pressure is reduced, and the polymer is usually extruded in spaghetti-like strands, cooled, and cut into chips. In some cases, the molten polymer is conveyed directly to melt-spinning equipment.

In spinning, the polymer chips are re-melted, either on a grid or in the melting zone of a screw extruder, metered via a gear pump, passed through a filter (usually sand) to remove solid impurities, and extruded vertically downwards through spin-nerettes. The filaments are cooled and solidified, lubricated with a material designed to eliminate static and provide desired frictional characteristics, and wound on a ro-tating tube to form a "package." At this stage, the fiber is essentially unoriented, weak, and unsuitable for textile use.

Fiber strength is developed by subjecting the as-spun yarn to a drawing step, in which orientation and crystallization of the polymer molecules occur. This stretching is achieved by passing the yarn between two sets of rolls, the second of which operates at a surface speed of up to ca 5 times that of the feed roll, so that the yarn is drawn to several times the original length. In the draw zone, the yarn is usually passed over a pin to localize the draw point and enhance uniformity. Drawing is usually effected at ambient temperature, heat actually being generated in the individual filaments by the stretching process. Textile nylon is usually stretched and twisted simulta-neously, the multifilament yarn being given a degree of twist to aid in subsequent processing. For certain uses, hot drawing (at temperatures, however, well below the polymer's crystalline melting point of about 263°C) or multiple drawing, or both, may be used to produce high tenacity with reduced extensibility.

Nylon-6,6 is a strong, tough, abrasion-resistant fiber that is relatively stable and relatively readily dyeable. It has good knot strength (ie, high percentage retention of tenacity when knotted) and good elasticity. By heat setting after being drawn, nylon fiber can be dimensionally stabilized at temperatures up to that of the annealing process, permitted the "boarding" of hosiery to preset shapes, pleat setting in garments, and importantly, providing ease-of-care characteristics.

Because of its broad range of characteristics, nylon-6,6 has a wide spectrum of uses. Textile nylon is used in hosiery, apparel, and home furnishings. Most of this fiber is multifilament, but monofilament nylon is used in sheer women's stockings. The high strength of nylon permits manufacture of lightweight and very sheer fabrics. Because of good abrasion resistance and resilience, nylon dominates the U.S. carpet market (8). A particular product for this area is BCF, or bulked continuous filament nylon. The bulking is accomplished by mechanically crimping multifilament yarn while it is heated, then cooling it in the bulky configuration. Nylon staple is also important in carpets.

Strong, durable nylon monofilaments are used in fishing lines and fishnets; high-strength nylon multifilament has a variety of automotive and industrial uses ranging from belting to filter fabrics. Tire cord (qv) is a major use, nylon having dis-placed high strength viscose rayon in this application.

Nylon-6.   Although Carother's research in the late 1920s and 1930s touched on fiber-forming polymers from aminocaproic acid, commercialization of nylon-6 [25038-54-4] fibers, so named because of the six carbons in the caprolactam monomer, first occurred in Europe shortly after commercialization in the United States of nylon-6,6. Manufacture is based on the use of caprolactam, the internal cyclic amide of aminocaproic acid, derived petrochemically (see Polyamides).

Unlike nylon-6,6, caprolactam is polymerized by ring opening, followed by re-petitive addition until the desired molecular weight is achieved. This reaction is also a melt polymerization in which an equilibrium content of monomer exists, the latter being removed after polymerization, usually by water washing. Nylon-6 polymer is also melt-spun, and the manufacture, fiber properties, and uses are very similar to those of nylon-6,6. Schlack at I.G. Farbenindustrie is credited with establishing ap-propriate conditions of converting caprolactam to a useful fiber-forming polymer.

Nylon-6 production has grown steadily, and has long since spread from Europe to the United States and elsewhere.

Qiana. In 1968 DuPont introduced a polyamide textile fiber different from either nylon-6 or 6,6. This fiber, named Qiana, is a condensation polymer containing alicyclic rings (9). As with other nylons, melt polymerization and melt spinning are used. The fiber is very silk-like in appearance and behavior, particularly with respect to fabric resilience after crushing. Qiana fiber is used mainly in apparel, particularly dresses, blouses, and shirts.

Aromatic Polyamides. Also introduced in the 1960s was a series of polyamides (10) in which the intermediates are wholly aromatic (see Aramid fibers). By virtue of the aromatic ring structures coupled with the relative regularity of the polymolecular chain, they are very high melting, very high in thermal stability, and possessed of generally high performance properties.

These polymers do not melt, for all practical purposes, other than at temperatures involving decomposition, and they are nearly insoluble. Consequently, conventional melt polymerization and melt spinning are not feasible. Polymers can be prepared from the dihalides of the dibasic acid by interfacial condensation with diamines at relatively low (<100°C) temperature or by reaction in suitable solvents (see Polymerization). In some cases sulfuric acid may be used as the solvent for these rather intractable polymers.

The intermediates, the polymerization, the fiber spinning and aftertreatment (drawing to orient, crystallizing and heat treatment) are all more costly than for the aliphatic nylons such as 6,6 or 6. The aramid fibers thus are higher in price, and find use in applications where exceptionally high strength, very high modulus (stiffness) or high resistance to heat, or both, are required.

Commercial aramid fibers (qv) are trademarked Nomex and Kevlar by DuPont. Nomex, based on meta-linked isophthalic acid and m-phenylenediamine, is used in specialty papers (11) for electrical insulation and aircraft structures, in protective garments, and in other applications requiring high thermal stability. Kevlar is an aromatic polyamide containing para-oriented linkages. It melts at over 500°C, is exceptionally high in strength with a tenacity more than twice that of high strength nylon or polyester, and a very high modulus. It finds use in heavy duty conveyor belts, and in composite structures with casting resins such as epoxies (see Embedding). It is contending against steel in radial tire reinforcement (12).

Polyester Fibers. Although Carothers had demonstrated that fiber-forming polyesters could be synthesized from glycols and dibasic acids, or from hydroxy acids, the laboratory work focused on polyamides for development. The original, and still growing, commercial success of polyester fibers is attributable to the work of Whinfield and Dickson in the United Kingdom (13). They demonstrated that the presence of aromatic rings in a linear polyester of sufficient molecular weight led to high melting points, good stability, and strong, tough fibers. Commercial production was initiated in the late 1940s and early 1950s (see Polyester fibers).

Polyester, now the largest-volume synthetic fiber in the world, surpassed nylon worldwide in 1972. The polymer, like nylon, is made by melt polymerization. Ethylene glycol reacts with either dimethyl terephthalate (DMT) or terephthalic acid (TPA) under heat and high vacuum. In the former case, methanol is eliminated in the condensation/transesterification reaction; in the latter, water is evolved. The intermediates are again petrochemicals-based. The polymer has a crystalline melting point of ca 250°C, similar to that of nylon-6,6.

Melt-spinning of polyester fibers is also similar to that of nylon. The as-spun fiber is amorphous and comparatively weak. Drawing, carried out hot because the glass transition temperature is higher than that of nylon, causes orientation of the polymer chains along with crystallization, and fiber strength is developed. The fiber may be heat-set (annealed) under tension to relax strains in the molecule and induce some additional crystallization. Like nylon, the extent and nature of the drawing can be used to control tenacity and elongation, these essentially being interchangeable. In more recent years, fiber production processes involving a combined spin-draw, or a combined drawing and texturing step, have proliferated.

Eastman Chemical Products, Inc. produces not only poly(ethylene terephthalate) [25038-59-9] under the trade name Kodel, but also a polyester based on condensing terephthalic acid with 1,4-dimethylolcyclohexane (14). This fiber is in many respects similar to the ethylene glycol-based polyester but melts higher and has a lower specific gravity and excellent recovery from stretch. It is designated Kodel II.

Polyester fibers are produced in both staple and continuous filament form. The former predominated in earlier years, but by 1977 polyester filament volume was 80% that of staple. The fiber is remarkably versatile, probably more so than any other chemical fiber. It is strong, abrasion resistant, relatively stable, higher in modulus than nylon, and of lower moisture regain. Its structure is such that polyester is more difficult to dye than most textile fibers, requiring somewhat specialized techniques.

Uses range widely over the apparel, home furnishings, automotive, and industrial fields. Enormous quantities of polyester staple are blended with cotton, rayon, or wool in spun yarns used for apparel. Good abrasion resistance also suits polyester to use in carpeting. Continuous filament yarn finds wide use in apparel; much of it is textured to produce bulk and opacity for this use.

In the automotive field, just as nylon tire cord (qv) to a great extent displaced rayon, high-tenacity polyester tire cord has made inroads against nylon, particularly in passenger car tires (15). Other industrial uses include belts, ropes, and filter fabrics.

Acrylic Fibers. Synthetic fibers based on acrylonitrile (qv) appeared in the United States in the 1940s (16). Subsequently, fibers of lower acrylonitrile unit content were introduced. These modacrylics present some use problems because they are lower in melting point and less resistant to the effects of heat than are acrylics, but they are relatively flame-resistant and useful in applications where this property is desirable (17) (see Flame retardant textiles; Acrylic and modacrylic fibers).

Acrylic and modacrylic fibers combined are now produced worldwide and represent the third largest in volume of the synthetic fibers produced. They are derived from petrochemical-based intermediates. Because acrylonitrile polymers cannot be melted without decomposition, melt spinning is not applicable. Highly polar solvents, such as dimethylacetamide, have been found to dissolve the acrylic polymers for either dry or wet spinning, both of which are used commercially (see Acetic acid). Virtually all acrylic fiber is produced in staple form. Earlier, filament acrylics were produced, but they proved very difficult to dye uniformly. A small amount of textile filament yarn is still produced in Japan, and acrylonitrile homopolymer filament is produced for carbonization in the manufacture of carbon fibers (see Carbon).

The fiber-forming acrylic polymers are high in molecular weight and are produced primarily in aqueous medium by free radical-initiated addition polymerization. The regular acrylics are usually copolymers with minor amounts of one or more comonomers, such as methyl acrylate, used to provide accessibility or sites for dyestuffs. Like other synthetics, acrylic fibers are stretched to develop orientation and fiber strength.

Acrylic fibers have good tenacity, although less than polyester or polyamides, excellent stability to sunlight, good dye acceptance as a result of the copolymer system used for this purpose, and a soft, pleasing hand of wool-like characteristics. Abrasion resistance, although well below that of nylon or polyester, is nevertheless good, and superior to that of wool. For these reasons, the acrylics find widespread use in both indoor and outdoor furnishings, including awnings and draperies, and in blankets, sweaters, and carpets (see Textiles). Blending low- and high-shrinkage yarns and shrinking the blend produces a bulky yarn. The levels of shrinkage are produced by different degrees of stretching in fiber production.

A variety of modacrylic fibers are produced in the United States and abroad. These are acrylonitrile copolymers or terpolymers with significant comonomer contents. Among the comonomers used are vinylidene chloride, vinyl chloride, and apparently acrylamides (qv). The fiber-forming modacrylic polymers are more soluble than the acrylics, some of them in acetone, for example, and are made commercially both by dry and wet spinning. The modacrylic fibers have various textile uses, including fake fur pile fabrics (see Furs, synthetic) and protective garments, among others. Modacrylics, because of their self-extinguishing or flame-resistant characteristics are also used in wigs (qv) and doll's hair.

Polypropylene. The discovery that catalyst systems could convert certain olefins, notably ethylene and propylene, to crystalline, ordered, or stereoregular polymers provided a basis for ordered polyolefin fibers of defined melting points (see Olefin fibers). Large-scale availability of low cost monomers from petrochemical operations, polymer specific gravities less than one, and melt-spinnability were key factors fostering the major development and commercialization of polyolefin fibers.

World polyolefin fiber production, most of it in the United States and Western Europe, is fourth in volume among the synthetics. Polypropylene [9003-07-0] is thought to constitute the majority of commercial polyolefin fiber. In stereoregular form, polypropylene has a crystalline melting point of about 105°C, compared to ca 125°C for low-pressure (regular) polyethylene [9002-88-4] and ca 110–120°C for high-pressure polyethylene (see Olefin polymers).

These fibers are melt-spun and produced in monofilament, multifilament, and staple forms. Both conventional melt spinning and a type of film extrusion are used. In the latter the film is mechanically slit, split, or fibrillated to produce coarser yarns where tex (denier) uniformity is not critical (18). In all cases the as-spun or extruded material is stretched to develop orientation and strength.

Polypropylene fiber is relatively low melting for a chemical fiber, limiting its uses. Offsetting this and other relative deficiencies is low cost, high strength, very great chemical inertness, and, because of the low density, high yardage of fiber of a given tex per kilogram (den/lb).

Because of the chemical inertness, dye acceptance is inadequate. For uses where color is requisite, either the base polymer is modified to provide dyeability or the fiber is spun-colored; that is, produced with pigments incorporated in the polymer and fiber during melting. Because of oxygen and light sensitivity, suitable stabilizers are also incorporated (19).

Textile uses include upholstery and carpeting, particularly indoor-outdoor carpets. In addition to face yarns for carpeting, polypropylene fiber, particularly that produced from film, is used for woven carpet backings. The low melting point, which prevents easy ironing, and the fabric hand have largely precluded polyolefins from application in apparel fabrics. Other important uses include rope and cordage, fishnets, and filter media. Negligible moisture absorption, resistance to decay by organisms, and low density, which causes the fiber to float make polyolefins particularly suited to such uses.

Vinyon and Other Vinyl Fibers. In addition to vinyl addition polymers based on acrylonitrile (ie, vinyl cyanide) others have been commercialized. Indeed, fibers known as Vinyon were commercialized in 1939 closely on the heels of the introduction of nylon-6,6, as a result of the joint efforts of the Union Carbide and American Viscose Corporations (now Avtex Fibers, Inc.) (see Vinyl polymers).

Poly(vinyl chloride) [9002-86-2] by itself tends to decompose on heating and is difficultly soluble in common solvents. However, a small amount of vinyl acetate co-polymerized with it acts as an internal plasticizer, promotes solubility in solvents such as acetone, and yields a copolymer capable of producing useful fibers. Commercially the copolymers are about 85–90% vinyl chloride and 10–15% vinyl acetate units. Although textile filament was produced earlier, U.S. Vinyon production as of 1978 is entirely in staple form by Avtex. It is made by a dry-spinning process like that for cellulose acetate fiber. The fiber is temperature sensitive, starting to soften, shrink, and become tacky below the boiling point of water. It finds specialty use in applications requiring bonding and heat sealing, in conjunction with other fibers.

Other Vinyon fibers from vinyl chloride homopolymers, rather than copolymers, are produced abroad, notably in France, Italy, and Japan (20). They are produced primarily by spinning from solution, carbon disulfide being one of the solvents used. A variety of products are made but all share relatively limited thermal stability, tending to shrink at temperatures in the neighborhood of boiling water. They are, however, resistant to moisture and rotting, and inherently nonflammable, suiting them to specialty uses such as filter cloths, nonflammable garments, fishnetting, and felts (qv) for insulating purposes. Strength and residual shrinkage can be varied to fit the application.

Saran. Saran fibers based on vinylidene chloride polymers (21) are produced in several countries under various trade names. They are based on vinylidene chloride copolymerized with small amounts of vinyl chloride, and still smaller amounts of acrylonitrile.

Saran fibers are melt-spun and then stretched. They are characterized by a pale straw color and by resistance to water, fire, and light, and bacterial and insect attack. Their relatively low melting points require excessively low ironing temperature. They are difficult to dye, requiring pigmentation in the melt. The fibers find specialty use in certain types of upholstery, filter cloths, and fishnets. Both staple fibers and filament yarns are produced, the latter in both heavy tex (denier) monofilaments and in multifilament form. The relatively low cost of these fibers is one of their attractions (see Vinylidene polymers).

PeCe. PeCe fiber, a post-chlorinated vinyl chloride polymer, was developed in Germany and was of some significance there during World War II. The post-chlorination raises the chlorine content of the polymer from 57 to 64% and confers acetone solubility, permitting ready wet spinning. Like many other vinyl fibers, it is low melting and restricted to special applications.

Vinal. Vinal is the U.S. term applied to poly(vinyl alcohol) [9002-89-5] fibers. These fibers originated in Japan and considerable production continues there. Although vinyl alcohol does not exist as a monomer, its polymer can be obtained by the hydrolysis of the acetyl groups from poly(vinyl acetate). Vinyl acetate, a relatively abundant and modest-cost monomer, is polymerized and then saponified to poly(vinyl alcohol), which is wet-spun from hot water. In this form, the fiber is water soluble. To make it commercially useful, the fiber is treated with formalin and heat, rendering it water insoluble. As with other fibers, the relative tensile strength and extensibility can be varied by varying the degree of stretch used. The fiber has reasonable strength, a moderately low melting point (222°C), limited elastic recovery, good chemical resistance, and resistance to degradation by organisms. Both staple and filament forms, including monofilament, are produced. Vinal is used in bristles, filter cloths, sewing thread, fishnets, and apparel. Although other producers have closely studied this fiber, production has remained largely confined to Japan. Elsewhere, it would appear, the overall cost-performance ratio has not justified commercialization.

Polychlal. Polychlal fibers (22), unique to Japan, are closely related both to the Vinyon and Vinal types. They are produced from an emulsion of poly(vinyl chloride) and a matrix of poly(vinyl alcohol). This is essentially wet-spun into aqueous sodium sulfate and formaldehyde-treated. Production on a modest scale continues in Japan, and some of this fiber has been exported to the United States for use in flame-retardant apparel.

Other Synthetic Fibers. Synthetic fibers other than those already described have been commercialized. Although production is, in most cases, relatively small, these products are frequently important in specific applications.

Benzoate Fibers. A melt-spun polyester fiber is produced in Japan from the self-condensation polymer derived from p-(β-hydroxyethoxy)benzoic acid, the aim being to provide a silk-like fiber (22). Although limited production of this fiber continues, it does not appear likely to grow significantly.

Poly(hydroxyacetic Acid Ester) Fibers. Although most of the aliphatic polyesters, as distinct from the aromatic acid-based type, are low melting, poly(glycolic acid) [26124-68-5, 26009-03-0], also known as poly(hydroxyacetic acid) or polyglycolide, is an exception, melting at over 200°C (23). Fibers can be melt spun from homopolymers derived from glycolide or from copolymers derived from glycolide and a minor amount of lactide. The fibers are crystalline, and can be oriented by stretching. Because they are absorbable in the body, they have become important in surgical suture applications, where they replace catgut sutures in many such applications (see Sutures). Compared to the textile fibers, these are relatively high-priced, low-volume specialty items.

Spandex Fibers. Synthetic elastomeric fibers, developed to compete with or improve on natural rubber fiber, differ from ordinary textile fibers in having very high extensibility to break (500–600%) and high recovery from such stretching. The segments of the polyurethane are based on low molecular weight polyethers, such as poly(ethylene glycol), or polyesters, which then react with an aromatic diisocyanate to give a three-dimensional, elastomeric product. Though the chemistry is complex, basically the elastomeric products depend on the presence of hard and soft segments and cross-linking. The fibers produced are white and dyeable and are stronger and lighter than rubber, making them particularly suited for use in foundation garments, bathing suits, support hose, and other elastic garments (see Fibers, elastomeric; Urethane polymers).

Fluorine-Containing Fibers.   Polymers based on polytetrafluorothylene [*9002-84-0*], DuPont's Teflon in the United States, have uniquely high chemical stability and inertness, no water absorption, low frictional characteristics, and high melting points with decomposition preceding and accompanying melting (see Fluorine compounds, organic). These properties clearly make conversion to fiber form difficult. Nevertheless, the problem was solved in the 1950s by emulsion spinning. This process appears basically to comprise sintering or fusing extremely small polymer particles that are themselves fibrous and have been aligned by passing a colloidal dispersion through a capillary. The fused fiber is subsequently drawn. Teflon fibers in both filament and staple forms are high priced and find, as would be expected, highly specialized applications, such as packings, special filtration fabrics, and other uses where corrosion resistance, lubricity, and temperature resistance are requisite.

Carbon Fibers.   Carbon fibers (24), the subject of a great deal of research and development in the past decade, are characterized by extremely high strength and modulus along with high temperature resistance. These characteristics suggest themselves for particular applicability to high performance, reinforced composite structures where strength, stiffness, and lightness of weight are at a premium, eg, in special aerospace, industrial, and recreational applications. The composite structure may include other fibers, such as glass, along with a matrix resin, such as polyesters, epoxies, or polyimides.

Carbon fibers are made by controlled carbonization of an already formed fibrous structure based on an appropriate organic polymer. In particular carbonization has been applied to filament yarns of homopolymers of acrylonitrile, to high tenacity viscose rayon multifilament yarns, and to pitch for lower strength carbon fibers (see Ablative materials; Carbon; Composite materials).

Phenolic-Type Fibers.   The Carborundum Company developed a fire-resistant fiber based on the technology of the well-known phenol–formaldehyde resin condensates. The latter, long used for moldings, laminates, and similar thermoset resin applications, are cross-linked, three-dimensional polymers made by curing a non-cross-linked precursor condensate of formaldehyde with phenolic compounds. Special processes were developed to produce fibers. The result, trade-named Kynol, is presently produced in Japan (Nippon Kynol Co., U.S. distributor, American Kynol Co.) and used for protective garments (see Novoloid fibers; Phenolic resins).

Polybenzimidazole Fibers.   Poly(benzimidazole) [*26986-65-9*] fibers (25) have been produced semicommercially from condensation polymers based on the reaction of 3,3'-diaminobenzidine and diphenyl isophthalate. A process in which the fibers are dry-spun and then drawn was developed by the Celanese Corporation, partially supported by the government. The fibers are of the high performance type, of very high melting point, good strength and extensibility, nonflammable, and surprisingly for a synthetic fiber, of high moisture regain: 13%. These fibers are of interest for aerospace applications and potentially for industrial applications as well subject to incompletely defined economics (see Heat resistant polymers; Polyimides).

Glass.   The world production of textile glass fiber listed at ca 800,000 metric tons for 1977 does not include substantial production of glass fiber batting materials. In many ways glass fiber is unique among the chemical fibers. It is an inorganic fiber, in contrast to the numerous organic, natural, and synthetic fibers, and the fibers are neither oriented nor crystalline. The fibers are strong, nonflammable, and rather heat-resistant, as well as highly resistant to chemicals, moisture, and attack by organisms. Not surprisingly, glass fibers are quite low in extensibility and higher in density than the organic fibers.

The fibers are produced by melt spinning. To offset their brittleness, they are spun into fine filaments. After lubrication, these are wound as a multifilament strand unless insulating batting is being produced; in the latter case the filaments are attenuated while still molten and are deposited in the form of a thick bat (see Insulation, thermal).

Glass fibers, in forms ranging from filament yarn to mats, woven rovings, and staple, find a variety of important uses. In addition to use as insulating bats, they are used in fabric form for draperies where the nonflammability, inertness, and resistance to the effects of sunlight are assets. Glass fiber cloth, as well as batting, is used in a variety of insulating applications and for filtration.

One of the most important uses for glass fiber is in reinforced plastics, particularly reinforced thermosetting polyester resins (26). Since the 1950s the construction of pleasure boats, both sail and power, has been shifted from hulls of wood to glass-reinforced polyester. Such composites are widely used in industrial and automotive applications. Glass-filled thermoplastic resins have also grown in volume. The glass fibers are used in nylon, polyacetal, and poly(butylene terephthalate) molding resins, for example (see Laminated and reinforced plastics). Glass fibers are also used in both radial and bias-ply automotive tire reinforcement (see Fiber optics; Glass).

Metal and Other Inorganic Fibers.   Fine steel wire is, of course, well-known for radial tire reinforcement. However, a number of metallic and other inorganic fibers have been investigated for special applications. Metallic fibers have been used in small quantities with organic fiber carpet yarns to minimize static problems (see Antistatic agents). Fibers of boron, boron–tungsten, steel, beryllium, boron and silicon carbides, boron and silicon nitrides, alumina, zirconia, and other inorganics have received attention (27), particularly for high performance specialty uses. Most are quite costly, and the quantities employed quite small.

Fiber Modifications

It is axiomatic that the applicability of a given fiber is a function of the overall price-performance considerations for that fiber. Although certain fibers are barred from a particular use by some inherent limitation, there is much overlapping and inter-fiber competition. Moreover, as the chemical fiber industry has grown, modifications of individual fiber types have been developed to broaden uses.

These modifications may be chemical, physical, or a combination of both, to improve performance in a given use or to adapt it to uses for which it was not previously satisfactory. Since the preceding sections have only alluded briefly to such modifications, it is in order to review them in somewhat greater depth.

Physical Modifications.   As noted, when chemical fibers are stretched, their tensile strength generally increases and their extensibility decreases. Use is made of this in producing highly drawn, high tenacity nylon and polyester for tire cord and industrial uses. By adjusting the draw ratio the fiber modulus can be altered, and polyester staple of a modulus suitable for blending with cotton is made this way.

By decreasing the filament tex, and therefore increasing the number of filaments in a yarn of given total tex, the suppleness of the yarn and the softness of hand in fabrics made from it are increased. This is the trend in polyester filament yarns for apparel (28).

Texturing or bulking of filament yarns is used to produce yarn with more of the characteristics of yarn spun from staple fibers while eliminating several steps involved in producing the latter. Texturing can also be used to produce stretch yarns, as from nylon. Most texturing of chemical fibers is based on the fact that if the yarn is distorted while heated to render it more plastic, then cooled, the yarn "remembers" its distorted structure and seeks to restore it. Several methods are briefly summarized in Table 3.

False-twist texturing has been used extensively particularly with polyester filament yarns. Basically, a running thread line is temporarily highly twisted between two points and heat-set. It is then untwisted, whereupon it kinks or bulks. Nylon of a high degree of stretch is thereby produced. Nonstretch bulked polyester yarn is also made this way; a second heater is used to set the bulked configuration. Initially, a spindle rotating at very high speeds and through which the yarn passes was used to insert the false twist. In recent years, the use of friction false-twist texturing has been growing. In this method, the running thread line proceeds over the edge of several disks in series, the twist being imparted by friction.

False-twist texturized polyester filament yarn has become a major item of commerce, with huge quantities being used in apparel. Originally, fully drawn filament yarn was textured. In recent years, it has been found advantageous in many cases for the fiber producer to make a partially drawn yarn with only some degree of orientation. This is known as POY (partially oriented or preoriented yarn). The texturer, who may also be the fiber producer, then simultaneously draws and textures this yarn.

Staple fiber bulk is enhanced by crimping before cutting into staple lengths and by using differential shrinkage fibers as described for acrylics. Rather elegant use of this principle is made in bicomponent acrylic staple fibers in which each fiber consists of two components fused together. These components are formulated to have different shrinkage potential. When subjected to hot water, a helical crimp develops.

Changing the cross section of the individual filaments in a yarn may affect fabric appearance and hand. In melt spinning the use of spinneret holes of special shapes

Table 3.   Filament Yarn Texturing/Bulking Methods for Fibers

Means of yarn distortion	Comments
knitting, fabric heat-setting, de-knitting	called knit-de-knit; cumbersome, currently used little or not at all
passage over hot knife edge	Agilon process; spiral crimp; nylon stretch yarn an example
passage between gear teeth	gear-crimping process; zig-zag configuration; used for BCF carpet yarn
passage into tightly enclosed space	stuffer-box crimping; saw-toothed configuration; Banlon yarns; tows for crimped staple fiber
twisting, particularly false-twisting	stretch and bulky nonstretch yarns; textured polyester apparel yarn so produced
air turbulence	Taslan yarns; filaments looped randomly by air jet while over-feeding; not dependent on heat-setting; used for blends of nylon and polyester with acetates

can produce multilobal filament cross sections. Polyester yarn, for example, that is so spun produces a fabric without the sometimes objectionable shiny or plastic appearance. Alternatively, special cross sections can be used to provide "sparkle" yarns.

Chemical Modifications.   Modified fibers can be produced by chemical alteration of the polymer, fiber, or fabric, or by the use of nonreacting additives. Improved dyeability is usually effected by use of a suitable comonomer. Polyester fibers ordinarily dyed with disperse dyestuffs are made cationically dyeable by use of a comonomer providing acidic sites in the base polymer.

Spun-colored fibers can be produced by incorporating a pigment before spinning. Acetate, triacetate, polyester, polyolefin, and acrylic fibers have used this approach for certain applications. In related processes, white pigments, particularly titanium dioxide, are used to dull the natural luster of chemical fibers, or optical brighteners may be added, such as to polyester, to increase the perceived whiteness of the fiber (see Brighteners, fluorescent).

Chemical approaches were of particular importance in seeking to meet U.S. Government regulations regarding flame-retardant garments for such uses as children's sleepwear. A halogenated phosphate compound known as Tris, tris(2,3-dibromopropyl) phosphate, used as an additive in acetate and triacetate spinning, and as a fabric additive with polyester, enabled fabrics from these fibers to meet federal requirements, but its use was dropped in light of work suggesting possible carcinogenicity. More recently, Hoechst Fiber Industries has introduced in the United States a flame-resistant polyester variant trademarked Trevira 271 based on copolymerization involving a phosphinic acid derivative (29) (see Flame retardants, phosphorus compounds).

The introduction of permanent-press treatment of polyester cotton staple fiber blends in fabric form was a major factor in the growth of polyester usage. Wash-and-wear characteristics can be imparted to fabrics by treatment with suitable thermosetting (cross-linking) resin systems, but an accompanying strength loss made this approach unsuitable for all-cellulosic fibers such as cotton or rayon. Blending the strong polyester fiber with the cellulosic overcomes this problem and yields a blend widely used in shirting and slacks fabrics.

Economic Aspects

Chemical fiber economics are complex, varying not only with the specific fiber and process, but also with the item (ie, staple or filament, tex, etc) of a given fiber, and with geography, scale, capacity utilization, and other factors. Certain generalities, however, provide guideposts. Of the noncapital-related costs, raw materials are the single largest item, followed by labor and energy. Both capital and production costs tend to be higher for filament yarn than for staple-fiber plants, and higher for low tex (denier) filament than for higher tex (denier) material. Because solvent and solution handling tend to increase energy consumption, labor, and environmental problems, both capital and operating costs for wet- and dry-spun fiber tend to be higher than those for melt-spun fibers.

High capital costs have for some years prevented construction of new rayon plants, except in special circumstances in certain Communist and developing countries. Rayon textile filament yarn has virtually disappeared because of high costs of wet-spinning fine tex items, despite the fact that cellulose is a renewable resource. Among the major synthetics, those with most favorable raw materials costs are polyester, polyolefins, and acrylics, with nylon-6 and 6,6 being significantly higher. Capital and energy costs, however, are less favorable for the acrylics. All these factors, plus versatility in use, have conspired to favor polyester in both staple and filament form.

Though highly profitable in much of their history, chemical fibers have been severely buffeted in recent years. In addition to the traditional cyclicality of the textile industry, during the 1970s overcapacity in both the fiber and petrochemicals industries and increases in energy and raw materials costs have significantly affected profitability for both U.S. and European fiber producers.

Table 4 lists typical market prices for several important fibers for 1970 and 1978, with 1973 and 1974 also included to reflect sharp increases in petroleum prices in that period. It must be emphasized that these prices reflect external factors as well as actual underlying production costs, that there are scores of other items at different prices, and that certain markets do not command these prices. All these economic pressures combine to place emphasis on the research and development of process improvements and cost reduction.

Present Status and Future Possibilities

Today there is one or more fibers to fit virtually any use, and there is an enormous range of properties in the broad spectrum of commercially available fibers. In addition, the direct conversion of synthetic polymers to nonwoven fabric form (30) has grown over the last few years. Polyethylene, polypropylene, and polyester resins, for example, can be converted to what are called spun-bonded fabrics in processes that combine melt-extrusion with a means of laying the filaments in a web and bonding the thermoplastic fibers at cross-over points to form fabrics useful in a variety of applications. Polypropylene carpet backing is just one example (see Nonwoven textiles).

Concurrent with these product developments has been a steady, continuing increase in uniformity of fiber products regarding properties, dyeability, and the like.

Scientific and technological possibilities are still plentiful, but in the future their exploitation is likely to be overshadowed by economic and political developments,

Table 4. Typical U.S. Market Prices Selected Chemical Fibers, 1970–1978[a], $/kg

Fiber and type	1970	1973	1974	1978
rayon staple, regular[b]	0.59	0.72	1.06	1.25
acetate filament, 8.3 tex (75 den)	1.61	1.47	2.02	2.42
16.7 tex (150 den)	1.23	1.10	1.63	2.11
polyester staple	0.90	0.84	0.97	1.19
polyester filament, 16.7 tex (150 den), drawn	2.99	1.94	1.87	1.43
nylon filament, 4.4 tex (40 den)	2.31	2.77	3.12	3.52
95.5 tex (860 den)	1.76	1.71	1.98	2.55
acrylic textile staple[c], low tex (denier)	1.96	1.71	1.76	1.63
polypropylene textile staple, natural[d]	1.01–1.08	0.79–0.86	1.03–1.10	1.14–1.21
177.8 tex (1600 den) BCF, colored filament	2.09	2.20	2.20	2.20

[a] Data from various industry sources, 1978.
[b] High wet modulus rayon staple ranges from $0.05–0.09 higher.
[c] Modacrylics tend to be higher, as do specialty items such as bicomponent acrylics.
[d] Natural refers to nonpigmented fiber.

particularly the increasing shortage of oil, increased cost of energy, environmental regulations, and other regulatory restrictions.

Overall, the chemical fiber industry can be expected to continue to grow, but at a rate lower than that of the 1950s and 1960s. Growth in the cellulosic sector is unlikely except for acetate cigarette filter tow. Growth will be in the synthetic fibers sector, and further inroads against the natural fibers are likely. Many chemical possibilities are in principle capable of exploitation. In the condensation polymers, eg, nylon or polyesters, the dibasic acid, the diamine, or the glycol can be varied widely to affect the properties of the fiber. For example, use of longer-chain glycols than ethylene glycol can impart greater resilience to polyester fibers based on terephthalic acid. Copolymerization in general, including block and graft copolymerization, offers seemingly endless possibilities for both addition and condensation polymers. Certain monomers not presently used commercially are known to produce interesting fibers; a notable case is pyrrolidone from which, by ring-opening polymerization, is obtainable nylon 4, a fiber of higher moisture regain than other synthetic fibers. Apparently, technical problems associated with polymer stability in melt-spinning and/or economic factors have prevented commercialization of nylon 4 fibers despite much work over several decades (see Polyamides).

Nevertheless, for the remainder of this century and perhaps indefinitely, it seems unlikely there will be any new, broad-based chemical fibers or that those relatively small-volume textile fibers already in existence will grow rapidly or become more widespread than at present. Thus, among the synthetics, the predominance of polyester, nylon, acrylic, and olefin fibers is likely to continue with continuing development of very specific modifications for specific uses.

BIBLIOGRAPHY

"Fibers" in *ECT* 1st ed., Vol. 6, pp. 453–467, by H. F. Mark, Polytechnic Institute of Brooklyn; "Fibers, Man-Made" in *ECT* 2nd ed., Vol. 9, pp. 151–170, by H. F. Mark, Polytechnic Institute of Brooklyn and S. M. Atlas, Bronx Community College.

1. H. Mark and G. Whitby, *Collected Papers of W. H. Carothers*, Interscience Publishers, Inc., New York, 1940.
2. *Textile Organon* **39**(6), 89 (June 1968).
3. *Ibid.*, **49**(6), 73 (June 1978).
4. *Mod. Text.* **52**(4), 19 (1976).
5. J. W. Schappel, F. R. Smith, and E. A. Zawistowski, *Textile Research Institute 48th Annual Research and Technology Conference, Atlanta Georgia, April 5–6, 1978.*
6. *Mod. Text.* **58**(6), 18 (1977).
7. *Ibid.*, **58**(3), 6 (1977).
8. *Chem. Eng. News.* **56**, 11 (Dec. 4, 1978).
9. G. Clayton and co-workers, *Text. Prog.* **8**, 1, 27 (1976).
10. *Ibid.*, pp. 98–115.
11. *Properties of Nomex Types E-54 and E-55 Aramid Paper*, DuPont Preliminary Information Memo 401, E. I. du Pont de Nemours & Co., Inc., April 1978.
12. R. E. Wilfong and J. Zimmerman, *J. Appl. Polym. Sci.* **31**, 1 (1977).
13. J. R. Whinfield, *Text. Res. J.* **23**, 289 (1953).
14. E. V. Martin, *Text. Res. J.* **32**, 619 (1962).
15. M. E. Denahm, *Mod. Text.* **59**(11), 13 (1978).
16. R. Meredith, *Text. Prog.* **7**, 21 (1975).
17. Ref. 9, pp. 61–69.
18. Ref. 9, p. 81.
19. Ref. 9, pp. 78–79.

20. R. W. Moncrieff, *Man-Made Fibers,* 6th ed., John Wiley & Sons, Inc., New York, 1975, pp. 522–532.
21. H. Mark, S. M. Atlas, and E. Cernia, *Man-Made Fibers Science and Technology,* Vol. 3, Wiley-Interscience, New York, 1968, pp. 303–326.
22. *Chemical Fibers of Japan 1970/1971,* Japan Chemical Fibers Association, Tokyo, 1970, p. 50.
23. Ref. 9, p. 50.
24. E. Fitzer and M. Heym, *Chem. Ind.* (16), 663 (1976).
25. R. H. Jackson, *Text. Res. J.* **48,** 314 (1978).
26. M. E. Dunham, *Mod. Text.* **59**(11), 16 (1978).
27. R. S. Goy and J. A. Jenkins, *Text. Prog.* **2,** 1, 31 (1970).
28. Ref. 8, p. 10.
29. W. S. Wagner in ref. 5.
30. R. A. A. Hentschel, *Chem. Technol.* **4,** 32 (1974).

General References

Refs. 9, 20, 21, and 27.
R. Hill, *Fibers from Synthetic Polymers,* Elsevier Scientific Publishing Company, Amsterdam, 1953.
J. J. Press, ed., *Man-Made Textile Encyclopedia,* Textile Book Publishers, New York, 1959.
E. M. Hicks and co-workers, "The Production of Synthetic Fiber," *Text. Prog.* **8,** 1 (1976).

WILLIAM J. ROBERTS
Consultant

FIBERS, ELASTOMERIC

An elastomeric fiber can be made from any natural or synthetic polymeric material that has high elongation and good recovery properties; however, at present only natural rubber (qv) and urethane polymers (qv) are raw materials for commercially successful elastomeric fibers. Previously, attempts have been made to introduce commercially fibers based on elastomeric polyamides (qv), polyesters (qv), and polyacrylonitriles (see Acrylonitrile polymers) (1–3). These efforts did not meet with success, owing primarily to economic problems and mechanical deficiencies in the experimental fibers. During times of natural rubber shortage, eg, World War II, neoprene (polychloroprene) has been used as a raw material for elastomeric fibers but, because of higher cost and less desirable performance characteristics, it is not used under normal supply conditions (see Elastomers, synthetic).

Elastomeric fibers based on urethane polymers and techniques for their experimental production were first reported in the early 1950s by Farbenfabriken Bayer, a pioneer in urethane and diisocyanate chemistry (4–5). It was not until the late 1950s, however, that elastomeric polyurethane fibers from pilot plant production facilities of the U.S. Rubber Co. (now Uniroyal, Inc.) and E. I. du Pont de Nemours & Co., Inc. were first introduced to the trade on a semicommercial scale. The fibers were given the generic designation spandex (6). The term spandex has come into common trade usage throughout the world in referring to fibers based on elastomeric urethane polymers.

Natural rubber thread has been in wide use for many years. If made by slitting rubber sheets, it is referred to as cut rubber thread. If made by extruding a natural rubber latex into an acid coagulation bath followed by washing, drying, and curing, it is referred to as extruded latex thread or fiber. Both production methods result in a thread or fiber that is a continuous monofilament. Cut rubber thread has a square or rectangular cross section, and extruded latex thread is substantially round.

Thermoplastic, hard (ie, inelastic) fibers, such as nylon and polyester, may be provided with a springlike, helical or zigzag structure by special processing procedures. Such fibers exhibit high elongations as the helical or zigzag structure is stretched but have very low recovery power. This apparent elasticity results from the geometric form of the filaments and not from polymer structure. Only fibers with elastic properties resulting primarily from entropy forces inherent within the basic polymer structure are described below.

Hard fibers, such as polyolefins, polyamides, and polyesters, can be converted into a state exhibiting relatively high degrees of elasticity by the use of critically controlled drawing and annealing techniques (7). Such fibers have been prepared with elongations in excess of 100% and good recoveries. The elastic recovery forces, however, result primarily from energetic effects (ie, ΔE) and not from entropy changes (ie, ΔS) as is the case in the rubberlike elasticity exhibited by natural rubber fibers and spandex. Hard, elastic fibers have not been produced commercially.

In the mid 1970s Monsanto Co. attempted to introduce commercially a nylon–spandex bicomponent fiber trade named Monvelle (8–10). Work on the development of nylon–spandex bicomponent fibers has also been reported by Toray Industries (11–14). These side-by-side (Monsanto) and sheath-core (Toray) bicomponent fibers were produced by extruding both nylon-6 and a polyurethane simultaneously through the same spinnerette hole. The resulting bicomponent fiber was then drawn under very carefully controlled conditions so that the nylon component was permanently deformed into a new, more highly oriented structure whereas the spandex was only elongated. On relaxation, recovery forces in the spandex component pulled the bicomponent monofilament back into a very tightly wound helical structure. An applied tensile force would tend to pull the helix and be resisted by the spandex component. Monvelle met with some customer acceptance as a replacement for nylon-covered spandex in stretch hosiery for women but was withdrawn from the market in 1976. Toray Industries has apparently made no attempts to commercialize its product.

Properties

In both rubber thread and spandex, mechanical properties may be varied over a relatively broad range. In rubber, variations are made by changes in the degree of cross-linking or vulcanization which is accomplished by changing the amount of vulcanizing agent, generally sulfur, and accelerants used. In spandex, many more possibilities for variation are available. By definition spandex is made from a urethane polymer (6). Again by definition, however, any polymer containing urethane linkages in the repeat structure:

$$\underset{}{-\!\!\left(\!R\!-\!\overset{\overset{\textstyle O}{\|}}{O}CNH\!-\!R'\!-\!NH\overset{\overset{\textstyle O}{\|}}{C}O\!\right)_{\!n}}$$

may be classified as a urethane polymer. The number of polymers in this classification is obviously very large. Most urethane polymers in current use for the manufacture of spandex are made by the reaction of a ca 2000 molecular weight polyester or polyether glycol with a diisocyanate at a molar ratio of ca 1:2, followed by reaction of the resulting isocyanate-terminated prepolymer with a diamine to produce a high molecular weight urethane polymer. The average repeat unit of the polymer molecule contains a polyester or a polyether chain of ca 2000 molecular weight, two urethane linkages, two residues from the diisocyanate, and a urea linkage. Mechanical properties may be affected by changing the particular polyester or polyether glycol, diisocyanate, and diamine used, and they also may be markedly modified by changing the molecular weight of the glycol and by changing the glycol–diisocyanate molar ratio used (15–18).

The long-chain urethane polymer molecules in spandex fibers are substantially linear block copolymers comprising relatively long blocks in which molecular interactions are weak, interconnected by shorter blocks in which interactions are strong (see Copolymers). The weakly interacting blocks, commonly referred to as soft segments, are from the polyester or polyether glycol component whereas the blocks having strong interactions result from the diisocyanate and chain extender and are referred to as hard segments. Hard segments from neighboring chains interact strongly, principally through hydrogen bonding, to form tie points between chains. As there are many hard segments in each chain, a network structure results. Any deformation of the network, either extension (stretching) or compression, results in an increase in the degree of order (ie, more uniform molecular alignment) and is opposed by entropy forces within the soft segments (19). Although in rubber the effect is similar, the network structure is formed not by strongly interacting hard segments but by covalent bonds between chain molecules that result from vulcanization with sulfur. For both spandex and rubber, modulus is directly, but not linearly, related to tie-point density, eg, in the case of spandex, hard-segment density. Similarly, the relationship for maximum elongation is an inverse function of tie-point density. In rubber threads, tie-point density is controlled by the amount of vulcanizing agent (sulfur), accelerants, and the reaction conditions used. In spandex, tie-point (hard segment) density is controlled by soft segment (glycol) molecular weight and by glycol–diisocyanate molar ratio.

Although rubber and spandex fibers can be produced with widely varying sets of mechanical properties, there is a relatively narrow range of properties for all commercial elastomeric fibers owing to common end-use requirements, practical limitations in raw materials availability and equipment limitations. Typical properties for spandex and rubber fiber are shown in Table 1. In Figure 1, stress–strain curves are shown for spandex and rubber fiber in comparison to hard fibers and hard fibers with mechanical stretch properties.

Manufacture

Cut Rubber.   In producing cut rubber thread, smoked rubber sheet or crepe rubber is milled with vulcanizing agents along with stabilizers and pigments. The milled stock is calendered into sheets of thicknesses of 0.3–1.3 mm, depending on the final size of rubber thread desired. Multiple sheets (usually 20 or 40) are layered, heat-treated to vulcanize, then slit to form ribbons containing the same number of threads as layered sheets (see Fig. 2). Individual filaments for textile uses are square in cross section and may be readily separated from the ribbon.

Table 1.   Physical Properties of Elastomeric Fibers

	Spandex	Extruded latex	Cut rubber
size range available[a]	2.2–360 tex (20–3,200 den)	16–610 tex (140–5,500 den)	2.5–21 μm dia (1,200–10,000 gauge)
tenacity, N/tex[b]	0.05–0.09	0.02–0.03	0.01–0.02
elongation, %	500–600	600–700	600–700
modulus (first cycle stress at 300% elongation), N/tex[b]	0.013–0.022	0.004–0.005	0.002–0.004
uv, ozone, and oxides of nitrogen resistance[c]	good (slowly yellows on uv exposure)	degrades rapidly	degrades rapidly
active chlorine resistance[c]	good (yellows on continual exposure)	poor	poor
oil (body oils, cosmetics, etc) resistance	good	poor	poor
dyeability	readily dyeable to broad range of colors	not dyeable	not dyeable
abrasion resistance (eg, nylon on elastomer)	very good	poor (fine sizes must be protected by covering yarns)	poor

[a] Spandex size is usually expressed in denier which is the weight in grams of a 9000 m length. However, the SI units is tex, the weight in g/1000 m. Rubber size is expressed as gauge which is the reciprocal of the diameter or side in inches (eg, extruded latex thread with a diameter of 0.01 in. is 100 gauge).

[b] To convert N/tex to g/den, multiply by 11.33.

[c] Both spandex and rubber threads are normally compounded with antioxidants and other stabilizers.

Extruded Latex Thread.   In the manufacture of extruded latex thread, a concentrated (ie, containing up to ca 50% solids) natural rubber latex from the *Hevea Brasiliensis* tree is blended with aqueous dispersions of vulcanizing agents, pigments, and stabilizing agents. The compounded latex is held under carefully controlled temperature conditions for 8–36 h in order to allow some initial vulcanization to occur. This has the effect of increasing wet strength and thus processability of the extruded threads. The matured latex is extruded at constant pressure through precision-bore glass capillaries into an acetic acid bath where coagulation into thread form occurs. Threads are removed from the coagulation bath by transfer rollers, washed free of excess acid, and conducted through a dryer, after which a finish is applied and the threads are wound on individual bobbins or formed into multi-end ribbons and wound on ribbon spools. The thread is finally vulcanized by heating the ribbon spools or bobbins in vulcanizing ovens for up to 18 h. Recently, many producers have been changing from batch to continuous, in-line vulcanization. After vulcanizing, the multi-end ribbons are packed without support in boxes for shipment to the customer. Individual ends are shipped on bobbins. A typical extruded latex thread production line is illustrated in Figure 3.

The basic method for producing extruded latex thread has remained virtually unchanged since the process was first developed in the 1930s. In the past twenty-five years, however, there have been three modifications to the process that have significantly improved economics and convenience of use:

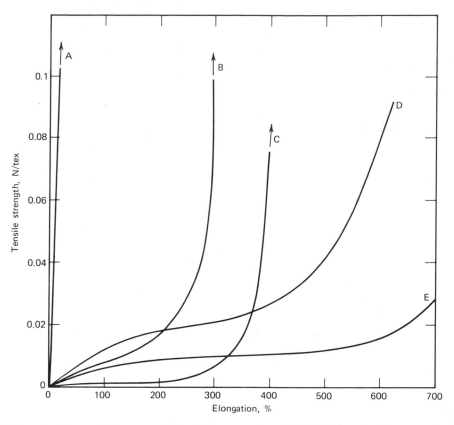

Figure 1. Stress–strain curves: A, hard fiber, eg, nylon, 0.35–0.79 N/tex; B, bicomponent nylon–spandex, 0.26–0.35 N/tex; C, mechanical stretch nylon, 0.26–0.44 N/tex; D, spandex; E, extruded latex. To convert N/tex to g/den, multiply by 11.33.

Ribbon-Forming Techniques for Latex Thread in the 1950s.   Previously all latex thread was wound and sold on single-end bobbins. Most producers now form the threads into multi-end ribbons continuously on the production line by bringing groups of the dried and finished but unvulcanized ends into contiguous side-by-side config-uration and applying carefully controlled pressure. The use of ribbons greatly reduces packaging expense and eliminates the need for returnable bobbins. It also simplifies operations in the customer's plant.

Talc-Free Rubber Thread.   Until recent years the finish used for all extruded latex thread was talc powder (a problem in the pollution of the atmosphere of the working area and, because of its abrasive nature, acceleration of wear in production and processing machinery). Most producers have shifted from talc to newly developed silicone oil-based finishes.

Continuous Cure (Vulcanization).   Until the early 1970s, the vulcanization process was conducted as a separate batch operation. Improvements in compounding and in oven design permit major producers to change to continuous in-line curing with savings in manpower and space as well as improved uniformities.

Extruded latex thread production lines vary widely. A typical line extrudes 100–200 ends (threads) into the coagulation bath in closely spaced side-by-side rela-

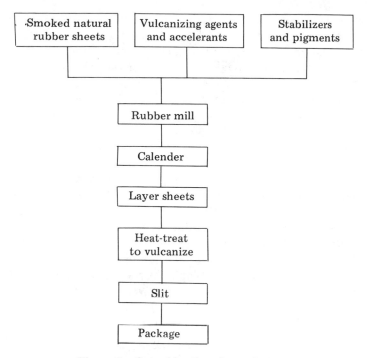

Figure 2. Cut-rubber thread manufacture.

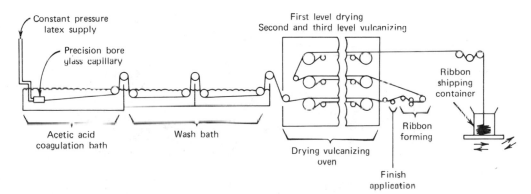

Figure 3. Extruded-latex thread production.

tionship. One or more wash baths follow the coagulation bath. In modern lines the oven is multipass with drying in the first pass followed by finish application and ribbon formation. The ribbons are continuously returned to other dryer levels for curing. Production speeds vary with thread size and equipment but, owing to hydrodynamic drag and the weak nature of the coagulating thread, maximum line take-up speeds are < 30 m/min.

Spandex. Spandex is produced commercially by four very different processes—melt extrusion, reaction spinning, solution dry spinning, and solution wet spinning. These are illustrated in Figure 4. Other approaches, such as emulsion spinning and sheet slitting, have been proposed but are not now in any commercial use (20). As is obvious from Figure 4, all processes involve different practical applications of basically similar chemistry. A block copolymer is formed by the reaction of a 1,000–3,000 molecular weight polyester glycol, polyether glycol, mixed polyester–polyether glycol, hydroxyl-terminated polycaprolactone, or any mixture of the preceding with a diisocyanate at a glycol–diisocyanate molar ratio of ca 1:2 to form a prepolymer which subsequently reacts with a low molecular weight diol or diamine at near stoichiometry. If the diol or diamine reaction with the prepolymer is carried out in a solvent, the resulting urethane block copolymer solution may be wet- or dry-spun into fiber. Alternatively, the prepolymer may be reaction-spun by extrusion into an aqueous or nonaqueous diamine bath to form a fiber and begin polymerization simultaneously, or the prepolymer may be permitted to react in bulk with a diol and the resulting block copolymer melt-extruded in fiber form.

Melt Spinning.   Although considerable development work has been done in melt-extrudable urethane polymer formulation, particularly by Monsanto Co. and Toray Industries, there is currently only one commercial spandex (Mobilon) made by this process. Mobilon is produced in Japan by the Nisshin Spinning Co., and used primarily in waist bands and hosiery. Although many special techniques are required

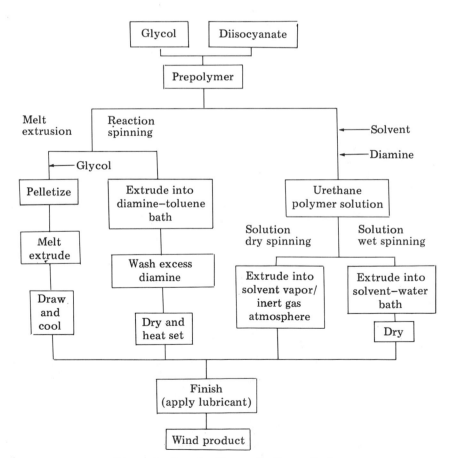

Figure 4. Spandex fiber production methods.

to accommodate heat sensitivity of the polyurethane, low modulus of the filaments, and inherent elastomeric tackiness, the process is similar to the melt extrusion of polyolefins.

Reaction Spinning.   One of the earliest commercial spandex fibers, Vyrene, was produced by the U.S. Rubber Co. (now Uniroyal, Inc.) from a reaction-spinning process (21). Subsequently, both Globe Manufacturing Co. and Firestone Rubber, Inc. introduced spandex fibers, tradenamed Glospan and Spandelle, respectively, that were produced by reaction-spinning techniques. U.S. Rubber Co. and Firestone discontinued spandex production in the mid 1960s. The Firestone process and product was licensed to Courtaulds Ltd. which has developed the process further and produces a spandex trade named Spanzelle. Globe Manufacturing continues to produce Glospan and an unpigmented spandex trade-named Clearspan.

In the manufacture of spandex by reaction-spinning, a 1000–3000 molecular weight polyester or polyether glycol reacts with a diisocyanate at a molar ratio of ca 1:2 to produce an isocyanate-terminated prepolymer. After adjusting viscosity by the addition of minor amounts (<10%) of inert solvent, the prepolymer is extruded into a diamine bath where filament and polymer formation occur simultaneously. The filaments are washed free of excess diamine, dried, cured, finished, and wound on tubes or bobbins. To minimize sticking filaments, the take-up tube, or bobbins are generally rewound before shipment to customer mills.

In general, some trifunctional hydroxy compound (eg, glycerol or trimethylolpropane) is added with the polyester or polyether glycol to produce covalent cross-links in the reaction-spun spandex. Additional covalent cross-links may result from allophanate and biuret formation during curing by the reaction of free isocyanate end groups with urethane or urea hydrogen along polymer chains. The early reaction-spun spandex fibers were extruded as monofilaments. Currently, a multiplicity of filaments are extruded from each spinnerette at individual filament tex values of 1.1–3.3 (10–30 den), then collected in bundles of the desired tex (den) at the exit of the reaction bath. This approach makes the surface area to mass ratio and diamine diffusion into the prepolymer cross section substantially constant in the diamine bath irrespective of final tex produced, thus minimizing condition changes required in changing tex. Since the individual filaments have reacted incompletely and are in a semiplastic state at the exit of the diamine bath, they interbond quite tightly into coherent bundles. The spandex fiber resulting from such production techniques is generally referred to in the trade as a fused multifilament.

As is the case with extruded latex thread, in reaction spinning production speeds are limited by filament weakness in the diamine bath and hydrodynamic drag. By combining multiples of relatively fine tex monofilaments to form one filament bundle, however, thread line strength develops much faster than with latex thread so that take-up speeds of as much as 100 m/min are possible, especially for the fine tex.

A master batch of delusterants (eg, TiO_2) and stabilizers usually is prepared by grinding the additives with the same polymeric glycol to be used in the prepolymer reaction. After grinding, the master batch is vacuum-dried and added to the polymeric glycol charge, prior to diisocyanate addition and the prepolymer reaction, in sufficient amount to provide the desired level in the finished fiber. Additional diisocyanate is added to the charge to account for the polymeric glycol introduced with the master batch. In selecting additives, care must be taken to avoid compounds that react significantly with diisocyanates or diamines under the conditions used in the process.

Further, additives chosen should have no adverse catalytic effect on the prepolymer or chain-extension reaction.

Reaction-spinning equipment is typically quite similar to that of solution wet spinning (see Fig. 6, p. 177). It differs principally in the use of fewer wash baths (only one or two as extruded latex) and in the use of a belt-type dryer rather than heated cans.

Solution Spinning. As in the case for both melt and reaction spinning, the first step in the preparation of a urethane polymer for wet or dry solution spinning involves reaction of a 1000–3000 molecular weight polyester or polyether glycol and a diisocyanate at a molar ratio of ca 1:2. In order to avoid solubility problems, great care must be taken to avoid any trifunctionality (or higher) in any of the reactants. Reaction conditions must be carefully selected and controlled so as to minimize side reactions, eg, allophanate and biuret formation, which can result in covalent cross-linking on chain extension (ie, reaction of the isocyanate prepolymer with diamine or diol to produce a high molecular weight urethane polymer). For the prepolymer reaction, most current commercial producers of solution-spun spandex fiber choose poly(tetramethylene ether) glycol (PTMEG) and methylenebis(4-phenyl isocyanate) (MDI) as the polymeric glycol and diisocyanate, respectively. Polyester glycols are lower in cost and, if properly selected, can yield substantially equivalent properties in the finished fibers. Solutions based on polyester glycols, however, are generally more difficult to handle in dry-spinning equipment, primarily owing to lower solvent removal rates.

In the polymerization reaction, commonly referred to as chain extension, the prepolymer, dissolved in a solvent, reacts with diamine or diol to form a urethane polymer in solution. In all commercial processes, the solvent of choice is dimethylformamide (DMF) or dimethylacetamide (DMAc). The chain extender is generally hydrazine or a diamine containing two primary amine groups. The diamine may be added to the prepolymer in solution or a solution of prepolymer may be added to a solution of the diamine. Stoichiometry is normally adjusted to provide a urethane polymer solution at 18–30 wt % solids and viscosity of 10–80 Pa.s (100–800 P) at 30°C. For dry spinning, viscosities and the percentages of solid are generally higher.

Additives, including pigments and stabilizers, are milled in the solvent of choice, generally with small amounts of the urethane polymer to improve dispersion stability, in a ball or sand mill, then blended to the desired concentration with the polymer solution after chain extension. Polymerization as well as additive preparation and addition of additives may be in either batch or continuous equipment. Some producers combine prepolymerization, chain extension, and additive addition and blending into one, integrated, continuous production line.

Dry Spinning. A majority of spandex manufacturers have chosen solution dry spinning as their production method. On a worldwide basis, ca 80% of all spandex produced is by various adaptations of this method. Although special mechanical techniques are required to handle dynamically the very low modulus, tacky filaments, dry spinning of spandex is very similar to the production of acrylic fiber by dry spinning. In both cases, DMF and DMAc are the solvents of choice (see Formic acid derivatives; Acetic acid derivatives).

The solution dry spinning process is illustrated by Figure 5. The spinning solution of urethane polymer is metered at a constant temperature by a precision gear pump through a spinnerette into a spinning cell of 30–50 cm dia and length of 3–6 m. Heated

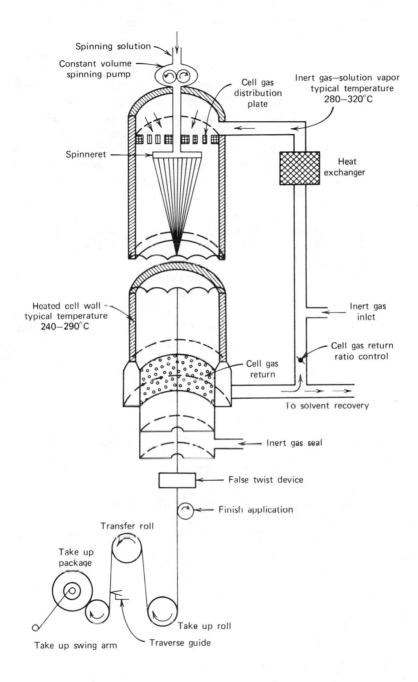

Spinning solution

Constant volume
spinning pump

Cell gas
distribution
plate

Inert gas—solution vapor
typical temperature
280—320°C

Spinneret

Heat
exchanger

Heated cell wall
typical temperature
240—290°C

Inert gas
inlet

Cell gas return
ratio control

Cell gas
return

To solvent recovery

Inert gas seal

False twist device

Finish application

Transfer roll

Take up
package

Take up swing arm

Traverse guide

Take up roll

Figure 5. Spandex production, solution dry spinning process.

145

cell gas, made up of solvent vapor and an inert gas (eg, N_2) is introduced at the top of the cell and passes through a distribution plate behind the spinnerette pack. Because both cell gas and cell walls are maintained at high temperatures, solvent evaporates rapidly from the filaments as they travel down the spinning cell. Individual filament tex is normally maintained in the range of 0.6–1.7 (5–15 den) to maximize, within operable limits, surface to mass ratio and solvent removal rate. Individual filaments are grouped into bundles of the desired final tex at the exit of the spinning cell by a coalescence guide. A commonly used guide employs compressed air to create a mini-vortex which imparts a false twist and rounded cross section to the filament bundle. Depending on the tex produced and the manufacturer, each spinning cell may produce 1–8 filament bundles. Solution dry-spun spandex is normally referred to as continuous multifilament or coalesced multifilament. The terms coalescence and coalesced are misused, as the individual monofilaments remain discrete, ie, they adhere to one another only by a natural elastomeric tack and do not coalesce into larger structures.

After coalescence, a finish is applied to the multifilament bundle before it is taken up onto a tube. In addition to serving as a lubricant in subsequent textile processing, the finish must prevent bundle-to-bundle adhesion on the package. The white mineral oil normally used is modified with silicone oils and antistatic agents. Take-up speeds are ca 300–1000 m/min, depending on tex and producer.

Solvent is condensed from the cell gases, purified by distillation, and returned to polymer making for reuse. In the spinning cell and on distillation at atmospheric pressures, DMF hydrolyzes to a small extent to dimethylamine and formic acid. These compounds must be substantially removed (eg, by an ion exchanger) before the recovered DMF may be used in polymer making, because both dimethylamine and formic acid act as chain terminators. Some producers use vacuum distillation in order to minimize such hydrolysis and avoid the need for ion exchange. DMAc is more stable at atmospheric pressure distillation conditions and requires no further purification to remove hydrolysis products.

Wet Spinning.   In general, any urethane polymer that may be solution dry-spun may also be wet-spun. In contrast, however, many wet-spinnable formulations may not be successfully spun on high-speed commercial dry-spinning equipment. Urethane polymer formulations utilizing polyester glycols are often used in wet spinning but only rarely in dry spinning.

A typical wet-spinning line is illustrated by Figure 6. Spinning solution is pumped by precision gear pumps through spinnerettes into a solvent–water coagulation bath. As in dry spinning, individual filament tex is maintained at ca 0.6–1.7 (5–15 den) in order to optimize solvent removal rates. At the exit of the coagulation bath, filaments are collected in bundles of the desired tex. A false twist may be imposed at the bath exit to give the multifilament bundles a more rounded cross section. After the coagulation bath, the multifilament bundles are countercurrently washed in successive extraction baths to remove residual solvent, then dried and heat-relaxed, generally on heated cans. Finally, as in dry spinning, a finish is applied and the multifilaments wound on individual tubes. A typical spinning line may produce 100–300 multifilaments at side-by-side filament spacings of less than 5 mm.

Water is continuously added to the last extraction bath and flows countercurrently to filament travel from bath to bath. Maximum solvent concentration of 15–30% is reached in the coagulation bath and maintained constant by continuously removing the solvent–water mixture for solvent recovery by distillation. Solvent recovery from

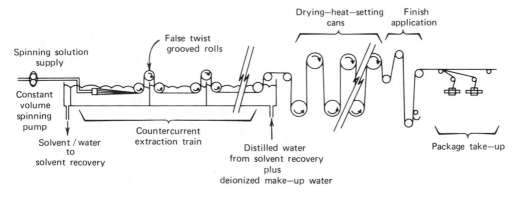

Figure 6. Wet-process spandex production.

wet spinning is generally a two-stage process, since three to five parts of water must be removed for every part of DMF or DMAc recovered. In the first stage, water vapor from the top of the column is condensed and either returned to the extraction train or sent to waste. DMF or DMAc is removed from the column reboiler as a liquid. The crude liquid DMF or DMAc, containing minor amounts of low molecular weight ure-thane polymer, extracted stabilizers, and other high or nonboiling contaminants, is transferred to a second column, vaporized, condensed, and stored for reuse in polymer making. In order to avoid accumulation of contaminants, the DMF or DMAc column reboiler must be purged periodically or continuously.

In wet-process synthetic fiber production, speeds are generally limited by hy-drodynamic drag of the bath medium. For wet-spun spandex, such a consideration limits bath exit speeds to ca 100–150 m/min. It is, perhaps, because of this apparent limitation of production rate that most spandex producers have chosen dry-spinning techniques. More recent work, however, has substantially minimized this limitation by subjecting the spandex filament to drawing as much as three to four times after the spinning bath (22–23). Temperatures and residence times are selected so that the filaments are brought to temperatures above their second-order transition points (ie, the hard-segment melting points). This allows the molecular chains to move freely to relieve stresses and results in filaments of fine tex but with the same mechanical properties as the heavier tex feed. Thus it is possible to take up from a wet spinning process at speeds in excess of 300 m/min by continuously drawing and heat-relaxing the filaments after drying.

Producers

Since spandex was introduced commercially in the late 1950s, many companies have attempted production. Most were unsuccessful owing to a variety of technological and economic problems, and the list of producers on a worldwide basis, shown in Table 2, is now relatively small. With the exception of Bayer's Dorlastan, original work on all spandex processes in current use was done in the United States or Japan. A large percentage of worldwide capacity is controlled by DuPont, either directly or through subsidiaries and joint ventures. These include two plants in North America; one in South America, two in western Europe; and one in Japan.

Table 2.   Spandex Fiber Producers

Company	Process	Trade name
North America		
Canada		
DuPont of Canada, Ltd.	solution dry spinning	Lycra
United States		
E. I. du Pont de Nemours & Co., Inc.	solution dry spinning	Lycra
Globe Manufacturing Co.	reaction spinning	Glospan and Clearspan
Western Europe		
Germany, Federal Republic of		
Farbenfabriken Bayer AG	solution dry spinning	Dorlastan
Italy		
Fillattice, S.p.A. (process licensed from Union Carbide Corp.)	solution wet spinning	Lynel
The Netherlands		
DuPont de Nemours, N.V.	solution dry spinning	Lycra
Spain		
Le Seda de Barcelona, S.A. (process originally licensed from Globe Manufacturing Co.)	reaction spinning	Enkaswing
United Kingdom		
Courtaulds, Ltd.	reaction spinning	Spanzelle
DuPont Company, Ltd.	solution dry spinning	Lycra
Eastern Europe		
USSR		
one plant licensed from and constructed by Toyobo Co., Ltd. of Japan	solution dry spinning	
Asia		
Japan		
Asahi Chemical Industry Co., Ltd.	solution dry spinning	Asahi Kasei Spandex
Fuji Spinning Co., Ltd. (process originally licensed from Ameliotex, Inc.)	solution wet spinning	Fujibo Spandex
Nisshin Spinning Co., Ltd	melt extrusion	Mobilon
Toyo Products Co., Ltd. (joint venture between E. I. du Pont de Nemours & Co., Inc. and Toray Industries, Inc.)	solution dry spinning	Opelon
Toyobo Co., Ltd.	solution dry spinning	Espa
South America		
Brazil		
DuPont Brazil, S.A.	solution dry spinning	Lycra

Economic Aspects

Except for some special purpose braids, no textile constructions consist entirely of elastomeric fibers. Use levels vary from a low of ca 5% in filling stretch cotton or cotton–polyester shirting fabrics (spandex core spun yarn in the fill) to a high of ca 40% for some warp-knitted tricot fabrics. The most commonly used raschel fabrics in the principal market area, foundation garments, contain ca 20–30% spandex.

Table 3 provides a summary of elastomeric fiber pricing in 1978 for the popular

Table 3. Elastomeric Fiber Pricing in the United States, 1978

		Extruded latex thread		
Tex[a]	Spandex Price[b], $/kg	Diameter, μm (gauge)	Approximate tex[a] equiv.	Price[c], $/kg
2.2	40.00	750 (34)	480	3.02
4.4	28.30	690 (37)	400	3.15
7.8	22.45	580 (44)	290	3.31
15.6	18.65	510 (50)	220	3.53
23.3	16.85	420 (60)	160	3.68
31.1	15.50	340 (75)	100	4.34
46.7	14.60			
62.2	12.35			
93.3	7.75			
124	5.95			
187	5.62			
249	5.40			

[a] To convert tex to den, divide by 0.111.

[b] On nonreturnable tubes.

[c] Unsupported multiend ribbons.

sizes. Cut rubber is not shown but prices, if adjusted for weight, are comparable to extruded latex thread. Raw material costs for spandex are on the order of $2.20–2.90/kg of finished product depending on the particular raw materials (eg, polyester or polyether glycol) used and conversion efficiencies. Similarly, rubber thread raw material costs are about $1.40–1.70/kg for both the cut and extruded processes. In comparing raw material costs to selling prices summarized in Table 3, it is obvious that the differences are much higher, especially for spandex, than for other chemical fibers such as nylon, polyester, acrylic, and acetate (see Fibers, chemical). Principal reasons for the higher differences are: the relatively small size of the total spandex market, which now totals worldwide ca 14,000 metric tons as compared to the multimillion ton levels enjoyed by the other chemical fibers, limits scale and thus efficiency of production units; and remaining technological problems limit maximum take-up speeds to much less than 1000 m/min and conversion efficiencies to under about 90%; and the relatively small number of producers continues to limit competition and allows higher gross profit margins.

In the early years of spandex commercial usage, many well-established, experienced synthetic fiber manufacturers projected near term markets on the order of 45,000 t. Such predictions were based on the following assumptions: (1) spandex would completely replace natural rubber; (2) major market areas for elastomeric fibers would continue to grow with the economy and population; (3) the superior performance characteristics of spandex would lead to many large volume end-use areas, eg, in outerwear, and shirtings, and sportswear; and (4) end-use areas for spandex staple fibers would develop rapidly.

Experience has shown that the first three assumptions were only partially correct and the fourth completely erroneous. Cost factors, improvements in rubber thread (eg, ribbon packaging and talc-free finishes) and adequate performance for specific areas have left considerable business to natural rubber thread. Dress styling and a freer life style have had a great negative effect on the use of foundation garments to the detriment of spandex growth. The rapid growth of lower cost and performance-adequate mechanical stretch yarns (eg, texturized nylons) have, instead of spandex, assumed a large volume potential in outerwear and sportswear areas, and the shift of consumer interest to wash-and-wear shirts retarded entry into shirting areas. After encountering many practical processing and construction problems in spandex staple research and development, all manufacturers have delayed introduction of a spandex staple fiber.

For the future, the original assumptions shown above may still be considered. Spandex raw material costs are not significantly higher than natural rubber fiber raw material cost. Cost differences between rubber and spandex can be expected to narrow, therefore, to the point where superior spandex properties will eventually mandate use of spandex over rubber in substantially all end-use areas. Despite the slow start, spandex is now beginning to penetrate sports and general outerwear areas and, as performance characteristics become more generally known, can be expected to grow at an increasing rate. Spandex staple is still commercially unknown but research and development work continues. Potentially, this area shows promise of large volume, as a staple spandex offers the possibility of making any staple yarn and staple yarn-based fabric elastomeric without change in the basic aesthetic qualities. In addition to the above, the Inmont Corporation has had a spandex spun-bonded fabric in semiworks-scale production for about three years that shows promise of broadening spandex use areas.

Uses

Cut Rubber and Extruded Latex. The manufacturing technology for cut and extruded rubber thread is much older and more widely known than that for spandex. Further production lines may be installed with a relatively modest capital investment. As a result, the manufacture of rubber thread is fragmented and more widely distributed with relatively few major and many minor producers. On a worldwide basis, Fillattice, S.p.A., of Italy with modern extruded latex plants in Italy, Spain, Malaysia, and the United States, is the largest producer of rubber thread. Second in production capacity is the Globe Manufacturing Co., Fall River, Mass., with production operations in the United States and England. Both of these firms also produce spandex (see Table 2).

Finer size extruded latex thread is generally wrapped with two covering yarns: one in a clockwise (S) direction and one in a counterclockwise (Z) direction. This covering yarn serves to protect the easily abraided rubber thread and to limit its performance to a selected area of the stress–strain curve.

Spandex. Both spandex and natural rubber fibers may be used in any area where extendability with good recovery is required. Initially, spandex was introduced as a replacement for natural cut rubber and extruded latex thread. Although spandex can indeed replace rubber with at least equivalent performance at lighter weight in almost all applications (excepting, primarily, golf balls where spandex is at a disadvantage owing to a slower recovery rate), such has not occurred to the degree originally expected mainly because of rubber's lower cost. Natural rubber fibers still find extensive use in narrow fabrics, braids, surgical hosiery, strip lace, and some circular knits. Spandex, however, has replaced almost all rubber in foundation garments and swimwear.

Spandex is processed into fabrics in three basic forms. These are: (*1*) bare (ie, as

Table 4.   Spandex Uses

Spandex form	Fabric types	Uses
bare	warp knits, circular knits, narrow fabrics (wovens, knits, and braids), and hosiery (knit)	foundation garments, swimwear, control tops of panty hose, elastic gloves, waist and leg bands, and upholstery
covered	warp knits, circular knits, and hosiery (knits)	support hosiery, foundation garments, elastic bandages, sportswear, and upholstery
core-spun	wovens, circular knits, and men's hosiery (knits)	shirting, brassieres, and sportswear

supplied by the maker on tubes or beams without further processing), (2) covered (ie, with one or, generally, two covering yarns, and (3) core-spun yarns in which a staple fiber such as cotton or cotton-polyester is spun around a fine tex spandex core. Use areas for the various forms are not clearly defined and overlap in many cases. A general summary is provided in Table 4 (see Textiles).

Spandex is sold in tex as low as 2.2 (20 den), whereas the finest diameter (gauge) at which extruded latex thread has been supplied is 140 μm (180) equivalent to ca 16 tex (140 den). The availability of spandex in such fine size and its unique properties compared to rubber (eg, dyeability, high modulus, abrasion resistance, and resistance to degradation) have led to new use areas (eg, support hosiery, lightweight circular and warp-knitted fabrics, core-spun yarns for woven and knitted fabrics, and elastic surgical dressings).

For textiles, cut rubber thread is used primarily in braids and narrow fabrics and as cut ribbons for waist, leg, and arm bands. Cut rubber is also used in all wound-core the same areas as cut rubber.

BIBLIOGRAPHY

"Spandex" in *ECT* 2nd ed., Vol. 18, pp. 614–633, by Robert A. Gregg, Uniroyal, Inc.

1. E. W. H. Becker, R. Hontz, and W. Watkins, *Ind. Eng. Chem.* **40**, 875 (1948).
2. A. Nishimura and H. Komagata, *J. Macromol. Sci. A.* **1**, 617 (1967).
3. S. Melamed and R. Minton, "Anidex Fibers" in N. M. Bikales, ed., *Encyclopedia of Polymer Science and Technology*, Vol. 15, Wiley-Interscience, New York, 1971, pp. 161–168.
4. Ger. Pat. 826,641 (Jan. 3, 1952), E. Windemuth (to Farbenfabriken Bayer).
5. Ger. Pat. 886,766 (1951), W. Brenschede (to Farbenfabriken Bayer).
6. *Textile Fibers Products Identification Act U.S. Public Law 85-897*, U.S. Federal Trade Commission, Washington, D.C., effective Mar. 3, 1960.
7. S. Cannon, G. McKenna, and W. Statton *J. Polym. Sci. Macromol. Rev.* **11**, 209 (1976).
8. R. Dunbar, W. Nunning, and D. Martin *Lenzinger Ber.* **43**, 96 (1977).
9. A. Bruner, N. Boe, and P. Byrne, *J. Elastoplast.* **5**, 201 (Oct. 1973).
10. J. Saunders and co-workers, *J. Appl. Polym. Sci.* **19**, 1387 (1975).
11. Jpn. Pat. 75 95,516 (July 30, 1975), T. Hidaka, S. Mizutani, and A. Tsuchimoto (to Toray Industries, Inc.).

12. Jpn. Pat. 75 138,124 (Nov. 4, 1975), T. Hidaka, K. Ikawa, and S. Mizutani (to Toray Industries, Inc.).
13. Jpn. Pat. 78 49,121 (May 4, 1978), T. Takeka, S. Mizutani, and M. Shirido (to Toray Industries, Inc.).
14. Jpn. Pat. 78 52,718 (May 13, 1978), T. Hidaka and co-workers (to Toray Industries, Inc.).
15. D. Allport and A. Mohajer in D. Allport and W. H. Janes, eds., *Block Copolymers,* John Wiley & Sons, Inc., New York, 1973, pp. 443–492.
16. R. Bonart, *Angew. Makromol. Chem.* **58–59**(1), 259 (1977).
17. H. Oertel, *Chem. Ztg.* **98**, 344 (1974); *Chemiefasern Text. Anwendungstech. Text. Ind.* **27**, 1090, 1095 (1977); *ibid.,* **28**, 44 (1978).
18. A. Bleijenberg and co-workers, *Br. Polym. J.* **4**(2), 125 (1972).
19. E. Hicks, Jr., A. Ultee, and J. Drougas, *Science* **147**, 373 (1965).
20. Jpn. Pat. 72 39,717 (Dec. 8, 1972), K. Kosonoi, K. Ichikawa, and Y. Nakahara (to Asahi Chemical Industry Co., Ltd.).
21. U.S. Pat. 2,953,839 (Sept. 27, 1960), R. Kohan, D. Slovin, and F. Bliven (to U.S. Rubber Co.).
22. U.S. Pat. 4,002,711 (Jan 11, 1977), T. V. Peters.
23. Jpn. Pat. 76 04,313 (Jan. 14, 1976), Y. Ikeda, T. Hirukawa, and Y. Ishiki (to Fuji Spinning Co., Ltd.).

TIMOTHY V. PETERS
Consultant

FIBERS, MULTICOMPONENT

Bicomponent fibers, conjugated, or composite fibers or heterofils have been known for some time. Glass bicomponent fibers were produced in the latter half of the nineteenth century and side–side viscose bicomponent fibers were patented in 1937 (1). Wool (qv), however, is a natural bicomponent fiber. Several acrylic bicomponent fibers were manufactured as self-crimping products in the early 1960s and nylon bicomponent fibers were introduced about the same time for use in hosiery and in nonwoven fabrics (qv) (see Acrylic and modacrylic fibers; Polyamides; Fibers, elastomeric; Fibers, chemical).

These products were all relatively simple with only a single interface between the two components. Today's fibers are more complex; they are manufactured with multiple interfaces and material–air boundaries. Interest in all types of multicomponent, though generally bicomponent, fibers has been revived because of the intense competition of the more conventional synthetic fabrics with the unacceptably high cost of introducing a fiber based on a totally new polymer; the need for some of the unique properties which can only be obtained via a bicomponent route; and improvements in technology which have allowed the manufacture of such sophisticated products.

The most commonly used designation for fibers composed of two polymers is bicomponent fibers, sometimes shortened to bico fibers. Since there are few examples, as yet, of three or more fiber-forming polymer components, bicomponent tends to be used as a general term to include these products. Synthetic fibers always contain materials such as stabilizers, catalyst residues, processing aids, pigments, and other additives. Alternative expresssions are bilaminar filaments, biconstituent fibers, composite fibers, heterofilaments (or heterofils), and sea–island fibers. It is important to distinguish between composite fibers (where every individual fiber or filament has two or more components) and composite yarns, sometimes called heteroyarns. The latter are yarns composed of a mixture of different types of filaments, usually single-component filaments, although a composite yarn may contain a mixture of single-component and bicomponent filaments (or any other combination).

It has been found possible to introduce controlled-composition variations into the cross sections of textile filaments of ca 0.1–0.5 tex (1–5 den), or only a few micrometers in diameter, especially in simpler bicomponent fibers with only a single interface, such as side–side and core–sheath bicomponents. Coupled with an increasing understanding of the viscous flow of fiber-forming polymer melts, and improved spinnerette manufacturing methods, this has led to the development of new bicon-stituent fibers including multiinterface or matrix–filament fibers (sea–island fibers). Japan, western Europe, and North America have all been involved in this development. In many cases these products are never marketed as bulk fiber but are converted by the producers into materials of higher value.

Bicomponent fibers are classified on the basis of the cross-sectional and longitudinal arrangement of the components (2–3). The simplest class, consisting of fibers with a single interface and with both components possessing an external boundary, is known as side–side fibers irrespective of the cross-sectional shape, which was not necessarily circular. The second class comprised fibers with a single interface but in which only one component had an external boundary, known as sheath–core fibers. These two classes exhibit special properties because of the presence of only a single interface. For example, side–side fibers usually tend to crimp when treated in various ways. There is no generally accepted term for fibers with more than one component–component interface and one or more external boundaries, although they are sometimes termed matrix–filament fibers. This term is used here in preference to sea–island fibers, which implies only a single external boundary, ie, multiple totally enclosed cores.

In the above examples, the cross sections of the components do not vary along the lengths of the fiber. Another class of bicomponent fibers, known as matrix–fibril fibers, exhibits cross-sectional variation with length. The second (minor) component is present as separate fibrils or short lengths of fiber embedded in a matrix of the first (principal) component. The main axis of the fibrils is aligned with the main axis of the fiber. Matrix–fibril fibers are not prepared by the controlled combination of polymer streams as are other bicomponent fibers; they are of little commercial importance.

A great variety of transverse and longitudinal fiber cross sections exists. The scanning electron microscope photograph of a transverse cross section of Diolen ultra matrix–filament fibers is shown in Figure 1. The center contains a complete filament comprising a star-shaped matrix of nylon into which are fitted six triangular filaments of poly(ethylene terephthalate) polyester. This whole cross section is surrounded by split matrix–and–filament components because this particular fiber is designed to split to give very fine fibers.

Materials and Processes

In many respects, the manufacture of bicomponent fibers differs very little from that of single-component fibers (see Acrylic and modacrylic fibers; Olefin fibers;

Figure 1.   Transverse cross section of Diolen Ultra fibers (magnification 2500×).

Polyamides; Polyester fibers). Many pairs of fiber-forming polymers may be employed in the manufacture of bicomponent fibers. In fact, it is not necessary for both components to be capable of forming fibers if spun alone. For example, polystyrene, which is not normally used in fiber manufacture, may be used as a temporary component in the manufacture of certain matrix–filament fibers. Clearly, it must be possible to handle both components by the same fiber-forming route, eg, melt-spinning. Thus it would be impossible to manufacture a polyester–acrylic bicomponent fiber because the acrylic component could not be spun by the melt-spinning route; bicomponent fibers with an acrylic component can only include components which can be wet or dry-spun. Similarly, materials which can only be wet-spun cannot be combined with materials that can only be melt-spun or dry-spun. Within these limitations, there is still a large number of possible combinations, especially in the melt-spun group, which today is the most important in technical terms.

The bicomponent fibers or their derived products are given in Table 1. The only products that are not melt-spun are the acrylic fibers (qv), which dominate production in volume, but in value are only a fraction of the unit cost of the most expensive products. Some of these products, such as the synthetic suedes, can cost up to $150/kg. Bicomponent rayon fibers have been reported experimentally, but are not produced commercially by a bicomponent route. The normal wet-spinning process, however, can form nonuniform fibers with self-crimp characteristics (4). Melt-spinning is a better process for bicomponent fibers than wet-spinning because of the reduced hole density and consequent loss of spinning capacity introduced by the assemblies behind the spinnerette required for bicomponent formation. This is particularly important in the case of the very high hole densities normally employed in wet-spinning.

The components employed in bicomponent-fiber manufacture must be compatible in a number of other respects as well as in their spinning route. Most important, their viscosities under the conditions of spinning must not be too dissimilar and high enough to prevent turbulence after the spinnerette. Even if turbulence does not occur,

Table 1. Manufacturers and Trade Names of Multicomponent Fibers or Multicomponent Fiber-Based Products

Manufacturer	Country	Trade name	Type	Composition (if known)
Allied Chemical Company	United States	Source[a]	staple, FY[b]	matrix–fibril nylon 6–polyester
ANIC	Italy	Euroacril	staple	side–side acrylic
Asahi Chemical Industry Co., Ltd.	Japan	Cashmilon GW and Cashmilon H	staple	side–side acrylic
		Lammus	suede	
Akzo N.V.	Netherlands	Diolen Ultra	FY[b]	matrix–filament nylon–polyester
		Diolen Biko	staple	matrix–filament nylon–polyester
		Diolen	staple	sheath–core polyethylene–polyester
		Diolen	staple	sheath–core nylon 6–nylon 6,6
		Enkatron	FY[b]	matrix–fibril nylon 6,6–polyester
American Cyanamid Company	United States	Creslan 68	staple	side–side acrylic
Bayer AG	Federal Republic of Germany	Dralon K/120, 140, 820	staple, FY[b]	side–side acrylic
Courtaulds Ltd.	United Kingdom	Courtelle LC	staple	side–side acrylic
Chisso Corporation	Japan	ES Fiber	staple	side–side polyethylene–polypropylene
		EA Fiber	staple	side–side EVA[c]–polypropylene
Colbond b.v.	Netherlands	Colbond, Colback	spunbonded	sheath–core nylon–polyester
Cyanenka	Spain	Crilenka Bico	staple	side–side acrylic
E. I. du Pont de Nemours & Co., Inc.	United States	Cantrece	FY[b]	side–side nylon 6,6–modified nylon 6,6
		Orlon Types 21, 24; Sayelle	staple	side–side acrylic
Hercules	United States	Herculon 404	staple, FY[b]	side–side polyolefin
ICI Fibers	United Kingdom	Cambrelle PBS3	staple nonwoven	sheath–core nylon 6–nylon 6,6
		Cambrelle PBS5	staple nonwoven	sheath–core modified polyester–polyester
		Epitropic fiber	staple blend	sheath–core modified polyester–polyester
		Bonafill	staple	sheath–core polypropylene–polyester
		Terram	spunbonded nonwoven	sheath–core polyethylene–polypropylene[d]
Japan Exlan	Japan	Exlan	staple	side–side acrylic
Kanebo Ltd.	Japan	Sideria[a]	FY[b]	sheath–core nylon–polyester
		Savina	suede	matrix–filament nylon–polyester
		Belleseime[e]	suede	matrix–filament nylon–polyester

Table 1 (continued)

Manufacturer	Country	Trade Name	Type	Composition (if known)
Kanebo Ltd.	Japan	Belina	staple	sheath–core nylon–polyester
		Kanebo Nylon 22	staple	side–side nylon 6–modified nylon 6
Kuraray Ltd.	Japan	Sofrina[f]	apparel leather	matrix–filament
		Amara	suede	matrix–filament
Mitsubishi	Japan	Vonnel V-57	staple	side–side acrylic
Monsanto Textile Company	United States	Monvelle[a]	FY[b]	side–side nylon–polyurethane
		Pa–Quel	staple	side–side acrylic
		Antistatic fibre	FY[b]	matrix–filament polyester–polyethylene
Montefibre	Italy	Leacril BC	staple	side–side acrylic
Rhone-Poulenc-textile	France	Tergal X-403	staple	side–side polyester–modified polyester
Snia	Italy	Velicren	staple	side–side acrylic
Teijin Ltd	Japan	Hilake	suede	matrix–filament nylon–polyester
		Tetoron 88	staple	side–side polyester
		S-28[a]	staple	matrix–fibril nylon 6–polyester
Toho Beslon	Japan	Beslon	staple	side–side acrylic
Toray Ltd.	Japan	Ecsaine[g]	suede	matrix–filament polystyrene–polyester
		Toraylina	suede	matrix–filament polystyrene–polyester
		Crestfil	raised fabric	
		SBII	leather	
		Tapilon	FY[b]	sheath–core nylon 6–nylon 6,6
		Metalian	FY[b] (antistatic)	sheath–core polyester
Unitika	Japan	Lauvest	suede	

[a] Production of these materials is now believed to have been discontinued.
[b] FY = filament yarn.
[c] Ethylene–vinyl acetate copolymer.
[d] Previously sheath–core nylon copolymer–polypropylene.
[e] Or Suede 21, Suede 21L.
[f] Napara for export.
[g] Ultrasuede or Alcantara for export.

if one component is less viscous than the other, the flow patterns designed to give the required filament cross section may not be stable. Even if the filament geometry is correct, if it is without a center of symmetry, and if both components have an external boundary, unstable extrusion can occur leading to flickering filaments and interfilament coalescence in a molten thread line. This can lead to severe problems in a simple side–side geometry even where both polymers are apparently similar; differences in frictional properties, flow behavior, and polymer-to-spinnerette adhesion play a part. As a rule, the viscosities of the two components should not differ by more than a factor of ca four, but ideally should be considerably closer. This can be accomplished by modifying the dope concentrations or the polymer molecular weights, although the effect on downstream processing and fiber properties must be considered. The choice of components depends on the fiber application. If possible, however, a polymer and one of its copolymers are selected because this choice can remove many of the compatibility problems, not only the primary one of viscosity balance, but also the subsequent ones of component adhesion and drawing ratio compatibility.

Frequently, component pairs do not adhere sufficiently to behave as a single fiber through various textile processes or in the final application. Splitting and breakage of components may occur on rolls, guides, or texturing devices resulting in heavy, sticky, or dusty deposits, particularly with matrix–filament fibers. Some matrix–filament fibers, however, are designed to separate during fabric finishing to confer certain properties on the fabrics. In other cases, splitting can be detrimental, for example, in carpets where it may result in matting (interlocking and compaction of the pile). Splitting may be alleviated by mechanically interlocking the components; the ultimate interlocking is a core–sheath fiber where the core component cannot escape. However, if the sheath is very thin or is uneven, it may be stripped or split away from the core. Mutual diffusion of the polymer components at the interface by slight flow disturbance in the spinnerette hole has been suggested for better adhesion in side–side fibers (5). Noninterlocked components may separate at the drawing stage, especially if the natural draw ratio of one component is exceeded.

The orientations, as indicated by the birefringences, of the two components, are always affected by the component structures, which are, by design, different. In some cases, optimal overall fiber strength is not needed because one component is removed at a later stage, or used as an adhesive. This simplifies drawing since the draw ratio can be selected to be suitable for the remaining component. If both components contribute to the fiber properties, a weighted average of the properties of either component results. Thus each property value must be below the value which would have been obtained for that property in a monocomponent fiber prepared under the same conditions from whichever component gives the better level of that property. Bicomponent fibers thus tend to have inferior properties compared to the equivalent monocomponent fiber, but have properties that the monocomponent material does not have. Typical properties of an acrylic bicomponent fiber are compared in Table 2 with those of the equivalent single-component fiber (6). In the case of acrylic fibers, the bicomponent technology is applied only to achieve self-crimping in order to obtain properties similar to those of wool. For self-crimping, the fiber must be asymmetric and the components must possess a differential shrinkage (7). As a rough approximation, the radius r of the helical crimp generated for a fiber of thickness h is given (3) by:

$$\frac{1}{r} \approx 1.6 \frac{\Delta L}{L} \cdot \frac{1}{h}$$

where $\Delta L/L$ is the differential change in length of the two components during the crimp development stage. The crimp may develop spontaneously as drawing tension is removed or it may be latent, ie, it develops as some condition of the fiber is changed. Some bicomponent fibers develop crimp when they are heated in a relaxed state. Some acrylic fibers develop crimp when they are dried, and lose it when they are immersed in water. Such a reversible crimp offers an advantage in that garments made from these fibers regain their original softness and bulk after washing.

No similar simple treatment is available for predicting the extension-recovery behavior or crimping or retraction forces that are associated with bicomponent fiber

Table 2.   Properties of Cashmilon Acrylic Fibers[a]

Property	Cashmilon GW, bicomponent	Cashmilon FW, monocomponent
strength, N/tex[b]		
dry	0.25	0.25–0.32
wet	0.21	0.23–0.28
elongation, %		
dry	40–45	28–40
wet	45–50	30–45
loop strength, N/tex[b]	0.31	0.18–0.27
loop elongation, %	20–25	5–20
modulus of tensile elasticity, %	99 ± 2	(90–92) ± 3
specific gravity	1.18	1.188
moisture regain at 20°C, %	1.8–1.6	1.6
shrinkage in hot water, %	4	2–4

[a] Ref. 6.
[b] To convert N/tex to gf/den, multiply by 11.33. N/tex = GPa/density.

crimping (3). Although self-crimping products are the most obvious application for bicomponent fibers, there has been no successful commercial application of the principle in the continuous-filament field, even though producers have tried to enter the textured yarn market. The crimping and retraction forces are too low to be of value in comparison with those obtained by the false-twist texturing or draw-texturing routes. In the case of staple fibers, the relatively weak forces can still be of value where the separated short fibers are concerned. Stronger retractive forces can be obtained if one of the components is elastomeric, but such products are not now commercialized. A number of self-crimping products have been obtained with a crimp potential generated by asymmetric treatment of the filaments after extrusion. Since part of the outside skin of each filament is treated, the maximum possible asymmetry results. Crimping can be achieved by water quenching or hot knife-edge treatments (8–9).

The processes by which multicomponent fibers are produced are largely independent of the polymers used except for the formation of a self-skinning fiber in the wet-spinning process (4). Production processes rely on the stability of the interfaces between the components to maintain the filament cross-sectional geometry as the cross-section area is reduced, perhaps several hundredfold, during extrusion and drawdown. Production methods for multicomponent fibers are given in Figure 2. To obtain the required properties under the best process economics, high filament spinning density is needed with high plant utilization and conversion efficiency, clean startup, and low production of waste fiber. In most production methods the components are combined after they have entered the spinning pack. The design of the spinning pack is more complex than for single-component fibers and this overcrowding reduces the potential filament density. This may not be important in the case of a low-count filament yarn (such as a hosiery yarn), where only a few filaments are required. It is more important in feedstock for staple fiber because the greater the number of filaments produced, the better are the process economics.

Multicomponent fibers can be produced by spinning, under suitable conditions, a blend of immiscible polymers, eg, nylon and polyester. A matrix–fibril fiber is produced in which the minor component forms short individual fibrils dispersed in a fiber

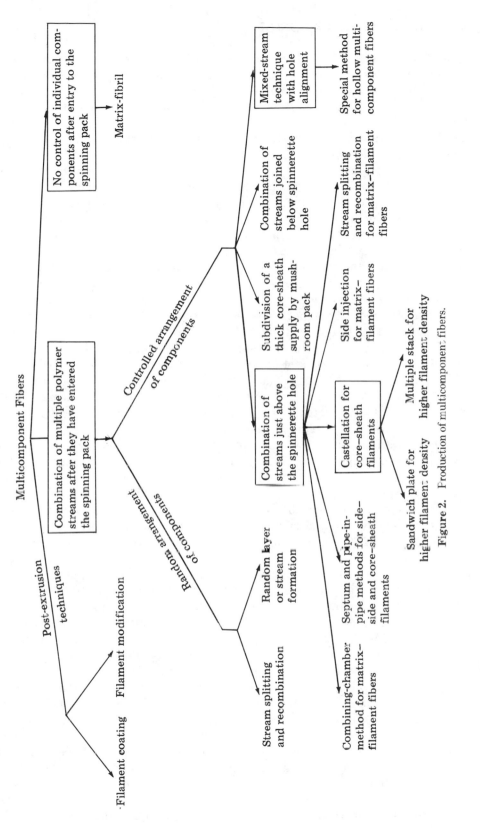

Figure 2. Production of multicomponent fibers.

159

composed of the principal component. New products were expected from this technique (10), eg, for low flat-spotting tire cords, but commercialization has not materialized. Changes in the ratio of the two components affect the structure of the fibrils. Because of orientation at spinning and drawing, the fibrils always orient parallel to the fiber axes. The components may be blended as chips or powders before extrusion. These fibers are relatively easy to produce and it is unfortunate that they are not of greater interest. However, it is possible that the fibril size and, therefore, the properties may depend upon the degree of mixing during extrusion, which is difficult to control. In order to obtain a fibrous structure, the polymers must be incompatible. For example, polyester is not sufficiently incompatible with polycarbonate to generate fibrils, but polypropylene forms fibrils in a polycarbonate matrix. The properties of the resulting fibers are superior to those of monolaminar polycarbonate filaments. The preparation of multicomponent fibers by post extrusion techniques is not easy. For example, it is difficult to coat fine filaments with a second component. It is possible that Teijin's Metalian antistatic fiber is prepared by vapor deposition of a layer of metal on the fibers (11). Antisoil coatings have been applied by filament-coating techniques to textile fibers, especially carpets (see Textiles, finishing). A perfluoroalkyl ester may be applied in a thin, durable coating to synthetic textile filaments as a spin-finish (12). In this case the sheath is extremely thin. Typical techniques include treatment of the hot filaments with cooling water (8), the passage of filaments over a heated knife edge (9), or the asymmetric heat treatment of the filaments in the spinnerette hole (13). Since the filaments are initially formed from a single component, the modifications change part of it to generate the crimp potential. This may be a change of orientation, crystallinity, molecular weight, or filament cross section.

True multicomponent fibers are prepared by the controlled combination of liquid polymer streams after the individual streams have entered the spinning pack. If only side–side fibers are required and a degree of uncertainty regarding the position of the interface can be tolerated, a conventional spinnerette may be fed with a polymer stream containing alternating layers of the two components (13). Since the mixed-stream flow is not turbulent, the alternating layer pattern is maintained as the stream reaches the spinnerette face and a bicomponent filament is formed whenever a spinnerette hole intersects a layer junction. This intersection takes place randomly and there is no control over the position of the interface in the filament. Depending on the number of spinnerette holes and the number of layers, the interfaces in any filament may vary, resulting in variation in the properties of the individual filaments. Alternatively, the streams can be mixed in one or more static or dichotomic mixers (14). Such mixers have four to nine alternate left- and right-hand helical elements in series. After the polymer components are fed to the opposite sides of the leading edge of the first element of each mixer they are split and recombined to form at the mixer exit a multilaminar stream which is then conducted to the spinnerette. Such mixers may be combined in a suitable configuration in a spinning pack. The principle of mixed-stream formation may also be used to form matrix–filament fibers (15–17). A greater number of streams, ie, fine cores, are normally formed than the number of spinnerette holes and, therefore, the number of interfaces per hole is high and less variable from filament to filament. Furthermore, because a simple spinnerette is employed, the number of filaments which may be produced from a spinnerette by the mixed-stream technique is not limited. In all mixed-stream methods, the number of layers or cores formed in the mixed stream or streams is, in theory, controlled because the streams are formed by the splitting and recombination of the polymers. A truly random number of fine cores is formed in the mixed stream by passage of the laminar stream from a mixer through a layer of gauze screens (18). The gauzes, of carefully selected opening size, thus break up the laminations (of which there may be >1000) to give as many as 150 cores per filament.

Other methods of preparing multicomponent fibers require the controlled flow of the components to the spinnerette holes or beyond. Most of these methods require the components to be combined and formed into the required cross-sectional configuration just above the spinnerette holes. However, three techniques require the controlled arrangement of components away from the spinnerette holes. In the first, the polymer streams are combined before they enter the spinning pack (19). A thick core–sheath supply stream passes through the sand filter and is distributed to a mushroom plate just above the spinnerette (Fig. 3). (Note: letters A and B arbitrarily indicate the positions of the two components. These letters do not necessarily correspond to the sheath and the core, respectively, in all figures.) The distribution channels above the plate are arranged in such a way that their position and areas lead the core and sheath polymers to flow between the underside of the mushroom plate and the upper surface of the spinnerette as an upper layer of core polymer and a lower layer of sheath polymer. Since both polymers are supplied at a fixed rate to the spinnerette pack, fixed quantities of each leave through each spinnerette hole as a concentric core–sheath filament. Despite the simplicity of this technique, it has disadvantages including the difficulty of establishing the correct flow at startup, the almost inevitable presence of some core-only and some sheath-only monocomponent filaments, a limitation to component ratios of between ca 2:1 and 1:1, and the need to design the pack internals for the specific core:sheath ratio needed. The technique is only used for concentric core–sheath spinning; more recent techniques are superior.

Another method is even more specialized; both components must be of the same material but with different orientation. Each filament is formed by the combination of two polymer streams below the spinnerette hole (see Fig. 4). The process is primarily applicable to combined melt spinning and drawing at high speeds (greater than 3500 m/min winding speed). Advantages include simplicity of process and equipment, except for the spinnerette. Disadvantages include limitations in polymer types and cross

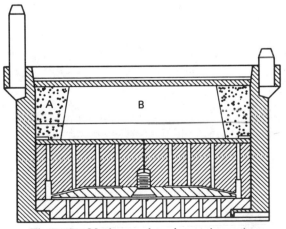

Figure 3.   Mushroom plate above spinnerette.

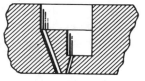

Spinnerette

Figure 4.   Combination of two polymer streams below spinnerette hole.

sections and low crimping forces. In addition, some spinning conditions can lead to filament oscillation and cross-section variability.

Side–side bicomponent fibers are also produced by feeding a mixed polymer stream to a simple spinnerette with the holes aligned with the stream interfaces. The mixed polymer streams are generated as concentric rings or parallel laminae (7). The pipe–in–pipe method is the simplest for mixed-stream formation (Fig. 5a); the concentric plate mixer (Fig. 5b) is actually a compact pipe–in–pipe system. At plate 1, polymer A forms a cylindrical stream which is narrowed at plate 2. Plate 3 adds a concentric layer of component B and the core–sheath stream is narrowed at the next plate 2. As many concentric layers may be added as there are pairs of plates. The proportion of the two polymers delivered to each spinnerette hole is determined by the supply rates to the spinning pack. The rings or lines of holes must be carefully aligned with the interfaces and the volume flow of each stream must be in proportion to the number of holes it supplies. If the number of interfaces is large, alignment becomes difficult and random layers form. Special spinnerettes are not needed, but there

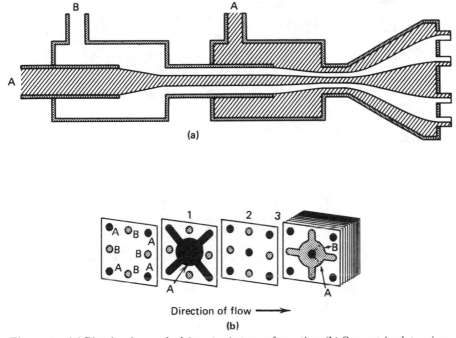

Figure 5.   (a) Pipe-in-pipe method for mixed-stream formation; (b) Concentric plate mixer.

is some loss of control over the components' ratio. The aligned mixed-stream method is not thought to be currently in commercial use. A variant is used in the formation of hollow matrix–filament fibers (20). For each filament, an interrupted, elongated slot meanders back and forth across the interface of a single core–sheath stream. This generates a number of separately spun interfaces, which combine below the spinnerette to form the hollow filament with alternate segments of the two constituents.

   In other methods, the polymers are combined just before their passage through the spinnerette hole. The pack designs are more complex than the methods described previously; they are more costly and may be limited in output. On the other hand, the filament cross sections are better controlled and parameters, such as the core:sheath ratio, may be changed without modifying the pack or spinnerette design. In the septum method for side–side fibers the components are kept apart until the very last moment by a partition above the spinnerette holes. The form of the partition, or septum, is not standardized (see Fig. 6); it may be a relatively thin plate (21) or somewhat thicker (22). It may even be a protrusion passing into the spinnerette hole (23). Since the septum generally covers a whole row or ring of holes, the pressure drop to each hole must be the same to obtain fiber uniformity. The ratio of components is determined by the pumping rates. Unless a condensed system is employed (21), the channel system behind the spinnerette severely limits the hole density. The condensed system is less suitable for highly viscous polymer melts because of the high pressures developing. In certain cases (24) a metering plate ensures uniform polymer flows to the spinnerette holes, and the same spinnerette is used to form some monocomponent fibers to yield a composite yarn. To prevent the splitting of filaments, the two components may be combined by a hole through the septum or some other modification (25); this distorts the interface in a controllable fashion. Another method of combining the components is by intermingling at the interface, for example, by the use of a gauze (5).

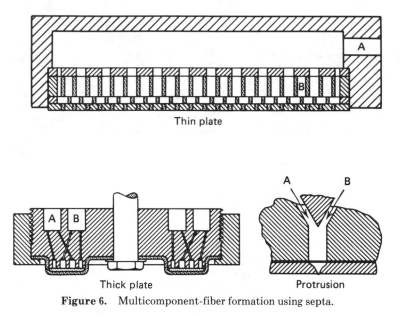

Figure 6.   Multicomponent-fiber formation using septa.

The pipe-in-pipe technique is analogous to the septum method. Here, the core stream is inserted directly into the sheath stream via a fine tube. The wall of this tube may be regarded as a circular septum (26). In principle, more than one core could be introduced into a filament by employing a number of injection tubes. In practice, the production of matrix–filament fibers by the pipe-in-pipe technique uses the so-called combining-chamber method in which a number of core–sheath filaments are combined and narrowed in a conical chamber below the spinnerette holes before final extrusion, as shown in Figure 7 (15,27). Various techniques are used to produce tricomponent filaments (28). For example, pipe-in-pipe core–sheath filaments are formed, which pass through a second pipe for addition of a second sheath before a number of such filaments pass into a combining chamber. All pipe-in-pipe designs suffer from the fragility of the injection pipes and the general complexity leading to high cost and low output. Strength may be improved by eliminating the delicate injection tube and merely extruding the core from a hole above the spinnerette hole (29). Larger diameter tubes may be fitted into the spinnerette-hole counterbores in order to receive the core

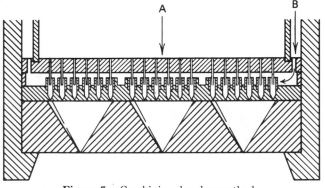

Figure 7.   Combining chamber method.

streams from simple holes (30). In still simpler designs, the core is extruded from an internal spinnerette placed above the main spinnerette without fine tubes or passageways. This modification may be used for multisegmented hollow filaments (31) or side–side filaments (32) by suitably offsetting the core holes. This simple hole–above–hole method requires an equal pressure of sheath polymer at each spinnerette hole and an even flow into each hole to ensure uniformity.

The need for control over the sheath polymer has led to the castellation method (Fig. 8), in which the sheath polymer flow is controlled by passage through a fine gap between the upper and lower spinnerettes (33). The upper spinnerette is generally called the distribution plate. The gap is ca 100 μm and extends evenly over the spinnerette; consequently highly uniform fibers are produced. The thickness of the metering gap is influenced by the viscosity of the sheath polymer. If the spinnerette or the distributor plate holes are of noncircular cross section, the filaments, cores, or both can have noncircular cross sections (34–35). The core can also be placed eccentrically by creating an uneven sheath polymer flow (3) by setting it eccentrically relative to the spinnerette hole; making it noncircular; sloping the face; or recessing the distributor plate at one side of the hole. Because of its versatility, reliability, and ruggedness, the castellation method seems likely to increase in importance, especially if the spinnerette area can be reduced. To achieve this, chambers containing the castellations are arranged in more than one plane in an overlapping manner resulting in closer hole spacing (36). Alternatively, an orifice plate containing small holes is aligned over the spinnerette counterbores (37). This permits smaller castellations on the distributor plate, which are closer together than if they meet the counterbore directly. Both techniques approximately double the number of holes per unit area.

Matrix–filament fibers are manufactured by the side-injection technique with a septum device in which a tube connects the chamber containing one component to the spinnerette hole. Fine holes in the sides of this tube communicate with a second chamber containing the second polymer component. By the usual control of supply rates, a segmented matrix–filament is formed. By varying the number of side holes, or introducing fine tubes into the side holes to lead the second component away from the edge of the filament cross section, or introducing a coaxial pin into the tube, complex cross sections may be achieved.

Similar matrix–filament cross sections may be obtained by a version of the plate mixer technique described above. In this case, each spinnerette hole has its own plate stack and thus each filament is a composite stream of the required complexity (38). In practice, many foil plates are connected by webs in such a manner that an assembly for many spinnerette holes can be built up by stacking a few webbed plates. This technique offers the promise of considerable development. As in all matrix–filament spinning, excessive die-swell must be kept in mind.

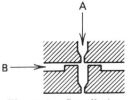

Figure 8.   Castellation.

Economic Aspects

The total production of multicomponent fibers throughout the world is difficult to estimate, because the fiber may be used directly by the manufacturer or may not be distinguished in the production statistics from monocomponent fiber of a similar type. It is estimated that world production of bicomponent fibers, excluding acrylics and those for synthetic suede manufacture, amounted to ca 25,000 metric tons in 1981, mainly manufactured in western Europe with a small production in Japan. Japanese synthetic suede capacity was ca 8.4×10^6 m^2 per year (39) by the end of 1981. A small licensed production is based in Italy. Average bicomponent fiber consumption is estimated at 100 g/m^2 of suede. Thus the consumption of bicomponent fiber is 840 metric tons. Suede fabric weights vary between about 150 g/m^2 and 235 g/m^2 of which between 10 and 35% is nonfibrous polyurethane and the remainder may be microdenier (ca 0.1 tex or less per filament) fibers generated from matrix–filament fibers plus woven, knitted, or nonwoven base fabric composed of monocomponent or bicomponent fibers. Up to 80% of the production of the various synthetic suedes may be exported to the United States and western Europe.

Health and Safety Factors

In multicomponent fibers, the individual components contribute to the fibers' properties. Where the components are generally inert, nontoxic, and nonallergenic, the multicomponent fibers behave similarly. Where one or more of the components is not an established textile polymer (eg, a physiologically active agent), special considerations may apply, and each case must be studied individually. Information is usually available from the manufacturers.

Uses

By far the most important application for multicomponent fibers is the creation of desirable textile properties or aesthetic effects. Although initially the aim was to achieve a steric crimp similar to that of wool, much more sophisticated effects can now be achieved. This is often done by using the techniques described above to generate microdenier fibers (<0.1 tex per filament), sometimes of sharply defined cross section. Such fibers are too fine to be spun directly. Other effects may be obtained by drastically modifying the surface of bicomponent fibers with one unmodified component to provide strength. There are hundreds of patents in this area, many from Japan; only a few can be mentioned here. Reversibly crimpable acrylic fibers are described in references 22 and 40–47; reference 3 also cites many earlier patents. Crimpable nylon for hosiery is discussed in references 3, 48, and 49. The use of self-crimping polyester filaments in a feed yarn for draw-texturing is described in reference 24, self-crimping poly(butylene terephthalate)/poly(ethylene terephthalate) filaments in reference 32, and self-crimping products with one elastomeric component in reference 50.

Microdenier (<0.1 tex per filament) fibers formed from matrix–filament fibers may be utilized in various ways. The simplest is in the production of knitted or woven fabrics with appearance and handle different from those made using fibers in the normal 0.1–0.5 tex per filament range. A more silklike appearance is often claimed (15,28,51–57). For textured microdenier yarns it may be necessary to treat the fabrics with a solvent to dissolve the matrix or to use special procedures to split the polymer junctions. In imitation fur fabric, tufts of microdenier (<0.1 tex per filament) fibers are formed from sea–island pile fabric (58). Flocked fabrics may also be produced (59). The most complicated application of bicomponent fibers is the preparation of washable synthetic suedes which have been very successful commercially (39,60). Such fabrics are based either on conventionally produced textiles or on nonwoven fabrics. The base material is impregnated with a resin, usually a polyurethane, and the matrix material is removed. Typical methods of producing artificial suedes or pile-surfaced fabrics are given in Figure 9 (16,27,31,62–65). An electron microscope photograph of the impregnated suede, Ecsaine, introduced by Toray in 1970, is shown in Figure 10.

In other applications one component serves as an adhesive. Core–sheath fibers are generally used in which the sheath has a melting point ca 20°C lower than the core. The products may be nonwoven fabrics. Applications range from textiles to three-dimensional fillings and foams for aircraft protection (66), to cigarette filters or fiber-tipped pens (67). The fiber–fiber bond may also be employed in the high speed manufacture of staple yarns (68) or in sizing false-twist yarns. If the adhesive action is employed to bond solid particles to fibers, so-called epitropic products are formed. For example, carbon black forms conducting filaments, which can be used for antistatic purposes or current carrying applications such as heaters (69). Silica particles impart water repellency (70).

In a third group of applications, multicomponent fibers are used to modify the physical properties of textiles, especially for antistatic applications (11,71–80). Many systems attempt to combine both antistatic and humectant or hygroscopic features. Many variations are intended to reduce body cling in women's apparel. Some improve the flat-spotting of nylon tire cords (18), rubber adhesion (81), abrasion resistance (82), and reduce abrasiveness (83). Others change density (with a metal core) (84) and modify tensile properties (85).

A recently developed area is the modification of the chemical fiber properties. For example, one component could impart desirable tensile properties and another, desirable chemical properties. Thus, chemical resistance is improved by coating a poly(ethylene naphthalene 2,6 dicarboxylate) with poly(tetramethylene naphthalene-2,6-dicarboxylate). Similarly, the use of two aromatic polyamides to form a sheath–core fiber with excellent heat-, dyeing-, and flame-resistant properties is claimed (86). Flame resistance may also be imparted by decabromobiphenyl and tin oxide (87). These additives may be incorporated into the sheath after fabric production to avoid spinning problems. Strength and easy dyeing properties may be obtained with cellulose acetate in the partially hydrolyzed sheath (88). A three-layer sheath–sheath–core fiber may possess dyeability and polyester properties by using a polyester core and a saponifiable polyester copolymer sheath (89). Good heat resistance may be achieved in a sheath–core polyamide fiber with a high concentration of metal ions (90).

The inclusion of more than one component in a fiber may confer various optical properties. For example, a bichromic shot effect may be obtained with a fiber of flat cross section with the second component located at the rounded end portions (91). Moire effects are produced with matrix–filament fibers which have been treated to expose the embedded filaments. Unusual color effects may also be obtained by dyeing the microdenier (<0.1 tex per filament) fibers formed by completely removing the matrix from matrix–filament fibers (92). A sheath–core structure in which the sheath

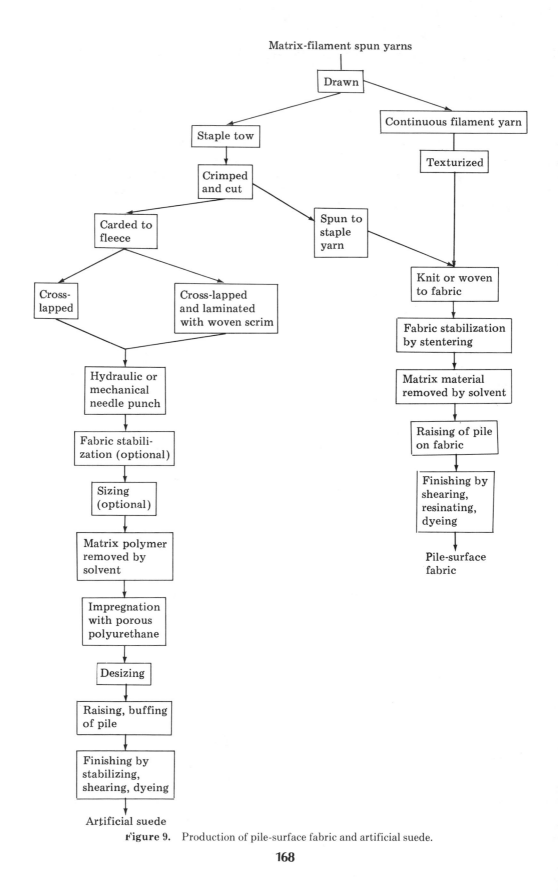

Figure 9. Production of pile-surface fabric and artificial suede.

Figure 10.   Impregnated suede (magnification 1000×).

possesses many hollow or uneven sections produces a silky luster (93). A nontransparent color effect is obtained from coloring material limited to the core of a sheath–core fiber (94).

More recently, multicomponent fibers have been proposed for use as containers or as diffusion barriers. For example, organic vapors and aerosols may be absorbed by filaments with a microporous sheath and an active sorbent core (95). In a reversal of this process, fibers with a pore-free core and a microporous sheath can be used to store and deliver detergents, wetting agents, bacteriocides, fungicides, etc (96). In a more sophisticated delivery system, sheath–core fibers with a core of progesterone and polycaprolactone and a sheath of polyethylene or polypropylene have been proposed as a long-term (5-year) female contraceptive device (97).

In a possible returning of interest to matrix–fibril fibers, another area of application for multicomponent fibers is in the rheological modification during melt-spinning. It has been recently observed that the presence of a small quantity of microfibers in the molten thread line modifies the drawing process with potential improvements in the processing, economics, or aesthetics of the fiber.

In addition, a glass–resin foam sheath–core fiber has been patented with enhanced aesthetic, chemical, and physical properties. Various three-component fibers are discussed in references 98–100. Bicomponent monofilaments are also proposed for various applications (101). Clearly, now that bicomponent fibers are readily accessible, more applications are likely to be developed using cheaper materials.

BIBLIOGRAPHY

1. P. A. Koch, *Ciba–Geigy Rev.* **1,** 1 (1974).
2. R. Jeffries, *Ciba–Geigy Rev.* **1,** 12 (1974).
3. R. Jeffries, *Bicomponent Fibres*, Merrow Publishing Co. Ltd., UK, 1971, pp. 2–4.
4. W. A. Sisson and F. F. Morehead, *Text. Res. J.* **23,** 152 (1953).
5. Brit. Pat. 1,004,251 (June 10, 1963), (to American Cyanamid Co.).
6. *Technical Information on Cashmilon GW Type (Bi-component)*, Asahi Chemical Industry Co. Ltd., Japan, 1976.
7. W. E. Fitzgerald and J. P. Knudsen, *Text. Res. J.* **37,** 447 (1967).
8. Brit. Pat. 1,425,705 (June 28, 1972), B. Boyes and A. Jones (to Imperial Chemical Industries Ltd.).
9. Brit. Pat. 808,213 (Nov. 30, 1955), R. H. Speakman and R. B. Macleod (to Imperial Chemical Industries Ltd.).
10. P. V. Papero, E. Kubu, and L. Roldan, *Text. Res. J.* **37,** 823 (1967).
11. U.S. Pat. 3,582,448 (Apr. 23, 1968), T. Okuhashi and K. Kumura (to Teijin Ltd.).
12. U.S. Pat. 4,325,857 (May 13, 1980), N. Jivanial and co-workers (to E. I. du Pont de Nemours & Co., Inc.).
13. Brit. Pat. 2,009,032 (Nov. 25, 1977), J. Venot and A. Mottet (to ASA S.A.).
14. Brit. Pat. 2,010,739 (Dec. 22, 1977), P. Chion and co-workers (to Rhone-Poulenc-Textile).
15. U.S. Pat. 3,531,368 (Jan. 7, 1966), M. Okamoto and co-workers (to Toyo Rayon Kabushiki Kaisha).
16. Brit. Pat. 1,514,430 (July 14, 1975), J. Nakagawa, T. Agaki, and K. Hiramatsu (to Kuraray Co., Ltd.).
17. Brit. Pat. 1,045,047 (Feb. 20, 1963), (to Mitsubishi Rayon Co., Ltd.).
18. U.S. Pat. 3,641,232 (Aug. 28, 1967), W. J. Fontijn and K. J. M. van Drunen (to American Enka Corp.).
19. Brit. Pat. 1,100,430 (Dec. 16, 1965), W. G. Parr and A. W. D. Hudgell (to Imperial Chemical Industries Ltd.).
20. Brit. Pat. 2,045,153 (Mar. 27, 1979), J. E. Bromley and J. P. Yu (to Monsanto Co.).
21. Brit. Pat. 1,004,251 (June 10, 1963), (to American Cyanamid Co.).
22. U.S. Pat. 3,038,237 (Nov. 3, 1958), R. B. Taylor, Jr. (to E. I. du Pont de Nemours & Co., Inc.).
23. Brit. Pat. 954,272 (July 14, 1961), (to Japan Exlan Co. Ltd.).
24. Eur. Pat. 0 001 880 (Sept. 16, 1977), M. E. Mirhej (to E. I. du Pont de Nemours & Co., Inc.).
25. Brit. Pat. 1,048,370 (Sept. 26, 1964), (to Kanegafuchi Boseki Kabushiki Kaisha).
26. Brit. Pat. 805,033 (Feb. 26, 1954), (to E. I. du Pont de Nemours & Co., Inc.).
27. Brit. Pat. 1,300,268 (June 4, 1970), (to Toray Industries, Inc.).
28. Brit. Pat. 2,057,344 (Sept. 7, 1979), M. Okamoto and M. Asada (to Toray Industries, Inc.).
29. Brit. Pat. 1,258,760 (Dec. 21, 1967), (to Kanegafuchi Boseki Kabushiki Kaisha).
30. Brit. Pat. 1,302,584 (June 23, 1970), M. Okamoto and co-workers (to Toray Industries, Inc.).
31. Brit. Pat. 1,502,360 (Dec. 12, 1974), (to Teijin Ltd.).
32. Brit. Pat. 1,546,021 (Feb. 27, 1976), G. Barbe and R. Habault (to Rhone-Poulenc-Textile).
33. Brit. Pat. 1,095,166 (Nov. 18, 1964), A. W. D. Hudgell (to Imperial Chemical Industries Ltd.).
34. Brit. Pat. 1,120,241 (July 27, 1965), M. Matsui (to Kanegafuchi Boseki Kabushiki Kaisha).
35. Brit. Pat. 1,066,418 (Nov. 27, 1964), (to E. I. du Pont de Nemours & Co., Inc.).
36. Brit. Pat. 1,565,155 (Nov. 25, 1977), (to Monsanto Co.).
37. Eur. Pat. 0 011 954 (Nov. 30, 1978), P. C. Parkin (to Imperial Chemical Industries Ltd.).
38. Brit. Pat. 2,024,096 (June 19, 1978), J. Bohler (to Akzo, N.V.).
39. *Jpn. Text. News* (327), 56 (Feb. 1982).
40. U.S. Pat. 3,038,236 (Nov. 3, 1958), A. L. Breen (to E. I. du Pont de Nemours & Co., Inc.).
41. U.S. Pat. 3,038,238 (Nov. 20, 1958), T. C. Wu (to E. I. du Pont de Nemours & Co., Inc.).
42. U.S. Pat. 3,038,239 (Mar. 16, 1959), G. M. Moulds (to E. I. du Pont de Nemours & Co., Inc.).
43. U.S. Pat. 3,038,240 (Feb. 2, 1960), F. J. Kovarik (to E. I. du Pont de Nemours & Co., Inc.).
44. U.S. Pat. 3,039,524 (Nov. 3, 1958), L. H. Belck and K. G. Siedschlag, Jr. (to E. I. du Pont de Nemours & Co., Inc.).
45. U.S. Pat. 3,092,892 (Apr. 10, 1961), J. F. Ryan, Jr. and R. L. Tichenor (to E. I. du Pont de Nemours & Co., Inc.).
46. U.S. Pat. 3,470,060 (Feb. 2, 1966), J. Szitz, A. Nogaj, and H. Marzolph (to Farbenfabriken Bayer Aktiengesellschaft).

47. U.S. Pat. 3,473,998 (Aug. 7, 1963), D. R. Spriestersbach and co-workers (to E. I. du Pont de Nemours & Co., Inc.).
48. Brit. Pat. 1,478,101 (July 31, 1974), W. B. Seagraves and K. L. Mullholland (to E. I. du Pont de Nemours & Co., Inc.).
49. Brit. Pat. 2,007,588 (Nov. 10, 1977), C-M. Cerutti and co-workers (to Rhone-Poulenc-Textile).
50. Brit. Pat. 1,518,500 (Nov. 26, 1974), C. M. Bowes (to Courtaulds Ltd.).
51. Brit. Pat. 1,584,313 (Oct. 6, 1976), (to Toray Industries, Inc.).
52. Brit. Pat. 2,015,421 (Dec. 23, 1978), K. Gerlach, N. Mathes, and F. Wechs (to Akzo N.V.).
53. Brit. Pat. 2,043,731 (Mar. 2, 1979), H. W. Bruecher, K-H. Hense, and R. Modtler (to Akzo N.V.).
54. Brit. Pat. 2,062,537 (Nov. 9, 1979), M. Okamoto (to Toray Industries, Inc.).
55. Eur. Pat. 0 44 221 (July 14, 1980), J. T. Summers (to E. I. du Pont de Nemours & Co., Inc.).
56. U.S. Pat. 4,122,658 (May 10, 1977), A. Morioka and T. Nakashima (to Toray Industries, Inc.).
57. Jpn. Pat. 55 116,873 (Feb. 26, 1979), (to Kanebo Kabushiki Kaisha).
58. Eur. Pat. 0 045 611 (Aug. 4, 1980), (to Toray Industries, Inc.).
59. Jpn. Pat. 56 107,080 (Jan. 24, 1980), (to Toray Industries, Inc.).
60. *Jpn. Text. News* (315), 18 (Feb. 1981).
61. Brit. Pat. 1,321,852 (July 1, 1969), (to Toray Industries, Inc.).
62. Brit. Pat. 1,496,369 (Dec. 25, 1974), (to Kabushiki Kaisha Kuraray).
63. Brit. Pat. 1,514,553 (Sept. 13, 1974), T. Kusunose and co-workers (to Asahi Kasei Kogyo Kabushiki Kaisha).
64. Brit. Pat. 1,541,774 (July 12, 1976), (to Mitsubishi Rayon Company Ltd.).
65. Brit. Pat. 2,004,496 (Sept. 6, 1977), K. Ozaki, M. Matsui, and N. Minemura (to Teijin Ltd.).
66. K. Porter, *Melded Fabrics—A Versatile And Unique New Family of Materials*, to be published, *Materials in Engineering*, Scientific and Technical Press Ltd., UK, 1982.
67. Brit. Pat. 2,036,115 (Nov. 15, 1978), T. Sugihara and H. Sonoda (to Chisso Corporation).
68. Brit. Pat. 1,495,546 (Mar. 21, 1974), P. L. Carr and P. M. Ellis (to Imperial Chemical Industries Ltd.).
69. Brit. Pat. 1,417,394 (Feb. 3, 1972), V. S. Ellis and K. W. Mieszkis (to Imperial Chemical Industries Ltd.).
70. Brit. Pat. 1,488,682 (Mar. 11, 1974), B. Walker and D. B. Chambers (to Imperial Chemical Industries Ltd.).
71. Brit. Pat. 1,502,719 (May 27, 1975), (to Monsanto Co.).
72. Brit. Pat. 2,001,901 (Aug. 8, 1977), T. Naruse and co-workers (to Kanebo Kabushiki Kaisha).
73. Brit. Pat. 1,393,234 (July 21, 1972), D. R. Hull (to E. I. du Pont de Nemours & Co., Inc.).
74. Brit. Pat. 1,527,192 (Jan. 12, 1976), (to Fiber Industries, Inc.).
75. Brit. Pat. 1,585,575 (Apr. 29, 1977), G. A. Paton, S. M. Nichols, and J. H. Sanders (to Dow Badische Co.).
76. Brit. Pat. 2,003,004 (July 7, 1977), M. Wandel and co-workers (to Bayer Aktiengesellschaft).
77. Brit. Pat. 2,036,638 (Nov. 23, 1978), S. J. Van Der Meer, W. Peschke, and D. Schilo (to Akzo N.V.).
78. U.S. Pat. 3,582,445 (Nov. 18, 1967), T. Okuhashi (to Teijin Ltd.).
79. U.S. Pat. 3,586,597 (Nov. 20, 1967), T. Okuhashi (to Teijin Ltd.).
80. U.S. Pat. 3,590,570 (Jan. 2, 1969), T. Okuhashi and K. Komura (to Teijin Ltd.).
81. Ger. Pat. 3,022,325 (June 15, 1979), (to Teijin Kabushiki Kaisha).
82. Jpn. Pat. 56 144,218 (Apr. 8, 1980), (to Kureha Chemical Industries Kabushiki Kaisha).
83. Jpn. Pat. 55 158,331 (May 25, 1979), (to Toray Industries, Inc.).
84. Jpn. Pat. 50 118,018 (Mar. 12, 1974), (to Toray Industries, Inc.).
85. Brit. Pat. 1,574,220 (Apr. 7, 1976), S. Yoshikawa, T. Sasaki, and H. Endo (to Kureha Kagaku Kabushiki Kaisha).
86. Eur. Pat. 0 018 523 (Apr. 24, 1979), (to Teijin Ltd.).
87. Brit. Pat. 2,077,310 (May 23, 1980), I. S. Fisher and S. R. Munks (to Imperial Chemical Industries Ltd.).
88. Brit. Pat. 1,601,585 (June 30, 1977), A. L. Heard, P. D. Randall, and D. J. Waters (to Courtaulds Ltd.).
89. Jpn. Pat. 55 132,716 (Apr. 2, 1979), (to Toray Industries, Inc.).
90. Jpn. Pat. 57 047,915 (Aug. 30, 1980), (to Asahi Chemical Industries Kabushiki Kaisha).
91. Jpn. Pat. 56 144,219 (Apr. 7, 1980), (to Kuraray Kabushiki Kaisha).
92. Jpn. Pat. 56 085,474 (Dec. 5, 1979), (to Toray Industries, Inc.).
93. Jpn. Pat. 56 031,015 (Aug. 14, 1979), (to Toray Industries, Inc.).

94. Jpn. Pat. 55 062,210 (Oct. 31, 1978), (to Teijin Kabushiki Kaisha).
95. U.S. Pat. 4,302,509 (June 6, 1980), M. J. Coplan and G. Lopatin (to Albany International Corp.).
96. Eur. Pat. 0 047 797 (Sept. 15, 1980), (to Carl Freudenberg F.A.).
97. R. L. Dunn and co-workers, *Controlled Release Pestic. Pharm.*, *Proceedings of International Symposium*, Plenum, New York, 1980, pp. 125–146.
98. Jpn. Pat. 55 093,813 (Jan. 1, 1979), (to Unitika Kabushiki Kaisha).
99. Jpn. Pat. 55 148,214 (May 1, 1979), (to Asahi Chemical Industries Kabushiki Kaisha).
100. Jpn. Pat. 56 015,416 (July 12, 1979), (to Unitika Kabushiki Kaisha).
101. Ger. Pat. 2,713,435 (Apr. 7, 1976), (to Kureha Kagaku Kogyo).

K. PORTER
ICI Fibres

FIBERS, VEGETABLE

Natural fibers of vegetable origin are constituted of cellulose (qv), a polymeric substance made from glucose molecules, bound to lignin (qv) and associated with varying amounts of other natural materials. A small number of a vast array of these vegetable fibers have industrial importance for use in textiles (qv), cordage, brushes and mats, and paper (qv) products. Cotton fiber makes up about three-quarters of the world vegetable-fiber tonnage (see Cotton).

Vegetable fibers are conveniently classified according to the part of the plant where they occur and from which they are extracted as shown in Table 1 (1–2): (*1*) leaf fibers are obtained from the leaves of monocotyledonous plants, mostly tropical, which are part of their fibro-vascular systems. Commercially important examples are abaca (Manila hemp), sisal, and henequen. These are the hard fibers used for cordage. The long, multicelled fibers do not readily split apart and are wire-like in texture. (*2*) Bast fibers are obtained from the bast tissue or bark of the plant stem. The group includes flax, hemp, jute, and ramie which are the so-called soft fibers that are converted into textiles, thread, yarn, and twine. The long, multicelled fibers in this case can be readily split into finer cells which are manufactured into textile and coarse yarns. (*3*) Seed-hair fibers, eg, cotton, kapok, and the flosses, are obtained from seeds, seedpods, and the inner walls of the fruit. These fibers are short and single-celled. (*4*) Miscellaneous

Table 1. Selected Vegetable Fibers of Commercial Interest [a]

Commercial name	Botanical name	Geographical source	Use
Leaf (hard) fibers			
abaca	*Musa textilis*	Borneo, Philippines, Sumatra	cordage
cantala	*Agave cantala*	Philippines, Indonesia	cordage
caroa	*Neoglaziovia variegata*	Brazil	cordage, coarse textiles
henequen	*Agave fourcroydes*	Australia, Cuba, Mexico	cordage, coarse textiles
istle (generic)	*Agave* (various species)	Mexico	cordage, coarse textiles
Mauritius	*Furcraea gigantea*	Brazil, Mauritius, Venezuela, tropics	cordage, coarse textiles
phormium	*Phormium tenax*	Argentina, Chile, New Zealand	cordage
bowstring hemp	*Sansevieria* (entire genus)	Africa, Asia, South America	cordage
sisal	*Agave sisalana*	Haiti, Java, Mexico, South Africa	cordage
Bast (soft) fibers			
China jute	*Abutilon theophrasti*	China	cordage, coarse textiles
flax	*Linum usitatissimum*	north and south temperate zones	textiles, threads
hemp	*Cannabis sativa*	all temperate zones	cordage, oakum
jute	*Corchorus capsularis; C. olitorius*	India	cordage, coarse textiles
kenaf	*Hibiscus cannabinus*	India, Iran, USSR, South America	coarse textiles
ramie	*Boehmeira nivea*	China, Japan, United States	textiles
roselle	*Hibiscus sabdarifa*	Brazil, Indonesia (Java)	cordage, coarse textiles
sunn	*Crotalaria juncea*	India	cordage, coarse textiles
cadillo	*Urena lobata*	Zaire, Brazil	cordage, coarse textiles
Seed-hair fibers			
cotton	*Gossypium* sp.	United States, Asia, Africa	all grades of textiles, cordage
kapok	*Ceiba pentranda*	tropics	stuffing
Miscellaneous fibers			
broom root (roots)	*Muhlenbergia macroura*	Mexico	brooms, brushes
coir (coconut husk fiber)	*Cocos nucifera*	tropics	cordage, brushes
crin vegetal (palm leaf segments)	*Chamaerops humilis*	North Africa	stuffing
piassava (palm leaf base fiber)	*Attalea funifera*	Brazil	cordage, brushes

[a] Ref. 2.

fibers are obtained from the sheathing leaf-stalks of palms, leaves, stem segments, stems, and fibrous husks, eg, piassava and coir. These strawlike, woody, and coarse fibers are used for brush and broom bristles, matting, and stuffing.

Other classifications are applied according to different points of view such as commercial (hard and soft fibers), practical (textile, cordage, brushes and mats, stuffing and upholstery, papermaking and baskets, etc) (3), bast fiber crops for fine textiles, packaging, and soft cordage (4), morphological (hairs, bast, leaves, woody, and others) (3), and crop cultivation (cotton, jute, sisal, flax and hemp; and other fiber crops) (5) (see also Pulp; Bagasse; Sugar; Wood). Leaf fibers are also classed as structural fibers (6).

Cordage and fabrics were important to ancient man for fishing and trapping and applying motive power (7–8). The making of ropes and cords started in the paleolithic age (ca 20,000 BC); a mesolithic cave drawing (Spain) shows ropes partway down a cliff to recover honey (9). Evidence of predynastic Egyptian (ca 4000 BC) use of ropes and cords from reeds, grasses, and flax has been found (10). The vegetable fibers mostly used for rope were date palm fibers and papyrus. Matting was another important use of vegetable fibers in ancient Egypt, rushes and reeds, as well as papyrus grasses bound with flax string were used. Brushes from vegetable fibers were also used in ancient Egypt.

Flax, jute, ramie and sedges, rushes, and grasses have a long history of use in textiles and basketing.

Properties

Chemical Composition. Chemical analyses or compositions of various vegetable fibers are given in Tables 2 and 3. Cotton is at the high end of the range in cellulose content with >90%; jute and kenaf (bast fibers) and abaca (leaf fiber) have a relatively low cellulose content of 63–65%. The chemical compositions vary greatly between plants and within specific fibers depending on genetic characteristics, the part of the plant used and growth, harvesting, and preparation conditions.

Fiber Dimensions. Leaf fibers are multicelled, and not readily split into the component cells, whereas the bast fibers are readily broken down, permitting spinning. Strand and cell dimensions are given in Table 4. The cells themselves are composed of microfibrils which are comprised of groups of parallel cellulose chains.

Physical Properties. Mechanical properties of certain vegetable fibers are shown in Tables 5 and 6. The vegetable fibers are stronger but less extensible than cotton, ie, they have a higher breaking length and elasticity modulus with a lower strain (extensibility) and work modulus. The vegetable fibers approach glass in stiffness (resistance to deformation) and are considerably stiffer than man-made fibers, but have lower toughness (ability to absorb work) (see Fibers, chemical) (2). Kapok and other seed-hair fibers are relatively low in strength but have great buoyancy.

Vegetable fibers have spiral molecules that are highly parallel to one another. The spiral angles are low in flax, ramie, hemp, and other bast fibers, considerably less than cotton, accounting for the former fibers' low extensibility (12).

Table 2. Chemical Analysis of Vegetable Fibers[a], wt %

Fiber	Cellulose[b]	Moisture	Ash	Lignin[c] and pectins	Extractives
Leaf fibers					
abaca	63.72	11.83	1.02	21.83	1.6
bowstring hemp	69.7	9.7	0.7	13.7	6.2
caroa	60	10		12	18
cebu maguey (cantala)	75.8	5.5	1.4	14.1	3.2
henequen	77.6	4.6	1.1	13.1	3.6
phormium	63	11.61	0.63	23.07	1.69
piteira (Mauritius)	75.6	5.2	1.6	17.2	0.4
sisal	77.2	6.2	1.0	14.5	1.1
tula istle	73.48	5.6	1.65	17.37	1.9
Bast fibers					
Congo jute (cadillo)	75.3	7.7	1.8	13.5	1.4
hemp	77.07	8.76	0.82	9.31	4.04
jute	63.24	9.93	0.68	24.41	1.42
kenaf	65.7	9.8	1.0	21.6	1.9
ramie	91			0.65	
sunn	80.4	9.6	0.6	6.4	3.0
Seed and hair fibers					
cotton	90	8.0	1.0	0.5	0.5

[a] Ref. 2.
[b] See Cellulose.
[c] See Lignin.

Table 3. Chemical Composition of Various Vegetable Fibers[a], wt %

Type of fiber	Cellulose	Hemi-cellulose	Pectins	Lignin	Water-soluble compounds	Fats[b] and waxes[c]
cotton	91.8	6.4			1.1	0.7
flax (retted)	71.2	18.6	2.0	2.2	4.3	1.7
flax (nonretted)	62.8	17.1	4.2	2.8	11.6	1.5
Italian hemp	74.4	17.9	0.9	3.7	2.3	0.8
jute	71.5	13.4	0.2	13.1	1.2	0.6
Manila	70.2	21.8	0.6	5.7	1.5	0.2
New Zealand cotton	50.1	33.4	0.8	12.4	2.4	0.1
ramie	76.2	14.6	2.1	0.7	6.1	0.3
sisal	73.1	13.3	0.9	11.0	1.4	0.3

[a] Ref. 11.
[b] See Fats and fatty oils.
[c] See Waxes.

Plant Descriptions, Preparations, and Grades of Fibers

The vegetable fibers are grouped below partly according to their plant source and partly based on the importance of use within source groups (3–5,13).

Table 4.   Dimensions of Fibers and Cells of Vegetable Fibers[a]

Fiber	Cells or ultimate fibers						Fibers or strands	
	Length, mm			Diameter, mm			Length, cm Range	Width, mm Range
	Min	Max	Av	Min	Max	Av		
cotton	10	50	25	0.014	0.021	0.019	1.5–5.6	0.012–0.025
flax	8	69	32	0.008	0.031	0.019	20–140	0.04–0.62
hemp	5	55	25	0.013	0.041	0.025	100–300	
ramie	60	250	120	0.017	0.064	0.040	10–180	0.06–9.04
sunn	2	11	7	0.013	0.061	0.031		
jute	0.75	6	2.5	0.005	0.025	0.018	150–360	0.03–0.14
sisal	0.8	7.5	3	0.007	0.047	0.021	75–120	0.01–0.28
Manila (abaca)	2	12	6	0.010	0.032	0.024	180–340	0.01–0.28
phormium	2	11	6	0.005	0.025	0.013		
Mauritius	2	6		0.015	0.024			
kapok	15	30	19	0.010	0.030	0.018		
coir	0.3	1	0.7	0.010	0.024	0.020		
kenaf	2	11	3.3	0.013	0.034	0.023		
pineapple fiber	2	10	5.5	0.003	0.013	0.006		
nettle	4	70	38	0.020	0.070	0.042		
sansevieria	1	7	4	0.013	0.040	0.022		

[a] Ref 2.

Table 5.   Mechanical Properties of Vegetable Fibers[a]

Fiber	Breaking length, km[b]	Ultimate strain at break, %[c]	Work of rupture modulus, kN·m/N	Initial linear elasticity modulus, N/tex[d]
kapok	16–30	1.2	0.1	1300
jute	27–53	1.5	0.3	1700–1800
hemp	38–62	2–4	0.6–0.9	180
flax	24–70	2–3	0.9	1800–2000
ramie	32–67	2–7	1.1	1400–1600
henequen	27–34	3.5–5		
manila hemp	32–69	2–4.5	0.6	
sisal	30–45	2–3	0.7–0.8	2500–2600
coconut	18	16	1.6	430

[a] Ref. 11.

[b] Tensile strength is reported on the basis of breaking length which measures strength per unit area.

[c] Elongation (stretch) at rupture.

[d] To convert N/tex to g/den, multiply by 11.33.

LEAF (HARD) FIBERS

Abaca.   This fiber is extracted from the plant *Musa textilis*, native to the Republic of the Philippines. The fiber is also known as Manila fiber or Manila hemp. When mature, the individual abaca plant consists of 12 to 30 stalks radiating from a central root system. In appearance it is very similar to the banana plant (*Musa sapientum*). Each of the stalks is 2.7–6.7-m tall with a trunk 10–20-cm wide at the base. The sheaths forming the stalk expand into the overhanging leaf structure. The sheaths

Table 6.   Coarseness and Breaking Length of Leaf Fibers[a]

Fiber	Coarseness, km/kg	Breaking length, km
Manila hemp		
mean	32	41.5
sisal	40	35.0
cantala	58	20.5
henequen	32	20.0
New Zeland flax	38	26.0
istle	34	27.0

[a] Ref 11.

before expanding are 2–4-m long, 13–20-wide, and ca 10-mm thick at the center. The fibers run lengthwise in the sheaths. The sheaths vary in length and width, and also in color. The mature stalks are cut off at the roots and at a point just below where they begin to expand. In the Philippines, the stripping method is generally used, ie, the fiber layer from the cut sheath is separated by inserting a knife just under the layer and a fiber strip or tuxy, 5–7-cm wide, is pulled off the sheath. Modern plantations have replaced this hand-pulling operation with power-driven pulling machines to accelerate production. After stripping, the fiber is air-dried; rain and sunlight need to be avoided to preserve color and luster.

A small proportion of Philippine abaca fiber is produced by mechanical decortication. The sheaths are fed into a decorticating machine in which the pulpy material is scraped from the fiber, then the fiber is washed and dried. The abaca fiber so produced is lower in quality than hand-cleaned fiber, ie, it has less sheen and is harsher in texture. Abaca fiber grown in Sumatra (Indonesia) and Central America is machine-decorticated exclusively.

The Philippines is the main source of abaca, contributing ca 90% of the total supply. The remainder is produced in Saba (North Borneo), Indonesia, and Central America. Philippine abaca is carefully graded by the government of the Philippines. The long cordage fiber is classified into four groups, determined by the fineness of the fiber and degree of separation of the individual fibers. Each group is further subdivided according to the range of color and associated fiber length. There are 18 grades, designated AB, CD, E, F, S2, S3, I, J1, G, H, H2, K, L1, L2, M1, M2, DL, and DM. These grades range in color from almost white to dark brown. Philippine abaca is longer and stronger than the corresponding grades produced elsewhere, and is generally lighter in color but bolder in texture.

Abaca is produced in two grades in Central America (introduced during World War II): clear and streaky. In Indonesia it is supplied in superior, good, fair, fair X, and B grades.

Sisal. This fiber is extracted from the leaves of the plant *Agave sisalana* which is widely cultivated in the Western Hemisphere, Africa, Asia, and Oceania. Other *Agave* sp. that furnish commercial vegetable fibers are henequen, cantala and istle (described below). However, only *A. sisalana* provides true sisal, which is further identified as African, Indonesian, Brazilian, and Bahama sisal.

The agaves have rosettes of fleshy leaves, usually long and narrow, which grow out from a central bud. As the leaves mature they gradually spread out horizontally and are 1–2-m long, 10–15-cm wide, and ca 6 mm thick at the center. The fibers are embedded longitudinally in the leaves, and are most abundant near the leaf surfaces. The leaves contain ca 90% moisture, but the fleshy pulp is very firm and the leaves are rigid. The fiber is removed when the leaves are cut because dry fibers adhere to the pulp. The fiber is removed by scraping away the pulpy material, generally by a mechanical decortication process.

In decortication, the leaves are fed through sets of fluted crushing rollers. The crushed leaves are held firmly at their centers and both ends are passed between pairs of metal drums on which blades are mounted to scrape away the pulp, and the centers are scraped in the same way. The fiber strands are washed and either air- or artificially dried. In Indonesia and Africa, for choice grades, the dried fiber is held against a revolving metal drum to remove remnants of dry adhering pulp.

Sisal fiber is graded according to the country and district of growth and further subgraded according to color, cleanliness, and length. There are eight Indonesian grades, designated, A, B, C, X, Y, X, D, and L; seven East African grades designated 1, 2, 3 long, 3 short, A, UG, and SCWF; five Haitian grades, designated A, B, X, Y, and S; eight West African grades, designated Extra, 1, 2, 2SL, 3 3L, A, and R; three Philippine grades, designated SR-1, SR-2, SR-3; and five grades of Brazilian sisal, designated Tipo 1, Tipo 3, Tipo 5, Tipo 7, and Tipo 9.

Henequen.   This fiber is extracted from *Agave fourcryodes* and is also called Mexican henequen from its source, the state of Yucatan, Mexico. There are also El Salvador and Cuban henequens from *A. letonae*. These plants are harvested, decorticated, cleaned, and prepared for marketing by procedures similar to those used for sisal.

Mexican henequen is classified into seven grades according to color, cleanliness, and length. The grades are AA, A, B, B1, M, C, and M1.

Cantala.   This variety of agave fiber is derived from *Agave cantala* and is also called Manila maguey from its Philippines source (also Indonesia). The fiber is prepared in the Philippines by retting the leaves in sea water followed by cleaning by hand or by decorticating as described for sisal. The Indonesian industry employs mechanical decortication without the retting step.

Catalpa is graded according to the method of decortication, retting, or machine decortication, into grades 1, 2, and 3.

Istle.   This is the term for short, coarse fibers from the leaves of agaves and related plants growing wild in central and northern Mexico. There are three commercial varieties: tula istle from *A. lophanfu*, jaumave istle from *A. funhana*, and palma istle from *Samuela carnerosana*. The agave istle resembles small plants, whereas the palma istle looks like a small palm tree with leaves radiating from the top of the plant.

The istle fibers are extracted from the leaves by hand-scraping with subsequent sun-drying. The palma leaves are very gummy and are steamed before scraping.

Mauritius.   This fiber is obtained from the giant cabuya plant, *Furcraea gigantea*, native to Brazil where it is called piteira or pita. It is cultivated commercially on the island of Mauritius and is known as Mauritius hemp or aloe. The cabuya plant resembles the agaves except that the leaves are heavier and larger.

The Mauritius fiber is extracted from the leaves by simple mechanical decortication by which the crushed leaves are scraped free from pulpy material, washed, steeped in a soapy solution and sun-dried.

There are five standard grades of Mauritius fiber: extra prime, prime, very good, good, and fair. Brazilian pita is graded according to standards for Brazilian sisal.

Phormium. This fiber is extracted from the plant *Phormium tenax*, native to New Zealand. The fiber is commonly called New Zealand flax or hemp, though it has no bast fiber characteristics. The plant is a perennial with a fan-shaped cluster of leaves 1.6–4.3-m long, and 6–10-cm wide. The leaves are green with a red midrib and red margins. The fibers are obtained by mechanical decortication of the cut leaves in a similar manner to sisal processing and then washed and sun-dried.

The phormium fibers are graded by New Zealand hard fiber standards according to color, strength and cleanliness and then subgraded into six classes: A, superior or superfine; B, fine; C, good; D, fair high-point, fair low-point; E, common; and F, rejections.

Sansevieria. This fiber is extracted from the perennial plant *Sansevieria* native to Africa, Arabia, India, and Ceylon. The fiber is also called bowstring hemp because of its primitive bowstring application. The plant grows wild or is cultivated in many countries. It is propagated with cultivated leaf cuttings. Extraction by mechanical decortication is similar to sisal processing.

Caroa. This leaf fiber is obtained from the plant *Neoglaziovia variegata*, which belongs to the pineapple family and is native to eastern and northern Brazil. The leaves for fiber production are collected from plants growing wild and are sword-shaped, 1–1.3 m long, and 2.5–5-cm wide. The fiber is extracted by hand-scraping after beating the leaves to break up the pulpy tissue, or after a retting process that partially ferments and softens the leaves. Caroa is graded by standards similar to those specified for Brazilian sisal. Its color and texture resemble sisal although it is apt to be incompletely cleaned.

Piassava. This fiber is extracted from *Attalea funifera*, a palm indigenous to Bahia, Brazil, where it grows wild and under cultivation. Bahia piassava and bass fiber are other terms. The plant is a tall feather palm with erect leaves 12 m in length. The fiber is taken from the sheathing leaf vases. The leaf sheaths generally need to be treated in water to free the fibers from the pulp by scraping. The mature lower leaves also may be cut at the base, the leaves crushed, and the fibers combed out with a knife. The fibers are sometimes hand-pulled from the leaf stem. The piassava fiber is classified into premium and second grades.

Broomroot. This fiber is extracted from the roots of the bunchgrass, *Muhlenbergia macroura*, a plant growing wild from Texas to Central America but produced commercially in Mexico. The plant grows 1–2-m tall, although the fiber is obtained from the roots by beating or rubbing the chopped roots to remove the bark-like covering. The fibers are bleached with sulfur fumes at the processing factory and graded into four qualities.

BAST (SOFT) FIBERS

Fine Textile Fibers

Flax. This fiber is extracted from the plant *Linum usitatissimum* L. which is grown chiefly in the USSR, Poland, France, Rumania, Czechoslovakia, Belgium, and Ireland. The plant is cultivated mostly for its oil-bearing seed (linseed), although it is also an important source of a vegetable fiber (see Vegetable oils). Flax is an annual plant with a slender, greyish-green stem growing to a height of 90–120 cm.

Several methods of harvesting are used and the correct time for harvesting for highest fiber quality and seed production is important. The plants are pulled by hand or machine for highest yield and quality, although mowing is practiced for some grades. After deseeding, the straw is retted, ie, the fibers are liberated through enzymatic action on pectinous binding material in the stem; dew- or water-retting or variations are employed. In the commonly used dew-retting, the straw is spread thinly on the ground and subjected to atmospheric precipitation with turning at intervals. Fungi enter the stem to cause the retting action. Water-retting is performed on bundles of dried straw which are immersed in rivers, pools, or ditches. A modern variation is a controlled warm-water procedure and, more recently, aerated retting is being used for reducing pollution of the effluents, water usage, and odors. The retted straw is dried before the scutching or fiber separation and cleaning step. This breaking of the straw was first done by hand-beating, and eventually in turbine-type machines consisting of multiple pairs of corrugated roll breakers and scutching wheels for cleansing and polishing. The fiber is then hackled for alignment and final cleaning. The short fibers from these operations (tow) and the longer fibers (line) are used for linen.

The qualities considered in grading flax are fineness, softness, strength, density, color, uniformity, luster, length, handle, and cleanliness. The line fiber from sources in different countries varies in grades and buyers have their own grading systems.

Hemp. This fiber is extracted from the annual plant *Cannabis sativa* which originated in Central Asia. The plant grows readily in temperate and tropical climates, but its commercial production for fiber is chiefly in the People's Republic of China and eastern Europe.

The hemp plant is grown for fiber from the stem, for oil from the seeds or for drugs from the flowers or leaves. Marijuana (see Hypnotics, sedatives, and anticonvulsants) is the narcotic alkaloid (cannabin) derived from a related hemp plant, which is not a commercial fiber-producing plant. The Marijuana Tax Act of 1937, nevertheless, includes *C. sativa*.

When mature, the stalks are 5–7-m tall and 6–16-mm thick. They are generally smooth and without branches or foliage except at the top. The stem structure is hollow. Thin-walled tissue adjoins the hollow center and outside of this is a layer of woody substance. The next layer consists of gummy tissue that cements the fiber layer to the woody layer. The fiber or bast layer is enveloped by a thin bark that constitutes the outside of the stalk.

The mature hemp stalk is harvested at the proper time to ensure highest quality and yield by hand-cutting and spreading or by a harvester-spreader for dew-retting. When water-retting is used, bundles of straw are dried in the field. Subsequently the leaves and seeds are removed by beating the tops on the ground, and they are then stacked before the water-retting. Water-retting is done in Italy, Spain, Hungary, Poland, and to a small extent in the USSR. Dew-retting is used mostly in the USSR. The retted and dried straw is further treated by either hand-breaking and hackling to remove the woody stem portion (hurd) or by mechanical breaking followed by turbine scutching. A wooden breaking device augments the hand-breaking.

The grading is by color, luster, density, spinning quality, cleanliness, and strength. Quality varies with country and regions within countries. The national classifications may include a number of grades and subgrades.

Ramie. This fiber is produced from the stems of *Boehmeria nivea*, a nettle native to central and western China but growing in regions varying from temperate to tropical, including the People's Republic of China (largest grower), the Philippines, Brazil,

Taiwan, Japan, and others. Another name for the raw fiber is China grass. The plant grows 1–2-m high or higher with stems 8–16 mm in diameter. It has perennial roots and yields three crops annually.

The ramie fiber is contained in the bark and is usually extracted by hand stripping and scraping. The plant is cut green and defoliated manually and bast ribbons are stripped from the woody stem. These are scraped in the field or stored in water for scraping in a central location. In Brazil, the Republic of China (Taiwan), and other countries the ramie ribbons are decorticated mechanically in small operations. Sometimes the ribbons are bleached by sulfur fumes.

Ramie fibers are graded according to their length, color, and cleanliness. Each country has its own grading system.

The raw ramie fibers contain 25–35% plant gums (xylan and L-arabinan) and small quantities of parenchyma cells which must be removed before the fibers can be spun. Degumming by boiling in aqueous alkaline solutions is sometimes done under pressure and at times with additives, followed by washing, bleaching with an oxidizing agent (eg, a hypochlorite), washing, neutralizing, and oiling to facilitate combing.

Packaging Fibers

Jute. This fiber is obtained from two species of the annual, herbaceous plant *Corchorus, C. capsularis* of Indo-Burma origin and *C. olitorius* from Africa. The major jute production area is India and Bangladesh and secondarily in other Asian countries and Brazil. The two sources of jute are differentiated by their seeds and seed pods. The *C. capsularis* is round-podded and is called white jute; the *C. olitorius* is long-podded and is known as tossa or daisee. The jute plants of both species have cylindrical stalks 2–4-m tall and 10–20 mm in diameter.

The plants are harvested by hand at an early seed stage using knives and the stems are left on the ground several days to promote defoliation. The defoliated stems are bundled and taken for wet-retting in canals, ditches, or ponds (slowly running water to carry off colored and acidic products) for periods of 10–20 d. The retting time for *C. insularis* is shorter than for *C. olitorius*. The retting can also be conducted on ribbons in concrete tanks. Stripping of the fibers follows immediately after stem-retting with sun-drying; the ribbon-retted material also is sun-dried.

The jute is made into crude bales for transport to a baling and grading center. Here, the fiber is designated as to species (white, tossa, or daisee), and source and grade according to the grades being used. The jute is then baled for shipment to domestic users or for export.

Kenaf and Roselle. These fibers are extracted from the stems of two closely related plants, *Hibiscus cannabinus* L. and *H. sabdariffa* L. var. *altissima*, known respectively as kenaf, Deccan hemp or mesta, and roselle. The two plants are native to Africa but are grown commercially in greatest quantity in the People's Republic of China, the USSR and Egypt for the kenaf, and India and Thailand for the roselle. The plants are herbaceous annuals growing in single stems to heights of 1–4 m for kenaf and 1–5 m for roselle with stem diameters of 10 to 20 mm. Kenaf is harvested as flowering begins whereas with roselle this takes place somewhat earlier. The kenaf can be hand-cut or mowed or pulled; the roselle may be hand-pulled. Both species may be field-dried for defoliation. Hand or mechanical ribboning followed by tank-retting is used on occasion, but the kenaf is often stem-retted followed by hand or mechanical stripping and washing. The washed fiber is finally dried on racks.

Kenaf is graded into three grades: A, good; B, medium; C, poor. The grading is based on color, uniformity, strength, and cleanliness. Roselle may be hackled to improve quality by removing dirt and increasing softness and luster. In Thailand roselle is classified into seven grades as to softness, color, and impurities.

Urena. This jutelike fiber is obtained from *Urena lobata* L., commonly called urena with many local names because of its widespread growth, although it is grown commercially mostly in Brazil and Zaire. This perennial shrub grows under cultivation as an annual with few branches and it grows to a height of 4–5 m with stems 12–18 mm in diameter. The stems are cut 20 cm above ground level because of the highly lignified base. The plants are defoliated green or after piling in the field; retting is similar to that for jute or kenaf and requires 8–10 d. The retted material is then stripped, washed, and sometimes hand-rubbed to remove impurities.

Urena fiber is officially graded in Brazil and Zaire into four or five grades according to color, uniformity of color, luster, strength, and cleanliness.

China Jute. This jute-type fiber is extracted from the stem of a mallow, *Abutilon theophrasti* Medic or *A. avecennae Gaetn.*, growing chiefly commercially in the People's Republic of China and the USSR, although also in Japan, Korea, and Argentina, and known as China jute or Tientsin jute, and by many other local names. The plant is a herbaceous annual growing to a height of 3–6 m with stems 10–18 mm in diameter. China jute is harvested by hand in the early flowering stage, the leaves are removed, and bundles are water-retted similar to jute operations. The harvesting, retting and fiber extraction may be mechanical in certain regions. In the hand operation the fiber is hackled, baled, and graded into two classes.

Soft Cordage Fiber

Sunn Hemp. This fiber is extracted from the stem of the legume, *Crotolaria juncea* L., indigenous to India and known as sunn hemp, brown hemp, Indian hemp, Benares hemp, and other national references. India is the largest producer followed by Bangladesh and Brazil. Sunn hemp is an erect herbaceous shrub growing to a height of 1–5 m with a thin, cylindrical, branch-free stem. The plant is harvested in the seed-pod stage by hand-cutting or pulling and left in the field for defoliation. The stems are water-retted, washed, stripped, and dried by methods similar to those for stem fibers described previously. For export, sunn-hemp fibers are hackled or dressed and graded by a system with many classifications according to length, strength, firmness, color, uniformity, and percentage of extraneous matter.

Seed and Fruit-Hair Fibers

Coir. This fiber is contained in the fruit or husk of the coconut palm *Cocos nucifera* L., which is positioned beneath the outer covering of the fruit and envelops the kernel or coconut. Sri Lanka(Ceylon) and India are important producers. The fruits are gathered just short of ripeness and the husks are broken by hand or by use of a bursting machine. The main supply of fiber is obtained as a residue for copra production. The extraction of the fiber involves retting at the edges of rivers and also in pits or, in modern operations, in concrete tanks. The retted husks are beaten with sticks to remove extraneous matter and the dried fiber is suitable for spinning. Rougher fibers require less retting and the fibers are extracted from the husks mechanically. The fibers may be washed before drying and may be hackled before direct use or baling for shipment. The fiber can also be removed from the husk by a decorticating machine in connection with copra production.

Kapok.   This fiber is obtained from the seed pod of a tree, *Ceiba pentranda*. The tree grows to a height of ca 35 m, and is indigenous to Africa and Southeast Asia; it is also known as the silk-cotton tree. The fiber is exported from Thailand, India, and Indonesia. The pods are picked short of opening, hulled or broken open, and the floss is dried in an area enclosed with cloth or wire netting to contain the floss. The seeds are removed by hand or separated mechanically and further cleaned by an air separation method. The fiber is finally baled for shipment. The seeds contain up to 25% of a nondrying oil resembling cottonseed oil in properties and uses.

Miscellaneous.   A number of seed-hair fibers have some minor uses, including East Indian balsa fiber (*Ochroma pyramdale*), Indian kumbi (*Cochlospermum gossypium*), American milkweed floss (*Asclepias* sp.) and cattail fiber (*Typha* sp.).

Production

The world production of jute and hard fibers is given in Table 7 for the period 1963 to 1976 (14).

The growth since 1970 has been at a zero or negative rate.

The distribution for the production of the major hard cordage fibers in 1976 is shown in Table 8.

U.S. imports of unmanufactured vegetable fibers have shown serious decreases in the period of 1963 to 1976 (Table 9).

The consumption of paper pulps made from vegetable fibers in the United States in 1972 is summarized in Table 10 with sources of pulp and types of paper products.

In addition, 216,000 metric tons of U.S. sugar cane bagasse pulp were consumed in making wall board and fine papers in 1972.

Vegetable Fiber Yields.   The yields of prepared vegetable fibers are generally low, 1–6%, as shown in Table 11. Linen flax fiber and kapok are exceptions.

Table 7.   World Production of Jute and Hard Fibers in Thousands of Metric Tons

	1963	1970	1976
jute	2421	2051	2034
abaca	119	93	67
sisal-henequen	826	763	628

Table 8.   Percentage Distribution of Production of Major Hard Cordage, 1976[a]

	Thousands of metric tons	%
abaca	84.0	13
sisal	421.3	66
henequen	130.0	21
Total	*635.3*	*100*

[a] Ref. 15.

Table 9.   U.S. Imports of Unmanufactured Vegetable Fibers [a]

	1963	1970	1975
	Metric tons		
flax (Belgium, Luxembourg)	2,246	1,581	411
hemp (Canada)	35	0	0
jute and jute butts (Thailand, Bangladesh)	78,811	20,625	20,953
kapok (Thailand)	12,157	11,703	7,499
abaca (Philippines, Ecuador)	28,174	21,492	22,239
sisal and henequen (Brazil, Tanzania, Kenya, Mozambique, Mexico)	92,448	59,677	14,465
istle or tampico (Mexico)	7,074	29	0
others [b]	4,753	7,551	12,655
Total	225,698	122,658	78,222

[a] Ref. 14.

[b] Crin vegetal, phormium, broom root, rice straw, coir, sunn.

Table 10.   U.S. Consumption of Nonwood Paper Pulps in 1978 [a]

Vegetable fiber	Source	Pulp consumption, t	Paper products
cotton linters	United States	100,000	rayon and fine papers
rags, textile residues	United States	70,000	absorbent felt and fine papers
seed flax straw	United States, Europe	60,000	cigarette and lightweight papers
abaca	United States (old ropes), Philippines	25,000	specialty papers, including tea bags
miscellaneous		30,000	
Total		285,000	

[a] Ref. 16.

Table 11.   Yields of Prepared Vegetable Fibers [a]

Plant	Fiber form	Yield [b], %	Plant	Fiber form	Yield [b], %
Leaf fibers					
abaca	decorticated	2–3	hemp	dry-line fiber	3.5
sisal	decorticated	3 [c]		dry tow	1.0
henequen	decorticated	2–3	ramie	decorticated	3.5
Seed-hair fiber			jute	dry-retted fiber	6.0
kapok		17 [d]	kenaf	dry-retted fiber	4.8
Bast fibers			roselle	dry-retted fiber	4.4
flax	scutched line fiber	12–16	sunn hemp	dry-retted fiber	3.4
	scutched clean tow	4–8			

[a] Ref. 3–5.

[b] Green plant basis.

[c] Leaf basis.

[d] Pod basis (600 pods per tree).

Economic Aspects

Prices.   The prices of vegetable fibers for cordage and noncordage uses have varied widely over the years depending on the supply and demand, world economic conditions, weather, and starting in the 1960s, competition from man-made fibers particularly polypropylene fibers (see Olefin fibers). In 1970, polypropylene had a price of ca $400/t on the U.S. market and was projected to fall by 1980 to $275/t, making vegetable fibers uncompetitive. In 1978, however, polypropylene had risen to $1200/t because of the oil situation (17–18) (see also Olefin polymers).

The 1961–62 prices for unprocessed fibers imported into the United States for the hard fibers, abaca and sisal, were $33.00 and $17.82/100 kg, respectively. Recent prices for these fibers and a synthetic fiber are shown in Table 12.

Projections. Vegetable fibers, excluding cotton and flax, have been relatively unimportant in the total fiber supply and will become less important in relation to converted cellulosic and noncellulosic fibers as projected in Table 13.

Estimated world demands for hard fibers, as shown in Table 14, is expected to remain static (zero growth) for ropes, cables, nets and other cordage, paper and padding, and other uses.

Cordage goods from vegetable fibers accounted for 87% and synthetics for 13% of the total of 21,000 t estimated for the U.S. market in 1964–1966. The vegetable fiber

Table 12. Recent Prices for Various Fibers [a]

Fiber	Grade	Date	Price, U.S. $/100 kg	Market
abaca	Davao I	av 1977	44.80	ex (export) ship, New York
		Dec. 1978	37.90	
sisal	British East Africa No. 1	av 1977	102.30	landed, New York
		Oct. 1978	102.80	
jute	raw Bangladesh White C	av 1977	46.90	CIF (cost, insurance, and freight),
		Oct. 1978	48.90	United Kingdom
flax	water-retted	av 1977	185.30	United Kingdom
		Sept. 1978	215.80	
polyester		av 1977	123.50	
fiber[b]		July 1978	116.90	wholesale, United States

[a] Ref. 18.
[b] See Polyester fibers.

Table 13. Percentage of Total Fiber Supply in the United States [a]

	Cotton	Wool	Rayon and acetate	Non-cellulosic	Other[b]
1950	64	6.7	25.3	3.5	0.2
1960	55.9	4.4	19.5	19.5	0.3
1980 (estd)	46	2.5	10.8	40.5	0.2
2000 (estd)	44	1.9	5.4	48.6	0.1

[a] Ref. 19.
[b] Including sisal, jute, kapok, abaca.

Table 14.   Estimated World Demand for Hard Fibers, 1000 Metric Tons[a]

	1964–66	1968–70	1980
sisal-henequen	772	765	695
abaca	120	95	95
Total	*892*	*860*	*790*

[a] Ref. 20.

share slipped to 71% of a total of 927,000 t in 1968–1970 and is projected to drop to 60% of a total of 965,000 t in 1980. The influence of higher oil prices on the cost of the polypropylene resins used in the synthetic cordages may well change the percentages and trend shown above.

Although there was a zero or negative annual growth rate for nonwood plant fibers in U.S. papermaking in the 1970s, it is expected that these fibers will hold their small percentage of under 1% of the total U.S. pulp consumption through the year 2000 (21).

Uses

Information on the broad end uses of vegetable fibers is given in Table 1 and consumptions by general products are discussed subsequently. Uses by certain categories are described as follows (1–3,5,22).

Textiles and Woven Goods.   Clothing, sacks and bags, canvas and sailcloth, and fabrics are made from the bast fibers flax, hemp, ramie, jute, kenaf, roselle, and nettle. Coarse sacks, coffee and sugar bags, floor coverings, and webbing are made from the leaf fibers sisal, henequen, abaca, and Mauritius hemp. Phormium, nettle, and cantala also are used in coarse bagging.

Cordage and Twines.   Industrial ropes, hoisting and drilling cables, nets, and agricultural twines are made from abaca, sisal, and henequen. Abaca serves particularly for hawsers and ship cables. Cantala, phormium, and Mauritius hemp are also used for cordage. Hemp, flax, and jute are used for string and yarns, and ramie is used for thread. Hemp and kenaf are used for nets as well.

Brushes.   Fibers suitable for bristles in scrubbing and scraping brushes and brooms include coir, piassava, istle and broomwort.

Stuffing and Upholstery Materials.   Stuffing for mattresses and pillows and furniture is made from kapok, crin vegetal, sisal tow, and flosses. Kapok is also used in life preservers and coir for door mats.

Paper.   In addition to the production of bond and fine writing papers from cotton and linen textile residues and roofing felt from low-grade rags, heavy-duty, multiwalled bags for industrial packaging of flour, cement, chemicals, fertilizers, and hardware are made containing salvaged vegetable fibers or residues from primary product manufacturing of sisal, jute, abaca hemp, henequen, phormium, caroa, and Mauritius hemp. Other specialty papers using these fibers are abrasive papers, and gaskets. Abaca fibers are used specifically in tea bag paper and flax tow and sunn in cigarette paper. Esparto (*Stipa tenacissima*) fibers are made into highest-grade printing papers. Kenaf is proposed for newsprint (see Paper).

Miscellaneous.   Tie material, basketry and furniture are made from raffia (*Raphia raffia*), a palm leaf segment, and rattan (*Calamus* sp.), a stem fiber.

BIBLIOGRAPHY

"Fibers, Vegetable" in *ECT* 1st ed., Vol. 6, pp. 467–476, by David Himmelfarb, Boston Naval Shipyard; "Fibers, Vegetable" in *ECT* 2nd ed., Vol. 9, pp. 171–185 by David Himmelfarb, Boston Naval Shipyard.

1. J. Cook, *Handbook of Textile Fibers,* 4th ed., Morrow Publishing Co., Ltd. Watford, Hertz, England, 1968.
2. M. Harris, *Handbook of Textile Fibers,* Harris Research Laboratories, Inc., Washington, D.C., 1954.
3. R. H. Kirby, *Vegetable Fibers,* Interscience Publishers, Inc., New York, 1943.
4. J. M. Dempsey, *Fiber Crops,* The University Presses of Florida, Gainesville, 1975.
5. J. Berger, *The World's Major Fibre Crops: Their Cultivation and Manuring,* Centre d'Etude de l'Azote, Conzett and Huber, Zurich, 1969.
6. W. Von Bergen and W. Krauss, *Textile Fiber Atlas,* American Wool Handbook Company, New York, 1942.
7. K. R. Gilbert, "Rope Making" in C. Singer, ed., *History of Technology,* Vol. I, Oxford University Press at Clarendon, 1954.
8. J. Grant, "A Note on the Materials of Ancient Textiles and Baskets," in ref. 7.
9. G. E. Linton, *Natural and Man-made Textile Fibers,* Duel, Sloan and Pearce, New York, 1966.
10. J. H. Harris, *Ancient Egyptian Materials and Industry,* Edward Arnold Publishers Ltd., London, 1962.
11. T. Zylinski, *Fiber Science,* Office of Technical Services, U.S. Dept. of Commerce, Washington, D.C., 1964.
12. W. E. Morton and J. W. S. Hearle, *Physical Properties of Textile Fibers,* 2nd ed., John Wiley & Sons, Inc., New York, 1975.
13. "Wood, Paper, Textiles, Plastics and Photographic Materials" in *Chemical Technology: An Encyclopedic Treatment,* Vol. VI, Barnes and Noble Books, 1973.
14. *U.S.D.A. Agricultural Statistics,* U.S. Government Printing Office, Washington, D.C., 1977.
15. H. Jiler, ed., *1977 Commodity Yearbook, Hard Fibers,* Commodity Research Bureaus, Inc., New York, 1977.
16. J. E. Atchison, *An Update on Utilization of Nonwood Plant Fibers, Recent Developments and Some Observations on a Visit to China,* paper presented at the 1979 Pulp and Fiber Fall Seminar, American Paper Institute, Charleston, S.C., Nov. 4, 1979.
17. "General Outlook" in *FAO Agricultural Commodity Projections 1970–1980,* Vol. I, Rome, 1971, pp. 264–2822.
18. *FAO Monthly Bulletin of Agricultural Economics and Statistics* 1(12), 45, 47, 49 (Dec. 1978).
19. G. W. Thomas, S. E. Curl, and W. F. Bennet, *Food and Fiber for a Changing World,* The Interstate Printers and Publishers, Inc., Danville, Ill. 1976.
20. R. Grilli, *The Future of Hard Fibers and Competition from Synthetics,* World Bank Staff, The John Hopkins University Press, Baltimore, 1975.
21. J. N. McGovern, *Non-Wood Plant Fiber Pulping,* TAPPI, CAR Report No. 67, 1976.
22. T. F. Clark in R. G. Macdonald, ed., *Pulp and Paper Manufacture,* Vol. II, McGraw-Hill Book Company, New York, 1969.

General References

J. E. Atchison, "Agricultural Residues and Other Nonwood Plant Fibers," *Science* **191,** 768 (1976).

D. Lapedes, ed., *Encyclopedia of Science and Technology,* Vol. 5, McGraw-Hill Book Company, New York, 1977.

G. E. Linton, *THE Modern Textile and Apparel Dictionary,* 4th ed., Textile Book Service, Plainfield, N.J., 1973.

JOHN N. MCGOVERN
The University of Wisconsin

FLAME RETARDANTS FOR TEXTILES

Hazards associated with the ready combustibility of cellulosic materials were recognized as early as the 4th century BC, when Aeneas is said to have recommended treatment of wood with vinegar to impart fire resistance (1). The annals of Claudius record that wooden storming towers used in the siege of Piraeus in 83 BC were treated with a solution of alum to protect them against fire.

The technique of imparting flame resistance to textile fabrics is relatively new. Among the earliest references is an article by Sabattini published in 1638. Recognizing a need to prevent fire, he suggested that clay or gypsum pigments be added to the paint used for theater scenery to impart some flame resistance. Perhaps the first noteworthy recorded attempt to impart flame resistance to cellulose was made in England in 1735 (2) when Obadiah Wyld was granted a patent for a flame-retardant mixture containing alum, ferrous sulfate, and borax.

In France in 1821, Gay-Lussac (3) developed a flame-resistant finish by treating linen and jute fabrics with a mixture of ammonium phosphate, ammonium chloride, and borax.

The first successful, launder-resistant, flame-retardant finish for fabric was based on the work of Perkin (4) who precipitated stannic oxide within the fiber. This fabric was flame resistant but afterglow was severe and persistent enough to completely consume the fabric.

Flame retardants are mainly used on cottons and rayons. Fabrics made from wool (qv), silk (qv), and protein-like synthetic polymers are not considered sufficiently combustible, for the most part, to warrant the need for flame-retardant finishes (see Biopolymers; Textiles).

Since World War II the flammability of textiles of all types has received greatly increased attention, spurred by the Conference on Burns and Flame Retardant Fabric in 1966 (5), and by the 1967 amendment to the Flammable Fabrics Act of 1953 (6). Flammability standards were established by the Department of Commerce and enforced by the Federal Trade Commission. This responsibility was taken over by the Consumer Product Safety Commission when it was created in 1973.

The terms used in connection with flame-resistant fabrics are sometimes confusing. Fire resistance and flame resistance are often used in the same context as the terms fireproof or flameproof. A textile that is flame resistant or fire resistant does not continue to burn or glow once the source of ignition has been removed, although there is some change in the physical and chemical characteristics. Fireproof or flameproof, on the other hand, refer to material that is totally resistant to fire or flame. No appreciable change in the physical or chemical properties is noted. Asbestos is an example of a fireproof material.

Most organic fibers undergo a glowing action after the flame has been extinguished, and flame-resistant fabrics should also be glow resistant. Afterglow may cause as much damage as the flaming itself since it can completely consume the fabric. The burning (decomposition) temperature of cellulose is about 230°C, whereas afterglow temperature is approximately 345°C.

Chemical modification of cellulose with fire retardants gives products whose resistance to laundering and weathering is superior to that of finishes based on the physical deposition of the flame retardant within the fabric, yarn, or fiber. The reactions involved are either esterification or etherification. The latter is preferred because ether linkages are more stable to hydrolysis.

Flame Resistance

The flame resistance of a textile fiber is affected by the chemical nature of the fiber; ease of combustion; fabric weight and construction; efficiency of the flame retardant; environment; and laundering conditions.

Fire-resistant characteristics can change significantly when treated fabric is exposed to sunlight, followed by laundering, even though repeated washing and tumble drying of samples of the same specimen did not indicate any significant changes, especially in the durability of the finish (7). Dry heat alone, followed by laundering or autoclaving can also have a deleterious effect (8).

Fibers are classified into natural fibers, eg, cotton, flax, silk, or wool; regenerated fibers, eg, rayon; synthetic fibers, eg, nylons, vinyls, polyester, acrylics; and inorganic fibers, eg, glass or asbestos. Combustibility depends on chemical makeup and whether the fiber is inorganic, organic, or a mixture of both (see Fibers, chemical).

The weight and construction of the fabric affect its burning rate and ease of ignition. Lightweight, loose-weave fabrics usually burn much faster than heavier-weight fabrics; therefore, a higher weight add-on of fire retardant is needed to impart adequate flame resistance.

Phosphorus-containing materials are by far the most important class of compounds used to impart durable flame resistance to cellulose (see also Flame retardants, phosphorus compounds). They usually contain either nitrogen or bromine and sometimes both. A combination of urea and phosphoric acid imparts flame resistance to cotton fabrics at a lower add-on than when the acid or urea is used alone (9). Other nitrogenous compounds, such as guanidine, or guanylurea, could be used instead of urea. Amide and amine nitrogen generally increase flame resistance, whereas nitrile nitrogen can detract from the flame resistance contributed by phosphorus. The most efficient flame-retardant systems contain two retardants, one acting in the solid and the other in the vapor phase.

Bromine in flame-resistant fabric escapes from the tar to the vapor phase during pyrolysis in air. It appears to have little or no effect on the amount of phosphorus remaining in the char. Bromine contributes flame resistance almost completely in the vapor phase.

Nitrogen when used in conjunction with phosphorus compounds has synergistic effects (10–12). Phosphorus content can be reduced without changing the efficiency of the flame retardant (13).

The temperature of the environment influences the burning characteristics of fabric as measured by the oxygen index (OI). This is true for untreated as well as flame-retardant fabrics. For example, the OI value of untreated fabric is 0.18 when burned at 25°C and 0.14 when burned at 150°C. For flame-retardant fabric, an OI value of 0.35 at 25°C may be reduced to 0.27 when burned at 150°C. Sunlight and heat can also destroy some flame retardancy (13–15), especially when followed by laundering or autoclaving (7–8). The moisture content of fabric can also affect flame retardancy (16).

Fire retardancy of a treated cellulosic fabric is reduced when the retardant contains acid groups and the treated fabric is soaked or laundered in water containing calcium, magnesium, or alkali metal ions. Phosphate- and carbonate-based detergents affect durability of fire retardants (17). Soap-based detergents can result in a substantial loss of fire resistance because of deposit of fatty acid salts (18). Phosphorus-based flame retardants are adversely affected by water hardness and sodium hypochlorite (see Drycleaning and laundering).

Mechanisms.   Cellulose, as such, has no appreciable vapor pressure and does not burn. However, on exposure to high temperatures it decomposes exothermically into flammable compounds causing further degradation and decomposition until complete disintegration has taken place. Numerous studies have been made on burning of untreated and flame-retardant-treated cellulose (19–24). Decomposition takes place in two stages. First, thermal decomposition causes cellulose to decompose heterogeneously into gaseous, liquid, tarry, and solid products (25,26). The flammable gases thus produced ignite, causing the liquids and tars to volatilize to some extent. This produces additional volatile fractions which ignite and produce a carbonized residue which does not burn readily. This process continues until only carbonaceous material remains. After the flame has subsided, the second stage begins: the residual carbonized residue slowly oxidizes and glowing continues until carbonaceous char is consumed.

Cotton (qv) treated with an effective flame retardant forms, in general, the same decomposition products upon burning as untreated cotton; however, the amount of tar is greatly reduced with a corresponding increase in the solid char. Consequently, as decomposition takes place, smaller amounts of the flammable gases are available from the tar, and greater amounts of nonflammable gases from the decomposition of the char fraction.

Char is essentially carbon and its oxidation causes afterglow. Phosphorus-containing compounds (in some cases, polymers) are particularly effective in inhibiting char oxidation. The oxidation of carbon takes place through either of the following reactions:

$$C + \tfrac{1}{2}\,O_2 \rightarrow CO \quad \Delta H = 110.45 \text{ kJ (26.4 kcal)}$$

$$C + O_2 \rightarrow CO_2 \quad \Delta H = 394.96 \text{ kJ (94.4 kcal)}$$

Effective glow-retardant chemicals, such as compounds containing phosphorus, cause the first reaction to be prevalent. Oxidation of carbon monoxide is not sufficiently exothermic to maintain afterglow of the char.

Imparting flame resistance to cellulose has been explained by the following theories:

Coating theory.   As early as 1821, Gay-Lussac (3) suggested that fire resistance was due to formation of a layer of fusible material which melted and formed a coating thereby excluding the air necessary for the propagation of a flame. This was based on the efficiency of some easily fusible salts as flame retardants. Carbonates, borates, and ammonium salts are good examples of coating materials that produce a foam on the fiber by liberation of gases such as carbon dioxide, water vapor, ammonia, etc.

Gas theory.   The flame retardant produces noncombustible gases at burning temperature which dilute the flammable gases produced by decomposition of the cellulose to a concentration below the flaming limit.

Thermal theory.   Heat input from a source is dissipated by an endothermic change in the retardant and the heat supplied from the source is conducted away from the fibers so rapidly that the fabric never reaches temperature of combustion.

Chemical theory. Strong acids, bases, metal oxides, and oxidants that tend to degrade cellulose, especially under the influence of heat, usually impart some degree of flame resistance to cellulose (27). This is also true of the more efficient flame retardants, such as phosphoric and sulfuric acid, which are good dehydrating agents. When this happens, cellulose on combustion produces mainly carbon and water rather than carbon dioxide and water.

Flame retardants for cotton may possibly act through a dehydration process by Lewis acid or base formation through a carbonium ion or carbanion mechanism (21). This theory is being further investigated.

Earlier theories suggested that flame-retarded cellulose decomposed at high temperatures to l-glucosan which in turn broke down to form other volatile products which were highly flammable. However, if bases are present in the fabric during burning, dehydrocellulose is formed by a base-catalyzed dehydration followed by char formation (28). Base-catalyzed dehydration prevents formation of l-glucosan by propagating structural changes at an energy level below that required to convert the coformers of the glucopyranose ring (29).

Durability

Nondurable Finishes. Flame-retardant finishes that are not durable to laundering and leaching are, in general, relatively inexpensive and efficient (30). In some cases, mixtures of two or more salts are much more effective than any one of the components alone. For example, an add-on of 60% of borax $Na_2B_4O_7.10H_2O$ is required to prevent fabric from burning. Boric acid, H_3BO_3, by itself, is ineffective as a flame retardant even in amounts that equal the weight of the fabric. However, a mixture of seven parts borax and three parts boric acid imparts flame resistance to a fabric with as little as $6\frac{1}{2}\%$ add-on.

The water-soluble flame retardants are most easily applied by impregnating the fabric with a water solution of the retardant, followed by drying. Adjustment of the concentration and regulation of the fabric wet pickup controls the amount of retardant deposited in the fabric. Fabric can be processed on a finishing range consisting of any convenient means of wetting the fabric with the solution, such as a padder or dip tank, followed by a drying on cans, in an oven, on a tenter frame, or merely by tumbling in a mechanical dryer. Water-soluble flame retardants also may be applied by spraying or brushing, or dipping fabrics, or as a final rinse in a commercial or home laundry (31). The water-soluble flame retardants most widely used for textiles are listed in Table 1. Less commonly used are sulfamates of urea or other amides and amines; aliphatic amine phosphates, such as triethanolamine phosphate [*10017-56-8*], phosphamic acid [*2817-45-0*] (monoimido phosphoric acid), $H_2PO_3NH_2$, and its salts; and alkylamine bromides, phosphates, and borates.

Semidurable Finishes. Semidurable fire retardants are those that resist removal for one to about 15 launderings. Such retardants are adequate for applications such as drapes, upholstery, and mattress ticking. If they are sufficiently resistant to sunlight or can be easily protected from actinic degradation, they can also be applied to outdoor textile products.

The principal disadvantage of the water-soluble flame retardants is their lack of durability. It can be overcome by precipitating inorganic oxides on the fabric; for example, $WO_3.xH_2O$ and $SnO_2.yH_2O$:

$$2\,Na_2WO_4 + SnCl_4 + (2x + y)H_2O \rightarrow 4\,NaCl + 2\,WO_3.xH_2O + SnO_2.yH_2O$$

Table 1.   Water-Soluble Flame Retardants for Textiles

Component	CAS Registry Number	Formula	Composition
borax–	[1330-43-4]	$Na_2B_4O_7.10H_2O$	70
boric acid	[10043-35-3]	H_3BO_3	30
borax–			47
boric acid–			20
diammonium phosphate	[7783-28-0]	$(NH_4)_2HPO_4$	33
sodium phosphate dodecahydrate–	[10101-89-0]	$Na_3PO_4.12H_2O$	50
boric acid			50
boric acid–			50
diammonium phosphate			50
borax–			50
boric acid–			35
sodium phosphate dodecahydrate			15
ammonium sulfamate–	[7773-06-0]	$NH_4OSO_2NH_2$	75
diammonium phosphate			25
ammonium bromide	[12124-97-9]	NH_4Br	100
borax–			15
boric acid–			47
sodium phosphate–	[7601-54-9]	Na_3PO_4	18
sodium tungstate dihydrate	[10213-10-2]	$Na_2WO_4.2H_2O$	20

These codeposits add flame- and glow-resistance properties to textile fabrics. However, some insoluble deposits may also degrade the fabrics. Codeposits quite frequently improve glow resistance. They are usually more soluble than the deposit responsible for flame resistance and are more easily removed during the laundering process.

There are several methods for introducing the insoluble deposits into the fabric structure. Most generally used is the multiple-bath method, in which the fabric is first impregnated with a water-soluble salt or salts in one bath, and is then passed into a second bath which contains the precipitant.

Most semidurable retardants are used on cotton and are based on a combination of phosphorus and nitrogen compounds (32).

Durable Finishes.   Earlier studies to produce durable flame retardants for cellulose were based on treatment with inorganic compounds containing antimony and titanium (33). Numerous patents were issued based on these types of treatments, eg, DuPont's Erifon process (34) and the Titanox FR process of the Titanium Pigment Corporation (35).

In the Erifon process titanium and antimony oxychlorides were applied from acid solution (pH 4) to fabric, which was then neutralized by passing through a solution of sodium carbonate, followed by rinsing and drying. Fabrics thus finished exhibited good flame resistance but also considerable afterglow. Some fabric characteristics were changed by this treatment. A large amount of tent fabric was treated by this type of process for the military service. However, it has now been replaced by flame retardants based on phosphorus.

The basic chemicals used in the Titanox FR process were titanium acetate chloride and antimony oxychloride. As with the Erifon process, it was difficult to process the fabric without dulling its appearance.

Excellent fire-resistant fabric was obtained by treating fabric with a suspension or emulsion of insoluble fire-retardant salts or oxides, eg, antimony oxide, with a chlorinated organic vehicle, such as chlorinated paraffin (36). Antimony oxide alone is only a poor flame retardant. When used, however, in conjunction with a chlorinated compound, which can form hydrochloric acid on heating, a very good flame retardant is produced.

The abbreviation, FWWMR, for fire, water, weather, and mildew resistance has frequently been used to describe treatment with a chlorinated organic metal oxide. A plasticizer, coloring pigments, fillers, stabilizers, or fungicides are usually added. However, hand, drape, flexibility, and color of the fabric are more affected by this type of finish than by other flame retardants. Add-ons of up to 60% are required in many cases to obtain adequate flame resistance. Durability of this finish is good and fabric processed properly retains its flame resistance after four to five years of outdoor exposure. This type finish is well suited for very heavy fabrics, eg, tents, tarpaulins, or awnings, but not for clothing or interior decorating fabrics. The metal oxides can be fixed to cotton by use of resins, eg, vinyl acetate–chloride copolymers (vinylite VYHH) or PVC (37).

A flame retardant has been developed based on an oil–water emulsion containing a plasticizer (PVC latex) and antimony oxide (38). High add-ons are necessary to impart adequate flame resistance but the strength of the fabric is little affected.

Test Methods

The important test methods are listed in Table 2.

Fire Resistance. The most widely used test is the standard vertical flame test, although it is only one of the many tests in which the fabric to be tested is suspended vertically. A strip of fabric (30 cm × 6.3 cm) is suspended vertically with its lower edge hanging just 1.9 cm above the top of a bunsen or tirrell burner. The burner is adjusted to give a 3.8-cm luminous flame, extending 1.9 cm over the fabric. After an ignition time of 12 s the flame is removed and the duration of the afterglow and the length of the charred area are measured. Most of the specifications allow an afterflaming of 2 s and an average char length (10 strips) of 8.9 cm, with a maximum of 11 cm for any one strip. These values are, of course, based on weights of the fabric. This method has been incorporated into a great many specifications, including CCCT 191a and CCCD 746 (Federal), 6-345 (Army), 24-C-20 (Navy), DOC FF 3-71, DOC FF 5-74, and others.

Refinements of this method include standard draft-free cabinets, conditioning the specimens, special fabric clamps and holders, manometers to control flow of gas to the burner, and techniques for measuring char length.

Flammability. The position of a fabric usually influences its burning rate. A fabric suspended vertically and ignited on the bottom burns considerably more rapidly than one held in a horizontal position. The speed with which vertically suspended fabrics burn makes distinctions among fabrics difficult and for that reason many flammability testers mount the fabric specimen at some angle from the vertical position.

The AATCC (American Association of Textile Chemists and Colorists) or inclined flammability tester was developed after World War II and has been more widely used than any other rate-of-burning device: A strip of fabric (15.2 cm × 5 cm) is clamped in a rack inclined at 45°, and the surface of the fabric near the bottom is exposed to the flame of a microjet burner for 1 s. If the sample is ignited by this flame, the time

Table 2. Summary of Test Methods For Textiles

Test standard	Instrument	Materials tested	Specimen size, cm	Angle of specimen	Ignition source	Properties measured
FT MS 191-5910	burn rate of	fabrics	2.5 × 15.2	30°	paper match	burn rate
CS 191-53 (Rev.), AATCC 33-1962, ASTM D1230-61, NFPA 702-1975, California Tech. Bulletin 117	45° flammability tester for wearing apparel	fabrics used for wearing apparel, bonded and nonbonded upholstery, filling materials and fabrics	varies with procedure	45°	microburner (butane gas)	ease of ignition, flame spread time, self-extinguishing time
FT MS 191-5908.1	burn rate tester (Method 5908)	fabrics	2.5 × 15.2	45°	microburner (butane gas)	ease of ignition, flame spread time
DOC FF-1-70 DOC FF-2-70, ASTM D2859 CPAI-84 DDD-C-95	carpet flammability tester	carpets, rugs, tent flooring	22.8 × 22.8	horizontal	burning methenamine tablet	extent of char area
DOT FMVS 302, California Recreational Vehicles	horizontal flammability tester	motor vehicle interior materials	10.1 × 35.5	horizontal	bunsen burner	flame spread time
CPAI-75	flammability tester	multicomponent sleeping bags	30.4 × 35.5	horizontal	bunsen burner (mfg gas type B)	burning rate over 25.4 cm
FTMS 191-5906, FAA Reg. Para. 25.853	burn rate tester (Method 5906)	fabrics-aircraft interiors	11.4 × 31.7	horizontal	bunsen burner	flame spread time
FT MS 372, NFPA 253	radiant panel tester for floor covering systems	floor covering systems	25.4 × 107	horizontal	radiant panel—gas/air mixture gas flame	critical radiant flux at flame out
NFPA 254-77	furnace tester for floor covering systems	floor covering systems	60.9 × 244	horizontal	bunsen burner	flame propagation index
NFPA-701-1976 (large scale), State of California Title 19-1237.3	vertical flammability tester (NFPA 701 large scale)	fabrics and films	in sheets—12.7 × 213 in folds—63.5 × 213	vertical	bunsen burner	afterflame, afterglow, char length
ASTM D568-74, NFPA 701-1976, (small scale), City of Boston Fire Code 1959, City of New York Fire Code 1940, State of California Title 19-1237.1	flammability tester for fabrics and plastics	fabrics, plastic sheet and film	varies with procedure	vertical	bunsen burner	burning rate, afterflame, afterglow, char length

Test method	Apparatus	Materials	Specimen size (cm)	Orientation	Ignition source	Measurements
FT MS 191-5903, FAA Reg. Para. 25.853 CPAI-84, California Tech. Bulletin 117, Sections A, B and F	vertical flammability tester	aircraft interior materials tenting-walls and top, resilient cellular materials and plastics	varies with procedure	vertical	bunsen burner	burn length, afterglow, char length
DOC FF3-71 DOC FF5-74 FAA proposed std flight attendants' uniforms	children's sleepwear flammability tester	fabrics and garments children's sleepwear, flight attendants' uniforms	8.9 × 25.4	vertical	special methane gas burner	residual flame time, char length
J. C. Penney modification of DOC FF3-71, FF3-71 proposed, ASTM method	vertical flammability tester (semirestraint method)	wearing apparel fabrics	15.2 × 38.1	vertical	special methane gas burner	burning time, weight loss, char length, char area
TAPPI-T461-5U-72	TAPPI flammability tester	treated paper and paper board	6.9 × 20.9	vertical	bunsen burner (natural gas)	flame time, afterglow, char length
CPSC proposed standard for the flammability wearing apparel	mushroom apparel flammability	clothing items, fabrics and related materials used in apparel	31.7 × 60.9	vertical-clamped to heat sensor assembly	microburner, methane gas	ignition time, maximum heat time
proposed FAA regulation	heat-flux resistance apparatus	fabrics for flight attendants' uniforms	15.2 × 15.2	vertical	elec. heated radiant panel	heat flux transmission
ASTM D2863-74	oxygen-index flammability tester	plastics, textiles, paper and other materials	0.6 × 0.3 × 12.7	vertical	gas flame, burning in atmosphere of variable oxygen concentration	oxygen index (oxygen content required to support combustion)
FTMS-191 Method 5920	Mackey apparatus	cloth or related materials	5 × 30.4	not applicable	not applicable	tendency of material to undergo self-heating at moderate temperatures

of flame travel over 12.7 cm of the specimen is used as an indication of relative flammability of the fabric (39). In July 1954 this method was adopted for the first federal legislation relating to combustible products.

With the passage of the amendment to the 1953 Flammable Fabrics Act in 1967, other federal flammability standards have been issued by the Department of Commerce (DOC). The test methods and textile products relating to these standards are as follows:

DOC FF 1-70 (40) and DOC FF 2-70 (41) Standards for the Surface Flammability of Carpets and Rugs (April 16 and December 29, 1971, respectively).

DOC FF 3-71 (42) and DOC FF 5-74 (43) Standards for the Flammability of Children's Sleepwear are a modification of the vertical flame test (July 29, 1973 and May 1, 1975, respectively). These standards have been amended by FF3-71 and FF5-74 (February 6, 1978). The former requires that no single sample could have flaming material on the bottom of the test cabinet 10 s after the ignition source was removed.

DOC FF 4-72 (as amended) Standard for the Flammability of Mattresses (44).

Motor Vehicle Safety Standard No. 302—Flammability of Interior Materials; (passenger cars, multipurpose passenger vehicles, trucks), tested in a horizontal position.

AATCC Test Method 34-1969—Fire Resistance of Textile Fabrics (45), based on a modification of the vertical flame test.

Oxygen Index Test. This test (OI) is based on the minimum oxygen concentration that supports combustion of the material (46). The OI values of various fibers are listed in Table 3.

Types of Retardants

Mesylated and Tosylated Celluloses. A study on imparting flame resistance to cellulose by chemical modification (47) established that the flame resistance of cellulose is improved by oxidation of —CH_2OH groups to —COOH. To correct some of the shortcomings of this treatment, mesyl or tosyl cellulose was prepared and the mesyl (CH_3SO_2) or tosyl ($CH_3C_6H_4SO_2$) group was replaced with bromine or iodine (47):

$$R_{cell}OSO_2CH_3 + NaBr \rightarrow R_{cell}Br + CH_3SO_2Na$$

This treatment produced a fabric that had durable flame resistance, good strength retention, but suffered from an undesirable afterglow which was eliminated by phosphorylation with diethyl chlorophosphate.

Table 3. Oxygen Indexes of Various Fabrics

Fabric	OI	Fabric	OI
acrylic	0.182	nylon	0.201
acetate	0.186	polyester	0.206
polypropylene	0.186	wool	0.252
rayon	0.197	flame-retardant cotton	0.270
cotton	0.201	aramid	0.282

Urea-Phosphate Type. Phosphoric acid [7664-38-2] alone imparts flame resistance to cellulose (25–26) resulting, however, in degradation. This can be prevented by incorporation of urea [57-13-6]. Phosphorylating agents for cellulose include ammonium phosphate [7783-28-0], urea–phosphoric acid, phosphorus tri- and oxychloride [7719-12-2] and [10025-87-3], respectively, monophenyl phosphate [701-64-4], phosphorus pentoxide [1314-56-3], and chlorides of partially esterified phosphoric acids.

Cellulose phosphate esters are also produced by treatment with sodium hexametaphosphate [14550-21-1] by the pad-dry-cure technique with a high retention of breaking and tearing strength (48). Phosphorus contents up to 3.5% have been reported. The reaction products contain more than 1.6% phosphorus and were insoluble in cupriethylenediamine, indicating that the cellulose was cross-linked:

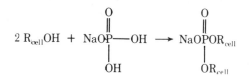

$$2\ R_{cell}OH\ +\ NaO\underset{\underset{OH}{|}}{\overset{\overset{O}{\|}}{P}}{-}OH\ \longrightarrow\ NaO\underset{\underset{OR_{cell}}{|}}{\overset{\overset{O}{\|}}{P}}OR_{cell}$$

Phosphorylated cottons are flame resistant in the form of the free acid or the ammonium salt. Since these fabrics have ion-exchange properties, conversion to the sodium salt takes place easily in laundering if acid sours and basic tap water are used. However, flame retardation can be restored if the fabric is treated with a solution of ammonium hydroxide after the acid sour and dried.

$$2\ R_{cell}OH\ +\ NaO\underset{\underset{OH}{|}}{\overset{\overset{O}{\|}}{P}}{-}OH\ \longrightarrow\ NaO\underset{\underset{OR_{cell}}{|}}{\overset{\overset{O}{\|}}{P}}R_{cell}$$

Phosphonomethylated Ethers. A phosphorus-containing ether of cellulose was prepared by the reaction of cotton cellulose with chloromethylphosphonic acid [2565-58-4] in the presence of sodium hydroxide by the pad-dry-cure technique (49). Phosphorus contents between 0.2 and 4% were obtained. This finish was durable but had high ion-exchange properties and was flame resistant only as the ammonium salt.

A durable flame-retardant cellulosic fabric with good hand is obtained by treating phosphorylated or phosphonomethylated cotton with titanium sulfate [13825-74-6] (50). The reactions are shown below.

Amide-Based Systems. *Cyanamide.* Pyroset CP (trade name for 50% cyanamide) solution produces a fire retardant that does not adversely affect the initial handle of treated fabric (51). Durability to laundering on industrial fabrics, blankets, draperies, and garments made from heavyweight fabrics is good; however, durability to laundering on lightweight fabrics remains limited (see Cyanamides). Treated fabrics also show an increase in wrinkle recovery. Drying and curing is accomplished in a single step, preferably at 149°C for 2–5 minutes. The finish is subject to ion exchange and, therefore, the hardness of the wash water has a decisive effect on durability of the flame retardancy. Fabric flammability fails after seven washings at a water hardness of 70

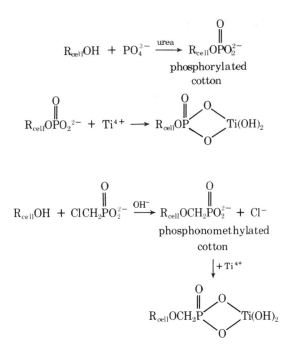

ppm whereas in soft water (0 ppm hardness) flame retardancy can withstand 50 wash cycles. Tensile strength losses of Pyroset CP treated fabric are 40–45%. The treated fabrics are dimensionally stable and have improved rot resistance.

Another system (52) is based on methylphosphoric acid [993-13-5] (MPA) and cyanamide [420-04-2]:

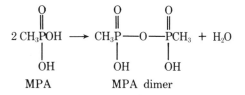

One or more of the hydroxyls in MPA or in its dimer react with cellulose:

$$
\underset{\substack{|\\ \text{OH}\\ \\}}{\overset{\overset{\displaystyle O}{\parallel}}{CH_3\overset{}{P}OH}} + R_{cell}OH \longrightarrow \underset{\substack{|\\ \text{OH}}}{\overset{\overset{\displaystyle O}{\parallel}}{CH_3\overset{}{P}OR_{cell}}} + H_2O
$$

The water is taken up by the cyanamide, forming urea:

$$
H_2NCN + H_2O \longrightarrow H_2N\overset{\overset{\displaystyle O}{\parallel}}{C}NH_2
$$

Fabrics are treated by a pad-dry-cure technique; however, smoke is evolved in the curing step. At an add-on of 10% this flame-retardant finish was durable to 40–50 laundry cycles, had dry wrinkle-recovery angles of 200–220°, and wet wrinkle-recovery angles of 220–270°. (The wrinkle-recovery angle reported is the sum of the recoveries in the warp and fill directions (W + F); thus, the maximum possible wrinkle recovery is 360°). The same finish gave tensile-strength retention of 60–80% and tear-strength retention of about 50%, and raised the moisture regain of the fabric about 3%. The system shows a high tolerance for calcium.

Acrylamide. Pyrovatex CP, is based on the reaction product of a dialkyl phosphite and acrylamide (qv) which is methylolated with formaldehyde (53).

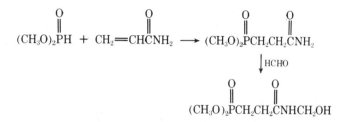

Excess formaldehyde reacts with cotton in the presence of an acid catalyst. A cross-linking agent, such as trimethylolmelamine [1017-56-7], TMM, improves the efficacy and durability of this finish. It is nonirritating to the skin, does not adversely affect the hand of the treated fabric, and is durable to 3–50 launderings and dry-cleanings. Tensile-strength losses are between 20 and 30% and tear-strength losses about 30%. Cotton textiles require add-ons of about 20–35%.

Dialkylphosphonopropionamides. Cellulosic derivatives very closely resembling those based on the dialkylphosphonopropionamides have also been prepared (54). The fabric was treated with N-hydroxymethylhaloacetamides (chloro, bromo, or iodo) in DMF solution by a pad-dry-cure technique with a zinc nitrate catalyst. It was then allowed to react in solution with trimethyl phosphite [121-45-9] at about 140–150°C; the reaction rates decreased in the order iodo > bromo > chloro.

With phosphorus contents above 1.5%, good flame resistance durable to laundering was obtained without noticeable loss in fabric strength properties.

Triazines. When the dialkoxyphosphinyl group is attached to the triazine ring rather than to an alkyl group, 2,4-diamino-6-diethoxyphosphinyl-1,3,5-triazine [4230-55-1] (DAPT) is obtained (41,55,56):

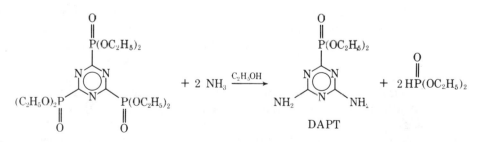

Formaldehyde and DAPT form a derivative that readily gives an insoluble cross-linked polymer. The fabric is treated by a pad-dry-cure procedure, with an optimum solution pH of 6.6 (no catalyst). Fabrics containing from 17.5–20% resin add-on passed the standard vertical flame tests even after 35 laundry cycles. Tear-strength losses were about 35% and tensile-strength losses about 18%. Although the treated fabrics did not yellow on chlorine bleaching and scorching, some strength loss resulted, indicating chlorine retention.

More recently a new effective flame retardant for cotton based on 2,4-diamino-6-(3,3,3-tribromopropyl)-1,3,5-triazine [62160-38-7] (DABT) was prepared from ethyl γ-tribromobutyrate and biguanide:

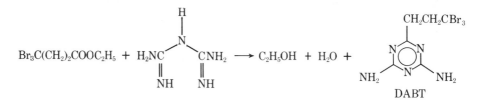

The tetramethylol derivative of DABT, prepared by reaction of DABT with alkaline aqueous formaldehyde, polymerized readily on cotton to impart excellent flame retardancy which was very durable to laundering with carbonate- or phosphate-based detergents and to hypochlorite bleach. This was accomplished at low add-on without use of phosphorus compounds or antimony oxide (57–59).

THPC Based.   A flame-retardant finish for cotton (60) is based on tetrakis(hydroxymethyl)phosphonium chloride [124-64-1] (THPC), $[HOCH_2]_4P^+Cl^-$. It is water soluble and is prepared by the reaction of phosphine with formaldehyde and HCl. The fact that THPC reacts with aminized cotton to impart flame resistance led to further studies with materials such as melamine and urea, in an effort to eliminate the costly aminization step. THPC reacts and polymerizes with methylolmelamine [937-35-9], as well as with many other materials containing active hydrogens, eg, amines, phenols, and polybasic acids (61), to form insoluble polymers in the cotton fiber and produce good flame- and glow-resistant finishes (see also Amino resins).

The methylol groups in THPC react with amines (or amides) in the following manner:

$$(HOCH_2)_4PCl + RNH_2 \rightarrow RNHCH_2PCl(CH_2OH)_3 + H_2O$$

One or more methylol groups are replaced by the $RNHCH_2$ groups and the ionic chlorine is hydrolyzed:

$$(RNHCH_2)_3PCl(CH_2OH) + H_2O \rightarrow HCl + CH_3OH + (RNHCH_2)_3PO$$

Thus, the product contains no chlorine and the phosphorus is in the form of phosphine oxide, a structure that is very stable to hydrolysis. If the nitrogen compounds are di- or polyfunctional, the product is a highly insoluble polymer. THPC alone also reacts with cellulose, but only very slowly under processing conditions associated with the finishing of cotton textiles.

A typical application procedure is as follows: a 268 g/m² cotton twill is treated by padding it through a solution containing 15.8% THPC, 9.5% methylolmelamine, 3% triethanolamine, and 9.9% urea, then dry cured for 4–5 minutes at 140°C, and washed. A dry add-on of 18–20% on the dry fabric imparts flame and glow resistance and also some crease and rot resistance.

A durable flame-retardant finish is obtained from bromoform [75-25-2] and triallyl phosphate [1623-19-4] (62). Although this finish (known as BAP) is durable to laundering, it causes some stiffness and imparts color to fabric. The char lengths obtained with the standard vertical flame test were longer than those obtained with other durable flame retardants.

Another durable flame retardant (63) is based on the telomerization of poly(allyl phosphonitrilate) with bromoform and its application to cellulose by the pad-dry-cure technique. This finish, known as PNE-bromoform, is more efficient than BAP but has some of the same shortcomings (see also Allyl monomers and polymers).

The THPC-amide flame-retardant finish penetrates the fibers whereas BAP and PNE-bromoform have a coating effect. To achieve penetration and coating, the THPC-amide is combined with either PNE-bromoform (64) or BAP (65). These finishes produced good results on many fabrics; however, again the treated fabrics were somewhat stiff because of deposited telomer.

A highly effective and durable flame retardant for cotton cellulose was obtained by treating cotton with the reaction product of THPC with tris(1-aziridinyl) phosphine oxide [545-55-1] (APO) (66). A thermosetting resin forms inside the fiber with a pad-dry-cure treatment. However, owing to the toxicity of APO, this product is no longer made (see Chemotherapeutics, antimitotic).

Flame resistance can also be imparted to fabric by chemical fixation of a water-soluble precondensate of THPC and amides with ammonia and/or ammonium hydroxide (67–69). This technique (the ammonia cure method) improves the degree of flame resistance by adding additional nitrogen to the finish. Thus the treated fabric does not lose strength as is the case with the heat cure method. Even though very little cross-linking is obtained by this technique, it greatly improves durability to laundering. The English flame retardant, Proban, is based on this technique (68).

A new, highly durable flame-retardant finish for cotton is based upon THPC, cyanamide, and phosphoric acid (70). Fabric is padded through a water solution of the three components, then dried and cured at elevated temperature. This formulation imparts a high degree of flame resistance to fabric. Substituting Na_2HPO_4 [13708-85-5] for H_3PO_4 maintains fabric strength.

Modified THPC-amide formulations are available containing antimony oxide and organic chlorine compounds (71–72), generally PVC or a chlorinated paraffin. Although these compounds are cheaper than THPC, they are less effective on a weight basis (see Flame retardants, antimony and other inorganic compounds).

A formulation based on THPC, thiourea [62-56-6], and small amounts of sodium dihydrogen orthophosphate [7558-80-7] has been reported (73). A pad-dry cure procedure was used. The treated fabric retained its flame retardancy for 30 laundry cycles.

Before 1976, all flame retardants based on THP salts were made from THPC. After this date, however, both Hooker Chemical Co. and American Cyanamid Company replaced THPC with the corresponding sulfate, (THPS) because of the possible formation of the carcinogen bischloromethyl ether (BCME) from THPC. No evidence has been found, however, showing that such a compound is formed. Although neither THPC or THPS are currently manufactured in the United States, Albright and Wilson, Co. Ltd. in England produces both of these compounds, and markets them in many countries including the United States.

Flame retardants based upon THPC are predominantly used for apparel and household goods, and for certain military items, eg, tent-liner fabrics. Processes based upon THPC are employed in the United States, Europe, and other countries (74,75).

THPOH Based.   Reactive methylol phosphorus compounds are made from 1 mole of THPC with 0.8–1 mole of sodium hydroxide. The aqueous solution of the product, referred to as THPOH (76), is not yet adequately characterized. Some researchers consider it to be mainly an equilibrium mixture of tetrakis (hydroxymethyl)phosphonium hydroxide [512-82-3], THPOH, and tris(hydroxymethyl)phosphine [2767-80-8], THP (78):

$$(HOCH_2)_4PCl + NaOH \rightarrow (HOCH_2)_4POH + NaCl$$

$$(HOCH_2)_4POH \rightleftharpoons (HOCH_2)_3P + HCHO + H_2O$$

Others have evidence that the solution consists of a mixture of THP and a hemiacetal of THP (79):

$$(HOCH_2)_4PCl + NaOH \rightarrow (HOCH_2)_3P + HCHO + NaCl + H_2O$$

$$(HOCH_2)_3P + HCHO \rightleftharpoons HOCH_2OCH_2P(CH_2OH)_2$$

In the reaction of THPC with sodium hydroxide (79) or other bases to produce THPOH, the pH of the solution must not exceed 7.1–7.5. In this range THPOH solution is stable. Like most other phosphines, THP can be oxidized at pH above 7.8 to tris(hydroxymethyl)phosphine oxide [1067-12-5] (THPO), which reacts only slowly with amides and amines.

Amides with two or more NH groups react with THPOH to produce insoluble polymers containing nitrogen and phosphorus. The phosphorus is in the form of a phosphine oxide; a methylol phosphine structure is probably an intermediate in these and similar reactions (see Phosphorus compounds).

THPOH–NH₃ Flame Retardants.   This flame retardant is applied by padding fabric with a solution of THPOH and various auxiliaries, then partially drying the fabric and finally exposing the partially dried fabric to ammonia gas (80). A typical formulation for flameproofing a 268 g/m² cotton fabric contains 30% THPOH (5.4% P) and wetting and softening agents. The fabric is dried to about 10–20% moisture, exposed to ammonia, and passed through a bath of H_2O_2 solution, which extends the durability of the finish by oxidizing phosphorus to the most stable phosphorus oxide (P=O) form, and removes odors and discoloration. The process can be modified by the addition of copper salts (81) to produce a system stable in the presence of ammonium hydroxide.

Since THPOH–NH₃ flame retardant does not impart stiffness, the process is applicable to most cotton fabrics. Fabrics weighing as little as 67 g/m² (56 g/yd²) have been satisfactorily flameproofed by this process. The treated fabric has high breaking strength, however tear strength is frequently reduced. With the proper amount of softening, tear strength can be restored. The phosphorus-to-nitrogen ratio in fabric treated by this process is about 3.5 to 1.

THPOH–TMM–NH₃ Flame Retardant.   A third system contains typically 18.5% THPOH, 11.5% trimethylolmelamine (TMM), plus wetting and softening agents (82). Such a formulation is adequate for a 268 g/m² cotton fabric, and enables it to pass the standard vertical flame test. TMM can replace about one-third of the THPOH required in the THPOH–NH₃ process at a considerable savings (83).

THPOH–Amide Flame Retardant.   This flame retardant is applicable to a wide variety of fabrics (84). The reagents are padded onto the fabric, dried, and then cured at about 150°C for 3 min. Fabrics retain 80–90% of their original breaking strength. Stiffness imparted is less than that imparted by the THPC-amide process, and like

THPC-amide, it gives a high degree of rot resistance. The treated fabrics have less tendency to yellow when exposed to sodium hypochlorite bleach.

   Miscellaneous.   Highly efficient flame retardants have been developed with reactive water-soluble compounds of low molecular weight and high phosphorus content. These compounds can be insolubilized by reaction with cellulose hydroxyls rather than *in situ* polymerization or copolymerizations (85). Included in this group are bis(aziridinyl)chloromethylphosphine oxide [*13846-34-9*] (86) and chloromethylphosphonic diamide [*6326-70-1*] (87–90). Cotton fabric treated with methylphosphonic diamide [*4759-30-2*] (MPDA) containing 12–14% add-on passes the children's sleepwear flammability standard FF3-71 (91).

   Flame and crease resistance can be imparted to cotton textiles in one operation (52), using guanidine [*133-00-8*], dimethyl phosphite [*868-85-9*], and formaldehyde [*50-00-0*].

$$
\underset{\underset{\displaystyle NH_2}{\overset{\displaystyle NH}{\|}}{NH_2-C-NH_2}} \;+\; 4\ HCHO \;+\; 2\ HPO(OCH_3)_2 \;\longrightarrow\;
$$

$$
HOCH_2 \underset{O \atop \overset{\|}{(CH_3O)_2P-CH_2}}{\overset{\displaystyle NH \atop \overset{\displaystyle \|}{C}}{\diagdown N \diagup \diagdown N \diagup}} CH_2OH \;\; \underset{CH_2-P(OCH_3)_2}{\overset{O}{\|}}
$$

A durable finish is obtained by using methanol in place of water as the solvent; an application time of 45 min is required for cotton. Excessive strength losses and poor reproducibility diminish the usefulness of this finish.

   Fyre-Fix.   The Fyre-Fix finish is based on a fiber-reactive material (92), and is a stable water-soluble material which is easy to handle and compatible with most resins and softeners. It is applied by a pad-dry-cure technique. The treated fabrics have good flame resistance, a soft hand, and no ion-exchange properties.

   Fyrol 76.   This flame retardant is a reactive vinyl phosphorus ester containing 22.5% phosphorus. It can be used alone or with methylolacrylamide (NMA) and a free-radical catalyst, eg, potassium persulfate. Fabric is treated with the conventional pad-dry-cure technique; rapid steam curing or radiation curing also can be applied. Sleepwear fabric requires approximately 25–30% add-on to pass the DOC FF3-71 test. The fabric has an excellent hand and some permanent press properties. Approximately 80–100% tensile strength and about 65% tearing strength are retained. Chlorine bleach and some carbonate detergents reduce the durability of this finish, whereas perborate bleaches do not affect it. As with the THPC-type finishes, an oxidation step is generally used.

   Pentamethylphosphorotriamide.   Of the phosphoramide derivatives, pentamethylphosphorotriamide [*10159-46-3*] is the most effective finish when applied to fabric in conjunction with dimethylolmelamine and when using an amine hydrochloride as catalyst. The finished fabric passes the FF3-71 flammability test. Its main application is for use on heavyweight clothes since the finish imparts a harsh hand to lightweight fabrics (93).

Application Techniques

   Use of radiation to effect fixation of some flame retardants is being investigated (94). Electron-beam fixation requires the selection of compounds that can be insolubilized inside or outside of the fiber with high yield in a short time. Polyunsaturated compounds, eg, Fyrol 76, are showing promise (see Radiation curing).

Flame retardants suitable for cotton are also suitable for rayon. A much better product is obtained by incorporating flame retardants in the viscose dope before fiber formation. The major classes of flame retardants used in the dope are listed in Table 4.

Thermoplastic Fibers. Thermoplastic fibers, eg, polyester and nylon, are considered less flammable than natural fibers because they possess a relatively low melting point and the melt drips rather than remaining to propagate the flame when the source of ignition is removed. Most common synthetic fibers have low melting points. Reported values for polyester and nylon are 255–290°C and 210–260°C, respectively.

Flame retardancy of the synthetic fibers is obtained by either mechanically building the retardant with the polymer before it is drawn into a fiber, or chemically modifying the polymer itself. Incorporation of chemicals in the dope before spinning the fiber has not been very successful. The most widely used technique for flame-retarding polyester acetate and triacetate is the application of Tris, tris(2,3-dibromopropyl) phosphate [126-72-7] (TDBP) [(BrCH$_2$CHBrCH$_2$O)$_3$P=O] by a Thermosol diffusion or an exhaustion technique (130–133). Polyester fiber can only retain about 4–5% TDBP calculated on the weight of the fiber. This was banned in 1977 by the Consumer Product Safety Commission (CPSC) as a potential carcinogen (see also Flame retardants, halogenated, and phosphorus compounds). Tris is not effective on cotton fabrics and was not used as a flame retardant for cotton textiles.

Table 4. Major Classes of Dope-Additive Flame Retardants for Viscose Rayon[a]

Active elements	Examples	Comments
phosphorus	alkyl and aryl phosphates, phosphonates and phosphites, polyphosphonates, Sandoflam 5060[b]	members of this class can be mixed with halogenated materials for synergistic effects
phosphorus–nitrogen	phosphazenes, phosphoryl or thionophosphoryl amides or ester amides, spirocyclotriphosphazenes, THPC-amines of condensates[c]	very efficient and effective; no unusual toxicological effects have been reported for a large number of these
phosphorus–halogen	halogenated alkyl or aryl phosphonates or polyphosphonates, halogenated alkyl or aryl phosphates, phosphites, or phosphazenes[d]	probably the largest class; severe toxicological problems for most useful members of this class
halogen	poly(vinyl or vinylidene halides) as latexes, poly(halogenated acrylate) latexes; emulsion or dispersions of alkyl or aryl halides, halogenated paraffins[e]	halogenated materials are not effective by themselves but are excellent synergists when combined with phosphorus compounds or colloidal antimony pentoxide; some latexes are not stable in viscose
antimony oxide	colloidal antimony pentoxide[f]	pigmentary antimony trioxide is not compatible with viscose; must be mixed with suitable halogenated synergist for necessary effectiveness

[a] Ref. 95.
[b] Ref. 96, 97.
[c] Ref. 98–109.
[d] Ref. 110–127.
[e] Ref. 127.
[f] Ref. 128, 129.

An alternative to the diffusion technique is the application of decabromodiphenyl oxide, on the surface of fabrics in conjunction with binders (134). Experimental finishes using graft polymerization, *in situ* polymerization of phosphorus-containing vinyl monomers or surface halogenation of the fibers also have been reported (132–133,135,136).

Mobil developed a flame retardant, Antiblaze 19, for polyester fibers (137). It is a nontoxic mixture of cyclic phosphonate esters. Antiblaze 19 is 100% active whereas Antiblaze 19T is a 93% active, low-viscosity formulation for textile use. Both are miscible with water and are compatible with wetting agents, thickeners, buffers, and most disperse dyes formulations.

Antiblaze 19/19T can be diffused into 100% polyester fabrics by the Thermosol process for disperse dyeing and printing. This requires heating at 170–220°C for 30–60 s.

Fire-Retardant Fiber Blends

The flammability behavior of thermoplastic and nonmelting polyester/cellulosic fiber blends cannot be predicted from the flammability of the single-fiber structures (138–141). The treatment of one of the fiber components with a flame-retarding agent specifically effective for that component does not necessarily render the two-component blend flame resistant (138,140) unless the treated component comprises at least 85% of the blend (74). An effective flame-retarding treatment must reduce the flammability of each component. It is sufficient to flameproof only one fiber component and the second one can be completely free of finish, if a flame-retarding agent is used that is effective on both types of fibers (142).

Considerable effort is being made to develop satisfactory flame retardants for blended fabrics. It is now feasible to produce flame-resistant blended fabrics that contain about 65% or more cellulosic fibers. Before additional significant advances can be made, it is necessary to develop flame retardants for thermoplastic fibers. Flame retardants are already available for the cellulosic fibers. The most practical approach may be to produce flame resistant thermoplastic fibers by altering the chemical structure of the polymers and, after blending these fibers with cotton or rayon, treating the finished textile with an appropriate flame retardant for the cellulose.

In 1972, the White Chemical Company developed Caliban F/R P-44, a nontoxic flame retardant for textiles and textile blends (143). It consists of decabromodiphenyl oxide, an acrylic latex, and antimony oxide in emulsion form, which is applied by the pad-dry-cure technique.

A flame retardant based on THPC-amide plus poly(vinyl bromide) [*25951-54-6*] (144) is suitable for use on 50/50 polyester-cotton blends. It is applied by the pad-dry-cure process, with curing at 150°C for about 3 min. A typical formulation contains 20% THPC, 3% disodium acid phosphate, 6% urea, 3% trimethylolglycouril [*496-46-8*], and 12% poly(vinyl bromide) solids. Approximately 20% add-on is required to impart flame retardancy to a 168 g/m² 50/50 polyester–cotton fabric. Treated fabrics pass the FF 3-71 test.

The THPOH-NH₃ finish, and the Fyrol 76 finish also impart flame retardancy to certain cotton–polyester blends if the blends contain at least 65% cotton.

The use of LRC-100 flame retardant for 50/50 polyester cotton blends has been reported recently (145). It is a condensation product of tetrakis(hydroxymethyl)-

Table 5. Flame-Resistant Synthetic Fibers [a]

Fiber	Company	Chemical nature	Composition of fabric	Applications
PFR Rayon	Avnet Fibers	rayon containing phosphazene ester	100% PFR, blends	safety clothing, aircraft upholstery and institutional drapery
Polyester Type G.H.	Toyobo Co.	polyester containing aromatic, sulfur-containing phosphonate	100% polyester, blends with certain fibers	apparel and interior furnishings
Extar FR [b]	Teijin	bromine-containing polyester	100% polyester, blends with certain fibers	developmental product; drapery
Trevira 271	Hoechst	polyester with comonomer		commercial product; apparel, interior furnishings
SEF [c]	Monsanto	modacrylic	100% SEF, blends with acrylic or polyester	apparel, drapery and industrial fabrics
Verel	Eastman	modacrylic	100% Verel, blends with rayon or polyester	interior furnishings
Kanekaron	Kanegafunci Chemical Co.	modacrylic		apparel, interior furnishings
Lufnen	Kanebo Acrylic Fibers Co.	modacrylic		apparel, interior furnishings
Treviron [d]	Teijin	vinyon	100% vinyon, blends	apparel, interior furnishing
Leavil [e]	Montedison	vinyon	100% vinyon, blends	apparel, interior furnishings
Clevyl T	Rhone-Poulenc	vinyon	100% vinyon, blends	apparel, interior furnishings
Cordelan I and II [f]	Kohjin Co.	polychlal	100% Cordelan, blends with up to 40% polyester	apparel, interior furnishings
Nomex 111	DuPont	aramid	100% Nomex, blends with Kynol, wool, etc	safety apparel, aircraft upholstery, specialty
Kevlar	DuPont	aramid	100% Kevlar, blends	safety apparel, industrial fabrics
Durette	Fire Safety Products, Inc.	modified aramid	100% Durette	space program, high oxygen areas, race drivers' gloves, specialty products
Fiberglass	Owens-Corning	glass	100% Fiberglass, blends	interior furnishing, safety apparel
Kynol	American Kynol	novoloid	100% Kynol, blends	safety apparel, specialty products

[a] From ref. 148, unless otherwise stated.
[b] Ref. 149.
[c] Ref. 150.
[d] Ref. 151.
[e] Ref. 152.
[f] Ref. 153.

phosphonium salt (THP salt) and *N,N',N"*-trimethylphosphoramide [*6326-72-3*] (TMPA). The precondensate is prepared by heating the THP salt and TMPA in a 2.3 to 1 mole ratio for one h at 60–65°C and is applied in conjunction with urea and trimethylol melamine in a pad-dry-cure oxidation wash procedure. To pass the FF 3-71 test, 3.5–4.0% phosphorus is needed.

In February 1978 CPSC approved changes in the FF-3 and FF-5 standards for childrens' sleepwear, eliminating the melt-drip time limit and coverage for sizes below 1, and revised the method of testing the trim. This permits the use of untreated 100% nylon and 100% polyester for children's sleepwear.

Halogens are less effective flame retardants on nylon than on polyester, and most of the flame retardants effective on cellulosics or on polyester substrates are not effective on nylon. Thiourea and thiourea formaldehyde appear to be the most effective treatments for imparting flame resistance to nylon (146–147).

The burning characteristics of synthetic and natural fibers are very different. Synthetic fibers tend to melt, drip, and shrink away when in contact with a flame, whereas cotton does not distort when subjected to a heat source. When synthetic and natural fibers are combined, the cotton serves as a grid for the polyester, thus preventing it from dripping, and the polyester, which has a higher heat of combustion, accelerates the pyrolysis of cotton. Thus, when natural and synthetic fibers are combined, many of the resulting fabrics are more hazardous than fabrics made from the individual components. Table 5 lists the commercially available inherent flame-retardant fibers.

Mutagenicity

Because Tris—a polyester flame-retardant chemical—is a potential carcinogen (154–156), workers in this field have tested a number of commonly used chemicals for potential mutagenicity.

The THPOH–NH$_3$ finish was tested recently (157) by the Ames mutagenicity test (158–161) and neither the finish nor its extracts caused a significant systematic increase in mutations.

The Hooker Chemical Co. has just released the results of tests conducted by an independent laboratory which indicated no significant mutagenic potential from any of the company's proprietary textile flame retardants.

Fyrol 76 also is reported to be nontoxic and nonmutagenic.

Stauffer's substitute for Tris, Fyrol FR2, was accused of mutagenic activity by the Environmental Defense Fund, and was withdrawn from the market by the company.

Because of such uncertainties, several major suppliers of flame-retardant chemicals have reduced their operations in this area or have abandoned the market altogether. This has forced the apparel industry to use inherently flame-resistant fibers.

BIBLIOGRAPHY

"Fire-Resistant Textiles" in *ECT* 1st ed., Vol. 6, pp. 544–558, by G. S. Buck, Jr., National Cotton Council of America; "Fire-Resistant Textiles" in *ECT* 2nd ed., Vol. 9, pp. 300–315, by George L. Drake, Jr., U.S. Department of Agriculture.

1. M. M. Sandholtzer, *Natl. Bur. Std. (U.S.), Circ.* **C455** (1946).
2. Br. Pat. 551 (1735), Obadiah Wyld.

3. J. L. Gay-Lussac, *Ann. Chim.* **18**(2), 211 (1821).

4. W. H. Perkin, *J. Ind. Eng. Chem.* **5,** 57 (1913); *Met. Chem. Eng.* **10,** 636 (1912).

5. *Bull. N.Y. Acad. Med.* **43**(8), 615 (1967).

6. *Public Law 90-189, S 1003, 90th Congr., Sect. 4, Dec. 14, 1967,* U.S. Government Printing Office, Washington, D.C., 1967.

7. D. A. Yeadon and co-workers, in *Proc. 10th Ctn. Util. Res. Conf. USDA, ARS 72-83,* SRRC, New Orleans, La., p. 70.

8. L. W. Mazzeno, Jr., and co-workers, *Text. Chem. Color* **5**(3), 43 (1973).

9. R. W. Little, *Flameproofing of Textile Fabrics,* Reinhold Publishing Corp., New York, 1947.

10. G. C. Tesoro, *Textilveredlung* **2,** 435 (1967).

11. G. C. Tesoro, S. B. Sello, and J. Willard, Jr., *Text. Res. J.* **38,** 245 (1968).

12. *Ibid.,* **39,** 180 (1969).

13. J. E. Hendrix and co-workers, *J. Appl. Sci.* **14**(7), 1701 (1970).

14. J. E. Hendrix and co-workers, *J. Fire Flammability* **3,** 2 (1972).

15. D. A. Yeadon and co-workers, *Proc. 10th Ctn. Util. Conf. USDA, ARS 72-83,* SRRC, New Orleans, La., 1970, p. 70.

16. Gulf Coast Section, *Text. Chem. Color.* **4**(12), 287 (1972); J. E. Hendrix and co-workers, *Text. Res. J.* **41**(10), 854 (1971).

17. R. J. Brysson and co-workers, *Information Council on Fabric Flammability, Proceedings of the 5th Annual Meeting,* Dec. 9, 1971, New York, p. 138.

18. R. M. Perkins, G. L. Drake, and W. A. Reeves, *J. Am. Oil Chem. Soc.* **48**(7), 303 (1971).

19. J. M. Church, "Fundamental Studies of Chemical Reactants for Fire-Resistant Treatment of Textiles" in *U.S.Q.N.C. Textile Series, Report* **38,** U.S.Q.N.C., Natick, Mass., 1952.

20. R. W. Little, *Flameproofing Textile Fabrics, American Chemical Society Monograph No. 104,* Reinhold Publishing Corp., New York, 1947, p. 41.

21. H. A. Schuyten, J. W. Weaver, and J. D. Reid, *Adv. Chem. Ser.* **9,** 7 (June, 1954).

22. "A Fundamental Study of the Pyrolysis of Cotton Cellulose To Provide Information Needed for Improvement of Flame Resistant Treatments For Cotton," *Final Report 1959/1964 on PL 480 Project No URE-29-(20)-9; Grant No. FG-UK-108-59,* The Cotton, Silk and Man-Made Fibers Research Association, Didsbury, England, 1964.

23. J. B. Berkowitz-Mattuck and T. Naguchi, *J. Appl. Polym. Sci.* **7,** 709 (1963).

24. E. Heuser, *The Chemistry of Cellulose,* John Wiley & Sons, Inc., New York, 1944, p. 546.

25. R. W. Little, *Flameproofing Textile Fabrics, American Chemical Society Monograph No. 104,* Reinhold Publishing Corp., New York, 1947; U.S. Pat. 2,482,755 (Sept. 27, 1949), F. M. Ford and W. P. Hall (to Joseph Bancroft and Sons Co.).

26. J. D. Reid and L. W. Mazzeno, Jr., *Ind. Eng. Chem.* **41,** 2828 (1949).

27. M. Leatherman, *U.S. Dep. Agric. Circ.* **466** Washington, D.C., 1938.

28. F. J. Kilzer and A. Broido, *Pyrodynamics* **2,** 151 (1965).

29. C. H. Mack and D. J. Donaldson, *Text. Res. J.* **37,** 1063 (1967).

30. W. A. Reeves, G. L. Drake, Jr., and R. M. Perkins, *Fire Resistant Textiles Handbook,* Technomic Publishing Co., Inc., Westport, Conn., 1974, p. 45.

31. J. D. Reid, W. A. Reeves, and J. G. Frick, *U.S. Dep. Agric. Leaflet No. 454,* USDA, Washington, D.C., 1959.

32. J. R. W. Perfect, *J. Soc. Dyers and Colour.* **74,** 829 (1958).

33. U.S. Pat. 2,570,566 (Oct. 9, 1951), F. W. Lane and W. L. Dills; U.S. Pat. 2,607,729 (Aug. 19, 1952), W. L. Dills; U.S. Pat. 2,785,041 (March 12, 1957), W. W. Riches (to E. I. du Pont de Nemours & Co., Inc.); U.S. Pat. 2,658,000 (Nov. 3, 1953), W. F. Sullivan and I. M. Panik; U.S. Pat. 2,691,594 (Oct. 12, 1954), J. P. Wadington; U.S. Pat. 2,728,680 (Dec. 27, 1955), D. Duane (to National Lead Co.).

34. U.S. Pat. 2,570,566 (Oct. 9, 1951), F. W. Lane and W. L. Dills (to E. I. du Pont de Nemours & Co., Inc.); H. C. Gulledge and G. R. Seidel, *Ind. Eng. Chem.* **42,** 440 (1950); *J. Text. Inst.* **41**(7), 357 (July, 1950).

35. U.S. Pat. 2,658,000 (Nov. 3, 1953), W. F. Sullivan and I. M. Panik (to National Lead Corp.).

36. U.S. Pat. 2,229,612 (Oct. 20, 1942), E. C. Clayton and L. L. Heffner (to Wm. E. Hooper and Sons Co.).

37. R. Van Tuyle, *Am. Dyest. Rep.* **32,** 297 (1943); *J. Text. Inst.* **34**(10), 587 (Oct. 1943); N. J. Read and E. G. Heighway-bury, *J. Soc. Dyers Colour.* **74,** 823 (1958).

38. U.S. Pat. 2,518,241 (Aug. 8, 1950), J. F. McCarthy (to Treesdale Laboratories); U.S. Pat. 2,591,368 (April 1, 1952), S. H. McAllister (to Shell Development Co.).

39. U.S. Dept. of Commerce, *Commercial Standards 191-53* (revised), U.S. Government Printing Office, Washington, D.C., 1954.
40. *Fed. Regist.* **35**(74), 6211 (1970).
41. *Fed. Regist.* **35**(251), 19702 (1970).
42. *Fed. Regist.* **37**(141), 14624 (1972).
43. *Fed. Regist.* **39**(85), 15210 (1974).
44. *Fed. Regist.* **37**(110), 11362 (1972); *Fed. Regist.* **38**(78), 10111 (1973); *Fed. Regist.* **38**(81), 10482 (1973); *Fed. Regist.* **38**(110), 15095 (1973); *Fed. Regist.* **38**(117), 15990 (1973).
45. *Text. Chem. Color.* **2**(3), 49 (1970).
46. C. P. Fenimore and F. J. Martin, *Mod. Plast.* **43,** 141, 146, 148, 192 (1966); J. S. Isaacs, *J. Fire Flammability* **1,** 36 (1970).
47. E. Pacsu and R. F. Schwenker, *Text. Res. J.* **27,** 173 (1957); R. F. Schwenker and E. Pacsu, *Ind. Eng. Chem.* **50,** 91 (1958); U.S. Pat. 2,990,232–2,990,233 (June 27, 1961), E. Pacsu and R. F. Schwenker, Jr. (to Textile Research Institute).
48. D. M. Gallagher, *Am. Dyest. Rep.* **53,** 23 (1964).
49. G. L. Drake, Jr., W. A. Reeves, and J. D. Guthrie, *Text. Res. J.* **29,** 270 (1959).
50. R. B. LeBlanc and D. A. LeBlanc, *Proceedings of the Symposium on Textile Flammability,* LeBlanc Research Corp., East Greenwich, R.I., 1974, p. 1.
51. S. J. O'Brien, *Text. Res. J.* **38,** 256 (1968); *Pyroset DO Fire Retardant, Text. Fin. Bull. No. 130,* 2nd ed., American Cyanamid Co., Bound Brook, N.J., 1959.
52. W. A. Sanderson, W. A. Muller, and R. Swidler, *Text. Res. J.* **40**(3), 217 (1970).
53. Fr. Pats. 1,395,178 (April 9, 1965) and 1,466,744 (Jan. 20, 1967) (to Ciba Ltd.); U.S. Pat. 3,374,292 (March 19, 1968), A. C. Zahir (to Ciba Ltd.); R. Aenishanslin and co-workers, *Text. Res. J.* **39**(4), 375 (1969); *Technical Bulletin 081-4B (270M),* Ciba Chemical and Dye Co., Toms River, N.J.
54. G. C. Tesoro, S. B. Sello, and J. J. Willard, *Text. Res. J.* **38**(3), 245 (1968).
55. J. P. Moreau and L. H. Chance, *Am. Dyest. Rep.* **59**(5), 37, 64 (1970).
56. L. H. Chance and J. P. Moreau, Paper presented at the *American Chemical Society 159th Meeting, Division of Cellulose, Wood, and Fiber Chemistry, Feb. 22–27, 1970, ACS,* Houston, Texas, 1970.
57. L. H. Chance and J. D. Timpa, *J. Chem. Eng. Data* **22**(1), 116 (1977).
58. L. H. Chance and J. D. Timpa, *Text. Res. J.* **47**(6), 418 (1977).
59. J. D. Timpa and L. H. Chance, *J. Fire Retard. Chem.* **5,** 93 (1978).
60. W. A. Reeves and J. D. Guthrie, *Text. World* **104**(2), 176, 178, 180, 182 (Feb. 1954).
61. W. A. Reeves and J. D. Guthrie, *Ind. Eng. Chem.* **4,** 64 (1956).
62. J. G. Frick, Jr., J. W. Weaver, and J. D. Reid, *Text. Res. J.* **25,** 100 (1955).
63. C. Hamalainen and J. D. Guthrie, *Text. Res. J.* **26,** 141 (1956).
64. C. Hamalainen, W. A. Reeves, and J. D. Guthrie, *Text. Res. J.* **26,** 145 (1956).
65. J. D. Reid, J. G. Frick, Jr., and R. L. Arceneaux, *Text. Res. J.* **26,** 137 (1956).
66. W. A. Reeves and co-workers, *Text. Res. J.* **27,** 260 (1957); G. L. Drake, Jr., J. V. Beninate, and J. D. Guthrie, *Am. Dyest. Rep.* **50,** 129 (1961).
67. U.S. Pat. 2,722,188 (Nov. 27, 1956), W. A. Reeves and J. D. Guthrie (to U.S. Secy Agr.).
68. U.S. Pat. 2,983,623 (May 9, 1961), H. Coates (to Albright and Wilson Ltd.); Brit. Pat. 906,314 (Sept. 19, 1962), H. Coates (to Albright and Wilson Ltd.); Brit. Pat. 938,989–938,990 (Oct. 9, 1963), H. Coates and B. Chalkley (to Albright and Wilson Ltd.).
69. G. L. Drake, Jr., W. A. Reeves, and R. M. Perkins, *Am. Dyest. Rep.* **52,** 608 (1963).
70. F. L. Normand, D. J. Donaldson, and G. L. Drake, Jr., *Text. Ind.* **134**(6), 169, 176, 186, 188, Atlanta (1970).
71. R. B. LeBlanc, *Text. Ind.* **132**(10), 274, 294, 296, 298, Atlanta (1968).
72. U.S. Pat. 3,243,391 (March 29, 1966), G. M. Wagner (to Hooker Chemical Corp.).
73. J. K. Sharma, L. LaL, and H. I. Bhatnagar, *Indian J. Text. Res.* **2,** 116 (1977).
74. W. A. Reeves and J. D. Guthrie, *U.S. Dep. Agric. Ind. Chem. Bull.* (364), (1953).
75. J. D. Guthrie, G. L. Drake, Jr., and W. A. Reeves, *Am. Dyest. Rep.* **44**(10), 328 (1955).
76. J. V. Beninate and co-workers, *Text. Ind.* **131**(11), 110, Atlanta (1967).
77. N. Filipescu and co-workers, *Can. J. Chem.* **41,** 821 (1963).
78. W. J. Vullo, *J. Org. Chem.* **33**(9), 3665 (1968).
79. S. E. Ellzey, Jr., and co-workers, private communication, 1970.
80. J. V. Beninate and co-workers, *Am. Dyest. Rep.* **57**(25), 74 (1968).
81. D. J. Donaldson and D. J. Daigle, *Text. Res. J.* **39**(4), 363 (1969).
82. J. V. Beninate and co-workers, *Text. Res. J.* **39**(4), 368 (1969).

83. D. J. Daigle and D. J. Donaldson, *Text. Chem. Colour.* 1(24), 34 (1969).

84. J. V. Beninate and co-workers, *Text. Res. J.* 38(3), 267 (1968).

85. G. C. Tesoro, *Pure Appl. Chem.* 46, 239 (1976).

86. G. C. Tesoro, W. F. Olds, and R. M. Babb, *Text. Chem. Color.* 6, 148 (1974).

87. Dutch Pat. 661 5460 (May 3, 1967), (to Badische Anilin & Soda Fabrik Co.); Brit. Pat. 1,126,259 (Sept. 5, 1968).

88. C. E. Morris and L. H. Chance, *Text. Res. J.* 43, 336 (1973).

89. G. Tesoro, E. I. Valko, and W. Olds in V. M. Bhatnagar, ed., *Proceedings of the Second International Conference on Flammability and Flame Retardants Montreal, Canada, May 22–23, 1975,* Technomic Publishing Co. Inc., Westport, Conn.

90. G. Tesoro, E. I. Valko, and W. Olds, *Text. Res. J.* 46, 152 (1976).

91. G. Tesoro, *J. Appl. Polym. Sci.* 21, 1073 (1977).

92. *Fyre-Fix,* Syntron, Inc., Pawtucket, R.I.

93. R. B. LeBlanc, J. H. Badger, and K. P. Clark, *AATCC 1974 National Technical Conference, Book of Papers,* AATCC, 1974, p. 462.

94. B. Court and co-workers, *Am. Dyest. Rep.* 67(1), 32 (1978).

95. N. A. Portnoy and G. C. Daul, *AATCC Book of Papers, AATCC 1978 Nat. Tech. Conf.,* AATCC, Research Triangle Park, N.C., 1978, p. 269.

96. A. Granzow, *Acc. Chem. Res.* 11(5), 177 (1978).

97. Brit. Pat. 1,156,588 (Oct. 24, 1967), H. Watanabe, K. Yagomi, and T. Tumori (to Kokoku Rayon & Pulp Co.).

98. Gen. Offen. 2,242,681 (March 15, 1973), H. Noehbur and A. Maeder (to Ciba Geigy A.-G.).

99. U.S. Pat. 3,947,276 (March 30, 1976), F. Sidan and P. Rossi (to Snia Viscosa Societa Nazionale Industria Applicapori Viscosa S.p.A.).

100. Fr. Pat. 2,310,376 (Dec. 3, 1976), (to Ethyl Corp.).

101. Jpn. Pat. 76,139,197 (Dec. 1, 1976), (to Ethyl Corp.).

102. U.S. Pat. 3,994,996 (Nov. 30, 1976), R. Borivo; and co-workers (to FMC).

103. U.S. Pat. 4,006,125 (Feb. 1, 1977), K. A. Regnard and A. H. Guber (to Horizons, Inc.).

104. U.S. Pat. 3,990,900 (Nov. 9, 1976), R. Borivoj, E. F. Orwoll, and J. F. Start, (to FMC).

105. U.S. Pat. 4,011,089 (March 8, 1977), J. T. Kao (to Ethyl Corp.).

106. Jpn. Pat. 74 124,323 (Nov. 28, 1974), I. Masuda and co-workers (to Omi Kenshi Spinning Co. Ltd.).

107. Jpn. Kokai 76 17,321 (Feb. 12, 1976), A. Kawai and co-workers (to Mitsubishi Rayon Co. Ltd.).

108. Jpn. Pat. 76 26,319 (March 4, 1976), H. Terasawa, T. Kondo, and I. Saito (to Dondo, Ltd.).

109. R. Wolf, "Flame Retardant Dope Additives for Regenerated Cellulose Fibers" in R. Bruce LeBlanc, ed., *Proceedings of the Symposium on Textile Flammability,* LeBlanc Research Corporation, East Greenwich, R.I., 1978, p. 248.

110. U.S. Pat. 3,266,918 (Aug. 16, 1966), J. W. Schappel and A. I. Bates (to FMC Corp.); U.S. Pat. 3,455,713 (July 15, 1969), J. W. Schappel and A. I. Bates (to FMC Corp.).

111. J. W. Shappel, *Mod. Text.* 49, 54 (1968).

112. L. E. A. Godfrey, *Text. Res. J.* 40, 116 (Feb. 1970).

113. Can. Pat. 769,630 (Oct. 17, 1967), M. P. Godsay (to Courtaulds, Canada, Ltd.).

114. Neth. Appl. 6,914,207 (Dec. 29, 1969), N. V. Aku.

115. Jpn. Kokai 76 17,325 (Feb. 12, 1976), K. Mimura, A. Kawai, and H. Ito (to Mitsubishi Rayon Co., Ltd.).

116. Jpn. Kokai 76 17,324 (Feb. 12, 1976), M. Inoue, K. Mimura, and K. Kagewa (to Mitsubishi Rayon Co., Ltd.).

117. Jpn. Kokai 76 17,322 (Feb. 12, 1976), M. Inoue, M. Mimura, and K. Kagewa (to Mitsubishi Rayon Co., Ltd.).

118. Jpn. Kokai 76 17,323 (Feb. 12, 1976), K. Mimura, A. Kawai, and T. Kafsuyama (to Mitsubishi Rayon Co., Ltd.).

119. Jpn. Pat. 72 23,612 (Oct. 13, 1962), Y. Oskubo, M. Ohki, and H. Terayawa (to Kaneyafuchi Co., Ltd.).

120. Jpn. Pat. 74 72,415 (July 12, 1974), Y. Kato and co-workers (to Toyobo Co. Ltd.).

121. Jpn. Pat. 74 125,625 (Dec. 2, 1974), K. Nagai and co-workers (to Toyo Rayon Co., Ltd.).

122. Jpn. Pat. 74 75,821 (July 22, 1974), Y. Koto and co-workers (to Toyobo Co., Ltd.).

123. Jpn. Pat. 75 59,512 (May 22, 1975), S. Nagai and co-workers (to Mitsubishi Rayon Co. Ltd.).

124. Jpn. Pat. 73 75,817 (Oct. 12, 1973), H. Kishida and co-workers (to Diawa Spinning Co., Ltd.).

125. Jpn. Pat. 73 91,312 (Nov. 28, 1973), K. Nagai, H. Okada, and I. Takequichi (to Toyo Rayon Co., Ltd.).
126. Jpn. Pat. 73 91,313 (Nov. 28, 1973), N. Yamomoto.
127. Jpn. Pat. 75 12,318 (Feb. 7, 1975), K. Nagai, H. Okoda, and I. Takeuchi (to Toyo Rayon Co. Ltd.).
128. U.S. Pat. 3,575,898 (April 20, 1971), R. L. McClure (to Beaunit Corp.).
129. U.S. Pat. allowed but unissued (Ser. No. 800,186), D. J. Unrau, N. A. Portnoy, and P. J. Hartmann.
130. E. Baer, in E. Greenwich, ed., *Proceedings of the Symposium on Textile Flammability*, LeBlanc Research Corp., East Greenwich, R.I., 1973, p. 117.
131. T. J. McGreehan and J. T. Maddock, *A Study of Flame Retardants for Textiles Report No. PB 251-441/AS*, National Information Service, Springfield, Va., 1976.
132. H. E. Stepniczka, *Textilveredlung* **10,** 188 (1975).
133. M. Lewin, S. M. Atlas, and E. M. Pearce, *Flame Retardant Polymeric Materials*, Plenum Publishing Corp., New York, 1975.
134. V. Mischutin, in R. B. Leblanc, ed., *Proceedings of the Symposium on Textile Flammability*, LeBlanc Research Corp., East Greenwich, R.I., 1975, p. 211.
135. R. Liepins and co-workers, *J. Appl. Polym. Sci.* **21,** 2529 (1977).
136. *Ibid.*, 2403 (1977).
137. B. E. Johnston, A. T. Jurewicz, and T. Ellison, *Proceedings of the Symposium on Textile Flammability*, LeBlanc Research Corp., East Greenwich, R.I., 1977, p. 209.
138. G. C. Tesoro and C. Meiser, Jr., *Text. Res. J.* **40,** 430 (1970).
139. W. Kruse and K. Filipp, *Mell.* **49,** 203 (1968).
140. W. Kruse, *Mell.* **50,** 460 (1969).
141. G. C. Tesoro and J. Rivlin, *Text. Chem. Col.* **3,** 156 (1971).
142. P. Linden and co-workers, *Textilveredlung* **6,** 651 (1971).
143. V. Mischutin, *J. Coated Fabr.* **7,** 308 (1978).
144. D. J. Donaldson and co-workers, *Am. Dyest. Rep.* **64**(9), 30 (1975).
145. R. B. LeBlanc, J. P. Dicarlo, and D. A. LeBlanc in ref. 138, p. 177.
146. D. Douglas, *J. Soc. Dyers Colour.* **73,** 258 (1957).
147. H. Stepniczka, *Ind. Eng. Chem. Prod. Res. Devel.* **12**(1), 29 (1973).
148. R. B. LeBlanc, *Text. Ind.* **142**(2), 37, 44, 50, 53, 55, 57, 59 (1978).
149. H. Mori in ref. 135, p. 124.
150. A. A. Dunham in ref. 50, p. 181.
151. R. Tsuzuki, A. Hashizume, and H. Mori in ref. 150, p. 283.
152. P. L. Susani in ref. 50, p. 268.
153. W. L. Morrison, *Proceedings of the Symposium on Textile Flammability*, R. I. LeBlanc Corp., 1976, p. 144.
154. R. W. Morrow and co-workers, *Am. Ind. Hyg. Assoc. J.* **37,** 192 (1976).
155. F. A. Daniher, *Information Council on Fabric Flammability, Tenth Annual Meeting, New York, 10 December 1976.*
156. N. K. Hooper and B. N. Ames in *Regulation of Cancer-Causing Flame-Retardant Chemicals and Governmental Coordination of Testing of Toxic Chemicals, Serial No. 95-33,* Government Printing Office, Washington, D.C., 1977, p. 42–45.
157. L. W. Mazzeno, Jr. and N. Greuner, *Text. Chem. Color.* **9**(8), 38 (1977).
158. B. N. Ames, J. McCann, and E. Yamasaki, *Mutat. Res.* **31**(6), 347 (1975).
159. A. F. Kerst, *JFF/Fire Retard. Chem.* **1,** 205 (1974).
160. J. McCann and co-workers, *Proc. Natl. Acad. Sci. U.S.A.* **72,** 5135 (1975); J. McCann and B. N. Ames, *Proc. Natl. Acad. Sci.* **73,** 950 (1976).
161. B. N. Ames, J. McCann and E. Yamasaki, *Mutat. Res.* **31,** 347 (1975).

General References

W. C. Kuryla and A. J. Papa, eds. (1973–75), *Flame Retardancy of Polymeric Materials,* Vol. 5, Marcel Dekker, Inc., New York, 1979.
J. W. Lyons, *The Chemistry and Uses of Fire Retardants,* Wiley-Interscience, New York, 1970.
W. A. Reeves, G. L. Drake, Jr., and R. M. Perkins, *Fire Resistant Textiles Handbook,* Technomic Publishing Co., Inc., Westport, Conn. 1974.

Textile Flammability, A Handbook of Regulations, Standards and Test Methods, American Association of Textile Chemists and Colorists, Research Triangle Park, N.C., 1975.

GEORGE L. DRAKE, JR.
United States Department of Agriculture

LEATHERLIKE MATERIALS

Leatherlike material normally refers to synthetic materials that are used as substitutes for leather in those applications where the latter traditionally has been used. In footwear applications, these products generally are referred to as man mades, synthetics, or poromerics. In seating and upholstery, they usually are called coated fabrics or vinyls (see Coated fabrics). In apparel and accessory applications, they usually are given a name that is suggestive of the currently popular styling, eg, the wet look in the early 1970s, supersuede in the mid 1970s, and totes or disco in the late 1970s.

Leather is the material of choice; synthetics are materials of economic necessity. The labor shortage during World War II caused the substitution of neoprene for leather in shoe soles and insoles (see Elastomers, synthetic). The profit squeeze in the U.S. shoe industry during the 1950s led to the adoption of vinyl-coated fabrics for women's shoes. Periodic shortages caused by the cyclical nature of leather supplies cause continued inroads by new products into traditional leather markets. These include the use of urethane-coated fabrics in shoes and apparel and the development of poromerics for men's shoes (see Urethane polymers). Each of these products is designed to simulate certain properties of leather that are relevant to a specific application (see Leather). The suitability of synthetics as leather substitutes depends upon their being able to duplicate certain desirable characteristics of leather. These characteristics arise from the physico-chemical structure of the leather (1). The high permeability for air and moisture vapor is one of leather's most desirable properties in its main use area, ie, footwear.

It has been shown that high moisture absorption is necessary for a feeling of coolness and comfort in the uppers of closed shoes. The modulus of leather is low for the first 10–15% of elongation, after which point it increases abruptly. This type of modulus characterizes leather as highly adaptable to the lasting process in the manufacture of shoes.

Most tanned hides are not used in their original thicknesses but are split into two or more layers. Leather can be split easily to yield whatever thickness is needed for a given application. The various splits usually are sold as weight grades (kg/m^2 or oz/ft^2)

rather than by gauge (mm or mils). Leather also is readily skived to allow smooth folds in certain applications. However, it is difficult to incorporate the ability to be split or skived into many types of synthetics. Splitting is eliminated by manufacturing various weight grades, but skiving often remains a problem.

Any artificial, leatherlike material ideally should have mechanical properties that are similar to those of leather, as well as breathability and moisture vapor permeability. However, these ideal properties often are compromised in favor of economic considerations, and many products that have been marketed successfully merely look like leather but have very few of its physical properties. Therefore, it can be argued that the most important characteristics of leatherlike synthetics are esthetic ones.

Types

Leatherlike materials have one common characteristic; they look like leather in the goods in which they are used. For purposes of a technological description, however, they may be divided into two general categories: coated fabrics and poromerics. Poromerics are manufactured so as to resemble leather closely in both its barrier and transition properties, eg, for heat or moisture, and its workability and machinability. The barrier or permeability properties normally are obtained by manufacturing a controlled microporous structure.

The modulus characteristics of synthetics should match those of leather at 80°C rather than at room temperature. Although the latter temperature commonly is used as the standard in the industry, the former more closely matches the lasting conditions.

Coated Fabrics. All commercial nonporomerics are coated fabrics. Several types of fabrics have been used and the coating may be either vinyl or urethane, depending upon the desired balance between physical properties and economic considerations.

The earliest important coated fabric was solid vinyl on cotton sateen. Solid vinyl has the economic advantage of being relatively inexpensive. It can be embossed readily with a leatherlike grain and it fulfills, at least to a first approximation, the visual esthetic requirements for a leather substitute. However, it does not acquire a leatherlike crease or break with usage. This break, which can be observed in any leather shoe, garment, or chair is the formation of tiny permanent wrinkles in areas that are subjected to repeated flexing. Solid vinyl also feels cold and hard when touched, and it does not allow the transmission of air or moisture vapor. The latter deficiency is particularly disadvantageous in footwear applications.

A second type of coated fabric is expanded vinyl, which is of more recent origin than solid vinyl and gives a distinct improvement over the latter in tactile esthetics. Expanded vinyl is manufactured so as to have internal air bubbles which give the material enhanced flexibility and cause it to feel warmer and softer to the touch than solid vinyl. Figure 1 is a photomicrograph of a solid-vinyl coated fabric, and Figures 2 and 3 are examples of expanded vinyls; Figure 2 is a footwear-grade product and Figure 3 is an upholstery-grade product. The former is constructed of a woven fabric, and the latter has a knit backing and a thinner skin layer. The bubble structure, which is a closed-cell type of foam, does not confer any appreciable degree of permeability on the material. Expanded vinyl has the same shortcomings in the area of moisture-vapor transmission as does the solid product (see also Foamed plastics).

Figure 1.   Cross-sectional view of solid vinyl (100×).

The most recent development in coated fabrics is the use of urethane coating. Urethane-coated fabrics can be constructed so as to give natural break characteristics and, thus, have a leatherlike appearance. Since urethane is a hydrophilic polymer, urethane-coated fabrics show a certain amount of moisture permeability. Values from 5–18 $g/(m^2 \cdot h)$ have been reported (4). Although this represents some improvement over the total impermeability of vinyl, it is far below that of the better poromerics. Studies indicate that a minimum of 40 $g/(m^2 \cdot h)$ and, preferably, 60–80 $g/(m^2 \cdot h)$ is needed for comfort in a closed shoe. On the other hand, 15–20 $g/(m^2 \cdot h)$ sometimes is sufficient permeability for semiopen shoes. Figure 4 is a photomicrograph of a typical urethane-coated fabric.

Poromerics.   Poromerics have been developed in order to match the moisture vapor permeability, skivability, and nonfraying characteristics of leather; the products have microscopic, open-celled structures. The structure of a poromeric is more homogeneous than that of a coated fabric, since it is either all polymer or polymer reinforced with nonwoven fibers. The size and configuration of the cells is such that the passage of moisture vapor but not liquid water can occur. Although the vapor transmission rates of commercial poromerics vary considerably, the better ones can equal that of leather.

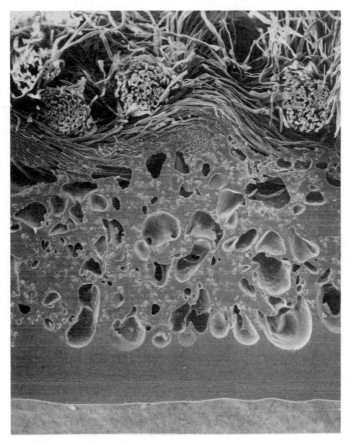

Figure 2.   Cross-sectional view of expanded vinyl (50×).

The attempt to obtain a skivable material generally involves using a nonwoven fibrous structure. One product, Porvair, is skivable because it has a homogeneous, unreinforced, all-urethane structure. Skivable products that are based on a woven/nonwoven combination structure have been introduced. The nonwoven approach has not been popular in the United States, but it has been very successful in Japan (eg, Clarino). The homogeneous product has been quite successful in Europe, and the woven/nonwoven approach is preferred in the United States (see Nonwoven textiles).

Figure 5 is a photomicrograph of Porvair. An unreinforced structure has the advantage of relative simplicity and flexibility in manufacture. Its principal drawback is that its modulus characteristics are the least leatherlike of any of the commercial synthetics. The nonreinforced poromeric has a rubberlike modulus curve. The pronounced modulus difference from leather requires considerable modification of lasting machinery in shoe applications. This disadvantage, however, must be balanced against the ability of a homogeneous product to be split or skived in any manner. Splitability allows simple adjustment of gauge for a wide variety of applications, and skivability allows application in areas requiring tight folding of the material.

A woven, reinforced product, eg, DuPont's Corfam, gives a more leatherlike

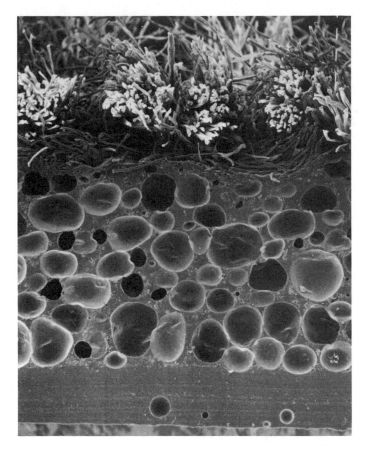

Figure 3.   Cross-sectional view of expanded vinyl (50×).

modulus curve than Porvair. However, the Corfam type of poromeric cannot be skived. One highly successful reinforced poromeric that can be skived is Clarino. This product is based on a loosely meshed nonwoven, about which a poromeric urethane is formed so that the fibers are not solidly bonded to the urethane but lie within the micropores. The different structures are distinguished in Figures 6, 7, and 8 of Aztran, Clarino, and Corfam, respectively.

Manufacture and Processing

Coating of Fabrics.   The type of fabric used as a coating substrate is determined by the intended application of the product. Woven fabrics are used for most footwear products, whereas knits are used for upholstery, apparel, or accessories where greater extensibility is required. The first vinyl-coated fabrics had cotton as the substrate. Recently, however, polyester and polyester/cotton-blend textiles also have been used; the choice depends upon a balance of factors, eg, cost, availability, and physical properties (see Polyesters). Fabric coating is reviewed in ref. 5.

The weight of solid vinyl coatings varies from 200–270 g/m^2 (6 to 8 oz/yd^2) and the thickness of the vinyl from 0.30–0.75 mm (12–30 mils), depending upon the specific

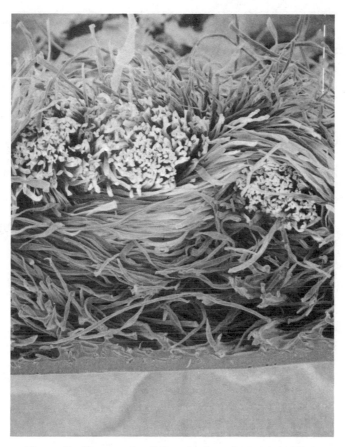

Figure 4.   Cross-sectional view of polyurethane fabric (70×).

use. The vinyl usually is a plasticized vinyl chloride homopolymer that is compounded with various stabilizers and pigments as desired. The ester-type plasticizer normally is 30–40 wt % of the vinyl, depending upon the required degree of softness and flexibility. The formulated vinyl is applied to the continuously moving textile web by coating, calendering, or extrusion (see Coating processes).

The form of vinyl polymer that is used depends upon the coating method. The first form is general-purpose resin, which is made using suspension polymerization and is processed as a melt. The second form is vinyl plastisol which is a suspension of fine vinyl particles in a plasticizer in which the vinyl is not readily soluble at room temperature. Additional nonsolvent may be added to reduce the viscosity in which case the mixture is called an organosol. The fine particles usually are obtained by spray drying emulsion-polymerized vinyl. This procedure yields particles from 0.5–5.0 μm in diameter. Plastisols are easier to handle, formulate, and pigment than are general purpose resins; however, the former are more expensive. Vinyl is used as plastisol in casting which may be done by either a direct or a transfer method. In the former, the plastisol is knife-coated directly onto the fabric which then is passed through an oven for fusion of the coating. Often, more than one coat is needed to obtain a smooth coating of the desired weight. In such cases, the multiple coats usually are applied sequentially

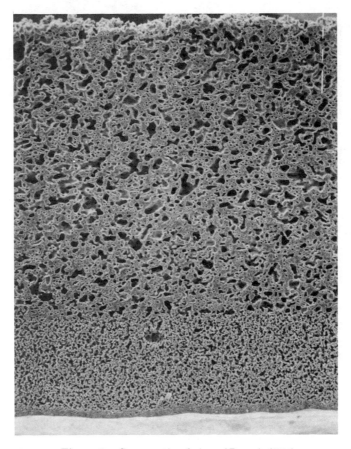

Figure 5.   Cross-sectional view of Porvair (65×).

on a single line. Although the direct-coating method is simple, it tends to produce a product that is relatively stiff as a result of the filling of the interstices in the fabric by the vinyl.

The loss of textile flexibility can be avoided by using the transfer-coating technique. The plastisol is deposited, usually by reverse-roll coating, on a web of release paper. An adhesive layer is applied to the plastisol after it has been passed through an oven. The fabric backing is laminated against the adhesive layer, and the sandwich is passed through another oven for fusion of the adhesive layer. Finally, the web is cooled, and the release paper is removed. As the adhesive layer does not penetrate very deeply into the fabric, a fairly flexible product can be obtained. Transfer casting is applicable to a wider variety of plastisol formulations than is direct casting.

In calender coating of solid vinyl, compounded vinyl first is formed into a thin uniform sheet on a calender and then is combined with the fabric by one of two methods. The two may be combined by fusing the vinyl to the fabric with heat and pressure, or they may be laminated with an adhesive layer. Calender coating lacks the processing flexibility of cast coating and generally is used for long runs of a single product.

Extrusion coating is similar to calender coating, except that the molten vinyl

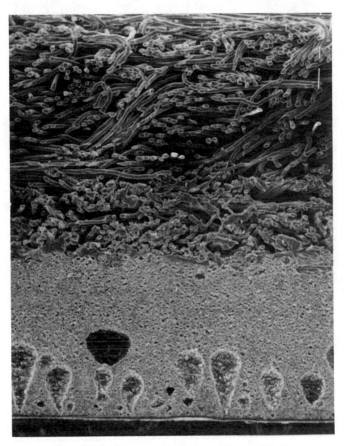

Figure 6.   Cross-sectional view of Aztran (55×).

compound is extruded through a sheet die directly into a nip where it is fused to the fabric. The products resemble those produced by heat-laminated calender coating. Like calender coating, extrusion coating is only suitable for long runs of a single product. Both of these techniques, however, have the advantage over cast coating of involving less-expensive, general-purpose grades of vinyl.

Expanded vinyl coatings are prepared by either the transfer coating or calender coating technique (6). In the case of transfer coating, at least two layers are necessary, ie, a skin layer and a foam layer. In U.S. manufacture, the foam layer acts as the adhesive layer. It is partially gelled, but not yet foamed, prior to its contact with the fabric. The usual practice in Europe is to use a three-layer coating, which includes a separate adhesive layer in addition to the skin and foam layers. The latter process requires a production line with three ovens, whereas the U.S. method involves only two ovens.

Foaming or blowing the interlayers is accomplished by including a blowing agent in the formulation of the vinyl. Upon being heated, the blowing agent decomposes to yield a gas, eg, N_2 or CO_2. The gas, in turn, produces a closed-cell foam structure. When expanded vinyl is made by calender coating, two layers of vinyl must be applied, ie, an expandable middle layer and a solid skin layer; the skin layer may be applied either before or after the foam layer is blown.

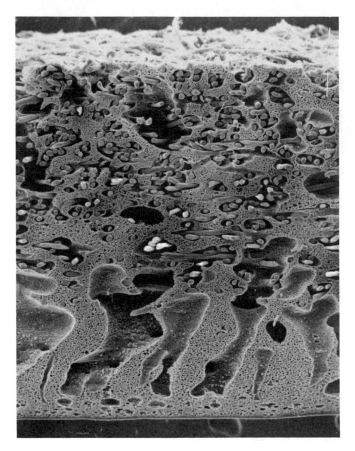

Figure 7.   Cross-sectional view of Clarino (70×).

The Oakes process, a mechanical alternative to chemical blowing, has been adopted by a number of European manufacturers. By this process, a foam is produced by beating air into the vinyl coating which then is cast onto the fabric. Problems of odor and bleeding have been caused by the surfactants used to produce the foams; however, unlike other vinyl products, these coatings are somewhat permeable. They are used in shoe linings and counters, whereas blown foams predominate in applications not requiring permeability, eg, open shoe uppers, handbags, and upholstery.

Urethane-coated fabrics first were made by direct coating of two-component systems. The components, a polyol and a polyisocyanate, were mixed as liquids and they reacted on the fabric to form the solid urethane (7) (see Isocyanates, organic). Because of problems regarding adhesion to the fabric and controlling softness, production changed to cast coating of thermoplastic urethane solutions. Urethanes have an advantage over vinyl in that their molecular structure renders them inherently flexible. The commercial products are block, or segmented, ester–urethane or ether–urethane copolymers. Whereas vinyl must be softened by the addition of a plasticizer, the urethanes are softened by decreasing the urethane segment lengths relative to the ester or ether segments (see Plasticizers). This segmented structure gives urethanes the widest span between glass transition temperature T_g and melting point T_m of any

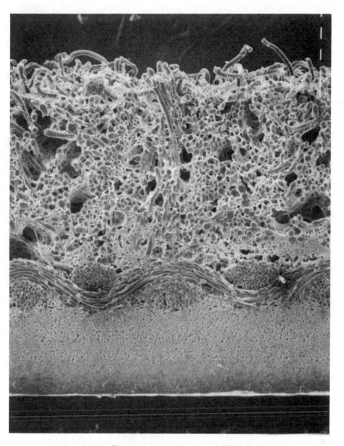

Figure 8.   Cross-sectional view of Corfam (70×).

commercial polymer. The temperature spread is wide enough to produce a product that can be subjected to very high temperatures in shoe lasting (eg, when used with hot-melt adhesives) but remain sufficiently flexible for use in outdoor winter temperatures.

Urethanes also have tensile strengths that are as much as ten times greater than those of vinyls. Because of their strength and their higher cost, urethanes are coated at much lower thicknesses than vinyl. Dry coating weights from 50–100 g/m² (1.5–3.0 oz/yd²) commonly are used. These low coat weights give the product a low heat capacity which gives the coated fabric a warm or cool hand under temperature conditions where vinyl-coated fabrics feel hot or cold; this heat capacity effect is of particular importance in upholstery applications. The thin coating also enhances the break characteristics of the product. To make the break more leatherlike, urethanes usually are coated onto fabrics that are napped on the coating side. The nap provides volume into which the coating can buckle to produce the creases of the break.

Urethanes do not require the stabilization against thermal degradation that vinyl does, but they do require stabilization against hydrolytic decomposition. The level of stabilization depends upon the type of urethane, with polyether-based types being considerably more stable to hydrolysis than polyester-based ones.

One of the major disadvantages of urethanes, with respect to vinyls, is that urethane solutions require strong solvents, eg, dimethylformamide (DMF) or tetrahydrofuran, both of which are expensive and toxic. Solvent recovery systems are available, but a large-volume line is needed to justify their cost.

Formation of Poromeric Structures. The manufacture of poromeric structures, particularly those with sufficient permeability for footwear applications, has presented the largest technological problem in the leather-substitute field. In the mid 1960s, over 20 chemical companies in the United States had projects directed toward this objective. The failure of DuPont's Corfam and Goodrich's Aztran were symptomatic of the problems in this area.

The simplest poromeric products are those that contain no reinforcing fiber, eg, Porvair (8). The most direct method of forming such a homogeneous microporous web is by polymer coagulation which involves extruding a web of urethane solution onto a suitable support. The web is immersed in a nonsolvent (for the polymer) which is miscible with the solvent used. The usual solvent/nonsolvent combination is DMF/water. When the urethane web is immersed in the water, DMF migrates to the aqueous phase and water begins to migrate into the web; simultaneously, the polymer precipitates from solution. The net effect of this precipitation/migration is the formation of a solid urethane that contains microscopic channels or pores.

Figure 9 is a flow diagram for the production of a typical poromeric, and Figure 10 is a diagram for the production of Porvair. Close control of the pore size is necessary in order to produce a product that has high permeability without having pores so large that they are visible to the naked eye. Pore size can be controlled by a number of factors, eg, urethane composition and solids content or time and temperature of the dunking cycle. Other process variations include adding small, measured amounts of nonsolvent to the urethane solution before extrusion and blowing steam onto the extruded web just before dunking.

One method of controlling pore size involves the use of a finely ground inert filler which is removed later in the process (see Fillers). For example, finely ground sodium

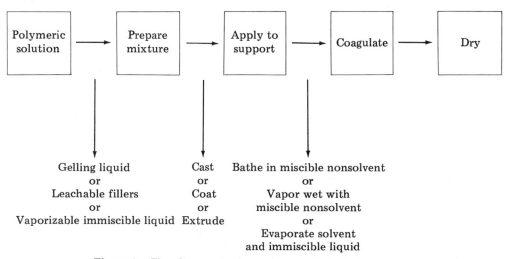

Figure 9.   Flow diagram for the production of a typical poromeric.

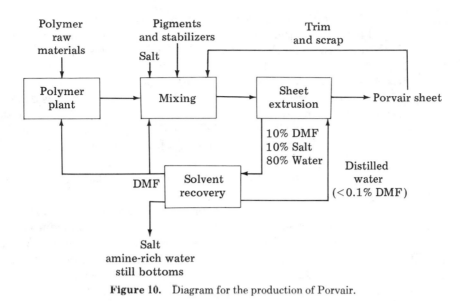

Figure 10.   Diagram for the production of Porvair.

chloride or ammonium sulfate is mixed with the urethane/DMF solution. When the solution is coagulated by dunking in water, the salt slowly leaches into the aqueous phase leaving cells in the web. The final solid structure from this process consists of large cells of ca 13 μm in diameter that are interconnected with smaller channels ca 1 μm in diameter.

The various techniques for producing fiber-reinforced poromerics result in a large variety of structures (9). The major variations include the type of fiber structure used and the degree to which it is impregnated with polymer. Although some poromerics, eg, Corfam, include a woven-fabric reinforcement, recent practice has tended toward using only nonwoven materials; the nonwoven-based products tend to match leather more closely in modulus. On the other hand, poromerics that are based upon woven or knitted fabrics generally have better drape characteristics. The nonwoven fibers usually are polyester or nylon; both confer strength and resiliency. A proportion of rayon (qv) commonly is used to give a degree of moisture absorption, and a small amount of polypropylene also is used (see Olefin polymers).

In general, these fibers are formed into a web by air-laying them to form a batt which is needle-punched to give the web density and integrity; needle punching is the most critical step in determining the web properties. As for general-purpose nonwoven textiles, a punching density of 60–90 punches per square centimeter is sufficient to yield a web that has planar integrity. However, needle punching results in a web of relatively low cohesive strength and of a uniform density on a macro scale but not on a micro scale. To achieve microscale uniformity, a punching density of 900–1500 punches per square centimeter would be needed. This would give a fabric with the same surface uniformity as a 90 × 60- or 100 × 100-count woven fabric. Such high punching densities are impractical by traditional needling methods. DuPont tried to avoid this problem first by using shrinkable fibers and then by putting a fine percale woven fabric between the nonwoven and the urethane layer (Corfam). This solution had drawbacks both in the high cost of the percale and in its total lack of extensibility.

The Japanese have developed an elegant nonwoven approach in the production of Clarino by producing fibers that are close in properties to the collagen fibers in leather; the method that was developed yields an islands-in-the-sea fiber. Two immiscible molten polymers are extruded to give a fiber with a cross-section, as shown in Figure 11. By subsequent selective solvent extraction of either the island or the sea phase, a hollow supple compressible fiber or a bundle of very fine fibers is obtained (10). Clarino is based upon a nonwoven web of the former type of fiber made of nylon. The latter type of fiber is used in Ultrasuede. The use of fibers of complex morphology or of bundles of fine fibers allows the formation of a high integrity microscopically uniform, nonwoven substrate at a relatively low needle-punching density. In addition to improving the process economics, this leads to a product of considerably enhanced suppleness.

After needle punching, the nonwoven is impregnated with a reinforcing resin and is coated with a urethane solution. The coating step is followed immediately by a coagulation step similar to that described for nonreinforced poromerics. Various products differ in the degree of impregnation of the fabric. The degree of impregnation has a pronounced effect on moisture absorption and permeability. Some products may be only lightly treated with a polymer, usually as a latex, to bind the fibers at their points of intersection (see Latex technology). This binding gives an open structure which, together with the coagulated coating layer, gives high permeability. In other cases, the nonwoven web may be heavily saturated with polymer solution, which subsequently is coagulated with the coating layer; this gives a structure which is microporous throughout. An extreme case of a polymer-saturated web is Clarino; the heavy impregnation, unique fiber structure, and carefully controlled coagulation conditions yields a structure that contains micropores which contain individual unbonded fibers.

A technique that has gained considerable commercial importance is coagulation coating (11) by which urethane/DMF solutions are cast onto a fabric similarly as for normal coated fabrics. However, instead of being dried by evaporation, they are coagulated in a water bath. This procedure produces a structure that has a poromeric layer on top of a fabric. These products are intermediate in properties between reinforced poromerics and urethane-coated fabrics and have the air permeability of the former and the physical properties of the latter.

Composites are manufactured by laminating a fabric backer, whether woven or nonwoven, to a previously prepared, unreinforced poromeric (see Composite materials).

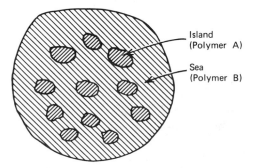

Figure 11.   Cross-sectional view of islands-in-the-sea fiber.

The difficulty with this manufacturing technique is in designing a bonding method that does not greatly reduce the permeability of the final structure or unduly increase its stiffness; however, this problem can be overcome by applying a discontinuous adhesive layer. The percentage of adhesive coverage must be closely controlled so that a good balance between permeability and adhesion is achieved. The microporous poromeric layer contributes elastic recovery and barrier properties, the fabric contributes leatherlike modulus, and the combination yields the product's tear strength. Composites represent one of the few instances where the addition of a plastic coating to a fibrous layer increases, rather than decreases, the tear strength of the fibrous substrate. However, a problem that is encountered with composites is that of achieving adhesion that is resistant to the moist, high temperature conditions that are encountered in shoe lasting. Traditionally, leather shoe uppers are steamed to soften them before lasting, and shoe manufacturers tend to steam synthetics, even though this is not necessarily needed.

The best overall properties in laminated structures have been obtained with a combination woven/nonwoven structure. The structure is prepared by heavily napping a sateen fabric, shearing the nap, and then impregnating the fabric with a reinforcing resin. The process results in a backing that is skivable and that has a leatherlike hand.

Finishing.   In addition to having suitable physical properties, it is necessary for a synthetic leather to have a leatherlike appearance. The appearance of a piece of leather results principally from two factors: color and texture. The finished surface of the material must have some resistance to abrasion and appropriate tactile properties. Three types of surfaces are commercially important: grained, patent, and suede, and each requires different finishing techniques. The method used for producing grain on a surface affects the manufacturing process for the basic material, eg, with a transfer-coated fabric, grain can be formed during the casting of the coating. This is done by casting onto a release paper that has had a negative version of the grain embossed into it (see Abherents). When the coating is laminated to the fabric, a product with the desired grain is obtained. With most poromerics, grain is obtained by embossing the complete structure. Embossing usually is preceded by one or more printing, spraying, or coating steps. Coating may be in a very thin layer that is just sufficient to give the product the desired color. Printing may be used where it is desired to have a color variation that matches the grain.

Embossing is a process of thermoforming, ie, heating the material, deforming it, and cooling to set it in its new shape. The web is passed over heated rolls that are engraved with a negative version of the desired grain. The grain can vary from very light (eg, calf) to very heavy (eg, reptile). In order that the grain be permanently impressed upon the elastomeric urethane, the embossing temperature must be very close to the material's melting point. However, at such high temperatures there is a risk that the pore structure may collapse (12). Even a very thin, impermeable layer on the surface can drastically reduce the overall moisture vapor transmission rate (MVT) of poromerics. It can be shown that for a structure of n layers,

$$K2_T = \left(\sum_{i=1}^{n} \frac{1}{K_i} \right)^{-1},$$

where K_T is the MVT of the total structure and K_i is the MVT of the ith layer in g/(m^2·h). Even one low value of K_i can markedly reduce K_T.

Maintaining permeability is not possible in a patent finish. However, as patent leather also has negligible permeability, this is not really a problem. Producing a patent effect requires a highly smooth surface which is coated with a clear layer (13); the latter gives the patent surface its appearance of depth. Producing a smooth, dust-free surface for coating is the critical step. One technique is to heat-fuse the surface in contact with a Mylar sheet by passing the Mylar and product web over a heated drum. Subsequent stripping of the Mylar then leaves a smooth, glossy surface. The lacquer used in the coating step may be urethane, vinyl, or acrylic. The urethanes may be either thermoplastic or two-component reactive systems; the reactive systems, although more difficult to handle, give a smoother, glossier surface than the thermoplastics. The viscosity increase resulting from the reaction occurs at a slower rate than that resulting from solvent evaporation. This slower rate allows the surface to become more level before it solidifies. Knife, spray, and roller-coating techniques have been used to apply patent lacquers.

A drawback of patent leather has been its tendency to crack during lasting; however, recent studies (14) indicate that synthetic patent leathers are not as prone to this problem. The difference is attributed to the greater extensibility at lasting temperature of the coatings used on the synthetics.

Inexpensive coated fabrics can be given a suedelike surface by floccing (15) which involves the bonding of short fibers more or less perpendicularly to the surface. The appearance mimics that of suede, which is made by napping leather. Suedes also have been produced from the type of heavily napped fabric described above (16). Rather than being laminated to poromeric layers, these fibers are sanded to give them a suedelike surface. Impregnation is controlled so as to provide the material with a high level of porosity.

A process that has been developed for vinyls involves coating an expanded vinyl on a fabric substrate and then tearing the cellular layer while it is hot. The tearing leaves a rough, irregular cell edge structure that has a fibrous appearance (17). A similar edge structure can be produced by sanding the surface of an expanded vinyl (18). The aesthetic qualities of the sanded surface can be improved if butadiene–acrylonitrile copolymer is included in the vinyl formulation (19). These products, which are based upon expanded vinyls, continue to hold a large share of the U.S. synthetic suede market.

Sueded, urethane-coated fabrics usually are produced by molding or hot embossing. A silicone rubber mold having cavities that closely resemble suede filaments is made; normally, the master mold is a flocked material. The thermoplastic urethane film is pressed into the mold, heated until it flows into the cavities, cooled, and removed. A semicontinuous version of this process was practiced in the U.S. in the early 1970s using flat-bed heated presses (20) which allowed the production of sheets only as large as the press size, ie, 127 × 137 cm. A continuous version of the process has been reported (21).

Sueded poromerics normally are produced from columnar cellular structures. The coagulation of the urethane solution is achieved under compositional and temperature conditions such that cylindrical, rather than spherical, cells are produced. The surface is abraded with rotary sanding machines to produce a suedelike effect (22). The columnar technique also has been applied to expanded vinyls (23).

None of the above methods can truly match the tactile qualities of suede, which result from the extremely fine diameter of the collagen fibers in leather. Conventional synthetic-fiber spinning techniques cannot produce fibers of sufficiently low tex (denier). Fine fibers have been made, however, using the islands-in-the-sea technique developed for Clarino (10). Commercial products, eg, Ultrasuede, have been produced by forming the islands-in-the-sea fibers into a nonwoven fabric and then dissolving the sea phase. This leaves a very highly entangled, nonwoven web of very fine fibers which then are impregnated with urethane. The unique nature of the nonwoven yields excellent drape characteristics, making it an exception to the general rule that woven-based products have better drape. The very low tex of the fibers also gives the product a feel that is almost indistinguishable from that of genuine suede.

Economic Aspects

In almost all cases, synthetic materials are considered to be less expensive substitutes for leather. Therefore, the economics of leatherlike synthetics are dominated by the state of the leather market. The leather market, in turn, is largely determined by the market for meat and dairy products, as hides are by-products of the cattle industry. As this latter business tends to be cyclical, so do hide prices. Because leather prices are only a fraction of the cost of finished leather products (10–20% in shoes), a rather large upward swing in leather prices is required to produce a significant advantage by switching to synthetics.

Substitution of synthetics for leather can be explained by the cyclical shortage of hides rather than by their cost. Hides must be processed within a short time after their production to avoid spoilage. The availability of hides is controlled by the slaughter rate which is controlled by the supply of and demand for meat. It has been observed that the demand side behaves normally, ie, it increases with increasing population and standard of living. On the other hand, the supply side behaves quite erratically, with a severe dip occurring approximately every seven years (eg, 1965, 1972, and 1979). This dip seems to be tied to the life cycle of cattle. Increasing future cattle herds requires an increase in calf production which requires that some heifers be bred as cows, thus removing them from the pool of slaughterable cattle, ie, current production must be sacrificed for the benefit of future increases. It has been observed that synthetics have an increased share of the footwear market during the dips in leather availability. However, once a synthetic does gain use in a market area, its use continues even when leather prices enter a subsequent down cycle. This phenomenon can be attributed to the properties of the synthetics. As they are uniform in composition and mostly thermoplastic in nature, they are well suited to quicker and more efficient manufacturing techniques. Cutting can be achieved in multiple layers from rolls; setting of uppers can be carried out rapidly by dry heat rather than by long steaming as with leather; and soling can be done by direct injection molding instead of the time-consuming sewing or cementing processes used with leather.

The future competitive position of synthetics may be seriously hampered by rising petroleum costs. With the single exception of cotton substrates for coated fabrics, all of the components of synthetics are derived from petrochemical feedstocks (qv). For example, poly(vinyl chloride) (PVC) is produced by way of ethylene from naphtha. Urethanes are based on aromatic isocyanates which are derived from benzene or toluene. Prices of these petrochemicals have increased dramatically since 1973 and probably will continue to do so. Eventually, these materials could be made from coal; however, although this would assure supply, it would do little to reduce cost. It should

Table 1.   Worldwide Production of Coated Fabrics in 1978, 10^6 m²[a,b]

	Urethane	PVC
Far East	80	400
Western Europe	150	550
Eastern Europe	30	300
North America	55	450
South America	15	100
Africa	10	25
Total	*340*	*1825*

[a] Ref. 24.
[b] To convert m² to ft², multiply by 10.76.

Table 2.   Estimated Consumption of Poromerics in 1978, 10^6 m²[a,b]

	All uses	Footwear only
Western Europe	2.6	1.9
Eastern Europe	3.7	2.3
North America	1.4	0.9
Far East	3.1	1.3
Total	*10.8*	*6.4*

[a] Ref. 24.
[b] To convert m² to ft², multiply by 10.76.

be noted that leather prices are not entirely independent of the petroleum market. Cattle ranching, particularly in the U.S. where cattle are grain fed rather than range fed, is an energy-intensive industry. Steeply increased energy costs could discourage the production and consumption of beef and, thereby, decrease the hide supply. Prices (1979) for synthetics range from ca \$4.30/m² (\$0.40/ft²) for vinyls to ca \$13.45/m²

Table 3.   Worldwide Production of Poromerics in 1978, 10^6 m² [a,b]

Company	Product	Total capacity	Estimated production
Far East			
Kuraray	Clarino		
Kanebo	Patora	8.8	6.9
Toray	Alcantara		
Teijin	Cordley		
Eastern Europe			
Pronit	Polcorfam (Poland)		
Exico	Barex (Czechoslovakia)	3.3	1.9
Graboplast	Graboxan (Hungary)		
VEB Vogtlandische	Ekraled (GDR)		
Western Europe			
Porvair	Porvair and Vantel		
Yagi	Patora	2.3	2.0
Societe Belge	Kabipor		
North America			
Uniroyal	Capilair	na	na
Scott-Chatham	Tanera		

[a] Ref. 24.
[b] To convert m² to ft², multiply by 10.76.

($1.25/ft^2) for poromerics, as compared to leather prices which range from $13.45/m^2 ($1.25/ft^2) for side leather to $75.00/m^2 ($7.00/ft^2) or more for high-grade, calf leather.

The worldwide production of coated fabrics in 1978 is given in Table 1. Note that vinyls are produced in considerably larger quantity than are urethanes, as might be expected from their lower cost. The worldwide consumption pattern by market for urethane-coated fabrics in 1978 was clothing, 50%; footwear, 15%; bags, 25%; upholstery, 5%; and industrial, 5% (24). Clothing and accessories are the principal markets for these products; whereas shoes comprise the principal market for poromerics, as shown in Table 2. This is to be expected, as poromerics originally were developed specifically for footwear applications. The poromerics that were produced commercially in 1978 are listed in Table 3.

The 1978 U.S. consumption of various upper materials for footwear is listed in Table 4. These figures exclude the 250 × 10^6 pairs of rubber and fabric shoes produced that year. The trend of U.S. footwear consumption and manufacturing is shown in Table 5. The large increase in imports is approaching 50% of domestic consumption. The shift to foreign production, especially to developing nations, results from the labor-intensive nature of shoe manufacturing. Producing a pair of shoes requires 0.5–1.0 h of labor, depending upon style, and ca 0.15 m^2 (1.6 ft^2) of upper material. Another trend shown in Table 4 is the relatively flat consumption rate, except for a dip during the 1974–1975 recession, which can be explained by a change in consumer tastes. It is estimated that consumption of rubber and fabric shoes (eg, sneakers) increased from (70–250) × 10^6 pairs from 1965–1978. Rubber and fabric shoes are manufactured using nonleatherlike products: the soles normally are of rubber, eg, plasticized PVC, urethane, neoprene, or SBR (styrene–butadiene rubber); and the uppers vary from injection-molded PVC in ski boots to nylon fabric in tennis shoes. Thus, these shoes can offer serious competition to those made of leatherlike synthetics in some low priced markets.

Table 4. Footwear Upper Materials Consumed in the U.S. in 1978, 10^6 pairs[a]

	Leather	Vinyl	Other	Total
production	235	104	80	419
imported	174	146	54	374
Total	*409*	*250*	*134*	*793*

[a] Data from American Footwear Industries Association.

Table 5. The U.S. Market for Upper Materials in Shoes, 10^6 pairs[a]

	1965	1970	1975	1978
U.S. production	626	562	413	419
imported	88	242	286	374
Total	*714*	*804*	*699*	*793*

[a] Data from American Footwear Industries Association.

BIBLIOGRAPHY

"Poromeric Materials" in *ECT* 2nd ed., Vol. 16, pp. 345–360, by J. L. Hollowell, E. I. du Pont de Nemours & Co., Inc.

1. F. O'Flaherty, W. T. Roddy, and R. M. Loolar, eds., *The Chemistry and Technology of Leather, ACS Monograph Series, No. 134*, Reinhold, New York, 1965.
2. W. Riess, *Proceedings SATRA North American Int. Conf., Ontario*, 1974, pp. 56–68.
3. H. Herfeld, *Index 78 Programme*, Session 6, paper 4.
4. L. C. Hole and B. Keech, *SATRA Bull.*, 37 (March 1968); *Proc. Xth Congr. International Union Leather Chem. Soc. London, 1969.*
5. H. R. Lasman in ref. 2, pp. 80–89.
6. U.S. Pat. 2,964,799 (Dec. 20, 1960), P. E. Roggi and R. A. Chartier (to U.S. Rubber).
7. H. J. Koch, *Melliand Textilbe.* **51,** 1313 (1970); R. R. Grant, *Urethane Plast. Prod.* **1,** 1 (1971).
8. U.S. Pat. 3,696,180 (Oct. 3, 1972), V. R. Cunningham and T. S. Dodson (to Porous Plastics Ltd.); U.S. Pat. 3,729,536 (April 24, 1973), E. A. Warwicker (to Porvair); U.S. Pat. 3,860,680 (Jan. 14, 1975), E. A. Warwicker and R. Hogkinson (to Porvair); U.S. Pat. 3,968,292 (July 6, 1976), A. W. Pearman and S. J. Wright (to Porvair); U.S. Pat. 4,157,424 (June 5, 1979), D. L. Bontle (to Porvair).
9. T. Hayaski, *Chem. Tech.*, 28 (Jan. 1975); L. C. Hole, *Rubber J.*, 152, 72 (1970); R. C. Hole and R. E. Whittaker, *J. Mater. Sci.* **6,** 1 (1971).
10. U.S. Pat. 3,865,678 (Feb. 11, 1975), M. Okamoto and co-workers (to Toray); U.S. Pat. 3,873,406 (March 25, 1975), K. Okazaki, A. Higuchi, and N. Imaeda (to Toray); U.S. Pat. 3,908,060 (Sept. 23, 1975), K. Okazaki, A. Higuchi, and N. Imaeda (to Toray); U.S. Pat. 4,103,054 (July 25, 1978), M. Okamoto and S. Yoshida (to Toray); U.S. Pat. 4,136,221 (Jan. 23, 1979), M. Okamoto and Y. Yoshida (to Toray).
11. *Mod. Plant. Int.* 12 (Dec. 1975).
12. U.S. Pat. 3,764,363 (Oct. 9, 1973), F. P. Civardi and H. G. Kuenstler (to Inmont); U.S. Pat. 3,931,437 (Jan. 6, 1976), F. P. Civardi and H. G. Kuenstler (to Inmont).
13. U.S. Pat. 2,801,949 (Aug. 6, 1957), A. W. Bateman (to E. I. du Pont de Nemours & Co., Inc.).
14. G. Hole and D. Hill, *SATRA Bull.* 18(9), 85 (1978).
15. U.S. Pat. 3,222,208 (Dec. 7, 1965), N. J. Bertollo (to Interchemical Corp.).
16. U.S. Pat. 3,998,488 (Oct. 26, 1976), F. P. Civardi (to Inmont); U.S. Pat. 4,055,693 (Oct. 25, 1977), F. P. Civardi (to Inmont); U.S. Pat. 4,122,223 (Oct. 25, 1978), F. P. Civardi and F. C. Loew (to Inmont).
17. U.S. Pat. 3,709,752 (Jan. 9, 1973), R. Wisotzky and R. E. Patersen (to Pandel-Bradford); U.S. Pat. 4,048,269 (Sept. 13, 1977), R. Wisotzky and J. C. Bolger (to Pandel-Bradford); U.S. Pat. 4,052,236 (Oct. 4, 1977), V. C. Kapasi and co-workers (to Pandel-Bradford).
18. U.S. Pat. 3,041,193 (June 26, 1962), E. Hamway and B. Edwards (to General Tire and Rubber Co.).
19. U.S. Pat. 3,949,123 (April 6, 1976), R. N. Steel (to Uniroyal).
20. U.S. Pat. 3,533,895 (Oct. 13, 1970), K. Norcross (to Nairn-Williamson); U.S. Pat. 3,632,727 (Jan. 4, 1972), K. Norcross (to Nairn-Williamson).
21. U.S. Pat. 4,044,183 (Aug. 23, 1977), N. Forrest; U.S. Pat. 4,124,428 (Nov. 7, 1978), N. Forrest.
22. U.S. Pat. 3,284,274 (Nov. 8, 1966), D. G. Hulsander and W. F. Manwaring (to E. I. du Pont de Nemours & Co., Inc.); U.S. Pat. 3,912,834 (Oct. 14, 1975), S. Imai, T. Eguchi, and M. Shimokawa (to Kanegafuchi).
23. U.S. Pat. 3,776,790 (Dec. 4, 1973), G. N. Harrington and F. P. Civardi (to Inmont).
24. S. D. Cleaver, Senior Information Officer, SATRA, private communication, March 7, 1980.

F. P. CIVARDI
G. FREDERICK HUTTER
Inmont Corporation

METAL FIBERS

A wide range of products, eg, textile products, paper, floor covering, insulation products, and many composites, depend on fiber manipulation technology for their economical manufacture and on the inherent fiber characteristics for their end-use properties. However, not many products have embodied the use of metal fibers. Until fairly recently, metals and alloys generally have been available at reasonable cost only in the form of wire of diameters significantly greater than natural fibers. The large diameter and the inherently high elastic modulus of metals and alloys has limited the use of many well-developed and economical fiber-manipulation processes; exceptions are woven wire products for which special looms and knitting equipment have been developed. Also, processes for producing metals and alloys with acceptable properties and dimensions have been developed slowly and are incomparable in scale and volume to those used in the synthetic organic fiber or the glass-fiber industry (see Fibers, chemical; Glass). The most significant reason for the slow development of economical processes is the difficulty of forming metal filament directly from the liquid phase. This arises from the unusually high ratio of surface tension to viscosity, which results in liquid-jet instability, and from the inability to attenuate significantly the stream outside the spinnerette as is the usual practice in the synthetic-fiber industry. Despite these shortcomings, a metal-fiber industry is emerging and will have a large impact on product and process design and performance.

Properties

Fiber. Fiber properties result from a combination of the material properties, the effect of processing the material into fiber form and, in some cases, the geometry of the final fiber. The mechanical, physical, and chemical characteristics of metals and alloys are high modulus and high strength, high density, high hardness, good electrical and thermal conductivity, can be magnetic, high temperature stability, good oxidation resistance, and corrosion resistance to varied chemical groups.

It may be that only one of the characteristics determines the choice of a particular metal or alloy for a specific application. For example, where a high specific modulus, ie, modulus-to-weight ratio, is desirable, beryllium with its very low density but high modulus may be the preferred material. In the generic material group, a metal or alloy usually can be formed and have properties that are functionally and economically acceptable for a particular application. However, for metal fibers, the choice of material is restricted since only a small number of metals and alloys are available in fiber form, particularly those with diameters <50 μm.

Most of the metal-fiber industry is directed to markets where one or more of the following properties predominate: strength–stiffness, corrosion resistance, high temperature oxidation resistance, and electrical conduction. These markets generally are satisfied by the following materials: carbon and low alloy steels, stainless steels, iron, nickel, and cobalt-based superalloys.

Property modifications can be made. For example, commercial stainless steels made according to standard melting practice contain small quantities of impurities which manifest themselves in the solidified ingot as small, nonmetallic inclusions. These inclusions are relatively unimportant in the majority of applications of the

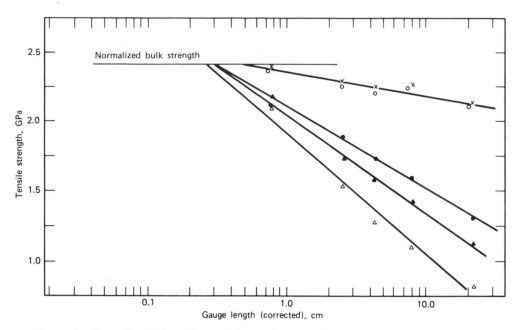

Figure 1. Gauge length dependence of the tensile strength of stainless steel filaments (diameter = 12 μm): X, heat E; O, heat F; ●, heat G; ▲, heat H; △, heat I. To convert GPa to psi, multiply by 145,000.

material because of their small size. However, in the case of stainless steel fiber, the diameter of the fiber can be of the same order as that of the inclusion and, consequently, the inclusion can significantly affect the strength of the fiber. The apparent strength of the fiber tends to decrease and the probability of finding an inclusion in the test sample increases with an increase in the length of the test fixture. The breaking strength of 12-μm fibers processed from five different heats of steel are plotted in Figure 1 as a function of the length of the fiber in the test machine. The gauge length effect is significantly more pronounced for heats G, H, and I as compared to heats E and F. The inclusion count for the five steels is presented in Table 1 and, as the data

Table 1. Inclusion-Count Data for Stainless Steel Heats

Mean inclusion diameters, μm	Number of inclusions for various heats				
	E	F	G	H	I
1	5	16	170	224	219
2		3	21	41	39
3		1	11	22	22
4			2	9	14
5			3	2	11
6				1	10
7					5
8				1	
9					
10					2

indicate, steels with low inclusion counts show the least sensitivity to fiber length. The normalized bulk strength value identified in Figure 1 corresponds to the intrinsic value for the specific stainless steel as measured in sufficient diameters, ie, >50 μm, to be uninfluenced by inclusion defects. The normalization is to a base-line chemistry since each heat shows small composition variations within the type specification. Even with the ultraclean, ie, low inclusion-count steels, breaking-load discrepancies within a fiber may exist resulting from small variations in fiber cross-section area. These cross-section variations, which are inherent to the bundle-drawing process, typically may result in a coefficient of variation in breaking load of about 6–8% for 12-μm diameter fiber, but only 3–4% for 25-μm fiber. Thus, fine-filament manufacturers using, eg, the bundle-drawing process must not only carefully control the process parameters but must establish strict specifications for the starting material in order to provide a quality product. Similar requirements are necessary for other fiber-forming techniques.

In a few cases, smallness improves material properties, eg, the mechanical properties of metal whiskers (1). The preparation of metal whiskers by vapor deposition or decomposition of a gaseous compound results in a slow buildup of material at high temperature. The resulting fiber is essentially free from dislocations and, consequently, exhibits mechanical properties which approach theoretical, ie, values related to the actual breaking of metal-to-metal bonds rather than the sliding of atoms or atom planes past each other as is the usual deformation mechanism.

Physical and chemical properties of metal fibers also may tend to be modified or exaggerated because of the small fiber diameter which results in high surface-to-volume ratios. For example, in electrical conduction, direct current is carried uniformly through a fiber cross section, whereas at high frequencies, electricity is carried close to the conductor (fiber) surface; the surface current density increasing with increasing frequency. Thus, the ratio of alternating-current to direct-current resistance of a small-diameter fiber is close to unity to significantly high frequency ranges; consequently, the fine fiber is an efficient high frequency conductor compared to large diameter wire.

In terms of chemical characteristics, the high surface-to-volume ratio is advantageous where the fiber serves as a catalyst but is disadvantageous where it is desired to minimize reaction of the fiber with the environment. The latter situation becomes important in many applications of fibers, eg, filtration (qv), seals, and acoustic treatment (see Insulation, acoustic). Fibers are chosen carefully to provide long life for the product.

Assembly.   The properties of any assemblage of fibers often are determined by the particular arrangement of the fibers in that structure. For example, the mechanical tensile strength of a number of fibers or filaments in a yarn depends on the degree of twist imparted to the group of fibers during the spinning of the yarn. Too low a twist results in the fibers being loaded nonuniformly with application of tension and in progressive fiber failure which produces a low yarn strength. At optimum twist, the load applied to the yarn is equally distributed on all fibers and the yarn exhibits strength which is the sum of the breaking loads of all of the fibers.

The mechanical properties of sintered-fiber, randomly oriented structures, as exemplified by elastic modulus and tensile strength, have been reported by a number of investigators and there is considerable variation even for the same alloy fiber. A composite of tensile data obtained from a variety of samples of 304-type stainless steel fiber mat is shown in Figure 2. The samples include porous structures prepared by press and sinter techniques using 12-, 25-, and 125-μm fibers with aspect ratios of 25,

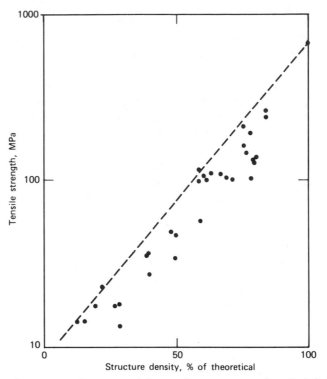

Figure 2. Tensile strengtn of porous, stainless steel structures as a function of structure density. To convert MPa to psi, multiply by 145.

62, and 187, as well as sintered air-laid web of 12-μm fiber calendered to a density of about 20% of theoretical. The curve in Figure 2 signifies the probable maximum attainable strength for 304-type stainless steel sintered-fiber structures as a function of structure density for the particular fiber types employed, ie, chopped conventionally drawn wire and bundle-drawn fiber.

Measurement of the elastic modulus of sintered-fiber porous structures, again of 304-type stainless steel, have been reported over a density range of about 40–85% (2). The data was generated from samples composed of 50- and 100-μm diameter fiber, 4 mm in length. The data is reasonably consistent with the theoretical relation between elastic modulus and structure density that was established for porous materials (3):

$$E_m = E_0 \left(1 - 1.9\,P + 0.9\,P^2\right)$$

where E_0 = modulus of elasticity (in tension) of 100% dense material, E_m = derived structure modulus (in tension) of the porous structure, and P = fractional porosity. The compressive modulus of elasticity is approximately one order of magnitude lower than the tension data. When compressed the fibers tend to buckle at low stress levels resulting in low structure modulus.

Measurement of the thermal conductivity of various types of porous materials has been reported (4). The samples range from sintered spherical powder of OFHC copper and 304L stainless steel in the 10–30% porosity range to fibrous stainless steel mats composed of 12-μm diameter filament with a porosity of 78%. The empirical relationship that best fits the data (5) is

$$\frac{\lambda}{\lambda_0} = \frac{1 - \epsilon}{1 + 11} \epsilon^2$$

where λ = sample thermal conductivity, λ_0 = solid material thermal conductivity, and ϵ = fractional porosity. The relationship is reliable for porosities up to about 80%. An attempt to derive thermal conductivity values from the measurement of electrical conductivity using a modified Wiedemann-Franz relationship is not successful in the case of stainless steels of >40% porosity (6).

The complications arising from changes in flow velocity; media geometry, including pore size and tortuosity; and fluid properties often require empirical approaches in order to satisfactorily describe observed results. All media present a finite resistance to this flow of fluids. In many cases, it is the magnitude of this resistance that is a determining economic factor in the choice of a particular porous material to perform a specific function, whether it be a filtration application or a lubricating device.

The resistance to flow is expressed in terms of the pressure drop across the medium per unit of length and the flow rate per unit area. The simplest relationship involving only viscous flow is Darcy's Law:

$$\frac{\Delta p}{L} = \frac{\mu}{K} \frac{Q}{A}$$

where Δp = pressure drop (101.3 kPa = 1 atm); L = media thickness, cm; μ = viscosity, mPa·s (= cP); Q = flow rate, cm^3/s; A = cross-section area, cm^2; K = permeability, darcys. The relationship defines the common permeability unit, the darcy (units of cm^2). When fluid flow involves inertial energy losses and viscous drag, the expression is modified to account for the increased energy loss:

$$\frac{\Delta p}{L} = \frac{\mu}{K} \cdot \frac{Q}{A} + \frac{\rho}{K^1} \cdot \frac{Q^2}{A^2}$$

where ρ = fluid density, and K^1 = permeability factor associated with kinetic energy loss. Compressible fluids require additional energy-loss terms to account for the work performed in fluid compression.

Thus for each porous structure and each specific fluid, there is a unique relationship defining the permeability of the medium under a particular set of conditions; however, in many cases the relationship cannot be derived from first principles. For the same medium type but with differing porosity values, the permeability is different. A relationship for laminar-flow conditions has been derived and satisfactorily predicts the change of permeability with change in porosity for the same fiber and structure geometry (fiber orientation):

$$K \propto \frac{\epsilon^3}{(1 - \epsilon)^2}$$

where ϵ = fractional porosity. The permeability of randomly oriented fiber structures can be very high when compared to other porous materials primarily because low density structures of good mechanical integrity can be fabricated. The use of fine fibers reduces viscous drag and, thus, compliments the special structure features.

The high permeability properties of randomly oriented fiber structures are particularly attractive in filtration applications. Comparative data for three nominal 20-μm rated filter media composed of sintered powder metal, woven wire cloth, and a randomly oriented fiber structure illustrate the advantage of the latter: the measured nitrogen permeability, normalized to the fiber structure data, are 0.12 cm, 0.42 cm, and 1.00 cm, respectively.

In addition to the high permeability values, the randomly oriented fiber structure exhibits another desirable property with regard to filtration, ie, the high dirt-holding capacity or on-stream life. This property can be quantified by the time taken to develop a particular cut-off Δp and can be as much as a factor of two compared to sintered powder media and a factor of 1.5 compared to the woven-wire structure.

Manufacture and Processing

Often there are two distinct elements in the fabrication of fiber products, ie, the formation of the fiber and the assembly of the fibers into a useful structure or form. The majority of commercial applications involve a large degree of secondary processing by various fiber-manipulation techniques.

Certain fiber-forming processes yield free fiber and others tend to produce a primitive fiber assembly, eg, a tow or mechanically interlocked bundles comparable to the bale of natural fiber. Thus, the commercially available form depends on the type of forming process employed. It also may depend on the business strategy of the manufacturer, who may limit the availability of the primitive form in order to attain the benefits of the value added by further in-house processing.

Fiber dimensions, as defined by the natural- and synthetic-fiber industry, tend to be restricted to diameters or equivalent diameters for noncircular cross sections of less than 250 μm. This would correspond to normal limitation for further processing by common textile-fiber manipulation techniques. This limitation also applies to metal fibers.

Fiber Forming. The principal methods that have been developed for metal-fiber forming relate to the basic starting material form. Mechanical processing incorporates processes that rely on plastic deformation to produce a fiber from a solid precursor. In liquid-metal processing or casting, the fiber is formed directly from the liquid phase.

Mechanical Processing. The mechanical processes involve material attenuation by gross deformation or they involve the parting of material from a source, eg, a strip or rod. The first group encompasses wire-drawing techniques and solid-state extrusion, and the second group consists of cutting or scraping-type operations, ie, slitting, broaching, shaving, and grinding.

The various processes are identified in Figure 3 in relation to their source material; both current and potential commercial processes are given. For processes, eg, conventional wire drawing, the manufacturing cost is highly dependent on the diameter of the final product. Material attenuation must take place in a series of small, usually equal steps of ca 20% area reduction per step. Since drawing speed cannot be continually increased to compensate for the lower through-put per step, succeeding reducing steps involve lower and lower efficiency, ie, quantity produced per unit time. The modified wire-drawing process, ie, bundle drawing, circumvents this strong diameter dependency by drawing many wires or filaments simultaneously through a reduction die. Consequently, the mass flow (kg/h) of the wire-drawing machines is increased dramatically and results in a large reduction in manufacturing cost. Some of this ad-

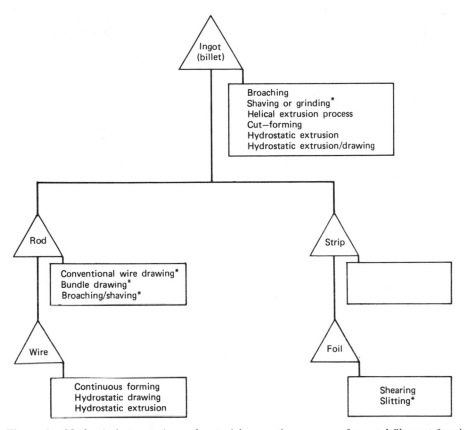

Figure 3. Mechanical attenuation and material separation processes for metal filament forming: △, stock material form; □, fiber-forming processes; *, commercially available product.

vantage, however, is offset by the equipment and processing that are necessary to form the initial bundle of wires and that are required to separate the fibers at the conclusion of the process. On balance, the overall economy improvement factor is highly favorable and can be as much as 40 or 50 to one for fibers of ca 12 μm in diameter. The bundle-drawing process (BDP) is an important source of quality fiber. A schematic process flow diagram comparing conventional wire drawing is shown in Figure 4.

The wire drawing, modified wire drawing, and extrusion processes produce continuous filaments of basically circular cross section and of unique properties that are partly a consequence of the processing history. Attenuation or constraining the material in the reducing die produces changes in the internal structure or morphology of the metal, eg, reduction in grain size, development of preferred orientation and, in some cases, induced phase transformation. In multiple-step reduction processes, the effect of such changes in material properties causes succeeding steps to become increasingly more difficult and usually it is necessary to eradicate the work-hardening effect by annealing the material. The drawing–annealing schedule can be arranged to provide efficient processing and desirable mechanical properties in the final fiber product.

The slitting, broaching, shaving, and grinding machines and devices designed

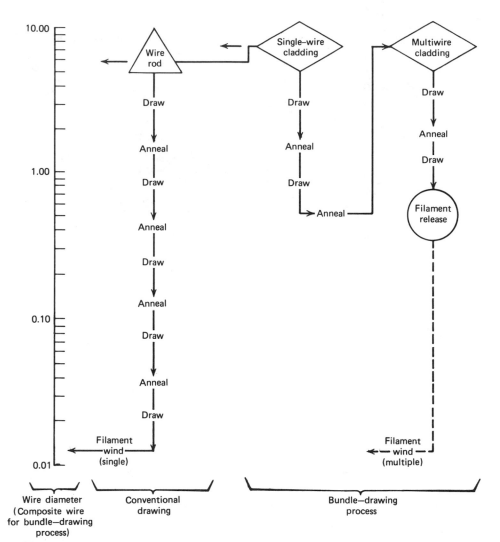

Figure 4. Flow diagram comparing conventional and bundle-drawing processes: △, starting material; ◇, ○, special operation.

Table 2. Mechanical Fiber-Forming Processes and Related Fiber Characteristics

Process	Typical fiber diameter, μm	Typical length	Materials	Cross-section shape	Economics
conventional wire drawing	≥12	continuous	all ductile metals and alloys	round (other sections possible)	304 stainless steel ca $3.00/kg at 250 μm; ca $3000/kg at 12 μm
bundle drawing	≥4; typically 8 or 12	continuous	ductile metals and alloys	rough surface	304 stainless steel ca $100/kg at 12 μm
broaching or shaving, eg, wire rod, and billet	≥8	short to continuous	most ductile metals and alloys	generally triangular	low carbon steel $2–4/kg depending on grade
slitting and shaving, eg, foil and sheet	ca 25 and greater	short (0.0004–4 cm) or continuous	most ductile metals and alloys	ductile square or rectangle	shaving fiber <$10/kg (not available currently)

to convert standard mill forms of metal to fiber have been described (7–9). The wire-shaving technique produces a fiber product at low cost and with wide application (10). Modern machines operate automatically on a multiplicity of wires and can produce metal wool with an equivalent diameter of <25 μm. The process involves the shaving of wire, which may be ca 12 mm in diameter, in a series of steps by serrated, chisel-type tools; the size and spacing of the serrations determining the cross-section dimensions of the individual fibers. The industry standards define fiber-wool grades as grade 3 (largest) through 0 to the very fine 0000 grade, with the extremes corresponding to a mean fiber width of 178–241 μm and 15–25 μm, respectively (11). The cross section of the shaved fiber generally is triangular with the apices defining sharp edges along the fiber length. The coarser grades contain very long fiber and, even in the finer grades, the fiber is of sufficient length to cling together without unraveling in the rolls or pads, which are the usual form for shipping. Material compositions are mainly ferrous alloys, either low carbon steel, or 400 series ferrite stainless steel.

Many variations of the wire-shaving process have been developed. A group of devices or machines have been designed for the shearing or shaving of stacks of thin metal foil, although no such commercial fiber product is on the market (8). The product produced by these processes usually is rectangular or square in cross section and the length depends on the process employed: for edge-shearing or slitting the fiber length can be equal to the foil length, whereas for end-shearing the width of the foil is the limiting factor. From the material standpoint, the versatility of mechanical processing is limited only by the required availability of the foil form. The slitting of metal foil in single thicknesses to provide continuous filament usually is unattractive because of the associated low production rate. For some specialized applications, however, eg, for decorative textile thread, the cost may be justified especially if the foil costs are reduced by substitution of metal coated plastics.

Other mechanical processing techniques identified in Figure 3 either are not used for fiber production or are emerging technology. Hydrostatic extrusion and extrusion-drawing techniques offer opportunities for fiber-forming materials which are difficult to work or deform (12). These processes also allow for comparatively high reduction ratios, ie, the ratio of the diameter of the metal wire or rod stock to the diameter of the final wire or filament. Helical extrusion and continuous forming offer potential for a low cost product but it is too early to predict the lower limit of product dimensions consistent with economic production (13–14). A 1 mm diameter copper wire can be produced directly from a 150 mm diameter and 1 m long billet by helical extrusion and a 1 mm diameter aluminum wire can be converted directly from 10 mm diameter wire stock by continuous forming; the cut-form process is a combination of the two processes (15). The first step is the formation of a metal chip from a billet; the second step involves feeding the chip into a forming die. The advantages of this process are the attractive economics of step one and the improved product quality associated with the second step. The process is in the early developmental stage.

The fiber characteristics and relative economics of the commercially significant fiber-forming techniques by mechanical processing are summarized in Table 2. Scanning electron micrographs of representative fiber products are presented in Figure 5.

Liquid-Metal or Casting Processes. Melt spinning of glass and certain polymers is an established technique for mass production of fine filaments or fibers. The liquid material is forced through a carefully designed orifice or spinerette and solidifies in

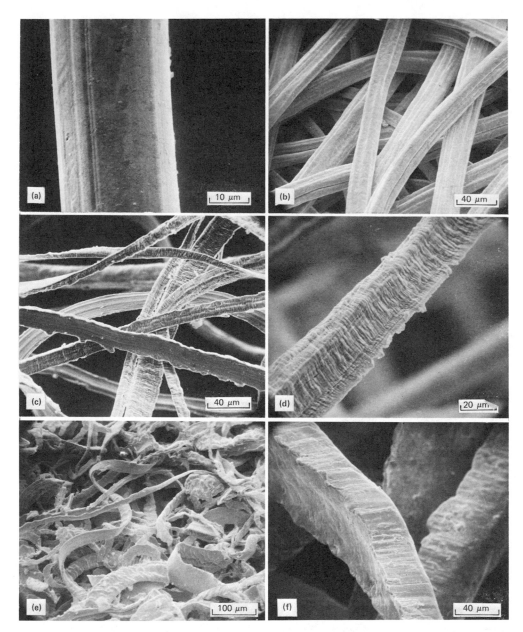

Figure 5. Scanning electron micrographs of fibers produced by mechanical fiber-forming processes: (a) 25 μm stainless steel, conventional wire drawing; (b) 19 μm 304-type stainless steel, bundle-drawing process; (c) 0000-grade steel wool, shaving process; (d) transverse shaving marks of 0000-grade steel wool, shaving process; (e) felt metal (FM), 1100 series, 347 stainless steel; (f) sheared, low carbon, steel fiber edge.

a cooled environment, usually after considerable attenuation, before being wound on a spool. However, this process cannot be adapted easily to metals. The development of liquid-metal fiber-forming processes has revolved around overcoming the inherently low viscosity of molten metals. The low viscosity and the high surface tension of liquid metals make it extremely difficult to establish free-liquid jet stability over a length sufficient to allow freezing of the metal into a fiber before the jet separates into droplets. The problem has been solved with varying degrees of success by one of the following approaches: altering the surface of the liquid jet by chemical reaction; promoting jet stabilization by indirect physical means, eg, an electrostatic field; or accelerating the removal of heat from the jet to promote solidification before breakup occurs. An alternative approach which has not been commercially exploited is that of placing a glass envelope around the molten metal to control the formability and thus circumvent the problem of jet stabilization (16–17). However, the final application of the product may require the removal of the glass envelope, which would add considerably to the product cost since it may account for 75 vol % of the material produced. The various liquid-metal fiber-forming processes of historical and commercial interest are identified in Figure 6.

Melt spinning involving a free-liquid jet is used for a variety of low-melting-point metals or alloys including Pb, Sn, Zn, and Al. Various techniques are involved in the cooling and quenching process including co-current gas flow, mists, and liquid media. The processes permit production of fiber of 25–250 μm in diameter and of continuous length. Melt-spin processes applied to higher-melting-point metals, particularly the ferrous alloys, requires the removal of very large quantities of heat from the liquid-metal jet in a very short time. Generally, those techniques that are successful for the low melting alloys cannot be used for the high melting alloys. An alternative approach involves the reaction of the liquid jet and the cooling medium, or some addition to that medium which results in the formation of a case or envelope of sufficient strength to prevent breakup of the liquid-metal jet.

In some cases, the chemical composition of the melt must be adjusted prior to spinning in order to prevent dissolution of the stabilizing film in the molten jet. A typical process flow diagram for the melt spinning of steel wire is shown in Figure 7 (18). The process is designed for two steel compositions, an aluminum steel (1.00 wt % Al, 0.36 wt % C) and a silicon steel (2.40 wt % Si, 0.32 wt % C). Carbon monoxide is the stabilizing medium and surface films of aluminum oxide or silicon oxide are formed. A continuous filament of 75–200 μm in diameter is spun at 240–120 m/min with continuous operation to 100 h. A useful degree of attenuation of the molten-metal jet outside the orifice is achieved by using a radially directed helium gas flow. Consequently, the final filament diameter can be controlled independently of the spinning orifice or spinerette geometry.

The alternative approach to fiber forming from the melt is by greatly accelerated heat transfer (19–20). The processes essentially are based on the quenching of the molten-metal jet on a chill plate positioned close to the orifice or at a point prior to the onset of jet instability and breakup. A number of different configurations for the chill plate have been proposed, ie, from rotating drums to concave disks. The quench rate of the molten jet on the chill plate is as high as 10^6 °C/s. Quench rates of this magnitude can produce nonequilibrium conditions in certain alloys and a characteristic grain structure associated with the impressed unidirectional heat flow and the kinetics of the nucleation–growth process. The fibers produced by these chill-plate processes generally are ribbonlike but can be continuous.

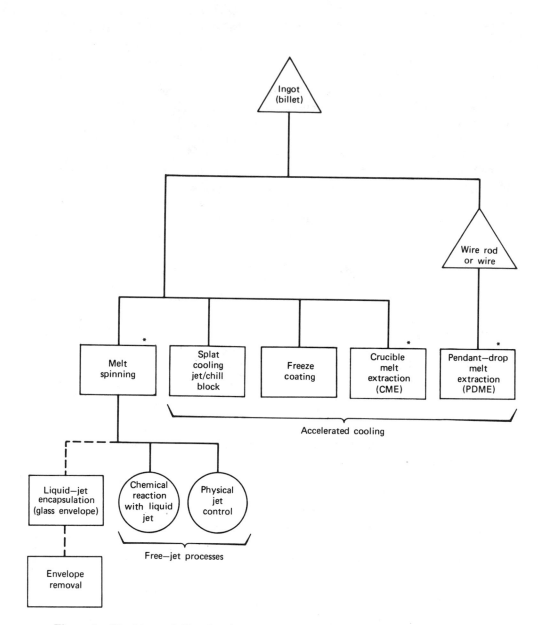

Figure 6. Liquid-metal, fiber-forming processes: △, stock material form; □, basic fiber process; ○, process modifications; *, commercially available fiber.

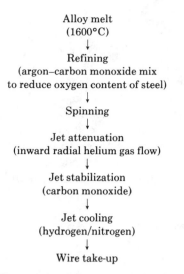

Alloy melt
(1600°C)
↓
Refining
(argon–carbon monoxide mix
to reduce oxygen content of steel)
↓
Spinning
↓
Jet attenuation
(inward radial helium gas flow)
↓
Jet stabilization
(carbon monoxide)
↓
Jet cooling
(hydrogen/nitrogen)
↓
Wire take-up

Figure 7. Flow diagram for silicon steel and aluminum steel melt spinning with jet stabilization.

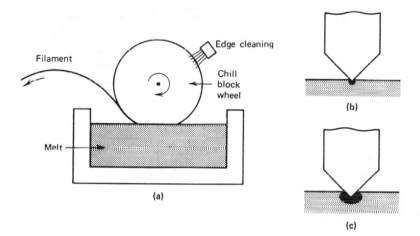

Figure 8. (**a**) Crucible melt-extraction process (CME); (**b**) and (**c**) crescent shaped, larger filaments.

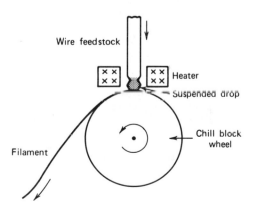

Figure 9. Pendant-drop, melt-extraction process (PDME)

243

The crucible melt extraction (CME) process is an extension of the chill-plate processes and is illustrated in Figure 8a (21). There is no spin orifice and, thus, no need for forming a liquid stream. The chill surface, the shaped rim of a revolving disk, is dipped into the crucible so that the shaped edge just contacts the liquid-metal surface. The cool disk edge immediately causes solidification of a small volume of liquid metal, which is carried out of the crucible and is ejected from the wheel by centrifugal action. The shape of the fiber cross section is dependent on the wheel-edge geometry and the depth of immersion, but it can be made circular for small diameter filaments (25–75 μm); larger filaments tend to be crescent shaped, as indicated in Figures 8b–8c.

A further evolution in the chill-plate concept is the pendant-drop melt-extraction process (PDME) (21). In this process, the orifice for jet forming is eliminated and the crucible is replaced by a suspended drop held by surface tension to the feedstock, which is a wire or wire rod. The PDME process conveniently circumvents jet stability problems and solves the basic material compatibility problems associated with the molten-metal-orifice and the molten-metal-crucible interfaces. The basic elements of this process are schematically represented in Figure 9. The process can be applied to a wide range of metals and alloys and can produce fibers with equivalent diameters as small as ca 25 μm.

The PDME and CME processes have been varied to provide discrete fiber lengths as opposed to continuous filament. The introduction of discontinuities on the chill-wheel rim at fixed intervals results in the casting of short fibers with the discontinuity spacing defining the fiber length. The short fibers tend to have a dog-bone configuration which, for certain applications, is advantageous. A summary of the liquid-metal fiber processes in terms of fiber characteristics and comparative economics is presented in Table 3. Scanning electron micrographs of representative fibers produced by the various casting techniques are reproduced in Figure 10.

Table 3. Liquid-Metal Fiber-Forming Processes and Related Fiber Characteristics

Process	Typical fiber diameter	Typical length	Materials	Cross-section shape	Economics
melt spin	$\geq$25 μm	continuous	Pb low melting point metals and alloys	circular	plumbers wool (lead) <$2/kg
melt spin with jet stabilization (chemical)	$\geq$75 μm	continuous	most metals and alloys (special compositions)	circular	
crucible melt extraction (CME)	$\geq$25 μm	continuous or controlled length	most metals and alloys	small diameter, circular; large diameter, crescent shaped	stainless steel $2–7/kg
pendant-drop melt extraction (PDME)	$\geq$25 μm	continuous or controlled length	most metals and alloys	small diameter, circular; large diameter, crescent shaped	

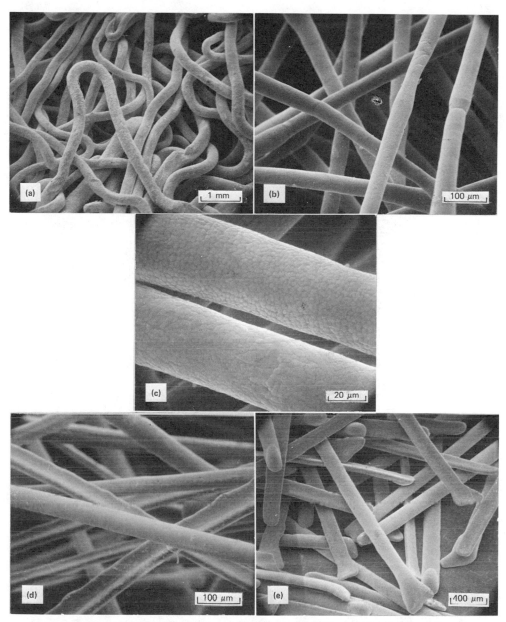

Figure 10. Scanning electron micrographs of fibers produced by liquid-metal or casting processes: (a) 250-μm lead filament, melt spinning; (b) 40-μm aluminum filament, melt spinning; (c) 35-μm melt spin aluminum filament showing surface structure; (d) 50 × 100-μm stainless steel filament, PDME; (e) 150-μm stainless steel filament (3 mm long), PDME.

Miscellaneous Processes. Production of low cost metal filaments or products with unique properties has been reviewed (7,22–26). A sampling of significant technology other than mechanical and liquid-metal processes is presented in Table 4. Many of these processes could, with additional development and the use of recent advances in material science and related technology, lead to attractive processes.

Assembly. The various assembly processes of metal fibers into structures are similar to those developed for natural and synthetic fibers. The main fiber characteristics and the process parameters most likely to be affected by those characteristics,

Table 4. Miscellaneous Metal-Fiber Forming Processes

Filament material source	Process	Product	Status
metal powder	slurry forming, extrusion with binder followed by sintering	potentially most metals and alloys; specifically W–Ni (250 μm in dia)	experimental
any metal form	vapor deposition on glass fiber or plastic strip which is subsequently slit[a]	aluminum-coated product for conductive or decorative applications	commercially available
	freeze coating on glass or other fiber	aluminum on glass	experimental
inorganic chemical or mineral	electroplating on a helical mandrel[b]	50 × 12-μm Ni continuous filament	experimental
	chemical decomposition in an organic fiber precursor[c]	Ni–Cr alloy fiber with dimensions equal to precursor	experimental
	halide reduction by hydrogen[d]	whiskers of many metals, eg, Cu, Ag, Fe, Ni, and Co; length, 3–20 mm	mostly experimental

[a] Ref. 22.
[b] Ref. 26.
[c] Ref. 25.
[d] Ref. 23–24.

Table 5. Special Metal-Fiber Characteristics and Processability

Fiber characteristics	Process parameter	Remedy
surface roughness high hardness	friction–wear wear	surface treatment: lubrication; change material of rubbing surfaces in equipment
irregular shape high density	fluid dynamics	modify process media conditions, eg, flow velocity, density, viscosity
high modulus	general processability	use fiber with smaller cross-section (lower section modulus)
high yield	resistance to take permanent set (spring back)	anneal to reduce yield strength

and suggested solutions or remedies to alleviate processing problems are given in Table 5. Applications of metal-fiber products may require a combination of fiber types, eg, metal fibers and organic fibers. Thus, processes must be adapted to accomodate blends of fibers of vastly differing properties and processing characteristics.

The basic fiber-processing techniques in relation to the preferred fiber form for each process and the resulting products are summarized in Table 6. In textile processing particularly, the product of one process often becomes the preferred fiber form for a second process; thus, sequential processing is common for the more advanced products. In addition to the principal fiber-structure processing techniques enumerated in Table 5, a variety of secondary or special processes are important in modifying or optimizing the fiber-structure properties; these are listed in Table 7 with their typical applications.

Table 6.   Fiber-Structure Fabrication Processes

Technology	Preferred fiber form	Process	Product structure
papermaking	free fiber; length >12 mm for fibers 12–250 μm	wet slurry casting on screen	green random-fiber mat
textile processing	continuous filament	twisting	continuous filament yarn
		breaking	sliver–roving
	sliver–roving	drafting–spinning	staple yarn (blends)
		carding	random web
		air laying	random web
	yarn (staple or continuous filament)	weaving	fabric sheet
		knitting	fabric sheet or tube
		braiding	ribbon or tube
		tufting	carpetlike structures
	random web	needle punching	feltlike products
miscellaneous	chopped fiber; length >4 mm	flocking	short-pile structure on substrate
	continuous filament yarn	filament winding	composite structures
	random web, woven or knitted structures	vacuum or gravity infiltration of matrix material	composite structures

Table 7.   Secondary or Special Processing Techniques

Process	Typical application or particular advantage
annealing	softens fiber, increases fiber ductility, assists in further processing
sintering	fuses fibers at contact points, provides increased mechanical strength in fibrous structures
fiber alignment	produces anisotropic characteristics in web structures or composite materials
plating	provides protection from corrosive environments
coatings	provides interfiber lubrication
calendering (rolling)	precise pore size and density control, particularly in nonwovens
pressing	precise pore size and density control, particularly in nonwovens
crimping	geometrical elasticity in fiber, yarn, or nonwovens

Uses

Applications for metal fibers are of two types. One is the substitution for other fiber types or for metal powder in the case of porous metals to improve performance or to provide a cost benefit, eg, high temperature oxidation resistance or improved permeability in porous structures. The other application is in the development of new products that are based on the unique fiber properties or property combinations of metals, eg, in high gradient magnetic separation (see Magnetic separations).

A representative listing of applications, with brief descriptions and main advantages associated with the use of metal fibers, is presented in Table 8. The fiber structure applications are exploitations of the structure and the inherent fiber properties. The composite applications benefit from both the fiber and fiber structure characteristics and their interaction with the matrix material (see Composite materials; Laminated and reinforced metals).

Table 8.   Metal-Fiber Applications

Application	Description	Status[a]	Special advantages and principal fiber function
Textile products and porous structures			
filters			
surface	screen or wire-mesh products in disk or cartridge configuration used for low contamination level fluids, eg, hydraulic fluids	C	mechanical strength, nonmigrating, corrosion resistant
depth	nonwoven or random web structures in disk or cartridge configuration; general industrial applications	C	high permeability, high dirt-holding capacity
electrostatic	particle-capture augmentation of charged particles using low density web structures	R	improved filtration efficiency, electrical conduction
	filter-cake density control using designed electric-field configuration in bag-house filters	D	improved cake permeability
magnetic (HGMS)	magnetic attraction of ferromagnetic or paramagnetic particles through field distortion associated with ferromagnetic fibers in a uniform magnetic field; used for decontamination of kaolin	C	high field distortion associated with fine filaments
seals			
abradable	zero-clearance seal formed by turbine blade tips mating with deformable, porous-mat engine-housing lining	C	improved engine performance (efficiency)
gaskets	rope-type structures and nonwoven mats which conform to surface irregularities under compression; general industrial high temperature applications	C	high temperature, resilient structures
abrasion	metal-wool abrasive products for material removal and polishing	C	sharp cutting edges, shaved fibers

Table 8 (*continued*)

Application	Description	Status[a]	Special advantages and principal fiber function
antistatic			
textiles	blended yarn (stainless steel–nylon fiber) up to 10 wt % steel woven into fabric for clothing applications	C	electrical conduction and flexibility
carpets	blended yarn introduced into tufting operation to provide <1% steel fiber in face yarns	C	electrical conduction and flexibility
filter bags	needle-punched blend of organic fiber and steel to provide static control in bag-house filtration applications	C	electrical conduction and flexibility
brush	steel-fiber brush for static control of paper in photocopying devices	C	electrical conduction and flexibility
insulation	low density web (<5%) with suitable attachments used for aerospace applications, eg, rocket-engine nozzle insulation	C	high temperature stability, nonfriable
acoustics	acoustic impedance control for tuned resonator-duct liners in jet engines, etc	C	fine fiberweb structures provide improved linearity in absorption characteristics
catalysts	worn wire-mesh stacks for nitric acid production (75 μm diameter, Pt–Rh wire)	C	optimum surface-to-volume ratio
electromagnetic interference control and field production	Faraday suits of stainless steel/nylon fabric for high voltage transmission-line maintenance: heat shrinkable cable-termination shields embodying air-laid web	C	electrical conduction, flexibility
fluid flow	a family of products including, flow restrictors, snubbers, silencers, vents, demisters, homogenizer plugs, etc, using porous fiber structures	C	mechanical integrity, controlled permeability
heated fabrics	built-in flexible electrical conductors for heating, clothing, etc	C	electrical conduction, flexibility
electrodes	battery plaques, current collectors for high energy batteries, fuel cells; electrodes for electrolytic capacitors	D	high surface area, distributed conductor network
bearings	liquid lubricant or air bearings utilizing porous fiber structures	D	controlled porosity, mechanical integrity
wicks	capillary structure for liquid-phase transport in heat pipes	D	high temperature stability, controlled size, interconnected pores
shock mounts	friction damping in shock and vibration isolation mounts	C	high fiber-to-fiber friction, resilience
flame trap	protection of flammable fluids using sintered fiber-metal porous plugs	C	high thermal capacity and heat conductivity
transpirational cooling	surface cooling by controlled fluid flow through component structure, eg, turbine blades	D	structural integrity with designed permeability
fluidizer plate	support and diffuser plate for fluidized bed, eg, grain cars		porous-plate structure with load-bearing capabilities

Table 8 (*continued*)

Application	Description	Status[a]	Special advantages and principal fiber function
ceramic attachment	compliant layer of fibers for ceramic attachment to metal surfaces	C	resilient fiber structure accomodating differential thermal expansion coefficients of substrate and ceramic
yarn and cable products	yarn and cable applications demanding flexibility with high strength, eg, in instrument control cables and medical sutures	C	good hand, high knot strength, biocompatibility, high flexibility
Composite applications			
tire cord	steel tire cord containing 0.15–0.25-mm, brass-plated filament in designed cord configurations	C	strength, dimensional stability, flexibility, heat conduction, bonding to rubber
tire tread	tread impregnation with chopped wire for off-the-road vehicle and aircraft tires	C	abrasion and cut resistance
timing belts	reinforced, nonstretching belts	?	dimensional stability, strength
brake lining	high friction composite material with dispersed short metal fiber for heat conduction	C	improved operating performance by temperature control using conductive fibers
refractory bricks (linings)	furnace-lining reinforcement with castable refractories embodying chopped fiber or melt extraction process fiber to reduce friability	C	mechanical integrity, crack arrestor function
concrete	reinforcement with short fibers to increase load-bearing capability	C	crack arrestor
conductive plastics	plastic housings for electrical equipment to control electromagnetic interference (EMI)	C	electrical conduction
	fabric containing short metal fiber for shielding–reflecting microwave radiation	C	efficient coupling
superconductors	continuous filament, small-diameter superconductors embedded in high thermal conductivity matrix	C	more efficient use of superconductivity material and protection during conduction state transition

[a] C = commercial, D = development, and R = research.

BIBLIOGRAPHY

1. R. V. Coleman, *Metall. Rev.* **9**(35), 261 (1964).
2. P. Ducheyne, E. Aernoudt, and P. De Meester, *J. Mater. Sci.* **13**, 2650 (1978).
3. J. K. Mackenzie, *Proc. Phys. Soc. (London)* **B63**, 2 (1950).
4. R. P. Tye, *A.S.M.E. Publication No. 73-HT-47*, American Society of Mechanical Engineers, New York, 1973.
5. J. Y. C. Koh and A. Fortini, *NASA Publication No. CR-120854*, 1972.
6. R. W. Powell, *Iron and Steel Institute Special Report No. 43*, p. 315.
7. C. Z. Carroll-Porczynski, *Advanced Materials*, Chemical Publishing Co., Inc., New York, 1969.
8. U.S. Pat. 3,122,038 (Feb. 25, 1964), J. Juras.
9. W. M. Stocker, Jr., *Am. Mach.* (May 15, 1950).
10. L. E. Browne, *Steel* (Feb. 25, 1946).
11. *Federal Specification FF-S-740a*, U.S. Government Printing Office, Washington, D.C., Oct. 1965.
12. H. Ll. D. Pugh and A. H. Low, *J. Inst. Met.* **93**, 201 (1964–1965).
13. D. Green, *J. Inst. Met.* **99**, 76 (1971).
14. *Ibid.*, **100**, 295 (1972).
15. T. Hoshi and M. C. Shaw, *J. Eng. Ind. Trans. ASME*, 225 (Feb. 1977)
16. G. F. Taylor, *Phys. Rev.* **23**, 655 (1924).
17. U.S. Pat. 1,793,529 (Feb. 24, 1931), G. F. Taylor.
18. *AIChE Symp. Ser.* **74**(180), (1978).
19. U.S. Pat. 2,879,566 (1959), R. B. Pond.
20. U.S. Pat. 2,976,590 (1961), R. B. Pond.
21. R. E. Maringer and C. E. Mobley, *J. Vac. Sci. Technol.* **11**, 1067 (1974).
22. J. F. C. Morden, *Met. Ind.*, 495 (June 17, 1960).
23. S. S. Brenner, *Science*, **128**, 569 (1958)
24. S. S. Brenner, *Acta Met.* **4**, 62 (1956).
25. W. H. Dresher, *Technical Report AFML-TR-67-382*, WPAFB, Dec. 1967.
26. E. H. Newton and D. E. Johnson, *Technical Report AFML-TR-65-124*, WPAFB, April 1965.

JOHN A. ROBERTS
Arco Ventures Co.

NONWOVEN TEXTILE FABRICS

Spunbonded,   252
Staple fibers,   284

SPUNBONDED

Spunbondeds are a significant and growing area of the nonwovens industry and, in 1979, amounted to about one third by weight of the North American nonwovens market (1). Staple-fiber nonwovens result from shortening the textile chain in the preparation of textiles from fibers; spunbondeds result from the preparation of synthetic-fiber nonwovens to achieve chemical-to-fabric routes. The introduction of fabrics prepared from continuous filaments by such routes dates from the mid-1960s in Europe and North America (2–4). However, chemical-to-fabric routes had been described, patented, or commercialized earlier (5–7).

The development of spunbondeds has been confined to the Western world and Japan and their development to profitability has been reviewed (8). Spunbonded lines have a high unit cost, ca $(1.2–1.5) \times 10^7$ in 1977, but can produce vast quantities of product (5). Spunbondeds are no longer identified as textile substitutes but as materials in their own right. However, although limited penetration of the domestic textile field has been achieved, spunbondeds are deficient in esthetic properties, eg, drape, conformity, and textile appeal. These deficiencies result at least in part from their bonded structure which is of rigid fiber-to-fiber links. The major area for staple-fiber, nonwoven fabric sales, ie, cover stock, could represent a major growth area for spunbondeds. The nonwovens market, in general, is expected to grow rapidly but the direction of growth is unclear (9–14). However, it is likely that, in the short term, spunbondeds will grow in areas where they have a foothold, eg, in civil engineering and in the carpet, automotive, furniture, durable-paper, disposable-apparel, and coated-fabrics (qv) industries.

Properties

Appearance.   Spunbonded nonwoven textiles generally are manufactured as roll goods in widths up to ca 5.5 m with basis weights of 5–500 g/m^2 or greater. In some cases, shaped articles such as filters may be produced as unit items but these are the exception. The basis weight of the majority of fabrics are 50–180 g/m^2. Generally, they are white, although gray and colored versions are produced for deliberate application benefits and occasionally as a result of recovered or scrap raw material forming part of the feedstock.

The fabrics generally have a random fibrous texture but they can have an orderly arrangement of fibers or be filmlike or netlike. Thicknesses generally are 0.1–5 mm; the majority of fabrics are 0.1–2.00 mm thick. The fibers comprising the fabrics generally are continuous and circular but, depending on the particular fabric, they may be in short lengths or in a mixture of lengths and they may have relatively uniform, noncircular cross sections or nonuniform, circular or noncircular cross sections. Filament diameters are 0.1–50 μm, with the preferred range being 15–35 μm, except for microdenier spunbondeds which are 0.1–1.0 μm in diameter. Spunbonded nonwovens have a characteristic stiff or feltlike handle which is the result of their rigid structure in which the filaments lack shear and, therefore, have a restricted relative movement.

The fabrics may have a glazed, plasticlike finish or a fibrous look. In some cases, the surface texture is discontinuous and may consist of a pattern of regular semiglazed dots which occupy 10–20% of the surface area, with the rest of the material having a more fibrous appearance. Such fabrics are point-bonded fabrics and generally are less stiff than fabrics which are bonded throughout. Thick, feltlike fabrics often are characterized by the presence of holes up to 1.0 mm in diameter from surface to surface; this is the result of needle punching, which is a common bonding method. In thinner fabrics, the pattern in which filaments have been laid down may sometimes be visible, as may the existence of heavier and lighter areas, thus indicating the limitations of the filament distributing apparatus.

Physical and Chemical. Physical and chemical properties, eg, specific gravity, moisture absorption, electrical behavior, chemical and solvent resistance, temperature resistance, dyeability, resistance to microorganisms and insects, and light stability of spunbonded fabrics, are determined by the material composition. Reference to the equivalent textile or polymer composition generally gives a satisfactory indication of these properties. Light stability, where it is important in relation to the fabric application, invariably must be approved by the customer in a test relevant to the use situation. In general, spunbonded textiles are engineered to be of adequate stability for their design applications, and the availability of excellent stabilizers enables virtually any desired degree of stability to be achieved. Illustrative tensile strength-loss curves are shown in Figure 1.

Mechanical, Hydraulic, and Textile Properties. Properties of spunbonded textiles are related to the mechanical properties and physical form of their fiber constituents and to their geometrical construction and binding. These properties are described by a wide range of tests (see Analytical and Test Methods) (15). Important areas for fabric characterization are thickness, basis weight, opacity, compressibility, isotropy, uniformity, tensile and modulus properties, recovery, burst properties, tear properties, permeability, porosity, abrasion resistance, creep, flex life, dimensional stability, static properties, absorbency, fraying properties, wrinkle and crease resistance, seam strength, and drape/handle characterization. Some of these properties are listed in Table 1. The basis weight per unit area is often the most important fabric parameter to the customer, since it is related to the yield and, thus, the area cost. The fabric thickness may be of importance in some applications and is measured by simple techniques which take into account the fabric uniformity, compressibility, etc (15).

Properties which are common to conventional textiles and nonwovens often are compared when spunbondeds are used in textile applications. These include dimensional stability to washing or to dry heating, eg, for coating substrates, and typically, values as low as 1.0% area shrinkage from machine washing may be achieved by the appropriate selection of the raw material, eg, polyester, and of the process. The abrasion resistance of spunbondeds and their resistance to pilling, snagging, filamentation, or

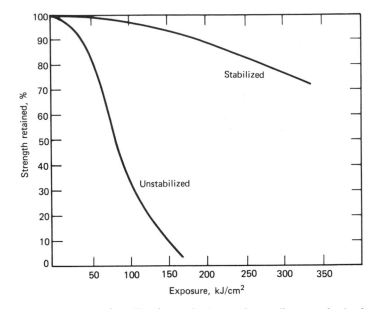

Figure 1. The effect of outdoor Florida weathering on the tensile strength of polyethylene/polypropylene, bicomponent, spunbonded fabrics. (Fabrics facing due south at 45°.) To convert kJ/cm^2 to langleys, multiply by 239.

cobwebbing also may be important and well-documented test procedures that are based on conventional textile testing have been described (15) (see Textiles, testing). Generally, the properties of spunbondeds are within a wide range of values. Spunbondeds excel in fray and crease resistance. The moisture absorption of most spunbondeds is very low, as is common with most synthetic-fiber textiles. However, the water holding of the fabrics can be quite high because of the high void content, especially with needled fabrics, where water may be retained by capillary action. The esthetic characteristics of spunbondeds are rarely quantified, except in the research environment.

Properties for objective comparison with conventional textiles include shear, drape, and nonrecoverable extension; numerous tests have been devised to assess fabric softness and hand (16). In some instances, shear, or the resistance of the fabric to in-plane shearing stress, may be the most important single factor determining fabric performance (17). The resistance of a woven fabric to in-plane shearing stresses is low because the yarns of which it is composed are free to move relative to each other in the fabric plane since they are interlaced. In a bonded nonwoven fabric, the filaments are not only locked together by rigid bonds, but the usually random arrangement of filaments produces the equivalent of triangulation in an engineering structure, thus presenting great resistance to in-plane movement of the filaments. Since the ability of a textile to accept shear deformation is a necessary condition for conformable fitting to a three-dimensional surface, spunbonded structures are excluded from many textile market sectors if they lack shear. Spunbondeds differ from most conventional textiles in their out-of-plane deformation behavior. This is usually quantified by some test, eg, drape or bending length (16).

Manufacture and Processing

A variety of raw materials and routes are used in the manufacture of spunbonded nonwovens. Virtually all of the spunbonded fabrics that are marketed are prepared from thermoplastic polymers (see Polymers, thermoplastic). A number of patents dealing with the manufacture of spunbonded fabrics based on dry-spun acetate fibers have been published (18–19); however, such work has not led to the marketing of acetate spunbonded fabrics. Trade names and manufacturers of spunbonded fabrics are listed alphabetically according to trade name in Table 2.

The basic polymers that are used are those that are common in fiber- and film-forming operations, ie, polyamide (nylon-6 and nylon-6,6), polyester, isotactic polypropylene, and polyethylene. Nylon-6,6 probably is used more commonly than nylon-6, although the latter is incorporated in a number of spunbonded fabrics. In some fabrics, both types of nylon are used, which allows thermal bonding based on the difference in melting points of the two materials (20). Until recently, certain versions of Terram and Mirafi were based on a nylon-6:6,6 copolymer which could be bonded thermally at a low temperature in the presence of saturated steam (21) (see Copolymers). Other nylon copolymers may be employed in the manufacture of fusible interlinings where a low melting point is needed. Nylons have a number of advantages, such as, easy bonding and dyeability, in nonwovens for certain applications. Their major disadvantage is their high energy content, which has led to their increasing cost and decreasing economic attractiveness as compared to polyesters and polyolefins. Generally, there is little need for precise end-group control for spunbondeds, and only medium tenacity and normal shrinkage filaments are required.

Poly(ethylene terephthalate), because of its desirable blend of useful properties for textiles, has become the major synthetic textile fiber. Its use in spunbonded nonwovens has not mirrored its penetration of the textile market, probably for economic reasons; although its high processing temperature and the necessity for moisture-free handling prior to extrusion lessens its attractiveness. However, it is used in a number of fabrics at normal textile molecular weights and mainly where its good dimensional stability at elevated temperatures is desirable (2–3). For example, poly(ethylene terephthalate) is used in carpet backings which undergo hot wet treatments under high tensions during carpet manufacture. As with other fibers, its properties and, hence, those of the fabrics, may be tailored, within limits, during the fiber-production process. For thermal-bonding processes, the melting point of a proportion of the fiber may be reduced either by leaving it undrawn or by employing a copolymer (22). The most usual technique is the replacement of 10–30 mol % of the terephthalic acid by isophthalic acid, since the latter compound is available relatively cheaply as a raw material for unsaturated polyester resin manufacture (see Polyesters, unsaturated).

In the polyolefin field, both isotactic polypropylene and polyethylene are used for spunbonded fabric manufacture (2); the factors dictating the choice are both economic and technical (see Olefin polymers). In recent years, polypropylene has been

Table 1. Typical Properties of Spunbonded Materials

Property	Cerex[a]	Corovin PP-S[a]	Fibertex 200[a]	Lillionette 506[a]	Lutradur H7210[a]	Net 909 P520[a]	Novaweb AB-17[a]	Terram 1000[a]	Trevira 30/150[a]	Typar 3351[a]	Tyvek 1073-B[a]
composition	nylon-6	polypropylene	polypropylene	nylon	polyester	high density polyethylene, polypropylene, or blend	polyethylene/polypropylene	low density polyethylene/polypropylene	polyester	polypropylene	high density polyethylene
bonding	autogenous area bonded	thermally point-bonded	needled	area resin bonded	area thermally bonded	fibrillated film	foamed film	area thermally bonded bicomponent	area resin bonded	area thermally bonded using undrawn segments	area thermally bonded (calendered)
basis weight, g/m²	10–68	20–100		30–100	50–250	9–54	10–30	70–300	85–170	54–142	39–110
typical basis weight, g/m²	34[b]	75[c]		60[b]	100[c]	27[d]	16–18[d]	101/170	142[c]	119[d]	75[d]
thickness, mm	0.12[e]		1.27[e]		0.38[f]	0.12[d]		0.7[d]	0.35[f]	0.33[d]	0.2[d]
breaking strength, N[l]	182/116[g]	130[h]	578[g]	103/98[g]	200/180[h]	9.8/7.8[d,i]	15/1.4[d,j]	850[g,k]	62/55[h]	512[g]	79/93[d,i]
breaking elongation, %		45[h]	125[g]	89/77[g]	40/40[h]		30/80[d]	80[g,k]	40/40[h]		26/33[d]
tear strength, N[l]	36/30[m]	15[n]	200[m]	34/32[o]	50/50[p]			250[q]		255[r]	4.5/4.5[s]
burst strength, kPa[t]	276[u]		1724[u]	245[v]				1100[w]			1180[x]

256

property							
opacity	intermediate	high	intermediate	intermediate	high	high	88%[y]
flex life, cycles		106[z]				high	>10[aa]
air permeability, L/(dm^2·min)	104[u]	70[bb]	3040[cc]	287[cc,dd]			304[ee]
water permeability, L/(dm^2·min)			30[ff]			24[gg]	641[hh]
pore size, µm			80–100[ii]		300[d]	100[jj]	

[a] For name of manufacturer, see Table 2.
[b] ASTM D 1910.
[c] DIN 53854.
[d] Test method not given.
[e] ASTM D 1777.
[f] DIN 53855.
[g] ASTM D 1117-1682 (grab test).
[h] DIN 53857 (strip test).
[i] N/10 mm.
[j] N/50 mm.
[k] Sample width 200 mm.
[l] To convert to dyne, multiply by 10⁵.
[m] ASTM D 2263.
[n] DIN 53356.
[o] ASTM D 2261.
[p] DIN 53859.
[q] BS (British Standard test) 4303-1963.
[r] Trapezoid test.
[s] Elmendorf test.
[t] To convert kPa to psi, multiply by 0.145.
[u] ASTM D 231.
[v] ASTM D 774-67.
[w] BS 4768-1972.
[x] Mullen test.
[y] Eddy opacity test.
[z] BS 3424; Method 11B.
[aa] MIT Flex test.
[bb] DIN 53887.
[cc] ASTM D 737.
[dd] L/min.
[ee] Frazier Air Porosity.
[ff] U.S Corp of Engineers, unit cm./s.
[gg] Manufacturer's test method.
[hh] Water vapor (g/m^2/24 h).
[ii] Equivalent opening sieve.
[jj] Pore size O_{90} µm.

Table 2. Trade Names and Manufacturers of Spunbonded Fabrics

Admel	MPC-Spunbond Inc., Japan
Asahi Spunbond	Asahi Chemical Industry Co., Japan
Axtar	Toray Industries, Inc., Japan
Bem-Liese	Asahi Chemical Industry Co., Japan
Bidim	Rhone-Poulenc Textile, France, South Africa, Brazil / Monsanto Textile Co., U.S.
Cerex	Monsanto Textiles Co., U.S.
Colback	Colbond bv, Netherlands
Conwed	Conwed Corporation Inc., U.S.
Colbond	Colbond bv, Netherlands
Corovin	J. H. Benecke GmbH, FRG
Celestra / Crowntex	Crown Zellerbach Corporation, U.S.
Delnet / Delweve	Hercules, U.S.
Dipryl	Sodoca S.a.r.l., France
Duon	Phillips Fibers Corporation, U.S.
Dynac	Toyobo Spun Bond, Japan
Enkamat	Enka Glanzstoff AG, FRG
Evolution/Evolution II	Kimberly Clark Corporation, U.S.
Fibretex	Crown Zellerbach Corporation, U.S.
Kimcloth	Kimberly Clark Corporation, U.S.
Kridee	VEB Textile Verpackungsmittel, GDR
Kyrel	J. P. Stevens, U.S. (imported from Asahi)
Lillionette	Snia Viscosa SpA, Italy
Lutrabond	Lutravil Spinnvlies, FRG
Lutradur	Lutravil Spinnvlies, FRG
Lutrasil	Lutravil Spinnvlies, FRG
Marix	Unitika Ltd., Japan
Mirafi	Fiber Industries Inc., U.S.
Novaweb	Bonded Fibre Fabrics, Ltd., UK
Netlon	Netlon Ltd., UK
Net 909	Smith and Nephew Ltd., UK
Petex	Juta, Czechoslovakia
Petromat	Phillips Fibers Corporation, U.S.
Polyfelt	Chemie Linz AG, Austria
Polyweb	Riegel Products Corporation, U.S.
Reemay	E. I. du Pont de Nemours & Co., Inc., U.S.
Sharnet	Inmont, U.S.
Silheim	Toray Ltd., Japan
Sodospun	Sodoca S.a.r.l., France
Sualen	VEB Chemiefaserkombinat Wilhelm Pieck, GDR
Supac	Phillips Fibers Corporation, U.S.
Syntex	MPC-Spunbond Inc., Japan
Tapyrus	Tonen Petrochemical Co. Ltd., Japan
Tafnel	MPC-Spunbond Inc., Japan
TCF	Mitsubishi Rayon Co., Ltd., Japan
Terram	ICI Fibres, UK
Texizol	Chemosvit np., Czechoslovakia
Thinsulate	Minnesota Mining & Manufacturing Co. Inc., U.S.
Toyobo Spun-Bond	Toyobo Co. Ltd., Japan

Table 2 (*continued*)

Trevira-Spunbond	Hoechst AG, FRG and U.S. (1980)
Typar	E. I. du Pont de Nemours & Co., Inc., U.S. and Luxembourg
Tyvek	E. I. du Pont de Nemours & Co., Inc., U.S.
Unisel	Teijin Ltd., Japan
Viledon-M	Carl Freudenberg, FRG
Vivelle	ICI Mond, UK, FRG

the least expensive fiber-forming polymer available and it is less dense than all of the polymers, particularly polyester whose fabric volume yield is less than 70% of that of polypropylene. Thus, unless major technical factors inhibit the choice of polypropylene, it is the favored raw material for spunbonded fabrics. Because of its low density, polypropylene fabrics float on water, which can be an advantage, for example, in the installation phase with civil-engineering fabrics. Standard predegraded fiber- or film-forming grades normally are used in this application, together with whatever additive masterbatches are dictated by processing or fabric application considerations. Conventional spinning and drawing technology usually is employed to yield medium tenacity filaments. Generally, it is not necessary to employ special techniques to yield low shrinkage filaments unless the fabrics are to be used in some critical high temperature application, eg, a coating substrate. Where thermal bonding techniques are used, the fabric is necessarily subjected to a heat treatment which tends to stabilize it against further shrinkage. In contrast to nylon and polyester, delustering pigments, eg, titania, normally are not included in polypropylene that is destined for fiber manufacture, so that polypropylene spunbonded fabrics have a fairly lustrous appearance. Colored pigments sometimes are employed where colored fabrics are required, since polypropylene cannot be dyed easily.

Polypropylene filaments have a further advantage in spunbondeds manufacture in that, under the influence of heat and pressure, quite strong bonds can be formed, even with drawn filaments, without destroying them. This is particularly useful in point bonding where high pressures are available (23). Bonding also may be achieved by employing undrawn and, hence, lower softening temperature filaments or by the use of a variety of binders (24). A further advantage of polypropylene, which also applies to polyethylene, is that scrap fabric or spinning waste can be reused and does not require elaborate recovery techniques. This is especially important in spunbondeds manufacture where the effectiveness of the operation may depend upon achieving a good conversion efficiency and, inevitably, up to 5% of the fabric has to be trimmed as edge waste.

Polyethylene is more expensive than polypropylene, it has a slightly lower fabric yield, and it cannot be made to produce filaments of the high strengths of polypropylene. However, its lower melting point permits less expensive and easier processing, and it is used in a number of materials where its lower service temperature is not a disadvantage. It is used in the manufacture of extruded nets, tack-spun items, film-fibril spunbondeds, and civil-engineering fabrics. The use of polypropylene and polyethylene in some types of spunbonded fabrics also results from their ability to fibrillate easily. As with polypropylene, fiber or film grades of polyethylene are used because their purity and melt viscosity are suitable for precision extrusion. Both low and high density polyethylenes are used in this area; the new linear, low pressure polyethylenes offer potential advantage of higher strength and lower cost. Polyethylene can be used in a self-bonding mode, either while still tacky after extrusion, or under the influence of heat and pressure (25). It also can be used as the binder with other fibers mentioned above (26). Several combinations of the four fiber-forming polymers are possible and a number are employed in various spunbondeds.

Spunbonded fabrics are produced by many routes which are combinations of alternative process steps (2). From these combinations, a small number of major processes can be identified by which the majority of spunbonded fabrics are produced. The predominant technology consists of continuous filament extrusion, followed by drawing, web formation by the use of some type of ejector, and bonding of the web.

Basic Spun-Filament Route. Spunbonded fabrics based on the extrusion of continuous filaments were introduced commercially as Reemay by DuPont in 1965 (4). The continuous-filament process may best be illustrated by reference to Figure 2 which shows diagrammatically the basic elements of typical mainstream, spunbonded fabric processes (2–3).

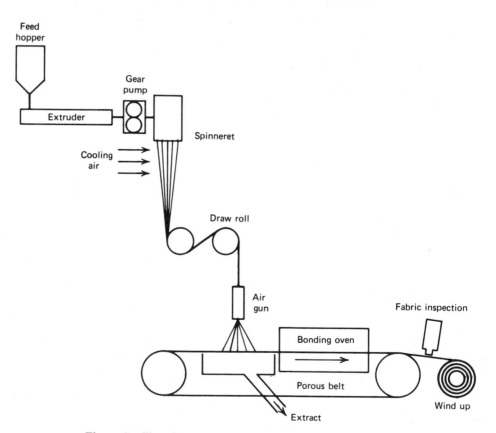

Figure 2. Flow sheet for typical spunbonded-fabric manufacture.

Typically, the polymer is melt spun in an identical manner to that used for the manufacture of continuous-filament textile or industrial yarns (27). In the case of polyester filaments, a homogenous granular polyester feedstock of suitable intrinsic viscosity is dried to a low moisture content and is fed to a suitable melter, either a screw pressure melter or a screw extruder, where it forms a pressurized (650–7000 kPa or 80–1000 psig), heated (265–280°C) stream from which it is fed to a metering pump which both meters the polymer stream and boosts its pressure to 7–35 MPa (1000–5000 psig). The molten polyester is extruded from a stainless steel spinneret containing from a few to several hundred holes ca 0.2 mm in diameter by way of a sand or other type of filter. A flow of cooling air solidifies the filaments below the extrusion point. If a polypropylene spunbonded fabric is to be prepared, similar melt-spinning principles apply, although the details differ.

In the early days of melt spinning, maximum wind-up speeds were limited, for mechanical reasons, to ca 1000–1500 m/min. Thus, in any combined spin-draw process, because of the effect of draw ratio, the spun yarn haul-off speed or the spinning speed could be as low as 200 m/min. This severely limited the throughput of the melt-spinning process, which, uncombined, could be run easily at the upper mechanical limit to the haul-off speed of 1000–1500 m/min which was more favorable economically. The development of improved winders capable of operating at up to 4000 m/min or higher and the knowledge that, as the spun yarn haul-off speeds increase, the required draw ratio is reduced to 2:1 or less, allowed the development of partially oriented yarns (POYs) (28). The more natural combination of drawing and texturing could be used for POYs, since false-twist texturing is limited in speed to below ca 1000 m/min because of the high levels of false twist which have to be inserted into the yarns.

This same limitation in haul-off speeds applies to spunbonded fabric manufacturing routes, and a number of techniques have been employed to circumvent it. However, since no wind-up unit is involved and, hence, no traverse, the mechanical limitations on speeds are not so low. If separate spinning and drawing processes are used, the attractions of an integrated process disappear; alternatively, if spin-draw is used, the melt-spinning process may not operate at maximum efficiency. This has been overcome either by hauling off the spun filaments at the highest possible speeds and accepting the reduced filament properties which result from the lack of a drawing stage or by tolerating the reduced extrusion rates employed with the spin-draw route (3,27). In the former route, haul-off is achieved by the use of a pneumatic jet or gun, which is capable of forwarding the filaments at speeds up to those for POYs, ie, ca 3000 m/min (27). This has led to filament tenacities that are ca 50–70% of those which would have been obtained by conventional drawing and to extensions that are considerably higher than those of textile filaments.

In the integrated spunbonded fabric process, the overall line speed may be limited by factors other than spinning-machine output. For example, the throughput of filaments per gun, the gun-scanning speed, and the web bonding speed all may be controlling elements for the overall manufacturing rate of the fabric. The gun is one of the key elements in the manufacture of spunbonded fabrics.

The alternative roll spin-draw route has much to recommend it, since it permits closer matching of the properties of the filaments and, therefore, of the fabrics to the requirements of the application. A diagrammatic version of this route is shown in

Figure 3; the diagram illustrates a number of features of a more sophisticated spun-bonding process. Extruded filaments are supplied from two spinnerets: those from one spinneret pass directly to the gun without drawing and those from the second spinneret pass over a guide roll 1 to a heated feed roll 2 and over a slot heater 3 to a draw roll 4 before entering the gun. Roll 4 is arranged to run faster than roll 2; thus, the filaments are elongated by an appropriate amount. The gun 5 traverses, from side to side above the moving porous belt 6, on which filaments are deposited in a layered arrangement. From belt 6, the mat of filaments passes to an oven which is maintained at a suitable temperature so that the undrawn filaments are preferentially melted and bond the drawn filaments from spinneret 7.

The electrode 8 and the grounded plate 9 improve the filament separation and, therefore, the cover and uniformity of the fabric. If an electrostatic charge is induced onto the filaments and as soon as they are released from the tension of the gun, they repel and are attracted to grounded surfaces. Thus, electrode 8 causes charging of the filaments either by direct charging from a high voltage source or by triboelectric charging, ie, the charge is generated by the friction of the filaments against a grounded, conductive surface (27). The mutually repelled, charged filaments are attracted to plate 9 and lose their charge after being precipitated in web form. In the absence of a grounded target, the charged filaments continue to repel each other after hitting

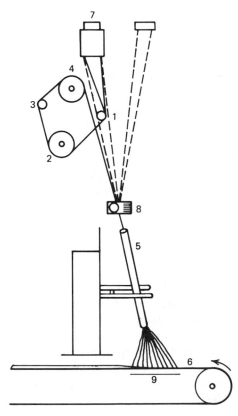

Figure 3.   The spin-draw route to spunbonded fabrics showing the optional use of a separate feed of bonding filaments (27).

the porous belt and form an unmanageable mass. For simplicity, the porous belt may form the target if, for example, it is made of wire mesh and is earthed. Static eliminators or high humidities may be employed to assist in the removal of charge, because it may take several seconds to leak to the grounded target since the filaments are nonconducting. If the filaments are charged triboelectrically, and since it is often convenient for the grounded surface to act as a yarn guide, ceramics such as chromium oxide often are used to resist abrasion. It is possible to modify the polymer composition or to select suitable polymer/guide combinations to enhance the triboelectric charging effect.

Because spin finishes contain an antistatic agent, the use of such a filament treatment would vitiate the charging process and would seriously hinder the formation of uniform filament webs. Since it is very difficult to wind packages of filament yarn without a spin finish because of static problems, split, spunbonded-fabric processes are not common. In principle, the winding of packages and the subsequent separate formation of the yarns into webs offers some advantages, eg, a better match between the spinning and the web-making processes (29); however, these are outweighed by process difficulties. The formation of webs from continuous filament tow has been described (30).

An alternative to charging the filaments to obtain controlled laydown in the form of a mat is to suck the air from below the porous belt (3). An overall downward air flow pins down the filaments and reduces disturbance by the air from the guns. This technique is used almost universally in mainstream spunbonded-fabric manufacture, although it is not without problems.

Another technique which is employed to prevent web disturbance after laying is spontaneous bonding or tacking. In the case of certain nylons or nylon copolymers, freshly spun filaments are tacky for a few seconds until they crystallize (3). Thus, webs of these filaments exhibit a significant coherence which can eliminate the need for suction. This process may be enhanced by treating the filament mats with water vapor or steam to increase the crystallization rate (31). This treatment also may cause spontaneous elongation of the filaments which results in some filament interlocking as well as tacking. The process can yield webs which are stable enough to wind and unwind prior to bonding, thus allowing a process break if web formation and bonding are mismatched in speed. Polypropylene and polyester do not exhibit spontaneous tackiness but, the webs may be temporarily tacked by slight heat, eg, 120°C, and pressure to give handleable webs prior to further bonding (32–33).

After web formation, either with or without temporary consolidation, the webs must be bonded to achieve their final strength. Bonding usually is achieved integrally with the web-formation process and at the web-formation speed of between a few and tens of meters per minute. Generally, for thermal area-bonding, some type of drum oven is used (34). From the oven, the web passes to a winder where it is rolled either for direct packaging or for later rewinding and breakdown into rolls. For thermal point-bonding, a heated textile calender is used and consists of two heated steel rolls; one is plain and the other bears a pattern of raised points (35). Control of filament orientation is necessary to concentrate the strength of a polypropylene carpet-backing fabric in the warp and weft directions to provide greater resistance to stretching and neckdown during carpet processing (4). This is achieved by arranging for the guns to scan at high speeds, some in the machine direction and some in the cross-machine direction (36). Since the scanning speeds are comparable to the filament-forwarding velocities, great control of the filament orientation is obtained on the porous belt.

The filament spray is scanned by deflecting it just as it emerges from the gun exit by means of oscillating air jets. This technique for achieving a measure of deliberate filament orientation gives a combination of randomly oriented filaments and a proportion of filaments of the desired orientation. Recently, for purposes of manufacturing spunbonded nonwovens for textile applications, a technique has been proposed for achieving 100% orientation of filaments in any chosen direction (37). The technique makes use of scanning jets that involve a very narrow filament spray and that confine the filament spray between pairs of plates extending from the gun exit to the porous belt. Thus, webs can be prepared for spunbonded fabrics in which the filament orientation is controlled as closely and as uniformly as in woven fabrics. The introduction of warp orientations depends on traversing a complete gun/plate assembly at right angles to the direction of the belt movement.

As illustrated in Figure 4, a gun for forwarding filaments consists essentially of a means for conducting the bundle of filaments to a point at the high velocity, low pressure zone of a venturi. Air is entrained with the filaments as a result of the low pressure generated in the jet. Once the filaments have reached the low pressure zone, the filament bundle and the entrained air move with the expanding supply air to form a cone or a fan of separated filaments; the exact shape depends on the detailed design of the gun (3,38–39). The pressure of the supply air may be from hundreds to thousands of kilopascals (several to tens of atmospheres) and air velocities in the low pressure

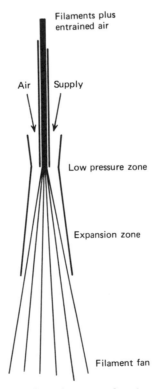

Figure 4.   The principal features of an air gun used to forward filaments in spunbonded-fabric manufacture.

zone may be several times supersonic. The air velocity must always be greater than the filament velocity in order to generate tension in the filament bundle. The filaments may be electrostatically charged prior to entering the gun to assist in separating them by mutual repulsion once they enter the expansion zone; tension generated in the low pressure zone holds the bundle together up to this point. Air flows of free air may be tens or hundreds of liters per minute and tensions of tens of grams are generated in the threadline.

The objective of good gun design is achieving the best separation of the filament bundle into individual filaments that are suitably distributed over the desired spray profile. If good filament separation is not achieved, then the fabric will have a rope appearance and consist of ropes of filaments and areas of low filament cover. The greater the quantity of supply air per filament, the better is the separation. However, equally important considerations in gun design are achieving maximum filament throughput at minimum air pressure and consumption, and the tension generated by the gun and the spray width. The greater the spray width, the fewer the number of guns required to cover a given width of fabric.

The tension requirements of the gun depend on the process configuration. If a roll-drawing route is employed, the gun needs only to generate sufficient tension to receive the filaments from the last godet, ca 0.003–0.01 N/tex (0.03–0.1 gf/den). If only gun drawdown is used, higher tensions of ca 0.01–0.1 N/tex (0.1–1.1 gf/den) are required at higher velocities (higher velocities because sufficient drawing tension must be generated by air drag below the spinneret). Thus, much more powerful guns with high supply pressures are required in order to maintain a favorable excess of supply air velocity relative to filament velocities of ≥ 2300 m/min (27,38). It is not possible to achieve the highest filament tenacities by gun drawdown because of the limited tensions which can be achieved at the highest gun-supply pressures, particularly with filament diameters at the upper end, ie, 35 μm, of the working region where surface/volume ratios for air friction purposes are reducing rapidly. For this reason, gun-drawdown processes are less in favor than roll-drawing processes.

The filaments emerging from the gun form a spray pattern which may be circular or elliptical; an elliptical profile, which approximates a line of filaments, is preferred In some types of spunbonding operations, the spinnerets are arranged in line, and this linear arrangement is continued through the guns and to the porous belt (3). This is difficult to arrange if the filaments are to be roll-drawn and, in general, throughput and technical considerations rule out this arrangement. Usually, it is necessary to generate uniform sheets of web from a number of fans of filaments of elliptical plan with the longer axis of a few to tens of centimeters in length.

The weight distribution of deposited fiber depends on the gun design but tends to be approximately triangular on both ellipse axes. Thus, if a single gun is held above the porous belt with the larger ellipse axis at right angles to the direction of belt travel, a strip of web of approximately constant density will be deposited along any line parallel to the direction of belt movement. A uniform web over a porous belt several or many times wider than the gun spray width can be formed by combining a number of such strips so that gaps and overlaps are not a problem and enough coherence is obtained between strips to ensure adequate cross-machine-direction fabric strength (40–42). The web can be achieved by arranging for several layers of strips to be laid from multiple guns which are overlapped like shingles. If the gun is traversed from side to side of the porous belt, the path traced out by the web is as shown in Figure 5.

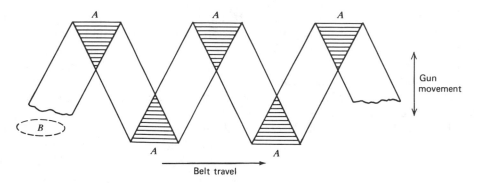

Figure 5.   Path traced on a moving porous belt by a traversing spray (B = spray pattern in plan view).

In the deeply hatched triangular areas A, the web thickness is twice that in the remainder of the strip. The size of areas A depend on the relative rates of movement of the gun and belt and on the spray width. If the belt advances by one spray width for every cycle of gun movement, then adjacent areas A touch; at this stage all of the porous belt becomes covered by a double web thickness. If the belt advances by less than one spray width per cycle of the gun, then the leading and trailing edges of the strip of web overlap, as do adjacent areas A on each edge, giving high density areas in the web.

The effect of moving guns also may be achieved by using stationary or swivelling deflector plates from which the spray is bounced onto the belt (43–46). Scans may be arranged along and at right angles to the direction of belt movement to achieve the desired orientation of the filaments in the web (36). Normally, in spunbonded-fabric manufacture, the filaments crumple as they hit the porous belt. Crumpling must occur unless the apparent relative velocities of the filaments and the belt are zero. Thus, if crumpling is to be avoided and if the filaments arrive at the belt at 2000 m/min, then the laydown speed also must be 2000 m/min. If a total scan width of one meter is desired, a scan rate of about 17 Hz is required to lay the filaments without crumpling. Such a rate is possible for small excursions obtained by scanning but not for the large excursions used with traversing guns. However, it is possible to superimpose a scanning motion onto a traversing motion.

Bonding.   As with staple-fiber nonwoven textiles, there are three major bonding techniques, ie, thermal, chemical/adhesive, and needling, that are used for spunbondeds. Except with needling, a choice may be made of point- or area-bonding. Point-bonding consists of cohering the filaments in small, discrete, and closely spaced areas of the fabric. Area-bonding involves using all available bond sites in the fabric; however, every filament contact is not necessarily bonded, since not every contact necessarily is capable of forming a bond (47). The three basic techniques of bonding do not lead to three completely different varieties of fabric physical properties. Thermal and chemical/adhesive bonding produce fabrics which are not radically different.

Needling is the less versatile technique since the only bonding variants are punch density, depth, and needle design; the products are thick fabrics of good flexibility, low initial modulus, high ultimate strength, high tear strength, and low fatigue life (2). Needling is not suitable for lightweight fabrics, ie, <100 g/m², because the needling action tends to concentrate the filaments in the already denser areas and, thus, destroys the fabric uniformity. Modern needle looms operate at high speed over high widths and, thus, the technique is perfectly applicable to spunbonded lines (48). It is less expensive to operate and requires much less sophisticated control than thermal bonding does. Many Docan process licensees use it and it tends to be popular for civil-engineering fabrics which are generally medium to heavyweight, ie, 120–500 g/m^2 (49).

Thermal and chemical/adhesive bonding offer the choice of point or area-bonding; the choice of either is dictated mainly by the ultimate fabric application. Because the bonded area usually is only 5–25% of the total fabric area, point-bonded fabrics tend to be more flexible and textilelike than area-bonded fabrics. The essential requirement for thermal bonding is the presence in the web of some material which holds together crossing or adjacent filaments and which is activated by heat. Usually the material is a polymer and often is the same material as that from which the web is manufactured. In the simplest form, thermal bonding makes use of the sticky nature of the extruded filaments emerging from the spinning machine (25). In the majority of thermally bonded fabrics, secondary heating is employed; the web is formed from fibers which have cooled to below their softening point and additional heat must be supplied to cause fusion.

In the simplest arrangement, the heat may be supplied by a bank of ir heaters, which are particularly applicable in the case of materials which have a wide softening range, eg, copolymers for fusible interlinings. Generally, fairly accurate temperature control is necessary. In the case of drawn polypropylene filaments, the useful bonding range is only 0.5°C, outside of which either no bonds are formed or the filament properties deteriorate (50). The use of hot-air drum ovens also is common (see Fig. 6). In hot air drum ovens, accurate temperature control is achieved by high air flows. The porous pinning belt provides a restraining pressure to prevent movement of the filaments during bonding and it helps to prevent fabric shrinkage. If the belt pressure

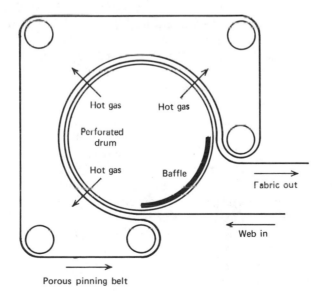

Figure 6.   Typical bonding oven used in the manufacture of heat-bonded nonwovens.

is insufficient to give the desired degree of web compression, then additional nip rolls may be employed at the fabric exit while the filaments are still soft. Highly accurate temperature control also may be obtained by the use of pressurized steam ovens. In the steam ovens, the web, which is trapped between two porous belts, is passed through suitable seals into a pressurized steam box where the heat of condensation plus specific heat of the steam almost instantaneously heat the belts and web to the steam-box temperature. After the fabric leaves the exit seal, which may provide substantial nip pressure on the web, any residual moisture is evaporated by the heat that is retained in the belts and web.

Since the temperature of the steam box is very sensitive to the steam pressure, which can be easily and accurately controlled, very good temperature control is possible. Operating temperatures are limited to about the softening point of polypropylene, because temperatures above this require very high steam pressures which are difficult to seal. A further advantage of steam-heated bonding is that discoloration of materials that are sensitive to oxidation, eg, nylon-6,6, does not occur.

For lightweight fabrics, where heat transfer through the fabric is rapid, thermal area-bonding may be achieved by hot calendering the web through plain rolls; this gives the highest pressures if maximum fabric strength is required. If controlled web compression is needed, then fixed-gap, hot calendering is possible. In principle, other types of heating, eg, dielectric or ultrasonic, could be used for thermal area-bonding.

The introduction of bonding sites for activation by thermal bonding may be accomplished in a variety of ways, many or all of which also are applicable to staple-fiber nonwovens. The simplest technique but the most difficult to control is softening the filaments of the web until they stick at the crossover points. A common technique is to combine filaments of the same material but of different softening points; an example is the use of undrawn polypropylene filaments as a binder for webs of drawn polypropylene (50). A softening point difference of 3°C is generated. The most sophisticated way of achieving this, eg, in Typar, is the introduction of undrawn segments at intervals along the polypropylene filaments of which the remainder is drawn (50). Undrawn polyester also may be used as a bonding agent (27). A similar technique is to include a different material of a lower softening point in the web filaments. However, this is complicated, since a separate extrusion system is required, for spunbonded fabrics. The different material may be introduced as a powder which is distributed throughout the web. With the powder technique, it is difficult to achieve uniform bonding throughout the width, length, and depth of the fabric.

Another sophisticated technique for achieving a uniform distribution of a polymeric binding agent is to use bicomponent filaments in which the binding agent forms a thin sheath around each web filament or is present as an integral sector of the filament cross section; such bicomponent filaments are called, respectively, core/sheath and side/side heterofilaments (see Polyamides, polyamide fibers). Combinations in spunbonded fabrics have been polyethylene/polypropylene (51), nylon-6/nylon-6,6 (21), nylon-6/polyester (52), and nylon-6:6,6 copolymer/polypropylene (21). All of these are of the core/sheath configuration; the sheath material is given first in each case.

Although, in principle, the manufacture of heterofilaments does introduce significant complications to the extrusion process, heterofilaments can be economically manufactured to the same precision as single-component filaments (53). The advantage of the heterofilament route to thermal bonding is the high degree of control that it confers on the properties of the spunbonded fabric. By variation of the characteristics of the sheath material, the core-to-sheath ratio and the proportion of heterofilaments present in the web, in combination with variations in the characteristics of the core material, the drawing conditions and the filament diameter, the balance of tensiles, and tear and other properties of the textile can be varied over a wide range to tailor the properties closely to the application.

Adhesive, or resin bonding also is common to staple-fiber nonwovens. The process usually consists of applying an acrylic or similar resin that is dispersed in water to the web, removing the surplus by mangling or other means, and passing the web over drying cans or rolls which evaporate the water and polymerize the resin. The resin accumulates at the filament contact points prior to water removal and polymerization as a result of surface tension. The web is bonded and, if a thinner product is required, also may be calendered which flattens the filaments and the bonds. Generally, resin bonding is not used in the thermoplastic spunbonded fabric area except for special applications, because it is a wet process which offers no particular advantages against a choice of alternative dry-bonding techniques (54).

Solvent bonding also has been disclosed for certain specialized applications (55), but chemical bonding is more important commercially. The latter is a system of bonding which is applicable to a number of nylon polymers (56). When the filaments are in close contact, the presence of anhydrous hydrogen chloride, or any of a number of other gaseous substances, initiates the formation of interfilament bonds, possibly by breaking interchain hydrogen bonds by forming an HCl complex with the amide group. When the HCl is desorbed, the hydrogen bonds immediately reform between molecular chains from different filaments. The bonding agent may be present in the gaseous phase or it may be dissolved in an inert organic carrier.

The relationship of fabric properties to the manufacturing variables is complex. Broadly, the tensile properties of the fabric are determined by the tensile properties of the filaments. High tenacity, low extension filaments lead to the formation of high tensile strength, low extensibility fabrics. However, the bonding medium properties also exercise an influence on the tensile properties. A high modulus binder tends to yield a high initial fabric modulus and, usually, the binder modulus is related to the binder tensile strength. Since needled fabrics have no binder, this corresponds to a very low modulus binder, and, hence, low initial-modulus fabrics. A high concentration of binder, which tends to lock the filaments together tightly, tends to effect high tensile strength. However, as the proportion of binder increases, the fabric becomes more filmlike and, therefore, less resistant to dynamic tearing. It also becomes stiffer, less flexible, and less permeable to gases and liquids. Conversely, with reduced binder proportions, fabrics are more feltlike or fibrous, less abrasion resistant, more flexible, more resistant to dynamic tearing, and more permeable. Again, needled fabrics are equivalent to zero binder levels. The influence of binder at levels of up to 30% or so, where it does not form a major proportion of the fabric, is summarized in Figure 7. The introduction of crimp into the filaments tends to increase the fabric flexibility. However, since the flexibility is a function of the average free filament length between bonds, a similar effect on fabric flexibility may be obtained by point-bonding, which results in a much greater average free filament length.

The most common form of point-bonding is thermal point-bonding, which is effected by the use of a heated calendar using virtually any of the techniques available

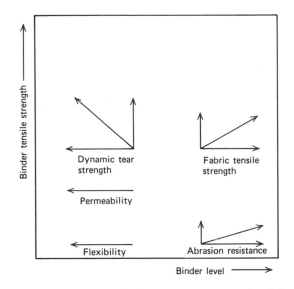

Figure 7. The qualitative influence of binder parameters on spunbonded textile properties. The arrows indicate increasing levels of the indicated property.

for area bonding, ie, bicomponent filaments, filament blends, undrawn filaments, or direct bonding of the drawn filaments. Since bonding is by both heat and pressure, the bonding of drawn filaments is easier to control than it is in area-bonding where temperature is the only variable. Based on this relative ease of bonding of single-component webs, point-bonded, drawn polypropylene filament fabrics are common because of the cost advantage of this polymer and its well-established filament-spinning route (32). The chemical and adhesive bonding routes can also be adapted for point-bonding, although in the case of adhesive bonding which is used for applying the adhesive spots, the process is termed print bonding. As far as is known, it is not used at all with spunbonded fabrics; one of the disadvantages is the inevitable relative coarseness of the print patterns since control over the bond size, adhesive penetration, etc, is imprecise. Chemical point-bonding is applicable to nylon fabrics and is based on the same technique as chemical bonding (57).

In thermal point-bonding, a heated calendar is used. Point-bonding calenders may be characterized by a 1–3-m working width and consist of heated steel work rolls between which the web is passed for point-bonding. The pattern may be cut on one or both rolls, the diameters of which are related to the working width of the calender for stiffness purposes. For a 3-m working width, rolls of 0.5-m diameter are not unusual. Of critical importance is the uniformity of pressure across the working width of the machine. Since the hydraulic or pneumatic pressure which is applied at the ends of the rolls to create the working nip pressure is several tons, it is impossible to prevent some bowing of the work rolls no matter how stiff the roll geometry is. Thus, some expedient must be adopted to obtain adequate pressure uniformity. Possibilities are backing the work rolls with crowned supporting rolls, crowning the work rolls, or setting the work rolls at a slight angle to each other. The most common solution is to employ crowned support rolls, since crowning the work rolls creates complications in pattern cutting.

Crowning consists merely in symmetrically reducing the roll diameter progressively from the center to the ends of the rolls; the total reduction is only a fraction of a millimeter. The pressure on the ends of the support rolls causes a balancing pressure to the center of the work rolls as a consequence of the crowning. One or both of the work rolls may be heated. Since accurate temperature control, typically $\pm1°C$ at up to $225°C$, is essential, oil heating is common although smaller calendar rolls may be electrically heated. With oil heating, the heated rolls are hollow and contain suitable oil inlet and outlet rotating seals and inner distribution channels to ensure high and uniform heat transfer. Calendering speeds depend on the fabric basis weight, the pattern design, the web material, and other factors. However, a range of 5–100 m/min probably is available; the higher speeds correspond to the lower basis weights.

Duplex bonding reduces the presence of area bonding by allowing the heat required to form the point bonds to penetrate from both sides of the web (53). In duplex bonding, part of the pattern is cut onto each roll and both rolls are heated to the bonding temperature. For example, opposing helical lands on the rolls generate a pattern of lozenge-shaped point bonds (58). Since the bonding heat has only half as far to penetrate into the web, the roll temperatures may be lower and the residence time in the nip zone may be shorter; thus, there is a reduction in the degree of adventitious area-bonding. Further, if the rolls are identical in land width and spacing, the fabric is identical on both faces. Any part of the web in contact with a heated surface undergoes bonding of one sort or another if the surface is at or above the bonding temperature. If the web which is in contact with the heated surface is under pressure, then primary bonds, ie, the lozenges, are formed; if the contact is without the bonding pressure, then secondary bonds result. In the case of the helical lands, each side of the fabric contains diagonal lines of secondary bonds interrupted by lozenge-shaped primary bonds which pass through the thickness of the fabric and which occur at the intersection of the secondary bonds. Only the primary bonds are common to both sides of the fabric; the helical lands of one work roll can create lines of secondary bonds only on one side of the fabric. The presence of secondary bond lines affects the fabric's flexibility. A fold can occur easily when the fold line is along a bond line. Folding at right angles to the bond line is less easy. Thus, fabrics with bond lines at right angles to each other have hindered flexibility.

The continuous secondary-bond lines reduce the fabric flexibility and, therefore, a further development of duplex bonding has evolved to overcome this problem (59–60). In principle, the use of identical point patterns on both rolls should eliminate all secondary bonding. However, mechanical considerations prevent the use of this solution, ie, such patterns can never be held in perfect register. Thus, the solution has been to use on the rolls point patterns of different pitch which can never mesh and damage the rolls but which, when interacting, generate a point-bond pattern of the required technical and esthetic characteristics. An example is shown in Figure 8 and illustrates the formation of a bark-weave design by the superimposition of two checkerboard designs. No continuous lines of secondary bonding can be formed from either of the two checkerboard rolls; therefore, fabric stiffness is minimized. However, isolated, secondary point bonds are associated with each primary point bond of the bark-weave pattern, and these isolated secondary point bonds are useful in improving the abrasion resistance of the fabric. An intermediate effect may be obtained by the combination of nonmeshing points and lands. Further esthetic improvements may be achieved by the duplex combination of nonmeshing basic point or line-point patterns with specialized half-tone point or line patterns (61).

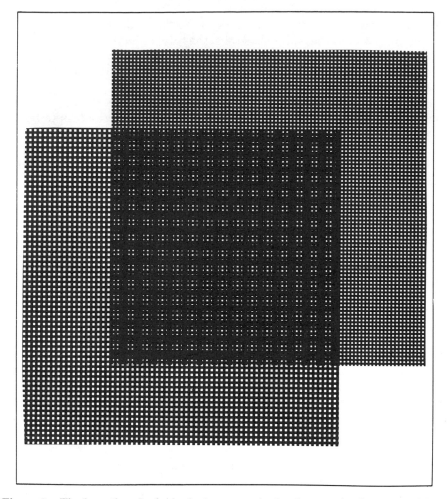

Figure 8.   The formation of a duplex bark-weave point-bond pattern by the superposition of two checkerboard designs (53).

The properties of point-bonded fabrics are closely related to those of area-bonded fabrics of the same composition. Point-bonding variables are total bond area, bond spacing, and bond shape; the bond size is determined if the three variables are fixed. The total primary bond area usually is 7–12%. In a duplex system, a primary bond area of 9% implies that the point area on each roll amounts to 30% of the roll area if the design is balanced, ie, equal proportions of secondary bonding form on each side of the fabric. In such a case, 21% of each side of the fabric area is secondary bonded. A primary bond area of 9% is obtained by the interaction of point areas of 20% and 45%. However, in the latter case, one side of the fabric has 11% of its area secondary bonded and the other side, 36%. Increasing the percentage primary bond area moves the fabric properties toward those of area-bonded fabrics, whereas changing the point-spacing or shape has a complex effect on the overall property balance.

Melt-Blown Textiles. Historically, the oldest chemical-to-fabric route is the melt-blown route. Work beginning in the late 1930s provided the basis for processes that are licensed to a number of companies (5–6,62). In this process, the fibers differ from those of the basic process both in diameter, ie, they are microdenier fibers with diameters of 0.1–1 μm, and in continuity, ie, they are staple fibers up to several centimeters long. Products include battery separators, hospital–medical products, ultrafine filters, insulating batting, and semidurable tablecloths.

The manufacturing technique for melt-blown textiles, which is inherently more expensive than that for conventional spunbondeds, consists of extruding the fiber-forming polymer through a single-extrusion orifice directly into a high velocity heated air stream (62–64). A diagram of the typical nozzle arrangement is shown in Figure 9. Typically, air at 300°C and 450 kPa (50 psig) is used. Virtually all of the fiber-forming thermoplastic polymers are suitable for melt blowing, but polypropylene and polyester are used commercially. The fiber diameter depends inversely on the air pressure but the use of higher pressures increases the cost.

After fiber formation, a web is collected by the interposition of a suitable screen conveyor. No binder is used; the fibers are held together by a combination of fiber interlacing and thermal bonding resulting from the residual heat of the extrusion and blowing process. Since the fibers are not drawn in the conventional textile sense, they are unoriented and have a low tensile strength. For some applications, they may be calendered to give thin, microporous sheets. They also may be embossed, vacuum- or thermo-formed, or laminated to other materials. Conventional or net-type spunbondeds may form the conveyor material to provide a high strength, low porosity laminate (62).

Flash-Spun Textiles. Flash-spinning is an alternative technique for the conversion of fiber-forming polymer into fibers. The fibers that are produced are in the form of a three-dimensional network of thin, continuous interconnected ribbons that are <4 μm thick that are termed film-fibrils or plexifilaments (65). Plexifilaments are produced by extruding the fiber-forming polymer through a single orifice as a high temperature, high pressure solution in an inert solvent with a boiling point 25°C or more below the melting point of the polymer. Thus, a yarn bundle is formed without the need for individual spinneret holes. The solvent must not act as a solvent for the

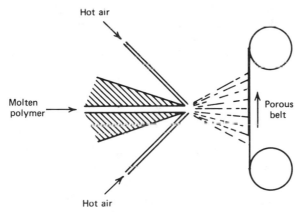

Figure 9. The manufacture of melt-blown spunbonded textiles.

polymer at temperatures below its normal boiling point, unlike solution spinning where solvents are used in which the polymers are stable at temperatures well below the solvent boiling point. Typically, high density polyethylene is used as the fiber-forming polymer and methylene chloride is used as a solvent. A preferred polymer extrusion temperature can be obtained which is within 45°C of the critical temperature of the solvent, ie, 193°C, with polymer concentrations being 2–20%. The presence of a dissolved inert gas, eg, CO_2, is used to increase the degree of fibrillation.

It is necessary to use a polymer or polymer mixture which is capable of crystallizing rapidly to a high degree of crystallization, if high strengths are to be obtained. Since the plexifilaments may be extruded at high speeds, up to nearly 16 km/min, they can have a high orientation and, therefore, a high strength, up to 0.37 N/tex (0.033 gf/den), in the as-spun state. The plexifilament networks may take several forms, depending on the conditions of manufacture. For the manufacture of spunbonded fabrics, the preferred form is one in which the fibrillated strands may be spread transversely into a lacey sheet, when they are hot calendered, to give a strong, paperlike product (4–5,8–9,66–67). Both area- and point-bonding may be used; the latter gives a more flexible product. As in solvent-spinning processes, the recovery of solvent influences the economics. A typical product which is based on this technique is Tyvek, which is used for book covers, maps, banners, signs, flags, tags, labels, packaging materials, and single-use garments for industrial, medical, and consumer applications.

Foam Spinning. In the continuing search for the simplest possible chemical-to-fabric process, several techniques have been proposed for converting extruded film into fabrics by slitting or fibrillating them (2). A technique which is similar to the flash-spinning process but which is related to film extrusion has been patented (68–69). In this process, fiber-forming polymers are extruded through a radial die. However, the molten polymer, typically polypropylene, contains a foaming agent, eg, azodicarbonamide, which evolves gas at the extrusion temperature (see Foamed plastics; Initiators). As the film is cooled immediately after extrusion by a blast of cold air, a foamed film forms and is reheated to above its glass-transition temperature and is stretched. The biaxial stretching or drawing process converts the foamed structure into a balanced fibrous tubular network which may be either collapsed or slit and wound up as textile roll goods. A diagrammatic indication of the equipment used in this process is given in Figure 10. The overall draw ratio, as given by the ratio of the diameters of the die and the pullring, may be 1.5–8; the upper limit is reached when fibers start to break out of the fiber network.

Integral Fibrillated Nets. A technique for the direct conversion of film to fabric or to a textilelike material has been commercialized as the Net 909 process (70–71). Extruded film is embossed while molten by being passed through a pair of heated nip rolls, one of which is engraved so as to form a pattern of usually hexagonal bosses on the sheet. When the embossed film is biaxially stretched, preferential drawing of the thin portions takes place, thereby causing splits to occur in the film and resulting in a sheet material that consists of a series of bosses connected by fibrillar strands.

Direct Extrusion. The possibility of forming useful textilelike materials by the direct extrusion of thick polymeric monofilaments has been exploited in such applications as land reclamation. The principle of direct extrusion is that filaments 0.2–1.5 mm in diameter of fiber-forming polymers are extruded from suitable spinneret orifices, usually in a multicurtain arrangement, and are allowed to drop ca 10 to 20 cm and to collect on a suitable moving collector, eg, a spiked roller (25). The technique

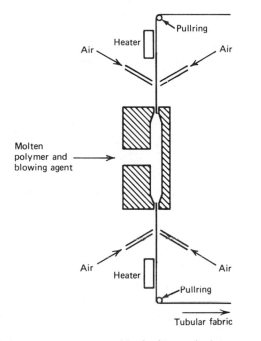

Figure 10. Die and drawing equipment used in the foam-spinning process for the manufacture of nonwoven textiles.

results in the formation of a coarse mat of thick, looped, undrawn monofilaments of any desired mat thickness from 5 mm to 20 mm or more which is suitable for upholstery material, for filtration (qv), as vertical and horizontal drainage, in soil stabilization and antierosion applications, as underwater mats in hydraulic engineering, and for numerous other purposes.

Integral Extruded Nets. The Netlon process for the manufacture of integral extruded plastic nets dates from 1955 (72). This process has been licensed worldwide and the products are familiar as polyethylene or polypropylene net packaging of fruit, agricultural nets, bird netting, plastic fencing, etc. Since the basic patent appeared, many similar but alternative techniques have appeared leading to a wide variety of integral net products (72–73). The basic process involves a spinneret composed of two rotating die members. The spinneret holes are situated at the annular intersection of the faces of the two members such that a proportion of each hole is in each member. As the die members rotate in opposite directions, strands emerging from the die orifices are full thickness only at the instant at which the portions of each hole are coincident (see Fig. 11). The tubular net that is produced is cooled, usually in a water bath, and is stretched slightly over a mandrel before being slit either lengthwise or on the bias to give substantially unoriented roll goods with either diamond or square mesh.

If full orientation of the strands is required, then monoaxial or biaxial stretching, usually with heating, is carried out. With suitable extrusion systems, the extruded product has the appearance of a perforated plastic sheet with 180 or more holes per square centimeter at the extrusion stage and 120 or more after the sheet is passed over the mandrel, or the strands may merge into a nonporous structure. When the material is biaxially oriented, the net structure is regenerated with well-defined strands (13)

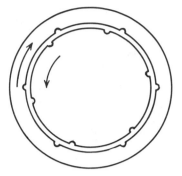

Figure 11.   Split die used in the manufacture of integral extruded nets by the Netlon process.

and a relatively high number of holes per square centimeter, perhaps 8 or more, depending on the draw ratio.

Tack-Spun Textiles.   The tack-spinning process for the preparation of pile-surfaced products is sometimes used with spunbonded textile substrates (74–75). Fiber-forming polymer that is in the form of an extruded film is fed with a carrier layer, which may be paper (qv), nonwoven textile, etc, into the nip of a pair of rolls which are heated to above the melting point of the film but well below the melting point, if any, of the carrier. When the film contacts the hot roll, it melts and is carried through the nip with the carrier material. As the heated roll and carrier surfaces part, fibers are drawn out of the melt and are immediately cooled by an air blast (76). The pile density and height may be controlled by varying the operating parameters of the process. The fibers that are formed are basically unoriented. Typically, either natural or pigmented polyethylene is used as the fiber-forming polymer and paper is used as the carrier. The product may be used for decorative packaging, hospital disposable applications, etc.

Fabrics From Yarns.   In principle, it is possible to prepare nonwoven fabrics from continuous filament yarns or tows, and such split processes could have some advantages although they do not have chemical-to-fabric appeal. The advantages of the process accrue from potentially better matching of the fiber and fabric production rates and improved overall utilization of the plant. A number of papers and patents have appeared for such processes, but no commercial products are available based on such techniques, with the possible exception of a Japanese spunbonded textile (29–30).

Rayon Textiles.   The technical knowledge for the manufacture of spunbonded rayon textiles has been available at least since 1972 in the UK and Japan (37,77–80) (see Rayon). However, the relatively low spinning speed (ca 125 m/min) and the increasingly expensive energy requirements relative to moisture removal in a wet process such as for spunbonded rayons lead to unfavorable economics, especially at a time when manufacturers of competitive dry-laid staple nonwovens are seeking binderless or thermal bonding processes to increase process speeds and reduce energy costs.

Cellulose Esters.   Like spunbonded rayon fabrics, a process for the manufacture of spunbonded nonwoven textiles from cellulose esters (qv) is available, but is not in major commercial use (18–19). The route was developed for the improved manufacture of cigarette filter rods by obviating the use of additional plasticizers or a curing stage. The production of cellulose ester nonwovens involves the dry-spinning process.

Economic Aspects

In 1979, spunbonded fabrics represented about one third by weight of all non-wovens production in the United States (1). In Europe in 1978, spunbondeds comprised ca 29% of all nonwovens. Their development and manufacture has occurred primarily in the United States and Western Europe. Only small tonnages are produced in Japan and virtually none is exported (81). Some spunbonded fabrics are produced in Eastern Europe but no figures are available and production is thought to be small. Because spunbondeds and nonwovens are new materials, data are not easy to locate nor necessarily reliable, especially as definitions for the various types of nonwovens are not universally agreed upon and cash volumes may be related to roll goods or finished consumer products. However, estimates of world spunbonded fabric production in recent years are summarized in Table 3. The weight of fabric is composed of a number of different basis weights. In the United States in 1977 the average weight was ca 60 g/m^2. Thus, the U.S. market in 1977 was ca 10^9 m^2 of fabric out of a total of approximately 6.0×10^9 m^2 of all types of nonwovens (82). In 1979, over one half of the U.S. spunbonded capacity was based on polypropylene, about one fifth each on polyester and polyethylene, and under a tenth on polyamide (1). Spunbondeds are expected to continue their growth in the United States at a true rate in value terms of 12%/yr during 1979–1982 (1).

Analytical and Test Methods

The simplest test for characterizing the strength of spunbonded textiles is the strip tensile test in which a strip of the material, which usually is 25 or 50 mm wide, is clamped in jaws of the same width in a conventional machine for textile testing. A gauge length of 200 mm is used and the jaws are separated every minute to provide a stress-strain curve of the material (84). An alternative tensile test is the grab tensile test in which a specimen that is considerably wider than the jaws is used (15). Special sample widths may be necessary to achieve reproducible results (51). Another type of tensile test is the plane-strain test (51,85). Grab tensile test stress-strain curves for a number of spunbonded fabrics are illustrated in Figure 12, and the isotropy of a

Table 3. World Production of Spunbonded Fabrics (1973–1979), thousand metric tons

Year	U.S.	Europe[a]	Japan	Total
1973		11		
1974	42[b]	17	1.8[e]	60.8
1975		20	1.7[e]	
1976		28	2.3[e]	
1977	57[d]	39	3.3[e]	99.3
1978	64[c]	44	3.7[f]	111.7
1979	84[b]			

[a] European Disposables and Nonwovens Association (EDANA).
[b] Ref. 82.
[c] Ref. 83.
[d] International Nonwovens and Disposables Association (INDA).
[e] Ref. 81.
[f] Ref. 79.

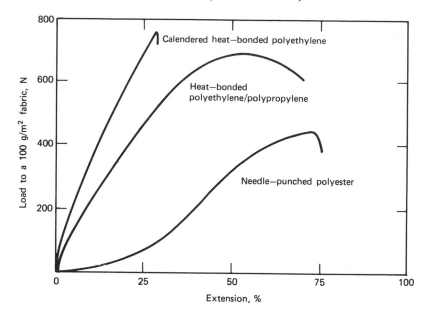

Figure 12. Typical stress-strain curves for spunbonded textiles of various types (ASTM D 1682; 200-mm sample width used).

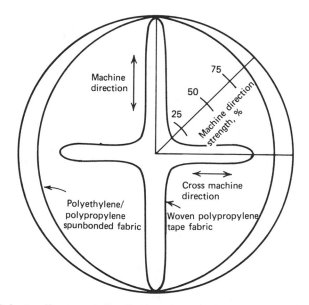

Figure 13. Polar tensile-strength distribution of a typical spunbonded fabric compared with that of a woven fabric (DIN 53857).

typical spunbonded fabric as compared with a woven fabric is illustrated in Figure 13. The work to rupture and the modulus of the fabrics under different types of test situations may be relevant to certain applications. A multidirectional tensile test, eg, the French cylinder test, sometimes may be used when considering the overall fabric resistance to multidirectional stress (86).

Typical measures of tear strength are the trapezoid and wing tear tests; both are carried out at low rates of tearing (15). A test which is commonly used for high tear rates is the falling pendulum or Elmendorf tear test. Although this test is noted as being generally unsuitable for nonwovens, it often is used as the only well-established dynamic tear test (87). Less well known but becoming increasingly adopted in Europe in the civil-engineering field is the cone-drop test, in which the diameter of the puncture that is produced by a heavy pointed cone falling from a specific height is recorded (88).

The burst strengths of spunbondeds may be measured by the Mullen test (89). In civil engineering, a plunger test, which is known as the CBR test because of its relationship to the California Bearing Ratio test and which gives both the strength and deformation to failure, is being used increasingly (86).

Hydraulic tests include measurement of the permeability of the fabric by air and water perpendicular to and in its plane and measurement of the pore size distribution, often by a graded sieving technique (15,51). The void ratio of the fabric, or its porosity, may be calculated from the thickness and the basis weight (51).

Health and Safety Factors

Considerations relevant to textiles of similar composition, which generally are inert, nontoxic, and nonallergenic, probably are broadly applicable. Flammability, allergic reactivity, food contact considerations, linting in surgical applications, etc, should be considered for health and safety. Specialized information usually is available from manufacturers and from the literature.

Uses

The applications for nonwoven textiles often are categorized as disposable or durable; disposable may be equated with dry or wet-laid staple-fiber nonwovens and durable applications often involve spunbonded fabrics. In the United States, disposable applications accounted for ca 4.4×10^9 m^2 of fabric and durable applications accounted for 1.6×10^9 m^2 of fabric in 1977 (12,82). About 60% of the total volume of disposable textiles was used in the cover-stock field for items, eg, diapers, sanitary napkins, tampons, underpads, etc. Until recently, very little spunbonded fabric was used as cover stock although, because of its dominance in volume terms, it is a market that is attractive to the spunbonded producer (90). The lightweight (15–25 g/m^2) fabrics for this market have well-established, demanding specifications and are highly price competitive and sell at only a few cents per square meter. At a total market value of 1.55×10^8, the average price in 1977 was ca $0.05/m^2 compared to an average price for fabrics for all durable applications of ca $0.27/m^2. In Europe, sanitary and medical applications accounted for approximately 16% by weight of total nonwovens fiber consumption compared to ca 32% in the United States (91).

A detailed breakdown of nonwoven-fabric applications is given by market sector in Table 4. In order of importance, the spunbonded market sectors are carpet, bedding/home furnishing, disposable apparel and durable paper, coated fabrics, furniture components, interlinings, and civil-engineering fabrics. Spunbondeds also are marketed in smaller proportions in a number of other sectors and in many miscellaneous applications.

Table 4.   Estimated U.S. Consumption of Nonwoven Textiles by Market in 1977 [a]

Application	Consumption				Spunbonded content
	10^6 m^2	10^3 t	10^6 $	%	
coverstock	2600	64	155	17.7	very small
surgical packs	460	34	90	10.3	small
interlinings	195	24	72	8.2	small
coated fabrics	155	22	62	7.1	intermediate
wipes	376	22	59	6.8	very small
filters	343	18	55	6.3	intermediate
carpet components	343	23	55	6.3	large
bedding/furnishings	272	19	52	5.9	intermediate
disposable apparel	125	8	21	2.4	large
durable papers	134	7	20	2.3	large
furniture	46	5.4	14	1.6	intermediate
fabric softeners	176	3.6	13	1.5	very small
civil engineering	13	2.3	7	0.8	large
other disposable	269	16	42	4.8	intermediate
other durable	403[b]	42	157	18.0	intermediate
Total	*5910*[b]	*310.3*	*874*	*100.0*	

[a] Ref. 12.
[b] Ref. 82.

The first spunbonded fabrics were developed for use in the carpet sector to replace jute which was used as a primary carpet backing; however, a number of factors prevented the market penetration that had been planned (62). Spunbonded primary backing entered segments of the carpet market where it offered unbeatable technical advantages, eg, a backing for printed carpets because of its superior dimensional stability and for fine gauge tufting for which its finer filament structure is superior to woven tape fabrics. Spunbondeds also are used in the carpet field as secondary backings, where the esthetics of melt-pigmented spunbonded fabrics are advantageous, and as an underlay carrier, where strength and lightness, hence, cost competitiveness, are beneficial. The carpet sector annual growth rate is leveling and 5% growth in weight of product consumed was forecast for the period 1977–1980 (82).

An area of much greater potential growth is in furnishings and automotive components. The former, which is termed hidden furniture fabrics, was expected to show a growth of 80% in value from 1977 to 1980 (12); another report forecast an annual 18% volume increase in coated and laminated fabrics for the latter outlets from 1977 to 1981 (82). Furniture applications are basically structural and include spring pocketings, dust covers, cushion backings, bottoming cloths, skirt liners, and pull strips (92). In the automotive area, there are eight main applications: paneling, visors, seat backing, seat listing, insulators, substrates for coated or laminated upholstery, landau tops, and trunk linings. Spunbonded fabrics comprise 36% of the nonwovens market in furniture applications and 43% of the nonwovens automotive market. The penetration of nonwovens into the automotive market is expected to be accelerated by the energy crisis since weight reductions may be obtained by their use.

Many durable nonwoven components are used in apparel which were made from woven materials 10–15 yr ago (93). These include metal fly-clasp reinforcements, belt-loop linings, pocket stays, waistbands, shirt collars, center placket and cuff interlinings. Probably only a small proportion of these nonwovens is spunbonded. One use for spunbondeds is as a fusible binder. Some spunbondeds are used in shoes as internal components and in luggage and accessories (92) (see Leatherlike materials). The advantages of nonwovens in these areas are improved stability, ease of handling, isotropy and uniformity, and suitability to automation with consequent savings in labor process steps and energy. Growth in these areas is not expected to be dramatic (82). Very significant growth is anticipated, however, in the disposable apparel area which is based on spunbonded fabrics and which is expected to double in value by 1981 compared to the 1977 level (12). The advantages of disposable garments are appreciated in the nuclear field, hospitals, laboratories, and many fields where smooth-surfaced overalls, smocks, laboratory coats, etc, resist contamination and, when contaminated, may be inexpensively and safely disposed of.

A 26 vol % annual growth rate in the bedding and home-furnishing area has been forecast in the United States (82). A significant proportion of nonwoven fabric used in this field is spunbonded and much of this may be point bonded. Typical uses are mattress pads, quilts, bedspreads, draperies and headers, and other accessory fabrics (12).

Civil-engineering fabrics form the final and major growth market for spunbonded fabrics; as a proportion of the total U.S. market for nonwovens, this sector is still quite small and was thought at ca 2.3×10^7 m^2 to be about half the size of the European civil-engineering fabric market in 1978 (94). However, it is expected to grow to ca $(1.0-1.5) \times 10^8$ m^2 in North America by 1985 when it will equal the European market. Use of nonwovens in civil engineering in the rest of the world is expected to equal that in Europe and North America, producing a total world market of $(3-4) \times 10^8$ m^2 in 1985. Virtually all of the fabrics used in this application are spunbonded. The application considerations for nonwoven fabrics in civil-engineering uses have been described (51). The fabric performs three functions when installed in a civil engineering or a geotechnical application, ie, separation, filtration, and reinforcement. Specifically, these functions provide embankment stabilization, drainage, and improved soil/fabric composite strength.

Spunbonded civil-engineering fabrics have expanded into a multitude of related civil-engineering and agricultural uses, eg, erosion control, revetment protection, railroads, canal and reservoir lining protection, highway and airfield blacktop cracking prevention, etc. National and international specifications are being agreed upon and established (10). Largely as a result of cooperation in producing fabrics of the appropriate specification with the civil-engineering industry by leading spunbondeds producers, geotextiles are being included in civil-engineering projects to an increasing extent at the project-design stage (10). The particular properties of spunbondeds which are responsible for this revolution are chemical and physical stability, high strength/cost ratio, and their unique and highly controllable structure which can be tailored to provide the desired characteristics, especially with respect to water and soil permeabilities. Spunbonded fabrics also are used for agricultural shading and insulation and other agricultural uses, wall coverings, medical and surgical disposable reinforcements, decorative packaging, capillary mattings, reinforced plastics, textile linings, electrical and building applications, wipes, scrims, blinds, battery separators, sound absorbents, gaskets, abrasives, composites, etc (see Insulation, acoustic).

Outlook

There is no clear agreement on the detailed direction of the growth of the non-wovens industry. Disposables may outgrow durables from 1978–1988 (83), although durables may lead disposables in short-term (1977–1981) growth. All authorities agree that nonwovens will grow at a significant rate in the next few years and that a real term sales-value growth figure of ca 8% will apply to which will be added a price inflation of ca 7%/yr (1,10–12,82–83,95–96). If durables do lead, it is likely that the newer techniques, eg, spunbondeds, will spearhead the growth and it is considered that this technology and spunlaid fabrics will grow at 19%/yr from 1978 to 1982 (1,12). Spunbonded lines are expensive and, as the output may be $(1-4) \times 10^7$ m^2/yr per line, very firm and profitable outlets must be anticipated for investment to be made. Thus, both technical and economic factors are retarding the technical progress of spunbondeds, and market growth in the near future is likely to occur by expansion of the applications for fabrics produced by existing technology.

BIBLIOGRAPHY

1. J. R. Starr, *European Disposables and Nonwovens Association*, AGM Paper, Munich, June 12, 1980.
2. A. Newton and J. E. Ford, *Text. Prog.* **5**(3), 1 (1973).
3. L. Hartman, *Text. Manuf.* **101**, 26 (Sept. 1974).
4. O. L. Shealey, *Text. Inst. Ind.* **9,** 10 (Jan. 1971).
5. F. F. Hand, Sr., *Text. Ind.* **143**, 86 (July 1979).
6. R. G. Mansfield, *Text. World* **129**, 83 (Feb. 1979).
7. Brit. Pat. 836,555 (Nov. 9, 1955), F. B. Mercer (to Plastic Textile Accessories Limited).
8. R. G. Mansfield, *Text. World* **127**, 81 (Sept. 1977).
9. W. W. Powell, *Mod. Text.* **LVII**, 39 (July 1977).
10. *Text. Manuf.* **79**(2), 19 and **79**(3), 26 (1979).
11. *Text. World* **129**, 57 (Aug. 1979).
12. J. R. Starr, *Am. Dyest. Rep.* **68**, 45 (March 1979).
13. M. E. Denham, *Mod. Text.* **LX**, 8 (June 1979).
14. *Pulp and Pap. Int.*, 59 (Mar. 1978).
15. *INDA Standard Test Methods*, International Nonwovens and Disposables Association, New York.
16. *Softness—Hand Research Study*, International Nonwovens and Disposables Association, New York, 1974.
17. J. Skelton, *Text. Res. J.* **46**, 862 (1976).
18. U.S. Pat. 3,148,101 (Sept. 8, 1964), W. T. Allman and co-workers (to Celanese Corporation).
19. U.S. Pat. 3,669,788 (June 13, 1972), W. T. Allman and co-workers (to Celanese Corporation).
20. A. E. Pedder and S. J. D. Hay, *Nonwovens Year Book 1976*, Texpress, Stockport, UK, pp. 35, 62.
21. Brit. Pat. 1,157,437 (July 9, 1969), B. L. Davies (to Imperial Chemical Industries Limited).
22. U.S. Pat. 2,836,576 (May 27, 1958), J. A. Piccard and F. K. Signaigo (to E. I. du Pont de Nemours & Co., Inc.).
23. U.S. Pat. 3,855,045 (Dec. 17, 1974), R. J. Brock (to Kimberley Clark Corporation).
24. Brit. Pat. 993,920 (June 2, 1965), P. J. Couzens (to Imperial Chemical Industries Limited).
25. Neth. Pat. 7,607,664 (Jan. 11, 1977), A. Rasen and co-workers (to Akzo NV).
26. Ger. Pat. 2,812,429 (Sept. 14, 1978), H. A. Booker, B. L. Davies, A. J. Hughes, and C. Shimalla (to Fiber Industries Inc.).
27. U.S. Pat. 3,338,992 (Aug. 29, 1967), G. A. Kinney (to E. I. du Pont de Nemours & Co., Inc.).
28. U.S. Pat. 3,771,307 (Nov. 13, 1973), D. G. Petrille (to E. I. du Pont de Nemours & Co., Inc.).
29. R. Krčma, *Textiltechnik* **28**, 775 (1978).
30. Jpn. Pat. 48 056,967 (Aug. 10, 1973), (to Teijin Ltd.).
31. Brit. Pat. 1,288,802 (Sept. 13, 1972), W. G. Parr (to Imperial Chemical Industries Limited).
32. U.S. Pat. 3,855,046 (Dec. 17, 1974), P. B. Hansen and B. Pennings (to Kimberly Clark Corporation).

33. U.S. Pat. 3,989,788 (Nov. 2, 1978), L. L. Estes, Jr., A. F. Fridrichsen, and V. S. Koshkin (to E. I. du Pont de Nemours & Co., Inc.).
34. *Nonwovens Rep. Int.*, 1 (Oct. 1979).
35. V. D. Freedland, *Nonwovens Ind.* **10,** 33 (Aug. 1979).
36. U.S. Pat. 3,991,244 (Nov. 9, 1976), S. C. Debbas (to E. I. du Pont de Nemours & Co., Inc.).
37. Brit. Pat. 2,006,844 (Oct. 26, 1977), P. M. Ellis and R. D. Gibb (to Imperial Chemical Industries Limited).
38. Ger. Pat. 2,785,158 (July 6, 1972), H. O. Dorschner and co-workers (to Metallgesellschaft AG).
39. Brit. Pat. 1,436,545 (May 19, 1976), J. Brock (to Imperial Chemical Industries Limited).
40. Ger. Pat. 2,706,976 (Dec. 7, 1978), V. Semjonow (to Hoechst AG).
41. Ger. Pat. 1,560,790 (Mar. 27, 1975), L. Hartmann (to Lutravil Spinnvlies GmbH and Co.).
42. Brit. Pat. 1,231,066 (May 5, 1971), C. H. Weightman (to Imperial Chemical Industries Limited).
43. U.S. Pat. 4,163,305 (Aug. 7, 1979), V. Semjonov and J. Foedrowitz (to Hoechst AG).
44. Ger. Pat. 2,421,401 (Mar. 22, 1979), K. Mente, W. Raddatz, and G. Knitsch (to J. H. Benecke GmbH).
45. U.S. Pat. 4,017,580 (Apr. 12, 1977), J. Barbey (to Rhone-Poulenc Textile).
46. Neth. Pat. 7,710,470 (Mar. 28, 1979), D. J. Viezee, P. J. M. Mekkelholt, and J. A. Juijn (to Akzo NV).
47. K. West, *Chem. Br.* **7,** 333 (Aug. 1971).
48. D. Ward, *Nonwovens Yearbook 1979*, Texpress, Stockport, UK, pp. 19–26.
49. Ger. Pat. 1,966,031 (May 19, 1971), H. O. Dorschner, F. J. Carduck, and C. Storkebaum (to Metallgesellschaft AG).
50. U.S. Pat. 3,322,607 (May 30, 1967), S. L. Jung (to E. I. du Pont de Nemours & Co., Inc.).
51. *Designing with TERRAM*, ICI Fibres, 'Terram Group,' Pontypool, Gwent, UK, Aug. 1978.
52. *Nonwovens Rep. Int.*, 1 (Jan. 1980).
53. K. Porter, *Phys. Technol.* **8,** 204 (Sept. 1977).
54. U.S. Pat. 4,125,663 (Nov. 14, 1978), P. Eckhardt (to Hoechst AG).
55. U.S. Pat. 4,181,640 (Jan. 1, 1980), G. P. Morie, C. H. Sloan, W. J. Jackson, Jr., and H. F. Kuhfuss (to Eastman Kodak Company).
56. U.S. Pat. 3,542,615 (June 16, 1967), E. J. Dobo, D. W. Kim, and W. C. Mallonee (to Monsanto Company).
57. U.S. Pat. 4,075,383 (Feb. 21, 1978), R. M. Anderson and co-workers (to Monsanto Company).
58. L. G. Kaseho, *International Nonwoven and Disposable Association, Technical Symposium Paper*, Washington, Mar. 1974.
59. Brit. Pat. 1,474,101 (May 18, 1977), D. C. Cumbers and D. Williams (to Imperial Chemical Industries Limited).
60. Brit. Pat. 1,474,102 (May 18, 1977), D. C. Cumbers and D. Williams (to Imperial Chemical Industries Limited).
61. Brit. Pat. 1,499,178 (Jan. 25, 1978), K. Porter (to Imperial Chemical Industries Limited).
62. *Nonwovens Ind.* **10,** 10 (Sept. 1979).
63. U.S. Pat. 3,972,759 (Aug. 3, 1976), R. R. Buntin (to Exxon Research and Engineering Co.).
64. U.S. Pat. 4,168,138 (Sept. 18, 1979), D. McNally (to Celanese Corporation).
65. U.S. Pat. 3,081,519 (Mar. 19, 1963), H. Blades and J. R. White (to E. I. du Pont de Nemours & Co., Inc.).
66. U.S. Pat. 3,442,740 (May 6, 1969), J. C. David (to E. I. du Pont de Nemours & Co., Inc.).
67. U.S. Pat. 4,069,078 (Jan. 17, 1978), M. D. Marder, R. Osmalov, and G. P. Pfeiffer (to E. I. du Pont de Nemours & Co., Inc.).
68. U.S. Pat. 4,085,175 (Apr. 18, 1978), H. W. Keuchel (to PNC Corporation).
69. Jpn. Pat. 53 078,374 (Dec. 20, 1976), (to Teijin Ltd.).
70. Brit. Pat. 914,489 (July 19, 1960), D. E. Seymour and D. J. Ketteridge (to Smith and Nephew Research Limited).
71. Brit. Pat. 1,548,865 (July 18, 1979), A. G. Patchell, W. O. Murphy, and R. Lloyd (to Smith and Nephew Research Limited).
72. U.S. Pat. 3,959,057 (May 25, 1976), J. J. Smith.
73. U.S. Pat. 4,123,491 (Oct. 31, 1978), R. L. Larsen (to Conwed Corporation).
74. Ger. Pat. 1,902,880 (Aug. 20, 1970), K. Seiffert.
75. Brit. Pat. 1,503,669 (Mar. 15, 1978), A. A. A. Giovanelli and E. W. Schmidt (to Imperial Chemical Industries Limited).

76. Brit. Pat. 1,378,638 (Dec. 27, 1974), D. M. Fisher and co-workers (to Imperial Chemical Industries Limited).
77. M. J. Welch and J. A. McCombes, *Nonwovens Ind.* **11,** 10 (May 1980).
78. *Nonwovens Rep. Int.*, 5 (Sept. 1978).
79. *Nonwovens Rep. Int.*, 3 (Jan. 1980).
80. *Textile World* **129,** 101 (Nov. 1979).
81. S. Tauchibayashi, *Nonwovens Ind.*, 14 (Mar. 1980).
82. L. E. Seidel, *Textile Ind.* **142,** 41 (Jan. 1978).
83. *Mod. Text.* **LX,** 7 (Oct. 1979).
84. *DIN 53857, B1.2*, DIN Deutsches Institut für Normung eV., Berlin, FRG.
85. C. R. Sissons, *C.R. Coll. Int. Sols. Text.* **II,** 287 (1977).
86. W.Wilmers, *Str. Autobahn* **31**(2), 69 (1980).
87. *ASTM D 1424-63 (Reapproved 1975)*, Part 32, ASTM Standards, American Society for Testing and Materials, Philadelphia, Pa., 1979, p. 352.
88. S. L. Alfheim and A. Sorlie, *C.R. Coll. Int. Sols. Text.* **II,** 333 (1977).
89. *ASTM D231-62 (Reapproved 1975)*, Part 32, ASTM Standards, American Society for Testing and Materials, Philadelphia, Pa., 1979, p. 119.
90. D. K. George, *Nonwovens Ind.* **11,** 22, 43, 49 (Mar. 1980).
91. *Nonwovens Rep. Int.*, 4 (Apr. 1979).
92. *Text. Ind.* **143,** 66 (Feb. 1979).
93. F. F. Hand, Sr., *Text. Ind.* **144,** 92 (Feb. 1980).
94. R. G. Mansfield, *Text. World* **128,** 52 (Dec. 1978).
95. *Mod. Text.* **LX,** 11 (May 1979).
96. *Mod. Text.* **LX,** 27 (July 1979).

K. PORTER
ICI Fibres

STAPLE FIBERS

Nonwoven textile fabrics are porous, textilelike materials, which usually are in sheet form, are composed primarily of fibers, and are manufactured by processes other than spinning, weaving, knitting, or knotting (see Textiles). These materials also have been called bonded fabrics, formed fabrics, or engineered fabrics. The thickness of the sheets may vary from 25 μm to several centimeters. The weight may vary from ca 5 g/m^2 to ca 1 kg/m^2. The texture or feel may range from soft to harsh. The sheet may be compact and crisp as paper, supple and drapable as a fine conventional textile, or high loft like natural down; also, it may be highly resilient or limp. Its tensile properties may range from barely self-sustaining to impossible to tear, abrade, or damage by hand. The fiber components may be natural or synthetic, from 1–3 mm to essentially endless. The total composition may consist only of one type or a mixture of fibers within which frictional forces provide the basis for the product's tensile properties. However, the finished product may include a lesser or greater proportion of film-forming polymeric additive functioning, which is an adhesive binder, to impart these properties. All or some of the fibers may be welded chemically, ie, with a solvent, or physically, ie, with heat. A scrim, gauze, netting, yarn, or other conventional sheet material may be added to one or both faces or embedded within the nonwoven as reinforcement.

The above description does not exhaust the possible variations in composition and in physical properties which are or can be made. A wide range of products exist, with paperlike materials at one extreme to woven fabrics at the other.

There have been many attempts to formulate a definition of nonwoven fabrics as well as a more affirmative name (1–3). Felted fabrics from animal hairs, eg, wool (qv), are excluded from the class, even though their structure fulfills descriptions of nonwoven textile fabrics.

Conventional textile fabrics are based on yarns and, in special cases, monofilaments. Yarns are composed of fibers which have been parallelized and twisted by a process called spinning to form cohesive and strong one-dimensional elements. In making textile fabrics, the yarns (or the monofilaments) are interlaced, looped or knotted together in a highly regular repetitive design in any of many well-known ways to form a fabric (see Fig. 1). The fabric strength and other physical properties are derived from friction of individual fibers against each other in each yarn, and friction between adjacent yarns.

The present nonwoven technology started ca 1938. Nonwovens are derived from conventional textile technology where long, ie, staple, fibers are used (see Fig. 2), and conventional paper (qv) technology where short fibers are used. In contrast to conventional textiles (4–5), the base structure of all nonwovens is a fibrous web; thus, the basic element is the single fiber. The individual fibers are arranged randomly. Tensile and other, eg, stress–strain and tactile, properties are imparted to the web by adhesive or other chemical and physical bonding of fiber to fiber; fiber-to-fiber friction which, primarily, is created by entanglement; and/or reinforcement of the web by added structures, eg, woven and knitted fabrics, yarns, scrims, nettings, films, and foams.

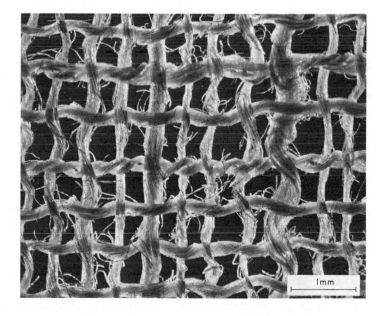

Figure 1.   An ancient Egyptian woven linen fabric which closely resembles a modern woven gauze, except for the improved uniformity of weight and twist in modern yarns.

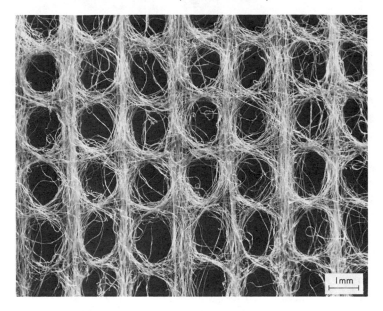

Figure 2. A rearranged card web of rayon fibers also resembles a woven gauze. Billions (10^9) of meters of nonwovens, such as shown here, have been produced.

The modern history of staple nonwoven textile fabrics began with the publication of patents representing two different approaches for the manufacture of flexible bonded sheets. The first, issued in 1936, describes an intermittently bonded nonwoven fabric made by printing an adhesive in a predetermined repetitive pattern onto an oriented fiber web (6). The bonded areas provide strength and the unbonded areas, where the fibers are free to move relative to each other, ensure a degree of softness (see Fig. 3). The second approach was patented in 1942 and describes a method for making a flexible bonded sheet using a random mixture of potentially fusible and nonfusible fibers (7). A sheet which is held together by a system of random intermittent adhesive bonds is produced when the web is subjected to physical or chemical conditions to selectively soften the fusible fibers while appropriate pressure is applied. Nonwovens also were made by saturating webs with water-based latexes which then were heat-dried.

There are a few basic elements which must be varied and controlled to produce the great range of nonwoven fabric types which are available. These include: the fibers, including chemical types and physical variations; the web and the average geometric arrangement of its fibers as predetermined by its method of forming and subsequent processing; the bonding of the fibers within the web, including the properties and geometric disposition of the adhesive binder or, alternatively, the frictional forces between fibers, primarily as created by close fiber contact or entanglement; and re-inforcements, eg, yarns, scrims, films, and nettings. In practice, each element can be varied and, thus, can exert a powerful influence, alone and in combination, on the final fabric properties.

Staple fibers originally were defined as the approximately longest or functionally the most important length of natural fibers, eg, cotton (qv) or wool (8). In the present context, as applied to regenerated cellulose and synthetic fibers, staple fibers are of

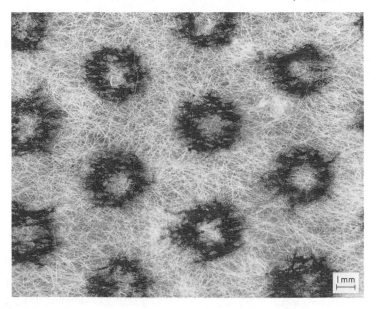

Figure 3. This is a typical print-bonded nonwoven fabric, made from card web and bonded by the rotogravure process with poly(vinyl acetate) latex. The binder, in a regular pattern of annular circles, is stained differentially by a solution of $I_2.KI$.

relatively uniform length, ca 1.3–10.2 cm, and can be processed on conventional textile machinery. Since regenerated and other extruded fibers essentially are endless as formed, they are cut during the manufacturing process to a specific length to meet a processing or market need. Extruded fibers also are produced as continuous filaments. Woven or nonwoven fabrics that are made from staple fibers appear fuzzier and feel softer and fuller than fabrics that are made from endless filaments. The processes for forming webs from staple fibers are entirely different than those in use for continuous filaments. Staple and filament fiber webs lead to products which differ substantially in their properties.

Components and Processes

Fibers. The fibers, as defined by their chemical composition and as a result of their physical–mechanical properties, determine the ultimate fabric properties. Other determinants, such as web structure and bonding maximize inherent fiber characteristics, eg, strength, resilience, abrasion resistance, chemical properties, and absorbency or hydrophobicity.

Any of the natural fibers which are available in commercial quantities can be used. In practice, wood pulp, which is far shorter than staple length, is the only one used in large amounts because of its high water absorbency, bulk, and low cost. Occasionally it is used in mixtures with staple-length rayon (qv) or polyester and, less often, with nylon (see Polyamides). Cotton and manila hemp also are used but in small amounts. Cotton has excellent inherent properties for use in nonwovens (9). Recent simplifications in cotton-fiber cleaning and bleaching should increase its commercial usage (10). Fibers such as wool, silk (qv), and linen are far too costly for serious consideration.

Regenerated fibers include viscose rayon and cellulose acetate (see Cellulose acetate and triacetate fibers). Viscose rayon was the overwhelmingly predominant fiber in use until recent years, when its cost rapidly escalated and new uses were developed for which rayon is not suitable. It is an easy fiber to process into webs and is bonded readily by several different kinds of adhesives, notably latexes, and by mechanical entanglement. It is available as regular and high wet modulus types; both, but primarily regular modulus, are used. However, it is not a strong fiber and this deficiency is evident in products made from it. Rayon is almost ideal for surgical and sanitary products, filters for food and industrial solvents, wiping cloths, certain lightweight facings, hand towels, and hospital garments. Common denominators for these uses are biodegradability, compatibility with body tissues, inertness to common solvents, good water absorbency, and moderate cost; therefore, they are well-suited for disposability. In recent years, viscose-rayon manufacturing plants in the United States and Western Europe have been closing because of increasing cost of the fiber. Considerable research is underway to replace the viscose process and to improve rayon fiber properties (11–12). Products comprised of wood pulp are gaining in importance because of the latter's low cost and equal purity–absorbency properties. However, the short fiber length, 1–3 mm of common commercial grades, creates problems in bonding, if product softness is needed. Polyester fibers (qv) are now (1981) the fibers used in the greatest quantity in staple form.

All synthetic fibers are potentially available for use. However, only polyester poly(ethylene terephthalate), nylon (types 6,6 and 6), vinyon, polypropylene (see Olefin polymers), and poly(vinyl alcohol) are significant commercially (see Vinyl polymers); polyester fibers represent by far the largest volume usage. Polyester cost is slightly lower than viscose rayon, and its strength and resilience are superior. Polyester fibers are hydrophobic which is desirable in lightweight facing fabrics, eg, in disposable diapers. They impart an easily perceptible dry feel to the facing even when the inner absorbent wood pulp is saturated. As practical methods are developed to process and bond polyester fibers, they are displacing rayon fiber facings in the marketplace.

Nylon fibers have been used in nonwovens, eg, in garment interlinings, because of their strength and resilience. However, their cost is substantially higher than polyester and they do not have sufficient compensating properties for many uses; therefore, they are used in staple form in moderate amounts.

Polypropylene fibers are strong, low in cost, and highly hydrophobic. Because of their low density, their yields, ie, high polymer volume, large diameter, is better than other fibers. Even though they are derived ultimately from petroleum, their projected future cost is favorable (see Petroleum, products). Polypropylene is difficult to bond with the widely used latexes, but effective binders are being developed. When the bonding problem is solved, staple polypropylene fibers probably will be used in large amounts (see Fibers, chemical; Olefin fibers).

Poly(vinyl alcohol) fibers, which are cross-linked using formaldehyde, are manufactured in substantial commercial quantities and only in Japan. They are strong, water insoluble, and hydrophilic. They are comparable to rayon in their response to latex binders and produce nonwovens which are slightly stiffer, slightly less absorbent but substantially stronger (13). The comparison of original fiber strengths of rayon with poly(vinyl alcohol) and the tensile properties of the respective finished fabrics illustrate the retention of fiber properties in the finished nonwoven. Poly(vinyl alcohol) fibers are not widely used because of unfavorable availability and cost.

Cellulose acetate fibers, although moderately inexpensive, are weak and difficult to bond. Hence, these fibers are not used in significant amounts.

In the early stages of the nonwoven industry, the staple fibers which were used were, with minor exceptions, those available for the conventional textile industry. The exceptions were the small amounts of thermoplastic fibers made primarily as nonwoven binders in the 1940s and 1950s, ie, plasticized cellulose acetate for fusible–nonfusible fiber mixtures (7) and vinyon HH fibers of poly(vinyl chloride)/acetate for wet-formed nonwovens used for tea bags. In recent years, the fiber industry has provided polypropylene fibers and undrawn or only slightly drawn polyester fibers as more effective thermoplastic-fiber binders (14).

Manufacturers in England (ICI), Japan (Chisso), and the U.S. (Enka) have developed quite effective binder fibers especially for the nonwoven industry. These are bi-component fibers characterized by a low melting skin and a high melting core. They appear to be, by far, the best thermoplastic binder fibers made to date. When the binder fibers are heated, the skin melts or softens to bond to adjacent fibers but the core is not affected. The fibers can be used alone or blended with conventional fibers. Bonding can be effected by even heat and pressure or in a pattern.

Several manufacturers have been developing modified cellulose fibers that are made with a microcrimp superimposed on a conventional fiber crimp (11), that are produced from cellulose solvent processes other than viscose (11,15), that have a round or irregular crosssection with a correspondingly shaped lumen to increase absorbency (16), and that are alloyed with sodium polyacrylate to increase absorbency (American Enka, Avtex) (11) (see Acrylic and modacrylic fibers; Acrylic acid).

In addition to the substantial numbers and variety of natural, regenerated, and synthetic fibers which are readily available, mechanical differences are introduced intentionally in extruded fibers to affect processing conditions and web and final product properties. The variations which may be produced include fiber length, diameter, crimp, cross-section, spin finish, draw ratio, and inclusion of delustering agent.

Length. Synthetic and regenerated fibers are made by extrusion; therefore, they can be prepared in any length by appropriate cutting operations. Extruded fibers are furnished in precision cut, uniform lengths of ca ±0.3 cm. Staple extruded fibers are provided in lengths of ca 2–6 cm. Fibers that are intended for wet-processing into webs are cut much shorter.

Diameter. Regenerated and synthetic fibers are produced in a wide range of diameters for conventional textile uses; these diameters also are accepted and used in making nonwovens. Fiber diameter is described by tex, which is the weight of a 1000 meters of filament. The practical metric unit is decitex (= 0.9 den), ie, the weight of 10 km of filament. Fiber diameter is directly proportional to the square root of its tex and industry proportional to the specific gravity of the fiber polymer.

The fiber staple length usually is chosen for convenience in the nonwoven manufacturing process; when longer than ca 25 mm, length has a small effect on product properties. Decitex, by contrast, has a great effect on properties of the finished fabric and is carefully selected. Fiber decitex, hence diameter, affects the hand of the fabric; bonding properties, whether by adhesive or by friction; fabric cover or opacity, especially in light weights; fabric resilience and web thickness; fabric liquid or gas filtering properties; and machine processing requirements.

The diameter of typical elliptically shaped fine United States cotton fibers is ca

8 μm (minor axis) and 20 μm (major axis), which is comparable to viscose rayon fibers of 1.67 dtex (1.5 den) (17). Such a decitex yields satisfactory products and is convenient for processing. However, the proliferation of end uses has required the use of higher decitex (den) fibers for many products; thus, there is no standard decitex (den).

Crimp.   Crimp refers to a fine, periodic, three-dimensional sine-wave or saw-tooth shape along its length. The scale of this wave is crimps per centimeter. Natural fibers, eg, cotton or wool, have an inherent, irregular crimp. Synthetic staple fibers virtually always are crimped deliberately during manufacture by a mechanical stuffing or gear crimping process. Rayon staple fibers also are crimped either mechanically or chemically.

In mechanical processing, the crimp is necessary so that the teeth of the cards and the licker-in cylinder can grip and separate the fibers. The crimp enhances interfiber friction which is necessary for the web so that it can sustain itself during the manufacturing process. In the finished nonwoven, the presence of residual crimp in the fibers gives the product a third dimension and, therefore, a degree of resilience and a degree of textile appearance and feel. Fibers which are intended to be made into webs by wet or paper-making processes customarily are not crimped, since this would increase uncontrolled fiber entanglement in the water dispersion.

Cross Section.   Conventional viscose rayon fibers are roughly circular with characteristic crenulations. High wet-modulus viscose rayon fibers do not show crenulations and are either round or bilobal. Special, highly water-absorbent rayon fibers have been made with a round or cross section and a central lumen (16). Cotton fibers have a distinctive, irregular cross-sectional shape which is like a kidney bean but with a central, frequently flattened lumen.

The common synthetic fibers, ie, polyester, nylon (6 or 6,6), and polypropylene, normally are circular in cross section, although they have been extruded in a variety of shapes. It has been proposed that specific, noncircular shapes are better for bonding efficiency and fabric softness; however, customarily, round cross-section fibers are used (18).

The bi-component fibers, which were developed specifically for the nonwoven industry as combination structural and binder fibers, have an approximately spherical cross section. There are two important varieties: in one, a constituent is distributed around a higher melting core; in the other, the two components are present more or less side by side.

Spin Finish.   Spin finish refers to the lubricant which is added in small amounts (ca 0.2 wt %) to the surface of extruded fibers to enhance their processability in the mechanical operations, up to and including spinning, by controlling lubrication and friction and subduing static electricity. A fiber with no finish cannot be made into yarn. The effectiveness of the spin-finish is critical for the commercial acceptance of a fiber and, as a result, its chemical identity frequently is proprietary.

Finishes permit dry processing at high speeds and, for synthetic fibers, frequently are hydrophilic to enhance the wettability of the fibers by water-based latex binders. Where the fibers are intended to be dispersed in water or wet-laid for the primary web-forming step, rayon fibers have been provided with no spin-finish called hard finish, whereas the normally hydrophobic synthetic fibers, eg, polyester, have a hydrophilic finish. The proper finish on a synthetic fiber helps dispersability in the wet web-forming process so that longer fibers can be used.

Delustered Fibers. Most of the research effort to improve rayon is to achieve the properties of cotton (11). Delustering is accomplished by dispersing a small amount (ca 1–2 wt %) of TiO_2 pigment in the viscose dope prior to extrusion. The use of delustered fibers increases the opacity and, therefore, the covering power of lightweight nonwovens and gives these, as well as heavier-weight products, an attractive white matte appearance. A special, highly delustered rayon fiber with a higher TiO_2 content has been made for the nonwoven industry. The incorporation of delustering agents does not have a significant effect on the tensile properties of the fiber.

Delustering agents also are used in synthetic fibers. Colored pigments can be dispersed in extruded fibers, and the resultant dope-dyed fibers have some minor use in the nonwoven industry.

Draw Ratio. Synthetic fibers for nonwoven use are drawn in the same manner as for general textile use, except in special cases, eg, in making bonding fibers. Poly-(ethylene terephthalate) fibers, either undrawn or drawn much less than standard, have a substantially lower softening temperature and are used widely as bonding fibers; they are activated by heat and pressure.

Fiber Webs. A web is the common constituent of all nonwoven fabrics. Polymer nettings, foams, and scrims which are made from cross-laid yarns incorrectly have been claimed to be nonwoven fabrics. The characteristic properties of the base web are determined by fiber geometry, largely as determined by mode of formation; fiber characteristics, both chemical and mechanical; web weight; and further processing, including compression, fiber rearrangement, and fiber entanglement. The mode of web formation is a primary determinant of fiber morphology, including fiber orientation and frictional engagement. Therefore, it has a strong influence on the properties of the finished product. Among the important aspects of fiber morphology are the average directional fiber orientation, whether isotropic or anisotropic; the geometric shape of the fibers, predominantly stretched-out, hooked, or curled; interfiber engagement or entanglement; and residual crimp. Of these, fiber orientation properties are the most important.

There are only a few ways to form fiber webs and most are derived from classical textile or paper-making processes. The methods are dry formation, including carding and air-laying (19); and wet-laying in large volumes of water by modified paper-making methods and in high viscosity, low volume foam. Development of fundamental new methods for forming webs or major modifications of old methods is difficult.

Dry Formation. *Carding.* Carding is a mechanical process whereby clumps of staple fibers are separated into individual fibers and simultaneously made into a coherent web (20). The operation is carried out on a machine which utilizes opposed moving beds of fine, angled, closely spaced needles or their equivalent to pull and tease the clumps apart. Card clothing is the structure comprised of needles, wires, or fine metallic teeth embedded in a heavy cloth or in a metal foundation. The carding machine may be considered as a convenient frame to support the clothing so that it may operate at maximum efficiency (20). The opposing, moving beds of needles are wrapped on a large cast-iron main cylinder, which is the heart of the card, and a large number of narrow, cast-iron flats, which are held on an endless belt that is placed over the top of the main cylinder.

The needles of the two opposing surfaces must be inclined in opposite directions and must be moving at different speeds relative to each other. The main cylinder moves at a higher surface speed than the flats. The clumps between the two beds of needles are separated into fibers and are statistically aligned in the machine direction as each fiber is theoretically held at each end by individual needles from the two beds. The individualized fibers engage each other randomly, and, with the help of their crimp, form a coherent web at and below the surface of the needles of the clothing on the main cylinder.

All of the mechanisms are built around the main cylinder and, since the various settings must be precise, the cylinder is massive. In addition to the flats, the carding machine includes means to carry a crude lap or batting onto the cylinder where the carding action takes place. It contains other mechanical means to strip or doff the web off the cylinder after completion of the process. The doffed web is deposited gently onto a moving belt where it is plied with other webs and carried to the next step which, presumably, is rearranging or bonding. Most nonwovens, including the most light-weight products, are made of webs combined from several cards which are aligned in tandem in order to average nonuniformities in the webs.

The orientation ratio of webs at the doffer of a conventional cotton-system card is ca 3:1 in the machine direction. This web is highly sensitive to stretching or drafting. Drafting as little as ca 5% greatly increases the web anisotropy ultimately to 10:1 or more. The physical properties of nonwoven fabrics are highly dependent on fiber orientation; even saturation with binder cannot overcome the effects of fiber orientation on tensile properties. Consequently, nonwovens that are made from webs from conventional cards characteristically have high machine direction and low cross-machine direction strengths. Fabrics made from card webs can be torn readily along the machine direction, ie, parallel to fiber orientation direction, and can hardly be torn across the predominant fiber orientation axis.

In the preparation of fiber assemblies for spinning into yarns, it is desirable that the fibers are oriented in a parallel manner; in making nonwoven fabrics, it usually is desirable that fibers are not highly oriented. Cards are available which are very wide, ie, as much as 355 cm; can be run at high speeds, eg, as much as 137 m/min; produce webs with low long-to-cross orientation ratios; and are less susceptible to becoming highly oriented by drafting than conventional textile cards.

A sizable portion of the industry has solved the low cross-tensile strength dilemma by producing webs which have been cross-laid. Cross-laying is a process by which oriented webs are laid down at or near alternating 45° angles on another oriented web on a moving belt. Cross-laying has not been successful with lightweight webs because they cannot be placed accurately enough to avoid unsightly edge lines; the process is quite successful for heavy weights. A major use is for preparing webs for subsequent needle-bonding. Garnett cards, or modified cotton cards, can be used in this process.

Air-Laying.   The earliest and most successful means to overcome the severe anisotropy of webs that are made by carding is the formation of webs by capturing fibers on a screen from an air stream. The fibers in the air stream are individualized and totally randomized so that there should be no preferential orientation when they are collected on the screen. Actually, this ideal situation is not fully attained, and there is a slight orientation in the machine direction and imparted to the web by the air stream and the forward motion of the screen. Fibers that are deposited from an air stream appear to be more curled, thereby having a shorter effective length than fibers in a card web. Drafting the web in handling also orients the fibers to a degree but less so than in card webs.

The length of fibers used in air-laying varies from 1.9 to 6.4 cm (21). The shorter lengths are used more commonly because they allow higher production speeds with better web uniformity. Long fibers require a higher air volume, ie, a lower fiber density in the air stream, to avoid tangling. Use of high production speeds requires the use of high air velocity which promotes nonuniform air flow which, in turn, has a disastrous effect on web quality.

Air-laying cannot be carried out at speeds as high as those used in carding; this problem is especially true for lightweight webs. The quality (primarily, the uniformity) of air-laid webs is enhanced if preformed webs, eg, card webs, are fed into the air-laying equipment; however, this adds to the cost of the process. The adjustment and maintenance of the equipment must be rigorously optimized for best results in speed and web quality. Nevertheless, the advantages of an isotropic or nearly isotropic base web for nonwovens are so considerable that air-laying is an important commercial practice.

A typical example of a specialized variation of web forming is the dual rotor process and apparatus (22). This is an air-laying process which produces a web containing two different fibers, eg, wood pulp and staple rayon or polyester. Supplies of fibers are fed to oppositely rotating licker-ins that are rotated at speeds which are optimum for the fibers being individualized by the licker-ins. The fibers from each supply are entrained in their respective air streams which are impelled at high speed toward each other. The degree of mixing of the dissimilar fibers can be controlled to a considerable degree. Another significant innovation is the development of a card which has a short air-scrambling process at the doffer, thereby producing a card web with a nearly isotropic fiber web (23).

The dry mechanical processes commonly have metal needles or teeth which open clumps and individualize fibers by strong, tearing forces. The tearing action unavoidably breaks fibers, so that the true fiber lengths span a range and, on average, are shorter than the starting fibers. Fiber damage varies from process to process, fiber to fiber and, perhaps, day to day on a single production line. Analysis of fiber-length distribution originally was tedious but is greatly simplified and can be used to quantify fiber breakage, which may be a source of product variation that has not been adequately recognized (24–25).

Wet Formation.   Wet formation is a variation of the methods by which paper is made. Fibers are dispersed in water. The water is filtered through a moving screen, leaving a fiber web on the screen. The web is transferred to appropriate belts and/or felts and is dried. Wood pulp fibers disperse readily in water, but longer fibers present many practical difficulties.

High decitex (den), ie, high diameter, high wet-modulus fibers, are dispersible at lengths up to ≥1.9 cm or more. Long, fine diameter fibers, ie, 1.7–3.3 dtex (1.5–3 den), have been dispersed successfully on laboratory equipment and in Japanese handmade paper processes by using very high molecular weight hydrocolloids as dispersing aids (26) (see Dispersants). Without dispersing aids, the long fibers entangle and form a nonuniform web. It appears that 0.95 cm is the upper feasible length and 0.64 cm is the upper practical commercial length for fine decitex hydrophilic fibers, eg, 1.67 dtex (1.5 den) viscose rayon. Fine decitex fibers, ie, 1.67 dtex (1.5 den), of staple length, 1.3 cm or longer, are not in commercial use as ingredients of wet-formed webs.

An unconventional wet-forming process for the dispersion of relatively long, 1.67

dtex (1.5 den), 1 cm fibers has been developed. It is based on the dispersion of the fibers in a low volume viscous foam (27).

It appears that wet formation should produce isotropic webs. At the moderate process speeds of nonwoven manufacture, web orientation is slightly anisotropic. However, high water volumes are needed, and this is adequately handled by the Rotoformer or by an inclined wire (28–29) (see also Paper).

Bonding. Bonding implies the use of an adhesive ingredient or heat, including embossing, or chemical treatment, but bonding also can be effected by interfiber friction from mechanical entanglement. The adhesive methods are numerous and varied. Binder can be applied in many different practical ways, with a choice of numerous kinds of adhesives; greater or lesser amounts; and in uniform, nonuniform, random, or patterned distribution in the web. In choosing and applying binder, the developer of a nonwoven fabric has the largest scope in selection and application of any of the principal elements, ie, fiber, web, binder, and reinforcement. The selected binder may be added in the form of a synthetic water-based emulsion or dispersion, which commonly is called a latex; a solution; a foam; thermoplastic particles, including fibers among others; monomer or oligomer for polymerization in place; plastisols; or polymers that react with the fibers. Any film-forming polymeric substance, which can be dissolved or dispersed and/or partially or fully fused while in contact with the fibers of a web, may be used as an additive binder.

In addition, methods have been devised where most or all of the base fibers of the web have been chemically or thermally softened or fused to give them adhesive properties. The activated fibers subsequently are resolidified to connect fibers to fibers and, therefore, function as a binder. This method, in which the fibers of the web provide the bonding substance, is autogenic bonding.

The most common commercial bonding system is based on synthetic, aqueous latexes. Latexes have many practical advantages, including availability of numerous varieties with different chemical and physical properties, ease of handling, and low cost. They can be modified easily or tailor-made for special needs. The use of latexes is so widespread that they comprise a substantial industry in their own right (30). Present latex or water-dispersed binders, with rare exceptions, are vinyl polymers which are emulsion polymerized by free-radical mechanisms (31).

In additive adhesive bonding, the web is formed first, usually is compressed, and the binder substance is added to it. Thermoplastic binder fibers are incorporated in the web as it is formed. In its final functional form, the binder is a film-forming polymer. For the formation of a continuous chain of bridging throughout the web, each fiber must be bonded at least two times (preferably more) to different fibers. Fibers which escape any bonding are likely to fall out of the web during use, causing undesirable dusting. Fibers which are bonded only once do not contribute to tensile properties but may enhance bulk, softness, and absorbency.

The mechanism of adhesive bonding is not definitively understood. The customary explanation is that bonding is directly related to adhesion, ie, shear or peel adhesion or a combination of the two. The alternative to adhesion is the mechanical entrapment of microfibers within encompassing bonds. At the Chicopee laboratory, corona-discharge treatments were applied to polymer films in a reactive environment; the result was greatly enhanced peel adhesion of latex to the film. The same treatment, as applied to fiber webs of nominally the same chemical composition as the film and bonded by the same latex, produced nonwoven fabrics with very little improved strength (32). The peel adhesion properties of a latex on nylon-6,6 film were improved five-fold by modifying the zeta potential or electrophoretic mobility of the latex. Applying this modified latex to a web of nylon-6,6 fibers, however, did not produce a measurably stronger nonwoven fabric (33).

Experiments were carried out at North Carolina State University where polyester film was treated with diacetyl, 2,3-butanedione, an adhesion assistant, and coated with an acrylic monomer which was polymerized by electron-beam radiation. The diacetyl pretreatment greatly improved the peel adhesion of the polymer to the polyester film. A parallel experiment was carried out using a web of poly(ethylene terephthalate) fibers in place of the poly(ethylene terephthalate) film. They were pretreated with diacetyl, and bonded with monomer that was polymerized in place by electron-beam radiation. The resultant fabric showed no improvement in tensile properties over a fabric that had not been pretreated with the adhesion enhancer (34).

However, there are contrary data which indicate that specific adhesion can be important. An example is a commercial, lightweight nonwoven fabric composed of viscose rayon fibers bonded with ca 2 wt % cellulose, which is applied as a viscose solution and is regenerated. Microscopic examination of fabric cross sections show no interface between viscose binder and viscose fibers, indicating fusion of the two. Viscose rayon fiber fabrics require 10–15 wt % added latex binders to achieve comparable tensile properties. If adhesion is very high, it can be a controlling factor in bonding. In addition, where the binder content is so low that fiber embedment is not possible, eg, in lightly bonded wood-pulp webs, adhesion is important (see Pulp).

Experiments have been carried out to investigate bonding by chemical reaction, ie, formation of covalent bonds between binder and fibers, based on the chemistry of fiber-reactive dyes and ionic bonds (35 36) (see Dyes, reactive). Some latex bonding of rayon, as carried out in large commercial volume, is enhanced by the reaction of excess latex cross-linking agent, eg, N-methylolacrylamide, with the polymer and active sites on the fiber (37).

Latex. Synthetic latexes have the most extensive history as binders and are the most widely used. The estimated total usage of latexes in 1979 in nonwoven fabrics was 54,000 metric tons (dry basis) (30). Although it is possible to make a reasonably complete listing of the major categories of commercial latexes, the number of minor variations in use is large. The main types include polyacrylates, eg, poly(ethyl acrylate) and copolymers; vinyl acetate–acrylic acid copolymers; poly(vinyl acetate)s; poly(vinyl chloride)s; ethylene–vinyl acetates; styrene–butadiene carboxylates; and polyacrylonitriles (see Latex technology).

Most of the latex binders include a small amount of a self-cross-linking agent, typically N-methylolacrylamide (see Acrylamide). Other, nonformaldehyde cross-linking systems are being developed. The advent of the self-cross-linking latexes at the end of the 1950s made durable, washable, wet-abrasion resistant nonwovens practical and, possibly, was the most important factor in the subsequent rapid growth of nonwovens made from staple fibers. The self-cross-linking monomer does not increase the modulus of the cured binders to a significant degree but greatly improves water and solvent resistance.

All of the above latexes are effective binders for rayon webs. Except for a few obvious latex properties, eg, film-forming ability, modulus of the cast film, solvent or water resistance, and wettability, the relationship of latex properties to final fabric properties is not reliably predictable, although a substantial effort has been expended to establish such relationships (38–40). However, manufacturers of latexes and of fabrics, have developed rapid screening procedures and can readily search for latexes which satisfy commercial needs.

The latexes which are, for the most part, universally good binders for rayon and other hydrophilic fibers are not successful, ie, they do not lead to the formation of strong fabrics with reasonable amounts of added binder, when used on synthetic fibers, in particular, polyester and polypropylene. Even though the synthetic fibers have tenacity values (N/tex or gf/den) two to four times greater than that of rayon, the synthetic fiber webs which are bonded with conventional latexes are much weaker, ie, typically, one half to one fourth as strong.

Latex binders can be efficient impregnants on 100% polyester webs if the following parameters are optimized: a hydrophilic finish on the binder and a binder formulation which wets the fibers readily (37). Presumably, the binder effectiveness is dependent on its micromorphology in the web and on the fibers, as well as on its specific adhesion. In addition, effective binders for polyester fibers have been developed from water-dispersable polyester polymers (31).

Needs of the Manufacturing Process.   Latex binders may be applied to dry-formed webs by overall saturation, application in pattern by the rotogravure process or a modified silk-screen process, spraying on one or both web faces, and application as a foam. However, these methods subject the latex to severe shear stress. Therefore, a high degree of mechanical stability at either full or diluted concentration is required. The latex binder must be able to accept miscellaneous additives without becoming unstable. It should be stable in storage for many months but, once applied, it should dry and cross-link under reasonable time and temperature conditions. Consecutive production batches must not vary in composition, viscosity, and other rheological properties; surface tension; latex particle size; and size distribution. It should be a good film former, but it should be readily dispersable for clean-up. This list can be expanded by processors with specific needs.

Needs of the Nonwoven User.   Each commercial product has a complex of properties which are directly dependent on the binder. These may include softness, strength, water absorbency or repellency, wet–dry abrasion resistance, and water or solvent resistance. The binder must have no toxic, allergenic, or odorous ingredients. Its film properties, including water-white color, strength, and modulus, must be stable for a reasonable period of time. In addition, each user may have specific needs.

Binder Formulation.   It is not customary for a latex to be used exactly as received from the supplier. The latexes usually are manufactured at ca 50 wt % solids; the user often dilutes the latex with water. An acid or latent acid catalyst, a defoaming agent, a viscosity modifier, usually a thickener, and/or a surfactant or a water repellent may be added (see Defoamers; Surfactants and detersive systems). Less often a pigment, either white or colored, a rust-inhibitor, a fire-retardant salt but only if the latex is salt-tolerant, and a binder-migration control agent are added.

The amount of binder which is added to the web varies from ca 5 wt % to ca 50 wt %, depending upon desired fabric properties, commonly 10–35 wt %. The higher ranges of binder content tend to obscure fiber properties; one measure of binder and process efficiency is the minimum binder content required for needed tensile and other physical properties.

Application of Latex Binder. Saturation. Saturation or impregnation involves immersing the fiber web in the latex and removing the excess latex. The web can be fed directly into the bath under an idler roll and exit between the mangle rolls. An unbonded web is not strong enough for this simple process and requires support on a carrier or, more likely, is carried between two open-mesh screens. There are many variations in web handling and threading practices.

As the web enters the nip of the mangle, it is exposed to the excess liquid which simultaneously is moving in the reverse direction. The liquid can severely damage the fragile web which, in practice, means that the speed at which this kind of saturation process can be run is limited. The problem is especially severe for lightweight webs. One approach to solving the problem is the use of mechanical means to divert the exiting stream from contact with the web. However, the customary method for impregnation is to circumvent the saturation process by metering latex onto the web using an engraved roll and a very fine, close pattern. The close pattern allows the latex to diffuse, and the web is effectively saturated with a controlled binder content.

In concept, if not execution, saturation bonding with latex is the simplest way to make a nonwoven fabric. Variations in properties and amount of binder added, the method of drying and curing, and the degree of compacting the web enable the formation of a great diversity in products.

Print Bonding. One of the frequently desired attributes in a nonwoven fabric is a textilelike flexibility, ie, hand or drape. This can be accomplished by saturating a web with a relatively small amount of soft binder, ie, one that has a low film modulus or low second-order transition temperature (38–40). One deficiency of such a saturated staple fiber web is that all fibers and fiber ends are tied down, giving the product a thin, flat dimension and a slick surface. Another method which is used to optimize fabric hand is intermittent bonding: binder is distributed intermittently so that the resultant product has alternating areas with and without binder. The bonded areas provide the strength and the unbonded areas provide a degree of bulk, flexibility, and surface texture.

Random intermittent bonding has been achieved by use of a mixture of fusible and nonfusible fibers; the former are activated by heat and pressure. Another approach was the printing of latex binders on the web in a pattern using an engraved roll. Billions (10^9) of meters of nonwovens have been made by the second method, which is particularly suitable for products as lightweight as 15 g/m² but which can be efficiently applied to webs as heavy as ca 100 g/m² (ca 4 oz/yd²). Print bonding is applicable readily to card webs of any hydrophilic fiber, eg, rayon, both standard and high wet modulus; cotton; poly(vinyl alcohol); and their blends (13). Research is being carried out to optimize print bonding of hydrophobic fibers, including polyesters and polyolefins. Print bonding is a widely used and mature technology.

Commercial equipment has been developed to apply latex binder by a patterned-screen method. Binder formulations are modified to greatly increase viscosity so that, according to machine design, the latex is forced through the screen pattern into the web. Line conditions and binder compositions are different from those customarily used in rotogravure printing. Screen printing characteristically deposits the binder more precisely and with less sidewise diffusion or migration than its older counterpart. Rotogravure printing is most conveniently applied to webs that are composed of long fibers, are prewet with water, and are produced by carding so that fibers are oriented preferably in the machine direction. All of the above conditions are not required in screen printing.

Print-Pattern Design.   The earliest print-pattern design is a series of parallel lines running transverse to the machine direction and is used when the fibers are oriented predominantly in the machine direction (6). The rationale for this pattern was that it maximizes cross-strength; as if each transverse binder line is continuous and acts as a filling yarn in a woven fabric. A cursory examination using a low power microscope shows that the line is discontinuous and that the binder is deposited preferentially at fiber crossovers and other high surface energy locations. This simple line pattern is fairly effective and a slight variation, ie, a pattern of horizontal, parallel wavy lines, is in wide use. It was soon recognized that significant improvements in fabric properties, other than simple tensile strength, can be effected by the design of the print pattern.

Other designs were developed to increase fabric cross-directional toughness, ie, the area under the stress–strain tensile curve; resilience; and wet-abrasion resistance. Cross-directional toughness is improved substantially by patterns based on rectangular or oval units (41–42). Cross-resilience is improved by diagonal line patterns, and wet-abrasion resistance is improved by a miniaturization of the pattern (43). Since the latex binders frequently are pigmented, the pattern design becomes a decorative design. Consequently, the design of a new pattern becomes a combination of technology and esthetics.

In the rotogravure system, the latex binder usually is printed onto a wet web. The water-based latex penetrates through the web and simultaneously diffuses sidewise; this phenomenon is migration. Excess migration blurs the carefully designed pattern, sometimes unpredictably, and generally is undesirable. The degree of migration can be controlled by coagulating the latex in place as it contacts the web by using physicochemical methods (44) or, when the web reaches the heat of the dryers, by using heat-unstable surfactants or other heat-reactive additives in the latex. Heat-unstable additives also are used in saturated webs to prevent migration of binders to the fabric faces during drying (46–47). The ability of patterns to substantially improve cross-tensile properties is limited. Fiber orientation has a far greater effect; eg, no print-pattern or other bonding method can upgrade cross-tensile strength nearly as much as changing the long–cross ratio of fiber orientation (see also Printing processes).

Spray Bonding.   Binders, particularly latexes, can be applied by direct spraying onto a web. Several requirements must be fulfilled for this process to be practical, including low viscosity binders which can tolerate the extreme shear conditions at the spray nozzle, nonplugging nozzles, and an airless spray to avoid disrupting the web. The spray usually is applied to a dry web which may be composed of short and/or long fibers, including wood pulp alone and mixtures with staple fibers. The spray tends to penetrate only partly into the web and is used primarily where high loft properties are desired in the final product. The spray can be applied to one or both faces of the web. In heavy applications, the sprayed binder penetrates fully.

Foam Bonding.   Binder may be applied to the web as a water–air foam which implies that less water is used and increased energy savings are achieved in the drying process. Foam characteristically is deposited on the web surface. Binder that is applied as foam which collapses during drying is called froth. It is also possible to apply foam using formulations permitting the foam structure to survive the drying process.

Thermoplastic Bonding.   A thermoplastic bond is a physical or adhesive bond made from an added thermoplastic polymer which has been softened or fused as it is in close contact with the fibers. Upon solidification and while still in close contact, the polymer binds the structural fibers. Thermoplastic bonding is one of the oldest and most effective types of bonding. The thermoplastic material may be in the form of fibers, rods (short-cut, high denier filaments), powder, granules, netting, film, or particles in irregular shapes (7,47–48). Each system has economical or functional merits for an intended use. Bonding with thermoplastic particles is a versatile technique because of the great number of variations which may be practiced, including choice of polymer for strength, melting temperature and rheological properties of the melts, and adhesive and chemical properties; amount of polymer added; shape and size of activable particles; and heat and/or solvent vapor, pressure, and time conditions of activation.

Fibers are the most widely used form. Thermoplastic fibers have the added advantage of simplicity of mixing with base-web fibers. The fibers customarily are intermixed at random and activated by heat and pressure or an intermittently patterned heat and pressure. Until recently, polypropylene fibers were the most effective and practical type. Recently, ICI (UK), Chisso (Japan), and Enka (U.S.) have introduced bimodal, staple fibers which have promising properties.

Autogenic Bonding. Autogenic bonding is the bonding of a web by partially solubilizing the base-web fibers in close contact using heat, pressure, or solvents and resolidification to form polymer bonds. The concept has been popular but not often commercially successful. Typical examples include: immersion of a compressed web of cellulose fibers in concentrated H_2SO_4 or other cellulose solvent followed by quenching and washing in water; immersion of a compressed viscose rayon fiber web in 8–10 wt % NaOH, within which concentration range viscose rayon becomes gelatinous, followed by quenching and washing with water; even or patterned exposure of a 100% thermoplastic polyolefin fiber web to heat and pressure, followed by cooling; exposure of nylon fiber-web to HCl gas, followed by washing to remove HCl; and addition of aqueous solutions of certain salts, ie, LiCl, KCNS, and others, to nylon, acrylic, or polyester fiber webs, removal of water by heat to locate the resultant concentrated salt solution at fiber crossovers (thereby dissolving fiber polymer), followed by washing to solidify the bond and to remove salt (49).

Although autogenic bonding appears attractive, it has flaws in practice. Since the bond substance comes from the fibers, the latter are subject to damage. Furthermore, experience has shown that the range of useful bonding is limited; there is a strong tendency toward only a narrow range of practical process variability. Too little treatment produces underbonding; too much treatment damages fibers and embrittles the resultant product.

Solution Bonding. Bonding with organic solutions of polymers is an efficient process by the criterion of strength attained binder added, especially at low binder content (50). The practical problems of solvent hazards and cost and the difficulties of handling and applying viscous but low solids solutions have negated the widespread use of solution bonding. However, water solutions of binders have specialized uses. Poly(vinyl alcohol), which usually is cross-linked with a formaldehyde donor, is practical as a binder and has a commercial history (52). Another interesting water solution system is viscose, ie, cellulose xanthate dissolved in aqueous NaOH. This has been used on a small but successful commercial scale for many years on rayon and cotton-fiber webs. The product is effectively bonded with much lower binder added than is customary and is a classic example of and evidences excellent binder efficiency when specific adhesion is very high.

Polymerization of Monomer. Only recently, with the advent of electron-beam polymerization, has polymerization of added monomer or oligomer *in situ* in a fiber web been feasible (52–53). The process is notable for its low power requirements. Nonwoven fabrics have been bonded experimentally by this process, but results have not been good enough to justify the high cost of the special monomers required and difficulties in application (see Radiation curing).

Reactive Bonding. In one experiment, a binder is applied to a web of cellulose fibers on which it reacts to form covalent bonds (54). The chemical that is used is based on fiber-reactive dye systems using cyanuric chloride, which reacts with the hydroxyl groups of cellulose fibers and polymeric diamines of the binder. This method should be equal or superior to those of systems with high specific adhesion.

An inexorable problem associated with adhesive bonding is that maximum strength and softness are mutually irreconcilable. Strength may be increased only by the sacrifice of suppleness and vice versa:

$$\text{strength} + \text{softness} = K$$

where K is an arbitrary constant. Exceptionally effective adhesion, softer and stronger binders, or optimization of binder morphology may increase the value of K, but only to a limited degree.

Bonding by Fiber-to-Fiber Friction. Needle-punching describes a process which is in large-scale use and by which a base web is reinforced by passing it under banks of rapidly reciprocating barbed needles (55). The barbs entrap bundles of fibers of the web, forcing them downward through the web, thereby forming numerous small loci of fiber entanglement. The process is relatively slow, is most useful for heavy webs, and is capable of making very strong, supple products.

Laboratories supported by the governments of the GDR and of Czechoslovakia have developed a family of nonwoven fabrics that are made by extensive overstitching or knitting of fiber webs with yarns. The resultant products are unique, although they have a superficial resemblance to soft woven or knitted textiles. Tradenames associated with these nonwovens include Arachne, Malivlies, Schusspol, Maliwatt, Malimo, Malipol, and Voltex (56). Similar products are also available in the United States. These products can be used as manufactured or as a base for plastic coating.

An ingenious method for making a soft, highly absorbent all-cotton pad consists of exposing a web of unbonded cotton fibers to a mercerizing-strength NaOH solution. A card web of relatively short cotton fibers is carefully floated without tension on the cold, aqueous NaOH solution. Mercerization without tension occurs and is accompanied by vigorous curling and tangling of the fibers. After thorough washing and drying, the resultant material is bulky, resilient, and strong enough for uses as a specialty absorbent and as a carrier of reagents (57).

When a web of staple fibers is exposed to fluid forces, eg, multiple water jets, the fibers can be rearranged to form yarnlike bundles of fibers, thereby altering the web appearance and function. A web which is textureless in appearance can be made to resemble woven, knitted, or embroidered fabrics or can be given an entirely novel appearance by this process. From the inception of fiber rearrangement, it was recognized that some fiber entanglement takes place, and the patents describe these products as self-sustaining (58–60). However, they do not have adequate strength, without additional bonding, for broad usage. Webs with fibers that are rearranged as described and bonded with latexes have been manufactured in large amounts since the late 1950s.

Spunlace Bonding.   Fibers in a web can be tangled to a substantial degree by the application of fine, columnar water jets at very high pressures (61). As the jets penetrate the web and deflect from a wire screen or other permeable backing, some of the fluid splashes back into the web with considerable force. Fiber segments are carried by the turbulent fluid and become entangled on a semimicro scale. These products represent a new species of fabric where there is an irregular interlocking, hence, entanglement on a fiber-to-fiber scale. Manufacture of spunlace products presents difficult engineering problems and is highly energy-intensive. The products have excellent strength properties, except in very light weights. A major fault is their lack of stretch recovery. In addition, they tend to have low abrasion resistance, unless they are extensively processed (62). They have a fundamental advantage over chemically bonded counterparts in that strength and suppleness are not mutually exclusive. Both can be maximized at the same time. Spunlace products have not become the universal nonwovens but they occupy an important and expanding section in the product spectrum.

Reinforcements.   Fiber webs can be reinforced by woven fabrics, plastic nettings, cross-laid yarn scrims, foams, and polymer films. The combination of webs with a reinforcement usually is enhanced by adhesive bonding, although mechanical entanglement also is feasible. However, it is desirable that adhesives are effective for both components and that the stress–strain characteristics of both are similar so that they efficiently complement each other.

Manufacture

The manufacture of nonwoven fabric roll goods is achieved by combining the raw materials, ie, fibers, reinforcing structures, binders, or fibers alone, by two or more of the processes of web forming, adhesive bonding, and fiber entanglement. Although there are few processes and components, each exists in so many variations that there are numerous possible methods.

In all methods, the web is formed first. The chemical type of fiber and its decitex, length, etc, are chosen for ultimate product properties. If staple fibers are used, the web-forming process is always dry, either involving carding or air-forming. If the final product is intended to be bonded by thermoplastic fibers, they are included in the original fiber blend that composes the web. Other thermoplastic particles are likely to be added as the web is formed or webs are plied. After the web is formed, it may be wetted for ease of handling, especially if it is to be bonded with a water-based adhesive, eg, a latex or an aqueous solution. The latex may be added evenly, ie, by saturation, by printing in a pattern, by spraying, or by application as a foam. The type, amount, and geometric placement of the binder can be varied, depending on the resultant desired properties. If the fibers are rearranged to give the product a characteristic appearance, rearrangement is carried out at the web-forming–bonding step.

After the binder is applied when using, it is activated by heat thermoplastic adhesive; dried; and self-cross-linked using latex, foam, or solution; or exposed to specific solvent or heat conditions by autogenic bonding. During the bonding, the web customarily is compressed to a desired degree. A reinforcing structure may be incorporated in the web during its formation, between plies or on one or both faces after formation and before application of binder. The reinforcing web may be saturated with binder

Table 1.   United States Consumption of Nonwoven Fabrics[a]

| | Roll-stock value, 10^6 | | Growth, |
	1978	1983[b]	%/yr
Type			
disposable	455	930	15
durable	405	780	14
Process			
dry (carding, air-forming, needle-punching)	490	950	14
wet (modified paper process)	90	160	12
composite	70	105	9
spunbonded (on-line polymer to nonwoven)	190	435	18
spunlaced (entangled-fiber bonding)	20	60	24
End Use			
diaper-cover stock	120	250	16
surgical packs and gowns	100	200	15
interlinings and interfacings	80	130	10
coated and laminated fabrics base	75	140	13
home furnishings			
carpets and bedding	135	280	16
wipes and towels	70	135	14
sanitary napkins, tampons, and underpads	40	70	12
filtration media	65	135	16
all others	175	370	16

[a] Ref. 64.
[b] Estimated.

Table 2.   Estimated Worldwide Consumption of Nonwoven Fabrics[a], 1978, 10^3 Metric Tons

North America	
dry-formed	195
wet-formed	36
composite	32
spunbonded	54
spunlaced	6.8
Far East	
dry-formed	23
spunbonded	4.5
wet-formed	5.4
Western Europe	
dry-formed	79
wet-formed	20
spunbonded	27
Other	
dry-formed	29
wet-formed	2.3
spunbonded	6.8

[a] Ref. 64.

and placed between plies of the web; in this way, it represents a specialized means to combine the bonding and reinforcing steps.

Where the bonding is by mechanical entanglement rather than by an adhesive or another physicochemical method, the web is transported to the entanglement zone.

There it is either passed under a bed of reciprocating, barbed needles, overstitched with yarns, mercerized, or subjected to fiber-scale entanglement by fine high pressure jets of water.

Economic Aspects

There is no ideal or prototype nonwoven fabric. There are approximately 150 important end uses and the products are made on more than 1000 production lines worldwide. In 1979, the value of nonwoven fabrics that were produced as roll goods in the United States was ca 10^9 and the converted value was ca 3×10^9. In that year, the amount of nonwovens manufactured approached 10% of the entire textile industry, whereas weaving accounted for ca 65%, and knitting comprised ca 25% (63).

A breakdown of consumption of nonwoven fabrics in the United States and worldwide is given in Tables 1 and 2.

BIBLIOGRAPHY

1. P. Coppin and co-workers, "The Definition of Nonwovens Discussed," *European Disposables and Nonwovens Association Symposium, Gothenberg, Sweden, June 6–7, 1974.*
2. *Guide to Nonwoven Fabrics*, INDA, The Association of the Nonwoven Fabrics Industry, New York, 1978.
3. *ASTM D 123-70*, ASTM, Philadelphia, Pa., 1970.
4. W. D. Freeston, Jr., and M. M. Platt, *Text. Res. J.* **35**, 48 (1965).
5. S. Backer and D. R. Petterson, *Text. Res. J.* **30**, 704 (1960).
6. U.S. Pat. 2,039,312 (May 5, 1936), J. H. Goldman.
7. U.S. Pat. 2,277,049 (March 24, 1942), R. Reed (to The Kendall Co.).
8. E. R. Schwarz, K. R. Fox, and N. V. Wiley in H. R. Mauersberger, ed., *Matthews Textile Fibers*, 5th ed., John Wiley & Sons, Inc., 1947, Chapt. XXIV.
9. A. R. Winch, *Eighth INDA Technical Symposium*, Orlando, Fla., March 19–21, 1980, p. 224.
10. A. R. Winch, *Text. Res. J.* **50**, 64 (1980).
11. R. Remirez, *Chem. Eng.* **86**(7), 113 (March 26, 1979).
12. *Chem. Week* **124**(23), 27 (June 6, 1979).
13. U.S. Pat. 3,930,086 (Dec. 30, 1974), C. Harmon (to Johnson & Johnson).
14. U.S. Pat. 3,507,943 (April 21, 1970), J. J. Such and A. R. Olson (to The Kendall Co.).
15. A. F. Turbak and co-workers, *ACS Symp. Ser.* **58**, (1977).
16. M. J. Welch and J. A. McCombes in ref. 9, p. 3.
17. R. F. Nickerson in ref. 8, Chapt. V, p. 178.
18. G. G. Allen and L. A. Smith, *Cellul. Chem. Technol.* **2**, 80 (1968).
19. F. M. Buresh, *Nonwoven Fabrics*, Reinhold Publishing Company, New York, 1962.
20. G. R. Merrill and co-workers, *American Cotton Handbook*, 2nd ed., Textile Book Publishers Inc., New York, 1949, Chapt. 7.
21. G. B. Harvey, *Formed Fabric Ind.* **6**(9), 10 (1975).
22. U.S. Pat. 3,740,797 (June 26, 1973), A. P. Farrington (to Johnson & Johnson).
23. Manufactured by E. Fehrer, *Textilmaschinenfabrik und Stahlbau*, Linz, Austria.
24. W. L. Balls, *A Method of Measuring the Length of Cotton Hairs*, London, Eng., 1921.
25. *Evaluation of the Motion Control Text System*, USDA Agricultural Marketing Service, Memphis, Tenn., Nov. 1974.
26. U.S. Pat. 3,794,557 (Feb. 26, 1974), C. Harmon (to Johnson & Johnson).
27. B. Radvan and A. P. J. Gatward, *Tappi* **55**, 748 (1972).
28. U.S. Pat. 2,781,699 (Feb. 19, 1957), W. J. Joslyn (to The Sandy Hill Iron and Brass Works).
29. U.S. Pats. 2,045,095; 2,045,096 (June 23, 1936), F. Osborne (to C. H. Dexter Co.).
30. C. H. Kline & Co., *INDA Newsletter* **79**, 3 (Dec. 1979).
31. K. R. Barton, *Nonwovens Ind.* **10**(5), 28 (1979).
32. U.S. Pat. 3,661,735 (May 9, 1972), A. Drelich (to Johnson & Johnson).

33. U.S. Pat. 3,639,327 (Feb. 1, 1972), A. Drelich and P. Condon (to Johnson & Johnson).
34. W. K. Walsh, "Radiation Processing of Textiles," *Symposium, North Carolina State University, Raleigh, May 26–27, 1976.*
35. G. G. Allen, G. Bullick, and A. N. Neogi, *J. Polym. Sci.* **11**, 1759 (1973).
36. M. L. Miller, "Ionic Bonding in Rayon Nonwovens," Ph.D. dissertation, University of Washington, 1972.
37. W. F. Schlauch, "Recent Developments in Nonwoven Binder Technology," *Nonwoven Fabrics Forum, Clemson University, Clemson, S.C., 1979.*
38. J. W. S. Hearle and P. J. Stevenson, *Text. Res. J.* **34**, 181 (1964).
39. J. W. S. Hearle, R. I. C. Michie, and P. J. Stevenson, *Tex. Res. J.* **34**, 275 (1964).
40. R. I. C. Michie, *Text. Res. J.* **36**, 501 (1966).
41. U.S. Pat. 2,705,687 (April 5, 1955), D. R. Petterson and I. S. Ness (to Chicopee Manufacturing Corp.).
42. U.S. Pat. 3,009,823 (Nov. 21, 1961), A. Drelich and V. T. Kao (to Chicopee Manufacturing Corp.).
43. U.S. Pat. 2,880,111 (March 31, 1959), A. Drelich and H. W. Griswold (to Chicopee Manufacturing Corp.).
44. U.S. Pat. 4,084,033 (April 11, 1978), A. Drelich (to Johnson & Johnson).
45. U.S. Pat. 3,944,690 (April 6, 1976), D. Distler and co-workers (to Badische Anilin- und Soda-Fabrik Aktiengesellschaft).
46. U.S. Pat. 4,119,600 (Oct. 10, 1978), R. Bakule (to Rohm and Haas Co.).
47. U.S. Pat. 2,880,112 (March 31, 1959), A. Drelich (to Chicopee Manufacturing Corp.).
48. U.S. Pat. 2,880,113 (March 31, 1959), A. Drelich (to Chicopee Manufacturing Corp.).
49. U.S. Pat. 3,542,615 (Nov. 24, 1970), E. Y. Dabo and co-workers (to Monsanto Co.).
50. A. Drelich and P. N. Britton, "Physical Combinations of Cellulose Fibers and Polymers," *149th Meeting of the American Chemical Society, Detroit, Mich., April, 1965.*
51. U.S. Pat. 3,253,715 (May 31, 1966), E. V. Painter and co-workers (to Johnson & Johnson).
52. C. Houng, E. Bittencourt, J. Ennia, and W. K. Walsh, *Tappi* **59**, 98 (1976).
53. U.S. Pat. 4,146,417 (March 27, 1979), A. Drelich and D. Oney (to Johnson & Johnson).
54. G. G. Allan and T. Mattila, *Tappi* **53**, 1458 (1970).
55. P. Lennox-Kerr, ed., *Needle Felted Fabrics*, The Textile Trade Press, Manchester, Eng., 1972.
56. J. D. Singelyn, "Principles of Stitch Through Technology," *Nonwoven Fabrics Forum, Clemson University, Clemson, S.C., 1978.*
57. U.S. Pat. 2,625,733 (Jan. 20, 1953), H. Secrist (to The Kendall Co.).
58. U.S. Pat. 3,081,514 (March 19, 1963), H. Griswold (to Johnson & Johnson).
59. U.S. Pat. 2,862,251 (Dec. 2, 1958), F. Kalwaites (to Chicopee Manufacturing Corp.).
60. U.S. Pat. 3,033,721 (May 8, 1962), F. Kalwaites (to Chicopee Manufacturing Corp.).
61. U.S. Pat. 3,485,706 (Dec. 23, 1969), F. J. Evans (to E. I. du Pont de Nemours & Co., Inc.).
62. M. M. Johns and L. A. Auspos, "The Measurement of the Resistance to Distanglement of Spunlaced Fabrics," *Seventh INDA Technical Symposium, New Orleans. La., March 1979.*
63. E. Vaughan, Clemson University, and D. K. Smith, Chicopee, private communication, Sept. 1979.
64. J. R. Starr, consultant, private communication, Boston, Mass., 1980.

General References

F. M. Buresh, *Nonwoven Fabrics*, Reinhold Publishing Company, New York, 1962.
R. Krcma, *Manual of Nonwovens*, Textile Trade Press, Manchester, Eng., 1971.
E. M. Passot, "Computer Evaluation of Fiber Crimp and Spot-Bonding for Drapable Nonwovens," Ph.D. dissertation, University of Washington, 1974.
M. S. Caspar, *Nonwoven Textiles*, Noyes Data Corporation, Park Ridge, N.J., 1975.
INDA (Association of the Nonwoven Fabrics Industry, New York) Technical Symposia, 1973 to present.
Annual Symposium Papers, Clemson Nonwoven Fabrics Forum, Clemson University, Clemson, S.C., 1969 to present.

ARTHUR DRELICH
Chicopee Division
Johnson and Johnson

NOVOLOID FIBERS

Novoloid fibers are cross-linked phenolic–aldehyde fibers typically prepared by acid-catalyzed cross-linking of a melt–spun novolac resin with formaldehyde (see Phenolic resins). Such fibers are generally infusible and insoluble, and possess physical and chemical properties that clearly distinguish them from other synthetic and natural fibers (see Fibers, chemical). Novoloid fibers are used in flame- and chemical-resistant textiles and papers, in a broad range of composite materials, and as precursors for carbon and activated-carbon fibers, textiles, and composites.

The generic term novoloid was recognized in 1974 by the United States Federal Trade Commission (FTC) as designating a manufactured fiber containing at least 85 wt % of a cross-linked novolac (1). In granting its official approval, the FTC noted that its criteria for new generics required that a candidate fiber have a chemical composition radically different from other fibers resulting in distinctive physical properties of significance to the general public.

The basic novoloid patents were issued to the Carborundum Company, Niagara Falls, New York, in the early 1970s (2). At present, Nippon Kynol, Inc. (Japan), and American Kynol, Inc., as exclusive licensees of Carborundum, manufacture and sell novoloid fibers under the trademark Kynol.

Fiber Properties

Except where specifically noted, the properties described here are those of commercial Kynol novoloid fibers produced by melt-spinning and aqueous curing. Fiber properties are strongly affected by the chemical and physical characteristics of the novolac resin, the fiber diameter, and the method and degree of curing (cross-linking). These variables may be adjusted to produce the balance of properties required for a particular application; the data given here represent typical values or ranges.

Novoloid fibers are generally elliptical in cross-section, with a ratio of diameters of approximately 5:4. They are light gold in color and darken gradually to deeper shades with age and exposure to heat or light, although there is no significant concomitant change in other fiber properties. The fibers have a very soft touch or hand, and are generally without appreciable crimp, although crimp may be imparted by thermomechanical means to facilitate spinning.

Thermal Properties. Novoloid fibers are highly flame resistant, but are not high temperature fibers in the usual sense of the term. A 290 g/m^2 woven fabric withstands an oxyacetylene flame at 2500°C for 12 s or more without breakthrough. However, the practical temperature limits for long-term application are 150°C in air and 200–250°C in the absence of oxygen. This apparent paradox must be understood in terms of the chemical structure of the fiber (see also Heat resistant polymers).

Flame Behavior and Resistance: Combustion.   A measure of the flame resistance of a given material is the limiting oxygen index (LOI), ie, the percent concentration of atmospheric oxygen required for self-supporting combustion (see Flame retardants). The LOI of novoloid fiber materials varies with the configuration of the fibers (as yarn, felt, fabric, etc), but is generally in the range of 30 to 34, ie, higher than that of natural organic textile fibers and of all but the most exotic synthetic organic fibers. (Measured values of LOI are significantly influenced by the specific test method and apparatus employed, and, therefore, even higher values have been reported in the literature.) Moreover, when exposed to flame, novoloid fibers do not melt but gradually char until completely carbonized without losing their initial fiber form and configuration. This behavior is attributable to the cross-linked, amorphous, infusible structure of the fiber and to its high (76%) carbon content.

Combustion of an organic polymer substance generally involves not only oxidative attack at the surface but, more important, decomposition and volatilization of the interior material leading to oxidation in the gaseous phase. The process may be accelerated by softening or melting, which hastens decomposition and permits escape of flammable volatiles. Some insight into flame behavior may be gained, therefore, by consideration of behavior on heating alone, in the absence of the final, oxidative combustion step.

Thermogravimetric and differential thermal analytic data (tga and dta, respectively) on novoloid fibers indicate that above 250°C in the absence of oxygen novoloid fibers undergo gradual weight loss until, close to 700°C, the fiber is fully carbonized, with a carbon yield of 55–60% (3–4). Thus the high initial carbon content results in limited production of volatiles (weight loss 40–45%), only a portion of which is, in fact, flammable. Melting does not occur, and shrinkage is small, suggesting that the cross-linked structure of the material promotes gradual coalescence of the aromatic units into a stable, amorphous char; the relatively low thermal conductivity of the material aids in the moderation of this process.

The foregoing observations are helpful in understanding the high flame resistance of novoloid fibers and fabrics, particularly since oxygen generally is depleted in the actual zone of combustion. Exposure to flame results in formation of an amorphous surface char that serves both to radiate heat from the fiber and to retard the evolution of flammable volatiles; the latter is limited by the high carbon content. The amorphous (glassy) nature of the char presents a minimum reactive surface to the flame. Moreover, in a textile-fabric structure the charred fibers at the fabric face provide a protective insulating barrier retarding penetration of both heat and oxygen into the interior of the fabric; minimal shrinkage enhances the mechanical stability of this barrier. Finally, the formation of H_2O and CO_2 as products of decomposition and combustion provides an ablative type of cooling effect (see Ablative materials).

Since novoloid fibers are composed only of carbon, hydrogen, and oxygen, the products of combustion are principally water vapor, carbon dioxide, and carbon char. Moderate amounts of carbon monoxide may be produced under certain conditions; but the HCN, HCl, and other toxic by-products typical of combustion of many flame-resistant organic fibers are absent. The toxicity of the combustion products is thus very low or negligible (5). Smoke emission is also minimal, less than that of virtually any other organic fiber.

Oxidative Degradation. The mechanism of oxidative degradation of novoloids is similar to that of phenolic resins in general (6), and appears to involve mainly oxidative attack on the methylene linkages between aromatic units, with initial formation of a peroxide (see Hydrocarbon oxidation). This peroxide is subject to further decomposition to a carbonyl which, in conjugation with the phenolic hydroxyl, apparently forms the chromophoric keto–enol group which is thought to contribute to the characteristic golden color of the fiber (4).

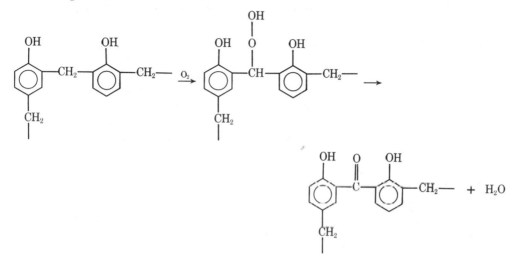

The subsequent, complex mechanism of oxidative degradation involves principally the same carbonyl group and leads to chain scission. Formation of quinoids and related compounds is probably responsible for further darkening.

Oxidative degradation is a comparatively slow process which, in the severe environment of a flame, is far less significant in determining fiber behavior than is the nonoxidative char formation described above. At ordinary temperatures it leads only to darkening of the fiber surface, with no significant effect on properties. However, oxidation is accelerated by increasing temperatures, leading to significant loss of weight and strength as temperatures approach 200°C. Therefore, the practical temperature limit for long-term applications in air is about 150°C.

Punking. Decomposition of the peroxide formed at the methylene linkage is an exothermic reaction. As a result this reaction may, under certain conditions, become self-sustaining, leading to the phenomenon known as punking.

Punking may occur in materials containing methylene bonds subject to peroxide formation, provided a large surface area is available for oxidation and the rate of heat removal is low. These conditions are met by large masses of tightly packed novoloid fibers, which are excellent heat insulators. Given sufficient oxygen for peroxide formation, sufficient fiber bulk for efficient insulation, and an initial temperature high enough for peroxide decomposition to generate more heat than is removed by ventilation or other means, internal temperatures may increase to 570–580°C, at which point ignition of the fiber may occur. The result is a smoldering combustion that continues until the fiber mass is consumed.

Punking may easily be prevented by appropriate control of storage and use. Thus, novoloid materials should not be stored in bulk at high temperatures or subjected to lengthy heat treatment over 120°C without adequate ventilation, and should be cooled below 60°C after heat treatment or high temperature use. Applications involving prolonged exposure to high temperatures with limited opportunity for heat escape—such as steam-pipe insulation—should be avoided. Punking does not occur in the absence of oxygen, eg, when the fibers are encapsulated by a matrix material (see also Peroxides, organic).

Chemical Resistance.   Novoloid fibers display excellent chemical and solvent resistance. They are attacked by concentrated or hot sulfuric and nitric acids and strong bases, but are virtually unaffected by nonoxidizing acids, including hydrofluoric and phosphoric acids; dilute bases; and organic solvents.

Other Properties.   Novoloid fiber materials display excellent thermal insulating characteristics. A nonwoven batting at a density of 10 kg/m³ showed a thermal conductivity of 0.04 W/(m·K) at 20°C and less than 0.03 W/(m·K) at −40°C (7) (see Insulation, thermal). Retention of properties at very low temperatures is excellent. Efficacy of sound absorption is high (see Insulation, acoustic). Ultraviolet radiation, although leading to deepening of color, has minimal effect on fiber properties; resistance to γ-radiation is also high.

Textile and Paper Properties

The textile properties of novoloid fibers are summarized in Table 1. The ranges shown for strength, elongation, and modulus reflect the dependence of these properties on fiber diameter, with the higher values corresponding to finer diameters.

Novoloid fibers are processed by suitably modified conventional textile techniques. The moderate tensile strength (comparable to that of cellulose acetate) and lack of inherent crimp require that suitable precautions be taken, particularly in carding staple fibers, to prevent excessive fiber breakage. Thorough opening is required, and carding speed is lower than that for most conventional fibers. A modified woolen system is probably best for spinning. Blending with other fibers such as aramids improves processing speeds and increases yarn tensile strength (see Aramid fibers).

Table 1.   Textile Properties of Novoloid Fibers

Property	Value
diameter, μm (tex[a])	14–33 (0.22–1.1)
specific gravity	1.27
tensile strength, GPa[b] (N/tex[c])	0.16–0.20 (0.12–0.16)
elongation, %	30–60
modulus, GPa[b] (N/tex[c])	3.4–4.5 (2.6–3.5)
loop strength, GPa[b] (N/tex[c])	0.24–0.35 (0.19–0.27)
knot strength, GPa[b] (N/tex[c])	0.12–0.17 (0.10–0.13)
elastic recovery[d], %	92–96
work-to-break, J/g (= mN/tex[e])	26–53
moisture regain at 20°C, 65% rh, %	6

[a] To convert tex to den, multiply by 9.

[b] To convert GPa to psi, multiply by 145,000.

[c] To convert N/tex to gf/den, multiply by 11.33; N/tex = GPa ÷ density.

[d] At 3% elongation.

[e] To convert mN/tex to gf/den, multiply by 0.01133.

Heavy yarns (300 tex) of 100% novoloid fiber are readily spun. With suitably modified equipment 100% novoloid yarns as fine as 30 tex and blended yarns of 20 tex are routinely produced.

Spun novoloid yarns are used in the production of woven fabrics in weights from 100 to >550 g/m^2, as well as in knitted products such as gloves. Blending with other fibers improves the low tensile strength and abrasion resistance of 100% novoloid materials. Novoloid fibers may be dyed with cationic or disperse dyes; however, color range and stability are limited by the inherent gold color of the fiber and its tendency to darken with exposure to heat and light.

Dry-formed webs and felts are produced by needle-punching and other felting techniques. Novoloid papers are produced by wet-laid paper methods. For paper-making, the physical characteristics of the fiber generally require admixture of a binder, such as PVC, polyesters, or epoxies. For 100% novoloid paper, a suitable proportion of partially cured fiber may be added. Hot-pressing or calendering of the web expresses the uncured novolac resin and promotes binding by cross-linking to the fiber surface (see Paper; Pulp, synthetic).

Manufacture

Novoloid fibers are infusible, insoluble, intractable three-dimensional network polymers. To produce fibers from such a material, fiberization must precede cross-linking; ie, preformed precursor fibers must undergo chemical modification.

Melt-Spinning. Novoloid fibers are typically produced by first melt-spinning a novolac resin. Novolac resins are manufactured from phenols and aldehydes using an acid catalyst and excess phenol. A suitable novolac for fiber production is that made from phenol and formaldehyde. Such a material may be melt-spun to produce an extremely friable, whitish fiber that is both readily fusible and soluble in organic solvents. Some molecular orientation parallel to the fiber axis may be observed at this stage.

Aqueous Cure. The melt-spun fibers are then cured by immersion in an acidic aqueous solution of formaldehyde, followed by gradual heating to promote the acid-catalyzed cross-linking reaction. Curing commences with the formation of a skin on the fiber surface, and proceeds inward as the acid (usually HCl) and formaldehyde diffuse into the material.

This process transforms the novolac resin into a three-dimensionally cross-linked network polymer through the formation of methylene and dimethyl ether linkages. The molecular orientation disappears, and the material becomes unoriented and

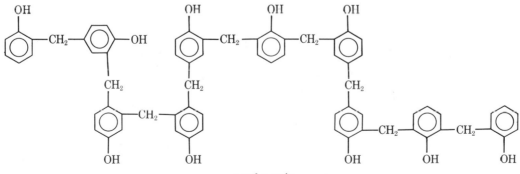

novolac resin

amorphous. As curing proceeds, the proportion of dimethyl ether linkages declines, and the cross-link density may approach that of cured phenolic resins. Generally, however, the density of cross-linking remains comparatively low. Although tensile strength increases by a factor of roughly ten, the cured fibers remain flexible and elastic.

The extent of cross-linking is limited by steric factors and by adjustment of curing conditions. As a result, methylol groups formed on sites where cross-links are not completed typically constitute 5–6 wt % of the cured fiber, and are available for further reaction with matrix materials at the fiber surface. These groups play a key role in the formation of novoloid-fiber composites. The proportion of these groups may be reduced by heating the cured fiber to 180°C.

The combination of melt-spinning followed by aqueous curing permits the large-scale production of textile-grade fibers with closely controlled and highly reproducible properties in diameters from <14 to >33 μm (0.2–1 tex or 2–10 den) and lengths from 100 μm to continuous filament. Such fibers are currently available for ca 10 \$/kg (staple fiber). Price reductions of 30–40% are expected as production increases.

Partial Cure.   Interruption of the curing process at an early stage gives partially cured fibers containing uncured novolac resin within an envelope of cross-linked material. The uncured core remains fusible and soluble, and may be extracted by organic solvents such as methanol to leave hollow novoloid fibers. Alternatively, the uncured material may be left in place to be removed later by thermomechanical means when it may serve, for instance, as a self-contained binder for novoloid-fiber papers.

Blown Fibers.   For finer fibers or when precise control of diameter and length are not required, other manufacturing methods may be used. Very fine precursor fibers can be produced by air attenuation of molten resin injected into a stream of hot air. When collected and cured as above, these blown fibers have a fuzzy or woolly appearance, random lengths, and random diameters in the range of 2–5 μm. Alternatively, fiberization may be carried out by centrifugal spinning in a cotton-candy-type machine.

Self-Curing Fibers.   Self-curing novoloid fibers also may be produced. Curing agents such as hexamethylenetetramine or paraformaldehyde are routinely blended into novolacs to provide bulk-molding resins which cure upon heating. A similar method has been demonstrated for the production of novoloid fibers (8). For successful fiber formation, the blend of resin and curing agent must be melted and fiberized quickly

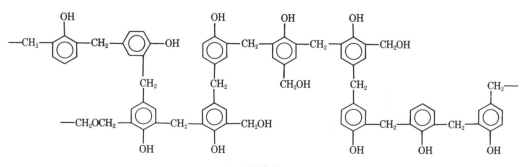

novoloid fiber

without curing, and then cured without remelting. Rapid fiberization is accomplished by hot air impingement, centrifugal spinning, or other suitable means, followed by curing in an acidic gas such as hydrogen chloride or boron trifluoride. This method permits very rapid curing, with cross-linking completed in minutes. Commercial feasibility of this method has not yet been established.

Acetylation.   Novoloid fibers produced by the above methods are typically golden, the coloration deepening with age and exposure to heat or light (reverse fading). Color development may be prevented by heating the cured fibers with acetic anhydride to produce bleached or white novoloids.

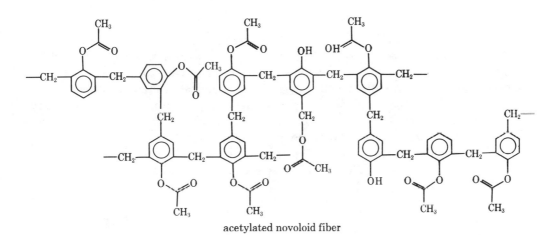

acetylated novoloid fiber

Acetylation of the phenolic hydroxyl groups inhibits peroxide formation at the methylene linkage, through both steric and inductive effects. Thus oxidative degradation and formation of chromophoric groups is prevented, and acetylated novoloid fibers are white and not subject to reverse fading. The limitation in dyeability is overcome and a broad spectrum of colors becomes available. Thermal stability is enhanced, and the possibility of punking is diminished. However, tensile strength and flame resistance are reduced.

Acetylation significantly increases fiber costs.

Analysis

Table 2 shows typical elemental analyses and approximate empirical formulas for novolac resin, cured novoloid fibers, and acetylated novoloid fibers.

Table 2.   Empirical Formulas and Elemental Analyses

Material	Empirical formula (approximate)	Elemental analysis, wt %		
		C	H	O
novolac resin	$C_{69}H_{59}O_{10}$	78.3	5.6	15.4
cured novoloid fiber	$C_{63}H_{55}O_{11}$	75.8	5.5	17.9
acetylated novoloid fiber	$C_{61}H_{56}O_{14}$	72.0	5.5	22.5

Health and Safety Factors

A toxicity evaluation of Kynol novoloid fibers in accordance with regulations under the *Federal Hazardous Substances Act* demonstrated that the fiber is neither toxic nor highly toxic under the oral and skin contact categories; is not an eye or primary skin irritant; and has a low order of acute inhalation hazard (9). The low toxicity of combustion products and limited smoke evolution are discussed above, as is the need for appropriate precautions in storage and use to forestall punking.

Uses

The combination of infusibility, excellent flame and chemical resistance, minimal evolution of smoke or toxic gases, and low specific gravity permit novoloid fibers (either alone or in combination with other materials) to replace asbestos in many uses (see Asbestos). Furthermore, the specific physical and chemical properties of these fibers lead to numerous other applications, particularly in new composite materials (see Composite materials). Novoloid fibers also are precursors of low modulus carbon and activated carbon fibers (see Carbon and artificial graphite).

Textile and Related Applications. Woven, knitted, and felted novoloid fibers, either alone or in combination with other materials, are used in protective apparel including fire suits, insulating gloves, and clothing for foundry workers (see Flame-retardant textiles). Fabrics may be aluminized or elastomer-coated for added protection and strength. Moderate tensile strength and abrasion resistance, and restricted dyeability, limit these applications.

Heavy woven novoloid fabrics, with or without aluminum or elastomeric coating, can be substituted for asbestos textiles in applications requiring resistance to flame and metal splash; they are lighter in weight and easier to manipulate. Because of its high chemical and solvent resistance, novoloid fiber also is utilized in gaskets and braided packings.

Novoloid spun yarns and papers are employed as filler yarns and wrapping tapes in high performance communications cables for installation where maximum circuit integrity must be maintained for safety reasons; eg, in nuclear-energy plants, highway tunnels, high-rise buildings, and underground concourses.

Composites. Novoloid fibers are used in combination with a wide variety of matrix materials such as thermosetting and thermoplastic resins, elastomers, and ceramics. Owing to a direct cross-linking reaction between fiber and matrix, in which the methylol groups formed during curing play the key role, novoloid composites frequently display unexpected synergistic improvements in properties. Typical examples are the composites of novoloid fiber with resole resin and chlorinated polyethylene (CPE).

As the fiber component of organic matrix composites, novoloid fibers are characterized by easy and uniform dispersion and generally excellent fiber-to-matrix wetting and adhesion. They improve properties such as heat resistance, impermeability, compressive strength, shock resistance, dimensional stability, and hardness. However, because of the comparatively low strength of the fiber itself, tensile strength generally is not markedly increased. The low specific gravity of novoloid fibers often leads to a reduction in weight of the composite material.

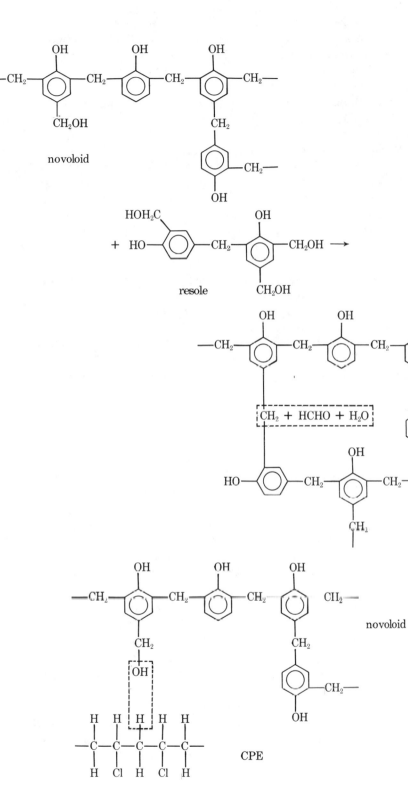

novoloid

resole

$\lceil CH_2 + HCHO + H_2O \rceil$

novoloid

CPE

313

Thermosetting Resins. Owing to the large degree of similarity in chemical structure, novoloid fibers are particularly suitable as fillers for phenolic resins. As much as 70 wt % fiber may be incorporated into a resole matrix. The composite has better thermal stability (up to 230°C), thermal insulation, electrical resistance (especially after boiling in water), high temperature shock resistance, chemical resistance, and machinability than resins filled with other fibers. Further evidence of the high compatibility and synergism of novoloid fiber and resole resin is the fact that, although the carbon yield on carbonization of novoloid fibers is 55–57% and that of resole alone 50–52%, a 50/50 composite of the two materials yields 65–70% carbon (see Fillers).

Similar results are obtained with epoxy and other reactive resins. Applications for such fiber-reinforced thermoset plastics include heat-resistant automotive and electrical parts, and chemical-resistant materials.

Thermoplastic Resins. Addition of novoloid fiber to PVC raises the melting point of the resin; at a fiber content of ca 15 wt %, the material does not melt. Similar results are seen with polyamides (qv). Potential applications include printed circuit boards and electrical panels. In combination with polyester resins, novoloid fibers provide a lightweight, readily machinable material with excellent cold resistance for piping and similar applications. Furthermore, woven novoloid fabrics and thin-paper surface veils combined with polyester resins provide long-lasting tank linings for corrosive materials such as HCl and HF.

Elastomers. It is difficult to reinforce cured CPE, a flame-retardant elastomer, with fibers because the fiber is degraded by HCl released on curing. Novoloid fibers, however, are unaffected by HCl; in fact, the fiber–matrix cross-linking reaction permits the fibers themselves to act as the curing agent for pure (unformulated) CPE, yielding a product with good hardness, dimensional stability, and resistance to flame, heat, and chemicals.

Similar results are achieved with numerous other elastomers including ethylene–propylene, acrylic, and nitrile rubbers, chlorosulfonated polyethylene, and polytetrafluoroethylene (PTFE). High ratios of uniform fiber incorporation, better than 50 wt %, may be obtained with conventional rubber-processing equipment. Because of the relatively low resistance of novoloid fibers to sulfuric acid, sulfur-curing systems should be avoided or applied with caution.

Applications for novoloid-filled elastomers include cushion boards and rollers; chemical, solvent, and fuel hoses; ducts; gaskets; packings; and the coating or saturation of novoloid fabrics for use in hoses and ducts and as protective sheets and curtains. A 550 g/m² novoloid fabric coated with 650 g/m² of formulated CPE, eg, has a tensile strength of 4–5 GPa (580,000–725,000 psi) and sheds molten steel.

Ceramics. Addition of small amounts of novoloid fibers to unfired ceramics stabilizes the material during rapid firing and increases the strength and integrity of the final product. Applications include daubable furnace linings and sprayable fire-protective cable coatings (see Ceramics).

Carbon Fibers. Heating in an inert atmosphere converts novoloid fibers gradually to pure carbon at yields of 55–57%. Neither pretreatment nor tension is required during carbonization. The rate at which carbonization can be carried out is governed mainly by oven configuration and the rate at which evolved gases are removed. Shrinkage of the fiber on carbonization is predictable and low, on the order of 20%.

As a consequence of this simplicity and stability of processing, novoloid fibers may readily be carbonized either as a continuous tow, or in a final product configuration (felt, fabric) pre-formed from the readily processable novoloid precursor fibers.

Novoloid-based carbon fibers are amorphous. Even treatment at the usual graphitization temperatures above 2000°C does not result in the formation of the typical well-ordered graphite molecular structure. Novoloid-based carbon is thus a low modulus, low strength material, at least in comparison to the high modulus fibers prepared from rayon and polyacrylonitrile (PAN) precursors.

Novoloid-based carbon fibers are obtainable in regular and high purity grades. Treatment of the precursor by controlled increase of temperature to about 800°C in an inert atmosphere gives highly stable carbon fibers with a carbon content of about 95%. Further heating to 2000°C yields fibers of 99.8% or greater carbon content.

Table 3 gives the typical properties of novoloid-based carbon fibers compared to PAN- (high modulus) and pitch- (low modulus) based carbon fibers. The decrease in specific gravity of novoloid-based carbon as temperature exceeds 1000°C is remarkable and has not been adequately explained. It does not appear related to any observable development of microporosity or voids. Despite the apparent decrease in interatomic spacing d_{002}, the x-ray peak remains broad and the material is essentially amorphous in nature. Of further interest is the fact that although fibers carbonized at 600°C retain the elliptical cross section of the novoloid precursor, those treated at 2000°C tend to become round.

Novoloid-based carbon fiber is soft and pliable, produces little dust or fly on processing, and has good lubricity. It retains much of the good dispersability and affinity for matrix materials of its novoloid precursor. In addition, its heat, chemical, and electrical characteristics are comparable to those of other carbon fibers. Thus,

Table 3. **Typical Properties of Carbon Fibers**

Property	Novoloid		Pitch		Polyacrylonitrile	
type	low modulus		low modulus		high modulus	
treatment temperature, °C	800	2000	1000	2000	1500	2000
specific gravity, g/cm³	1.55	1.37	1.63	1.55	1.8–1.9	1.9–2.0
carbon content, wt %	95	99.8+	95	99.5+	93	99.5+
x-ray diffraction profile, 002, 2θ, degrees	23.0[a]	25.0[a]	24.0[a]		25.0[b]	26.1[c]
interlayer spacing, d_{002}, pm	395	351				336
tensile strength, MPa[d]	500–700	400–600	500–1000		1500–3000	
elongation, %	2.0–3.0	1.5–2.5	1.5–2.5		1.0–1.5	
modulus, GPa[e]	20–30	15–20	30–50		150–300	
heat resistance, °C						
tga	436	541	416		519	
air	350	380	350		350	
specific resistivity, mΩ·cm	10–30	5–10	10–30		1–10	
affinity with PTFE, CPE, epoxies[f]	good		fair		poor	

[a] Broad.
[b] Medium.
[c] Sharp.
[d] To convert MPa to psi, multiply by 145.
[e] To convert GPa to psi, multiply by 145,000.
[f] PTFE = polytetrafluoroethylene; CPE = chlorinated polyethylene.

it is well suited for production of braided packings, either impregnated or unimpregnated (no coating is required for braiding even of the high purity filament yarn); bearings and other low friction materials; carbon-fiber-containing conductive elastomers with uniform, predictable, and high conductivity; and highly heat- and chemical-resistant gaskets and sheets. Paper made from chopped carbon fiber is suitable for resin impregnation to produce sheet electrodes for electrostatic precipitators (see Air pollution control methods). Yarns, fabrics, and felts of novoloid-based carbon are used in protective curtains, vacuum-furnace insulation, and static-elimination brushes.

Activated Carbon. Novoloid-based activated carbon is formed by a one-step process combining both carbonization and activation, in an oxygen-free atmosphere containing steam and/or CO_2, at ca 900°C. The material can be activated either as a continuous tow or after the formation of fabrics or felts. The degree of activation, as measured by the BET method, is controlled by time rather than temperature, and can approach 3000 m^2/g. Table 4 gives typical properties of novoloid-based activated-carbon fibers; data for PAN- and rayon-based activated-carbon fibers and for granular activated carbon are presented for comparison.

The pores of novoloid-based activated carbon are generally uniform in size and

Table 4. Typical Properties of Activated Carbon Fibers

Property	Novoloid	Novoloid	PAN[a]	Rayon[a]	Granules
yield, wt %	33	22	15	7	
diameter, μm	9.2	8.5	ca 5	ca 10	
pH	7.3	7.5	8.0		6.3
pore capacity					
specific surface area[b], m^2/g	1500	2000	870, max	1450, max	910, typical
pore volume[c], cm^3/g	0.63	0.75	0.28	0.53	0.42
mechanical properties					
tensile strength, MPa[d]	400	350	300	70	
elongation, %	2.8	2.7			
modulus, GPa[e]	14	12	50+		
gas adsorption capacity at 20°C[f], wt %					
benzene	53	67	34	46	33
toluene	57	80	32	47	35
trichloroethylene	83	104	53		61
moisture content at 20°C, 65% rh, wt %	37	20	30		27
methylene blue adsorption capacity[g], mL/g	200	300	40		100

In the header, "Precursor" spans the Novoloid, Novoloid, PAN[a], Rayon[a], Granules columns.

[a] Owing to their partially crystalline structure, these examples represent the practical limits of activation for these fibers. PAN = polyacrylonitrile.
[b] BET method (Brunauer, Emmett, Teller).
[c] Steam-adsorption method.
[d] To convert MPa to psi, multiply by 145.
[e] To convert GPa to psi, multiply by 145,000.
[f] JIS K-1474 Japan industry standard.
[g] JIS K-1470 Japan industry standard.

straight, rather than branched as in granules. Pore radius distribution shows a single peak at about 1.5–1.8 nm; radius and volume increase with increasing activation, and thus may be controlled for selective adsorption of polypeptides and similar large molecules. Pore configuration and the high surface-to-volume ratio of the fibers, compared to granular activated carbon, permit extremely rapid adsorption and desorption. A further advantage over granules is the reported potential for rapid and convenient reactivation of novoloid-based activated-carbon felts and fabrics by direct-resistance heating *in situ* with a moderate electrical current (10).

Owing to the strength and flexibility of novoloid-based activated carbon fibers, wet-laid paper webs can be formed from activated-carbon fiber and kraft pulp without any binder. Effective surface area of the finished sheet is therefore close to that calculated from the activated-carbon fiber content.

Applications for activated-carbon paper include medical supplies such as bandages for malodorous wounds; quick-acting air and liquid-purification filters; ozone eliminators for electrostatic copiers; and protective wrappings for electronic sensors. Activated-carbon fabrics hold promise for use in compact dialysis equipment.

Glassy Carbon. When a novoloid–resole composite is baked at high temperatures over an extended period of time, it completely carbonizes to form a uniform, amorphous, glassy or vitreous carbon in which the portions formerly comprising fiber and matrix are virtually indistinguishable. This material is harder than glass, has extremely low gas permeability, and is hardly affected by 5 h immersion in 50:50 sulfuric–nitric acid at 100°C. Applications are foreseen in extremely rigorous chemical, electrical, mechanical, and heat environments. Production cost is expected to be significantly lower than that of hitherto available glassy carbon products.

BIBLIOGRAPHY

"Phenolic Fibers" in *ECT* 2nd ed., Suppl. Vol., pp. 667–673, by James Economy and Luis C. Wohrer, The Carborundum Company.

1. *Fed. Reg.* **39,** 1833 (1974).
2. U.S. Pats. 3,650,102 (March 21, 1972); 3,723,588 (March 27, 1973), J. Economy and R. A. Clark (to the Carborundum Company).
3. J. Economy and L. Wohrer, "Phenolic Fibers" in N. M. Bikales, ed., *Encyclopedia of Polymer Science and Technology*, Vol. 15, Wiley-Interscience, New York, 1971, pp. 370–373.
4. J. Economy, L. C. Wohrer, F. J. Frechette, and G. Y. Lei, *Appl. Polym. Symp.* **21,** 81 (1973).
5. J. Economy, "Phenolic Fibers" in M. Lewin, S. M. Atlas, and E. M. Pearce, eds., *Flame-Retardant Polymer Materials*, Vol. 2, Plenum Press, New York, 1978, pp. 210–219.
6. R. T. Conley and D. F. Quinn, "Retardation of Combustion of Phenolic, Urea–Formaldehyde, Epoxy, and Related Resin Systems" in ref. 5, Vol. 1, 1975, pp. 339–344.
7. Dynatech R/D Company, report to American Kynol, Inc., April 26, 1976.
8. U.S. Pat. 4,076,692 (Feb. 28, 1978), H. D. Batha and G. J. Hazelet (to American Kynol, Inc.).
9. Industrial Health Foundation, Inc., report to the Carborundum Company, Sept. 1972.
10. J. Economy and R. Y. Lin, *Appl. Polym. Symp.* **29,** 199 (1976).

General References

References 3–5, and 10 are general references.
J. Economy, L. C. Wohrer, and F. J. Frechette, "Non-Flammable Phenolic Fibers," *J. Fire Flammability* **3,** 114 (April 1972).
R. Y. Lin and J. Economy, "Preparation and Properties of Activated Carbon Fibers Derived from Phenolic Precursors," *Appl. Polym. Symp.* **21,** 143 (1973).
J. Economy, "Kynol Novoloid Fibers," *Polym. News* **II** (7/8), 13 (1975).

H. D. Batha, "Resistance of Kynol Fabrics to [Molten] Metals" in V. M. Bhatnagar, ed., *Fire Retardants: Proceedings, 1976 International Symposium on Flammability and Fire Retardants*, Technomic, Westport, Conn., 1977, pp. 81–87.

K. Ashida, M. Ohtani, T. Yokoyama, K. Kosai, and S. Ohkubo, "Full Scale Investigation of the Fire Performance of Urethane Foam Cushions Using Novoloid Fiber Products as Interlayer," *J. Cellular Plast.*, 311 (Nov.–Dec. 1978).

J. Economy, "Now that's an interesting way to make a fiber," *Chemtech* **10**(4), 240 (April 1980).

JOSEPH S. HAYES, JR.
American Kynol, Inc.

OLEFIN FIBERS

Olefin fibers, which also are called polyolefin fibers, are manufactured fibers in which the fiber-forming substance is any long-chain, synthetic polymer of at least 85 wt % ethylene, propylene, or other olefin units (1). The olefin fibers of commercial importance are polypropylene and, to a lesser extent, polyethylene. Fibers have been produced on a laboratory scale from several other polyolefins, eg, poly(1-butene), poly(3-methyl-1-butene), and poly(4-methyl-1-pentene) (see Fibers, chemical).

The first polyolefin fibers to be extruded were made in England in the late 1930s from low density polyethylene (see Olefin polymers, high pressure (low and intermediate density) polyethylene). After World War II, polyethylene monofilaments were produced by several manufacturers; however, they lacked light and dimensional stability. Because their intended use was in automotive seat covers, this venture had only limited success (2). In 1957, high density polyethylene, which was produced by the Ziegler and Phillips organometallic-catalyst systems, became available for extrusion. Fibers from this polymer were substantially superior to the previous ones, especially in terms of their mechanical properties. They were used in certain industrial applications, eg, ropes and cables, and to a limited extent, as furniture fabric. Their general acceptance was hampered by their low melting temperature (130–138°C), lack of dyeability, and poor elastic properties or resilience. With few exceptions, these properties characterize modern polyethylene fibers (see Olefin polymers, linear (high density) polyethylene).

Polypropylene fibers resulted from work in stereospecific polymerization involving Ziegler catalysts (see Ziegler-Natta catalysts). The process yielded a polypropylene polymer that was nearly completely isotactic and, thus, crystallizable in contrast to the rubbery atactic polymer resulting from all prior polymerizations of propylene.

(Recently, Phillips was favored for possession of the U.S. patent; however, Montedison, DuPont, and Standard Oil of Indiana are contesting the claim (3).) Fibers from polypropylene have a significantly higher melting point, ie, 165–175°C; equal tenacity and abrasion resistance, and substantially better resilience than polyethylene fibers. Like polyethylene fibers, they are essentially undyeable in unmodified form. The advantage to using polypropylene was the expectation that it could be produced at a lower cost than any other synthetic fiber, since propylene is a common refinery by-product. This expectation has been realized only partially, owing to the cost of the various additive packages that are necessary to make polypropylene fibers that are technically competitive with other fibers. Olefin-fiber consumption in the United States is approaching 10% of the total synthetic-fiber consumption projected for 1987 (4) (see also Fibers, chemical).

Properties

Physical. Some physical properties of commercial polyethylene and polypropylene fibers are listed in Table 1 (5). Polyethylene and polypropylene fibers differ from other synthetic fibers in two important respects: their nearly total lack of water absorption ensures that wet properties are identical with their properties at standard conditions (65% rh, 21°C) and their low specific gravity leads to a higher covering power, ie, one kilogram of polypropylene can produce a fabric, carpet, etc, with as much as 45% more fiber per unit area than one kilogram of polyester.

Chemical. The general chemical characteristics of olefin fibers are extreme hydrophobicity and inertness to a wide variety of inorganic acids and bases and to organic solvents at room temperature. These properties derive from the hydrocarbon character of the fundamental unit and from the great molecular weights of commercial polymers. Typical equilibrium moisture regain at 65% rh and 21°C of some commercial fibers are compared with those of polyethylene and polypropylene fibers in Table 2. At room temperature, polyolefins are inert to a variety of industrial chemicals, including 6 N sulfuric acid, 6 N nitric acid, acetone, ethyl alcohol, and ammonium hydroxide. Sulfuric and nitric acids and other oxidizing acids degrade polyolefins at higher temperatures. At room temperature, there is no known solvent for isotactic polypropylene; although, at higher temperatures, some aromatic hydrocarbons and chlorinated hydrocarbons dissolve it. Polar solvents are ineffective at any temperature. However, in the case of extremely long immersion times, eg, ca 30–120 d, water causes a small, ie, 1–9% strength loss in polypropylene fibers (2).

Thermal. The thermal transitions of polyolefin fibers are compared with those of polyester and nylon in Table 3. Comparatively, polyethylene and polypropylene undergo all thermal transitions at substantially lower temperatures than either polyester or nylon.

The thermal degradation of olefin fibers is oxygen sensitive, and may be prob lematic in fiber formation and in use. In general, the absorption of oxygen by the polymer results in chain scission and, therefore, molecular weight degradation as a result of high temperature formation of hydroperoxides. The unprotected polymer cannot be melt spun. An enormous variety of antioxidant packages act as stabilizers and allow both melt spinning and reasonable high temperature performance (see Antioxidants and antiozonants) (9). In polyethylene, effective antioxidants may contain sterically hindered phenols and cresols, sometimes with thio or sulfide groups.

Table 1. Synthetic Fiber Properties[a]

Polymer	Fiber type	Tenacity, MPa (N/tex[b])		Modulus, GPa (N/tex)[b]	Breaking elongation, %		Elastic recovery, %[c]	Density, g/cm³
		Standard	Wet		Standard	Wet		
low density polyethylene	monofilament	92–280 (0.1–0.3)	92–280 (0.1–0.3)	0.18–1.0 (0.2–1.1)	20–80	20–80	95 (at 5%)	0.92
high density polyethylene	monofilament	290–570 (0.3–0.6)	290–570 (0.3–0.6)	1.7–4.2 (1.8–4.4)	10–45	10–45	100 (at 5%)	0.95–0.96
polypropylene	staple and tow	280–560 (0.3–0.6)	280–560 (0.3–0.6)	0.28–3.3 (0.3–3.5)	20–120	20–120	70–100 (at 5%)	0.90–0.96
	monofilament	270–540 (0.3–0.6)	270–540 (0.3–0.6)	1.6–4.8 (1.8–5.3)	14–30	14–30	98 (at 5%)	0.90–0.91
	multifilament	180–630 (0.2–0.7)	180–630 (0.2–0.7)	1.2–3.2 (1.3–3.5)	20–100	20–100	94–98 (at 5%)	0.90–0.91
polyester	staple and tow	280–830 (0.2–0.6)	280–830 (0.2–0.6)	1.2–2.1 (0.9–1.5)	12–55	12–55	81 (at 3%)	1.38
	multifilament	280–690 (0.2–0.5)	280–690 (0.2–0.5)	1.2–3.6 (0.9–2.6)	24–42	24–42	76 (at 3%)	1.38
nylon-6,6	staple and tow	230–680 (0.2–0.6)	230–570 (0.2–0.5)	1.0–4.5 (1.0–4.5)	16–75	18–28	82 (at 3%)	1.13–1.14
	multifilament	230–570 (0.2–0.5)	230–570 (0.2–0.5)	0.45–2.4 (0.4–2.1)	25–65	30–70	88 (at 3%)	1.13–1.14
acrylic	staple and tow	230 (0.2)	120–130 (0.1–0.2)	1.0 (0.9)	20–28	26–34	73 (at 3%)	1.16
rayon	staple, regular tenacity	150–450 (0.1–0.3)	150–300 (0.1–0.2)	0.75–2.3 (0.5–0.15)	15–30	20–40	82 (at 2%)	1.46–1.54
	staple, high tenacity	450–750 (0.3–0.5)	300–600 (0.2–0.4)	1.7–6.6 (1.1–4.4)	9–26	14–34	70–100 (at 2%)	1.46–1.54
cellulose acetate	staple, multi-filament	130 (0.1)	130 (0.1)	0.53–0.70 (0.4–0.53)	25–45	35–50	48–65 (at 4%)	1.32

[a] Ref. 29.
[b] To convert N/tex to gf/den, multiply by 11.33; N/tex = kJ/g; N/tex × density (g/cm³) = GPa. To convert GPa to psi, multiply by 145,000.
[c] Parenthetical number denotes the amount of initial extension.

Table 2.   Moisture Regain at 65% rh and 21°C of Commercial Fibers[a]

Fiber	Moisture regain, wt %
polyethylene	<0.1
polypropylene	<0.1
poly(ethylene terephthalate)	0.7
polyacrylonitrile	2.1
nylon-6,6	4.5
acetate, cellulose	6.5
cotton	8.0
rayon	11.0
wool	18.0

[a] Ref. 6.

Table 3.   Thermal Transition of Fibers[a]

Fiber	CAS Registry No.	Glass-transition temperature (T_g), °C[b]	Melting temperature (T_m), °C	Softening temperature (T_s), °C[c]	Thermal-degradation temperature, (T_d), °C
polyethylene (high density)	[9002-88-4]	−120	130	125	
polypropylene (isotactic)	[25085-53-4]	−18	170	165	290
poly(1-butene) (isotactic)	[25036-29-7]	−25	128		
poly(3-methyl-1-butene)	[26073-14-0]		315		
poly(4-methyl-1-pentene) (isotactic)	[24979-98-4]	18	250	244	
poly(ethylene terephthalate)	[25038-59-9]	70	265	235	400
nylon 6,6	[32131-17-2]	50	265	248	360

[a] Refs. 2, 7–8.
[b] For unoriented polymer.
[c] Or sticking temperature.

In polypropylene, thermal degradation is more difficult to control owing to the presence of tertiary carbons that are attacked easily. Usually several antioxidants, which are presumed to act synergistically, are necessary. Hindered phenols and cresols that are mixed with such compounds as dilauryl or distearyl thiodiproprionate are typical of the types of compounds used. An extremely large number of such mixtures is available.

Ultraviolet Degradation.   Sunlight affects olefin fibers, and linear polyethylene is somewhat less susceptible than other polyolefins. The effect is similar to that for thermal oxidation, ie, chain scission and molecular weight degradation, although the mechanism of photooxidation is different. At room temperature, there is no apparent influence of fiber size, molecular orientation, or crystallinity on the rate of photooxidation of polypropylene (10). However, the process does occur preferentially on surfaces, so that fibers and films tend to degrade more rapidly than thick objects. Classic uv stabilizers (qv) include finely divided carbon black and a number of the pigments that are used to color olefin fibers and which absorb in the uv region. An important property of these additives is that they are not lost by migration to the fiber surface during production and use. Since the rate of loss is related to diffusion of the additive through the polymer, fiber size, molecular orientation, and fiber morphology affect the practical stabilization of the fiber. The relative stabilization performance of several additives, which either absorb uv light or trap the radicals produced in the fiber during the photooxidative process, is given in Table 4 (11) (see also Hydrocarbon oxidation).

Flammability. One fundamental measure of the flammability of polymers is the oxygen index which defines the minimum oxygen concentration necessary to support combustion. Polyolefins are more combustible than many other common polymers (12). As a practical matter, however, the oxygen index does not predict accurately the fire performance of textiles, since the surface-to-volume ratio of the textile, the orientation in the test procedure, and the specimen mass and density interact to produce varied results in actual fire tests. If improved fire retardancy is required, additives containing antimony oxide or halogenated compounds are quite effective, although they often compromise other fiber properties (see Flame retardants; Flame retardants for textiles).

Dyeing. Olefin fibers are inherently difficult to dye, because there are no sites for the specific attraction of dye molecules, ie, no hydrogen-bonding or ionic groups, and dyeing only can take place by virtue of weak dye–fiber van der Waals forces. As a result, the most common method of coloring olefin fibers is to add pigments to the melt before extrusion. The pigments usually are predispersed in a polypropylene or low density polyethylene master batch to obtain efficient dispersion. The principal pigment types are titanium dioxide (white), cadmium salts (reds, oranges, yellows),

Table 4. Light Stabilization of Polypropylene Multifilaments[a]

Light stabilizer	Weatherometer hours at T_{50}[b], h
0.25% CR 141[c]	720
0.25% CR 144[d]	830
0.50% CR 141[c]	1200
0.50% CR 144[d]	1100
0.50% UV-1[e]	410
0.50% UV-2[f]	370
0.50% Ni-2[g]	380

[a] Ref. 11.

[b] The no. of hours in an artificial weathering instrument at which the original multifilament sample has lost 50% of its tensile strength.

[c] CR 141: mol wt = 1600, poly(2,2,6,6-tetramethyl-4-piperidylamino-1,3,5-triazine).

[d] CR 144: mol wt = 2500–3000, poly(2,2,6,6-tetramethyl-4-piperidylamino-1,3,5-triazine).

[e] UV-1 = 2-hydroxy-4-*n*-octyloxybenzophenone.

[f] UV-2 = 2-(2'-hydroxy-3',5'-*di-tert*-butylphenyl)-5-chlorobenzotriazole.

[g] Ni-2 = Ni-bis[O-monoethyl(3,5-*di-tert*-butyl-4-hydroxybenzyl)]phosphonate.

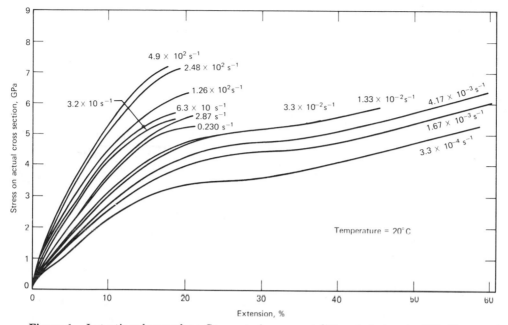

Figure 1. Isotactic polypropylene. Stress–strain curves at different strain rates (18). To convert GPa to psi, multiply by 145,000.

iron oxide (brown), and a variety of organic pigments (see Pigments) (13). Coloration through pigmentation is inflexible, requires continuous production changes, and involves large inventories. Two principal approaches toward dyeability have been taken, ie, to invent disperse dyes with high solubility in the fiber and to modify the fiber by incorporating polar dye sites in it (see Dyes and dye intermediates). Efforts to produce soluble disperse dyes have not been very successful, owing to poor lightfastness, poor fastness to dry cleaning, and generally low color build-up (14). On the other hand, fiber modification has been more successful and several are available commercially. Polypropylene fibers that can be acid-dyed contain poly(vinylpyridine) dispersions or vinylpyridine units grafted onto the polypropylene chain. A wide variety of acid dyes of suitable fastness is available, but they must be transported in water through the fiber, which is difficult (14). Nickel-modified fiber contains a small amount of an organic nickel compound which is capable of chelating and forming a metal dye complex with certain mordant disperse dyes (15–16). Each of these modified-fiber types is more costly than pigmented polypropylene fiber. An important use of the nickel-modified fibers and of those that can be acid-dyed is in printed carpets (14,17).

Stress–Strain Relationships. As is common for all polymeric fibers, the stress–strain curves for olefin fibers depend on strain rate, temperature, molecular weight, and fiber morphology, especially molecular orientation and crystallinity. The effect of varying strain rate from 490 s^{-1} to 0.00033 s^{-1} is represented graphically in Figure 1 for an isotactic polypropylene monofilament of average crystallinity, ie, 55%, and relatively high orientation (draw ratio ca 5.5:1, birefringence 350 $\times 10^{-4}$) at 20°C (18). For the same monofilament, the dependence of the stress–strain curve on temperature is represented in Figure 2 (18). There are similar effects on the breaking conditions of the fiber, ie, tenacity or stress at break, energy to rupture, and extension at break. The

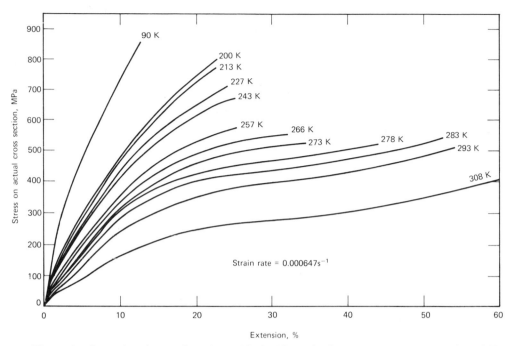

Figure 2.   Isotactic polypropylene. At 308 K, the fiber is broken at 74.8% extension and 484 MPa stress. Stress–strain curves at different temperatures (18). To convert MPa to psi, multiply by 145.

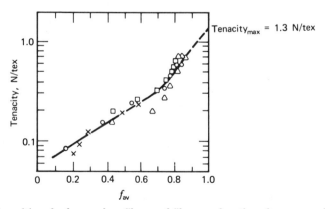

Figure 3.   Tenacities of polypropylene fibers and films as a function of average molecular orientation f_{av}. O, film drawn at 135°C; ×, film drawn at 110°C; □, fiber drawn at 90°C, △, heat-set fiber (19). To convert N/tex to gf/den, multiply by 11.33.

abrupt changes in breaking properties at 280 K, and at a strain rate of 10^{-3}–10^{-2} s^{-1}, are glass-transition effects. The effect of average molecular orientation on tenacity at 25°C and 300 s^{-1} is shown in Figure 3 (19), and that of crystallinity on tenacity is shown in ref. 20. In general, orientation constitutes the most important structural effect on breaking conditions.

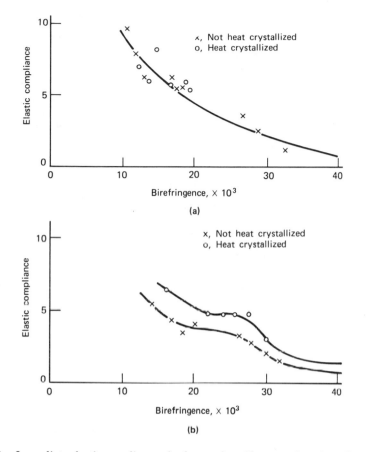

Figure 4. Immediate elastic compliance of polypropylene fibers as a function of birefringence. (**a**) Low molecular weight fiber; (**b**) high molecule weight fiber (21).

Creep and Stress Relaxation. Olefin fibers exhibit creep, ie, time-dependent deformation, of which only a portion is recoverable, under load; and the inverse phenomenon, ie, stress relaxation, which is the spontaneous relief of internal stress and occasionally is accompanied by small spontaneous length changes. Both of these behaviors depend on fiber molecular weight, molecular orientation, and crystallinity. Examples of these effects for polypropylene fibers are shown in Figure 4, which shows the relations between immediate elastic compliance, ie, a quantity deduced from a series of creep curves that are extrapolated to zero stress and 0.2 min, and birefringence, ie, a measure of molecular orientation, for low and high molecular weight fibers, both before and after heat crystallization (21). High orientation, high molecular weight fibers undergo less creep than low orientation, low molecular weight fibers; the effect of heat crystallization is less pronounced and depends on fiber molecular weight.

Elastic Recovery. The elastic recovery as a function of temperature for polyethylene and polypropylene fibers is shown in Figure 5 (7). Elastic recovery or resilience is the fractional length of the fiber recovered from the initial extension. Work recovery is the work energy recovered from that required for the initial extension, from the areas under the extension and recovery stress–strain curves; the product of elastic recovery

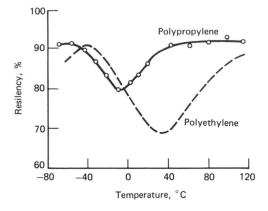

Figure 5. Resiliency of polypropylene and polyethylene multifilament yarns as a function of temperature (7).

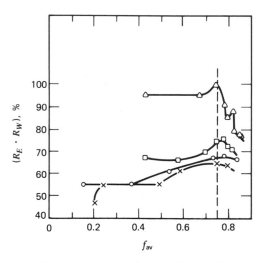

Figure 6. Recovery factor ($R_E \cdot R_W$) for polypropylene fibers and films as a function of average molecular orientation f_{AV}. O, film drawn at 135°; ×, film drawn at 110°C; □, fiber drawn at 90°C; △, heat-set fiber (19).

R_E and work recovery R_W corresponds reasonably with practical tests of resiliency, ie, in carpets. Figure 6 is an illustration of the relation between this product and the average molecular orientation for 5% extension of a series of polypropylene fibers and films (19). The dashed line in Figure 6 indicates the level of average molecular orientation for which all of these fibers and films show maximum recovery (f_{av} = 0.76) independent of fabrication history. Heat effects and, therefore, crystallinity or other morphological effects also are evident.

Dynamic Mechanical Properties. The response of a fiber to cyclic deformation is determined by the temperature and frequency of the deformation and by the structural characteristics of the molecule and fiber morphology. Both components of the complex modulus may be determined from the measure of the energy storage

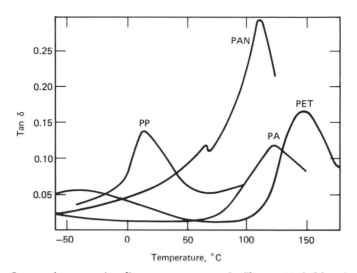

Figure 7. Loss-angle tangent (tan δ) versus temperature for fibers at 0% rh. Mean·frequency at 20°C: polypropylene (PP), 280 Hz; polyacrylonitrile (PAN), 278 Hz; nylon-6,6 (PA), 190 Hz; poly(ethylene terephthalate) (PET), 286 Hz (22).

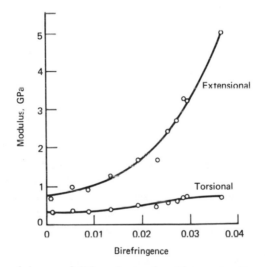

Figure 8. Tensile and shear moduli for polypropylene fibers as functions of birefringence (24). To convert GPa to psi, multiply by 145,000.

capacity of the fiber, ie, the tensile storage modulus E', and the tendency of the fiber to dissipate energy, ie, the tensile loss modulus E''. Energy dissipation also is determined by the ratio E''/E', which is related to tan δ or the phase angle between the stress and the strain. For low to medium values of tan δ, E' is the same as the usual Young's modulus of a material. Dynamic properties of fibers may be obtained from bending, torsion, or tensile experiments. The temperature dependences of tan δ for polypropylene (PP), polyacrylonitrile (PAN), nylon-6,6 (PA), and poly(ethylene terephthalate) (PET) fibers from bending data are compared in Figure 7 (22). The principal maximum in each curve is an approximate locator at the measurement frequency of the fiber glass-transition temperature at which fractional energy dissipation is maximized; ie, low values of tan δ confer good elasticity. The maximum is near room temperature for polypropylene and is significant for carpet resilience. Tan δ varies with varying mechanical and thermal history (23). The relationship of tensile and shear moduli as functions of birefringence, which largely is a function of molecular orientation, for polypropylene fibers is shown in Figure 8; the figure demonstrates the highly aniso-tropic nature of oriented fibers (24).

A detailed quantitative analysis of the relations among process conditions, fiber structure, and physical properties of polypropylene fibers has been published (19). However, correlations between processing conditions and fiber structure and between fiber structure and fiber physical properties are not fully understood.

Manufacture and Processing

Olefin fibers are fabricated commercially by one of several modifications of the melt-extrusion technique, of which the fundamental elements are the continuous expulsion of molten polymer through a die, ie, spinneret; the solidification of the ex-trudate by heat transfer to the surrounding fluid medium; and the winding of the solid extrudate onto packages. Further processing may include drawing the fiber to as much as six times its original length and a variety of heat treatments to relieve thermal stresses within the fiber. Texturizing processes, which are combinations of deforma-tions and heat treatments, also may be applied.

Extrusion. A generalized melt-spinning process for olefin fibers is depicted in Figure 9. Polymer pellets simultaneously are melted and forced through the system by a heated melting screw. The spinning pump meters the fluid polymer through a filter system to the spinneret where the fluid polymer is extruded under pressure through holes that, depending on the desired product, vary widely in size, shape, and number. For continuous-filament production, a spinneret may contain twenty to several hundred holes, ie, capillaries, measuring 0.3–0.5 mm in diameter. For staple-fiber production, a spinneret may contain as many as 500 holes in the same size range as for continuous-filament processes. Monofilaments are extruded from spinnerets with only a few relatively large holes, eg, 2.5 mm in diameter. The spinneret holes usually are round, but they may have other cross sections, eg, multilobal or rectangular. The length of the capillary in relation to its diameter must be appropriate to the melt viscosity of the polymer at the extrusion temperature and to the extrusion rate. The filters above the spinneret are designed to remove particles that otherwise would clog the capillaries. Both sand packs and metal screens may be used as filters.

Polyolefins are extruded at 100–150°C above their melting points. Thus, extrusion temperatures for both polyethylene and polypropylene are 225–300°C and are not very different from extrusion temperatures for higher melting condensation polymers, eg, nylon-6,6 and poly(ethylene terephthalate). The high molecular weight of polyo-lefins, ie, 200,000, requires temperatures substantially higher than the melting tem-perature to achieve sufficiently low melt viscosity for extrusion. Even so polyolefins typically must be extruded at higher melt viscosities than those associated with the extrusion of condensation polymers. Typical values of the shear-dependent viscosities of polypropylene and poly(ethylene terephthalate) are compared in Figure 10 and their temperature-dependent viscosities are compared in Figure 11. Typically, polyolefins

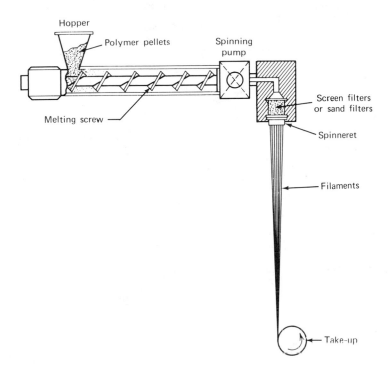

Figure 9.   Schematic melt-spinning process (25).

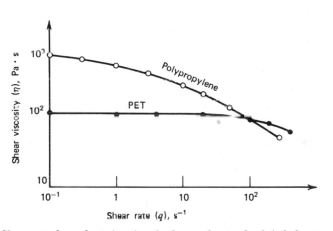

Figure 10.   Shear rate-dependent viscosity of polypropylene and poly(ethylene terephthalate) melts (25). To convert Pa·s to poise, multiply by 10.

are extruded at shear viscosities up to ≥ 1 kPa·s (10^4 P), which require higher extruder pressures than for condensation polymer extrusion. Typical shear rates are ca 10^3–10^4 s^{-1}; at typical extrusion temperatures, the upper portion of this range is not far from the melt-fracture point at which the threadline becomes unstable and produces a highly distorted and sometimes corkscrew-shaped fiber. Both distortion and die swell are consequences of the high degree of viscoelastic memory, which is characteristic of

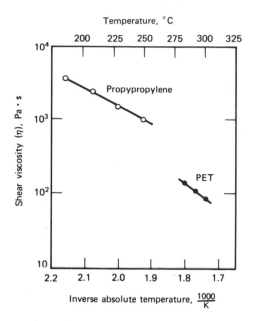

Figure 11. Temperature-dependent viscosity of polypropylene and poly(ethylene terephthalate) melts (25). To convert Pa·s to poise, multiply by 10.

molten polyolefins. Die swell is the formation of a bulge of molten polymer as it leaves the die or capillary; it is exaggerated by high melt viscosities and/or high shear rates, ie, high output rates. One consequence of die swell in polyolefin extrusion is that neither the extruded fiber diameter nor the degree of shear stress-induced molecular orientation in the fiber is influenced to any great extent by the capillary diameter; fiber formation occurs from the molten polymer mass or bulge immediately below the spinneret. Die-swell effects can be reduced by increasing the length-to-diameter ratio of the capillary, reducing the throughput rate, and decreasing the polymer melt viscosity by increasing the temperature or by choosing a different polymer. Minimizing die swell is especially important in the extrusion of fine-denier fibers.

Solidification. At some distance below the spinneret surface, enough heat has been transferred from the semimolten threadline to the surroundings that solidification of the fibers occurs. Because the important fiber-forming polyolefins are crystallizable and have glass-transition temperatures T_g below normal room temperature, solidification not only occurs by crystallization between the melting temperature T_m and T_g, but also can continue after the fiber is wound onto a package. Thus, solidification of these fibers depends both on extrusion conditions and on isothermal postcrystallization.

Values of T_g, T_m, T_{max}, and K, ie, the kinetic crystallizability for several fiber-forming polymers, are listed in Table 5 (26). Maximum crystallization rate is observed in isothermal experiments at T_{max}. The kinetic crystallizability is approximated from the width of the distribution of isothermal crystallization rates as a function of crystallization temperature multiplied by the maximum isothermal crystallization rate in the temperature range of appreciable crystallization rate. It characterizes the degree of crystallinity obtained upon cooling the fiber at the unit cooling rate from T_m to T_g

Table 5.   Thermal and Crystallization Characteristics for Several Polymers[a]

Polymer	T_m, °C	T_g, °C	T_{max}, °C	K, °C/s
natural rubber	30	−75	−24	3.4×10^{-3}
isotactic polypropylene	180	−20	65	35
poly(ethylene terephthalate)	267	67	190	1.1
nylon-6	228	45	146	6.8
nylon-6,6	264	45	150	140
isotactic polystyrene	240	100	170	0.16

[a] Ref. 26.

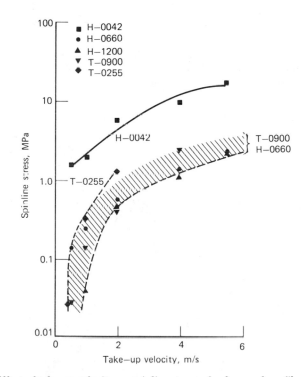

Figure 12.   Effect of take-up velocity on spinline stress of polypropylene fibers. Extrusion temperature, 230°C (30). To convert MPa to psi, multiply by 145. The fibers are composed of polymers of various molecular weights: H-0042, $\overline{M}_w = 3.14 \times 10^5$; H-0660, $\overline{M}_w = 2.74 \times 10^5$; H-1200, $\overline{M}_w = 1.94 \times 10^5$; T-0900, $\overline{M}_w = 2.13 \times 10^5$; and T-0255, $\overline{M}_w = 2.88 \times 10^5$.

(26). As evidenced in the Table 5 data, isotactic polypropylene, like most linear polyolefins, crystallizes relatively rapidly compared with most other polymers. Severe quenching may be used to prevent polyolefin fibers from crystallizing completely during the extrusion process.

One extrusion condition that is important in controlling solidification is the nature of the quenching medium. For example, the high thermal mass of monofilaments requires that an efficient heat-transfer medium be employed if they are to be solidified within a reasonable distance of the spinneret; thus, they are extruded into water baths. Water quenching also is used in compact spinning plants, which are designed to operate

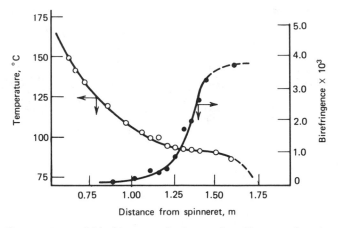

Figure 13. Temperature and birefringence of polypropylene fibers as a function of distance from the spinneret. Extrusion temperature, 230°C; take-up velocity, 0.8 m/s (30).

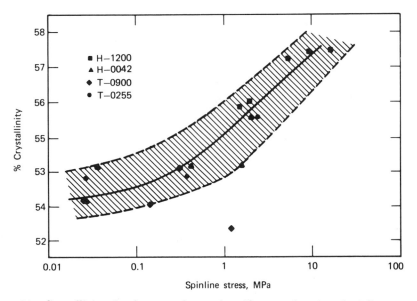

Figure 14. Crystallinity of melt-spun polypropylene fibers as a function of spinline stress (30). To convert MPa to psi, multiply by 145.

in minimum vertical space (27). If less efficient air quenching is used, the quench zone may be 2–10-m long, depending on the process extrusion velocity. Very rapid quenching, eg, in a water bath, leads to an as-spun smectic, paracrystalline, or hexagonal polypropylene fiber structure. The structure is characterized by very low crystallinity in the conventional sense, but it has substantial near-range order, ie, it is not completely amorphous (28). Substantial orientation of individual molecules can exist, however, in this state and this leads to different final fiber properties, which are achieved after additional orientation. The maximum obtainable draw ratio for hexagonal structure fibers is greater at any starting level of crystalline phase orientation

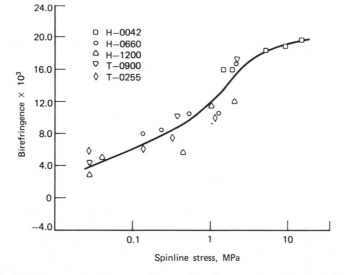

Figure 15. Birefringence of melt spun polypropylene fibers as a function of spinline stress (30). To convert MPa to psi, multiply by 145.

than for the normally obtained, highly crystalline, monoclinic structure (29). The reason for this difference is the mechanics of deforming crystallized monoclinic material during drawing compared with deforming ordered, but not crystallized, quenched structures.

The other extrusion condition that controls solidification is spinline stress, which is determined by the mass throughput at the spinneret and the linear take-up velocity and, therefore, the fiber-diameter reduction after solidification. The relation between spinline stress and take-up velocity depends on extrusion temperature and polymer molecular weight, as well as on spinneret throughput. Typical relationships between spinline stress and take-up velocity to 10 m/s for five polypropylene polymers with weight-average molecular weights of $(1.94-3.14) \times 10^5$ are represented in Figure 12 (30). Since significant spinline stress develops only in the partially solidified portion of the fiber, structure formation, eg, molecular orientation of individual molecular segments and of ordered regions, develops some distance away from the spinneret, as shown in Figure 13 for polypropylene (30). The orientation conditions in the fiber affect the rate and the kind of crystallization that occurs as shown for polypropylene polymers in Figures 14 and 15 (30).

Take-Up. Industrially, take-up velocities vary from ca 1–2 m/s for monofilaments to ca 10–20 m/s for fine-denier yarns. New developments are extending these limits in both directions. Small extrusion plants, the products of which can be changed rapidly with minimum cost, have been developed for polypropylene staple production that is characterized by take-up velocities of ca 3 m/s (27,31). Slowing the take-up velocity also allows in-line incorporation of subsequent drawing and/or texturing steps (32). High speed processes also have been described (33–35). Both crystallinity and birefringence reach a plateau at 33 m/s for isotactic polypropylene that is taken up at speeds as great as 117 m/s (35). Take-up speeds of 67 m/s are used in a polypropylene spunbonded process for producing nonwoven fabric directly from extruded fiber (see

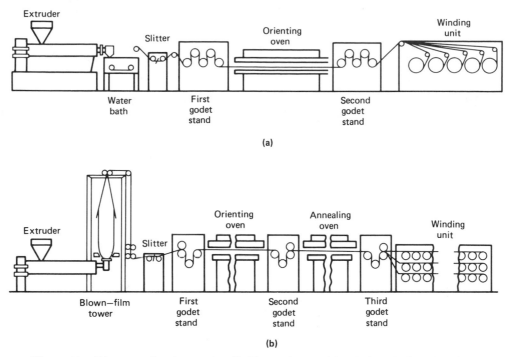

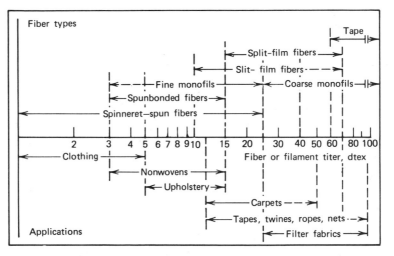

Figure 16.   Diagrams of equipment for slit-film production (6). (**a**) Flat-film line. (**b**) Blown-film line.

Figure 17.   Types and applications of fibers as functions of size (37). To convert dtex to den, multiply by 0.9.

Nonwoven textiles, spunbonded) (33). A continuous spin-draw-texture process, which involves air-jet texturing, can be operated at ca 42 m/s (34). It is particularly suitable for heavy-denier carpet yarns.

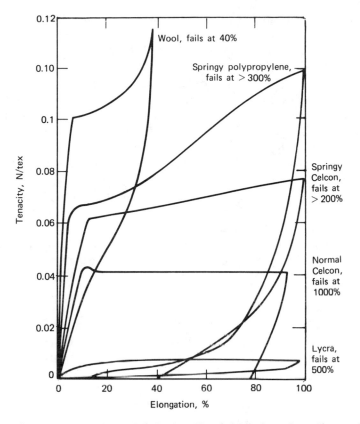

Figure 18. Comparative stress–strain behavior of hand elastic, ie, springy, fibers and normal fibers (39). To convert N/tex to gf/den, multiply by 11.33.

Drawing. Drawing or stretching the as-spun fiber orients the polymer molecules further and usually induces further crystallization. In monofilament processing, drawing usually is accomplished in-line but, in fine-denier processes, it usually is a separate processing step. In either event, it is necessary to raise the fiber temperature above its T_g, to stretch the fiber usually to 3–8 times its original length, and to stabilize the new structure by cooling. Since polyolefin fibers have T_gs that are below room temperature, they can be drawn slowly without heating, ie, they can be cold-drawn. In order to draw at commercially acceptable rates, ie, 1.7–6.7 m/s, they also are heated by being passed over a heated roll or a heated plate. If staple fiber is to be produced, drawing is performed on large bundles of as-spun fibers or tows that may contain up to 500,000 filaments. In such case, drawing usually is accomplished in several stages and is followed by in-line texturing and cutting to staple that is 10–150 mm long.

Texturizing. Conventional texturizing imparts bulk and/or elasticity to polyolefin fibers. The most important processes are the stuffer box process, which is used principally for staple and by which the heated tow is overfed into an enclosed box where crimping occurs and is cut to staple as it is removed; the air-jet process, which involves deforming the yarn by air turbulence in a nozzle and which is capable of high process speeds using heated or unheated air; and the steam-jet process, which is similar to the air-jet process but is characterized by pressurized or superheated steam. Both jet processes are popular for carpet yarns. False-twist processes occasionally are used for the small amount of polyolefin, ie, polypropylene, apparel yarn. Bicomponent yarns, usually of polypropylene/polyethylene, which are self-bulking on the application of heat, have been described (36).

Annealing.   Heat may be used to stabilize the final fiber structure and, therefore, its properties, eg, shrinkage and breaking elongation, by stress relief. Severe annealing treatments may be included in the production of hard elastic fibers. In the case of monofilament and film production, annealing treatments are conducted in-line.

Fibers from Film.   Fibers also are obtained from extruded film that has been mechanically slit and fibrillated. Although, prior to 1950, the general process had been used to obtain narrow tapes of regenerated cellulose, poly(vinyl chloride), and polystyrene, the superior properties of polyethylene and polypropylene have made the production of fibrous products from film nearly unique to these two polyolefins. The starting film may be a flat film that is extruded through a flat die and is either water-bath cooled or chill-roll cooled, or it may be a blown film that is extruded through an annular die and air-cooled. Schematic diagrams for these two processes are shown in Figure 16 (6). After the film is extruded, it is slit to widths of 1–10 mm for textile uses or 15–35 mm for rope production. Slitters essentially are bars that are equipped with razor blades which are set at a desired spacing (37). At the slitting stage, various fibrillation pretreatments may be imposed, eg, film perforation by knife, needle, or pin roller. The slit tapes then are drawn (oriented) in an oven to develop the molecular orientation and crystallinity necessary to produce acceptable properties in the final product. For many fibrillation techniques, a very high draw ratio, eg, 1:11, is necessary. Fibrillation may be followed by annealing primarily to control heat shrinkage. A certain amount of relaxation, to ca 15%, may be allowed in this step. The splitting or slitting of film tapes into interconnected networks of fibers usually is accomplished after tape stretching. Typical processes include purely mechanical shearing with rotating brushes

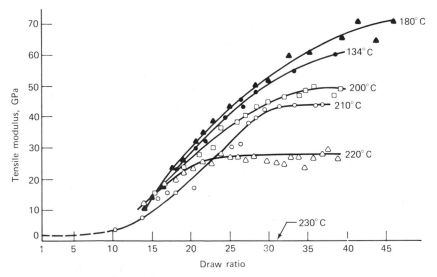

Figure 19.   Tensile moduli of cold-extruded polyethylene fibers as functions of total draw ratio at various crystallization temperatures. Fibers were crystallized at 0.49 GPa (7.1 × 10⁴ psi) and extruded at 120°C and 0.49 GPa (43). To convert GPa to psi, multiply by 145,000.

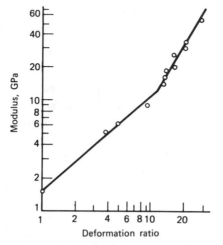

Figure 20. Tensile modulus as a function of deformation ratio for hydrostatically extruded poly-ethylene fibers (41). To convert GPa to psi, multiply by 145,000.

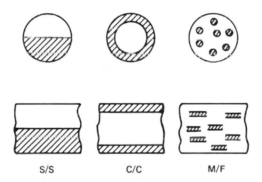

Figure 21. The major types of bicomponent and biconstituent fibers: S/S, side by side; C/C, cover/core; M/F, matrix/fibril (25).

or rubber-coated rollers, bending operations, and twist operations; all lead to largely uncontrolled fibrillation. Controlled fibrillation requires the profiling of the film at the extruder die or immediately thereafter, after which mechanical working is applied. Regular film perforation prior to drawing also tends to control fibrillation (37).

Split-film fibers tend to be made in sizes similar to those of coarse filament yarns or of fine monofilaments. Comparative fiber sizes and applications for spinneret spun fibers, monofilaments, slit-film fibers, and film tape are shown in Figure 17 (37). The major uses for slit-film fibers are in the upholstery, carpet, and industrial areas. Split-film fibers have lower production costs than spinneret fibers of >0.6 tex (>5 den), whereas the reverse is true for fibers of <0.6 tex (<5 den) (37).

The choice of polymer for split-film processes affects not only final mechanical properties, but also the tendency of the fibers to fibrillate uncontrollably. Polypropylenes with melt index values of 0.5–11.0 have been used successfully in a number of processes (38). However, fibrillation tends to be controlled most effectively by blending polypropylene and polyethylene. In addition, polystyrene, low density polyethylene, and various polyamides have been blended with polypropylene to alter physical properties, eg, tenacity, elongation, or fibrillation tendency (38).

Novel Olefin Fibers

Hard Elastic Fibers. Hard elastic fibers include but are not restricted to several polyolefins. They are characterized by reasonably high levels of tensile modulus and unusually high levels of elastic recovery from large deformations (39). Typically, 90% or more recovery from 50% or greater elongation is observed; this is contrasted with the usual case of 60–100% recovery from 2–5% elongation (see Table 1).

Typical process conditions include extrusion under conditions suitable to obtaining high molecular orientation in the as-spun fiber, minimal secondary drawing, and annealing treatment which sometimes is severe. This processing produces conditions for crystallization under stress that lead to a row-nucleated fiber morphology which, on annealing, becomes a lamellar structure. The lamellar structure is thought to allow bowing modes during deformation either through lamellar bending or shearing of individual molecular chains within the lamellae. The mechanism of elastic recovery is largely energetic, ie, driven by a minimization of system internal energy, rather than being entropic as is the usual case for recovery from deformation by the tendency of individual molecules to adopt the most random conformation. Among the polyolefins that have been reported to exhibit this behavior are polyethylene, polypropylene, poly(3-methyl-1-butene), and poly(4-methyl-1-pentene). Hard elastic fibers have been investigated since the mid-1960s. They are characterized by high extensibility with good recovery combined with at least a normal modulus that is suitable for a variety of textile applications.

Potential uses include those in stretch fabrics with substantial elastic force, fabrics having variable porosity, stretchable sewing thread, material for stuffing or padding cushions, nonwoven stretchable fabrics, and cardiovascular prosthetic devices (40). The stress–strain curves of several hard elastic fibers are compared with wool and Lycra in Figure 18 (see also Elastomers, synthetic) (39).

High Modulus Fibers. Extremely high values of tensile modulus and tensile strength may be obtained, particularly in polyethylene and polypropylene fibers, by cold drawing of the polymer at or well below the crystalline melting temperature; by cold extrusion of a plug of solid polymer through an orifice under pressure; or by hydrostatic extrusion, in which a solid plug is surrounded by a pressure-transmitting fluid that exerts a hydrostatic pressure on the plug as it is forced through an orifice. Although none of these processes are used commercially, they are subjects of intense research interest. The molecular models proposed for these solid-state deformation processes generally are based on the breakup of existing crystalline lamellae, their orientation in the deformation direction, and the pulling out of folded-chain molecules to form tie molecules within and between the disrupted lamellae. Fibril formation is observed for sufficiently large deformations; fibril formation leads to the extended chain molecular conformation that is responsible for the unusually high tensile properties of these materials (41). The solid-state deformation of polypropylene monofilaments has been studied extensively (42). Moduli as high as ca 11.5 N/tex (130 gf/den) and tenacities to ca 1.1 N/tex (13 gf/den) are possible and are much higher than the 1.8–5.3-N/tex (20–60-gf/den) and 0.3–0.6-N/tex (3.5–7.0-gf/den) moduli and tenacities, respectively, observed for normal monofilaments (see Table 1) (42). Cold

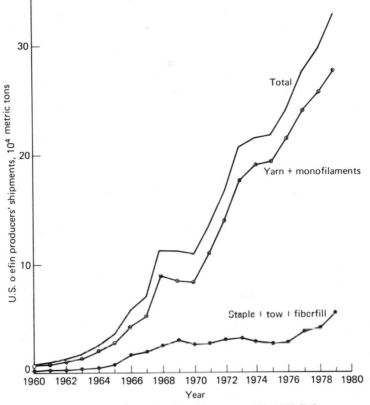

Figure 22. U.S. olefin-fiber shipments, 1960–1979 (54).

Table 6.   Polypropylene Consumption by Fiber Type, Thousand Metric Tons[a]

Type	Western Europe 1974	Western Europe 1980		U.S. 1974	U.S. 1980		Japan 1974	Japan 1980	
film fibers	190	350		83			44	111	
monofilament	26		65		83	642	3		3
spun bondeds	na								
bulked continuous-filament yarn		9	27	12		61	9		10
continuous-filament yarn				30					
staple, tow	45	78		26	45		41	42	
Total	*270*	*520*		*234*	*748*		*97*	*166*	

[a] Ref. 55.

extrusion of high density polyethylene, produces fibers for which the draw-ratio dependence of tensile modulus is very sensitive to the crystallization temperature that is imposed under pressure prior to extrusion. The draw ratios of predominantly extended-chain morphologies, which are characterized by crystallization temperatures of 134–180°C and which result in tensile moduli of 60–70 GPa ((8.7–10.2) × 10⁶ psi), are compared with those of predominantly folded-chain morphologies which are

Table 7. U.S. Mill-Fiber Consumption and Ten-Year Consumption Growth Averaged over Ten Years [a]

Type	Consumption, 10^3 t				Ten-year consumption growth rate, %		
	1957	1967	1977	1987 (estd)	1957–1967	1967–1977	1977–1987 (estd)
acrylic, modacrylic	381	1,932	2,994	3,561	17.6	4.5	1.8
nylon	1,433	5,076	10,678	15,014	13.5	7.9	3.4
olefin and others	91	857	3,085	5,511	25.5	16.6	6.0
polyester	354	3,193	16,370	28,282	25.2	18.2	5.6
rayon and acetate	5,479	6,963	3,992	3,266	2.4	−5.3	−2.0
textile glass	440	1,379	3,588	5,058	12.0	10.1	4.0
cotton	18,325	20,022	14,397	9,820	0.9	−3.2	−3.8
wool	1,674	1,447	608	454	−1.4	−8.2	−3.8
Total	*28,177*	*40,869*	*55,712*	*70,966*			
annual growth rate of fiber consumption (av)					3.8	3.2	2.4

[a] Ref. 4.

Table 8. Cost Summary for the Principal Synthetic Staple Fibers, $/kg [a]

Type	Cost in 1974	Cost in 1980	
		1974 dollars	1980 dollars
polyester	0.24	0.27	0.39
nylon-6,6	0.38	0.44	0.62
nylon 6	0.36	0.42	0.59
acrylics	0.29	0.32	0.46
polypropylene	0.22	0.25	0.35

[a] Ref. 25.

Table 9. Estimated Relative Fiber and Fabric Costs

Type	Density, g/cm^3	Fiber unit cost, $/kg	Relative fiber cost	Relative fabric cost
polyester	1.38	0.24	1.09	1.66
nylon-6,6	1.13	0.38	1.73	2.20
acrylics	1.16	0.29	1.32	1.72
polypropylene	0.90	0.22	1.00	1.00

characterized by crystallization temperatures of 200–220°C and which correspond to substantially lower tensile moduli are shown in Figure 19 (43). Not only do the fiber morphologies differ but the fiber birefringence approaches a value which is greater than the single-crystal limit ($\Delta n_c = 0.059$) for the extended-chain fibers (43). Thus, an essentially perfect alignment of molecular axes with the fiber axis is indicated. In hydrostatic extrusion, the presence of the pressurized fluid lubricates the polymer plug and the die, thereby allowing extrusion at considerably lower pressures than are common for cold extrusion. Extrusion ratios of ca 20 have been achieved with poly-

Table 10.   Relative Energy Costs of Fiber Production Based on Polypropylene[a]

Fiber	Per unit volume	Per unit mass
acrylics	3.2	2.5
polyamides	2.5	2.0
polyester	2.1	1.4
polypropylene	1.0	1.0

[a] Ref. 57.

ethylene with a marked increase in tensile modulus occurring at high ratios. Figure 20 indicates that the upper tensile modulus limit that can be achieved using this technique is comparable with that obtained by cold extrusion (41). The relation between birefringence and deformation ratio also is similar to that observed for cold extrusion. In all of these techniques, the major object is to achieve high alignment and high extension of the molecular chains; with such fiber morphologies, extremely high tensile moduli are possible.

Bicomponent and Biconstituent Fibers.   The three principal types of bicomponent and biconstituent fibers are shown in Figure 21 (26). The side-by-side (S/S) bicomponent consists of two different components which are fused along the fiber axis. Such a fiber is asymmetric with respect to its thermal properties, especially thermal shrinkage, for the proper choice of components and becomes self-crimping when heated. Good adhesion at the interface, which requires limited compatibility of the two components in the melt, also influences proper component choice. Two examples are a self-bulking polypropylene carpet yarn, in which the second component is high density polyethylene, and a polyethylene/polypropylene bicomponent designed for spunbonded nonwoven processes, in which heat not only produces fiber crimp but also bonds the fibers in the fabric at their crossover points through the polyethylene component (44–47).

The cover/core (C/C) bicomponent fiber makes possible the use of two components with radically different properties; however, suitable adhesion at the interface is a problem. Olefin fibers are used in this geometry usually in conjunction with a nonolefin fiber; the principal function of the olefin is as a carrier of some desirable substance in the fiber core. Thus, some antistatic carpet yarns and other conducting fibers utilize a carbon-loaded polyethylene core which is sheathed by a nylon or polyester cover (48–50). A usage that involves only olefin fiber is the reinforcement of polypropylene fibers by including a polypropylene core loaded with glass or asbestos fibers or graphite (51). A novel use for the C/C geometry involves the loading of a hollow polypropylene fiber with a wax-activated carbon mixture which absorbs gases and vapors (52).

Biconstituent fibers are of the matrix/fibril (M/F) geometry; fibrils or ellipsoids of one component are dispersed in another. The general properties of such fibers have been reviewed (53). An example of a commercial polyolefin of this type is a dyeable polypropylene fiber, in which the dispersed phase is a basic polycondensate or poly-(methyl sorbate) (53).

Economic Aspects

Production.   In the United States, olefin-fiber consumption has increased steadily since about 1966. Most of the increase (Fig. 22) has resulted from increased demand for continuous-filament yarn and monofilaments rather than from staple products (54). This demand distinction is not evident in western Europe and Japan, where staple products are more important. Consumption of polypropylene fibers in 1974 and estimated consumption in 1980 by fiber type for western Europe, the United States, and Japan are listed in Table 6 (55).

U.S. olefin-fiber consumption and ten-year average growth rates are compared with those of other fiber types in Table 7 (54). Olefin-fiber usage is greater than that for acrylic and rayon/acetate and is exceeded substantially only by nylon, polyester, and cotton consumption. The average annual growth rate of U.S. olefin-fiber consumption has been high and has been equaled only by that of polyester fibers. This trend is expected to continue even though the growth rate of per capita use of textile fibers is predicted to decline (4).

Relative Costs.   Since propylene is a by-product of the petroleum refining industry, it has had an historical cost advantage over the starting materials of other petroleum-based fibers. When the cost of the various stabilization packages is added, the

Table 11.   Production of Carpet from Polypropylene, 10^6 m^2 [a]

Carpet type	Face fiber	Backing type	U.S.			Western Europe		
			1979	1981	1983	1979	1981	1983
tufted	[b]	woven primary	603.0	658.0	721.0	380.9	383.6	413.0
tufted	[b]	nonwoven primary	72.0	77.0	81.0	72.9	88.3	97.3
tufted	[b]	woven secondary	151.0	395.0	461.0	22.3	41.3	71.8
tufted	staple	[c]	3.0	2.8	2.5	27.1	31.7	34.3
tufted	filament, slit film	[c]	61.0	71.2	74.6	15.0	21.0	27.8
needle-punched	staple	[c]	23.0	28.5	31.0	73.6	86.9	99.7
woven	staple and filament	[c]	1.4	2.6	3.5	5.1	5.1	5.2

[a] Ref. 58. To convert m^2 to yd^2, multiply by 1.20.
[b] All fiber types and yarn constructions.
[c] All backing types.

Table 12.   U.S. Polypropylene Usage in Upholstery, Thousand Metric Tons[a]

Year	Filament	Staple	Total
1969	35.8	13.1	48.9
1971	103.0	26.8	129.8
1973	167.4	27.2	194.6
1975	223.6	25.4	249.0
1977 (estd)	235.9	27.2	263.1

[a] Ref. 56.

cost per kilogram of polypropylene fiber still is lower than that of other synthetic fibers. Table 8 shows the cost per kilogram of polypropylene staple fiber compared with polyester, nylon, and acrylic staple fibers. In projecting 1980 costs, it was assumed that crude oil would cost $75/m^3 ($12/bbl), that an average annual inflation rate of 6% applied, and that a 15% return on investment would be obtained (56); these are grossly underestimated figures. The other cost factor of relevance is covering power, which is determined by the density of the fiber (see Table 1). An estimate of the true relative cost of producing, for example, a fabric with specific cover factor may be obtained from the normalized product of the density (see Table 1) and the unit cost of the fiber (see Table 8, 1974 data). Such an estimate is shown in Table 9. This estimate is slightly less favorable to polypropylene if the cost of pigmentation, ie, $0.02–0.08/kg, is added; pigmentation of polypropylene tends to be more expensive than conventional dyeing of the other fibers.

Energy Costs.   Polypropylene fiber production involves relatively lower consumption of energy than other synthetic fibers because of the former's low density, low melting temperature, and melt spinnability (see Table 10).

Uses

Olefin-fiber use patterns are concentrated in home furnishings and industrial areas, where good fiber-mechanical properties, relative chemical inertness, low moisture absorption and, sometimes, low density contribute to desirable product properties. Olefin fiber use in apparel has been restricted by low melting temperatures, which make the ironing of polyethylene and polypropylene fabrics impossible.

Home Furnishings.   Polypropylene fibers are used in every aspect of carpet construction, ie, from face fiber to primary and secondary backings. The use in 1979 and the predicted use in 1981 and 1983 of polypropylene in tufted carpets for Western Europe and for the United States are compared in Table 11.

Polypropylene's advantages over jute as carpet backing are dimensional stability, minimal moisture absorption, and stable price, availability, and improved product uniformity. Drawbacks include difficulty in dyeing and higher cost than jute fabrics. The use of face-fiber polypropylene has been accelerated in the United States by the introduction of bulked, continuous-filament (BCF) carpet yarns with improved crimp and elasticity. BCF carpet yarns are especially important in the contract carpet market, which is characterized by low, dense loops, where easy cleaning of polypropylene carpets is an advantage (see also Recreational surfaces).

Polypropylene in upholstery is estimated to comprise 30% of the total upholstery fabric production in the United States, and this share is expected to increase to 50% (56). The upholstery market is a major outlet for the dyeable polypropylene variants. The U.S. upholstery usage of polypropylene from 1969 to 1977 is given in Table 12. Other important home furnishing uses for olefin fibers include furniture structural uses, mostly for nonwoven fabrics, and wall-covering materials, including printable polyethylene-spunbonded fabrics.

Industrial.   Industrial uses for polyethylene fibers parallel those for polypropylene. Polyethylene and polypropylene tend to be used nearly interchangeably in many applications. Use of polypropylene monofilaments in 1977 is shown in Table 13. The increased usage of polypropylene fiber of all kinds for industrial applications is shown in Table 14 (59). Polyolefin usage in the industrial area has been growing steadily since 1970. Polyolefin rope and twine compete largely with the hard fibers, ie, jute, hemp and sisal and nylons. In 1976, polyolefins, hard fibers, and nylon comprised 34%, 38%, and 19%, respectively, of the rope and twine markets. The advantages of the polyolefins, ie, strength, lack of moisture effects on strength, and the ability to float on water, are the basis for predictions of their annual growth rate in these markets of 3%/yr (56). Polyolefin fibers accounted for 70% of the bag market in 1976 (59). Rope, twine, and bag applications are the only ones where olefin fibers predominate. However, other important areas include civil-engineering applications, which involves both woven and nonwoven fabric. Conservative growth rates of 35–40%/yr have been estimated (60). Typical applications include asphalt-pavement underlay, railroad-track bed support, drainage systems, and sedimentation control. Woven, needle-punched nonwoven, and spunbonded fabrics are sold for these purposes, with polyester, nylon, polypropylene, and polyethylene dominating. These same types of fabrics are also used largely in filtration systems, which is a rapidly growing market as a result of new environmental air- and water-quality laws.

Table 13.   1977 Industrial Markets for Polypropylene Monofilaments, Thousand Metric Tons [a]

Type	Used in 1977
rope	136
cordage	45
agricultural fabrics	23
filtration fabrics	14
civil-engineering fabrics	14
poultry-house fabrics	5
tarps and pool covers	5
rubber-liner fabrics	2
fish lines and nets	1
screens, bags, sutures, hose coverings	0.5

[a] Ref. 56.

Table 14.   U.S. Domestic Shipments of Polypropylene Fiber for Industrial Use, 10^3 t [a]

Application	1965	1970	1975	1978
bags	5	32	190	195
agricultural and industrial woven fabric	18	32	163	281
webbing	32	41	36	50
rope	45	77	154	177
twine	14	95	218	195
nonwoven fabrics	14	41	82	132
fish lines, nets, sewing thread, book bindings, pot cleaners, and others	14	41	36	177

[a] Ref. 59.

BIBLIOGRAPHY

"Polypropylene Fiber" in *ECT* 2nd ed., Supplement, pp. 808–836, by F. C. Cesare, M. Farber, and G. R. Cuthbertson, Uniroyal, Inc.

1. *The Textile Fiber Products Identification Act*, Public Law 85-897, Washington, D.C., Sept. 1958.
2. V. L. Erlich, "Olefin Fiber" in N. M. Bikales, ed., *Encyclopedia of Polymer Science and Technology*, Vol. 9, Interscience Publishers, a division of John Wiley & Sons, Inc., New York, 1968, pp. 403–440.
3. *Chem. Eng. News*, 10 (Jan. 21, 1980).
4. *Text. Ind.* **143**(2), 84 (1979).
5. N. C. Heimbold, *Text. World*, (insert) (Aug. 1978).
6. I. J. Satterfield, private communication 1, 1976.
7. G. M. Bryant, *Text. Res. J.* **37**, 552 (1967).
8. L. H. Nielsen, *Mechanical Properties of Polymers*, Reinhold Publishing Co., New York, 1962.
9. F. H. Winslow and W. L. Hawkins in R. A. V. Raff and K. W. Doak, eds., *Crystalline Olefin Polymers*, Wiley-Interscience, Part 2, New York, 1964.
10. A. Garton, D. J. Carlsson, and D. M. Wiles, *J. Polym. Sci. Polym. Chem. Ed.* **16**(1), 33 (1978).
11. A. Tozzi, G. Cantatore, and F. Masina, *Text. Res. J.* **48**, 433 (1978).
12. H. Ohe and K. Matsuura, *Text. Res. J.* **45**, 778 (1975).
13. C. Ripke, *Chemiefasern Textilind.* **30/82**, 110 (1980).
14. J. G. Lee, "Dyeing of Polypropylene Fibre Including The Printing of Nickel Modified Polypropylene," *Polypropylene Fibres in Textiles*, International Conference, University of York, Sept. 30 and Oct. 1, 1975, The Plastics and Rubber Institute, London, Eng., 1975.
15. A. F. Turbak, *Text. Res. J.* **37**, 350 (1967).
16. F. Vohwinkel and W. Langhausen, *Chemiefasern Textilind.* **28/80**, 42 (1978).
17. R. L. Baker, *Knitting Times* **46**, 18 (1977).
18. I. H. Hall, *J. Polym. Sci.* **54**, 505 (1961).
19. R. J. Samuels, *J. Macromol. Sci. Phys.* **B4**, 701 (1970).
20. W. C. Sheehan, R. E. Wellman, and T. B. Cole, *Text. Res. J.* **35**, 626 (1965).
21. D. W. Hadley and I. M. Ward, *J. Mech. Phys. Sol.* **13**, 397 (1965).
22. R. Meredith, *Proc. 5th Intl. Congr. Rheology* **1**, 43 (1969).
23. S. Nagou and K. Azuma, *J. Macromol. Sci. Phys.* **B16**, 435 (1979).
24. P. R. Pinnock and I. M. Ward, *Brit. J. Appl. Phys.* **17**, 575 (1966).
25. A. Valvassori, P. Longi, and P. Parrini in E. G. Hancock, ed., *Polypropylene and Its Industrial Derivatives*, Halsted Press, John Wiley & Sons, Inc., New York, 1973.
26. A. Ziabicki, *Fundamentals of Fibre Formation: The Science of Fibre Spinning and Drawing*, John Wiley & Sons, Inc., New York, 1976.
27. A. Schweitzer, *Chemiefasern Textilind.* **28/80**, 41 (1978).
28. R. L. Miller, *Polymer* **1**, 135 (1960).
29. M. Compostella, A. Coen, and F. Bertinotti, *Angew. Chem.* **74**, 618 (1962).
30. H. P. Nadella, H. M. Henson, J. E. Spruiell, and J. L. White, *J. Appl. Polym. Sci.* **21**, 3003 (1977).
31. F. Hensen, *Chemiefasern Textilind.* **28/80**, 36 (1978).
32. *Text. Month*, 60 (Nov. 1977).
33. R. Hoffmeister, *Fiber Prod.* **6**(5), 7 (1978).
34. R. Wiedermann, *Chemiefasern Textilind.* **28/80**, 888 (1978).
35. J. Shimizu, N. Okui, and Y. Imai, *paper presented at the ACS/CSJ Chem. Congress*, Honolulu, Hawaii, 1979, Part I (ISBN 8412-0487-x)CELL208.
36. I. Diacik and R. Simo, *Lenzinger Ber.* **47**, 213 (1979).
37. H. Krassig, *Macromol. Rev.* **12**, 321 (1977).
38. R. A. Gill, *Text. Mon.*, 58 (Mar. 1978).
39. S. L. Cannon, G. B. McKenna, and W. O. Statton, *J. Polym. Sci. Macromol. Rev.* **11**, 209 (1976).
40. Belg. Pat. 650,890 (Jan 23, 1965), (to Celanese Corp.).
41. W. G. Perkins and R. S. Porter, *J. Mater. Sci.* **12**, 2355 (1977).
42. W. C. Sheehan and T. B. Cole, *J. Appl. Polym. Sci.* **8**, 2359 (1964).
43. W. T. Mead and R. S. Porter in R. L. Miller, ed., *Flow-Induced Crystallization in Polymer Systems*, Gordon & Breach, New York, 1977.
44. G. Mackie and S. McMeekin, *Chemiefasern* **26**(1), 38 (1976).
45. G. Mackie and S. McMeekin, *Fiber Prod.* **5**, 48 (1977).
46. *Mod. Text.* **59**(5), 24 (1978).

47. Brit. Pat. 1,446,570 (Aug. 18, 1976), (to Chisso Corp.).
48. *Ind. Text.* (1053), 99 (1976).
49. Brit. Pat. 1,443,337 (July 21, 1976), (to E. I. du Pont de Nemours & Co., Inc.).
50. Brit. Pat. 1,527,192 (Oct. 4, 1978), (to Fiber Industries).
51. A. Rudin, H. L. Krein, and B. F. Hiscock, *J. Appl. Polym. Sci.* **22,** 299 (1978).
52. *Text. World* **128**(12), 65 (1978).
53. S. P. Hersh, *J. Appl. Polym. Sci. Appl. Polym. Symp.* **31,** 37 (1977).
54. *Text. Organon* **51**(2), 13 (1980).
55. E. Welfers, *Chemiefasern Textileind.* **28/80,** 25 (1978).
56. *Mod. Text.* **58**(10), 7 (1977).
57. V. D. Freedland, *Chemiefasern Textilind.* **30/82,** 35 (1980).
58. *Fiber Prod.* **8**(1), 15 (1980).
59. L. E. Seidel, *Text. Ind.* **141**(1), 74 (1977).
60. *Chem. Week*, 32 (June 6, 1979).

D. R. BUCHANAN
North Carolina State University

POLYAMIDES, FIBERS

Polyamide fibers have monomer units joined by amide groups $+\text{CONHRNHCOR}'+_n$ and are prepared from diamines and dicarboxylic acids, or, in the case of $+\text{RCONH}+_n$, from lactams. If R and R′ are aliphatic, alicyclic, or mixtures containing less than 85 wt % aromatic moieties, the polyamides usually are referred to as nylon. If more than 85 wt % of the repeating units are aromatic in structure, the fibers are called aramids (see Aramid fibers).

Nylon was the first significant fiber made from wholly synthetic polymer and probably was first characterized in 1899 (1). In 1929, Carothers initiated research that ultimately resulted in the transformation of a laboratory curiosity to a commercially practical fiber (2–5). This fundamental research resulted in the application for U.S. patents for polyamides made from aliphatic amino acids and lactams and included nylons-6 [25038-54-4], -7 [25035-01-2], -8 [25035-02-3], -9 [25035-03-4], -11 [25035-04-5], and -17 [79392-50-0] (6–7). Subsequent U.S. patents described diamine–diacid components of aliphatic aromatic, alicyclic, and heterocyclic structures in combination with each other or with other aliphatic monomers as one of the polyamide components (8–9). In 1938, a German patent for spinnable polycaproamide was applied for (10).

In the United States, DuPont began production of nylon in October, 1939 (11). The first nylon stockings were marketed in May, 1940. The second nylon-6,6 [32131-17-2] producer in the United States was Chemstrand Corp., now Monsanto Textiles Company, which began production of nylon-6,6 in 1952. Commercial production of nylon-6 in the United States began with the large-scale availability of caprolactam in 1955 (see Polyamides, caprolactam).

In Germany in 1939, I. G. Farbenindustrie, A.G. started production of coarse nylon monofilaments under the trade name Perluran (12). Large-scale production of nylon-6 was started in 1941 at Landsberg (now renamed Gorzow Wielkopolski). After World War II, production of nylon-6 was started by Vereinigte Glanzstoff Fabriken, A.G. at Obernburg, and of nylon-6,6 in 1949 by Deutsche Rhodiaceta, A.G. at Freiburg (13). In Italy, nylon-6,6 was made by Italian Rhodiaceta at Pallanza using a spinning machine that was imported from the United States in 1939 (14). In the United Kingdom, Courtaulds and Imperial Chemical Industries, Ltd. (ICI) formed British Nylon Spinners and began the production of nylon-6,6 in Covington in 1941. This was succeeded by a second plant built at the same location in 1942 (15). Other early producers of polyamide fibers were Canadian Industries Ltd. (DuPont and ICI) in Kingston, Ontario, in 1942; Argentina Ducilo (DuPont) in Buenos Aires in 1948; Société Rhodiaceta in Lyon-Vaise, France (nylon-6,6); Algemene Kunstzijde Unie in Arnhem, Netherlands, in 1949 (nylon-6); and Toyo Rayon in Nagoya, Japan, in 1949 (nylon-6,6). In 1945, I. G. Farbenindustrie, A.G. evaluated more than 3000 polyamide constituents; none showed any important improvement over nylon-6,6 and nylon-6 (16).

In 1950, fourteen plants in ten countries produced 55,000 metric tons of polyamide fibers (17). By 1980, worldwide production had expanded to 3.05×10^6 t and production in the United States was 1.05×10^6 t (18).

Nylon-6 and Nylon-6,6

Properties. One group of properties that often is difficult to measure quantitatively is that associated with esthetics, style, and perceived value. In apparel, these may include fabric softness, drape, any sound from movement, light reflection, and comfort. In carpets, they may include color brightness and firmness or plushness. Another group of properties that generally is easier to measure by conventional methods includes strength, abrasion resistance, recovery from deformation, creep, and resistance to environmental changes.

In some cases, it is difficult to determine whether a specific behavior results primarily from basic chemical and physical structure, or whether it is a consequence of the manufacturing method. For example, thermal stability in tire yarn is related to the heat stabilizer used by the producer, but adhesion and flex fatigue are affected by the fiber finish as well as by chemical structure (see Tire cords).

Comparisons of properties of the different yarns generally are based on tests of typical commercial products. The results may not always be representative of other examples of each nylon type. Usually yarn tex above 15.6 (>140 den) is considered medium and heavy, those below 15.6 tex are rated as light, and 15.6-tex yarns are included in both categories.

Tensile. The main tensile properties of representative nylon-6 and -6,6 yarns are listed in Table 1 and defined below.

Linear density: tex is the weight in grams of 1000 m of yarn; denier is the weight in grams of 9000 m of yarn.

Tenacity is the tensile stress at break. It is expressed as force per unit linear density of unstrained specimen, for example, N/tex or gf/den (gf/den = N/tex × 11.33).

Knot tenacity is the tensile stress required to rupture a single strand of yarn in which an overhead knot is tied in the portion between the testing clamps. It is expressed as force per unit linear density and is an approximate measure of brittleness of the yarn.

Table 1.   Physical Properties of Nylon-6 and -6,6 Continuous-Filament Yarns [a]

Property	Normal tenacity	High tenacity
tenacity, N/tex[b]	0.4–0.6	0.75–0.84
wet	0.36–0.51	0.64–0.7
loop	0.40–0.49	0.57–0.69
knot	0.35–0.42	0.52–0.72
tensile strength, MPa[c]	580–635	855–952
elongation at break (conditioned), %	23–43	12–17
wet	28–41	14–21
tensile modulus (conditioned), N/tex[b]	2.2–3.1	2.9–4.1
average toughness, N/tex[b]	0.07–0.14	0.06–0.14
moisture regain at 21°C, %		
65% rh	4.5	3.3–4.5
95% rh	7.1–7.8	7.8

[a] Conditioned at 65% rh and 21°C.

[b] To convert N/tex to gf/den, multiply by 11 33.

[c] To convert MPa to psi, multiply by 145.

Loop tenacity is the tensile stress required to rupture yarn when one strand of yarn is looped through another and then broken. It is expressed as force per unit linear density. It is an indication of brittleness but is not considered as sensitive as measurement of knot tenacity. Reported values are one-half actual test values.

Breaking strength is the maximum load in g (or lb) required to rupture a fiber.

Tensile strength is the maximum stress or load per unit area in units of Pa such as kPa (kPa/6.895 = psi) or MPa (MPa × 145 = psi). It is calculated as

tensile strength (MPa) = tenacity (N/tex) × specific gravity × 1005

(or psi = gf/den × specific gravity × 12860)

Elongation at break is the increase in sample length during a tensile test. It is expressed as a percentage of original length.

Tensile modulus (Young's modulus, initial modulus, or elastic modulus) is the load required to stretch a specimen of unit cross-section area by a unit amount. It is expressed as the ratio of change in strain in the initial straight-line portion of the stress–strain curve extrapolated to 100% elongation of the sample. Values for the tensile modulus of a polyamide fiber decrease at slower rates of extension, because primary and secondary creep contribute to the observed stress–strain curve. Modulus in semicrystalline fibers is dependent on the degree of crystallinity and orientation; thus, manufacturing and processing conditions affect the modulus. For example, high tenacity yarns drawn to relatively low elongation show a higher modulus than yarn with higher elongation. The temperature and humidity of the testing atmosphere also affect the modulus. *Stretch modulus* is the ratio of change in stress to change in strain in the initial straight-line portion of the stress–strain curve for 1% elongation, eg,

stretch modulus × 100 = tensile modulus

Work-to-break is the work required to rupture the material. It is proportional to the total area under the stress–strain curve.

Breaking toughness is the required work per unit linear density to rupture the material. It usually is calculated by dividing work to break by tex (or denier). *Average toughness* (toughness index) is an estimate of the breaking toughness, assuming a straight-line stress–strain curve, eg, for cotton. It is expressed as work per unit linear density of fiber that would cause rupture. Average toughness is not the measured area under the load-elongation curve.

average toughness = (N/tex or gf/den at break) (% elongation at break)/(2 × 100)

Yield point (elastic limit) is the point on the stress–strain curve where the load and elongation cease being directly proportional, ie, it is the point at which the stress–strain curve deviates from the tangent drawn to the initial straight-line portion of the curve, as in tensile modulus determinations.

Creep is the change in shape of a material while subject to a stress and is time-dependent. Primary creep is the recoverable component of creep, and secondary creep is the nonrecoverable component.

Elasticity is the ability of a material to recover its size and shape after deformation.

Stress–Strain Relationships.   Plots of the stress–strain behavior of representative nylon-6 and -6,6 yarns are shown in Figure 1. Under normal conditions, the stress–strain or load-elongation curves of a well-oriented nylon yarn or fiber show an initial straight-line portion in which the stress is proportional to the strain. This is followed by a yielding of the fiber structure and is indicated by an S-shaped curve: first, concave to the stress axis, then a curvature concave to the strain axis up to the break point. In the initial portion or Hookean region of the curve, ie, after removal of slack, crimp, or twist effects, stress is proportional to the strain. Extrapolation of the Hookean region provides data for calculating the tensile or Young's modulus.

When a nylon fiber is extended, the intermolecular forces that have kept the fiber from retracting oppose its extension. This short-range elasticity produces the initial straight-line portion of the curve. In this region, the fiber can return to its normal length upon removal of the stress. As portions of the chain molecules are extended further, the network of molecules becomes more oriented. The chains straighten and come in closer contact with each other so that more intermolecular bonding, ie, hydrogen bonding and van der Waals forces, becomes effective. Thus, the fiber passes through a yield region beyond which the molecular segments are unable to return to

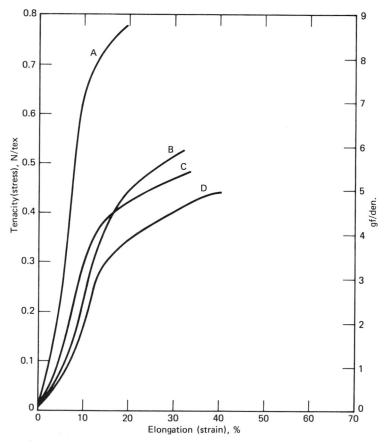

Figure 1.   Stress–strain curves of representative nylon-6 and -6,6 continuous-filament yarns. Yarns conditioned and tested at 65% rh and 21°C (19). A, 93.3-tex (840-den), 136-filament tire yarn; B, 22.2-tex (200-den), 16-filament automotive upholstery yarn; C, 7.8-tex (70-den), 12-filament apparel yarn; D, 117-tex (1050-den), 70-filament yarn for texturing. To convert tex to den, divide by 0.1111.

their original configuration upon release of the stress. The mechanism mainly responsible for the force, which resists strain beyond the yield point, is the tendency of the chain molecules in the amorphous region to assume a random configuration, ie, the tendency toward maximum entropy. The crystalline regions also hinder the movement of the chain segments. The sum of all these forces and the resistance to chain movement equals the total stress required to break the sample. It is unlikely that significant numbers of primary molecular bonds are broken when filaments rupture. Stress–strain curves are described in refs. 20–24.

Stress–strain curves from tests of adjacent portions of a strand generally are similar until just before rupture occurs. The confidence range along the stress–strain curve is no greater than that of the breaking point (25). The coefficient of variation of the breaking strength of nylon-6 filaments is 2–3% and its breaking elongation is 3.5% (25).

The stress–strain properties of nylons depend to a large extent on spinning speed and draw ratio (see Manufacture). More fundamentally, the properties are controlled by the intimate morphology of the fibers, especially the crystalline orientation and the amorphous orientation. Figure 2 shows the relationships among these properties in terms of the tenacity and the draw ratio for nylon-6. Figure 3 shows the relationships among crystalline orientation, amorphous orientation, and tenacity for nylon-6,6 (26).

Creep and Recovery. Time is one of the most important factors in any consideration of the mechanism of elastic recovery. When a load is applied to a fiber, it undergoes instantaneous extension and continues to extend or creep with time. On removal of the load, there is an instantaneous recovery of part of the extension followed by contraction or delayed recovery over a period of time. When recovery is incomplete, the residual is called permanent set (27). This behavior is represented graphically in Figure 4. Creep is the extension with time under an applied load and recovery is the reverse process. Creep and recovery with respect to time for nylon-6 yarns under different initial loads for different total elongations have been measured (28). Initially, recovery is rapid, and most of it occurs within a few minutes after release of the load.

The conditions of extension and recovery, ie, load, rate of extension and recovery, time of extension, etc, affect quantitative comparisons of nylon with other fibers.

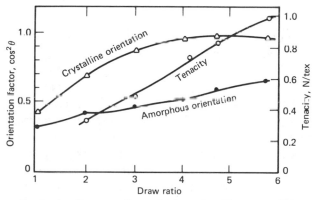

Figure 2. Strength of nylon-6 yarns: effects of orientation. To convert N/tex to gf/den, multiply by 11.33.

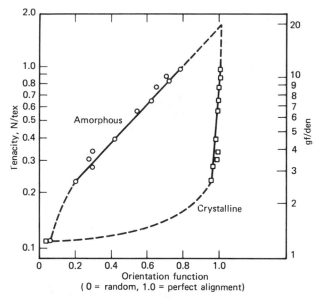

Figure 3. Nylon-6,6 yarn tenacity versus orientation. □ = f_c (crystalline orientation), σ = ca 0.004; ○ = f_{am} (amorphous orientation), error bars ±σ = ca 0.02. Max tenacity = Ca 1.8 N/Tex (20 f/den). To convert N/tex to gf/den, multiply by 11.33.

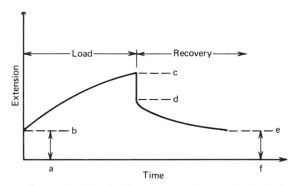

Figure 4. Creep under constant load and recovery under zero load: a–b, instantaneous extension; b–c, creep; c–d, instantaneous recovery; d–e, delayed recovery; and e–f, permanent set.

Contrasted with other fibers, however, the nylon yarns have an outstanding degree of elasticity and recover well from high loads and extensions. Nylon shows the best recovery properties of any of numerous fibers tested (29–30).

Elastic recovery is the ability of a material to regain its original form after being stretched. It enables fabrication of products that maintain their original shape or conform to specific contours of the body. The outstanding elastic recovery of the nylons is responsible for the cling or fit in women's hosiery and for their extensive market acceptance. Relaxation is an effect complementary to creep. In one case and as illustrated in Figure 5, an instantaneous stress occurs when a nylon fiber is stretched, but gradually decreases as time passes (27). The reduction of stress when a fiber shrinks is called relaxation. The elastic recovery, creep, and relaxation of nylon and other fibers is discussed at length in ref. 31.

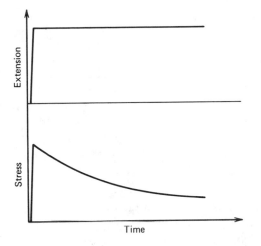

Figure 5. Relaxation of stress under constant extension.

Staple. The use of nylon staple rather than nylon continuous filament stems from a need to satisfy the demand for a soft and warm staple hand or sensation when fabric is touched with the hand. Best performance in many types of fabrics usually requires blends of fibers. Since intimate blending only can be achieved in staple form, the combining of the blend components takes place early in the textile spinning process, ie, either before or immediately after carding. Therefore, the properties of nylon staple must be compatible with each of the other blend components, and the fiber must also be adaptable to processing on the respective spinning systems. The outstanding contributions of nylon staple to carpet, apparel, and upholstery fabrics are its abrasion resistance and its low moisture absorption. The former property increases the wear life of the textile structure, the latter contributes to faster drying.

The strength of nylon continuous filament, which makes it so outstanding in industrial applications, eg, tire cord and webbing, is not desirable in the staple used in apparel. High fiber tenacity in conjunction with high abrasion resistance could result in unsightly pill formations and fabric-structure distortions. Furthermore, high tenacity and the corresponding low elongation do not always match natural-fiber properties in the respective blends. Nylon staple fiber is produced so that tenacities and extensions normally are 0.3–0.4 N/tex (3–5 gf/den) and 50–100% extension, respectively. Elongation and moduli are more important than strength considerations, since unmatched strain properties of blend components can easily overload one component to the detriment of the blend as a whole. The nylon fiber-blend component, when used in the proper blend ratio, matches and enhances the strength of wool, cotton, and rayon fibers.

Most other fiber properties of nylon staple differ little from those of the continuous filament. There is little difference in the property characteristics between nylon-6 and -6,6; distinctions resulting from degree of crimp, finish, heat set, and other properties of products furnished by producers of one type often are greater than those between the two types of nylons. The multiplicity of nylon blends, processing systems, and uses requires a large variety of staple types, which are classified below according to properties:

fiber tex	0.17–2.2 (1.5–20 den)
staple length	cotton system, 4 cm
	worsted system, up to 12 cm
	special applications, up to 15 cm
fiber cross section	round and modified
luster	bright and semidull
crimp	crimped and noncrimped
heat set	heat set and nonheat set

Thermal and Moisture. The thermal behavior of nylon fiber and the interrelated effects of moisture have a pronounced influence on the physical properties of the yarn and its products. They are of basic importance in fiber manufacture, in converting the yarn into fabric, and in ultimate use.

Thermal properties of fibers have not been extensively investigated, because much of the practical thermal behavior of textiles is strongly influenced by other factors. For example, temperature and heat changes are substantially affected by absorbed moisture, and thermal expansion of a nylon filament is much less than is its swelling caused by moisture. Nylon-6 is lower in melting point and in some other thermal properties than nylon-6,6 is (see Table 2).

The change in the moisture regain with relative humidity for nylon-6 and other fibers is shown in Figure 6. The values for moisture regain are not absolute for a particular fiber but vary, eg, with degree of orientation and extractable content. Some fibers, eg, nylon-6 and -6,6, have slightly different moisture regains, depending on whether equilibrium is established from a wet or dry state. The absorptions and desorptions of nylon-6 and wool are compared in Figure 7.

Table 2. Polyamide Temperature Index, °C[a]

	Temperature, °C	
Property	Nylon-6	Nylon-6,6
melting point	215; 220[b]	250
zero strength[c]	232[d]	240[d]
maximum setting temperature	190	225
softening point	170	235
starts to become plastic	160	220
critical temperature		
in air	163[e]	158[e,f]
	93[g]	130[g]
in steam	137	140
maximum ironing temperature	150	180
optimum setting (steam)	128	130
maximum wash temperature		
set by saturated steam	60	71
set by dry state	30	60

[a] Data from ref. 32, unless otherwise noted.
[b] Ref. 33.
[c] Heat-stabilized tire yarn, 1.6-kg load.
[d] Ref. 34.
[e] Ref. 35.
[f] Heat-stabilized tire yarn.
[g] Unstabilized yarn.

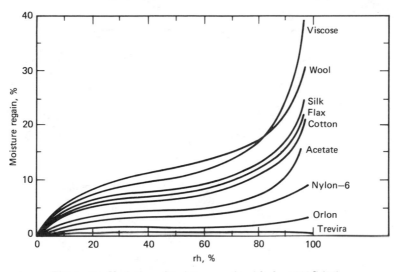

Figure 6. Variation of moisture regain with rh at 20°C (36).

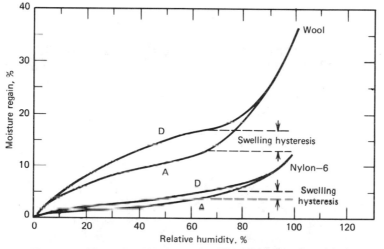

Figure 7. Absorption (A), desorption (D) of nylon-6 and wool (36).

Thermal and light-resistant properties are to a large extent related to the specific stabilizing substances used and to certain aspects of the processes by which the yarns are made. These vary from producer to producer and are changed from time to time by each producer. The thermal stability of nylon-6 and of nylon-6,6 are shown in Figure 8 (37).

Light and Heat Stabilization. *Light.* In the presence of light, titanium dioxide, which normally is used as a delusterant, reacts with oxygen to form peroxide, autocatalytically, and the peroxide degenerates polyamides in the absence of stabilizers, eg, manganese salts (see Heat stabilizers; Uv stabilizers). Additives, eg, hypophosphorous acids, phosphites, phosphates, etc, also are used (38).

Heat and Light. Water, carbon dioxide, and ammonia are the three principal gaseous products of nylon-6 pyrolysis (39–40). The preferred stabilizers in nondelustered fibers from degradation by heat and light are copper salts (41). The salts normally are added prior to polymerization and typically are present in filamentary

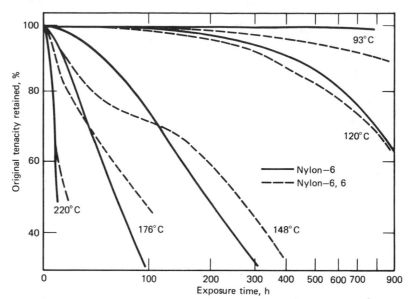

Figure 8. Effect of temperature and exposure time upon retained tenacity of nylon-6 and -6,6 (37).

yarn at 45–60 ppm Cu. The marked effect of copper salts on nylon-6 and -6,6 stabilization is illustrated in Figure 9 (42). Several organic or inorganic materials act as synergistic costabilizers with copper salts, eg, alkali metal iodides, stannous salts, 2-amino-3,5-diiodobenzoic acid, 2-hydroxybenzothiazole, hydroxybenzimidazoles, and 2-mercaptobenzomethylthiazole (38,43–46).

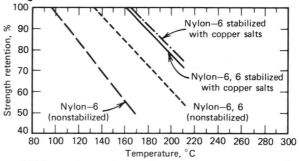

Figure 9. Effect of 24-h exposure to air at various temperatures on strength retention in stabilized and unstabilized polyamide fibers.

Stabilization Against High Temperature Strength Loss at High Loads. One of the significant factors that has made it possible to produce a nylon-6 tire yarn is the ability to increase the breaking temperature under high loads. 9,9-Dialkyldihydroacridine compounds are employed in polyamide stabilization at high temperature and high loads (47). These stabilizers are added at 0.4–1.0 wt % monomer before or

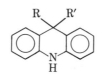

9,9-dialkyldihydroacridine

during polymerization. Certain Schiff bases also impart good thermal-load-index properties (48).

Stabilization Against Heat Disorientation in Liquids. The disorientation of polyamides as vulcanization temperatures of tires are raised becomes an increasingly significant problem. This disorientation is most pronounced during the time in which tire yarns are exposed to moisture at low load during vulcanization or when pressure is released after vulcanization. The problem can be alleviated partially by maintaining some degree of tension on the tire cord at the end of the pressure release. It also can be avoided partially by minimizing the moisture content of tire yarns prior to vulcanization.

Stabilization Against Hot–Wet Degradation. Undrawn nylon filaments of either nylon-6,6 or nylon-6 are susceptible to degradation in wet or humid conditions at 50–90°C (49). The very high, largely irreversible, extensibility of nylons with the absorption of large quantities of energy can be converted to a practical use in emergency arrestor gear in aircraft carriers. However, the undrawn nylon usually is prone to degradation by humidity. Exposure of these arrestor ropes to 8-hydroxyquinoline effectively protects against degradation for 500 d.

Electrical. Electrical conductance of nylon is very low. Conductivity increases as moisture content rises; its value for nylon-6 yarn increases by several orders of magnitude as rh increases from 0 to 100%. The absolute values depend upon the specific spin finish or after finish applied to the fiber and whether any of the finish is on the fiber at the time of testing.

The insulating properties of nylon are manifested in the readiness with which it accumulates static electrical charges. Positive or negative charges may be generated easily on the surface by rubbing or contact with other appropriate substances, followed by separation. These charges are not readily dissipated. The combination of ease of formation and difficulty in losing electrostatic charges is an unfavorable characteristic of nylon and other synthetic fibers with low moisture regain. Increasing the relative humidity, application of conductive finishes, or incorporation of certain substances in the polymer melt prior to spinning help to dissipate static charges (see Antistatic agents).

Some of the more important electrical constants for nylon-6 are given in Table 3. Values for nylon-6,6 are approximately equal except for the dielectric constant, which is lower than that for nylon-6.

Manufacture and Processing. Essentially all polyamides other than the aromatic (aramid) types are melt-spun. This process is more economical for processing polyamides that melt below about 280°C than is solution spinning and, therefore, it is preferred for nylon-6,6 and nylon-6. Molten polymer is delivered from an extruder, from a metal grid melter (an early commercial process), or directly from a polymerizer, through a filter, and to a meter pump. The pump accurately meters polymer to the pack, which is a combination of a small filter and spinneret. Upon leaving the spinneret,

Table 3. Some Important Electrical Constants for Nylon-6

Property	Value	Reference
surface resistance, Ω	2×10^{12}	50
specific resistance, $\Omega \cdot m$	2.6×10^{12}	50
dielectric strength, kV/m	9000	50
dielectric constant	3.4	51

the molten filaments pass into a vertical chimney, where they are air-cooled and simultaneously stretched to the desired diameter. Finish is applied and the yarn is wound on a bobbin. The fibers are then heated and stretched to 400–600% of their original length to effect oriented crystallization and greatly improved tensile strength. At this stage, the fiber or yarn, which consists of a bundle of fibers, may be suitable for some applications requiring straight or flat filaments. Where special properties, eg, very low shrinkage, may be required, as in tire yarn, a heat-relaxation step may be added. For many uses in apparel or carpeting, a textured, not a flat, yarn is preferred. Texturing introduces crimp, whereby the straight filaments are given a twisted, coiled, or randomly kinked structure. A yarn that is made of such filaments is more open in structure, softer, and provides fabrics that are more pleasing to the touch for many applications than the flat yarns.

In some cases, two or more of the steps are combined into consecutive processes, eg, spin–draw or spin–draw–texture, to reduce manufacturing costs. These manufacturing processes produce continuous filament yarn. A second important type of product is staple, which is obtained by cutting continuous filament yarn into 3–15-cm lengths. Staple can be processed and blended with natural fibers, eg, wool and cotton.

Spinning Continuous-Filament Yarn. In the first commercial process for nylon spinning, polymer chips were stored under nitrogen pressure in a sealed hopper from which they flowed by gravity to a pancake coil 17-cm in diameter heated by a central Dowtherm system (11). The nitrogen pressure moved the molten polymer to a gear pump, which forced the metered polymer stream through a sand-filled filter pack and a spinneret; temperatures of operation were maintained by a single heat-exchange system. The molten filaments were quenched by a cross-flow of ambient air in a chimney with side panels to prevent outside air disturbances, and the quenched filaments converged over a ceramic guide to form the single threadline per spinneret. Finish was applied with a roll and the threadline was wound on a friction-driven bobbin at a few hundred meters per minute. The spun yarn was stretched or drawn, given twist for coherence, and wound on shipping packages.

A continuous nylon-polymerization system was developed in the 1950s; it was coupled directly to the spinning units, thereby eliminating the need for polymer chips (11). Continuous polymerization is used for nylon apparel, carpet, industrial filament yarns, and staple. The capacity of a continuous polymerization–direct spinning machine can be as high as 70,000 metric tons per year. Another significant advance was the use of extruders, rather than grid melters, to melt polymer chips. Use of extruders is standard where scale and flexibility requirements argue against use of continuous polymerization and direct spinning. Extruders can be horizontal or vertical, vented or unvented, with capacities of 5000–7000 t/yr.

The molten polymer must be delivered as quickly as is practical to the spinneret in precisely metered amounts or the size of the filaments will vary and the final products will be unacceptable aesthetically or in performance. A metering pump is used subsequent to extrusion to provide exact flow rates of molten polymer per unit time against pressures as high as 70 MPa (10,000 psi) and at ca 300°C.

The polymer may contain catalyst residues, gel particles, precipitated additives, etc, all of which clog the spinneret holes. Thus, the polymer must be filtered and subjected to fairly high shear to achieve melt homogeneity. Filtration and shear are accomplished in the pack assembly, which consists of a filtration/shear device attached directly to the distribution plate and spinneret. One early pack consisted of a cylindrical cavity ca 3.7 cm in diameter, 3.7 cm deep, and filled with layers of different sizes of special sand with the finest on the bottom and the coarsest at the top. Fine mesh screens in the bottom and top of the cavity retained the sand in place (11). More recently, the layers of sand have been largely supplanted by specially designed screens and sintered metal. Pack designs must minimize the possibility of stagnant spots where polymer could be trapped and thermally degrade, thereby increasing the pressure drop through the pack and shortening its life.

Disk spinnerets are available with as many as 500 holes and, in rectangular spinnerets for staple, with 4000 holes. The molten filaments enter the top of a tower or chimney, where they are quenched or cooled by a cross-current of air. The air flow is controlled carefully to avoid turbulence. Toward the bottom of the chimney, the filaments converge to form the thread line in the V formed by crossed ceramic pins or other similar devices. This thread line passes to the floor, below which finish is applied, and it is wound on the spin bobbins. Winders wind 10–25-kg bobbins with two or more bobbins per winder and at speeds up to 6000 m/min. Automated winders change bobbins mechanically for continuous operation.

A conventional spinning apparatus, by which polymer is melted in a large extruder which feeds a manifold spinning line, is illustrated in Figure 10. Alternatively, the spinning manifold can be fed directly from a continuous polymerization process, thereby bypassing the extruder. The molten polymer feed lines are made as short as possible, and the polymer is distributed to a series of meter pumps and then to spinnerets where the yarn is formed. The yarn passes to quench chimneys, to tube conditioners, over finish rolls, over takeup godet rolls, and finally to the winders. In the apparatus in Figure 10, spare winders are used for rapid change to a new roll during doffing to avoid loss of yarn. The yarns then pass on monorail conveyors to the drawing areas (52). Of the various spinning steps, filtration, extrusion, quenching, and application of finish materials probably are the most significant in terms of their effect on the quality, strength, and uniformity of the fibers.

In a large polyamide plant, ie, one that produces 40,000–60,000 t/yr of industrial-type fibers, a large number of ends or strands of yarn are spun simultaneously. It is desirable and may be mandatory that each end of yarn be of uniform quality along any segment of its length in order to meet the specifications required by industry. There may be 140 filaments per yarn end. It is a general rule of thumb that any defect greater than 20% of the diameter of the filament may result in a filamentary break. A single end of yarn with different orientation, different heat treatment, or change in moisture can result in a streak in the final woven fabric. Thus, any portion of the polymer in the yarn must be as rheologically uniform as possible. Approximately 80% of the problems related to obtaining a uniformly drawn, high strength fiber are solved by absolute control over the processing operations up to a distance of ca 15 cm below the spinneret face.

As the speed of winding on the bobbin increases, the amorphous and crystalline regions become more oriented with respect to the fiber axis. As a result, the elongation and residual draw ratio decrease and the tenacity increases. For example, at a spinning speed of 500 m/min, the elongation may be 400–500%, depending upon other conditions, eg, polymer molecular weight, melt temperature, and tex (den) per filament. At 1000 m/min, the elongation may be 200–300% and declines almost linearly to 60–70% at 3500 m/min and more gradually after that as speed is increased. At ca 6000 m/min,

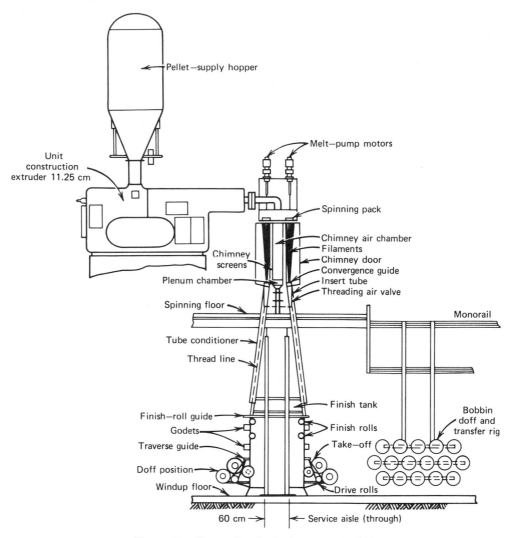

Figure 10. Conventional spinning apparatus (52).

the yarn requires no further drawing to develop properties useful for apparel applications.

The elongation of nylon-6,6 that is spun at ca 2500–3300 m/min is ca 70–100% (partially oriented nylon); the product is a suitable feedstock for draw-texturing at 700–800 m/min (53). Spinning nylon-6,6 at 2285–4500 m/min results in yarns with elongations of 35–80% (54).

Drawtwisting. Continuously spun yarn is drawn as needed to develop more useful properties. One of the early ways of doing this was to use a drawtwister, which stretched and twisted the yarn to make a more cohesive bundle. The spun bobbins were taken to the drawtwisters where the yarn was drawn or stretched by removing it from the bobbins at a fixed rate, snubbing it on a heated pin, and passing it over a usually hot draw roll, which was rotated faster than the feed rate by a factor depending on the amount of desired drawing. The drawn yarn was wound and twisted by a standard ring-and-traveler mechanism on a bobbin for shipment.

Spin–Drawing. The successful combination of spinning and drawing processes reduced manufacturing costs. In 1959, high speed winders necessary for cost reduction and fluid jets to tangle rather than to twist the yarn were developed. The first application was to polyester textile yarn and, in 1960, to nylon tire and apparel yarns (11). When combined with continuous polymerization, spin-drawing is the preferred process for filament yarns of nylon as well as for polyester. The spin-draw equipment is similar to that shown in Figure 10; it is generally a double-sided unit (55). The principal difference is the addition of stretch godets and relaxation equipment for spin-draw processes (56–58). The relaxation step is illustrated for tire-yarn manufacture in Figure 11 (59).

The draw ratio affects properties, eg, tenacity and elongation. As draw ratio increases, the tenacity generally increases and the elongation decreases. As a result, usually higher draw ratios are used for tire yarn than for carpet or apparel yarn.

Texturing. In general, textured yarns are filament yarns that have greater apparent volume or are made more extensible by mechanical distortion of the filaments (60). The distortion may be produced by buckling the filaments under endload compression, by either bending them over an edge of small radius or twisting the strands

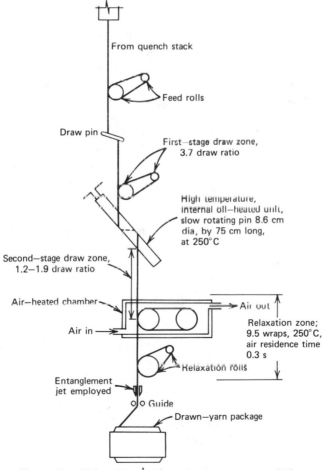

Figure 11. Nylon-6,6 spin–draw tire yarn apparatus (59).

as a whole. Bulked nylon yarns are either fine tex (1.7–22 tex (15–200 den)) for woven, knitted stretch, and textured fabrics in apparel applications or heavy tex (110–400 tex (1000–3600 den)) for carpet (60). Yarns are textured to obtain increased fabric cover, greater recovery from deformation, a more pleasing feel, and greater warmth. Requirements of texture uniformity for continuous-filament textured yarns are much more exacting than those for staple. The former are not characterized by the latter's thorough blending and leveling of nonuniformities (61).

Fine Tex. Stuffer-tube crimping of fine-tex yarns is described in a series of U.S. patents (62–63). Generally, commercial fine-tex crimping has been limited to one or a few ends, ie, yarns. However, in the Spunize system, a ball warper is used in conjunction with multiple end winding to crimp and wind 400 or more ends of fine-tex yarns simultaneously (64). Drawn yarn is accelerated by feed rolls and impacts against yarn in a stuffer tube where it bends, folds, and forms crimps as a result of heat softening from preheating or heat controllers in the stuffer tube. Crimped yarn is pulled from the stuffer tube at a constant rate. The amount of texture and texture permanence are controlled by yarn residence time and temperature in the stuffer tube. Increased residence time can be accomplished with increased weight or back pressure on the stuffer tube, resulting in increased crimp. The converse effect results from decreased weight. Yarns from the process frequently are plied to reduce nonuniformity of individual ends (65–66).

Edge crimping produces a bulked yarn and is effected by drawing a heated yarn over a blunt knife edge. In one process, two or more ends of drawn yarn are delivered to each bulking position (67–69). The yarns are fed to a rotating, heated cylinder, which stretches and heats them; the plasticized yarn bundle is then drawn over an edge, which can be heated. The acute angle through which the yarn travels causes the surface of each filament nearest the crimping edge to be compressed and the outside to be stretched. Ten to twenty percent or more of the filament surfaces can be flattened or, in some cases, there may be practically no visible deformation (69). The yarn appears untextured at this point, but the distortions and strains are fixed by passing the yarn over a cooling roll. The yarn then can be wound directly onto a package, or sent over another heated roll where the latent crimping energy can be stabilized and later released at different levels, thereby generating minute coils or crimps by varying tension and temperature. In other cases, the texture is not developed until the fabric is dyed and finished. One such marketed bulk yarn is Agilon.

False-twist texturing has largely replaced the tedius twist–set–detwist of the Helanca process (70). Special manufacturing techniques are used to produce a nylon feeder yarn with excellent false-twisting properties. The drawn yarn, which is heated by a radiant or contact heater to close to its melting point, is twisted to as high as 30–40 turns per centimeter, cooled, and untwisted to its original twist level (71). Machines with spindle speeds of up to 360,000 rpm have made this process much more economical than and more competitive with other texturing processes. The product is primarily a stretch rather than a bulk yarn. However, a modified false-twist process, in which stretch is reduced by subsequent heating and tension under controlled conditions, is used to produce a nylon bulked yarn by false twist. This is done either by following the heat–twist–untwist step with a second heat–twist–untwist step at a lower temperature or by a steam or hot-water treatment of the yarn on perforated tubes in a pressure dyeing machine (72).

Air-jet texturing or Taslan texturing, in which continuous-filament nylon yarns or other synthetics are passed into an air jet, controls tangling. The yarn is introduced into a venturi tube in the air jet, where turbulence from a stream of compressed air causes the development of loops and texture. Many variations of this process have been used to obtain different bulk and novelty effects (73–75). The product is characterized by substantial bulk and practically no stretch.

Draw-texturing involves a false-twist texturing process and a simultaneous drawing step for the production of textured hosiery and apparel yarn. Modern machinery consists of rotating disks for twisting of the yarn, and processing speeds are as high as 800–900 m/min. The draw ratio depends on the residual draw ratio in the feeder yarn. Partially oriented nylon feeder yarn with elongation of 80% usually requires a draw ratio of ca 1.3 to yield properties suitable for apparel use.

Bicomponent fibers, in which each filament is composed of two or more different polymers with different responses to environmental conditions, also can be used to obtain textured yarns.

Heavy Tex. Heavy-tex textured yarn is used primarily for tufting carpets.

A gear-crimping process for producing textured yarn is described in several patents (76–78). The texturing process can be adapted to a draw-twist or a spin-draw texture operation. Heat-plasticized drawn yarn is delivered to a pair of meshed gears. The crimp frequency is a function of the pitch of the gear teeth. Crimp permanency of the yarn depends on the yarn temperature during crimping and the setting temperature immediately thereafter.

In fluid-jet texturing, the yarn and hot or superheated steam pass at high speed into a small chamber (79–82). The yarn is first heat-plasticized by, eg, the steam, and then is textured by controlled turbulent flow of the steam. In some designs, the yarn is textured by impingement on a baffle plate or on the walls of a specially shaped chamber. Fluid-jet texturing differs from Taslan texturing in that texture in each filament of the former has a random, three-dimensional curvilinear configuration and the whole bundle is essentially free of loops, whereas the Taslan process bulks the yarn by forming loops and knots in the yarn bundle.

Stuffer-box crimping of heavy-tex yarns 90–560 tex (800–500 den) is similar to that described for the Spunize process for fine-tex yarns (84). Heavy-tex crimping differs from conventional fine-tex processes in that, generally, multiple ends are fed simultaneously into the stuffer box (83–87). The operation is similar to that used in crimping tow for staple, except for the separation of individual ends of textured yarns and the packaging as a parallel wound package or as cones. The basic false-twist process used in producing bulked fine-tex yarn also is used in texturing heavy-tex, nylon-6 yarns for carpet applications. The application of false-twist texturing to heavy-tex yarns has not been as widespread as some other processes, because of economic limitations resulting from the low rates attainable with the maximum available spindle speeds.

Initially, texturing processes were split processes, but in-line texturing was developed using fluid-jet techniques. Mechanical texturing is largely a split process, as most mechanical systems cannot operate at the speeds necessary for in-line processing. Modern carpet yarns are textured in-line by fluid-jet processes but texturing of textile denier yarns is overwhelmingly by mechanical means on separate machines (11).

Heat Setting. A heat-setting operation usually is employed for production of a bulked or stretch yarn. Heat setting is required to obtain a texture that is stable to dyeing and washing and that recovers from stresses applied during wearing or use of the finished goods. In nylon-6 bulking processes, yarn temperatures of 100–190°C usually are applied before the yarn is deformed by mechanical action. The yarn is cooled significantly before the deforming force is removed so as to set the crimp. The exact temperature used depends on whether steam or dry heat is employed and on the amount of mechanical deforming energy and the required degree of heat set. The optimum temperature for heat setting of nylon-6 fiber with wet steam is about 130°C and, with dry heat, ca 190°C (88). In the bulking process, crystallinity and orientation can be altered at points of deformation. The permanence of the deformation is controlled by the degree of heat setting.

Staple. The production of staple is much simpler than that of filament yarn, since the breaking of a filament in the spinning step usually is not problematic, and uniformity of filaments is not critical because the staple is blended. Several hundred to one thousand filaments can be spun from one spinneret, and many of the yarns can be combined into a large tow for drawing, crimping, and cutting. Polymerization is coupled with spinning to lower processing cost. As for filament, the polymer is metered to spinnerets, the filaments are quenched, and finish is applied. Yarns then are combined to form a tow, which is drawn by being passed over hot rolls. More finish can be applied at this step.

Crimping. Stuffer-box crimping generally is the most productive method for crimping nylon for staple. Multiple ends of tow are forced continuously into a constricted stuffer box by a precisely adjusted set of feed rolls. The box is sealed by means of an adjustable hinged and weighted gate. The strands filling the chamber are folded in uniform waves by compression. When the pressure in the stuffer box exceeds the pressure of the hinged gate, the gate rises, permitting discharge of the crimped tow. The operation is repetitive and proceeds continuously (62).

Tow Cutting. The tow usually is moistened prior to being cut to reduce the static charge, which otherwise may cause the fiber to stick to the cutters (89). Moistening the tow also increases the life of the knives. Since the toughness of polyamide fibers causes rapid wear of the cutting knives, special hardened steel must be used. After being cut, the staple is baled for shipment.

In staple spinning, the staple undergoes a series of processing steps which cause alignment of the fibers with the intended yarn axis and separation of the staple into bundles. These bundles are elongated to improve fiber orientation along the yarn axis. This series of steps is called carding and drawing. Finally the yarn is twisted to effect coherence and longitudinal strength (89–90). Nylon-6,6 staple was first produced by DuPont in 1947, and nylon-6 staple was introduced in the United States by Allied in 1955.

Modified Cross-Sections. Conventional spinneret orifices are circular. Improved machining methods, advances in fiber production technology, and the study of the effect of profiled fibers, ie, with noncircular-cross sections, on fabric characteristics, eg, luster, sparkle, opacity, air permeability, resistance to showing soil, and heat insulation, have given greater importance to the production of modified cross-section filaments (91–93). The spinning equipment and process for production of modified cross-section yarn are very similar to those used in the manufacture of round cross-section yarn, except for the shape of the spinneret orifice. Some of the process conditions that facilitate production of well-defined cross-sections include higher melt viscosity, lower polymer temperature, and rapid quench just below the spinneret (94–95).

In all nylon-spinneret manufacture, the final orifice is produced after counterboring all but the thin sections of the spinneret blank. Noncircular spinneret orifice shapes are made by electron-beam milling and electrodischarge machining. Both of these methods are used to produce a wide variety of noncircular orifices with extreme accuracy. Electrodischarge machining is the controlled erosion of electrical conductors by rapidly recurring electrospark discharges. During electron-beam milling, the beam is absorbed by a very thin surface layer of the material and its energy is converted into heat, causing the material below the shaped electrodes to melt and vaporize.

One spinning method depends upon the coalescence or fusing of melt streams below the spinneret with the consequent formation of a noncircular single filament (96–97). The other method consists of extrusion of the melt through profiled capillaries, which dictate the basic fiber shape. Various specific orifice shapes have been developed, and fiber cross-sections have been defined in terms of mathematical relationships (98–100). The simultaneous spinning of a combination of 50% round and 50% profiled filaments decreases the fiber-bundle density, thereby imparting warmth and moisture permeability to the textile fabric (101). Synthetic yarns also have been produced in which the cross-section shape varies, but the cross-section area remains constant along the filament; the resultant product is characterized by an apparent variable tex (den), random receptivity to dyeing, and novel feel, luster, and porosity (102). Some of the spinneret-hole arrangements and shapes which are used to produce solid profiled fibers are illustrated in Figure 12.

Hollow filaments are produced by the proper arrangement and design of orifices so that the melt streams coalesce below the spinneret (91,103) or through the design of spinning equipment to form a hollow cross-section within the capillary (91,103).

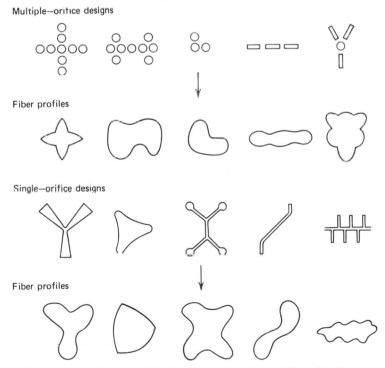

Figure 12.   Schematic orifice designs and resultant solid-profile fibers.

One method is injection of a gas into the filament during the melt stage (104–105). Some of the single- and multiple-orifice shapes used to produce hollow filaments are illustrated in Figure 13.

Finishes. The main functions of a fiber finish are to provide surface lubricity and yarn cohesion. At any point where a yarn or fiber contacts another surface, eg, draw pins, guides, or other fibers within the bundle, a lubricant is essential in order to maintain the fiber in useful form. From the time the filaments are formed until they are made into the final product, a finish must be applied to prevent damage to the fiber during processing. The principal causes of filament breakage during drawing are buildup of excessive tension and generation of a static charge on the yarn. These may be caused by interfilament friction and by the friction generated, eg, as the yarn passes over the draw pin and draw heater. These steps also generate heat, which makes operation impossible without adequate lubrication protection (106). The term lubricant as applied to a finish may be misleading, since to lubricate generally means to minimize friction. In textile processing, this is not always desirable; sometimes it is necessary to raise or otherwise adjust friction to meet specific conditions.

Spin finishes are applied to the fiber during spinning. Overfinishes sometimes are applied by the producer for special purposes; eg, the service life of nylon-6 as used in cordages is increased if finished with a compound with an oxidized polyethylene wax base (107). Yarn lubricant or coning oil defines all lubricants added subsequent to application of spin finish. Yarn lubricants are used in textile milling and processing, eg, picking, coning, slashing, weaving, and knitting.

The finish generally is an emulsion or a water-soluble mixture of one or more lubricants and an antistatic agent, to avoid static charge, which would cause the filaments to repel each other, giving a yarn with poor cohesion and, thus, flaring filaments

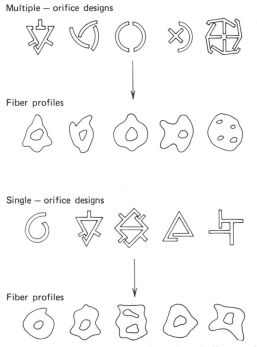

Figure 13. Schematic orifice designs and resultant hollow-profile fibers.

which causes uneven processing of the yarn. Wetting agents usually are added to aid the spreading of the finish on the yarn. The concentration of finish on the yarn and after evaporation of water usually is 0.3–0.8 wt %.

Lubricants and Antistatic Agents.   Mineral oils have long been used as lubricants. The fatty acids comprise the largest group; compounds with functional groups C_6–C_{18} provide excellent lubricity. These may be straight-chain esters, eg, methyl, ethyl, or butyl esters of fatty acids. Refined or synthesized polyglycols or their esters of fatty acids and fatty glycerides are widely used; the triglycerides are more stable to higher temperatures (108). More complex chemicals are used increasingly for greater stability. These include the esters of dibasic acids; pentaerythritol, neopentylene glycol, polyglycerols; aryl, alkyl, and alkyl aryl phosphates; and silicones, silicate esters, fluoroesters, and poly(phenyl ether)s (109). Modifications have been made by ester interchange, ethoxylation, or acetylation to effect changes in degree of lubricity, scroop, cohesiveness, and plasticity.

Antistatic agents (qv) generally are more complex than lubricants. On nonconductive surfaces, eg, of most textiles, charges remain isolated, and effective discharge cannot be accomplished through grounding techniques. Both positive and negative charges can be acquired by various fibers (110 111). Because water containing inorganic substances has a relatively high dielectric constant, its presence may reduce the intensity of the contact potential and result in reduced charge (112). In addition, water is a good conductor of electricity and can cause a nonconductive surface to exhibit some characteristic conductive properties. Therefore, hygroscopic chemicals are potentially good antistatic agents.

Organic compounds, eg, surface-active chemicals, generally are incorporated in finish formulations for their antistatic protection. Their action is attributed to their ability to absorb moisture and to become molecularly oriented on the fiber surface and, thus, exhibit some dielectric shielding and conductivity. Antistatic agents are nonionic, cationic, anionic, or ampholytic (112–114). Nonionics include the polyoxyethylene derivatives of fatty acids, alcohols, amines, and amides as well as derivatives of hydroxy amino compounds, eg, alkanolamines (see Alkanolamines). Quaternary ammonium compounds (qv), eg, the alkylmorpholiniums, are examples of cationics. Anionics are represented by soaps, esters of sulfuric and phosphoric acid, and alkane and alkylaromatic sulfonic acid derivatives. Ampholytes, eg, betaine, dissociate into either positively or negatively charged ions. Compounds of the betaine type are included in this category (see Surfactants and detersive systems).

Spin finish greatly influences the processing of fibers into yarns and fabrics. If, for example, the component fibers and yarns of a fabric do not have sufficient mobility to adjust because of improper lubrication, strains are introduced into the fabric and may produce uneven dyeing, decreased strength, or unpleasing esthetic qualities (115).

Evaluation and selection of a fiber finish depends on processability, economics, physical properties of the fiber, as well as the versatility of its ultimate use.

Application.   After the finish has been prepared in proper concentration and usually as an aqueous emulsion, it is pumped to a holding or storage tank from which it is circulated through the finish system in the trough or trays on the spinning machines. Finish concentration and level of pickup are balanced to give the final weight of lubricant, antistat, water, or other finish additive required for subsequent operations. Some of the various methods of application in industry are passing the bundle of fibers through the circulating finish (this method is used mostly in nonaqueous systems), passing the yarn bundle across a constantly revolving roll, or spraying the moving fiber bundle. More recently, fiber producers apply finish by metering a stream of finish through an orifice in a slotted pin or guide, which is positioned such that the yarn runs through the slot and is wetted by the finish.

Additives. A variety of substances usually are added to the reactive mixture before or during polymerization to change the properties and nature of the polymer.

Delustering. Delustering reduces transparency, increases whiteness, and prevents undesirable gloss in finished fabrics (116–121). It usually is accomplished by dispersing finely divided titanium dioxide particles 0.1–0.5 μm in diameter (avg) within the fiber (122–123). Semidull nylon yarns contain ca 0.3 wt % TiO_2; full-dull yarns, ca 2.0 wt % TiO_2. Improper incorporation of TiO_2 into molten monomer or polymer may result in one or more of the following problems (124): premature clogging of filter packs, leading to frequent replacement of packs and spinnerets; nubs or lumps in the filament during spinning; breaks in molten filament streams, poor drawing performance, poor yield, and variable tex (den); excessive yarn abrasiveness and rapid wear of processing machinery; poor dyeability and fabric streaking; and poor light stability and rapid loss of tensile properties upon exposure to sunlight or other uv radiation.

The anatase crystalline modification of titanium dioxide generally is used even though it catalyzes light degradation of nylon more than the rutile type (125). Moreover, it is not as effective a delusterant as rutile TiO_2. However, rutile TiO_2 is not quite as white and is much more abrasive than anatase TiO_2 (126) (see Pigments, inorganic).

Nylon-6 and nylon-6,6 delusterant technologies differ somewhat. In nylon-6,6 manufacture, the TiO_2 is added to the aqueous hexamethylenediamine–adipic acid salt. In nylon-6 manufacture, the TiO_2 is added to the caprolactam and frequently as a master batch containing up to 30 wt % TiO_2 in low molecular weight polymer (127).

Spin Dyeing. Spin dyeing or mass or dope dyeing is the coloring of yarn by incorporating dyes or pigments into the polymeric mass prior to spinning (128). Although pigments can be added during polymerization, they also can be added immediately prior to spinning in order to confine the colored material to as small a zone of the process as possible (129–130). The primary advantage of spin dyeing is the production of a stable, colorfast yarn requiring no subsequent dyeing. On the other hand, the variety of shades is limited, since each shade requires separate processing facilities and/or extensive flushing between runs which results in high waste.

Carbon black is the next most frequently used pigment after TiO_2. It can greatly enhance the fiber's resistance against light degradation.

Inorganic pigments often are used for spin coloration (131). Organic pigments usually are more intensely colored than the inorganic ones, but there are few that are available that do not degrade during spinning (see Pigments, organic pigments). The somewhat lower spinning temperature of nylon-6, ie, 255–270°C versus 280–290°C for nylon-6,6, makes the selection of stable pigments for the former somewhat easier. Some typical pigments used for spin-dyeing nylon-6 are titanium dioxide (white), carbon black (black), cadmium sulfide (yellow), cadmium selenide (yellow), Phthalocyanine Green (green), Phthalocyanine Blue (blue), Quindo Magenta (pink), and cadmium sulfoselenide (red) (130,132).

A number of optical brighteners (see Brighteners, fluorescent) also are available for incorporation in molten polymer (129–130). The portion of radiation absorbed in the uv portion of the spectrum is reemitted as visible, usually blue light.

Antistatic Agents.   Polyoxyethylene products at one time were added commercially to nylons to reduce the generation of static charges in fabrics in carpets and to improve moisture transport and soil release (131–132). These products were largely supplanted by more improved types (see Bicomponent fibers).

Flame-Retardants.   A variety of halogenated organic compounds, especially aromatics, and phosphorus derivatives have been added to nylons to enhance flame retardance (133) (see Flame retardants). Although some of these have appeared beneficial, based on laboratory test results, they usually introduce problems, eg, reduced polymer melt stability and reduced light resistance. In addition, there is concern about possible leaching of the additives from the fibers during laundering.

Dyeability.   *Fiber Characteristics.*   Chemical structure, crystallinity, molecular orientation, and fabric preparation affect the rate and extent of nylon dyeability. Nylon-6 and -6,6 differ in the arrangement of the carbonyl and amido groups between the hydrocarbon chains of the polymer. In nylon-6,6, their order is reversed alternately so that there is no difference in the order of occurrence from either end of the chain. Nylon-6 has a uniform order of these functional groups, and the chain has a right and a left end. When the molecules of the noncrystalline portions of the extruded fiber are aligned or oriented by drawing, nylon-6 molecules have a more random, open structure than nylon 6,6 molecules (134). Thus, the rate of diffusion of dyes into this open structure is more rapid, and the opportunity for dye fixation on the greater number of available hydrogen bonds is increased.

Some nylon chain ends of both types normally terminate with amine groups, which are the most active dye sites for acid dyes. The ratio of amine ends to other end groups varies with polymerization conditions and can be controlled by the introduction of other chain-terminating additives for specific uses.

As with other synthetic and natural fibers, nylon consists of a mixture of crystalline and noncrystalline regions. The molecules in crystalline arrangement are so closely packed and so much less responsive to chemical influences that they are probably not dyeable. The portion of the fiber made of such material, as determined by polymerization and spinning conditions, is a significant factor in determining the fiber's dyeability (135). The physical boundary between crystalline and amorphous regions in the fiber probably is characterized by a gradual transition. When the fiber is drawn to impart controlled tensile and elongation properties, a more regular spatial arrangement between molecules is established. The extent to which this orientation is complete also affects dyeability (135).

When the amido and carbonyl groups of adjacent molecules are in close proximity, the hydrogen atom of the amido group shows a strong electrostatic affinity for the oxygen atom of the carbonyl group. The formation of this hydrogen bond releases energy and makes the polymer less responsive and more impervious to dyeing. During drawing, many of these bonds are broken, others are formed, and some are strained. During subsequent fiber handling, the strains tend to return to a more relaxed state. Uncontrolled relaxation may introduce variations in these characteristics and, thus, in dyeability. Nonuniform application of heat, as from guides overheated by friction during conversion to fabric, may introduce streaks, which are evident in the fabric after dyeing. Heat-setting introduces energy under controlled conditions; the strained bonds are broken and reformed at more appropriate locations. Variations in heat-setting conditions from side-to-side or end-to-end of a piece may cause uneven dyeing (136).

Various studies have related dyeability to rate of diffusion, ie, the solubility of dye in the polymer; hydrogen bonding of dye to polymer; formation of ionic bonds between dye and polymer and the strength or affinity of these bonds; and the aggregation of dye molecules into particles trapped in interstices between the molecules. There is considerable uncertainty regarding the mechanism, because each dye responds differently to one or more of these mechanisms which, in turn, are influenced by variations in fiber and fabric preparation.

Dyebath Characteristics. In general, uniform dyeing is most easily achieved when dyeing conditions are selected so that dye, diffusing into and out of the fiber, establishes an equilibrium with an appreciable amount of dyestuff remaining in the bath. Under this condition, nonuniform fabrics eventually achieve a level shade. Disperse dyes are unexcelled for this purpose.

The nitrogen atoms in nylon absorb hydrogen ions with increasing avidity as the pH of the dyebath is lowered from ca pH 9 (137). The terminal amino groups are most active and fix the neutral-dyeing acid colors at high pH, either by reaction with the free dye acid or by addition of a hydrogen ion to form a cationic group, which absorbs the dye anion. The acid dyes that are fixed at high pH generally are slightly ionized, ie, weakly acidic. At below ca pH 6.5, the fiber becomes strongly cationic and rapidly absorbs dye anions. Under such conditions, the dye is exhausted quickly on the most readily accessible dye site and fiber variations appear as streaks in the fabric. Because these dyes are more firmly bonded, they do not tend to level by migration. A direct relationship can be demonstrated between the low ionizability of a dye and its level-dyeing capacity. The relationship between oligomers, ie, the low molecular weight constituents, their effect on fiber dyeability, and analysis has been described in refs. 138–140.

Many attempts have been made to develop dyebath additives that temporarily combine and inactivate either the cationic groups in the fiber or the anionic groups of the dyestuff for slower, more level dyeing. Generally, they provide only a partially satisfactory expedient. Cationic dyebath additives sometimes adversely affect the lightfastness of dyes.

Many dyes do not build deep shades, except from a dyebath at ca pH 2, because the available amino groups are not sufficient to fix more dye. At below ca pH 3.5, the amido groups absorb hydrogen ion and become effective dye sites for the acid-dyeing acid colors, which are polybasic and require many dye sites for fixing. If the neutral-dyeing colors, which usually are monobasic, are used with polybasic colors, the latter may be blocked or even displaced, because the monobasics require fewer strong dye sites for effective fixation.

Choice of Dye. Nylon has a marked affinity for virtually every class of dyestuff. Those classes with the widest applications are the disperse, acid premetalized, chrome, selected direct, and certain naphthol dye combinations (141). The choice usually is disperse dyes because of their easy application. Since they are fixed only by easily broken hydrogen bonding, they level well but also show poor washfastness. Brightness of shade and lightfastness usually are minimal. The acid dyes contribute brighter shades and some improvement in washfastness but require greater care in application. The neutral-dyeing, weakly acidic types can be used even in solid shades on well-prepared, uniform fabric. The acid-dyeing colors and the neutral-dyeing premetalized or strongly acidic colors provide the ultimate in performance. They generally are usable only where tone-on-tone effects or other devices obscure nonuniformity. To the dyer, chemical differences between the nylon-6 and -6,6 fibers are less significant than fiber and fabric uniformity.

Modified Nylon-6 and Nylon-6,6 Fibers

Bicomponent and Biconstituent Fibers. Bicomponent and biconstituent fibers are produced by melting separately two different components and extruding them through a common spinneret, as shown in Figure 14. The resulting fibers may have configurations as shown in Figure 15. The sheath-and-core arrangement recalls the structure of the cortex and cuticle of wool. Conventional drawing or spin-draw procedures are used. Subsequent to the drawing stage, the yarn passes through an essentially tensionless heating zone where the structure of the fiber causes helical crimping. Then the fiber is wound on the final sales package at very low tension.

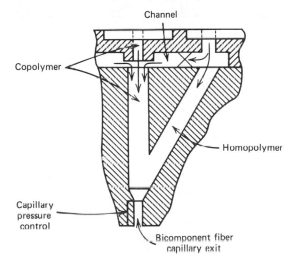

Figure 14. Bicomponent spinneret for sheath-and-core. If side-by-side fiber is desired, interconnecting channel X is not employed.

Components used in some early commercial bicomponent fibers, and their percentages, are shown in Table 4. Some properties of fabrics made from bicomponent fibers are compared with those of other polyamide fibers in Table 5.

Fabrics made from bicomponent yarn have unique properties of shape configuration as a consequence of their helical crimp. They have been used in hose and tricot fabrics. However, with improvement in texturing of nylon to give stretch yarns and with the lower cost of nylon, the bicomponents are no longer competitive. The most notable bicomponent nylon was DuPont's Cantrece (142).

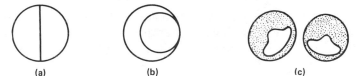

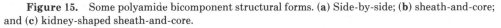

Figure 15. Some polyamide bicomponent structural forms. (**a**) Side-by-side; (**b**) sheath-and-core; and (**c**) kidney-shaped sheath-and-core.

Table 4.   Components Used in Some Commercial Bicomponent Fibers

Type	Component A copolymer, composition	Component B homopolymer
sheath-and-core	6,6–6,10, 50:50	6,6
side-by-side	6,6–6,I[a], 80:20	6,6
side-by-side	6,6–6, 80:20	6,6
sheath-and-core	6–6,T[b], 40:60	6

[a] 6,I = hexamethylene isophthalamide.
[b] 6,T = hexamethylene terephthalamide.

Table 5.   Properties of Some Polyamides in a 0.13-kg/m^2 [a] Fabric

	Bicomponent fiber	Conventional nylon	Stuffer-box crimped Y cross-section nylon	Spun staple nylon
bulk, cm^3/g	2.4	1.8		
compressibility thickness, ratio of thickness at 3 g/cm^3 to thickness at 230 g/cm^3	1.3	1.1	1.15	1.45
air permeability, m^2/(min·m^2) [b]	35	20		
covering power at equal tension, % surface uncovered in hosiery	4.0	9.0		

[a] To convert kg/m^2 to oz/yd^2, multiply by 29.5.
[b] ASTM D 737-46.

Bicomponent systems that include, as a minor component, a nylon heavily loaded with conductive carbon black are used commercially as conductive filaments. In some fibers of this type, the carbon black is a core component surrounded by a sheath of nylon-6,6; another is characterized by a side-by-side configuration so the carbon black component is seen as a very faint black stripe along the side of a nylon filament (143). The addition of 0.1–1 wt % of these conductive filaments into regular nylon yarns gives a static-resistant no-shock nylon carpet. Because of the fine size of the conductive filaments and the partial hiding of the carbon within the filaments and at the low percent of addition, the tiny amount of carbon does not interfere with dyeing. This is the preferred approach for producing shock-resistant carpets and has displaced the use of polyether additives in nylon for this purpose. Fibers with conductive materials in a sheath around a nylon core also have been used, but these generally cause a more apparent color contamination because of the larger area of conductive material that is visible (see Antistatic agents).

Biconstituent fibers are similar to bicomponents, except the two polymers are in generically different classes. Monsanto's Monvelle, which is no longer marketed because of unfavorable economics, was a biconstituent from nylon-6 and a melt-spinnable polyurethane (144–145). In the drawing step, the nylon component drew normally, whereas the polyurethane underwent elastic stretching. Cooling the fiber below the glass temperature of the nylon, as occurred on the bobbin, stabilized the fiber so that it appeared as a flat yarn. When knit into hosiery and then placed in a dye bath, the nylon softened, the polyurethane contracted, and a helical crimp formed. Monvelle hose combined the power of medium support hose with the sheerness of stretch hose.

Block Copolymers. There are four common methods of preparing nylon block copolymers: combination in the melt of two or more homopolyamides; use of a diamine or diacid monomer that contains an amide linkage and subsequent reaction with another diamine or diacid; reaction of a complex molecule, eg, a bisoxazolone, with a diamine to produce a wide range of multiple amide sequences along the chain; and reaction of a diisocyanate and a dicarboxylic acid (146).

Irrespective of the route selected to form block polyamides, and as a result of transamidation, the percentages of homopolymer, block copolymer, and random copolymer are functions of the time and temperature of the mixture of these polyamides when in the molten stage, as shown in Figure 16. In practice, however, time is much more critical than is indicated in Figure 16. In the production of a polyamide-block-copolymer tire yarn, there is considerable amide interchange after 13 min (147). It is this rapid rate of amide interchange that poses the principal problem in producing yarn of uniform properties from nylon-block copolymers.

A fundamental difference between block copolyamides and random copolyamides is illustrated in Figure 17 with respect to glass-transition temperature as a function of composition. The two curves are essentially mirror images of each other: the block copolymer has significantly improved properties and the curve for the random copolymer shows a degradation in properties (148).

Random Copolymers. The largest suggested use for random copolymers is in the bicomponent-fiber field, but bicomponents are not used commercially at this time (150). The principal limitations of most random copolymers are their high shrinkage, low softening points, reduced wet-strength properties, tackiness, low ironing temperatures, and rates at which creep failure occurs. Future commercial utility of this

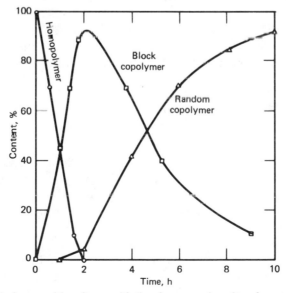

Figure 16. Typical composition change with time for conversion of two homopolymers first to a block copolymer and then to a random copolymer.

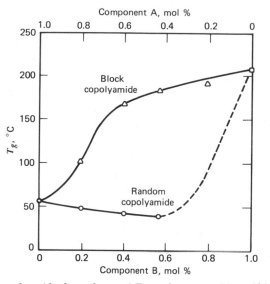

Figure 17. Block copolyamide dependence of T_g on the composition of block and random copolyamides versus the mole compositions of nylon-6 (component A) and 2,5-dimethylpiperazine-4,4'-biphenylene-3,3'-dimethyldicarboxamide (component B).

type of fiber may be in the fields of nonwovens, heat-molding of fabrics, and bonding of fabrics (see Nonwoven textiles). Eftrelin, 90 wt % nylon-6,6 and 10 wt % nylon-6, has good creasing or pleating properties and a soft hand; and Wetrelin, 43 wt % nylon-6 and 57 wt % nylon-6,T, has good compression elasticity, good creasing or pleating, better alkali resistance than nylon-6 or -6,6, and a woollike hand. Most other polyamide copolymer systems introduced to date have not presented sufficiently improved engineering properties to justify the additional cost of their production (149).

A random copolymer of terephthalic and isophthalic acids with hexamethylenediamine was studied in detail (150). At about 65 mol % terephthalic acid the fiber approached poly(ethylene terephthalate) in dimensional stability and could be dyed without the use of pressure or carriers. However, its advantages were not sufficient to result in its commercial use.

Other Nylons

Qiana. The only nylon of significant commercial use other than nylon-6 and -6,6 is Qiana. The polymer for this fiber is produced by condensation polymerization of bis(4-aminocyclohexyl)methane and dodecanedioic acid. The diamine occurs in several isomeric forms; ca 20 wt % cis–trans and 80 wt % trans–trans diamine is believed to be used in the polymer.

Commercial uses of Qiana, which was introduced by DuPont in 1968, have emphasized yarns with silklike appearances in luxury fabrics. Dimensional stability of the fabrics and their wrinkle resistance are similar to those of polyester. Some commercial yarns consist of fibers with different levels of shrinkage, so that self-texturing characteristics are obtained. Typical yarn properties are shown in Table 6 (152).

Table 6. Typical Properties of Qiana Nylon Yarns

Property	Value	Reference
melting point, °C	280	152
glass temp		
dry, °C	ca 190	
wet, °C	ca 85	
density, g/cm^3	1.03	153
tenacity, N/tex[a]	0.26–0.29	153
elongation, %	2.1–2.6	153
shrinkage in boiling water, %	4–10	153
moisture regain (at 65% rh), %	2.5	153

[a] To convert N/tex to gf/den, multiply by 11.33.

Dimensionally Stable and Miscellaneous. In an effort to improve the wrinkle resistance of nylon fabrics and the flat spotting performance of bias and bias-belted tires, a number of additional polyamides with high glass-transition temperatures, as compared to nylons-6 and -6,6, have been tested experimentally. Typical properties are shown in Table 7. As of 1981, none are produced commercially (see also Tire cord).

Another group of nylons, which are listed in Table 8, has been tested extensively and at 30–50% elongation for general use in apparel and carpets. None has gained commercial significance.

Table 7. Dimensionally Stable Experimental Nylons

Diamine[a]	Diacid	CAS Registry No.	mp, °C[b]	T_g, °C, Dry[b]	Tenacity, N/tex[c]	Elongation, %	Evaluation	Ref.
PACM	dodecanedioic	[25035-12-5]	298	~220	0.78	13	nonflatspotting in tires	151
CBMA	dodecanedioic	[53830-65-2]	280	175	0.8	13–16	nonflatspotting in tires	153
CBMA	sebacic	[53830-66-3]	291		0.6	14	nonflatspotting in tires	153
CBEA	dodecanedioic	[52277-94-8]	254	135	0.35–0.44	20–30		
MDA	sebacic		270	160	0.48	11	nonflatspotting in tires	154
PXD	sebacic	[31711-07-6]	290	158	0.44–0.53	12–15	tire flatspotting not satisfactory	153
HMD	terephthalic	[24938-70-3]	370	180	0.22–0.44	30–45	melting point is too high for melt spinning	155
HMD	terephthalic/ isophthalic, 65/35		310	165	0.3–0.66	8–40	nonflatspotting in tires	150

[a] PACM = bis(p-aminocyclohexyl)methane; CBMA = p-cyclohexanebis(methylamine); CBEA = p-cyclohexanebis(ethylamine); HMD = hexamethylenediamine; MDA = 4,4'-methylenedianiline, bis(p-aminophenyl)methane; PXD = p-xylylenediamine.

[b] Tg measured by Vibron analysis.

[c] To convert N/tex to gf/den, multiply by 11.33.

Table 8. Miscellaneous Nylons

Polyamide	Mp, °C	T_g, °C	Tenacity, N/tex[a]	Moisture regain, %	Evaluation	Reference
poly(m-xylylenedipamide) [25805-74-7]	243	90	0.26–0.44	4.0	hot–wet properties and light stability inferior to -6 and -6,6	156–157
nylon-4 [24938-56-2]	262	72	0.22–0.40	7.3	excessive property loss when wet	157–158
nylon-7 [25035-01-2]	235	60	0.26–0.40	2.8		157
nylon-11 [25035-04-5]	192	57	0.26–0.44	1.3		157
nylon-12 [24937-16-4]	175	40	0.22–0.40	2.0	melting point is too low	157

[a] To convert N/tex to gf/den, multiply by 11.33.

Uses

Principal commercial uses of nylon-6 and -6,6 are in carpets, apparel, tire rein-forcement, and other industrial applications. Qiana is used primarily in apparel. Nylon-6 and -6,6 are the most used fibers in carpets, because of their excellent wear resistance, retention of appearance, and low carpet cost. In addition, a full range of colors and luster are available. For a given level of carpet appearance and wear life, less nylon, in terms of kilograms per square meter, is required than for acrylic or polyester fibers. Additionally, soil-shedding and shock-free carpets are available and are based on the use of special spinning techniques in preparing the yarns and on the incorporation of a very low percentage of conducting filaments in the yarns.

Nylons also are the predominant fibers used in the carcasses of truck tires and in airplane tires, because of their excellent strength, adhesion to rubber, and fatigue resistance. To a lesser extent, nylons are used in the carcasses of radial tires for auto-mobiles and in replacement tires of bias and bias-belted construction. The nylons do flat-spot in these latter two designs in passenger tires and, hence, have lost the large part of this market to polyester. However, nylons are essentially nonflatspotting in the carcasses of radial tires.

In the general apparel area, nylon has lost its market share to polyester because of the easier care and better wrinkle resistance of polyester. Nylon has continued to be the principal fiber in women's hosiery, men's half hose, women's intimate apparel, and stretch fabrics for slacks. There continues to be some use of nylon, especially in blends with other fibers, in many other apparel applications.

Nylon is used in a range of industrial and military applications because of its good strength, toughness, and abrasion resistance. Other applications include ropes, some seatbelts, parachutes, fishing lines and nets, and substrates for coated fabrics (qv). Nylon is used in upholstery fabrics when long-wearing characteristics are desired.

BIBLIOGRAPHY

"Polyamides" in *ECT* 1st ed., Vol. 10, pp. 916–937, by Ferdinand Schulze, E. I. du Pont de Nemours & Co., Inc., and Harold Wittcoff, General Mills, Inc.; "Polyamide Fibers" in *ECT* 2nd ed., Vol. 16, pp. 46–87, by O. E. Snider and R. J. Richardson, Allied Chemical Corporation.

1. S. Gabriel and T. A. Maas, *Ber.* **32,** 1266 (1899).

2. D. G. Bannerman, *Synthetic Fibers in Papermaking*, Interscience Publishers, a division of John Wiley & Sons, Inc., New York, 1964, pp. 57–58.

3. J. G. Cook, *Handbook of Textile Fibers*, 3rd ed., Menon Publishing, Watford, Herts, Eng., 1964, pp. 265–271.

4. P. A. Koch, *Fibres Plast.* **22,** 196 (July 1961).

5. H. Mark and G. S. Whitby, eds., *High Polymers*, Vol. I, Interscience Publishers, Inc., New York, 1940.

6. U.S. Pat. 2,071,250 (Feb. 16, 1937), W. H. Carothers (to E. I. du Pont de Nemours & Co., Inc.).

7. U.S. Pat. 2,071,253 (Feb. 16, 1937), W. H. Carothers (to E. I. du Pont de Nemours & Co., Inc.).

8. U.S. Pat. 2,130,523 (Sept. 20, 1938), W. H. Carothers (to E. I. du Pont de Nemours & Co., Inc.).

9. U.S. Pat. 2,130,948 (Sept. 20, 1938), W. H. Carothers (to E. I. du Pont de Nemours & Co., Inc.).

10. P. A. Schlack, *Chemiefasern* **17,** 961 (Dec. 1967).

11. F. S. Riordan, Jr. and J. H. Saunders, "Forty Years of Melt Spinning," *American Chemical Society Symposium*, Washington, D.C., June 10, 1980.

12. H. Klare, *Technologie und Chemie der synthetischen Fasern aus Polyamiden*, Verlag Technik, Berlin, FRG, 1954, p. 13.

13. H. Hopff, *Synthetic Fiber Development in Germany*, Part II, H. M. Stationery Office, London, Eng., 1945, p. 719.

14. Ref. 13, pp. 458–459.

15. J. G. Cook, *Handbook of Textile Fibers*, 1st ed., Menon Publishing, Watford, Herts, Eng., 1959, p. 234.

16. Ref. 13, p. 101.

17. *Text. Organon* **24,** (June 1953).

18. *Chem. Eng. News*, 10 (Mar. 9, 1981).

19. Unpublished data from Allied Chemical (1964 production).

20. H. D. W. Smith, *Am. Soc. Test. Mater. Proc.* **44,** 543 (1944).

21. J. E. Booth, *Principles of Textile Testing*, Chemical Publishing Co., Inc., New York, 1961, Chapt. 8.

22. E. R. Kaswell, *Wellington Sears Handbook of Industrial Textiles*, Wellington Sears Co., New York, 1963.

23. R. Meredith, *Mechanical Properties of Textile Fibers*, Interscience Publishers, Inc., New York, 1956, Chapt. 16.

24. W. E. Morton and J. W. S. Hearle, *Physical Properties of Textile Fibers*, Butterworths, London, Eng., 1962, Chapt. 13.

25. H. Busch, *Z. Ges. Text. Ind.* **65,** 1014 (1963).

26. P. G. Simpson, J. H. Southern, and R. L. Ballman, *Text. Res. J.* **51,** 97 (1981).

27. Ref. 24, p. 333.

28. W. Wegener, *Reyon Zellwolle Chem. Fasern* **32,** 69 (1954).

29. R. Meredith, *Mechanical Properties of Textile Fibers*, Interscience Publishers, New York, 1956, p. 76.

30. L. F. Beste and R. M. Hoffman, *Text. Res. J.* **20,** 441 (1950).

31. W. E. Morton and J. W. S. Hearle, *Physical Properties of Textile Fibers*, Butterworths, London, Eng., 1962, Chapts. 15 and 16.

32. W. Grether, *SVF Fachorgan Textilveredlung* **15**(1), 29 (1960); H. U. Schmidlin, *Preparation and Dyeing of Synthetic Fibers*, Reinhold Publishing Corp., New York, 1960, p. 45.

33. *Caprolan Nylon Product Line*, Technical Bulletin C3, Allied Chemical Fibers Division, New York, 1963, p. 11.

34. *Caprolan Nylon, Industrial Rubber Uses*, Bulletin C15, Allied Chemical, New York, 1964.

35. F. Fourné, *Synthetische Fasern*, Wissenschaftlicher Verlag, Stuttgart, FRG, 1964, p. 276.

36. *Technical Bulletin A-2*, Perlon-Hoechst, pp. 7–8.

37. R. M. Moncrieff, *Man Made Text.* **41**(481), 34 (1964).

38. U.S. Pat. 3,242,134 (Mar. 22, 1966), P. V. Papero (to Allied Chemical).

39. S. Straus and L. A. Wall, *J. Res. Natl. Bur. Std.* **60,** 39 (1958).

40. S. Straus and L. A. Wall, *J. Res. Natl. Bur. Std.* **63A,** 269 (1959).

41. U.S. Pat. 3,113,120 (Dec. 3, 1963), P. V. Papero and R. L. Morter (to Allied Chemical).

42. F. Fourné, *Synthetische Fasern*, Wissenschaftlicher Verlag, Stuttgart, FRG, 1964, p. 276.

43. U.S. Pat. 2,705,227 (Mar. 29, 1955), G. Stamatoff (to E. I. du Pont de Nemours & Co., Inc.).

44. U.S. Pat. 3,280,053 (Oct. 18, 1966), I. C. Twilley and F. P. Poznik (to Allied Chemical).

45. U.S. Pat. 3,294,735 (Dec. 27, 1966), I. C. Twilley and F. P. Poznik (to Allied Chemical).
46. K. R. Osborn, *J. Polym. Sci.* **38**, 357 (1959).
47. U.S. Pat. 3,003,995 (Oct. 10, 1961), E. C. Schule (to Allied Chemical).
48. U.S. Pat. 3,321,436 (May 23, 1967), W. Stilz and co-workers (to Badische Anilin- und Soda-Fabrik).
49. E. Mikolajewski, J. E. Swallow, and M. W. Webb, *J. Appl. Polym. Sci.* **8**, 2067 (1964).
50. *Technical Bulletin NP-4*, American Enka, July 1960.
51. *Technical Bulletin A4-2*, Perlon-Hoechst.
52. A. Alexander, *Manmade Fiber Processing*, Noyes Development Corp., Parkridge, N.J., 1966, p. 82.
53. U.S. Pat. 3,994,121 (Nov. 30, 1976), E. B. Adams (to E. I. du Pont de Nemours & Co., Inc.).
54. U.S. Pat. 4,093,147 (June 6, 1978), J. E. Bromley, M. M. McNamara, and W. T. Mowe (to Monsanto).
55. Ref. 52, p. 111.
56. H. Klare, *Synthetic Fibers from Polyamides*, Akademie-Verlag, GmbH, Berlin, FRG, 1963.
57. F. Fourné, *Synthetische Fasern*, Wissenschaftlicher Verlag, Stuttgart, FRG, 1964, pp. 39–115.
58. H. H. Weinstock, ed., *Nylon 6 Fibers, Technology, Properties, and Applications*, Wiley-Interscience, New York, sections 10–16, in press.
59. U.S. Pat. 3,311,691 (Mar. 28, 1967), A. N. Good (to E. I. du Pont de Nemours & Co., Inc.).
60. B. L. Hathorne, *Woven, Stretch and Textured Fabrics*, Interscience Publishers, a division of John Wiley & Sons, Inc., New York, 1964.
61. E. A. Hutton and W. J. Morris, *A Survey of the Literature and Patents Relating to Bulked Continuous Filament Yarns*, Bulletin 81, Shirley Institute, Manchester, Eng., July 1963.
62. U.S. Pat. 2,575,781 (Nov. 20, 1951), J. L. Barach (to Alexander Smith).
63. U.S. Pats. 2,575,837, 2,575,838, and 2,575,839 (Nov. 20, 1951), L. W. Rainard (to Alexander Smith).
64. Spunize Dept., Allied Chemical Corp., New York.
65. J. S. Taylor and D. Bhattacharya, *Text. Rec.* **80**, 58 (1962).
66. *Mod. Text. Mag.* **42**, 38 (Oct. 1961).
67. U.S. Pat. 2,919,534 (Jan. 5, 1950), E. D. Bolinger and N. E. Klein (to Deering Milliken Research Corp.).
68. U.S. Pats. 3,021,588 (Feb. 13, 1962), and 3,035,328 (June 22, 1962), E. D. Bolinger (to Deering Milliken).
69. U.S. Pats. 3,025,584 (Mar. 20, 1962), and 3,931,089 (Apr. 5, 1960), C. G. Evans (to Deering Milliken).
70. U.S. Pat. 2,019,185 (Oct. 29, 1936), R. H. Kagi (to Herberlein Patent Corp.).
71. U.S. Pats. 2,803,105, 2,803,108, and 2,803,109 (Aug. 20, 1957), N. J. Stoddard and W. A. Seem (to Universal Winding Co.); U.S. Pat. 3,025,659 (Mar. 20, 1962), (to Lessona Corp.).
72. J. J. Press, ed., *Man-Made Textile Encyclopedia*, Interscience Publishers, Inc., New York, 1959, p. 241.
73. U.S. Pat. 2,958,112 (Nov. 1, 1960), J. N. Hall (to J. N. Hall).
74. U.S. Pat. 2,884,756 (May 5, 1959), W. I. Head (to Eastman Kodak Co.).
75. U.S. Pats. 2,783,609 (Mar. 5, 1957), 2,852,906 (Sept. 23, 1958), and 2,869,967 (Jan. 20, 1959), A. L. Breen (to E. I. du Pont de Nemours & Co., Inc.).
76. U.S. Pat. 3,140,525 (July 14, 1964), D. J. Lamb (to Monsanto).
77. U.S. Pats. 3,024,516 and 3,024,517 (Mar. 13, 1962), J. E. Bromley and W. H. Hills (to Chemstrand Corp.).
78. U.S. Pat. 3,137,911 (June 23, 1964), J. E. Bromley (to Monsanto).
79. U.S. Pat. 3,005,251 (Oct. 24, 1961), C. E. Hallden, Jr. and K. Murenbeeld (to E. I. du Pont de Nemours & Co., Inc.).
80. U.S. Pat. 2,942,402 (June 28, 1960), C. W. Palm (to Celanese).
81. U.S. Pat. 2,884,756 (May 5, 1959), W. I. Head (to Eastman Kodak).
82. U.S. Pat. 2,852,906 (June 23, 1958), A. L. Breen (to E. I. du Pont de Nemours & Co., Inc.).
83. U.S. Pat. 3,164,882 (Jan. 12, 1965), N. Rosenstein and A. J. Rosenstein (to Spunize Co. of America).
84. U.S. Pat. 3,101,521 (Aug. 27, 1963), A. J. Rosenstein, N. Rosenstein, and T. F. Suggs (to Spunize Co. of America).
85. U.S. Pats. 3,031,734 (May 1, 1962), and 3,037,260 (June 5, 1962), H. J. Pike (to Allied Chemical).
86. U.S. Pat. 2,862,879 (Dec. 2, 1958), H. D. Fardon and H. J. Pike (to Allied Chemical).

87. U.S. Pat. 2,933,771 (Apr. 26, 1960), H. H. Weinstock, Jr. (to Allied Chemical).

88. H. U. Schmidlin, *Preparation and Dyeing of Synthetic Fibers*, English ed., Reinhold Publishing Corp., New York, 1963, p. 177.

89. Ref. 72, Chapt. 5.

90. F. Fourné, *Synthetische Fasern*, Wissenschaftlicher Verlag, Stuttgart, 1964, Chapt. VI.

91. H. Boehringer and F. Bolland, *Faserforsch. Textiltech.* **9**, 405 (1958).

92. K. Greenwood, *Tex. Rec.* **78**(62), 113 (1960).

93. G. M. Richardson and H. Stanley, *Mod. Text. Mag.* **43**, 53 (Feb. 1962).

94. U.S. Pat. 2,891,277 (June 23, 1959), W. L. Sutor (to E. I. du Pont de Nemours & Co., Inc.).

95. Brit. Pat. 936,729 (Sept. 11, 1963), (to E. I. du Pont de Nemours & Co., Inc.).

96. Can. Pat. 591,686 (Feb. 2, 1960), R. B. Hayden (to E. I. du Pont de Nemours & Co., Inc.).

97. Fr. Pat. 1,358,092 (Mar. 2, 1964), (to Snia Viscosa).

98. U.S. Pats. 2,939,201 and 2,939,202 (June 7, 1960), M. C. Holland (to E. I. du Pont de Nemours & Co., Inc.).

99. U.S. Pat. 3,097,416 (July 16, 1963), A. H. McKinney (to E. I. du Pont de Nemours & Co., Inc.).

100. U.S. Pat. 3,109,220 (Nov. 5, 1963), A. H. McKinney and H. E. Stanley (to E. I. du Pont de Nemours & Co., Inc.).

101. Brit. Pat. 947,183 (Jan. 22, 1964), (to Farbwerke Hoechst, A.G.).

102. U.S. Pat. 3,138,516 (June 23, 1964), J. G. Sims (to Monsanto).

103. Brit. Pat. 843,179 (Aug. 4, 1960), G. Siemer, H. Laurioch, and W. Stockigt (to VEB Thüringisches Kunstfaserwerk).

104. U.S. Pat. 3,075,242 (Jan. 29, 1963), E. Grafried (to W. C. Heraeus).

105. U.S. Pat. 3,081,490 (Mar. 19, 1963), W. Heynen and W. Martin (to Vereinigte Glanzstoff-Fabriken, A.G.).

106. U.S. Pat. 3,113,369 (Dec. 10, 1963), H. D. Barrett, R. T. Estes, and G. C. Stow (to Monsanto).

107. U.S. Pat. 3,103,448 (Sept. 10, 1963), S. E. Ross (to Allied Chemical).

108. H. C. Speed and E. W. K. Schwarz, *Textile Chemicals and Auxiliaries*, 2nd ed., Reinhold Publishing Corp., New York, 1957, p. 222.

109. R. C. Gunderson and A. W. Hart, *Synthetic Lubricants*, Reinhold Publishing Corp., New York, 1962.

110. A. E. Henshall, *Text. Rundsch.* **14**(1), 28 (1959).

111. V. E. Shashoua, *Am. Chem. Soc. Div. Polym. Chem. Prepr.* **4**(1), 189 (1963).

112. R. D. Fine, *Am. Dyest. Rep.* **43**, 405 (June 21, 1954).

113. P. Senner, *Reyon Zellwolle Chem. Fasern* **8**, 666, 744, 807, 865 (1958).

114. J. Diemunsch and J. Chabert, *Bull. Inst. Text. Fr.* **102**, 887 (1962).

115. C. Schlatter, R. A. Olney, and B. N. Baer, *Text. Res. J.* **29**, 200 (Mar. 1959).

116. U.S. Pat. 1,692,372 (Nov. 11, 1929), H. A. Gardner.

117. U.S. Pat. 1,875,894 (Sept. 6, 1933), J. A. Singmaster.

118. U.S. Pat. 1,980,428 (Nov. 13, 1935), R. H. Parkinson (to Celanese Chemical Co.).

119. U.S. Pat. 2,205,722 (June 25, 1940), G. D. Graves (to E. I. du Pont de Nemours & Co., Inc.).

120. U.S. Pat. 2,278,878 (Apr. 7, 1942), G. P. Hoff (to E. I. du Pont de Nemours & Co., Inc.).

121. U.S. Pat. 3,002,947 (Oct. 3, 1961), D. E. Maple (to E. I. du Pont de Nemours & Co., Inc.).

122. U.S. Pat. 2,671,770 (Mar. 9, 1954), J. C. Lyons (to Societe Rhodiaceta).

123. Brit. Pat. 504,714 (Apr. 28, 1939), (to E. I. du Pont de Nemours & Co., Inc.).

124. U.S. Pat. 2,689,839 (Sept. 21, 1954), W. W. Heckert (to E. I. du Pont de Nemours & Co., Inc.).

125. A. Sippel, *Melliand Textilber.* **38**, 898 (1957).

126. R. J. Fahl, *Chem. Can.* **3**, 23 (1962).

127. U.S. Pat. 2,846,332 (Aug. 8, 1958), G. A. Nesty (to Allied Chemical).

128. Ger. Pat. 1,111,771 (Mar. 15, 1962), H. J. Twitchett and A. S. Weld (to Imperial Chemical Industries, Ltd.).

129. P. Schaeffer, *Chem. Tech.* **12**, 742 (1960).

130. U.S. Pat. 3,160,600 (Dec. 8, 1964), J. R. Holsten and J. S. Tapp (to Monsanto Co.).

131. U.S. Pat. 2,875,171 (Feb. 24, 1959), S. P. Foster and R. W. Peterson (to E. I. du Pont de Nemours & Co., Inc.).

132. Brit. Pat. 971,742 (Oct. 7, 1964), (to Allied Chemical).

133. U.S. Pat. 3,329,557 (July 4, 1967), E. E. Magat and D. Tanner (to E. I. du Pont de Nemours & Co., Inc.).

134. G. A. Nesty, *Text. Res. J.* **29**, 765 (1959).

135. W. C. Carter, *Chem. Eng. News* **22**, 44 (May 4, 1944).
136. H. U. Schmidlin, *Preparation and Dyeing of Synthetic Fibers*, English ed., Reinhold Publishing Corp., New York, 1963, p. 26.
137. T. Vickerstaff, *The Physical Chemistry of Dyeing*, 2nd ed., Interscience Publishers, Inc., New York, 1954, p. 451.
138. H. Zahn and H. Spoor, *Z. Anal. Chem.* **168**, 190 (1959).
139. W. E. Beier, V. A. Dorman-Smith, and G. C. Ongemach, *Anal. Chem.* **38**(1), 123 (1966).
140. H. H. Schenker, C. C. Casto, and P. W. Mullen, *Anal. Chem.* **29**, 825 (1957).
141. H. U. Schmidlin, *Preparation and Dyeing of Synthetic Fibers*, Reinhold Pub. Corp., New York, 1963, p. 177.
142. E. A. Tippetts, *Text. Res. J.* **37**, 524 (1967).
143. E. E. Magat and R. E. Morrison, *J. Polym. Sci. Symp.* **51**, 203 (1975).
144. A. H. Bruner, N. W. Boe, and P. Byrne, *J. Elastoplast.* **5**, 201 (1973).
145. J. H. Saunders and co-workers, *J. Appl. Polym. Sci.* **19**, 1387 (1975).
146. R. J. Ceresa "Block and Graft Copolymers" in N. Bikales, ed., *Encyclopedia of Polymer Science and Technology*, Vol. 2, Interscience Publishers, a division of John Wiley & Sons, Inc., New York, 1965, pp. 485–528.
147. Brit. Pat. 918,637 (Feb. 13, 1963), (to E. I. du Pont de Nemours & Co., Inc.).
148. V. V. Korshak and T. R. Frunze, *Synthetic Hetero-Chain Polyamides*, Davy, Daniel & Co., Inc., New York, 1964, pp. 232–233.
149. E. A. Tippetts, *Am. Dyest. Rep.* **54**, 141 (1965).
150. R. D. Chapman, D. A. Holmer, O. A. Pickett, K. R. Lea, and J. H. Saunders, *Text. Res. J.* **51**, 564 (1981).
151. J. W. Hannell, *Akron Rubber Group Meeting*, Jan., 1970; *Polym. News* **1**(1), 8 (1970).
152. J. S. Rumsey, *Mod. Text.* **51**(2), 48 (1970).
153. A. Bell, J. G. Smith, and C. J. Kibler, *J. Polym. Sci.* **A3**, 19 (1965).
154. D. A. Holmer, O. A. Pickett, Jr., and J. H. Saunders, *J. Polym. Sci.* **A1**, 10 (1972).
155. B. S. Sprague and R. W. Singleton, *Text. Res. J.* **35**, 999 (1965).
156. E. F. Carlston and F. G. Lum, *Ind. Eng. Chem.* **49**, 1239 (1957).
157. R. W. Longbottom, *Mod. Text.* **49**(12), 19 (1968).
158. E. M. Peters and J. A. Gervasi, *Chem. Technol.* **2**, 16 (1972).

General References

E. M. Hicks, Jr. and co-workers, "The Production of Synthetic-Polymer Fibers," *Text. Prog.* **3**(1), 1 (1971).
A. J. Hughes and J. E. McIntyre, "Nylon Fibers," *Text. Prog.* **8**, 18 (1976).
H. F. Mark, S. M. Atlas, and E. Cernia, eds., *Man-Made Fibers Science and Technology*, 3 Vols., Wiley-Interscience, New York, 1967–1968.
Z. K. Walczak, *Formation of Synthetic Fibers*, Gordon and Breach, New York, 1977.
A. Ziabicki, *Fundamentals of Fiber Formation*, John Wiley & Sons, Inc., New York, 1976.

J. H. SAUNDERS
Monsanto Company

POLYESTER FIBERS

A polyester fiber is a manufactured fiber in which the fiber-forming substance is any long-chain synthetic polymer composed of at least 85 wt % of an ester of a dihydric alcohol (HOROH) and terephthalic acid (p-HOOCC$_6$H$_4$COOH) (1). The most widely used polyester fiber is made from linear poly(ethylene terephthalate) [25038-59-9] (PET).

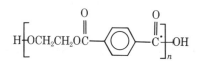

E. I. du Pont de Nemours & Co., Inc., acquired the U.S. patent rights for PET in 1948, and Imperial Chemical Industries, Ltd. (ICI) of the UK obtained patent rights for the rest of the world (2). Fiber-forming aliphatic polyesters were produced in the early 1930s, but these products had low melting points and were unsuitable for commercial use (3). Other aromatic polyesters have been investigated extensively (4). Polyester fibers became available commercially in the United States in 1953, and production expanded enormously in the 1960s and 1970s. Although the rate of increase has decreased somewhat, polyester fibers constitute the fastest-growing fiber market in the United States and worldwide (see also Fibers, chemical; Textiles).

Properties

The linear homopolymer PET, which is a condensation polymer of terephthalic acid (TA) or its dimethyl ester, dimethyl terephthalate (DMT), and ethylene glycol, is the dominant composition for commercial polyester fibers. Other homopolymers that have been produced commercially are poly(1,4-cyclohexylenedimethylene terephthalate) [24936-69-4] (Kodel II, Eastman Chemical Products, Inc.), poly(butylene terephthalate) [26062-94-2] (Fortrel II, Fiber Industries, Inc.), poly(ethylene 4-oxybenzoate) [25248-22-0] (A-Tell, Unitika, Ltd.), polyglycolide [26124-68-5] (Dexon, American Cyanamid Co.), and polypivalolactone [24937-51-7] (5–9). Only small quantities of these other homopolymers are manufactured (see Sutures).

Number average molecular weights of ca 15,000 are required for useful textile-fiber properties, but lower values give staple fibers of low tendency to pilling, ie, to forming unsightly, small fuzz balls on the fiber surface, and higher values provide high strength fibers for industrial uses (10–11). Most apparel products contain a delusterant in varying amounts, usually TiO$_2$ in quantities up to 2 wt %, which reduces fabric shine. Optically brightened polymers are common (12) (see Brighteners, fluorescent). There also are commercial fibers that contain minor amounts of copolymer that confer specific properties, eg, deep dyeability, cationic dyeability, and flame retardancy (13–15) (see Flame retardants in textiles).

Structural. Drawn polyester fibers may be considered to be composed of crystalline and noncrystalline regions. The structural unit or unit cell for PET has been deduced by means of x-ray diffraction techniques and is considered triclinic, with one repeating unit (16). The aromatic, carboxyl, and aliphatic molecular groups are nearly planar in configuration and exist in a side-by-side arrangement. The theoretical density of pure crystalline material can be determined mathematically from the dimensions of the unit cell. If the noncrystalline density can be deduced by, for example, quenching the amorphous polymer very rapidly, it is possible to calculate a percentage crystallinity, which is an important parameter in determining the physical properties of a fiber. Percentage crystallinity and molecular orientation relate to tensile strength and shrinkage; however, the various methods of measurement are problematic (17).

Mechanical.   A wide range of polyester properties is possible, depending on the method of manufacture. Generally, as the degree of stretch is increased, which yields higher crystallinity and greater molecular orientation, so are the properties, eg, tensile strength and initial Young's modulus. At the same time, ultimate extensibility or elongation normally decreases. An increase in molecular weight further increases tensile strength, modulus, and extensibility. Typical physical and mechanical properties of polyester fibers are summarized in Table 1, and some representative stress–strain curves for polyester fibers, including one for regular-tenacity nylon-6,6 filament, are given in Figure 1. Plots such as these are determined readily by use of a number of commercially available instruments (18).

Shrinkage of the fibers also varies with the mode of treatment. If relaxation of the stress and strain in the oriented fiber occurs, shrinkage decreases but the initial modulus also may be reduced. Yarns maintained at a fixed length and constant tension during heat treatment are less affected with respect to changes in modulus, etc, and reduced shrinkage values still are obtained. This latter aspect is important in fabric stabilization treatments (19).

Like most organic fibers, PET shows nonlinear and time-dependent elastic behavior. Creep occurs under load with subsequent delay in recovery on removal of the load, but compared to that of other melt-spun fibers, creep is small.

Chemical.   Polyesters show good resistance to most mineral acids, but dissolve with partial decomposition in concentrated sulfuric acid. Hydrolysis is highly dependent on temperature. On standing in water at 70°C for several weeks, conventional

Table 1.   Physical Properties of Poly(Ethylene Terephthalate)

Property	Filament yarn		Staple and tow	
	Regular tenacity[a]	High tenacity[b]	Regular tenacity[c]	High tenacity[d]
breaking tenacity[e], N/tex[f]	0.25–0.50	0.5–0.86	0.2–0.5	0.52–0.63
breaking elongation, %	19–40	10–34	25–65	18–40
elastic recovery, %	88–93 at 5%	90 at 5%	75–85 at 5%	75–85 at 5%
initial modulus, N/tex[f]	6.6–8.8	10.2–10.6	2.2–3.5	4.0–4.9
specific gravity	1.38	1.39	1.38	1.38
moisture regain[g], %	0.4	0.4	0.4	0.4
melting temperature, °C	258–263	258–263	258–263	258–263

[a] Textile filament yarns for woven and knit fabrics.
[b] Tire cord and high strength, high modulus industrial yarns.
[c] Regular staple for 100 wt % polyester fabrics, carpet yarn, fiberfill, and blending with cellulosics or wool.
[d] High strength, high modulus staple for industrial sewing thread and blending with cellulosics.
[e] Standard measurements are conducted in air at 65% rh and 22°C.
[f] To convert N/tex to g/den, multiply by 11.33.
[g] The equilibrium moisture content of the fibers at 21°C and 65% rh.

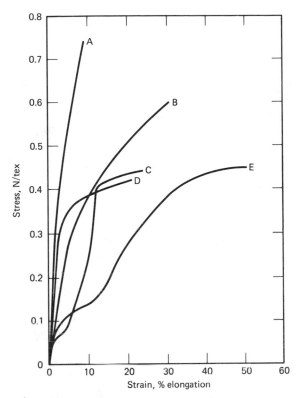

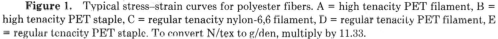

Figure 1. Typical stress–strain curves for polyester fibers. A = high tenacity PET filament, B = high tenacity PET staple, C = regular tenacity nylon-6,6 filament, D = regular tenacity PET filament, E = regular tenacity PET staple. To convert N/tex to g/den, multiply by 11.33.

polyester fibers show no measurable loss in strength; however, after one week at 100°C, the strength decreases by ca 20% (20). Basic substances attack the fiber in two ways. Strong alkalies, eg, caustic soda, etch the surface of the fiber and reduce its strength. Ammonia and other organic bases, eg, methylamine, penetrate the structure initially at the noncrystalline regions and cause degradation of the ester linkages and a general loss in physical properties (21). The polyesters display excellent resistance to conventional textile bleaching agents and are resistant to cleaning solvents and surfactants. Most of these effects are conditioned by the structural morphology of the specimens.

Other. Poly(ethylene terephthalate) fibers display good resistance toward sunlight and, in this respect, are surpassed only by the acrylics (see Acrylic and modacrylic fibers). Addition of modern light stabilizers and antioxidants enhances the polymer's performance (see Uv stabilizers; Antioxidants and antiozonants). The resistance to mildew, aging, and abrasion generally is excellent.

Manufacture and Processing

Terephthalic acid or its dimethyl ester reacts with ethylene glycol to form a diester monomer, which polymerizes to the homopolymer PET. The polymer is melted and extruded or spun through a spinneret, forming filaments which are solidified by cooling in a current of air. The spun fiber is drawn by heating and stretching the filaments to several times their original length to form a somewhat oriented crystalline structure with the desired physical properties.

Raw Materials. Prior to 1965, only DMT was used in the large-scale production of polyester fibers. Extensive development of direct *p*-xylene oxidation followed by purification made pure TA an acceptable alternative to DMT (22). Terephthalic acid accounts for ca 40% of the polyester raw-material market in the United States and worldwide. In Japan, a separate purification step is not used, and a medium purity TA is used with satisfactory results if a blueing agent, eg, divalent cobalt or optical brightener, is added during polymerization (23) (see Phthalic acids).

The most widely used TA process is based on catalytic air oxidation of *p*-xylene by means of bromine and other promoters or activators, followed by aqueous hydrogenation of the crude TA and crystallization (24). Dimethyl terephthalate is obtained by esterification of the crude TA derived from *p*-xylene oxidation or by the Witten process, which proceeds by concurrent oxidation and esterification of *p*-xylene. Crude DMT is purified by crystallization and distillation (25).

Ethylene glycol is derived from ethylene by oxidation to ethylene oxide, followed by acid hydrolysis. An alternative route based on acetoxylation of ethylene was attempted, but it failed because of corrosion-related problems (26) (see Glycols).

Polymerization. Polyester polymer is produced commercially in a two-step polymerization process, ie, monomer formation by ester interchange of DMT with glycol or esterification of TA with glycol, followed by polycondensation by removing excess glycol:

Step 1

Ester interchange:

Esterification:

Step 2

Polycondensation:

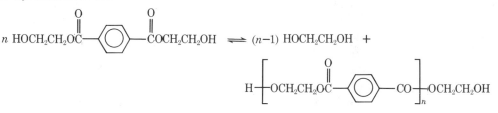

Monomer formation (step 1) by the catalyzed ester-interchange reaction between molten DMT and glycol takes place at ca 200°C. The product is a mixture of monomer, very low molecular weight polymer, and by-product methanol, which distills at ca 150°C. Ester-interchange catalysts are divalent salts of manganese, cobalt, magnesium, zinc, or calcium (see Esterification). An alternative monomer-formation system involves TA rather than DMT and an uncatalyzed direct esterification rather than ester interchange. High temperatures and pressure increase the reaction rate. The monomer, which is the same from both methods except for some end groups, usually is polymerized (step 2) in the presence of an antimony catalyst. Additives, eg, TiO_2 as a delusterant, can be added at the early stage of polymerization. Chain extension is promoted by removal of excess glycol from the very viscous melt at ca 280°C, with carefully controlled agitation and a progressive reduction of pressure to ca 200 Pa (ca 1.5 mm Hg). Heating is continued at ca 280°C until the desired degree of condensation is obtained (27). Control of molecular weight usually is by determinations of melt viscosity, and, frequently, the power input to the agitator is employed as an indicator. Side reactions also occur; the etherification of ethylene glycol leads to the formation of diethylene glycol, which is incorporated into the polymer chain as $—OCC_6H_4COO-$ $(CH_2)_2O(CH_2)_2O—$ and lowers the softening point of the polymer to an extent depending on the amount of this product generated. By-product formation of cyclic oligomers, eg, trimer and tetramer of TA and glycol, also occurs in minor amounts, depending on reaction conditions.

Older plants are based on the simpler batch process in which monomer is formed in one vessel by ester interchange or esterification and then is transferred to a second vessel for polymerization. The batch process is employed where small-scale plant economics and flexibility for manufacturing many specialty products are important. Most large-scale operations involve continuous polymerization systems, which often are coupled directly to spinning operations. As illustrated in Figure 2, a typical continuous polymerization system consists of a DMT melter, an ester-interchange column, a glycol flash vessel, a low polymerizer, a high polymerizer, and a molten-polymer manifold system that feeds into several banks of direct fiber melt spinning heads or into a solid-polymer chipping system. In some continuous polymerization units, a TA direct-esterification vessel replaces the ester-interchange column; these units are

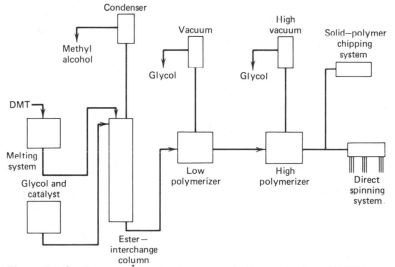

Figure 2. Continuous polymerization system for the production of PET fibers.

becoming more prominent because of favorable raw-material economics. On the large scale, continuous systems provide cost savings over batch operations because less equipment is needed to produce large quantities of polymer. Furthermore, continuous melt feed for direct spinning can be employed to eliminate the steps of forming polymer chips, followed by blending, drying, and remelting, which are required for chip spinning (see Spinning). High molecular weight products also are produced more easily by continuous processing (28).

The design of the high polymerizer is critical in determining final polymer properties and performance. It usually consists of a high vacuum, horizontal cylindrical vessel with a horizontal rotating shaft to which disks, cages, or shallow flight screws are attached; these stir the polymerizing mass to provide large surface areas so that glycol escapes more easily (29–31). The agitator also moves the highly viscous material slowly through the vessel from the inlet to the screw or pump that extracts the polymer from the high vacuum vessel. High-polymerizer vessel designs vary from falling strands, to disk-ring film generators, to twin-screw agitators (32–35). Control of the degree of polymerization is by on-line measurement of melt viscosity; deviations from a set value are corrected automatically by the degree of applied vacuum. Complex instrumentation is required, particularly in controlling the flow of material through the various states. Owing to the considerable inertia of the system, it is difficult to make quick, substantial changes in throughputs.

High molecular weight polyester polymer also can be prepared directly by employing a longer melt residence time, or by solid-state polymerization in which low molecular weight polymer chip is heated under an inert atmosphere at ca 200°C; the generated glycol is removed continuously (36). This method yields very high molecular weights and low polymer degradation, but a remelting stage is necessary for fiber formation.

Waste Recovery. Glycol and antimony catalysts are recycled into the polymerization system. Polymer and fiber waste can be separated into two chemical components, usually by methanolysis; the DMT that forms is purified and repolymerized. The waste also can be melted directly and converted to a lower quality fiber.

Drying. Molten polyester is extremely prone to hydrolysis; therefore, polymer chip must be dried to very low moisture levels, usually <0.005 wt %, before melting. This is done with hot, dry air or inert gas at ca 180°C in tumble, column, or fluidized-bed dryers (37) (see Drying).

Spinning. Poly(ethylene terephthalate), a thermoplastic, is converted to the fibrous form by melt-spinning from chip or directly from the polymerizer; melt-spinning allows great versatility in control of fiber diameter, structure, and cross-sectional shape (38–39). Filaments typically are formed by forcing molten polymer at ca 290°C through a pack containing a sandbed or metal filter, followed by extrusion through a stainless-steel spinneret containing many tiny holes which, traditionally, are round and 0.23 mm in dia. Control of polymer throughput normally is achieved by use of a special, low slip gear pump, which generates the required pressure, ie, ca 10–14 MPa (1500–2000 psi), to force the polymer through the filter–spinneret pack assembly. Hydrolytic, oxidative, and thermal degradation of molten PET is high; therefore, polymer must be conveyed to the spinneret under extremely dry, oxygen-free conditions and at carefully controlled temperatures. The extruded filaments cool mainly by air convection, and uniform cooling is achieved with a carefully controlled, turbulence-free forced-air quench system (38). Solidification of the filaments from the spinneret occurs at ca 0.6 m below the extrusion point, and the threadline converges lower down, passes over a spin-finish applicator, and then is collected at speeds of 600–4500 m/min for subsequent processing (see Fig. 3). Exact control of the process variable at each of the melt-spinning steps is necessary to achieve the desired highly uniform and high quality spun-yarn properties.

During the spinning operation, the filaments attenuate to ca 0.025 mm in dia; these fibers are amorphous but show a small degree of orientation along the fiber axis, as measured by optical birefringence (39). This contrasts with nylon-6,6, in which the filaments are crystalline but the crystallites are not well oriented. This molecular orientation in polyester is an important parameter as, together with the stretch applied at drawing, it defines the physical properties of the final yarn. Polymer molecular weight, melt viscosity, polymer throughput, extrusion temperature, spinneret geometry, threadline tension, quench rate, and drawdown or stretching affect the orientation of the extruded filament.

For staple fiber, several spinning threadlines, each of which contain 250–3000 filaments, are brought together, passed over capstans (a set of rolls) through an air ejector, and are coiled in a large can at windup speeds of 600–2000 m/min in preparation for subsequent drawing. High productivity spinning-machine designs include a large filter and spinneret pack assembly, many filaments per spinneret, a maximum number of spinning positions per machine, and high spinning speeds.

For continuous-filament yarns, the spinning threadline, which normally consists of 15–70 filaments, either is wound at 1000–3500 m/min on bobbins as feedstock for subsequent drawtwisting or drawtexturing, or it is drawn directly at high speed (2500–4000 m/min), usually over hot rolls, and wound on the final salable package. In textile yarns, the most significant commercial advancement has been the development of high speed spinning processes that yield partially oriented yarn (POY),

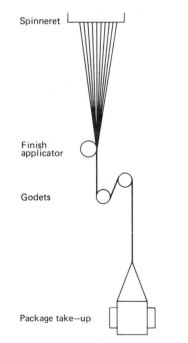

Figure 3.   Filaments extruding from the face of a spinneret.

which is the preferred feedstock for subsequent drawtexturing (40). In partially oriented yarn, the filaments have been stretched by air drag in the high speed spinning process, usually at 3000–3500 m/min, so that they show a birefringence of $(35-50) \times 10^{-3}$ but show no significant amount of crystallization (see Fig. 4). The partially oriented yarn possesses a high degree of stability and uniformity and, unlike lower birefringence spun yarn, does not deteriorate with time. This allows use of POY directly for texturing and eliminates an expensive draw-twisting step. Other advantages of the POY process are somewhat higher spinning throughput; better spun-yarn uniformity, which yields improved dye uniformity properties because of a more stable threadline at the higher spinning tension; and improved texturing properties at increased speeds. Recent advances in winder-equipment technology have made this high speed process economically feasible, and production of stable packages of increased size, ie, 10–20 kg, is achieved at lower costs. Further developments include winders with automatic handling systems and windup speeds of ca 7000 m/min (41).

In addition to the manufacture of normal circular cross-sectional fibers, spinnerets with noncircular holes yield filaments with multiple lobes, which reduce the undesirable glitter of some fabrics in bright light, as well as triangular filaments, which increase fabric sheen (42). Other specialty fiber properties are obtained by extrusion of hollow fibers; very fine, low tex/filament (den/filament) fibers; thick–thin conjugates

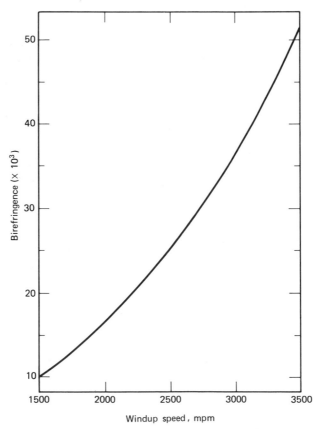

Figure 4. Typical effect of windup speed on the orientation or birefringence of polyester spun yarn.

of side-by-side composite fibers; multicore composite fibers; and core-sheath composite fibers (42–46). These composites can be biconstituent, ie, from two different polymer streams.

Drawing, Relaxing, and Stabilization. Drawing, relaxing, and stabilization produce the oriented crystalline structure that gives polyester fiber its characteristic properties. Drawing is the stretching of fibers of low molecular orientation to several times their original length; relaxing is the releasing of strains and stresses, which results in reduced shrinkage in the drawn fibers; and stabilization is the treatment of the fibers with heat to enable them to resist further changes in dimensions. In the drawing process, PET spun yarn is heated above its second-order (glass) transition temperature (T_g), ca 80°C. Normally, an ordered crystalline structure is produced, and the degree of order depends markedly on the temperature, rate of stretching, and draw ratio (39,47–49).

In staple-fiber manufacture, very large tows comprising ca $(1–3) \times 10^6$ filaments of undrawn yarn are made by combining several ends or lines of spun yarn and drawing them on a heavy-duty draw frame at ca 100–300 m/min. The draw is achieved by leading the tow between two sets of rolls operating at different speeds. To ensure that no slip occurs, the rolls are highly polished, which increases the coefficient of friction, and additional nip rolls, which grip the tow, may be used. Heating the spun yarn above the T_g is necessary to obtain uniform drawing and is achieved in a variety of ways, ie, with steam, hot water, hot rolls, ir heat, etc. The drawn tow is crimped, usually with a stuffer-box crimper; dried and heat-set; and either packaged as tow or cut into the required staple lengths, commonly 38–152 mm, and baled for sale.

Flat or hard yarn, ie, oriented, crystalline fiber, for direct weaving or knitting, or as a feedstock for subsequent texturing often is made on drawtwisters. This continuous-filament yarn, which consists of 3–100 filaments, is drawn by stretching the yarn between two rolls; the second or draw roll operates at approximately three and one-half times the speed of the feed roll. A typical drawtwister is characterized by about 200 individual positions. Slippage is minimized by making several turns around the highly polished rolls, which normally have associated with them smaller idler rolls that are canted so that the turns do not run together. The temperature of the spun yarn is raised to the drawing temperature with heated feed rolls, or by passing the yarn around a heated mat chrome pin or onto some other heated surface (50). The drawn yarn is wound at ca 1000 m/min on a cylindrical tube with a conventional ring and traveler mechanism, and the package, which contains ca 2–5 kg yarn and has a small amount of base twist, is either packed in cartons for sale or sent to a beaming operation. Higher stretch, which gives high strength industrial yarn, is achieved in a second stage of drawing at a high temperature (ca 200°C) (51). Additional heat can be applied by inserting a slightly curved hot plate or steam jet between the feed and draw rolls; the actual incremental draw ratio depends upon total draw ratio and temperature.

High speed drawing can be integrated with spinning (spin-draw). The polymer is melt-spun, heated, stretched, annealed, and wound at high speed (2500–4000 m/min) (52–53). Advantages of this process are lower labor costs and the elimination of several handling steps, but it involves much more complex machinery, and a low interruption rate is required.

Texturing. Continuous-filament, 5.6–33-tex (50–300-den) yarn can be textured easily by several, eg, false-twist, stuffer-box, knit-deknit, and air-jet techniques (54). In texturing, filaments are crimped or looped at random to give the yarn greater volume or bulk; then they are heat-set for dimensional stability. The most common textured yarns are the false-twist textured yarns made with aggregates of tiny, high speed spindles, stacked friction disks, or cross-belt nips, all of which impart a high degree of twist (55–56). The twist usually is 24–28 turns/cm and passes up the threadline and over an upstream hot plate at over 200°C. This treatment imparts a permanent helical deformation to the yarn. The yarn is cooled before detwisting and, as the yarn emerges from the aggregate, it is passed over another hot plate for stabilization, is lubricated, and is wound on a 2–5-kg package. The drawing, texturing, and heatsetting steps usually are carried out in an integrated manner by either a sequential, ie, drawn first, then textured, or a simultaneous process, ie, simultaneous drawing and texturing, on one piece of equipment (55,57). The draw-texture machine typically consists of ca 200 ends and can be operated at 150–900 m/min. With the advent of POY, draw-texturing technology has advanced rapidly and is now in common use. Another type of textured yarn that is increasing in importance can be made by air-jet treatment, which produces tiny loops along the length of the yarn (55).

Dyeing.   Polyester fibers are dyed almost exclusively with disperse dyes because of the lack of reactive dye sites in the fiber (see Dyes, application). The dyes act primarily through controlled dispersion with only weak hydrogen bonding of the dye molecule to the fiber (58–60). The choice of a disperse dye suitable for dyeing polyester fiber depends on its affinity to PET; leveling; colorfastness, particularly to light, dry cleaning, and to sublimation; size of the dye molecule; and the diffusion rate of the dye into the fiber. Other classes of dyes that have been used to a significant degree are azoic dyes for limited shades by a high temperature process, vat dyes mainly for Thermosol dyeing, leuco ester dyes for a few pale shades, resin-bonded pigments for pale-to-medium shades and printing, and cationic dyes for modified, ie, reactive, polyester semicolored fiber variants (see Azo dyes; Dyes and dye intermediates).

The rate of dyeing polyester depends on the temperature, time, and thermal history of the fiber, as is shown by the temperature–time plot for a typical disperse dyeing in Figure 5. Exhaustion of the dye, as shown in Figure 5, does not occur below 120°C within practical limits of time. Heat-setting polyester before dyeing affects dyeability markedly, as can be seen in Figure 6. The application of heat reorganizes or opens the polymer structure, thereby making it possible for the dye molecule to diffuse into the fiber (61–62). The temperature, time, and dye-uptake data in Figures 5 and 6 should not be considered absolute, because other factors, eg, yarn/fabric tension, molecular structure of the dye, and use of pressure dyeing, affect the relationships shown in these figures significantly.

Because of the hydrophobic nature of polyester and its lack of dye sites, dyeing from an aqueous bath requires temperatures above 100°C or the use of a dyeing accelerant or carrier (see Dye carriers). Carriers generally are aromatic organic compounds that swell the polyester fiber so that the monomolecular disperse dye can penetrate more rapidly into the fiber. Ideally, a carrier should not be volatile in steam, should be removed easily after dyeing, and cause no impairment of the lightfastness of the dye. A carrier increases dye uptake to an optimum concentration, above which stripping of the dye from the fiber occurs because of the solvent action on the dye. Therefore, carriers often are used at higher concentrations to achieve partial stripping of the dyed fabic (63–64).

The dyeing of polyester often is complicated by its combination with other fibers in fabric blends. Polyester–cotton and polyester–rayon blends can be piece-dyed with

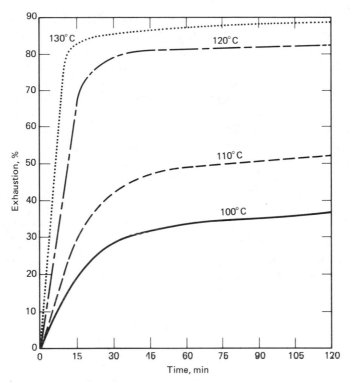

Figure 5. A temperature–time plot for a typical disperse dyeing of polyester fibers (the dye used was Palanil navy blue RE).

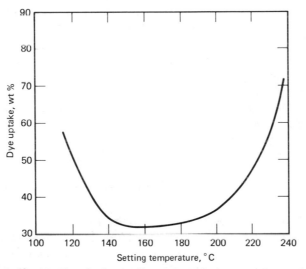

Figure 6. Effect of heat-setting of polyester fibers prior to dyeing on percentage dye uptake. (Terylene fiber treated with 2.0 wt % Dispersol Fast Scarlet B 150 fine powder at 100°C for 90 min).

disperse–vat, disperse–direct, and disperse–reactive dye systems (see Dyes, reactive). The important consideration in these systems is to minimize the staining of disperse dye on the cotton or rayon since cellulosic dyes do not stain polyester. For dyeing with heavy shades, it often is necessary to include an intermediate clearing step after dyeing with disperse dye and before dyeing the cellulosics. In the dyeing of polyester–wool yarns, staining on the wool is reduced by using a disperse dye. A one-bath method involving carriers at 100°C is favored because of the damage that results to the wool at high temperatures and over long dyeing times. Since rapid exhaustion of disperse dye from the bath reduces staining of wool, a type of polyester fiber that dyes faster than wool frequently is used. Nylon or cellulose triacetate sometimes are combined with polyester and can be disperse-dyed with the polyester. Polyester fiber can be modified with a comonomer for more rapid and deeper dyeing with disperse dyes; when it is blended with regular polyester, tone-on-tone color effects are achieved (65). Modified polyester allows dying without a carrier where use of the carrier presents an environmental problem. Polyester fiber that contains sulfonic acid groups can be dyed with disperse and cationic dyes. When this fiber is used in blends with regular polyester, wool, or cellulosic fibers, unique multicolor effects can be obtained.

Several commercial methods are used for dyeing polyester and polyester-blend fabrics and yarns. Package dyeing of polyester-textured filament or polyester spun yarn and blends is done under pressure with a carrier to obtain good dye penetration and level dyeing throughout the package. Beck and jet dyeing of polyester fabrics usually involves the same techniques, but some spun fabrics are dyed with atmospheric becks or jets, which are less expensive than pressure equipment. Most polyester–cellulosic woven blends are Thermosol-dyed continuously, which involves the use of a disperse-dye pad, an ir predyeing unit, a Thermosol oven, a cotton dye pad, a steamer, wash boxes, and drying cans. For larger yardages of the same color, this process presents the most economical dye process for polyester-containing fabrics. Such fabrics usually are 50–50, 65–35, or 80–20% blends of polyester and cotton or rayon, respectively. The polyester portion of the blend is dyed by heat diffusion of the dye into the fiber for 40–60 s at 190–200°C in the Thermosol oven. The cotton or rayon portion is dyed by steaming the fabric for 40–60 s at 100–110°C in a steamer with vat, fiber-reactive, or sulfur dyes.

In another commonly used system, polyester blends for sheeting- and shirting-weight fabrics are dyed or printed with pigments, which are resin-bonded on the fabric surface. Light and medium shades can be dyed or printed by this system very economically. Other techniques that are used to dye polyester fibers include organic solvent dyeing and spun-dyeing with melt-colored pigments. The dyeing of polyester fibers and blends is reviewed in ref. 42.

Finishing. Poly(ethylene terephthalate) is a thermoplastic and, therefore, heat is applied to set the structural memory desired for maximum performance of PET fiber either alone or in blends. This important heat-setting technique known as no-cure-permanent-press, or durable press, provides the wrinkleproof, ease-of-care, and low laundering shrinkage properties for which polyester fabrics are noted. In the finishing of polyester blends with cotton, the most important process is cured permanent press, in which cotton in 50–50, 65–35, or 80–20% polyester–cotton blends is cross-linked to improve its recovery power and wrinkle resistance (66). The resin-treated fabrics are hydrophobic and, therefore, exhibit greater soiling tendency. For this reason, additional soil-release finishes are necessary to change the surface characteristics of the permanent-press fabrics (67) (see Textiles, finishes).

Other textile finishing methods are employed after dyeing to change the properties of polyester fabrics (42). Esthetic finishes modify the appearance or tactile property of the fabric, or both, and functional finishes are used to improve the performance of a fabric under specific conditions of use. Both chemical, ie, wet, and mechanical, ie, dry, methods are utilized commercially. Finishing for esthetics includes napping and shearing to change a flat smooth surface to a bulky flannel or fleece surface; sueding to obtain a sandpaper-surface effect; embossing, glazing, and chintzing to produce a shiny, hard surface; and hand building or softening to yield either a firmer, harsher feel or a softer feel, respectively. Finishing for performance includes singeing to remove surface fiber and, thereby, minimize pilling; waterproofing in combination with soil-repelling; adding antistatic agents (qv) to facilitate laundering; adding hygienic agents; and heat-setting and permanent-press treatment.

Economic Aspects

Since the advent of polyester fibers in the 1940s, when TA was a rare and expensive chemical, world production of polyester fibers rose by 1980 to 5×10^6 metric tons (68). Polyester accounts for almost half of all synthetic fibers and one seventh of all textile fibers. The growth of the polyester fibers market in the United States is illustrated in Figure 7 for the main product lines (69). Factors that increase polyester staple and filament growths to these high levels are favorable interfiber economics resulting from increased volumes, expanding uses and new products and, most im-

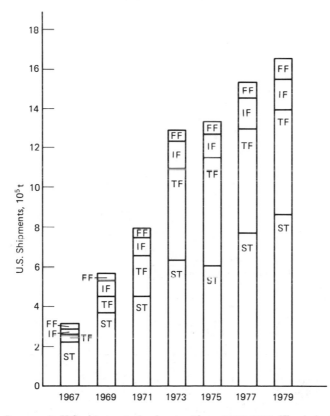

Figure 7. Increase in U.S. shipment of polyester fibers, 1967–1979. IF = industrial filament yarn, used in tire cord, V-belts, fire hose, and thread; TF = textile filament yarn, used in dresses, lingerie, uniforms, sport shirts, rainwear, home furnishings, and other items; ST = staple and tow, used in many of the apparel outlets listed above and in carpets, sheets, pillowcases, scatter rugs, and nonwovens; FF = fiberfill, a filling material used in pillows, sleeping bags, mattresses, cushions, and like products. Adapted from ref. 69.

portant, the substitution of polyester for cotton. Worldwide, these same factors have resulted in the emergence of many new producers, particularly in developing countries (70). U.S. polyester-fiber capacity in 1981 is listed in Table 2 according to manufacturer. World production of polyester fibers for 1980 by region is listed in Table 3.

Staple fiber represents the largest share (ca 60%) of the polyester market in the United States and worldwide. Staple capacity has remained in balance with a steadily increasing demand, and prices have risen along with raw-material and energy costs. However, the textile-filament yarn market increased tremendously in 1973, but demand has leveled off, and the excess capacity brought on-stream has resulted in a severe deterioration of prices in a period of rapid cost escalation.

Price and supply of raw-material feedstocks will be principal factors in the future economic position of polyester fiber. Manufacture of PET requires only a small portion of the petroleum stream but competes with gasoline octane boosters for the supply of p-xylene. Energy also is an important factor; production of 1 kg of polyester fiber requires a total energy consumption of ca 167 MJ (72,000 Btu).

Analytical and Test Methods

Chemical and physical methods that are used for most fibers can be applied to PET. Physical and mechanical methods are described in ref. 71. Poly(ethylene ter-

Table 2.   1981 U.S. Polyester Fiber Capacity, 10^3 t/yr[a]

Manufacturer	Trademark	Staple	Filament	Total
American Enka Corp.	Encron	27	50	77
Avtex Fibers, Inc.	Avlin		32	32
E. I. du Pont de Nemours & Co., Inc.	Dacron	431	318	749
Eastman Chemical Products, Inc.	Kodel	195	54	249
Fiber Industries, Inc.[b]	Fortrel	354	249[c]	603
Hoechst Fibers, Inc.	Trevira	159	48	207
Monsanto Co.	Blue C	45		
others		18	70	88
Total		*1229*	*821*	*2050*

[a] Adapted from ref. 70.

[b] Jointly owned by Celanese (62.5%) and ICI (37.5%).

[c] 45,000 t of polyester textile filament capacity is on standby.

Table 3.   1980 World Production of Polyester Fibers[a]

Region	10^6 t/yr
U.S.	1.8
other Western Hemisphere	0.4
West Europe	0.8
East Europe	0.4
Japan	0.6
East Asia	0.7
other	0.4
Total	*5.1*

[a] Ref. 68.

ephthalate) fibers can be characterized by microscopic tests, eg, the measurement of birefringence, specific gravity, and softening point. Many of the producers identify their own products by the use of tracer elements or by the catalyst residues remaining in the polymer. Catalyst residues usually are determined using x-ray fluorescence methods. Recourse to spectroscopic data also can be invaluable. Detailed structures of the ir spectra of PET have been reported (72–73). Nuclear magnetic resonance is useful for determining second-order transition temperatures of the polymer and oriented fibers (74). Differential scanning calorimetry and differential thermal analysis are other methods of detecting heat-energy transitions (75). Mass spectrometry, x-ray diffraction, and scanning electron microscopy also are employed in PET morphology studies.

With respect to wet methods, solution viscosity is used widely to measure molecular weight. Powerful solvents, eg, phenol, o-chlorophenol, tetrachloroethane, and dichloroacetic acid, are needed to dissolve polyester fibers. A number of correlations have been made between number-average molecular weight ($\overline{M}_n$) and the intrinsic viscosity (η); the general (Mark-Houwink) equation is (76)

$$(\eta) = KM_n^a$$

where K and a are measurable constants. For quality-control purposes, relative viscosities, ie, the ratio of solution viscosity to solvent viscosity at fixed concentration, are measured. Chromatography is used to separate and analyze PET components and by-products. Diethylene glycol content, oligomer distribution, comonomer content, spin finish formulations, etc, can be determined.

Health and Safety Factors

Poly(ethylene terephthalate) is chemically stable and virtually nonbiodegradable. In comparisons between *in vitro* toxicity and *in vivo* fibrinogenicity for various polymer and mineral dusts, it has been shown that asbestos exhibits the highest effects, and PET the lowest (77–78). Trevira polyester fiber is nonmutagenic in the Ames test (79).

Uses

Staple. Blends with cotton form polyester's principal use in the United States and worldwide. Normally, the material is sold in bales of ca 230–360 kg; the tex per filament of the textile and industrial fibers is ca 0.17–0.33 (1.5–3 den/filament); the cut length is 38–76 mm; the crimp frequency is ca 0.4–0.6 crimps/mm; and the tenacity is ca 0.5–0.6 N/tex (5–7 g/den). Cotton blends, which commonly are 65 wt % polyester for lightweight and 50 wt % polyester for heavier-weight fabrics, are particularly useful for permanent-press garments. Blends with rayon also are important. High modulus staple [0.35 N/tex (4 g/den)] yields lightweight fabrics of high strength (80) and has been a principal factor in the application of blends to other markets, eg, sheetings, home furnishings, and industrial uses. Heavier weight fabrics are made from lower-modulus [0.13 N/tex (1.5 g/den)] fibers. Wool blends containing 65 or 50 wt % polyester commonly are used for men's suiting. Staple length is 76–102 mm and tex per filament usually is 0.3 or 0.4 (3 or 4 den/filament). Low modulus fibers normally are used.

Use of polyester staple in sewing threads is well established. The tenacity and weight of the fiber is greater than 0.4 N/tex (6 g/den) and 0.14 tex/filament (1.25

den/filament), respectively. Also, 0.7–1.8-tex/filament (6–16-den/filament) staple is being used as a carpet fiber. The staple of lower tex per filament is used in throw rugs and tufted area rugs; the staple of higher tex per filament is used in broadloom carpeting. Other staple products have been engineered for uses as fiberfill for mattresses, sleeping bags, pillows, etc, and in the nonwoven area (see Nonwoven textile fabrics, staple fibers).

Filament. Flat, fully drawn textile yarns are offered in weights-per-length of 3–28 tex (30–250 den) with individual filament weight per length typically at 0.17–0.56 tex (1.5–5 den). Tenacities are ca 0.35 N/tex (4 g/den). These yarns are used directly in weaving, tricot knitting, and some texturing operations for apparel applications. Flat yarn has been almost entirely replaced by POY as a feedstock for texturing. Use of flat polyester filament yarn is increasing in the tricot-fabric market.

Draw-textured yarns are applied primarily in apparel knit, woven, and tricot fabrics. Properties and weights are similar to the flat yarns. These heat-set bulked-yarn products offer good dimensional stability and excellent garment wear properties, and the unstabilized yarns give a higher degree of stretch.

Industrial, high strength, continuous-filament yarns are made for automobile tire reinforcements as tire cord as well as for seat belts, fire hose, V-belts, etc, where strength, high modulus, dimensional stability, and low creep are important (see Tire cord). Yarn weights per length are 93–187 tex (840–1680 den), and the tex per filament is ca 0.7 (6 den/filament). Tenacities are ca 0.7–0.84 N/tex (8.0–9.5 g/den). Lower tex fibers [5–49 tex (45–440 den)] are used in corespun sewing threads.

Spun-Bonded. The use of spun-bonded nonwoven fabrics made from filament yarns is increasing steadily (see Nonwoven textile fabrics, spun-bonded). DuPont's Reemay is a spun-bonded product that makes use of polyester filaments held together with a polyester binder (81). Uses are in interlinings, electrical insulation, and carpet underlay reinforcement (see Insulation, electric). Needled spun-bonded products have been engineered for road reinforcement (82).

Other. Extensive effort has been made to widen the range of applications for polyester fibers by modifying the PET structure so as to improve properties. In addition to deep-disperse dyeability, cationic dyeability, and flame retardancy, modifications have introduced antipill, high shrinkage, antistatic, antisoiling, melt-colored, bicomponent, modified cross-section, low tex per filament, spunlike, and silklike properties (83–84).

BIBLIOGRAPHY

"Textile Fibers (Polyester)" in *ECT* 1st ed., Vol. 13, pp. 840–847, by S. A. Rossmassler, E. I. du Pont de Nemours & Co., Inc.; "Polyester Fibers" in *ECT* 2nd ed., Vol. 16, pp. 143–159, by G. Farrow and E. S. Hill, Fibers Industries, Inc.

1. G. Farrow and E. S. Hill in N. M. Bikales, ed., *Encyclopedia of Polymer Science and Technology*, Vol. 11, Interscience Publishers, a division of John Wiley & Sons, Inc., New York, 1969, p. 1.
2. U.S. Pat. 2,465,319 (March 22, 1949), J. R. Whinfield and J. T. Dickson (to E. I. du Pont de Nemours & Co., Inc.); Brit. Pat. 578,079 (Aug. 29, 1946), J. R. Whinfield and J. T. Dickson (to Imperial Chemical Industries, Ltd.).
3. W. H. Carothers and J. W. Hill, *J. Am. Chem. Soc.* **54,** 1577, 1579 (1932).
4. I. Goodman and J. A. Rhys, *Polyesters*, Vol. I, Iliffe Books Ltd., London, UK, 1965.
5. E. V. Martin and C. J. Kibler in H. F. Mark and co-workers, eds., *Man-Made Fibers, Science and Technology*, Vol. 3, Interscience Publishers, a division of John Wiley & Sons, Inc., New York, 1968, p. 83.

6. H. Sattler, *Text. Prax. Int.* **10**(II–IV), 1175 (1978).
7. K. Mihara, *Angew. Makromol. Chem.* **40–41**, 41 (1974).
8. U.S. Pat. 3,297,033 (Oct. 31, 1963), E. E. Schmitt (to American Cyanamid Co.).
9. H. A. Oosterhof, *Polymer* **15**, 49 (1974).
10. Brit. Pat. 840,796 (July 13, 1960), P. W. Eggleston and co-workers (to Imperial Chemical Industries, Ltd.).
11. U.S. Pat. 3,051,212 (Aug. 28, 1962), W. W. Daniels (to E. I. du Pont de Nemours & Co., Inc.).
12. P. Quensel and co-eds., *Ciba Geigy Rev.* **1**, 26 (1973).
13. Brit. Pat. 938,055 (Sept. 25, 1963), G. Richards and co-workers (to Imperial Chemical Industries, Ltd.).
14. U.S. Pat. 3,018,272 (Jan. 23, 1962), J. M. Griffing and W. R. Remington (to E. I. du Pont de Nemours & Co., Inc.).
15. U.S. Pat. 3,794,617 (Oct. 30, 1972), H. E. Mains and co-workers (to Emery Industries).
16. R. Daubney and co-workers, *Proc. R. Soc. London Ser. A* **226**, 531 (1954).
17. G. Farrow and I. M. Ward, *Polymer* **1**, 330 (1949).
18. H. Hindman and G. S. Burr, *J. Eng. Ind.* **71**, 789 (1960).
19. D. N. Marvin, *J. Soc. Dyers Colour.* **70**, 16 (1954).
20. *Chemical Properties of "Terylene": The Effect of Water and Steam*, Technical Bulletin, ICI Fibres, Ltd., Harrogate, UK.
21. G. Farrow and co-workers, *Polymer* **3**, 17 (1962).
22. W. S. Witts, *Chem. Process Eng.* **51**, 55 (1970).
23. Jpn. Pat. 1,091,991 (Aug. 12, 1976), (to Kuraray KK); Jpn. Pat. 54-120699 (Sept. 19, 1979), (to Toyobo KK).
24. Brit. Pat. 994,769 (March 29, 1963), (to Standard Oil Co.).
25. H. Ludewig, *Polyester Fibers, Chemistry and Technology*, Wiley Interscience, New York, 1979, p. 45.
26. U.S. Pat. 3,689,535 (March 24, 1969), J. Kollar (to Halcon International Inc.).
27. R. Hill, *Fibers From Synthetic Polymers*, Elsevier Publishing Co., Amsterdam, The Netherlands, 1953, p. 149.
28. U.S. Pat. 2,933,476 (Apr. 14, 1960), W. F. Fisher (to E. I. du Pont de Nemours & Co., Inc.).
29. U.S. Pat. 2,758,915 (Aug. 14, 1956), J. L. Vodonik (to E. I. du Pont de Nemours & Co., Inc.).
30. U.S. Pat. 3,118,739 (Jan. 21, 1964), W. G. Atkinson and J. L. Thomas (to E. I. du Pont de Nemours & Co., Inc.).
31. M. Sittig, *Chem. Process Rev.* (56), 118 (1971).
32. U.S. Pat. 3,110,547 (Nov. 12, 1963), R. E. Emmert (to E. I. du Pont de Nemours & Co., Inc.).
33. U.S. Pat. 3,499,873 (March 10, 1970), H. Kuchne and co workers (to Zimmer AG).
34. U.S. Pat. 3,964,874 (June 22, 1976), M. Maruko and co-workers (to Hitachi Ltd.).
35. U.S. Pat. 3,553,171 (Jan. 5, 1971), H. Ocker (to Werner & Pfleiderer).
36. F. C. Chen and co-workers, *J. Amer. Inst. Chem. Engrs.* **15**, 680 (1969).
37. Ref. 25, p. 119.
38. U.S. Pat. 3,135,811 (June 2, 1964), T. R. Barrett and H. E. Warner (to Imperial Chemical Industries, Ltd.).
39. A. Ziabichi, *Fundamentals of Fibre Formation, The Science of Spinning and Drawing*, Wiley-Interscience, New York, 1976.
40. G. Clayton and co-workers, *Text. Prog.* **8**(1), 31 (1976).
41. G. Schubert, *Internat. Text. Bull.* (World Edition Spinning 3), 229 (1980).
42. O. Pajgrt and B. Reichstadter, *Textile Science and Technology, Processing of Polyester Fibers*, Vol. 2, Elsevier Scientific Publishing Co., Amsterdam, The Netherlands, 1979, p. 38.
43. M. Okamoto, *Chemiefasern Text. Angewendungstech. Text. Ind.* **29**(81), 175 (1970).
44. U.S. Pat. 4,157,419 (June 5, 1979), M. E. Mirhej (to E. I. du Pont de Nemours & Co., Inc.).
45. Brit. Pat. 1,550,734 (Aug. 22, 1979), M. Okamoto (to Toray Industries, Inc.).
46. U.S. Pat. 3,760,052 (Sept. 18, 1973), N. Fukuma (to Asahi Chemical Industry Co., Ltd.).
47. K. Walczak, *Formation of Synthetic Fibers*, Gordon and Breach Science Publishers, New York, 1977.
48. I. Marshall and A. B. Thompson, *Proc. R. Soc. London Ser. A* **221**, 541 (1954).
49. A. B. Thompson, *J. Poly. Sci.* **34**, 741 (1959).
50. Ref. 27, p. 375.
51. U.S. Pat. 2,942,325 (Sept. 15, 1964), F. E. Spellman (to E. I. du Pont de Nemours & Co., Inc.).

52. U.S. Pat. 3,216,187 (Dec. 27, 1963), W. A. Chantry and A. E. Molini (to E. I. du Pont de Nemours & Co., Inc.).
53. Brit. Pat. 874,652 (Aug. 10, 1961), J. G. Gillet and co-workers (to Imperial Chemical Industries, Ltd.).
54. Ref. 42, p. 176.
55. M. J. Denton, *Polyester Textiles: Papers Presented at the Ninth Shirley International Seminar*, Shirley Institute Publication S26, Manchester, UK, 1977, p. 31.
56. Brit. Pat. 1,083,052 (Apr. 7, 1966), Heberlein; U.S. Pat. 4,047,373 (Sept. 13, 1977), I. Takai (to Oda Gosen Kogyo KK).
57. Ref. 40, p. 37.
58. H. U. Schmidlin, *Preparation and Dyeing of Synthetic Fibers*, Reinhold Publishing Corp., New York, 1963.
59. Ref. 42, p. 258.
60. *The Dyeing of Terylene Polyester Fiber*, Imperial Chemical Industries, Ltd., 1955.
61. W. J. Wygard, *Am. Dyestuff Rep.* **54,** 545 (1965).
62. W. Beckmann, *Can. Text. J.* **83,** 43 (March 1966).
63. E. Balmforth and co-workers, *J. Soc. Dyers Colour.* **82,** 405 (1966).
64. C. R. Jin and D. M. Cates, *Am. Dyestuff Rep.* **53,** 64 (1964).
65. F. S. Mousalli and T. L. Holmes, *1979 National Technical Conference, American Association of Textile Chemists and Colorists*, Philadelphia, Pa., Oct. 3–5, 1979, p. 144.
66. U.S. Pat. 2,974,432 (March 14, 1961), W. K. Warnock and F. G. Hubener (to Koret of California).
67. Ref. 42, p. 370.
68. *Textile Organon* **52,** 83 (1981).
69. C. A. Whitehead, ed., *Textile Organon,* Textile Economics Bureau, Inc., New York, 1967–1979.
70. *Chem. Week*, 22 (Apr. 22, 1981).
71. J. S. Hearle and R. Meredith, eds., *Physical Methods of Investigating Textiles*, Interscience Publishers, Inc., New York, 1959.
72. D. Grime and I. M. Ward, *Trans. Faraday Soc.* **54,** 959 (1958).
73. A. Miyake, *J. Poly. Sci.* **38,** 479, 497 (1959).
74. C. A. Boye and V. W. Goodlett, *J. Appl. Phys.* **34,** 59 (1963).
75. B. Ke, ed., *Newer Methods of Polymer Characterization*, Interscience Publishers, Inc., a division of John Wiley & Sons, Inc., New York, 1964, Chapt. IX.
76. P. J. Flory, *J. Am. Chem. Soc.* **65,** 372 (1943).
77. D. M. Conning and co-workers, *Proc. 3rd Int. Symp. Inhaled Particles* **1,** 499 (1971).
78. J. A. Styles and J. Wilson, *Ann. Occup. Hyg.* **16**(3), 241 (1973).
79. S. M. Suchecki, *Text. Ind.* **142**(8), 68 (1978).
80. K. C. McAlister, *Mod. Text. Mag.* **48,** 77 (1967).
81. R. A. A. Hentschel, *Chem. Technol.* 32, (Jan. 1974).
82. K. L. Floyd, *paper presented at the Ninth Shirley International Seminar*, 1977, Shirley Institute Publication S26, Manchester, UK, p. 171.
83. E. M. Hicks and co-workers, *Text. Prog.* **3**(1), 38 (1971).
84. Ref. 42, p. 30.

General References

References 1, 4, 5, 25, 27, 31, 39, 40, 42, 47, 55, 82 and 83 are all excellent for general review of polyester fibers.
R. W. Moncrieff, ed., *Man-Made Fibers*, 6th ed., Butterworth & Co., Ltd., London, UK, 1975, p. 434.
J. S. Robinson, ed., *Fiber-Forming Polymers, Recent Advances*, Noyes Data Corporation, Park Ridge, N.J., 1980, p. 3.
J. S. Robinson, ed., *Spinning, Extruding and Processing of Fibers, Recent Advances*, Noyes Data Corporation, Park Ridge, N.J., 1980, pp. 56, 143, 272, and 344.
F. Happey, ed., *Applied Fibre Science*, Vols. I–III, Academic Press, London, UK, 1977, 1978, and 1979.

GERALD W. DAVIS
ERIC S. HILL
Fiber Industries, Inc.

RAYON

The history of the inception and early development of rayon is well-documented (1–2). In the late 1950s and early 1960s rayon manufacturers, realizing that their regular rayon fibers lacked resistance to alkali, ie, regular rayon fibers exhibited loss of strength and lack of dimensional stability, developed high wet-modulus-type rayons (HWM) that largely overcame these deficiencies.

The first high performance (HP) rayons were marketed in the United States by American Viscose (Avril) and American Enka (Fiber 700) and, in the UK, by Courtaulds (Vincel). Other HWM fiber producers in Europe were Snia Viscosa (Koplon) and Lenzing (Hoch Modul 333). These rayons were prepared by the addition of various amines and polyglycols to viscose and of zinc to the acidic coagulation baths, and modification of the overall spinning conditions. During this period, the Japanese at the Tachikawa Research Institute took a somewhat different approach and used cellulose of very high degrees of polymerization (DP) and high CS_2 concentrations combined with low acid-bath concentrations (without zinc) and high stretching to produce so-called polynosic fibers (3). These fibers had high wet modulus and low caustic solubility. Fabrics produced from these polynosic fibers displayed the cover and crisp hand of cotton. As originally produced, polynosics had low elongation, low loop-and-knot strength, and poor abrasion resistance. Subsequently, they were prepared by some of the HWM-type technology, and today's fibers have good overall properties.

In the 1970s, further improvements in HP rayons were introduced to make these fibers more competitive with cotton. Among these were Avtex Fibers' Avril III, a multilobal fiber, and ITT Rayonier's Prima, a HWM crimped fiber (4–6). Courtaulds' Viloft is produced by adding sodium carbonate to the viscose to give a hollow fiber of higher water-holding capacity (7). Unfortunately, Viloft fibers and many of the other modified fibers are not high wet-modulus types. Fibers of higher water-holding capacity are American Enka's Absorbit and Avtex Fibers' Maxisorb, which are produced by addition of hydrophilic colloids, eg, carboxylated cellulose, methyl cellulose, or various polyacrylic acids, to the viscose prior to spinning (8–11).

During the past decade, the quest for flame-retardant fibers resulted in considerable work on viscose additives (12). Among the fire-retardant rayons are Avtex Fibers' PFR Fiber in the United States, Lenzing's Flamgard in Europe, and Daiwabo's HFG Fiber and Kanebo's Bell Flam in Japan. At present, the commercial demand for such products appears to be small (see Flame retardants in textiles).

As this work to improve the viscose process was going on, other significant developments relative to future rayon manufacture began to emerge in the mid- and late 1970s. In 1972, S. Kwolek at DuPont patented the use of anisotropic spinning wherein properly constructed polymers having the correct aspect ratio to form liquid-crystal solutions are spun to give fibers of exceptional strength (13). This patent discloses examples of nylon and describes how to make liquid-crystal solutions of cellulose acetate as well as how to spin these to obtain acetate fibers having up to three times the strength of the usual acetate fibers. It was not until 1980 that a liquid-crystal solution of cellulose was reported (14). However, no one has reported production of a rayon-type of fiber from a liquid-crystal cellulose solution (see Liquid crystals; Aramid fibers).

During the late 1970s, a new phase of rayon research emerged. By recognizing that the viscose system was almost 90 years old and that investment and pollution problems were becoming prohibitive in cost, Enka, Rayonier, Snia Viscosa, Rhone Poulenc, Courtaulds, and others began to examine the possibility of making solvent-spun rayons. At a major American Chemical Society symposium in 1977, Rayonier described its efforts to make solvent-spun high wet-modulus-type rayon fibers from cellulose solutions in dimethylformamide–nitrogen tetroxide (DMF/N_2O_4) and from dimethyl sulfoxide (DMSO)-paraformaldehyde solutions (15). In 1979, American Enka announced its Newcell process for making reconstituted cellulose fibers. This process is disclosed in patents that deal with solutions of cellulose in N-methylmorpholine N-oxide with the use of cellulose concentrations of ca 17–20 wt % (16). This process appears to have good potential for economical manufacture of reconstituted cellulose fibers of quality by a nonpolluting closed-loop method. In 1981, patents issued to Rayonier describe preparation of up to 16% cellulose solutions in dimethylacetamide (DMAc) or N-methylpyrrolidinone containing dissolved lithium chloride (17). These novel cellulose solutions were spun to give HWM-type reconstituted cellulose fibers. The Newcell and the LiCl–DMAc systems dissolve cellulose directly without forming derivatives; therefore, to call fibers from these systems rayons may be improper because all rayons have been produced by regenerating cellulose from a chemical derivative.

By far the main production of rayon today is by the viscose process; therefore, a large part of this review is dedicated to viscose technology. However, significant effort is being invested in search of other routes to make rayons more efficiently. These involve reexamination of the cuprammonium process and new solvent systems for cellulose.

Viscose Rayon

Most of the world's rayon is made by the viscose process. The term rayon designates a range of products that have widely varying properties, for the most part owing to the versatility of the viscose process. This versatility, which is the result of the many process steps and spinning changes that can be made, is a mixed blessing, since each stage of processing and spinning requires close attention to guarantee the desired product properties. The viscose process is a demanding process that requires continuous, year-long operation to prevent gelling of the system and to yield high quality products.

Process.   The stages in the preparation of a satisfactory cellulose xanthate solution are given in Figure 1. A few of the critical aspects of each stage are reviewed for both batch and continuous processes.

Steeping.   In order to have cellulose react with carbon disulfide to form the corresponding cellulose xanthate, cellulose must be converted to alkali cellulose (see Cellulose). Normally, this is done by placing cellulose-pulp sheets in a steeping press and filling the press with a closely controlled (±0.2 wt %) concentration of sodium hydroxide at a desired level of 18–20 wt %, depending on the type of cellulose used. With wood pulp, ca 18 wt % NaOH is used; cotton linters usually require higher NaOH concentrations of about 19.5–20% because linters are more difficult to penetrate than wood pulp. The rate of filling of the press is critical for obtaining an adequate alkali cellulose; wood pulp and cotton linters require different filling rates. The rate of filling

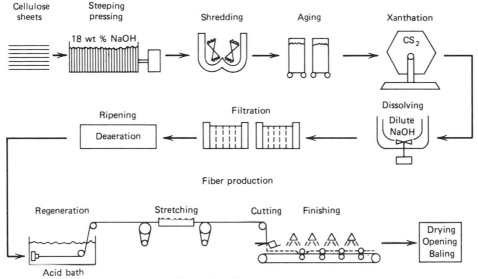

Cellulose sheets

Steeping pressing

18 wt % NaOH

Shredding

Aging

Xanthation

CS₂

Ripening

Deaeration

Filtration

Dissolving

Dilute NaOH

Fiber production

Regeneration

Stretching

Cutting   Finishing

Acid bath

Drying Opening Baling

Figure 1. Viscose process.

should correspond as nearly as possible with the rate of sheet wetting by the NaOH. If too slow a fill rate is used, a wicking action is observed where NaOH of a more dilute concentration wicks and wets the sheet ahead of the rising caustic liquid level. Such dilute wicking caustic does not open the cellulose crystallites adequately and results in incomplete conversion of cellulose I structure to the desired cellulose II found in alkali cellulose. Conversely, if the level of concentrated NaOH solution rises too rapidly, air bubbles are trapped in unwetted regions, which causes untreated cellulose fibers to remain in the unwetted regions; these subsequently create severe filtration problems. Although most steeps are carried out at 25°C, careful experiments have shown that the sheet temperature at the advancing caustic level is actually about 28–32°C because of exothermic reactions. The criticality of a good steep to viscose dissolution cannot be overemphasized. If the initial exposure of cellulose to caustic solution in the first 5 min is not sufficient, the damage is done; soaking for hours will not improve the situation.

The swelling of cellulose by water decreases with increasing temperature; this inverse swelling behavior is more pronounced when NaOH is present. Maximum swelling occurs with cold caustic solution (18); under proper conditions, some of the low molecular weight cellulose may dissolve along with the hemicelluloses that normally are removed. The hemicellulose concentration in the steep caustic must be controlled since it affects ultimate fiber properties. For regular rayon, this concentration is 2%, whereas for HWM fibers it is ca 0.7%, and for tire-cord fibers ca 0.5%. Therefore, pure dissolving pulps must be used or the alkali must be dialyzed if low hemicellulose concentrations are desired. High hemicellulose concentrations lead to reduced physical properties and increased brittleness in the final product. The effect of temperature on swelling is further illustrated in continuous slurry-steeping operations where higher caustic concentrations and higher temperatures (≤45°C) are used to reduce swelling of dispersed 5% cellulose slurry to ensure that the resulting alkali cellulose has a reasonable pressed-weight ratio.

After steeping, the cellulose is pressed under high pressure to a pressed-weight ratio of 2.6–3.0 to give an alkali cellulose with ca 34 wt % cellulose, 15.3 wt % caustic, and the remainder as water. As much excess caustic as possible is removed, because any excess will react with CS_2 in later steps to give undesirable by-products. Many steeping presses are arranged so that the caustic squeezing during the last phases of pressing are shunted to a separate receiver, because these final squeezings are always high in hemicellulose concentrations. This fact was recognized by Sihtola, who proposed the use of double steeping (19). In the Sini process, about 18% caustic is used to make the alkali cellulose, which is treated with more dilute caustic of ca 12%. This decreases the overall excess alkali and reduces subsequent by-product formation while also removing extra hemicellulose; the latter occurs because the more dilute caustic swells the cellulose to a greater extent, thereby facilitating removal of more hemicellulose.

Shredding. In addition to converting the alkali cellulose to a crumblike material, shredding serves two additional functions: first, the squeezing action of the shredder blades distributes the caustic more uniformly in the cellulose. Most alkali cellulose sheets are wetter at the edges than in the centers and, when the caustic solution drains, the bottoms of the sheets have more alkali. Thus, shredding gives a more uniform distribution of caustic in the alkali-cellulose crumb. Second, the shredder heats the crumb to a temperature favorable for aging, usually about 30–32°C. Typical shredders turn at about 30–60 rpm and have a blade clearance of about 0.08 cm. Improper blade clearance leads to the formation of undesirable ball-like aggregates. Too much shredding can give a low alkali-cellulose crumb density; the resulting crumb takes up too much space in the churn during xanthation.

Aging. Aging is used to decrease and control the cellulose DP, ie, the alkali cellulose stands in covered containers in a temperature-controlled room. The alkali crumb should never become dry. In the continuous process, the alkali cellulose may be dried while conveyed in an air stream; therefore, humidified air must be recirculated to the conveying section. In the batch process, drying can occur during aging if the cans are not covered properly. Formation of excess sodium carbonate at this stage is equally undesirable, since a 1% increase in carbonate concentration reduces filterability by 10%. Normally, sufficient oxygen is present in the alkali cellulose crumb, ie, added air exposure is unnecessary. Although the main DP decrease occurs during aging, it begins immediately upon exposure to caustic in steeping and continues during shredding. DP loss occurs during xanthation, where typically 75–125 additional monomer units are lost.

If the aging conditions are altered, the effect on final pulp DP can be predicted by use of two equations. To estimate the effect of aging time on DP at constant temperature, the rate constant k is obtained from the DP in the operation

$$k = \frac{\left(\dfrac{10^4}{DP_t} - \dfrac{10^4}{DP_o}\right)}{t}$$

where k = reaction-rate constant; t = aging time in hours; DP_t = DP at end of aging; and DP_o = DP at start of aging. If this value of k is used, DP at a certain time can be calculated.

To estimate the change of aging rate with temperature, a linear approximation is sufficient because the range of temperatures is small; thus

$$\log k_2 = \log k_1 + 0.05\, \Delta T(°C)$$

If these two equations are used judiciously, plant production cycles may be altered with minimum change in the product.

Xanthation. The four main types of xanthation units are the dry churn, the wet churn, the continuous belt xanthator (CBX), and a reciprocal screw of the Ko-Kneader-type called the Maurer unit. The latter two are essentially continuous xanthation units. Most viscose is made by the dry-churn method operating with addition of CS_2 at atmospheric pressure or at reduced pressure into evacuated churns containing alkali-cellulose crumb. The churn is rotated very slowly at ca 2.5 rev/min; thus, the crumb undergoes tumbling action and centrifuging action is avoided. A churn should be loaded to not more than 0.15 g/cm^3 density with alkali crumb. Usually, about 32 wt % CS_2 is used, and the temperature is kept at ca 32–33°C, ie, below the CS_2 bp of 46°C. The exothermic reaction is usually completed in ca 75–90 min. Complete reaction of CS_2 results in a partial vacuum. A final high vacuum is applied to the churn to remove traces of CS_2 before the churn is opened and the xanthate crumb is dumped. If the churn was dry when the alkali cellulose was added, the xanthate emerges as a free-flowing crumb. If water condensed inside the churn before the alkali-cellulose crumb was added, large balls will be present. In commercial practice, an economic balance is reached between operational conditions and by-product formation.

Xanthation is a critical part of the viscose process; many reactions proceed simultaneously during this step. Pure cellulose xanthate is white; any yellow color is owing to by-product formation, usually in the form of trithiocarbonate. The lower the xanthation temperature and the better the alkali-cellulose preparation, the lighter the color of the xanthate crumb and the less the hydrogen sulfide (H_2S) evolution during spinning, because H_2S comes only from the yellow trithiocarbonate (see eqs. 1–5).

$$CS_2 + H_2O \longrightarrow HS\overset{\overset{\textstyle S}{\|}}{C}OH \longrightarrow H_2S + COS \tag{1}$$

$$COS + H_2O \longrightarrow HO\overset{\overset{\textstyle O}{\|}}{C}SH \longrightarrow H_2S + CO_2 \tag{2}$$

$$CO_2 + 2\ NaOH \longrightarrow Na_2CO_3 + H_2O \tag{3}$$

$$H_2S + CS_2 \longrightarrow HS\overset{\overset{\textstyle S}{\|}}{C}SH \xrightarrow{\ 2\ NaOH\ } NaS\overset{\overset{\textstyle S}{\|}}{C}SNa + 2\ H_2O \tag{4}$$

$$NaS\overset{\overset{\textstyle S}{\|}}{C}SNa + H_2SO_4 \longrightarrow Na_2SO_4 + CS_2 + H_2S \tag{5}$$

In equations 1–4, sodium compounds such as Na_2S and $NaOH$ or their respective ions could be used in place of H_2S and H_2O, but the reactions are depicted more simply as shown.

Analysis of xanthate crumb corresponds closely to the theoretical ratios of expected carbonate to trithiocarbonate. There is evidence to suggest that COS reaction with cellulose is much faster than CS_2 and may precede CS_2 addition to cellulose.

Typically, about 25 wt % of the CS_2 is consumed in forming by-product trithiocarbonate through the side reaction, which leaves 75 wt % to react with the cellulose to give

However, these ratios can be changed significantly if improper steeping produces a poor alkali crumb or if xanthation temperatures are not controlled properly, in which cases far more by-product formation occurs.

The amount of cellulose xanthate actually formed often is reported as the extent of reaction, called the gamma (γ) number for xanthate formation. Thus, a γ number of 1.0 means that each anhydroglucose residue on the average has a 1.0 xanthate group. Correspondingly, a 0.5 γ indicates that every other anhydroglucose unit has a xanthate group attached to it. If 32 wt % CS_2 based on cellulose for xanthation is used and 75 wt % of this CS_2 gives cellulose xanthate, the xanthate crumb has a γ of ca 0.5. Xanthate sulfur was determined by iodine titration in the past (20); more rapid methods are available (21–22).

Cellulose has three hydroxyl groups: the C-2 and C-3 positions, which are secondary alcohols, and a C-6 position, which is a primary alcohol. During xanthation each of these hydroxyl groups reacts at a different rate mainly because of the nature of the primary or secondary alcohols, ie, its steric availability and the effect of adjacent hydroxyl groups. In freshly prepared xanthate crumb, more than half of the xanthate groups are on the 2 and 3 positions ($\gamma_{2,3}$) as compared to the 6 position (γ_6); the 2 and 3 positions are favored kinetically during xanthation even though reaction at the 6 position gives a thermodynamically more stable product. This is discussed in more detail under ripening.

Dissolving. Xanthate crumb is dumped into large, stirred tanks containing dilute caustic solution to dissolve the cellulose xanthate into a clear, honeylike, viscous dope known as viscose. The caustic and water in the dissolver are measured to give the desired cellulose:alkali ratios. Many dissolvers have standpipes that force the mixtures upwards in a cascading fashion while the moving blades propel the mixtures in a circular manner.

The most important factor in the dissolving cycle is the use of as cold a dilute caustic as possible. The initial dissolver temperature should be as low as possible, ie, preferably ca 5°C and definitely below 10°C. When the xanthate crumb is added, the temperature should not be permitted to rise above 10–12°C; it should be held <15–18°C during the 2–3 h required for dissolution to occur. Only during the last part of the dissolving cycle should the temperature be allowed to rise above 20°C to the temperature for ripening. Cellulose differs from almost all other polymers in that it and its derivatives dissolve more easily cold than hot. Faster stirring introduces more mechanical energy per unit time; this is converted to heat and warms the solution. Although fast stirring is not desirable, high shear does speed up dissolving. No one has successfully beaten cellulose into solution; it must be coaxed into solution with high shearing action at low temperatures. Any other approach produces too many undissolved fibers and gels that result in very poor filtrations.

Ripening. When the cellulose xanthate is first dissolved, it is not ready for spinning because it will not coagulate readily. The xanthate groups attached to the cellulose molecule are distributed over the three hydroxyl positions of the anhydroglucose monomer units; more xanthate groups are on the kinetically favored 2 and 3 positions ($\gamma_{2,3}$) than on the thermodynamically favored 6 position (γ_6). In essence, the xanthate groups in the 2 and 3 positions act as wedges to keep the cellulose chains from approaching one another; these must be removed or relocated to the 6 positions before closer chain packing can be achieved. This redistribution of xanthate groups is the main function of the ripening stage.

Under the alkaline conditions in the dissolved viscose, the 2- and 3-xanthate groups hydrolyze 15–20 times faster than the C-6 xanthate groups. The released xanthate groups can transxanthate with C-6 hydroxyl groups or react with the ever-present excess caustic to form the inorganic by-product sodium trithiocarbonate. The higher the ripening temperature, the more trithiocarbonate is formed. Therefore, ripening usually is done at rather low temperatures for long times and in vacuum to remove dissolved air.

During ripening, the xanthate concentration drops significantly. For example, a fresh viscose may have a 2.2% total sulfur and a concentration of about 1.6% xanthate sulfur (XS), which drops to about 0.9–1.1% XS after ripening. A significant fraction of the xanthate sulfur goes to by-product sulfur, mostly at the expense of the 2- and 3-xanthate groups. Several methods for measuring the readiness for spinning of ripened viscose have been developed. Since xanthate groups solubilize the cellulose, the number of such residual xanthate groups will have the greatest influence on viscose coagulability; a direct relationship between XS and ripening level exists. In the past, xanthate sulfur was determined by a tedious iodine titration.

Xanthate concentration alone does not determine coagulation rate; cellulose concentration, DP, caustic level, etc, influence coagulation rate. Therefore, the rayon industry adopted salt-index tests to measure coagulability. Two methods are used; for the sodium chloride index, saline solutions of varying concentrations are used to determine incipient coagulation when a single drop of viscose is added, whereas in the ammonium chloride or Hottenroth test, the viscose is diluted in a standard manner and titrated with 10% NH_4Cl to incipient precipitation.

A drop of fresh viscose may require about an 8% solution of NaCl to cause coagulation, whereas a 4% NaCl solution will coagulate the ripened viscose after 40 h, and 2% NaCl will work after 70 h. Similarly, it may be necessary to use 30 cm^3 of 10% NH_4Cl for fresh viscose and 10 cm^3 and 6 cm^3 after 40 h and 70 h, respectively. Each viscose plant uses its own standard methods, and great care must be taken when trying to compare salt indexes among various locations. Some use an acetic acid index. An approximate correlation of the three methods for a single viscose composition is given in Figure 2.

Spinning. Once the viscose is ripened to the proper xanthate level and the 2- and 3-xanthate wedges are removed so that proper rates of coagulation can be obtained, the viscose is ready for spinning. Chief goals in spinning are control of coagulation versus regeneration rates and the maximal use of the differences in these rates to obtain maximum responses to stretching. Stretching a soft gel does not give permanent alignment of molecules; when the gel hardens too much, further stretching is virtually impossible. In the spinning process, viscose is extruded into a bath containing both salt and acid. The salt usually is sodium sulfate at sufficiently high concentrations

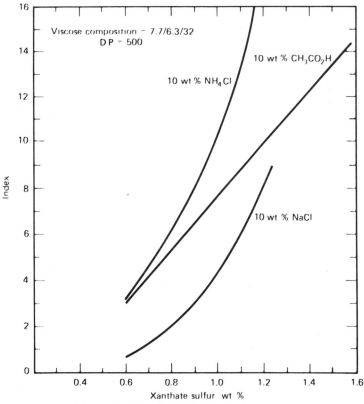

Figure 2.   Maturity index comparison curves.

to ensure that the sulfate solution behaves as a dehydrating system that removes water from the entering 7–10%-cellulose xanthate solution. Sodium sulfate has an RT solubility of 280 g/L; normally, it is used at 120–200 g/L in the coagulation baths. If greater dehydration action is desired, ammonium sulfate is used as the salt in the coagulation bath; its solubility is 440 g/L at RT, and it can be used at 230–350 g/L as a coagulant. Sulfuric acid is present to regenerate cellulose from the coagulated viscose. The concentration of sulfuric acid relative to the concentration of salt is critical in controlling the extent of water removal from the extruded viscose (by the salt) before the system is rigidly fixed by the acid removing the solubilizing xanthate groups. The control of densification of the extruded viscose by removing water with salt and stretching the densified mass versus the rate of acid hardening is the gist of the spinning process. The denser the viscose, the higher the resulting wet modulus. A high wet-modulus fiber with minimal stretch has been made in this manner. The higher the density, the more effective the stretching in aligning cellulose molecules; the final, dry and wet tenacities of the product will also be higher. In practice, the interplay of simultaneous reactions and other processes leads to significant interactions among the various factors governing spinning.

A comparison of the stress-strain performance of rayon and other fibers is given in Figure 3; the cross-sectional shapes of several typical rayons are given in Figure 4.

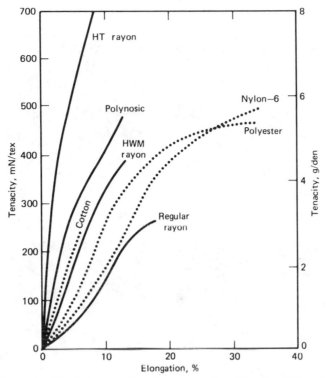

Figure 3.   Physical properties of rayons vs other fibers (conditioned).

Spinning Additives.   Rayon manufacturers became acutely aware of the need for controlling the complex interactions of coagulation, regeneration, and stretching when they first tried to prepare HWM fibers to improve rayon's competitive position relative to cotton. The simple approaches of using only Na_2SO_4 and H_2SO_4 were altered by the use of additives in both the viscose dope and the spin bath. The viscose ripeness level is an important factor in any study of additives. Several methods were developed to determine the rates of xanthate decomposition in the presence of various additives, eg, adding indicator-type dyes to the viscose and noting the distance from the spinneret where color changes occur; following the loss of the xanthate absorption peak at 303 nm when a cast viscose film is placed in a spin bath in a spectrophotometer cell; and dipping pH electrodes first in viscose and then in spin-bath liquor and measuring the pH changes with time. Hundreds of combinations of additives to the viscose and the spin bath were evaluated with these techniques. Controversy as to why certain combinations are effective exists; two main schools of thought seem to have developed—one believes that the modifying action takes place over the whole cross-section of the fiber, and the other feels that a surface layer of some type forms and controls further rates of transport into the fiber (23–24). Theories and mechanisms are in doubt, but it is certain that fiber properties are controlled by the use of additives. The best systems make use of some type of amine, eg, dimethylamine, and a polyetherglycol of ca 1500 mol wt added to the viscose as well as zinc in the coagulation bath (25). Proper combinations of these ingredients can increase regeneration times as much as 400–500% over unmodified systems.

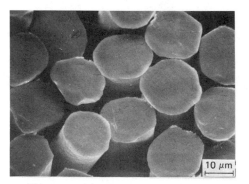

High wet modulus

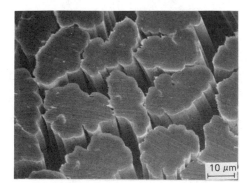

Regular rayon

Crimped HWM

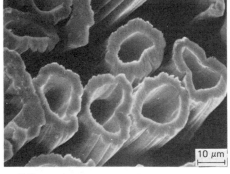

Hollow

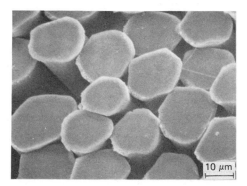

Cuprammonium

Figure 4.   Cross-sections of typical rayons.

Stretching.   An advantage of retarding regeneration is to make use of the densification of the extruded dope so as to obtain better molecular alignment on stretching. A fixed amount of stretching can occur; this is balanced between what is called jet stretch and godet stretch. Jet stretch is the ratio of the linear speed at the first godet roll to the average linear flow rate in the spinneret; it is expressed as follows:

$$\text{Jet stretch} = \text{linear speed of first godet} \cdot \left[\frac{\pi/4 \text{ (hole dia)}^2 \text{ (no. of holes)}}{\text{flow rate of dope}}\right]$$

Godet stretch is the stretching between the first and last godet rolls. The higher the godet stretch, the less residual elongation is left in the final fiber. Therefore, the manufacturer balances fiber strength and residual elongation.

Spinneret Design.   There probably are as many individual spinneret designs as there are fiber producers; yet in all cases, certain basic design parameters are dictated by the diffusion-controlled nature of the wet-spinning process. Each spinneret hole has a defined length to diameter (L/D) ratio and a defined conical entrance angle (α) (see Fig. 5).

For the most part, spinnerets have been made from a gold–platinum alloy to provide long-term resistance to corrosion. However, less expensive materials are now under investigation.

Some rayon producers favor one large spinneret (ca 7.5-cm dia) with various radial or circular arrangements of holes. In this design, several thousand holes can be drilled, but the regenerant reaches the center holes with difficulty. Another design uses small units with ca 2000 holes in a 1.9-cm-dia spinneret. A number of these small units are arranged in a circular cluster of ca 10-cm dia. In this fashion, 30,000 or more filaments can be spun from a single cluster head; the spaces among the individual spinnerets allow more complete acid penetration to the center filaments.

Finishing.   After the fibers are spun, cut, and washed, finishes are applied to give the proper frictional performance in subsequent textile processing. Slipperiness presents problems in spinning where physical properties of yarn are closely related to surface characteristics and friction, whereas stickiness results in improper fiber movement during processes such as carding and can lead to nep formation and fiber damage. Over the years, a wide variety of finishes have been proposed; one of the original and still one of the best is prepared from oleic and sulfonated oleic acids (see Textiles, finishes).

Modified Rayons.   Control of the regeneration conditions coupled with specific additives gives fiber manufacturers the opportunity to produce a host of modified fibers (see Fibers, chemical). An excellent review is given in ref. 26.

High Wet Modulus.   The production of HWM fibers by controlling regeneration kinetics through the use of additives was explained under the section on spinning. ISO permits any fiber with a wet strength of at least 220 mN/tex (2.5 g/den) and an elongation of less than 15% at this load to be called a Modal fiber. A wet modulus (tenacity at 5% stretch) of at least 45 mN/tex (0.5 g/den) is required for good performance.

Crimped Fibers.   In the early 1970s, Rayonier patented a chemically crimped rayon trademarked Prima (5–6). The crimp is produced by selective control of the conditions in the primary and secondary regeneration baths wherein a gel-core fiber is allowed to leave the primary bath and is exposed to hot-acid conditions in a secondary bath. Under these conditions, significant tensions are established throughout the fiber, which causes permanent chemical crimping. These fibers also have high wet modulus and

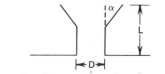

Figure 5.   Cross-section of a spinneret hole.

give cover, hand, bulk, and working performance in finished fabrics superior to that of ordinary rayon. When produced in higher deniers, Prima blends with polyester closely resemble wool fabrics.

Hollow Fibers.   Courtaulds has developed a hollow rayon fiber trademarked Viloft, (27). This fiber is much lighter than regular rayon and therefore gives significantly more coverage per unit weight. Moisture-absorbing capacity is about 50% greater than ordinary rayons; this makes Viloft a strong candidate for use in towels, sheets, pillow cases, and garments requiring improved absorbency. Although Viloft is not a high wet-modulus fiber, it reportedly behaves well if used up to 20 wt % in polyester or cotton blends. It is produced by adding sodium carbonate to the viscose and then controlling the regeneration conditions to capture the released CO_2 gases to give the hollow fiber. By increasing the carbonate content further, it is possible to produce a thinner-walled version of the hollow paper-making fiber, ie, PM fiber, which collapses and convolutes on drying. With further addition of carbonate, the fiber wall ruptures during regeneration and collapses from a superinflated state, ie, an SI fiber, to give a product resembling cotton in many respects (28) (see Hollow-fiber membranes).

Flame-Resistant Fibers.   A large body of literature exists on the various efforts to make an acceptable flame-resistant rayon. Some commercial successes have been reported, eg, Avtex's PFR fiber (29). Most such fibers are based on the addition of complex phosphorus-containing compounds to viscose. For example, various products based on phosphonitrilic acid and complex phosphorus esters are commercially available at this time (see Flame retardants). Although these materials give definite improvements in flame resistance, much more work will be required before wholly acceptable products are available. A good review on the present state of the art is available (12).

Test Methods.   A survey of tests is given in Table 1. In addition, ASTM and TAPPI standards should be consulted as well as the pulp suppliers.

Solvent-Spun Rayon

The viscose process is now 90 years old; current plant investment, labor, and operating and pollution costs make the construction of any new viscose plants unlikely in the United States. The present plants compete only because they have been written off for depreciation and new capital is not factored into the current selling prices. In a recent article in *Chemical Week*, an author summarizes the situation quite accurately by realizing the virtues of rayon as a fiber and simultaneously describing the many needs for modifying the overall viscose process (30). In spite of such candid realizations, some new HWM rayon capacity is being developed in Finland, Austria, and the USSR, although plant expansion may not be involved.

Several companies are exploring the possibility of using organic solvents with closed recycle–recovery loops to produce rayon-type fibers through solvent-spun processes because of the continued projected need for a fiber such as rayon. The first symposium on this subject was sponsored in 1977 by the American Chemical Society (15). In a subsequent publication, the past 130 years of cellulose-solvent technology was summarized, in which all cellulose solvents were categorized into four general classifications, and the considerations that would be required for a cellulose solvent to serve as a potential system for rayon manufacture were outlined (31). An expanded, companion review of the cellulose solvents and solubility is ref. 32. The American Enka

Table 1.   Test Methods for Viscose Rayon

Test	Parameter measured	Significance
$S_{18}{}^a$	hemicellulose content of starting pulp	high hemicellulose concentrations give brittle fibers
$S_{10}{}^b$	hemicellulose plus low DP-ends of cellulose	low DP cellulose adversely affects fiber strength
DP^c	av chain length of cellulose molecules	process control check and DP greatly influence fiber physical properties
plugging index	gels, fibers, etc, remaining undissolved	undissolved materials impair fiber quality; defines control efficiency
xanthate sulfur	xanthate groups on cellulose	xanthate concentration controls solubility and coagulability
total sulfur	organic plus inorganic sulfur	test for amount of CS_2 added and for inorganic by-product
salt index	coagulability by specified concentration of NaCl, NH_4Cl, or acetic acid	measures readiness for spinning
particle counts^d	particle-size distribution in viscose	too many particles above 15 μm impair fiber properties and suggest problems in viscose preparation
neutralization point^e	distance from spinneret at which viscose is neutralized by coagulating acid	spinning control; additives effectiveness
gel swell	water in fibers before drying; gel nature of fibers	effectiveness of regeneration coagulation stretch ratios; relates to ultimate fiber modulus
rewet swell	water retention or imbibition of dried and rewet fibers	relates to product field performance; also helps classify fiber type; HWM retains less water than regular rayon
$S_{6.5}{}^f$	concentration of nonoriented cellulose left in product	predicts how much product will be lost on repeated laundering

^{a} Soluble in 18% caustic.
^{b} Soluble in 10% caustic.
^{c} By cuen (copper ethylene diamine) viscosity.
^{d} By Coulter, HIAC, or laser scanners.
^{e} Diffusion value; film uv; electrode diffusion.
^{f} Soluble in 6.5% caustic.

Company has issued reports about the Newcell process, which appears to be based on spinning solutions of up to 20 wt % cellulose dissolved in hot aqueous N-methylmorpholine N-oxide (16,33). This system appears to be amenable to closed-loop recovery–recycle and relatively rapid spinning speeds. Although most previously investigated solvent-spun-rayon systems were based upon the use of unstable, easily regenerated cellulose derivatives, the Newcell process involves no discernable derivative formation but rather uses what appears to be a solution of cellulose; this leads to important economic advantages for the system.

Although the solubility parameter (δ) of the solvent must be nearly the same as that of cellulose, many liquids with the proper solubility-parameter values do not dissolve cellulose. For example, DMSO and DMF each have solubility parameters in the cellulose range, but neither dissolves cellulose. A solubility parameter is the square root of the vaporization energy per molar volume. It can be used for solutions of nonpolar liquids and amorphous polymers in nonpolar liquids to estimate or explain heats of mixing. For polar, crystalline polymers, other factors become significant, eg, with crystalline polymers, the heat of fusion or melting energy must be provided before solubility-parameter considerations can apply. In natural polymers such as cellulose and proteins, hydrogen bonding may be sufficient to prevent melting phenomena from occurring. These factors can be summarized in a thermodynamic approach that considers crystalline forces, derivative formation, and ultimate mixing. The change in free energy in dissolving the polymer may be written as

$$\Delta G_{\text{solution}} = \Delta G_{\text{fusion}} + \Delta G_{\text{derivatization}} + \Delta G_{\text{mixing}}$$

For dissolution to occur, the sum of all the changes in free energies for processes on the right side of the equation must be negative. Also, $\Delta G_{\text{derivatization}}$ will be negative if spontaneous derivative formation occurs. The reacted or derivatized material would be a new dissolved species and, to the extent that it formed, would alter the corresponding free energies of fusion and mixing. The change in free energy of mixing is

$$\Delta G_{\text{mixing}} = \Delta H_{\text{mix}} - T\Delta S_{\text{mix}}$$

The entropy of mixing ΔS_{mix} usually is positive; therefore, ΔG_{mix} is negative if the heat in mixing ΔH_{mix} has a sufficiently small positive value. For mixing nonpolar liquids in which the interactions among unlike molecules are geometric means of interactions among like molecules, $\Delta H_{\text{mix}} = K(\delta_1 - \delta_2)^2$ where (δ_1) and (δ_2) are the solubility parameters of the solvent and solute, respectively. Therefore, if the solubility parameters for both materials are as nearly equal as possible, the heat of mixing will be a small positive number which will be less than the negative $(-T\Delta S_{\text{mix}})$ term to give a negative ΔG_{mix}. For cellulose, the additional term ΔG_{fusion}, which is the free energy of melting the polymer, must be considered. In nonpolar polymers, the crystal forces can be overcome by heating, and the resultant melts dissolve readily in solvents having the proper boiling points and solubility parameters, ie, polypropylene dissolves readily in hot decalin or tetralin. In cellulose and proteins, the energies of intermolecular bonding that result from dispersion forces, hydrogen bonding, and other dipole–dipole interactions are too great for melting to occur at temperatures below those of decomposition; therefore, these polymers usually char or decompose rather than melt. This positive ΔG_{fusion} is significant and some type of physical work input or chemical change is required to overcome the effects of the relatively large bonding forces. Alternatives are to make $\Delta G_{\text{derivatization}}$ and ΔG_{mixing} sufficiently negative to overcome this positive ΔG_{fusion}.

A new solvent system for cellulose has been developed that can give up to 16% solutions of 500 DP cellulose. This system, which does not cause cellulose degradation even on extended heating or air exposure, utilizes lithium chloride in dimethylacetamide or in *N*-methylpyrrolidinone to form solutions of cellulose that are stable in extended storage (17). This system permits total recycle and recovery of all components to give a nonpolluting process. The economics of the process appear to be competitive with a new viscose plant installation.

Cuprammonium Rayon

Essentially all of the rayon made today is produced by the viscose process; nevertheless, there are some uses where the cuprammonium process has retained an advantage because of specific performance factors. Fibers from cuprammonium rayon are significantly more supple than viscose fibers and are used where a very soft hand is desired. The use of films and hollow fibers from the cuprammonium process for making artificial kidneys is critically important in this branch of medicine (see Dialysis; Prosthetic and biomedical devices). At present, almost all artificial kidney units use membranes prepared from such films and fibers. These films and fibers exhibit superior clearance performance for urea, creatinine, and metabolites, possess better dewatering characteristics, and cause less blood clotting as compared to any synthetics or corresponding products from the viscose process.

An analysis of the cuprammonium process suggests that it might be developed into a closed-loop system with the capability of complete recovery and recycle. Some workers in the USSR believe that the cuprammonium process should be reevaluated as a nonpolluting process to meet present-day ecological requirements. The following brief description is based on an updated review of the cuprammonium process practiced at the former I.G. Farbenindustrie Dormagen plant (34).

Dissolving. The ability of selected cupric ion, ammonia and alkali mixtures to dissolve cellulose was originally reported in 1847. Such solutions, known as Schweitzer's reagent, are used not only for making cuprammonium rayon, but also for measuring dilute-solution viscosities that can be related to degrees of polymerization of cellulose.

In the early 1900s, reports of how and why cellulose should dissolve in mixtures of cellulose and ammonia were published. In modern terms, the solution is described by using Werner-complex theory, in which four of the six d^3sp^2 orbitals of cupric ion are filled by NH_3 molecules as ligands, which leaves two orbitals available for ligand donation by 2- and 3-hydroxyl groups of cellulose. This is illustrated in Figure 6. Usually, some sodium hydroxide is added to help in the interaction of the $Cu(NH_3)_4^{2+}$ ions with cellulose. For some uses of cuprammonium rayon, the copper is precipitated as copper hydroxide, and the supernatant liquid containing sulfate ions (which are considered deleterious to good dissolution of cellulose) is removed, and the cupric hydroxide is added to the cellulose and ammonia and sodium hydroxide solution.

The foregoing brief, theoretical discussion establishes definite minimum ratios for concentrations of reagents that must be used in dissolving the cellulose. The minimum molar ratio that can be employed is 1 cellulose:1 copper:4 NH_3. In actual practice, excess NH_3 approaching 7 moles is used and NaOH is added to aid swelling and dissolution. The mole and weight ratios used by the Dormagen plant are shown in Table 2. I.G. Farbenindustrie used 0.53 pt of Na_2SO_3 in order to inhibit oxidative degradation.

Several methods have been tried to improve dissolution. These include mechanical

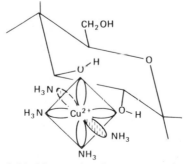

Figure 6. Orbital interactions in cuprammonium cellulose.

Table 2. Dormagen Plant Process Resulting in a 10% Cellulose Solution

Constituent	Mol ratio	Unit wt
cellulose	1.00	100
NH_3	7.52	79.1
copper	1.00	39.9
NaOH	0.655	16.2
Na_2SO_3[a]		0.53
water		764.27

[a] To inhibit oxidative degradation.

beating, ultrasonic dispersion, and use of air-dried slush pulp which, according to some investigators, dissolves faster than ordinary pulp.

The Dormagen plant used wood pulp containing only 89 wt % alpha cellulose for extensive year-by-year production of 40 metric tons per day of rayon staple and 13 t/d of filament. They employed both hot (>50°C) and cold (ca 20°C) purification processes. Pulps containing >99% alpha cellulose are available; therefore, wood pulp can be used in long-term commercial production through the cuprammonium process (see Pulp).

Cuprammonium rayon and the process compare favorably with viscose rayon and the viscose process. First, the DP of cuprammonium cellulose is 500–550 as compared to ca 350–400 for viscose rayon; second, the solutions are normally spun at 10% or more cellulose, which helps overall economics; third, the dissolution process for cuprammonium rayon is simpler than that for viscose; and fourth, the solutions are quite stable, a distinct advantage for the cuprammonium process.

Spinning. The spinning of cuprammonium yarns depends on the use of a funnel into which the coagulating–regenerating fluid flows and into which the dope is extruded.

As the flowing liquid travels down the funnel, its velocity increases and the entrained gelatinous cellulose fibers are stretched up to 400%. Subsequently, the fibers are washed as free as possible of occluded material, which is recycled, and the fibers enter a 5% H_2SO_4 bath where final removal of copper is achieved and where any remaining alkali or ammonia is converted to the corresponding sulfate.

The original stretch-spinning of cuprammonium rayon involved the use of reels. Filament on these reels had to be purified and rewound, which required considerable manpower and made the cuprammonium process noncompetitive relative to viscose. Since the mid 1940s, the cuprammonium process has been converted to a continuous process where yarn from 400–600 spinnerets is processed in line and wound directly onto a section beam. These section beams can be used directly for knitting, or several section beams can be combined to a larger warp beam for weaving. An excellent description of the continuous process is available in a patent issued to Beaunit Mills (35). A schematic presentation of both processes is given in Figure 7.

Fiber Properties. Cross-sectional views of fibers normally produced by the cuprammonium process are given in Figure 4; physical properties of cuprammonium and viscose rayons are given in Table 3. Conditioned and wet stress-strain properties of cuprammonium rayon and other rayons are compared in Figures 8 and 9. Cuprammonium rayon made to date is about equal to regular rayon in tensile properties and does not approach any of the improved viscose high wet modulus properties. If

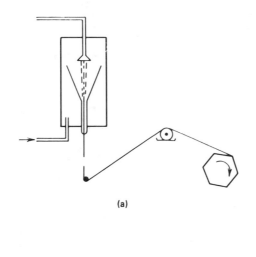

(a)

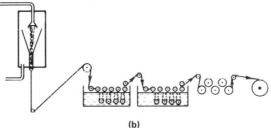

(b)

Figure 7. Cuprammonium process. (a) Reel spinning; (b) Continuous spinning.

this is the best that can be achieved in fiber properties, there is no hope for the cuprammonium process; to survive it must supply fibers that satisfy the varied commercial requirements. Recent patents suggest that improved fibers can be expected by spinning-process modifications.

Fiber properties are strongly dependent upon spinning conditions, which are related to recovery processes and conditions. Therefore, wide ranges of spinning conditions are found in the various references and patents. In the past, turbulent flow was regarded as harmful to fiber properties and the cause of filament sticking and breakage; consequently, many of the early patents were directed towards ways to cause smooth flow and prevent eddy currents. Later patents show how countercurrent flow is useful in preventing the deposit of copper precipitates that caused fiber breakage. Siphons were attached to the bottom of the funnels to help control flow and prevent precipitate buildup. Multicluster spinning into a partitioned funnel to increase production and various shapes for heads for multicluster spinning are reported. One enterprising inventor who used only water as the initial coagulation bath spun a highly ripened viscose dope in a cuprammonium unit and found no difficulty in obtaining yarns.

All patents agree that the spin coagulant liquid should be alkaline. The Dormagen plant used only 430 ppm of NH_3 (ca 0.025 M) in the spin water to help keep the funnels clean. Far more important than any of these factors is the comparison of flow rates in the spin funnel coupled with the new design of double-funnel spinning. Herein lies the possibility of achieving an improved wet modulus fiber through the cuprammonium process.

Table 3. Physical Properties of Commercial Rayons

Property	Cuprammonium	Viscose				Polynosic, unmodified
		Regular	LWM[a]	High tenacity	HWM	
				IWM[a]		
stress-strain behavior						
tenacity, mN/tex[b]						
conditioned	150–200	106–238	265–503	441–574	397–883	300–485
wet	84–120	62–159	132–380	291–353	309–706	238–353
standard loop	88–230	88–132	203–318	88–221	62–132	62–106
standard knot	62–150	62–124	194–309	141–282	106–247	106–256
elongation at break, %						
conditioned[c]	7–23	15–30	9–26	14–18	5–10	6.5–10
wet	16–43	20–40	14–34	17–22	6–11	7–12
specific gravity	1.52–1.54	1.50–1.53	1.52	1.53	1.53	1.51
moisture regain[d]	12.5	13	13	11–13	11–12.5	11.5–12.5
birefringence, by refractive index	0.026	0.018	0.036	0.039–0.044	0.046–0.057	0.040–0.045
water retention, %	100	90–100	65–80	65–75	60–70	55–70
solubility in 6.5% NaOH at 20°C, %	20–30	20–30	30–40	15–20	1.0–5.0	4–6
fiber DP[e]	450–550	300–350	300–500	350–600	500–800	550–650
wet modulus, mN/tex[b]	9–18	16–18	18–26	44–53	71–177	71–177

[a] Low and intermediate wet modulus.
[b] To convert N/tex to g/den, multiply by 11.33.
[c] Conditioned at 21°C and 65% rh.
[d] After centrifuging at 1000 g (ASTM D 2402-65T); commercial standard is 11%.
[e] Tenacity at 5% stretch.

416

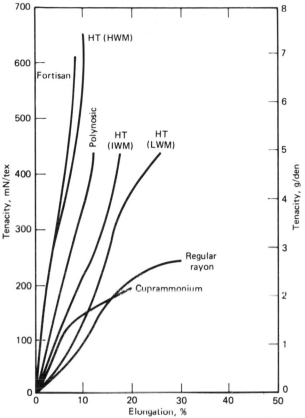

Figure 8.   Conditioned stress-strain curves of rayons at 21°C and 65% rh.

At the Dormagen plant, 1500 holes of 0.16 tex (1.4 den) fiber at 90 m/min were spun. Spinning water was introduced at 15 L/min (ca 4 gal/min), which means that it takes ca 720 L of spin bath flow to produce one kilogram of rayon. This translates to >650,000 L/t or 34.5 ML/d (9 × 10⁶ gal/d) for the production of 53 t/d. Although this is in line with water consumption for viscose production, it is still very high, since all of this flow must be recirculated or treated for recovery and recycle. A city of 72,000 people uses ca 34 ML water each day.

Spin water is only part of the plant's water flow in fiber spinning. For example, a viscose plant producing 50,000 t/yr capacity of rayon reportedly requires ca 81.8 ML (21.6 × 10⁶ gal) water usage per day. This usage, which is equivalent to the needs of a city of about 170,000 people, is broken down as follows:

12.5 ML (3.3 × 10⁶ gal) for soft water to process and to boilers.
27.3 ML (7.2 × 10⁶ gal) for filtered water for services.
42 ML (11.1 × 10⁶ gal) for refrigeration, cooling, and spin-bath recovery condensers.

Cuprammonium rayon plants have similar needs. Lowered spin flow directly diminishes all supporting energy and flow requirements. Much of this spin flow is considered necessary to obtain sufficient stretching of filaments; such high flow rates lead to considerable dilution of copper and ammonia and consequently, increase recovery

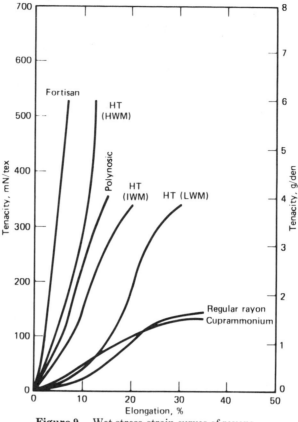

Figure 9. Wet stress-strain curves of rayons.

costs. The rayon industry realized these facts; most of the recent patents deal with improved methods for coagulating and stretching and the use of lower flow rates. Also, recent patents describe the use of various salts in the spin water to give coagulation followed by stretching, all of which should produce improved modulus fibers. For example, a British patent describes the addition of 0.8 mole $CaCl_2$ per liter spin water, which needs only 150% stretch to obtain 0.46 N/tex (5.2 g/den) conditioned tenacity and 0.39 N/tex (4.4 g/den) wet tenacity, ie, a high quality fiber (36). Bemberg SPA in Italy reports that precoagulated fibers are stretched 100% in the presence of 270 g/L $CaCl_2.2H_2O$ at 150 m/min to give rayon of 0.38 N/tex (4.3 g/den) and 0.30 N/tex (3.4 g/den) conditioned and wet tenacity, respectively (37).

An improved spinning funnel is described by Bemberg (38). A double funnel is used in which two different liquid flow rates can be controlled separately. Coagulation takes place in the upper funnel at a flow of 100 cm^3/min, whereas a faster flow of 500 cm^3/min imparts a high stretch in the free air space between the funnels. This design improves overall recovery.

Apparently, manufacturers who are still making cuprammonium rayon realize the need for lowering flow rates and the need for making improved-modulus fibers. Issued patents indicate that they are making efforts that appear to be properly directed for making such improved fibers.

Recovery. Recovery data are most difficult to obtain from the literature. The data for copper and for ammonia recovery are reviewed separately.

Copper Recovery. The amount of copper removed from the fiber in the spinning bath is low. Only 30% of the copper appears in the blue water; because of the high spin-water flow rates, it is in dilute solution. At the Dormagen plant, this gave 80 mg/L in the spin-bath effluent for yarn and 180 mg/L in the spin-bath effluent from the staple operation. This effluent is passed through ion-exchange resins, and the effluent from these resins is circulated back to the spinning funnels.

The 70% copper remaining in the yarns is removed in the acid-wash operation and exists as a solution containing 6.5% H_2SO_4 and 1.0% copper. This acid solution is used to regenerate the ion-exchange beds and then is adjusted to 0.5% acid and 1.2% copper. It passes to the sludging operation where Na_2CO_3 is used to precipitate the copper at pH of 5.0 to give a basic copper sulfate sludge; this copper sludge is recirculated to the cellulose-dissolving area. The overall recovery efficiency is 95% copper even at these high dilutions.

A method for reclaiming copper from the cuprammonium waste solution by precipitating it first as CuS and then reclaiming $Cu(OH)_2$ by use of alkali and air oxidation has been reported (39). Such a method might be advantageous in effecting almost quantitative copper removal to meet present-day environmental standards. The CuS could be concentrated either by sludging or by frothing methods that are well established. The extreme insolubility of CuS makes such thorough cleansing of effluents possible. A patent describes the use of mixing the waste streams to obtain a pH of 11–12 followed by evaporation of liberated NH_3 and 95% recovery of copper (40). A final example of copper recovery is reported at 98% for the Kustanai plant of the USSR (41). Ref. 41 quotes data issued by the Japan Organo Co. Ltd., according to which up to 99.9% of the copper can be recycled with modern techniques.

The use of lower spin-bath flow rates would significantly enhance overall copper recovery. In any case, the technology is available to achieve essentially quantitative copper recovery even if sulfide precipitation or electrolysis of the final effluent has to be employed.

Ammonia Recovery. The ammonia recovery differs from the copper recovery in that a much higher percentage of the ammonia can be recovered from the spin-bath effluent and by washing prior to the final acid bath. The use of a short hot-air exposure chamber between the spin bath and the final acid to achieve more complete NH_3 removal has been suggested (42). The ammonia that enters the acid wash and is converted to ammonium salt is recovered when the acid-wash liquor is made alkaline to precipitate the copper. However, the problem again is that the relative concentration of the ammonia is low owing to the rather high flow rates; thus, large liquid volumes must be processed. No recovery process removes the small levels that remain. The low residual concentrations present in the large water volumes signify losses in material and money. The only way to recover NH_3 at low concentrations is to process at reduced pressure. This means additional energy costs and higher investment costs, since more distillation vessels are required in vacuum processing as compared to processing at atmospheric pressure. Thus, the Dormagen plant had to process 25 ML (6.6×10^6 gal) of ammonia water at 1.27% concentration at reduced pressure in order to recover 30 t of NH_3 from the 38 t of NH_3 originally used to dissolve 48 t of cellulose. This required six recovery towers, which were 3.6-m dia and 19.8-m high. This amounts to handling ca 585 L (154 gal) of NH_3 recovery liquor per kilogram of rayon produced.

The cuprammonium process may be made a closed-loop system producing reasonably good quality rayon fibers if overall economics for handling recovery and recycling of dilute effluent streams can be improved.

Ecological and Pollution Considerations

In most processes developed many years ago, little consideration was given to pollution and ecological impact; the older rayon-manufacturing processes are not exceptions. The cuprammonium and the viscose processes consume huge amounts of water for each kilogram of product; ca 420–750 L (110–200 gal) of water per kilogram of rayon is needed directly for processing and 8–10 times that amount of water must be handled to provide the plant supplementary-service facilities. When water effluents did not require special treatments, overall process economics were favorable. However, treating ca 835 L (220 gal) of water per kilogram of fiber imposes on such processes new penalties that cannot be overlooked. Similar considerations exist for air handling in present-day plants. Each of the three different methods for making rayon reviewed in this article has its own pollution and ecological problems.

Cuprammonium Process.   About 40 kg copper and 80 kg ammonia are used for every 100 kg cellulose dissolved. Not only must this be recycled, but it must first be recovered from dilute solutions since ca 32,500 L (8600 gal) water are used to make this much fiber. The copper can be recovered with ion-exchange columns and by use of proper precipitation stages, but costs have become significant. Of more concern is the ammonia recovery in which dilute ammonia streams must be handled to give effective ammonia recovery. However, a group of rayon chemists in the USSR claim that the cuprammonium process is preferred to the viscose or solvent-spinning process from the recovery and overall economic viewpoints. Their data are needed for acceptance of such a position.

Viscose Process.   The viscose process also requires rather large amounts of water per kilogram of rayon. In this case, the aqueous effluent contains large amounts of sodium sulfate, much of which is recovered and sold. However, excess salt must be released in the effluent from the plant. Viscose plants must handle the H_2S gas and CS_2 vapor. Removing small amounts of these vapors in a plant by sweeping ca $2.8 \times (10^3–10^5)$ m^3/min [$(10^5–10^7)$ cfm] is a difficult engineering problem. H_2S and CS_2 in air are scrubbed with cascading caustic (if the air flow rate is less than ca 2.8×10^4 m^3/min (10^6 cfm), or adsorbed on zeolites or activated carbon for the higher flow rates where scrubbing is impractical. The presence of H_2S in the exhaust stream complicates the CS_2 recovery on activated-carbon beds since the H_2S is oxidized to sulfurous and sulfuric acids; this reduces the carbon-bed adsorptive capacity.

Most processes involve removal of from 50 to 300 ppm of H_2S present as HS^- and S^{2-} from exit streams, with the use of some type of alkaline medium in conjunction with some type of iron catalyst. The overall method based on the Claus reaction is

$$H_2S + \tfrac{1}{2} O_2 \xrightarrow{\text{Fe}_2\text{O}_3} S + H_2O$$

In the Ferrox process, simple iron salts are used, whereas in the Cataban process, the iron is present as a soluble iron chelate complex. In both cases, the iron is returned to the ferric state by passing air through the solution. The Cataban process probably has some advantage over the Ferrox process in that it has a higher reaction rate; this

Table 4. Energy Used in Fiber Production from Naphtha[a]

Product	Ratio of energy of raw materials to energy of naphtha combustion	Energy to process raw material, GJ[b]/t			Total processing energy plus energy of raw materials, GJ[b]	Ratio of the energies of raw materials plus processing to energy of naphtha consumption
		Monomer production	Polymerization	Fiber spinning		
Filament yarns						
nylon-6	2.09	65.1	16.3	31.1	204	4.97
nylon-6,6	2.15	54.5	14.0	34.5	198	4.71
cellulose acetate	0.69	93.6		68.2	192	4.58
polyester	1.41	55.1	22.2	40.4	174.8	4.17
rayon		28.6		85.6	114	2.72
Staple fiber						
nylon-6,6	2.15	54.5	12.0	18.4	180	4.28
polyester	1.41	55.1	18.8	21.3	157	3.75
acrylic	1.55	42.5		46.3	157	3.75
polypropylene	1.42	5.8	14.2	14.9	97.5	2.33
rayon		26.3		44.1	70.7	1.69

[a] Ref. 43.
[b] To convert J to Btu, divide by 1054; to convert J/t to Btu/lb, multiply by 0.4302.

421

Table 5.   Worldwide Rayon Production, 10^6 t

Year	Filament	Staple	Total
1965	1.040	1.905	2.945
1970	0.998	1.995	2.993
1975	0.816	1.814	2.630
1980	0.816	2.222	3.038

results in fewer undesirable sulfur-containing by-products. Processes that rely only on the use of alkaline scrubbing suffer from the use of large quantities of expensive caustic and from problems in disposing of the resulting alkaline sulfide liquors. Recent Federal regulations on effluent concentrations of Na_2SO_4, H_2S, and CS_2 place stringent restrictions on the viscose process.

Solvent-Spinning Process. In these cases, all recovery and recycle stages are considered to be closed loops with essentially none of the solvent being lost. This must be achieved because the solvent systems are too expensive to permit anything less than almost complete recycling. However, for solvent spinning, the biggest factors lie in the economics of recovery of sufficiently pure materials to allow for repeated recycle use. Small amounts of impurity can build up to cause some problems, but these should be resolvable. The discovery of a simple solvent that can be recycled easily will be the best approach to a long-range competitive rayon process.

Energy Requirements for Rayon vs Synthetics

As energy becomes more expensive, increasing emphasis is placed on the overall energy required to manufacture products. Energy considerations must include not only all aspects of fiber production but should also consider the energy involved in subsequent product performance. An approximate generalization of the production of an apparel garment is that it consumes between 7–8 kg of oil per kg of textiles. Much of this (1.5–2 kg oil/kg textile) is used in dyeing and finishing, whereas 20% is the energy content of the hydrocarbon starting materials, and the remainder is split among chemical processing, fiber spinning, and cloth production.

Since rayon is made from trees, no petroleum is used to make the original polymer; also, a significant part of the energy needed for separation and purification of cellulose is derived from pulping by-products as energy sources. These factors give rayon a favorable position relative to synthetic fibers with regard to total energy needed for fiber production, as demonstrated in Table 4 (43).

Economic Aspects

Over the last few years, the number of U.S. rayon plants in operation had decreased, whereas the number of rayon plants in the Far East and the USSR has increased. World rayon production since 1965 is given in Table 5.

Regardless of how rayon might be made in the future, whether by the viscose, cuprammonium, solvent-spun, or as yet some undiscovered process, fibers with the properties of rayon are and will be essential to the textile and related industries. A high demand for fibers with rayonlike properties will exist for many years to come. Rayon(s) or solvent-spun cellulose fibers will satisfy this demand.

BIBLIOGRAPHY

"Rayon and Acetate Fibers" in *ECT* 1st ed., Vol. 11, pp. 522–550, "Rayon," by Lionel A. Cox, American Viscose Corp., and "High-Tenacity Rayon," by P. M. Levin, E. I. du Pont de Nemours & Co., Inc.; "Rayon" in *ECT* 2nd ed., Vol. 17, pp. 168–209, by R. L. Mitchell and G. C. Daul, ITT Rayonier Inc.

1. G. E. Linton, *Natural and Man-Made Textile Fibers*, Duell Sloan & Pearce, a division of Meridith Press, New York, 1966.
2. R. L. Mitchell and G. C. Daul, "Rayon," in *Encyclopedia of Polymer Science and Technology*, Vol. 11, Interscience Publishers, a division of John Wiley & Sons, Inc., 1969, pp. 810–847.
3. U.S. Pat. 2,592,355 (Apr. 8, 1952), S. Tachikawa Sanjo and Higashiyama-Ku; U.S. Pat. 2,946,782 (July 26, 1960), S. Tachikawa, Higashiyama-Ku (to Tatsuji Tachikawa, heir).
4. I. H. Welch, *Am. Text. Rep. Bull. Edn.* **AT 8,** 49 (1978).
5. U.S. Pat. 3,632,468 (Jan. 4, 1972), and U.S. Pat. 3,793,136 (Feb. 19, 1974), G. C. Daul and F. B. Barch (to Rayonier, Inc.)
6. U.S. Pat. 3,720,743 (March 13, 1973), T. E. Muller and H. D. Stevens (to ITT Corp.).
7. J. S. Ward and R. Hill, *Text. Inst. Ind.* **7,** 5274 (1969).
8. U.S. Pats. 4,041,121 (Aug. 9, 1977); 4,063,558 (Dec. 20, 1977); 4,136,697 (Jan. 30, 1979), F. R. Smith (to Avtex Fibers, Inc.).
9. U.S. Pat. 4,066,584 (Jan. 3, 1978), T. C. Allen and D. B. Denning (to Akzona Inc.).
10. U.S. Pat. 4,104,214 (Aug. 1, 1978), A. W. Meierhoefer (to Akzona Inc.).
11. J. H. Welch and J. A. Combes, *Paper presented at the 8th Tech. Symp. of the International Nonwovens and Disposables Association*, 1980, p. 3.
12. N. A. Portnoy and G. C. Daul, *Paper presented at Nat. Tech. Conf. AATCC*, 1978, pp. 269–274.
13. U.S. Pat. 3,671,542 (June 20, 1972), S. Kwolek (to E. I. du Pont de Nemours & Co., Inc.).
14. H. Chanzy and A. Peguy, *J. Polym. Sci.* **18,** 1137 (1980); *Repr. 5th International Dissolving Pulps Conference*, Vienna, Austria, 1980.
15. A. F. Turbak, *Amer. Chem. Soc. Symp. Ser.* **58,** (1977).
16. U.S. Pat. 4,142,913 (March 6, 1979), C. C. McCorsley, III, and J. K. Varga (to Akzona Inc.); U.S. Pat. 4,144,080 (March 13, 1979), C. C. McCorsley, III (to Akzona Inc.); U.S. Pat. 4,145,532 (March 20, 1979), N. E. Franks and J. K. Varga (to Akzona Inc.).
17. U.S. Pat. 4,302,252 (Nov. 24, 1981), A. F. Turbak, A. El-Kafrawy, F. W. Snyder, Jr., and A. B. Auerbach (to ITT Corp.).
18. J. Marsh, *Mercerizing*, D. Van Nostrand Co., Inc., New York, 1942.
19. H. Sihtola and T. Rantanen, *Prepr. 4th International Dissolving Pulps Conference*, 1977, pp. 35–39.
20. W. H. Fock, *Kunstseide* **17,** 117 (1935).
21. D. Tunc, R. F. Bampton, and T. E. Muller, *Tappi* **52**(10), 1882 (1969).
22. M. Rahman, *Anal. Chem.* **43**(12), 1614 (1971).
23. D. K. Smith, *Text. Res. J.* **29,** 32 (1959).
24. F. R. Charles, *Can. Text. J.* **83**(16), 37 (1966).
25. U.S. Pat. 2,942,931 (June 28, 1960), R. L. Mitchell, J. W. Berry, and W. H. Wadman (to Rayonier, Inc.).
26. J. Dyer and G. C. Daul, *Ind. Eng. Chem. Prod. Res. Dev.* **20,** 222 (1981).
27. U.S. Pat. 3,626,045 (Dec. 7, 1971), C. Woodings (to Courtaulds Ltd.).
28. E. Attle, *Text. Mon.*, 24 (May 1977).
29. U.S. Pat. 4,040,483 (Aug. 9, 1977), B. R. Franko-Filipasc and J. F. Stuart (to FMC Corp.).
30. *Chem. Week*, 25 (July 29, 1981).
31. A. F. Turbak, R. B. Hammer, R. E. Davies, and H. L. Hergert, *Chemtech* **10,** 51 (Jan. 1980).
32. S. Hudson and J. Cuculo, *J. Macromol. Sci.* **C18**(1), 1 (1980).
33. R. Armstrong, J. K. Varga, and C. C. McCorsley, *Prepr. 5th International Dissolving Pulps Conf.*, Vienna, Austria, 1980.
34. *Synthetic Fiber Development in Germany*, compiled and edited by Leroy R. Smith, Textile Res. Inst., 10 East 40th St., New York 10016, 1946, based on the Technical Industrial Intelligence Committee Report on I.G. Farbenindustrie Plant, Dormagen, Germany, 1946.
35. U.S. Pat. 2,587,619 (March 4, 1952), H. Hoffman (to Beaunit Mills).
36. Brit. Pat. 815,189 (June 17, 1959) (to I. P. Bemberg Akt.-Ges.); H. Frind, *Kolloid Z.* **179,** 110 (1961).
37. U.S. Pat. 3,488,344 (Jan. 6, 1970), F. Biffi, E. Chesi, and O. Gallina (to Bemberg SpA).
38. U.S. Pat. 3,798,297 (March 19, 1974), O. Gallina (to Bemberg SpA).

39. Ger. Pat. 708,009 (June 5, 1941), G. T. Trout (to New Process Rayon, Inc.).
40. Ger. Pat. 1,006,147 (Apr. 11, 1957), P. Schubert (to I. P. Bemberg Akt. Ges.).
41. U. V. Grafov, G. S. Bykova and U. I. Malboroda, *Khim. Volokna* (3), 63 (May–June 1976).
42. U.S. Pat. 3,110,546 (Nov. 12, 1963), F. Hoelkeskamp (to J. P. Bemberg Akt.).
43. W. C. Firth, *Energy Balance of Man-Made Fiber Production*, Comité International de la Rayon et des Fibres Synthétiques, 1980.

JOHN LUNDBERG
ALBIN TURBAK
Georgia Institute of Technology

RECREATIONAL SURFACES

For the purposes of this article, the term recreational surfaces means man-made surfaces that provide a durable area of consistent properties for recreational activities. The characteristics of the playing surface may be selected to match natural surfaces under ideal conditions, or may provide special characteristics not otherwise available. In all cases, the intent is to provide desirable and durable playing characteristics. Recreational surfaces are used for football, soccer, field hockey, cricket, baseball, tennis, track, jumping, golf, wrestling, and general purposes. Included in the latter category are indoor–outdoor carpets, patio surfaces, and similar materials designed for low maintenance in light recreational service.

A grasslike artificial surface was installed commercially for the first time in 1966 in the Houston Astrodome. Since that time, other commercial fabrics of various constructions have become available.

Resilient surfacing compositions for recreational use were introduced in tennis courts in the early 1950s (1). In 1958, the first all-weather resilient athletic track was installed at the University of Florida using a composition similar to that employed for tennis courts. The polyurethane running track used in the Olympic games at Mexico City in 1968 started a new era in this highly competitive sport, and the uniformity and resilience of the synthetic track contributed to establishment of new speed records. Similar tracks soon became standard throughout the world, probably contributing to the more frequent occurrence of the four-minute mile. In addition to the performance properties, the all-weather aspects of a synthetic track offered advantages for scheduled events and practice. The original system was a rubber and clay-filled polyurethane with polyurethane chips on the surface. A wide variety of systems followed, including polyurethane with sand finish for an all-weather skidproof surface, vinyl chips, rubber chips, so-called sandwich tracks consisting of a bound-rubber base with polyurethane surface, bound-rubber chips alone with a painted surface, etc. A similar smooth-surface product was used for basketball. For tennis, a sand finish proved acceptable, but the polyurethanes were not competitive in cost with coated asphalt and molded vinyls and polyolefins.

These grasslike and resilient installations require substantial amounts of synthetic materials. A typical football field covered with artificial turf requires approximately 15,000 kg fabric, 15–30,000 kg shock-absorbing underpad, and 5–10,000 kg adhesive and seaming materials. The artificial surface for a 0.40-km running track typically might require 50–70,000 kg composition. Proper coatings for smooth basketball courts, and striping and marking of turf tracks and courts require additional material.

Types

Recreational surfaces must provide certain performance characteristics at acceptable cost, with a reasonable lifetime, and with acceptable appearance. For classification, arbitrary but useful distinctions depending on the primary function of the surface may be made: a covering intended primarily to provide an attractive surface for private leisure activities, a surface designed for service in a specific sport, or a surface designed for a broad range of heavy-duty recreational activities, including professional athletics. Examples of these three categories are, respectively: patio surfaces, track surfaces, and artificial turf for outdoor sports.

Light-Duty Recreational Surfaces. Artificial surfaces intended for incidental recreational use are designed primarily to provide a practical, durable, and attractive surface, eg, for swimming-pool decks, patios, and landscaping. Material cost is a prime consideration. Many contemporary surfaces in this category utilize polypropylene ribbon and a tufted fabric construction (see below).

Specific Athletic Surfaces. Included here are running-tracks, tennis courts, golf tees and putting greens, and other applications designed for a particular sport or recreational use. Specific performance criteria are important.

In the case of a tennis surface, for example, friction and resilience characteristics affect footing and behavior of the tennis ball. The common asphalt tennis courts have been improved significantly by all-weather coatings with superior appearance and characteristics. Typical coatings are vinyl or acrylic compositions in various colors. Poured-in-place and preformed systems of polyurethane, vinyl, and rubber, although often used, are more expensive than coated asphalt or concrete. More recently, open molded mats that are easily installed as interlocking tiles have been used to construct new and repair old tennis courts. These mats are less expensive than the poured-in-place and preformed systems that require glueing. They are easily removed if necessary, and thus are easily and quickly repaired. In addition, the open structure allows rapid drainage after rain. Such molded systems are usually made of polyolefins, vinyls, or polyolefin–rubber blends, with appropriate pigmentation and stabilizers (see Olefin polymers; Vinyl polymers).

Characteristics and design criteria differ, eg, for a putting green vs a wrestling mat. Another specific surface is a warning area or track adjacent to a sports surface, eg, the area between the fence and the playing surface of baseball fields. This area must feel different from the main area to warn the player that he or she is approaching the fence. A football field may be surrounded by a full-size running track or, where the space is limited, by a warning track.

Multi-Purpose Recreational Surfaces. The performance demands for artificial surfaces in this category control the design. A good example is the playing surface for professional football in the United States. The shock absorbency of the system affects player safety and long-term performance under very heavy use. The grasslike fabrics used for these applications are made from various pile materials, including polypropylene, nylon-6,6, nylon-6, and polyester (see Polyamides; Polyesters). The fabric may be woven, knitted, or tufted. The critical underpad is derived from various materials, representing a compromise of performance properties. Because of the importance of safety and performance, fabric and installation costs are higher than the lighter duty surfaces.

Performance Characteristics

User-Related Properties. Shoe traction for light-duty consumer purposes requires modest footing and reasonable surface uniformity. The frictional characteristics are obviously of much greater importance in surfaces designed for athletic use. They can be significantly affected by pile density, pile height, and other aspects of fabric construction.

Measuring the coefficient of static friction between the playing surface and the shoe or other contact surfaces determines traction. To test traction for grasslike surfaces, the force required to initiate movement in a weighted sports shoe resting on the artificial turf is measured (2). The coefficient of static friction is defined as the force pull in a direction parallel to the playing surface divided by vertical force loading. The magnitude of the vertical force loading must be sufficient to approximate actual penetration, but is not otherwise critical. The shoe characteristics significantly affect the traction.

Typical static friction coefficients are given in Table 1. These data demonstrate that the absolute traction values for synthetic surfaces are satisfactory in comparison

Table 1. Typical Traction Characteristics of Surfaces[a]

Surface	Static-friction coefficient[b]		Directionality index[c]
	Dry	Wet	
recreational surface			
tufted polypropylene	1.7–2.0	1.8–2.1	0.1–0.2
knitted nylon-6,6	1.8–2.0	1.6–1.8	0.10–0.25
tufted nylon-6	1.9–2.1		0.05–0.15
woven nylon-6	1.9–2.1		0.05–0.15
indoor-outdoor carpeting			
tufted polypropylene	0.4–1.5	0.4–1.5	
natural grass[d]	1.0–2.2	0.7–1.4	not applicable

[a] Ranges measured with appropriate sports shoes for the indicated surfaces.
[b] Defined as the average value measured in the four principal directions parallel to the fabric surface; two across the pile, one with, and one against the pile.
[c] Defined, from the same data under [b], as the average absolute deviation of the four traction values from the mean.
[d] The range is determined by the type and condition of grass.

with natural turf, provided that proper shoe surfaces are employed. Furthermore, synthetic surfaces, by virtue of their construction, are somewhat directional. This effect is evident in a measurement of shoe traction in various directions with respect to the turf-pile angle. Some traction characteristics are directly affected by the materials. Nylon pile fabrics, for example, show different traction characteristics under wet and dry conditions than polypropylene-based materials, since nylon exhibits higher moisture regain. Effects of artificial turf-fabric construction on shoe traction are given in Table 2.

For more specialized surfaces, shoe traction is equally critical. Gymnasium floors and running tracks must be uniform over their entire span. A balance of properties that gives good footing, traction and ball response is required. The surface must be tough and durable, provide long life, perform over a broad range of environmental conditions. A proper combination of energy and shock absorption enables the athlete to perform to maximum potential in relative comfort. For running tracks, resilience minimizes energy dissipation on the surface. A particular range is optimum for the track modulus of elasticity (3).

Spikes on track running shoes should be long enough to provide adequate traction with easy removal. Ideally, dull spikes afford traction by depression of the surface and gain additional energy for the runner as the resilient surface rebounds when the weight if lifted. As with turf, a specific shoe design for the surface is necessary.

Game-Related Properties. For some activities, like running and wrestling, the result of direct impact by the player is the only consideration. For others, like baseball or soccer, the system must also provide acceptable ball-to-surface contact properties. Important ball-response properties on the artificial surface are coefficients of restitution and friction, because these directly determine angle, speed, and spin of the ball.

The coefficient of restitution is defined as the ratio of the vertical components of the impact and rebound velocities resulting when a ball is dropped or thrown onto a playing surface. The velocities or related rebound heights are measured photographically. Criteria, such as ball inflation pressure, air temperature, and other details must be specified.

The coefficients of static friction between a sports ball and the playing surface are the ratios of the horizontal forces necessary to initiate sliding or rolling motion across the surface to the normal forces (wt) perpendicular to the surface. The sliding

Table 2.   The Effect of Fabric Properties on Traction Characteristics

Fabric[a] construction, density	Description	Static-friction coefficient[c]	Directionality index[c]
standard	height = 1.27 cm	1.8–2.0	0.2
standard	increased curl[b]	1.8–2.0	0.15
high	height = 1.02 cm	1.8–2.2	0.15
standard	texturized[d]	1.8–2.2	0.02–0.05

[a] 55.5 tex (500 den) nylon-6,6 pile ribbon.
[b] Curl is an index of filament modification imparted to the ribbon during processing.
[c] See Table 1 for definitions.
[d] Refers to a fiber-modification process.

and rolling coefficients of dynamic friction are similarly defined in terms of the forces necessary to sustain uniform motion across the playing surface. These friction coefficients determine slip or retention of inertial effects present upon impact. In golf, for example, the driven ball may bounce forward after the first impact with the surface, then bounce backward after the second. In this particular example, the combination of velocities and friction create a slipping condition on the first bounce; on the second, the rotational back spin imparted to the ball when first hit is activated by sufficiently large friction. In soccer, on the other hand, the ball in play rarely slips because a coefficient of friction ≥ 0.4, which is almost always achieved, is sufficient to transfer momentum.

Some illustrative values for ball-response parameters in various sports are given in Table 3 (4). Artificial surfaces can be designed to match certain desirable game-response parameters of natural grass surface and provide these properties consistently. Response is confined to a narrower range and is less affected by weather. As a general rule, however, artificial turf surfaces tend to be somewhat livelier in ball response, velocity, and distance of roll, with coefficients of friction lower than those for natural grass.

Shock Absorption.   Artificial playing surfaces for moderate to heavy use must provide a degree of shock absorbency for player comfort and safety. This requirement is achieved through incorporation of a resilient layer. For heavy recreational use, the shock-absorbing layer is a resilient underpad that provides a distinct layer between playing surface and substrate.

An ideal shock-absorbing medium, eg, for football, in the United States would provide a reasonable softness for player comfort in normal shoe contact combined with a high capacity for dissipation or distribution of kinetic energy involved in the impact of the player's fall. Various foamed elastomers are suitable for this purpose. The design

Table 3.   Physical Parameters[a] for Ball Response from Sports Surfaces

Surface	Coefficient of restitution	Friction coefficients		
		Static	Sliding	Rolling[b]
soccer				
nylon turf-pad	0.7	0.4	0.3–0.4	
polyester turf-pad	0.7	0.4	0.3–0.4	
polypropylene turf-pad	0.7	0.5	0.5	
natural grass				
long	0.6	0.8	0.8	
soft	0.5	0.8	0.8	
short	0.7	0.8	0.8	
baseball				
nylon turf-pad	0.6	0.5–0.6	0.6	0.1–0.2
natural grass	0.5	1.0	0.8	0.2
golf				
nylon turf-pad	0.5	0.3–0.5		0.1–0.2
natural grass	0.4	0.5		0.1–0.2
tennis				
nylon turf-pad	0.7	0.4–0.6		0.1–0.2
natural grass	0.8	0.7		0.1–0.2

[a] Approximate values.

[b] Rolling resistance increases markedly with rolling velocity.

criterion is the ability to dissipate energy of motion by reducing the deceleration and through hysteresis losses in the material. A useful device for characterizing the required properties is a dynamic-mechanical impact tester (5). It employs an instrumental missile that is allowed to fall freely from a specific height onto the resilient surface. Suitable sensing components record electronically the force- and displacement-time profiles of the missile throughout the interval of first penetration and rebound from the playing surface. The plots shown in Figure 1 illustrate the acceleration and displacement between initial and final contact of the missile with the playing surface, both vs time. The profiles show that the deceleration forces increase as the missile penetrates into the surface, reach a maximum, and then decrease as the missile rebounds from the surface. The effectiveness of the shock-absorbing medium is indicated by the height of the maximum or, more accurately, the integrated profile throughout the duration of impact, calculated according to the severity index, $\int g^{5/2}dt$, where g = acceleration due to gravity, and t = time. The more effective the shock-absorbing material, the less sharply peaked is the g_{max} curve, and the broader and shallower the g_{max} profile. Effective performance would be achieved by a material displaying large hysteresis in which the impact of the falling weight is progressively absorbed without rebound. Clearly, since a useful system also must be reversible, this extreme example is not practical. Useful materials for shock absorbency have the ability to gradually dissipate the impact of the falling object, with a moderate conversion of the total kinetic energy to heat through hysteresis losses.

The shock-absorbing characteristics of underpad materials or resilient surfaces are functions of material selection, physical composition, thickness, and temperature. The sensitivity of performance to physical characteristics and temperature of the shock-absorbing medium is illustrated in Table 4 and in Figure 2. Clearly, thicker materials offer better shock absorbency. However, in practice, an excessively thick underpad may result in unsure footing and increased cost. The effects of temperature must be anticipated, since pad temperatures can easily range from below freezing to 65°C in the sun. The ideal system would provide a relatively flat g_{max} response over this temperature range. A compromise is a design for g_{max} peaks no greater than about 250 and a severity index below 1000 within a reasonable range of temperatures.

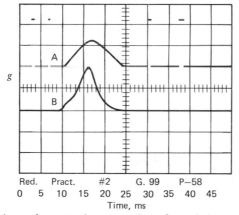

Figure 1. Deceleration and penetration curves from dynamic impact tester. Curve A is distance penetrated and curve B is deceleration plotted against time. Each square in curve A = 12.7 mm. Each square in curve B is equal to 50 g.

Table 4. Properties of Typical Underpad Materials

| Property | Closed-cell foam, % closed cells | | Poured elastomer |
	75–85	90–95	
thickness, cm	1.6	1.6	1.0
tensile strength, kPa[a]	620	655	2700
density, kg/m^3	96	256	1300–1400
g_{max} at °C			
21	85[b]	125[b]	105[c]
−12	105	150	
49	120	150	

[a] To convert kPa to psi, multiply by 0.145.
[b] 60-cm drop height of 9-kg flat-head missile (ASTM specifies 2 ft drop, 20 lb flat-head missile) (5).
[c] 22.8-cm (9-in.) drop height of hemispherical body (6).

Using the impact tester, values of g_{max} for grass-playing fields in late autumn range from about 75 for wet fields to 280 for frozen turf. The intermediate values observed depend upon soil type, moisture, condition, and other variables.

Durability. Grasslike surfaces intended for heavy-duty athletic use should have a service life of at least five years. The service life is approximately proportional to the amount of face ribbon available for wear. Pile density and height also affect the surface lifetime. In addition, different materials respond to abrasive wear to a greater or lesser extent. These effects cannot be measured except in simulated and controlled laboratory experiments, which do not necessarily reflect field conditions. Exposure to sunlight, uv light in particular, affects length of service. The Taber and Schiefer abrasion tests (7) evaluate fabrics and fabric constructions for potential wear properties. However, the data given in Table 5 indicate the unreliability of any specific accelerated wear test to predict longevity of fabrics, unless the tests are applied to closely related fabrics and for which actual wear-use data are available.

Other parameters provide indexes to surface durability including tuft bind (8), in which the force required to dislodge a surface element from its backing is measured; grab strength (9); and tensile strength. All are indications of strength and tear resistance. Artificial surfaces must be resistant to cigarette damage, vandalism, and the

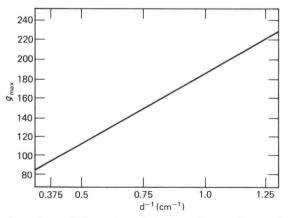

Figure 2. Shock absorption vs thickness for typical polyurethane resilient surface, measured according to Procedure B of ASTM F 355 using a 6.8-kg (ASTM specifies 15 lb) hemispherical missile dropped from a height of 30 cm (ASTM specifies 1 ft).

Table 5. Laboratory Simulations for Artificial Turf Wear

Surface	Effective pile loss, %	
	Taber method[a]	Schiefer method[b]
tufted polypropylene, 844 tex[c]	5	7
knitted nylon-6,6, 55.6 tex[c]	21	0.2
knitted polypropylene, 33.3 tex[c]		2.7

[a] ASTM D 1175, Rotary Platform, Double Head Method (5000 cycles).

[b] ASTM D 1175, Uniform Abrasion Method (5000 cycles).

[c] To convert tex to denier, divide by 0.1111.

like. A standard method for evaluating such resistance is the U.S. National Bureau of Standards Flooring Radiant Panel Test (10). In this test, a gas-fired panel maintains a heat flux, impinging on the sample to be tested, between 1.1 W/cm^2 at one end and 0.1 W/cm^2 at the other. The result of the burn is reported as the flux needed to sustain flame propagation in the sample. Higher values denote greater resistance to burning. The results obtained depend on both the material and surface construction. Polypropylene turf materials are characterized by critical radiant flux indexes that are considerably less than those for nylon, polyester, and acrylic polymers (11) (see also Flame retardants in textiles).

Materials and Components

The principal parts of a recreational surface system include the top surface material directly available for use and observation, backing materials that serve to hold together or reinforce the system, the fabric backing finish, the shock absorbing underpad system, if any, and adhesives or other materials.

Surface Materials. Pile materials used in grasslike surfaces are selected from fiber-forming synthetic polymers, such as polyolefins, polyamides, polyesters, polyacrylates, vinyl polymers, and many others (see Fibers, elastomeric). These polymers exhibit good mechanical strength in the necessary direction. The materials listed in Table 6 are thermoplastic polymers that may be suitably pigmented before extrusion.

For the relatively smooth recreational surfaces of running tracks and tennis courts, polyurethanes, polyolefins, and other flexible, durable elastomers or composites are employed (see Table 7).

Backing Materials. Any fiber-forming polymer of reasonable strength may be used in backing materials, including polyamides, polyesters, and polypropylenes. The backing provides strength and offers a medium to which the pile fibers may be attached. The backing usually is not visible in the finished product, nor does its presence contribute much to the characteristics of the playing surface. However, it provides dimensional stability and prolongs service life. Some properties of fabric backing materials may be inferred from the data in Table 6; however, more highly drawn fiber equivalents of greater strength are employed for backing uses.

Backing Finish. Usually, the backing material is consolidated with a pile ribbon (see Textiles). In tufting, for example, the tufts are locked to the backing medium. In weaving and knitting, the finish seals and stabilizes the product. Backing materials

Table 6.   Properties of Poured-in-Place Polyurethane Resilient Surfaces

Property	Test method	Desired range	Typical values[a]		
			A	B	C
impact resilience, %	ASTM D 2632	30–50	42	49	36
hardness, Shore A-2	ASTM D 2240	45–65	60	48	50
breaking strength, kPa[b]	ASTM D 412	>2,800	4,800	2,700	3,400
elongation to break, %	ASTM D 412	100–300	210	298	168
tear strength, N/m[c]	ASTM D 624	>11,000	17,200	15,200	11,200
10% compression, kPa[b]	ASTM D 575	590–860	720	500	520
compression recovery, %	ASTM D 395 (A)	>95	100	99.6	98.7

[a] Of laboratory samples prepared identically from different raw materials.
[b] To convert kPa to psi, multiply by 0.145.
[c] To convert N/m to lbf/in. (pli) sample thickness, divide by 171.1.

are usually applied as a coating that is subsequently heat-cured. For tufting, preferred choices are poly(vinyl acetate), poly(vinyl chloride), polyurethane resins, and various latex formulations. For knitted fabrics, poly(vinylidene chloride) and acrylics, or polystyrene–rubber latexes are used.

Underpads.   Shock-absorbing underpad material is usually made of foamed elastomer, which provides good energy absorption at reasonable cost (see Foamed plastics). The foamed materials may be poly(vinyl chloride), polyethylene, polyure-thanes, or combinations of these and other materials. Typical foam densities may range from 32–320 kg/m^3. Important criteria include a resistance to absorbing water, tensile strength and elongation, open-cell vs closed-cell construction, resistance to chemical attack, low cost, availability in continuous lengths, softness in energy-absorbing properties, and compression-set resistance (12). Water absorption is very important especially if the system is subjected to freezing temperatures. Generally, closed-cell materials are most resistant to water intrusion.

Some recreational surfaces, in particular the lighter-weight materials for patio surfaces, etc, either employ no shock-absorbing underpad system, or utilize a light coating that usually is joined directly to the turf during manufacture. This provides a certain amount of softness and grip adequate for the service intended. For running-track surfaces, on the other hand, the resiliency of the shock-absorbing medium and its relationship to athlete performance are obviously very important.

Adhesives and Joining Materials.   Grasslike surfaces are employed over substantial areas, and lengths of rolls must be joined, glued, or sewn together. A variety of adhesives ranging from low cost poly(vinyl acetate) materials to cross-linked epoxy cements are available (see Adhesives).

Table 7.   Typical Properties of Yarns Suitable for Pile Components of Artificial Surfaces

Property	Polypropylene	Poly(ethylene terephthalate)	Nylon-6,6
density, g/cm^3	0.91	1.38	1.14
melting point, °C	170	250	265
tenacity, N/tex[a]	0.22	0.18–0.35	0.31
elongation, %	25	30–100	33
moisture regain at 21°C and 65% rh, %	0.1	0.4	4

[a] To convert N/tex to g/den, multiply by 11.33.

Fabrication and Installation

Tufting. The tufting process is frequently employed in construction of grasslike surfaces (13). The techniques are essentially those developed for the carpet industry with economical high speed characteristics. In the tufting operation, pile yarn is inserted into one side of a woven or nonwoven fabric constituting the primary backing. Yarn is inserted by a series of needles, each creating a loop or tuft as the yarn penetrates the backing and forms the desired pattern on the other side. For artificial surfaces, the looped tufts that form in this process are cut to provide the desired individual blades in the playing surface. Cutting elements are incorporated in the tufting machine which sever the loops automatically in the process of forming the pile.

Depending upon the width of the fabric, a modern tufting machine may incorporate one to two thousand needles, simultaneously inserting the tufts along the fabric width. The needles may operate at speeds in excess of 500 strokes per minute, contributing to a highly efficient output of fabric yardage. The primary backing for tufted surfaces is usually a woven, synthetic filament fabric. After the tufts have been inserted, pile fiber and backing components are fused by application of the backing finish. The tufts are inserted in or looped through the backing fabric. Its quality determines the firmness of the attachment.

Knitting. The knitting process as applied to manufacture of artificial turf and related products provides a high strength, interlocked assembly of pile fibers and backing yarns. Pile yarn, stitch yarn, and stuffer yarn are assembled in the operation. The pile and stitch yarns run in the machine or warp direction, whereas the stuffer yarns interlock the wales, ie, rows, formed by the pile and stitch yarns, knotting the whole system together. Knitted fabrics typically possess high strength and high tuft bind.

A machine with approximately 1000 needles may be used to produce continuously a 5-m wide fabric. The assembly process is more complex than tufting. The pile yarn and stitch yarn are inserted into the knitting needle, whereas the stuffer yarn is interlocked with the others through a separate feed mechanism of the machine. As with tufting, the loops of pile fabric formed are slit, thus creating the desired individual blades.

The knitted fabric is subjected to a finishing operation in which a suitable backing material is applied to penetrate the yarn-contact points and stabilize the structure. This process usually is accompanied by a heat treatment that stabilizes the fabric and conditions the pile.

Weaving. As a general rule, weaving is slower than tufting or knitting. The process consists of a two or three-dimensional intermeshing of warp, pile, and fill yarns that may be of different types. In contrast to a knitted fabric, the yarns are not knotted together, but interwoven at right angles. The pile yarns are cut by a series of wires that are continuously assembled into and withdrawn through the fabric loops. A suitable finish further stabilizes the fabric.

Finishing. In each of the processes discussed above, the artificial turf fabric is subjected to a finishing operation in which a suitable adhesive is applied to the back side, thus bonding the components and stabilizing the material. The finish may be applied with a knife or brush, in foam or paste form, followed by a heating and drying stage. The temperatures also affect the pile-ribbon properties.

Underlay. An installed artificial turf system may or may not include components between the fabric and the subbase. As mentioned earlier, such components are not a requirement for light-duty use, but are essential in attaining the shock-absorbing properties required by heavy-duty surfaces. The foam underpads used in shock-absorbing systems are made by incorporating a chemical blowing agent into the foam latex or plastisol. Under the proper processing conditions, voids of controlled size and number are uniformly distributed throughout the foam material. Closed-cell foam structures are most desirable for outdoor use, because they resist water absorption.

Installation. In general, grass-like surfaces are glued down or otherwise affixed to a subbase. For light-duty purposes, it may suffice to tack the edges to the perimeter of the area to be covered. For heavy-duty systems, a solid subbase of asphalt or other material is first installed. The shock-absorbing underpad is glued to this surface, followed by glueing the turf on top of the pad. Other constructions with partial or complete glueing are also possible. However, it is essential to securely anchor the perimeter.

Additional fabric panels are bonded together by glueing onto a reinforcing tape, sewing, or some other technique. In Europe a secure stitch-seaming technique employing high-strength sewing yarns has an excellent durability record.

In another technique, common in Europe, the bonded turf–pad system is laid loosely onto permeable asphalt without glueing. This special subbase allows water to trickle through holes punched into the pad through the turf and to drain away laterally through the aggregate. Side-seam sewing is the preferred assembly technique.

For artificial surfaces in the athletic category, eg, running tracks, the installation techniques are different. A poured-in-place or interlocking-tile technique may be employed; the latter is used for tennis courts. Adequate provision for weathering and water drainage are essential. In general, the resilient surfaces are installed over a hard base that contains the necessary curbs to provide the proper finished level. Out-of-doors, asphalt is the most common base, and indoors, concrete. A poured-in-place polyurethane surface (14) is mixed on-site and cast from at least two components, an isocyanate and a filled polyol, of the polyether or polyester type. The latter usually contains an organic mercury catalyst (15) which provides a system with selective reaction toward organic hydroxyl groups, thus lowering the sensitivity toward moisture (see Urethane polymers). Amine-type catalysts, eg, Dabco 33 LV, may be used also. The isocyanate is of the polymeric type or a toluene diisocyanate or methylene bis-(phenyl isocyanate)-based prepolymer (see Isocyanates). Similar systems are used as binders for scrap-rubber granules. The surface properties can be varied by the type and amount of fillers and the size of the rubber granules.

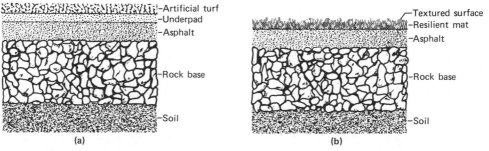

Figure 3. Cross-sections of typical artificial turf (**a**) and resilient track surfaces (**b**).

Table 8. Manufacturers and Trade Names for Artificial Turf Surfaces

Product	Manufacturer[a]	Description	Year
All-Pro Turf	All-Pro Turf	3.6-m width tufted fabric with polypropylene face yarn and synthetic backing yarns	current
AstroTurf Stadium Surface	Monsanto Company	4.5-m width fabric employing 55.5 tex (500 den) nylon-6,6 pile ribbon and high-strength polyester backing yarns	current
Clubturf	Clubturf, UK	woven polypropylene turf	current
Dunlop	Dunlop, UK	tufted nylon-6 fabric	current
Grand Turf 7000	Kureha Chemical Ind. Co., Ltd., Japan	a woven fabric with poly(vinylidene chloride) ribbon for the face pile and polyester warp and fill yarns	1977
Gräs	Fieldcrest/Karastan	woven fabric with textured nylon-6 face ribbon, synthetic yarns in backing	current
Grass Sport 500	Chevron	polypropylene turf–polyurethane pad system for sport use	current
Instant Turf	Instant Turf Industries, Inc.	polyolefin synthetic turf products for consumer and recreational use	current
Lancer	Lancer Enterprises, Inc.	light-duty recreational surfaces utilizing polypropylene or polyester face yarns, for consumer and marine applications	current
Playfield	Playfield Industries, Inc.	tufted product utilizing Chevron Polyloom II polypropylene yarn	current
Poligras	J. F. Adolff AG, FRG	knitted polypropylene pile fabric supplied with a bonded underpad	current
Poly-Turf	Sports Surfaces International Ltd., UK	nylon-6 tufted fabric installed with nitrile rubber–poly(vinyl chloride) pad	current
SuperTurf	SuperTurf, Inc.	tufted product utilizing polypropylene yarn and a synthetic backing	current
Tartan Turf	3M Company	tufted fabric with low-tex (den) nylon-6,6 filament	ca 1976
Marubeni/Toray GS-2	Mitsubishi Trading Company, Japan	tufted nylon-6 fabric utilizing woven polyester backing	current

[a] United States, unless otherwise indicated.

The mixed liquid is pumped into the area, where it levels and cures into the slab. It may be poured in two layers to eliminate imperfections in the base. The first layer may be a preformed rubber slab which is glued to the base, or a mixture of reground rubber and binders or rubber and polyurethane. A textured surface may be imparted to the second coat with sand or chips.

A permeable system may be strip-glued over a permeable asphalt base for drainage; this procedure is common in Europe. In some cases, colored binders impart the desired color, and a final protective coating of a urethane lacquer may be sprayed on the surface. Preformed slabs also may be used indoors, and covered by poured-in-place polyurethane. Whether solid or layered, indoor systems, particularly those used for basketball, require proper finish and maintenance coats to ensure adequate performance. The finish is selected according to the amount of use and the frequency of maintenance.

The general construction of artificial turf and resilient track surfaces is indicated in the cross-section drawings of Figure 3.

Table 9.   Manufacturers[a] and Trade Names for Resilient Surfaces

Product	Manufacturer	Description	Year
AstroTurf Track Surface	Monsanto Company	poured-in-place filled polyurethane with smooth, sand, or chip surfaces	to 1977
Cal Track	California Products Corporation	highly porous rubber-urethane mat installed over subbase	current
Chevron 440 track	Chevron	poured-in-place filled polyurethane base covered with a layer of colored granules adhered to the base to form the running surface	current
Chevron 400 track	Chevron	breathable surface over a 0.95-cm base mat of ground recycled rubber with a urethane textured coating containing resilient granules	current
Mateflex	Mateflex by Mele	patented molded polypropylene rubber interlocking modules 33-cm square tiles; installed over hard base; permeable; for tennis courts	current
Mondo Sport Surfaces	Robbins, Inc.	calendered and vulcanized surface with a base of polychloroprene rubber, mineral aggregates, stabilizing agents, pigments; preformed rolls; adhered to base; for track, tennis	current
Recaflex	C. Voight Söhne, FRG	sandwich-type track surface with ground rubber underlay and polyurethane top surface	current
Rekortan	C. Voight Söhne, FRG	single-layer, cast, filled polyurethane trace surface	current
RoyalDek	Uniroyal, Inc.	resilient, shock-absorbing artificial tennis surface formed by molded tiles	current
Rubaturf track	Rubaturf Sport Surfaces, Inc.	rubber granulates from ground scrap tires combined with rubber latex fillers, antioxidants, and hardening agents poured-in-place	current
Sportan resilient surfacing	Sportan Surfaces, Inc.	liquid poured-in-place polyurethane; smooth texture or granular topping for gymnasiums, tennis, track, or other resilient recreational surfaces	current
Sport-Tred	Pandel Chemical, Inc.	prefabricated, solid cast vinyl flooring; factory coating available; glued to subbase	current
Swiss Flex	Swiss Flex	modular molded surface square (645 cm^2); permeable; installed over hard surface on tennis courts	current
Tartan resilient surfacing	3M Company	liquid poured-in-place polyurethane; smooth on granule topping for gymnasium or track	current
Tracklite	Tracklite Systems	resilient track surface	current

[a] United States, unless otherwise indicated.

Paints and Striping.   Outdoor running tracks, indoor basketball courts, field house surfaces, and stadium turf require line markings and appropriate decorations. These are painted on with two-component epoxy paints or water-based acrylic latex, depending on the permanence desired. In multiple-use stadiums, the football striping is removed when the area is converted to baseball, and vice versa. Multiple markings in different colors for football, field hockey, lacrosse, and soccer may be desired in

Table 10.   Current U.S. Manufacturers and Trade Names of Various Recreational Surfaces and Components

Product	Manufacturer	Description
Ensolite	Uniroyal, Inc.	gym, wrestling, and judo mats
Interlock rubber floor systems	Pawling Rubber	interlocking tiles for exercise and recreational areas
Port a Pit	Ampro Corporation	pole-vault landing surface
Pro-Pit	Vantel Corporation	pole-vault landing surface
Quest	A. E. Quest & Sons, Inc.	pole-vault and high jump pits
Resilite	Resilite Sports Products, Inc.	wrestling mats, wall mats
Supreme Court	AllWeather Surfaces, Inc.	roll-down sports surfaces

community and school installations. Indoor paints usually are permanent, multiple color, and compatible with the surface, ie, deform with the resilient surface without cracking and accept any finish and maintenance coatings.

Economic Aspects

Manufacturers and trade names of commercial artificial surface products are given in Tables 8, 9, and 10.

Notice

Nothing contained herein should be construed as a recommendation to use any product, process, or apparatus in conflict with any patent.

BIBLIOGRAPHY

1. J. E. Nordale in N. M. Bikales, ed., *Encyclopedia of Polymer Science and Technology*, Vol. 15, Interscience Publishers, a division of John Wiley & Sons, Inc., New York, 1971, p. 490.
2. F. B. Roghelia and J. J. Burke, Monsanto Company Test Method, 1978.
3. T. A. McMahon and P. R. Greene, *J. Biomech.* **12**, 893 (1979); A. Chase, *Science* **81**, 90 (Apr. 1981).
4. J. J. Burke and F. B. Roghella (1978), G. Raumann (1967), J. Vinicki (1900), Monsanto Company Report.
5. *ASTM F 355*, *Procedure A*, American Society for Testing and Materials, Philadelphia, Pa., 1978.
6. Ref. 1, p. 493.
7. *ASTM D 1175*, American Society for Testing and Materials, Philadelphia, Pa., 1980.
8. *ASTM D 1335*, American Society for Testing and Materials, Philadelphia, Pa., 1972.
9. *ASTM D 1682*, American Society for Testing and Materials, Philadelphia, Pa., 1975.
10. I. A. Benjamin and S. Davis, *Final Report No. NBSIR 78-1436*, National Bureau of Standards, U.S. Department of Commerce, Apr. 1978; T. Kashiwagi, *JFF/Consumer Product Flammability* **1**, 267 (1974).
11. I. A. Benjamin and C. H. Adams, *Fire J.*, 63 (Mar. 1976).
12. *ASTM D 3574*, 1977; *ASTM D 1667*, 1976; *ASTM D 624*, 1973; *ASTM D 2856*, 1976, American Society for Testing and Materials, Philadelphia, Pa.
13. Ref. 1, Vol. 13, 1970, p. 692.
14. J. H. Saunders and K. C. Frisch, *Polyurethanes, Chemistry, and Technology*, Vol. 1, John Wiley & Sons, Inc., New York, 1963; Vol. 2, 1964.
15. U.S. Pat. 3,583,945 (June 8, 1971), J. Robins (to Minnesota Mining and Manufacturing Company).

W. F. HAMNER
T. A. OROFINO
Monsanto Company

REFRACTORY FIBERS

The term refractory fiber defines a wide range of amorphous and polycrystalline synthetic fibers used at temperatures generally above 1093°C (see also Fibers, chemical; Refractories). Chemically, these fibers can be separated into oxide and nonoxide fibers. The former include alumina–silica fibers and chemical modifications of the alumina–silica system, high silica fibers (>99% SiO_2), and polycrystalline zirconia, and alumina fibers. The diameters of these fibers are 0.5–10 μm (av ca 3 μm). Their length, as manufactured, ranges from 1 cm to continuous filaments, depending upon the chemical composition and manufacturing technique. Such fibers may contain up to ca 50 wt % unfiberized particles. Commonly referred to as shot, these particles are the result of melt fiberization usually associated with the manufacture of alumina–silica fibers. The presence of shot reduces the thermal efficiency of fibrous systems. Shot particles are not generated by the manufacturing techniques used for high silica and polycrystalline fibers, and consequently, these fibers usually contain <5 wt % unfiberized material. Refractory fibers are manufactured in the form of loose wool. From this state, they can be needled into flexible blanket form, combined with organic binders and pressed into flexible or rigid felts, fabricated into rope, textile, and paper forms, and vacuum-formed into a variety of intricate, rigid shapes.

The nonoxide fibers, silicon carbide, silicon nitride, boron nitride, carbon, or graphite [7782-42-5] have diameters of ca 0.5–50 μm. Generally, nonoxide fibers are much shorter than oxide fibers except for carbon, graphite, and boron fibers which are manufactured as continuous filaments. Carbon, graphite, and boron fibers are used for reinforcement in plastics in discontinuous form and for filament winding in continuous form. In addition to temperature resistance, these fibers have extremely high elastic modulus and tensile strength. Carbon and graphite fibers cannot be accurately classified as refractory fibers because they are oxidized above ca 400°C. This is also true of boron fibers which form liquid boron oxide at approximately 560°C. Most silicon carbide, silicon nitride, and boron nitride fibers are relatively short, ranging in length from single-crystal whiskers less than 1 mm to fibers as long as 5 cm. Diameters are ca 0.1–10 μm. These fibers are used mostly to reinforce composites of plastics, glass, metals, and ceramics. They also have limited application as insulation in and around rocket nozzles and in nuclear fusion technology requiring resistance to short-term temperatures above 2000°C. For inert atmosphere or vacuum applications, carbon-bonded carbon fiber composites provide effective thermal insulation up to 2500°C.

Until the early 1940s, natural asbestos was the principal source of insulating fibers (see Asbestos). In 1942, Owens-Corning Fiberglass Corporation developed a process for leaching and refiring E glass fibers to give a >99% pure silica fiber (1). This fiber was used widely in jet engines during World War II. Shortly after the war, the H.I. Thompson Company promoted insulating blankets for jet engines under the trade name Refrasil (2). The same material was manufactured in the UK by the Chemical and Insulating Company, Ltd. (3), and a similar material called Micro-Quartz was developed by Glass Fibers, Inc. (now Manville Corporation) from a glass composition specifically designed for the leaching process (4). Leached-glass fibers can be produced with extremely fine diameters using flame attenuation (see Glass).

Several techniques were developed by Engelhard Industries, Inc. in the early 1960s for manufacturing fused silica fibers by rod-drawing techniques (5). High thermal-efficiency insulating felts with very low densities are produced by rod-drawing. They are essentially free of shot, and were used as blankets for jet engines and space vehicles (see Ablative materials). High purity silica fiber was chosen for the tiles which form the first reusable thermal-protection system on the space shuttle Orbiter. Many of the leading edge and underside areas on the Orbiter are insulated with replaceable carbon-bonded carbon–fiber tile capable of protecting the metal skin beneath from the 1200–1800°C temperatures developed in those areas. In the early 1940s techniques for blowing molten kaolin [*1318-74-7*] clay ($Al_2O_3.2SO_2$) into fibers were also developed by Babcock & Wilcox Company (6) (see Clays). At about the same time, it was discovered that by blowing molten mullite [*1302-98-8*] ($3Al_2O_3.2SiO_2$) with air and steam jets, a portion of the melt is converted into glassy fibers (7).

The Johns-Manville Corporation became interested in alumina–silica fibers in 1946, establishing a cooperative development with Carborundum. Chemical modifications of the basic alumina–silica composition have resulted in three materials of interest. In the Carborundum Company process addition of ca 5 wt % zirconia to the basic 50 wt % alumina–50 wt % silica system produces a longer fiber for use at 1260°C. In 1965, Johns-Manville developed a chromia-modified, alumina–silica fiber with a temperature limit of 1427°C (8). More recently, the 3M Company has introduced a boria-modified alumina–silica fiber for continuous use at 1427°C. This fiber, trade named Nextel, is particularly suitable for fabrication into flexible high temperature textiles.

The next technological advance was the development of the precursor process by Union Carbide Corporation. An organic polymeric fiber is used as a precursor to absorb dissolved metal oxides. A subsequent heat treatment burns out the organic fiber, leaving a polycrystalline refractory metal oxide in the shape of the host polymeric material (9). With this technique Union Carbide developed the first commercially available 1650°C zirconia fiber, trade-named Zircar. At about the same time, Babcock & Wilcox, along with others, discovered that inorganic fibers could be made by spinning, blowing, or drawing a viscous, aqueous solution of metal salts into fibers followed by a heat treatment to convert the salts to the oxide form. The resulting fibers were polycrystalline and usually contained 5–10% porosity. This process allowed the fiberization of metallic oxides and combinations of metallic oxides, whose viscosity in the molten state would not allow fiberization by ordinary methods.

From this new technology, Babcock & Wilcox developed the first true mullite fiber, which was used experimentally in the development of tile insulation for the space shuttle Orbiter. In 1972, ICI introduced the first commercial fibers made by the sol process. Under the trade names Saffil alumina and Saffil zirconia, fibers were offered for continuous use at 1400 and 1600°C, respectively. The sol process eliminated the need for the relatively costly organic precursor fiber and thereby significantly reduced the cost of manufacturing relatively pure oxide fibers. The zirconia fiber, however, was dropped from production.

During the mid-1960s, ultralightweight high strength composites replaced traditional metallic structural members and other parts of the early space vehicles, and cast metals were reinforced with short fibers consisting of alumina, silicon carbide, silicon nitride, or boron nitride (see High temperature composites). Their strengths are 1.4–20 GPa (($2–29) \times 10^5$ psi) with elastic moduli of 70–700 GPa (ca $(1–10) \times 10^7$

psi). Silicon carbide whiskers (single-crystal fibers) are of special interest because they offer not only high strength and stiffness but are useful up to 1800°C. Numerous companies worldwide became involved in carbide–fiber research. In many manufacturing processes these fibers are grown as reaction products of gaseous silicon monoxide and carbon monoxide at high temperatures. In some cases, carbon fibers were allowed to react with silicon monoxide. By 1968, both Thermokinetic Fibers, Inc. and Carborundum sold short silicon carbide fibers at ca $500/kg. A recent innovation is the production of high quality short silicon carbide fibers by the pyrolysis of rice hulls in an inert or ammonia atmosphere.

Boron nitride exhibits exceptional resistance to thermal shock and can be used in inert or nonoxidizing atmospheres to 1650°C (11). In a Carborundum process, boron oxide fibers are converted to polycrystalline boron nitride at elevated temperatures in an ammonia atmosphere. A 1971 Lockheed Aircraft patent describes a similar process in which a boron filament is heated first in air and then converted to the nitride in a nitrogen atmosphere (12).

Properties

Refractory-fiber insulating materials are generally used in applications above 1063°C. Table 1 gives the maximum long-term use temperatures in both oxidizing and nonoxidizing atmospheres. For short exposures, some of these fibers can be used at temperatures much closer to their melting temperature without degradation.

The most important properties of these fibers are thermal conductivity, resistance to thermal and physical degradation at elevated temperatures, and tensile strength and elastic modulus. The thermal conductivity, in W/(cm·K), is actually the inverse of the R factor that expresses the insulating value of building insulations (see Insulation, thermal). Fibrous insulations with increasingly lower thermal conductivities require reliable test methods, and currently ASTM C 201 and ASTM C 177 are the standard methods. In addition, ASTM C 892-78 defines maximum thermal conductivities and unfiberized contents. Thermal conductivity is affected by the bulk density of the material, the fiber diameter, the amount of unfiberized material (shot), and the mean temperature of application. Typical silica fibers have a mean fiber diameter of

Table 1. Maximum Use Temperatures of Refractory Fibers

Fiber type	CAS Registry Number	Mp, °C	Maximum use temperature, °C	
			Oxidizing atmosphere	Nonoxidizing atmosphere
Al_2O_3	[1344-28-1]	2040	1540	1600
ZrO_2	[1314-23-4]	2650	1650	1650
SiO_2	[7631-86-9]	1660	1060	1060
Al_2O_3–SiO_2	[37287-16-4]	1760	1300	1300
Al_2O_3–SiO_2–Cr_2O_3	[65997-17-3]	1760	1427	1427
Al_2O_3–SiO_2–B_2O_3		1740	1427	1427
C	[7440-44-0]	3650	400	2500
B	[7440-42-8]	1260	560	1200
BN	[10043-11-5]	2980	700	1650
SiC	[409-21-2]	2690	1800	1800
Si_3N_4	[12033-89-5]	1900	1300	1800

1–2 µm and less than 5% unfiberized material. Alumina–silica fibers range in mean diameter from 2.5 to 5.5 µm and have 20–50 wt % shot, which significantly reduces their thermal efficiency (increases thermal conductivity). Alumina and zirconia fibers are manufactured with a uniform diameter of ca 3 µm and contain less than 5% unfiberized particles. With such a wide range of shot contents and fiber diameters, the thermal conductivity of equivalent bulk-density blankets or felts made from different fiber can differ significantly. A thermal conductivity-versus-mean-temperature plot is given for several fibers in Figure 1. The effect of density on the thermal conductivity of silica and alumina–silica fibers is shown in Figure 2. For reinforcing purposes, tensile strength measured at room temperature and stress-to-strain ratio measured as Youngs modulus are of primary importance (see Table 2). Alumina-silica fibers are not used as reinforcements, and consequently, strength and modulus measurements are not important.

At the relatively low densities of these materials, solid conduction of heat is negligible when considering total heat transfer (see Heat-exchange technology). Heat transfer is affected mainly by radiation, which at 1260°C, is reponsible for approximately 80% of all heat transfer. Heat transfer by convection is small, and gas conduction within the insulation is responsible for the balance of heat transfer through the material. Consequently, most efficient insulations are those that effectively block radiation. The temperature ratings given these fibers are generally the temperatures at which the linear shrinkage stabilizes within 24 h and does not exceed 5%. The 5% linear shrinkage value has become a design factor for furnace insulation and other high temperature applications.

Alumina–silica and leached-glass fibers have been converted from amorphous glass to their crystalline forms. The rate of crystallization, sometimes called devitrification, depends on time and temperature (see Crystallization). For some time, the degradation of mechanical strength of refractory fiber products was attributed to this crystallization. In alumina–silica fibers, crystallization is completed relatively quickly,

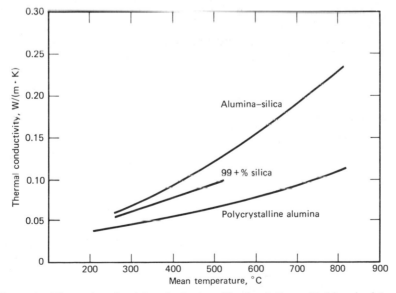

Figure 1.   Thermal conductivity of refractory fiber insulations with 96 mg/cm³ density.

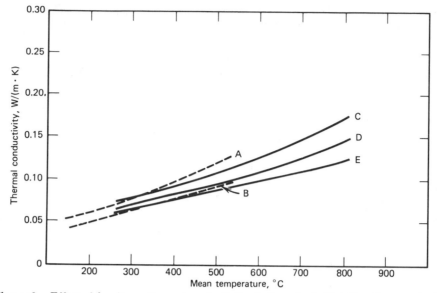

Figure 2.   Effect of density on thermal conductivity. A, 48 mg/cm³ silica fiber; B, 96 mg/cm³ silica fiber; C, 128 mg/cm³ alumina–silica fiber; D, 192 mg/cm³ alumina–silica fiber; E, 384 mg/cm³ alumina–silica fiber.

within 24 h, at rated use temperatures; further degradation of physical properties with time is due to grain growth in the fiber (13). Sintering, the bonding together of these fibers at temperatures well below their softening point, is also responsible for linear shrinkage. At contact points, solid-state diffusion of molecules has a bridging effect whereby previously flexible material is transformed into a rigid structure. A scanning electron micrograph of alumina–silica–chromia fiber after 120 h exposure to 1427°C is shown in Figure 3. Both the formation of mullite and cristobalite crystals in the originally amorphous fiber and the sintering at fiber intersections can be clearly seen. The diffusion of oxides in the fibers toward the contact point has a slight shortening effect upon the fiber that ultimately contributes to the overall shrinkage of the product. In general, with increased sintering the resistance to abrasion or mechanical vibration decreases.

Table 2.   Mechanical Properties of Oxide and Nonoxide Fibers

Fiber type	Density, g/cm³	Tensile strength, GPa[a]	Young's modulus, GPa[a]
SiO_2	2.19	5.9	72
Al_2O_3	3.15	2.1	170
ZrO_2	4.84	2.1	345
carbon	1.50	1.4	210
graphite[b]	1.66	1.8	700
BN	1.90	1.5	90
SiC[b]	3.21	2.0	480
Si_3N_4[b]	3.18	1.4	380

[a] To convert GPa to psi, multiply by 145,000.
[b] Single-crystal whiskers.

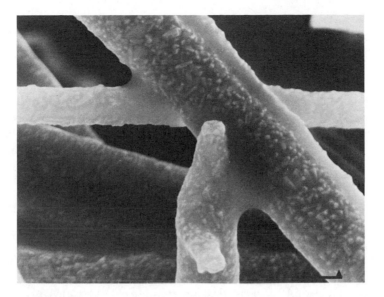

Figure 3. Alumina–silica–chromia fiber after 120 h at 1426°C showing crystallization and sintering at contact points. ×5000.

Alumina–Silica Fibers

Current refractory-fiber production consists mainly of melt-fiberized alumina–silica, and modified alumina–silica fibers. The 1260°C-alumina–silica fibers are produced by melting high purity alumina and silica or calcined kaolin in electric resistance furnaces. In either case, the resulting fiber contains ca 50 ± 2 wt % SiO_2 and Al_2O_3. The higher temperature grade, for use to 1427°C, is made by adding ca 3 wt % Cr_2O_3 to the basic composition or increasing the Al_2O_3 content to 55–60 wt %.

The raw materials are melted in a three-phase electric furnace that operates on the electrical resistance of the pool of molten material in which the electrodes are immersed. The initial melt is established by passing a current through graphite or coke granules that ultimately burn off, leaving the molten alumina–silica to serve as the conductive medium. Graphite is the common electrode material, although refractory metals, such as molybdenum and tungsten, are also used. The latter must be kept submerged in the molten pool to prevent oxidation. Some electrode loss occurs, but the refractory metal electrodes are generally not attacked by long-term use (14). The molten material is discharged through a temperature-controlled orifice at the bottom of the vessel (14–15), or the furnace is tilted to deliver the melt via a refractory trough to the fiberizing unit.

In the steam-blowing process, the liquid material is dropped in the path of a high velocity blast from a steam jet. Air jets are also used for fiberizing. Normally, the fiberizing pressures in the blowing processes are ca 700 kPa (ca 100 psi), but they can be much higher. The steam fiberizing of molten materials is an art long associated with the manufacture of mineral–wool insulations from low temperature melts. Ultra-high-speed photography has revealed that the blast initially shreds the molten stream into tiny droplets. As each droplet picks up momentum from the velocity of the steam blast, it elongates into a teardrop shape and attenuates into a fiber with a spherical globule of molten material as its head, called shot. Several examples of fiber with the spherical shot particles still attached are shown in Figure 4. Each fiber is therefore the tail from a droplet of nonfibrous material. However, the amount of fiber in the crude product on a volume basis is very large compared with that of nonfibrous material.

In the melt-fiberization or spinning process, the molten material is dropped on the periphery of a vertically oriented, rotating disk. The molten material bonds to the high-speed rotating disk surface from which the melt droplets are ejected. The fiber is attenuated both by centrifugal action on the ejected droplet and the fact that the fiber tail frequently is still attached to the disk's surface. Generally, spinning produces a longer fiber than steam blowing, and converts more of melt to fiber, depending on the melt-delivery rate.

High Purity Silica Fibers

Refractory fibers of essentially >99 wt % silica are not made by conventional melt-fiberizing techniques. Thus, to achieve the high purity required for operation at 1093–1260°C, a leaching process is employed (16). A glass of ca 75 wt % SiO_2 and 25 wt % Na_2O is melted in a typical glass furnace at ca 1100°C. Filaments with a diameter of ca 0.3 mm are drawn from orifices in the furnace bottom. These filaments are then passed in front of a gas flame and attenuated to a diameter of approximately 1.5 μm. The loose fibers are subjected twice to an acid-leaching process to remove the Na_2O. The fiber is thoroughly rinsed, and dried at ca 315°C. The resulting fiber has a maximum of 0.01 wt % Na_2O and K_2O, 0.20 wt % Al_2O_3, and 0.04 wt % MgO and CaO.

The direct manufacture of pure silica fibers requires either a fused or extruded silica particulate rod of 5–25-mm dia. The fused silica rods with diameters of 6–7 mm are formed from molten high purity silica and mounted in groups of 20 on a motor-driven carriage. The rods are reduced to ca 2 mm in dia by being forced through a graphite guide and over a vertically oriented oxyhydrogen burner operating at 1800°C.

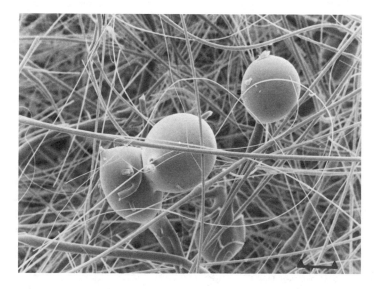

Figure 4.   Refractory fiber (1260°C) and unfiberized shot particles. ×1000.

These relatively large fibers are passed through another graphite block containing 20 holes. As the fibers exit, they are hit by an axial oxyhydrogen jet attenuating the fibers toward a rotating drum which collects them. Fibers with a diameter of 4–10 μm and a silica content of 99.95 wt % are produced (17).

Chemically Produced Oxide Fibers

Ceramic oxide fibers are difficult to produce by common melt technology because oxides such as zirconia and alumina have very high melting points and low viscosities (see also Ceramics). In 1969, Union Carbide marketed the first 1650°C zirconia refractory fiber manufactured by a chemical technique termed the precursor process. The precursor can be any organic fiber containing extremely small crystallites of polymer chains held together in a matrix of amorphous polymer. When immersed in a solvent such as water, the fiber swells, thus expanding the spaces between the crystallites. Although a number of organic fibers have this swelling characteristic, eg, wool, cotton, and cellulose acetate, rayon is preferred because of its structural uniformity and high purity. After an initial swelling and dewatering by centrifuge, the rayon fiber is immersed in a 2 M aqueous solution of zirconyl chloride containing a small amount of yttrium salt. The excess solution is centrifuged, and the fibers are dried and heated to ca 400°C in an atmosphere containing <10 vol % oxygen. This treatment pyrolyzes the rayon to a carbonaceous residue; the zirconyl chloride is converted to microcrystalline zirconia fiber containing yttrium oxide for phase stabilization and to prevent embrittlement.

Imperial Chemical Industries produces 1600–1650°C fibers by the sol process (18). Both silica-stabilized alumina and calcia-stabilized zirconia fibers are marketed as Saffil. First a metal salt, such as aluminum oxychloride in the case of alumina fiber, is mixed with a medium molecular weight polymer such as 2 wt % poly(vinyl alcohol). This solution is then slowly evaporated in a rotary evaporator to a viscosity of ca 80 Pa·s (800 P). This solution is then extruded through a 100-μm spinneret; the fibers are collected on a drum where they are fired to a temperature of 800°C. This action burns the organic component away and a fine-grained aluminum oxide fiber is formed with a porosity of 5–10 vol % and a fiber diameter of 3–5 μm. These fibers are used as filter media because of their inherent porosity. For refractory application, they are heated to 1400–1500°C, just long enough to eliminate the porosity, which would result in 3–4% linear shrinkage in application. The same process is followed for zirconium oxide fibers, starting with zirconium oxychloride, zirconium acetate, and calcium oxide. The only truly continuous alumina–silica-type fiber is made by the 3M Company employing a similar solution process (19). In this case, a 10-μm dia alumina–silica–boria fiber is manufactured for use to 1427°C. Basic aluminum acetate is dissolved in water and the solution is mixed with an aqueous dispersion of colloidal silica and dimethylformamide. The resulting solution is concentrated in a Rotavapor flask and centrifuged. The solution is then extruded through a 75-μm spinneret at 100 kPa (1 atm). The resulting fibers are collected on a conveyor chain and passed through a furnace at 870°C converting the filaments to metallic oxides. Heating in another furnace at 1000°C produces a glassy aluminum borosilicate [12794-54-5] with the calculated composition $3Al_2O_3 \cdot B_2O_3 \cdot 3SiO_2$.

Although fiber production by this technique is more expensive than melt fiberization, the sol process offers many other advantages. In melt fiberization, the effect upon viscosity and surface tension must be taken into account when adding even small amounts of other desirable oxides. In the sol process, however, the viscosity is controlled independently of the metals added and thus any number of metal salts can be easily added without adverse effect. The controlled addition of other metals can also serve as grain-growth inhibitors, sintering aids, phase stabilizers, or catalysts.

Nonoxide Refractory Fibers

The most important nonoxide refractory fibers are silicon carbide fibers and single-crystal whiskers. In the early work on silicon carbide, 1 part by volume SiO was mixed with 3 parts CO at 1300–1500°C in an inert or reducing carrier gas. The fibers formed on the colder parts of the reactor tube had 4–6-μm dia and were 50-mm long (20). In a similar but less complicated technique developed by Corning Glass Works, the SiO and CO gases are obtained by heating a mixture of carbon and silica in a molar ratio of 2:1 to 1300–1550°C in a hydrogen fluoride or hydrogen chloride atmosphere (21). The fibers grow more quickly than in an inert carrier gas and the process can be run continuously. These fibers are made up of beta silicon carbide crystals with a surface sheath of silica which prevents oxidation.

Since 1960, a number of processes for high quality silicon nitride fibers have been developed. These fibers are a by-product in the production of silicon nitride powder from silicon metal and nitrogen at high temperatures. Additions of a reducing agent greatly increased the fiber yield by converting silica first to silicon monoxide and then to silicon metal for the reaction with nitrogen (22).

$$SiO_2 + R \rightarrow SiO + RO$$

$$3\, SiO + R + 2\, N_2 \rightarrow 3\, RO + Si_3N_4$$

R is the reducing agent. Any silicate that forms thermally and chemically stable residual compounds as its SiO_2 content is reduced and provides a suitable source of silicon for the reaction. Alternate aluminum–silica and graphite plates are stacked in a graphite-lined alumina tube where the plates are separated by 2–4-cm graphite spacers. This tube is heated to 1400°C for 12 h in a nitrogen atmosphere in a silicon carbide-resistance furnace. After approximately 6 h the tube is cooled and the fibers are removed.

Boron nitride fibers are produced by nitriding boron filaments obtained by chemical vapor deposition of boron on heated tungsten wire. The tungsten wire is passed through a reactor containing boron trichloride in a carrier of hydrogen at 1100–1300°C. The boron trichloride is reduced and boron is deposited on the tungsten wire. The continuous boron filament exits the furnace and is wound upon a spool. In nonoxidizing atmospheres, boron filaments can be considered refractory fibers. When heated to 560°C, they develop a liquid boron oxide surface coating. For pure boron nitride fiber, the boron oxide filaments are further heated to 1000–1400°C in an ammonia atmosphere for ca 6 h. The following reactions are assumed to take place:

$$4\, B(s) + 3\, O_2(g) \rightarrow 2\, B_2O_3(l)$$

$$B_2O_3(l) + 2\, NH_3(g) \rightarrow 2\, BN(s) + 3\, H_2O(g)$$

Economic Aspects

Refractory fibers are 50–70% more efficient insulators than conventional brick linings at equivalent thicknesses. Although the initial cost of installing a fiber lining in a high temperature furnace is higher than that of installing brick lining, the difference can be quickly recovered in energy savings. As the cost of energy continues to increase, refractory-fiber furnace lining is becoming more and more attractive. In 1980, the average cost of 1260°C alumina–silica refractory fiber was ca $3.75/kg, whereas the 1970 price was ca $5/kg. The decrease in cost reflects the advancement in technology.

The 1981 selling price for the >99% silica fibers was approximately $55/kg in both bulk and felted forms. This price has remained relatively constant over the past ten years. The more refractory Saffil alumina fiber sells for approximately $33/kg. When first introduced in 1974, both the Saffil zirconia and alumina fibers sold for approximately $24/kg. Production quantities are proprietary.

A substantial reinforced-composites market for continuous filaments has developed in the past few years. Carbon, graphite, and boron fibers are produced commercially at prices of $40–400/kg, depending upon their properties. Silicon carbide whiskers have been produced commercially over the last 20 yr by several processes. Prices for these fibers have been based upon pilot plant quantities and have ranged from $2000/kg to as low as $150/kg in 1980. The price of silicon nitride fibers has been $200–500/kg. Boron nitride fibers have not been produced commercially until about 1975. Their price has remained at ca $50/kg since that time.

Health and Safety Factors

Synthetic or man-made refractory fibers do not appear to pose the health hazards of naturally occurring mineral fibers like asbestos. Because the diameter of most refractory fibers is <3.5 μm, they are considered respirable. Above 3.5 μm, fibers are not able to penetrate the functional components of the lung (23). Even though a portion of these fibers are respirable, they are considered a nuisance dust, because studies have shown they are not biologically active in living tissue. The Thermal Insulation Manufacturers Association is currently conducting animal inhalation studies on a number of man-made fibers including refractory fibers. Regardless of the results of these studies, proper respirators should be worn in environments of excessive exposure to refractory fibers.

Silica and alumina–silica refractory fibers, that have been in service above 1100°C, undergo partial conversion to cristobalite, a form of crystalline silica that can cause silicosis, a form of pneumoconiosis. The amount of cristobalite formed, the size of the individual crystallites, and the nature of the matrix in which they are embedded are time and temperature dependent. Under normal use conditions, refractory fibers are generally exposed to a temperature gradient. Consequently, it is most probable that only the fiber nearest the hot surface has an appreciable content of cristobalite. It is also possible that fiber containing devitrified cristobalite is more friable and therefore may generate a larger fraction of dust when it is removed from a high temperature furnace. Hence, removal of old furnace lining offers the greatest risk of exposure, and adequate protection against respiration should be provided. Fibers with diameters of ≥5 μm cause irritation to skin and mucous membranes. This is usually not a serious problem and can be avoided by wearing proper clothing and respirators. When heated in an oxidizing atmosphere, silicon carbide fibers form a surface coating of silica that converts to cristobalite. In addition, silicon carbide whiskers have been found to be biologically active in animal lung tissue and may cause pneumoconiosis. Adequate respiratory protection is advised when dealing with these fibers. The diameter of carbon and boron filaments places them above the respirable range, but they can cause skin irritation from penetration during handling. Silicon nitride and boron nitride fibers are not considered hazardous but may irritate the skin.

Uses

Blankets. Today, ca 60% of the refractory fiber production is used for 48–128-mg/cm^3 flexible needled blankets. The fiber is collected on a moving conveyor, run through compression rolls, and penetrated by barked needle boards. This needling has the effect of tying the fibers together; subsequent compression and heating increase the tensile strength of the product. Flexible needled blankets are commercially available in widths of 1.3 and 0.65 m, lengths ≤33 m, and thicknesses of 6 to 50 mm. They are primarily used as furnace-wall and roof insulations either as the exposed hot face or as back-up insulation behind refractory brick. Because of their lightweight flexible nature, blankets offer no structural support to the furnace wall and have to be anchored in place. Typically, the blankets are applied to furnace walls and roofs in overlapping layers by impaling them on metallic or ceramic studs fixed to the supporting metal framework. They can also be applied over existing brick walls using high temperature cement or mechanical anchoring systems providing a new, more insulating furnace interior (see Insulation, thermal; Furnaces). Because of their relatively low heat storage and thermal conductivity, they have replaced brick linings in most industrial kilns in order to reduce energy costs. Other applications for these blankets include insulation for automotive catalytic convertors and aircraft and space vehicle engines and a wide variety of uses in the steel-making industry.

Felts. Felts (qv) contain an organic binder, generally a phenolic resin or in some cases a latex material. As the fibers are collected after fiberization, they are mixed with a dry phenolic resin. The fiber is then passed through an oven and a constant pressure is applied by flight conveyors. Alternatively, the uncured fiber and resin are compressed between heated press plates to form felts with densities as high as 380 mg/cm^3. These high density felts are now used extensively in ingot-mold operations in steel foundries. Felts provide excellent expansion joints in high temperature applications because the fibers tend to expand after the organic binder has been burned out (see also Heat-resistant polymers).

Bulk Fibers. Bulk refractory fiber is used as a general-purpose high temperature filler for expansion joints, as stuffing wool, for furnace and oven construction, in steel mills and aluminum and brass foundries, in glass manufacturing operations, as a loose-fill insulation, and as a raw material for vacuum-formed shapes (see Fillers).

Vacuum-Formed Shapes. Approximately 20% of fiber production is converted to rigid shapes by a vacuum-forming process. The bulk fibers are mixed in aqueous suspension with clays, colloidal metal oxide particles, and organic binders. Molds with fine-mesh screen surfaces are used to accrete the solids into special shapes as the water is drawn through the screen by vacuum. These products are then dried at 100–200°C to obtain rigid shapes having densities generally between 200 and 300 mg/cm^3. In recent years, the more expensive 1650°C alumina and zirconia fibers have been mixed with small amounts of less expensive and lower temperature alumina–silica fibers in the vacuum-forming process to increase the use temperature. The uses of vacuum-formed refractory fiber products are extremely varied. The metal industry makes extensive use of vacuum-formed shapes from tap hole cones in furnaces and ladles to insulating risers in metal castings. The tiles of the space shuttle thermal protection system are made of 99.7 wt % silica fiber, trade named Q Fiber, made by a highly specialized vacuum-forming process. The largest application for vacuum-formed products is as lightweight board insulation for furnace linings. In the rigid vacuum-formed state, the fibers are less susceptible to abrasion from direct gas impingement emanating from burners or flue fans.

Modules. Lightweight refractory fiber modules are a recent innovation in furnace lining (24). The refractory fiber blanket is folded in an accordion fashion and then compressed and banded to form modules 0.3 by 0.3 m, and 0.2–0.3-m thick. Metallic hardware, attached to the back of each module as it is folded and compressed, is welded to the metal shells of furnace roofs and walls. The modules are snapped into place in a parquet fashion forming the complete insulation system for the furnace. This attachment technique offers distinct advantages over layered blanket construction by reducing installation time and labor and eliminating the need for metal or ceramic anchoring pins. The shrinkage effect is minimized because the resilient fibers expand somewhat when their compression banding is released during construction. This concept of furnace lining is gaining widespread acceptance in the industry because of the ease of installation and other technical advantages.

Fibers and Yarns. The continuous refractory oxide fibers, ie, silica and alumina–silica–boria can be twisted into yarns from which fabrics are woven. These fabrics are used extensively in heat-resistant clothing, flame curtains for furnace openings, thermocouple insulation, and electrical insulation. Coated with Teflon, these fibers are used for sewing threads for manufacturing speciality high temperature insulation shapes for air craft and space vehicles. The cloth is often used to encase other insulating fibers in flexible sheets of insulation. The spaces between the rigid tile on the space shuttle Orbiter are packed with this type of fiber formed into tape.

Nonoxide fibers in continuous form, such as carbon, graphite, and boron are primarily used in filament winding and the manufacture of high strength, high modulus fabrics. Lightweight, high strength pressure vessels are made by running these fibers through an epoxy impregnant and then winding them around plaster mandrels. After curing the resin, the plaster is dissolved leaving a thin-walled vessel. The drill stems used for the collection of lunar core samples were a combination of boron cloth with an epoxy–graphite filament. The carbon fibers in loose fill and cloth form are currently used in nonoxidizing processes as insulation up to 1800°C and higher temperatures for short duration. The shorter nonoxide fibers are primarily used as strength-enhancing reinforcements in resins, some ceramics, and metals. Silicon carbide and silicon nitride short fibers are dispersed in a number of resins that are then cast into various shapes. Applications for these cast parts are generally found in high technology areas such as the fabrication of specialty electrical parts, aircraft parts, and radomes (microwave windows). These fibers are also used as reinforcing inclusions in metals. Boron nitride is of particular value as a reinforcing fiber for cast aluminum parts because it is one of the few materials totally wet by molten aluminum. Silicon carbide and boron nitride fibers as well as short single-crystal alumina fibers are used to reinforce gold and silver castings.

The growth and commercialization of the nonoxide fiber industry parallels the growth of the high strength-composite industry. When these fibers can be produced in lengths of 2–10-cm in the \$10–20 per kilogram price range, a large >1600°C insulation market will open up.

BIBLIOGRAPHY

"Refractory Fibers" in *ECT* 2nd ed., Vol. 17, pp. 285–295, by T. R. Gould, Johns-Manville Corporation.

1. U.S. Pat. 2,461,841 (Feb. 15, 1949), M. E. Nordberg (to Corning Glass Works).
2. *Chem. Eng. News* **25**(18), 1290 (1947).
3. C. Z. Carroll-Porczynski, *Advanced Materials*, 1st ed., Astex Publishing Company, Ltd., Guildford, England, 1959.
4. U.S. Pat. 2,823,117 (Feb. 11, 1958), D. Labino (to Glass Fibers, Inc.).
5. U.S. Pat. 3,177,057 (Aug. 2, 1961), C. Potter and J. W. Lindenthal (to Engelhard Industries, Inc.).
6. U.S. Pat. 2,467,889 (Apr. 19, 1949), I. Harter, C. L. Horton, Jr., and L. D. Christie, Jr. (to Babcock & Wilcox Company).
7. *Ceram. Age* **78**(2), 37 (1962).
8. U.S. Pat. 3,449,137 (June 10, 1969), W. Ekdahl (to Johns-Manville Corporation).
9. U.S. Pat. 3,385,915 (May 28, 1968), B. Hamling (to Union Carbide Corporation).
10. H. W. Rauch, Sr., W. H. Sutton, and L. R. McCreight, *Ceramic Fibers and Fibrous Composite Materials*, Academic Press, New York, 1968, p. 24.
11. U.S. Pat. 3,429,722 (July 12, 1965), J. Economy and R. V. Anderson (to Carborundum Company).
12. U.S. Pat. 3,573,969 (Apr. 10, 1971), J. C. Millbrae and M. P. Gomez (to Lockheed Aircraft Corporation).
13. L. Olds, W. Miiller, and J. Pallo, *Am. Ceram. Soc. Bull.* **59**(7), 739 (1980).
14. U.S. Pat. 2,714,622 (Aug. 2, 1955), J. C. McMullen (to The Carborundum Company).
15. U.S. Pat. 3,066,504 (Dec. 4, 1962), F. J. Hartwig and F. H. Norton (to Babcock & Wilcox Company).
16. U.S. Pat. 4,200,485 (Apr. 29, 1980), G. Price and W. Kielmeyer (to Johns-Manville Corporation).
17. Brit. Pat. 507,951 (June 23, 1939).
18. Brit. Pat. 1,360,197 (July 17, 1974), M. Morton, J. Birchall, and J. Cassidy (to Imperial Chemical Industries, Ltd.).
19. U.S. Pat. 3,760,049 (Sept. 18, 1973), A. Borer and G. P. Krogseng (to Minnesota Mining and Manufacturing Company).
20. U.S. Pat. 3,246,950 (Apr. 19, 1966), B. A. Gruber (to Corning Glass Works).
21. U.S. Pat. 3,371,995 (March 5, 1968), W. W. Pultz (to Corning Glass Works).
22. U.S. Pat. 3,244,480 (Apr. 5, 1966), R. C. Johnson, J. K. Alley, and W. H. Warwick (to The United States of America).
23. J. Leineweber, *ASHRAE J.* **3,** 51 (1980).
24. U.S. Pat. 3,952,470 (Apr. 27, 1976), C. Byrd, Jr. (to J. T. Thorpe Company).

W. C. MIILLER
Manville Corporation

SILK

Silk is the solidified viscous fluid excreted from special glands (or orifices) by a number of insects and spiders. It is a polymer consisting of amino acids; glycine and alanine are the main components. The only significant source for textile usage is the silk-moth caterpillar, commonly referred to as the silkworm. Several varieties are known; the most valuable is the caterpillar of the moth *Bombyx mori*, which was domesticated centuries ago. A somewhat different product known as wild silk, sometimes called tussur, is produced by species of moth that are not domesticated.

The growth of domesticated silkworms is known as sericulture. The silk moth lays its eggs on leaves, where they are held by a secreted gummy substance. Upon hatching, the larvae feed on the leaf and grow rapidly from ca 0.63 to 8.9 cm in one month. The mature caterpillar prepares for metamorphosis by spinning the protective silk cocoon in ca three days. After ca two weeks, a fully developed silk moth emerges, invariably damaging the cocoon. In less than one week, the female mates, lays her eggs and, having accomplished her task of perpetuating the species, dies.

Although of no current commercial significance, silk is also produced by spiders in the class Arachnoidea. Considerable research has been done in recent years on spider silk (1–2).

The use of silk as a textile fiber predates written history. The earliest records indicate that silk was used in China over 4000 years ago. Silk and its production are mentioned in various Chinese legends (3). From China, commercial production of silk evolved and eventually spread throughout much of the world.

It is believed that several centuries passed before other countries, first India and Japan about 300 AD, discovered the secret of producing silk. In a comparatively short time, the methodology was known throughout Asia, Africa, and Europe. Attempts to establish silk in North America were first made by King James I of England in the 16th century. In the late 1700s, a significant amount of silk was produced in colonial America. Some production of silk was maintained in the United States until the early years of this century. Its decline in the United States but continued growth in China, Japan, and India are due to the low cost of labor in the latter countries.

Silk is available for textiles (qv) as a continuous filament and as staple yarn. Although not truly continuous, a filament may be 300–1200 m long. The strand produced by the silkworm consists of two filaments encased in a protein gum. The silkworm uses it to form the cocoon in which it encases itself for metamorphosis. If the moth is allowed to emerge alive, the filaments are ruptured. These and other damaged filaments, called silk waste, are used in producing staple yarns, which are formed by twisting short lengths of filament together.

Because of the enormous amount of manual labor required in the production of silk, it has always been expensive. Its properties have made it highly sought after and revered for apparel and furnishing fabrics for centuries. In recent years, man-made fibers have been produced that rival silk's desirable properties of luster, hand, and drape at a far lower cost (see Fibers, chemical). Consequently, the demand for silk has declined, although it is still used for specialty and high quality luxury items.

Physical Properties

Throughout history, silk has been sought after because of its unique fiber characteristics. It is both strong and elastic and has a tensile strength of 0.34–0.39 N/tex (3.9–4.5 g/den) and an elongation at break of 20–30%. Its natural crease resistance is due to good resilience and it recovers readily from deformation. A highly hygroscopic fiber (moisture regain of 10–15 percent), it is a poor conductor of heat and electricity. It has the desirable tactile properties associated with light weight, warmth, and good drapability. It is a smooth, translucent fiber with a triangular cross-section. Its cross-sectional dimensions correspond to ca 0.17 tex (1.5 den) per filament of fibroin. The continuous filament has a high luster or sheen that contributes to its aura of luxury.

In Table 1 the physical properties of silk are compared with those of a group of selected textile fibers. Data on cotton and polyester are included because of their dominant positions in the natural- and synthetic-fiber markets.

Different colors of silk are produced, depending upon the source of food for the silkworm. Although almost any color can be obtained, the principal ones are white when the caterpillar feeds on white mulberry (*Morus alba*) leaves, yellow from dwarf mulberry (*Morus nigra*) leaves, and green, which is from chlorophyll in the diet. Most wild silk is brown, probably a result of the variety of leaves consumed by the undomesticated caterpillar.

Table 1. Physical Properties of Selected Textile Fibers

Property	Cotton	Wool	Silk	Acetate	Nylon	Polyester
abrasion resistance	good	fair	fair	poor	excellent	good
absorbancy, % moisture regain	8.5	13.5	11	6.5	4.5	1
drapability	fair	good	excellent	excellent	good	fair
elasticity at 21°C, 65% rh						
elongation at break, %	3–10	20–40	20	23–45	26–40[a]	19–23[a]
recovery, %	75 45	99 65		94 23	100	97 80
from % strain	2 5	2 20	2[b]	2 20	8	2 8
environment						
mildew resistance	poor	good	good	excellent	excellent	excellent
renovation[c]	W, DC	DC	W, DC	DC	W, DC	W, DC
safe iron limit, °C	204	148	148	177	177	163
sunlight resistance	fair	good	poor	good	poor	good
hand	good	fair–excellent	excellent	excellent	fair	fair
pilling resistance	good	fair	good	good	poor	bad
resiliency	poor	good	fair	fair	good	excellent
specific gravity	1.54	1.32	1.30	1.32	1.14	1.22–1.38
static resistance	good	fair	fair	fair	poor	bad
strength, N/tex[d]	0.26–0.44	0.07–0.17	0.34–0.39	0.07–0.13	0.22–0.66	0.22–0.83
	good	fair	good	poor	excellent	excellent
strength loss when wet, %[e]	10	20	15	30	15	0
thermoplastic?	no	no	no	yes	yes	yes

[a] Regular.
[b] Poor if stretched beyond.
[c] W = wash; DC = dry clean.
[d] To convert N/tex to gf/den, multiply by 11.33.
[e] Approximate values.

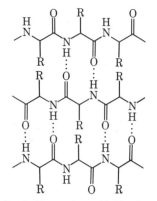

Figure 1.   Partial structure of fibroin; R = amino acid. The dotted lines indicate how hydrogen bonds might join the chains.

Chemical Properties

Raw silk consists primarily of the proteins sericin and fibroin. The bulk is fibroin or fiber, which is coated with 15–25% sericin or silk gum. The principal amino acid constituents are glycine, alanine, serine, and tyrosine. The percentage composition of these and other amino acids in the two proteins differ. A summary of the extensive studies on the chemical analysis of silk reports 18 different amino acids (4) (qv).

The fiber is composed of long-chain amino acid units joined by peptide links with hydrogen bonding between parallel chains. The structure shown in Figure 1 agrees with the silk-fibroin arrangement described by Pauling and Corey (5) as an antiparallel-chain pleated sheet structure. The single polypeptide chains probably pass through both the crystalline and amorphous regions; the amino acids with bulky side chains are in the latter areas. Since the two most common components in silk fibroin are glycine and alanine, which have small side chains, large sections of the chains can

Table 2.   Amino Acid Content of Silk Fibroin, *Bombyx Mori*

Amino acid	Mol/100 kg
glycine	567.2
alanine	385.7
leucine	6.2
isoleucine	6.9
valine	26.7
phenylalanine	8.0
serine	152.0
threonine	12.5
tyrosine	62.3
aspartic acid	17.6
glutamic acid	11.8
arginine	5.6
lysine	3.8
histidine	1.9
proline	5.1
tryptophan	2.5

Table 3.   Solubility of Silk

Chemical	Concentration, %	Temperature, °C	Time, min
soluble in			
sodium hypochlorite	5	20	20
sulfuric acid	59.5	20	20
insoluble in			
acetic acid	100	20	5
acetone	100	20	5
hydrochloric acid	20	20	10
formic acid	85	20	5
dimethylformamide	100	90	10

approach each other closely, which results in a mostly crystalline fiber structure (see Biopolymers).

A representative analysis of silk fibroin is given in Table 2. Small amounts of cystine and methionine occur in addition to those given in Table 2 (6). The presence of cystine in silk fibroin was first reported in 1955 (7); its composition was subsequently determined at ca 0.2 percent (8).

The chemistry of silk as related to its composition, structure, and reactivity is reviewed in ref. 9, which includes 124 literature references from 1901 to 1974.

Although not significantly affected by dilute acids, silk is readily hydrolyzed by concentrated sulfuric acid (see Table 3). It is highly sensitive to concentrated alkaline solutions and calcium thiocyanate but withstands treatment with hot dilute (1%) sodium hydroxide solutions.

Processing

Reeling.   Soon after the cocoon is completed by the silkworm, it is gathered, and the chrysalis, ie, the caterpillar undergoing metamorphis, is killed, usually by dry heat or steaming. The cocoons are sorted into standard grades based on quality and then converted to yarn by a process known as reeling. The gum coating is softened by immersion in soapy hot water, allowing the continuous filaments to be unwound. As the filaments are reeled, several are twisted together to produce a multifilament yarn.

Frequently, the filaments are combined further by an operation known as throwing, which produces a commercial yarn with the desired denier (the weight in grams of a 9-km (9000 m) length) and twist needed for the manufacture of a specific fabric by knitting or weaving.

Spinning.   Silk produced from damaged cocoons and fragments occurring in the reeling of continuous filaments is used for spun silk. Most of the gum is stripped from the comparatively short lengths of fiber by warm water and soap. After drying, these fibers are converted into yarn and then fabric by means of conventional equipment. Because of the delicate nature of the fiber, special handling techniques during manufacturing are required.

Preparation.   As with any fiber, it is necessary to remove all impurities present on the silk substrate before dyeing to ensure uniformity of penetration and surface coverage (levelness). These processes are known as scouring and bleaching. The four main impurities are sericin, lubricants and softeners added during throwing or in preparation for weaving or knitting, dirt and oils picked up incidentally during processing, and undesirable natural colors.

The scouring of silk presents special problems because the principal impurity, ie, sericin, is chemically very similar to fibroin. Thus, chemicals used to degum the silk, ie, remove the sericin, may also damage the fiber. Although the gum is soluble in boiling water, both sericin and fibroin are soluble in the alkaline soap solutions used to expedite cleansing. However, in dilute solutions at 82–87°C, fibroin is minimally affected, whereas sericin is dissolved. The chemicals and conditions used to remove the gum emulsify the other impurities (other than color) present, cleansing the fiber adequately for dyeing and finishing.

Treating domesticated silk with hydrogen peroxide removes unwanted color (10). Wild silk is much more difficult to strip of natural color and is therefore used frequently in its natural brownish hue. A fair white can be obtained on wild silk without excessive strength loss by special treatment with hydrogen peroxide followed by sodium hydrosulfite (10) (see Bleaching agents).

Weighting. During scouring and degumming, silk may lose 25 percent of its weight. This weight loss can be restored or even increased in the cleaned fabric by treatment with chemicals, such as iron compounds, tin compounds, or tannin. The amount of chemicals needed to return the silk to the original raw weight is known as par weight. In addition, these chemicals improve drapability. Fabrics have been produced with a weight increase of up to 400%.

Dyeing. Silk can be dyed with acid, basic, direct, mordant, reactive, and vat dyes. Since silk is amphoteric, it is readily dyed by the ionic dyes (acid, basic, direct) (see Dyes and dye intermediates). Special aftertreatments improve washfastness, which ranges from poor to satisfactory with these dyes. Vat dyes are noted for their excellent fastness to washing and light. However, they have to be solubilized in alkaline solution. At low dyeing temperatures, silk can withstand the necessary treatment. Both indigoid and anthraquinone vat dyes may be used on silk (11).

The mordant color most used today on silk is logwood, a natural dye. It gives a good black hue with excellent fastness. The dye forms a complex inside the fiber with metallic oxides (the mordant), usually chromium. Water solubility is thus diminished and washfastness enhanced. Certain acid colors also form such a complex and may be used on silk; they are either mordant acid or chrome types or premetallized acid types. The latter is manufactured as a dye–metal complex.

The reactive dyes form covalent bonds with the fiber and, therefore, excellent washfastness is obtained (12). The reaction between dye and fiber usually requires alkaline conditions, which silk can readily withstand. Among several types of reactive dyes, the dichlorotriazinyl types are most often used on silk. They are applied from an acid bath, which aids absorption. A subsequent alkaline treatment results in formation of the dye–fiber bond.

Silk is dyed in yarn, fabric, and garment form. The method used is determined by use and design considerations. Since yarn dyeing is somewhat more expensive and increases the risk of fiber damage, most coloration is applied to fabrics, both woven or knot. Hosiery is generally dyed after knitting.

As an alternative to yarn dyeing for design purposes, fabrics are often printed. Printing is a means of localizing dye in a pattern or design until fixation occurs. It is a versatile and comparatively inexpensive process. Fabrics may be printed with blocks, flat screens, rotary screens, or engraved rolls. Designs are produced by contact printing, where color is applied directly; discharge printing, where stripping chemicals are printed on the fabric that either destroy the color in the pattern area, leaving a white design, or replace the color with one that is unaffected by the discharge chemical; and resist printing, where a chemical that resists color uptake by the fiber is applied before dyeing.

Finishing. Finishing improves the fabric utility or enhances its properties. Because of their high price, silk products generally receive special care. Drape is improved by the addition of finishing chemicals that increase weighting. Organic acids impart a property known as scroop (a rustling sound), thermosetting and cross-linking resins improve wet and dry crease-recovery properties, and water-repellant, soil-resistant, and flame-retardant finishes provide the corresponding properties (see Flame retardants in textiles). Other chemicals improve tensile strength and abrasion resistance (13) (see Textiles, finishing).

Fiber Identification

Fibers and fiber blends may be identified by specific mechanical, chemical, or microscopical tests. All protein fibers upon burning impart the odor of burning hair or feathers. Silk, unlike other protein fibers, is soluble in 59.5 and 70% sulfuric acid. As the only protein fiber without surface scales, silk can be identified microscopically.

Fibers are sometimes blended to yield specific fabric characteristics or properties. Procedures are available for the quantitative analysis of fiber mixtures (14).

Economic Aspects and Uses

World production of raw silk increased steadily throughout the 1970s, although it had been higher in the late 1930s (see Table 4).

In the United States, silk import increased rapidly from 1900 to 1930; however, it declined appreciably during the 1930s and drastically after 1940 (see Table 5) (16–17), mainly because of the development of synthetic fibers.

Currently, only a modest amount of silk is imported into the United States. The *1980 Textile Blue Book* (18) lists 22 new silk importers and 42 raw-silk dealers, brokers, and agents. As shown in Table 6, total U.S. mill consumption of silk ranged from a low of 454,000 metric tons in 1975 to a high of 1.27×10^6 t in 1976 (19). The latest figures indicate that the U.S. raw-silk market has stabilized between 270,000 and 363,000 t per year. The principal suppliers are Japan and China, followed by Brazil (see Table 7) (20). Korea is the only other country currently exporting a significant amount of silk to the United States. These countries are the main producers of raw silk worldwide. The countries that produce raw silk are listed in the following decreasing order of importance (21): The People's Republic of China, Japan, Korea, Brazil, India.

The decrease in the use of silk in the United States is due to its very high price and the competition of synthetic fibers available at far lower prices. Silk prices cover

Table 4.   World Production of Raw Silk, Metric Tons[a]

1938	1970	1975	1978	1979
56,250	40,820	47,170	49,890	52,620

[a] From ref. 15.

Table 5.   U.S. Unmanufactured Silk Imports

Year	Metric tons[a]
1900	3675
1910	9750
1920	13,290
1930	36,560
1940	21,590
1950	4760
1960	3130
1970	816
1980	272[b]

[a] Ref. 16.
[b] Ref. 17.

Table 6.   U.S. Silk Imports, Metric Tons[a]

Year	Raw material[b]
1968	1814
1972	953
1975	454
1976	1270
1978	907
1980	454

[a] From ref. 19.
[b] Raw material means raw silk plus silk waste and noils.

Table 7.   U.S. Silk Imports, Metric Tons[a]

Year	United Kingdom	Italy	Japan	People's Rep. of China	Rep. of Korea	Brazil
1970	0.9	413	189		161	64
1975	0.45		175	160		80
1976	0.45	44	783	207		112
1977			558	121		48
1978		0.45	416	275	54	82

[a] From ref. 20.

a very large range of values, depending on raw material, manufacture, and use; no average price can be established. In any case, they are much higher than the prices of synthetic fibers of comparable quality such as polyester, nylon, or acetate.

In addition to raw silk, some 1580 t of finished goods were imported to the United States in 1978, primarily fabrics, handkerchiefs, toweling, napery, and dresses (22). Other uses of both domestic and imported products include underwear, nightwear, robes, loungewear, drapery, upholstery, narrow fabrics, slippers, sewing thread, electrical applications, rope, cordage and fishlines, and surgical sutures.

Serious competition to silk may arise from a new synthetic fiber, called Mitrelle, produced by ICI, which offers similar appearance and feel. Japanese manufacturers are also heavily committed to this market (23).

BIBLIOGRAPHY

"Silk" in *ECT* 1st ed., Vol. 12, pp. 414–452, by A. C. Hayes, North Carolina State College School of Textiles; "Silk" in *ECT* 2nd ed., Vol. 18, pp. 269–279, by A. C. Hayes, North Carolina State University School of Textiles.

1. R. W. Work, *Text. Res. J.* **46**(7), 485 (1976); **47**(10), 650 (1979).
2. F. Lucas and K. M. Rudall in M. Florkin and H. Stoltz, eds., *Comprehensive Biochemistry*, Elsevier Publishing Company, Amsterdam, The Netherlands, Vol. 26B, 1968, p. 475.
3. W. F. Leggett, *The Story of Silk*, Lifetime Editions, New York, 1949, pp. 70–73.
4. M. S. Otterburn in R. S. Asquith, ed., *Chemistry of Natural Protein Fibers*, Plenum Press, Inc., New York, 1977, p. 56.
5. L. Pauling and R. B. Corey, *Proc. Nat. Acad. Sci. U.S.A.* **39**, 253 (1953).
6. Ref. 2, p. 484.
7. W. A. Schroeder and L. M. Kay, *J. Am. Chem. Soc.* **77**, 3908 (1955).
8. Ref. 4, p. 62.
9. Ref. 4, pp. 55–73.
10. E. R. Trotman, *Dyeing and Chemical Technology of Textile Fibres*, 5th ed., Charles Griffin and Co., Ltd., London, UK, 1975, p. 261.
11. H. A. Rutherford, *The Application of Vat Dyes*, American Association of Textile Chemists and Colorists, Lowell, Mass., 1953, p. 169.
12. Ref. 4, p. 555.
13. U.S. Pat. 3,479,128 (Nov. 18, 1969), P. J. Borchert (to Miles Laboratories).
14. *AATCC Tech. Man.* **56**, 46 (1980).
15. *Text. Organon* **48**(7), 107 (July 1977); **51**(7), 105 (July 1980).
16. *U. S. Bureau of the Census, Historical Statistics of the U.S.—Colonial Times–1970*, Washington, D.C., 1975, p. 689.
17. *Text. Organon* **52**(11), 191 (Nov. 1981).
18. *Davison's Textile Blue Book*, 114th ed., Davison Publishing Co., Ridgewood, N.J., 1980, pp. 361–363.
19. *Text. Organon* **46**(11), 163 (Nov. 1975); **52**(11), 191 (Nov. 1981).
20. *Commodity Yearbook 1980*, Commodity Research Bureau, Inc., New York, 1980, p. 307.
21. H. Baumann, International Silk Association, Englewood Cliffs, N.J., private communication.
22. *Text. Organon* **50**(11), 171 (Nov. 1979).
23. *Chem. Week*, 41 (July 15, 1981).

CHARLES D. LIVENGOOD
North Carolina State University

TEXTILES

Survey,   459
Finishing,   466

SURVEY

Textile materials are among the most ubiquitous in society. They provide shelter and protection from the environment in the form of apparel, they provide comfort and decoration in the form of household textiles such as sheets, upholstery, carpeting, drapery, and wall covering, and they have a variety of industrial functions as tire reinforcement, tenting, filter media, conveyor belts, insulation, etc. Textile materials are produced from fibers (finite lengths) and filaments (continuous lengths) by a variety of processes to form woven, knitted, and nonwoven (felt-like) fabrics. In the case of woven and knitted fabrics, the fibers and filaments are formed into intermediate continuous-length structures known as yarns, which are then either interlaced by weaving or interlooped by knitting into planar flexible sheetlike structures known as fabrics. Nonwoven fabrics are formed directly from fibers and filaments by chemically or physically bonding or interlocking fibers that have been arranged in a planar configuration (see Nonwoven textile fabrics; Tire cords).

Textile fibers may be classified into two main categories and into a number of subcategories, as indicated in Table 1. The generic names of man-made fibers are defined and controlled by the Federal Trade Commission (1). With the exception of glass and asbestos fibers and the specialty metallic and ceramic fibers, textile fibers are formed from organic polymers. Cellulose (qv) and proteins (qv) are the only important natural polymers in naturally occurring fibers (see Biopolymers). Cellulose

Table 1. Classification of Textile Fibers

Naturally occurring fibers
    vegetable (based on cellulose), cotton, linen, hemp, jute, ramie
    animal (based on proteins), wool, mohair, vicuna, other animal hairs, silk
    mineral, asbestos
Man-made fibers
    based on natural organic polymers
        rayon, regenerated cellulose
        acetate, partially acetylated cellulose derivative
        triacetate, fully acetylated cellulose derivative
        azlon, regenerated protein
    based on synthetic organic polymers
        acrylic, based on polyacrylonitrile (also modacrylic)
        aramid, based on aromatic polyamides
        nylon, based on aliphatic polyamides
        olefin, based on polyolefins (polypropylene)
        polyester, based on polyester of an aromatic dicarboxylic acid and a dihydric alcohol
        spandex, based on segmented polyurethane
        vinyon, based on polyvinyl chloride
    based on inorganic substances
        glass
        metallic
        ceramic

Table 2. Trends in World Fiber Production, 10⁶ Metric Tons[a]

Fiber type	1974	1975	1976	1977	1978	1979	1980
natural fiber							
cotton	14.05	11.76	12.48	13.92	12.97	14.27	14.14
wool	1.53	1.54	1.49	1.49	1.53	1.57	1.58
silk	0.04	0.05	0.05	0.05	0.05	0.06	0.06
man-made fibers							
cellulosics	3.53	2.96	3.21	3.28	3.32	3.37	3.24
synthetic polymer	8.20	8.06	9.45	10.09	11.04	11.66	11.56

[a] Ref. 2.

is also the only polymer used in the production of regenerated or derivative man-made fibers. At this time, there are no commercial regenerated protein (azlon) fibers. Attempts at developing and commercializing azlon fibers from either vegetable or animal proteins (corn, peanuts, milk) have proven largely unsuccessful. Since currently used regenerated or derivative fibers (rayon, acetate, triacetate) are based on cellulose, this category of fibers is frequently referred to as "cellulosics." Synthetic polymers, as listed in Table 1, are widely used in fiber production (see Fibers, chemical; Fibers, elastomeric; Fibers, vegetable).

The 1980 total world production of textile fibers amounted to ca 30.6 × 10⁶ metric tons, distributed among the principal types of fibers as shown in Table 2. Cotton (qv) maintains its position of preeminence, although synthetic polymer fibers continue to grow in importance. Although the rate of growth of these fibers has been significantly lower in the late 1970s than in the 1960s, it is reasonable to expect a continuing growth in consumption of these fibers on a worldwide basis. Raw-cotton production is an important agricultural activity in many parts of the world, as indicated in Table 3.

Table 3. World Raw-Cotton Production by Region, 10⁶ Metric Tons[a]

Country	1978	1979	1980[b]	1981[b]
Americas				
United States	2.47	3.36	2.54	3.54
Brazil	0.58	0.60	0.65	0.61
Mexico	0.36	0.34	0.37	0.37
other	0.85	0.76	0.71	0.71
Western Europe	0.19	0.16	0.18	0.18
Asia				
USSR	2.74	2.96	3.24	3.08
People's Republic of China	2.27	2.32	2.83	2.95
India	1.41	1.43	1.40	1.47
Pakistan	0.50	0.79	0.75	0.82
other	1.05	1.03	1.02	1.12
Africa				
Egypt	0.46	0.51	0.55	0.53
other	0.70	0.71	0.68	0.73
Total	*13.58*	*14.97*	*14.92*	*16.11*

[a] Ref. 2.
[b] Preliminary data.

Table 4.   1980 World Man-Made Fiber Production by Region, 10^6 Metric Tons [a]

Country	Cellulosics	Synthetic polymer	Total (%)
Americas			
United States	0.37	3.63	4.00 (26.9)
other	0.15	0.83	0.98 (6.6)
Europe			
Western	0.74	2.58	3.32 (22.3)
Eastern	1.14	1.24	2.38 (16.0)
Japan	0.40	1.45	1.85 (12.5)
all other	0.45	1.88	2.33 (15.7)
Grand total	*3.25*	*11.61*	*14.86 (100.0)*

[a] Ref. 2.

Cotton production in any given year is dictated as much by weather conditions as by market demands. The United States, the USSR, and the People's Republic of China are the largest raw-cotton producers, although some cotton is grown in just about every country where rainfall and temperature are adequate. In the United States, ca 80% of the 1980 crop was grown in Arizona, California, Mississippi, and Texas. The high production of raw cotton in the western states is made possible by irrigation and highly mechanized farming practices.

Man-made fiber production is concentrated in the highly developed industrialized regions of the world, as shown in Table 4. The United States and Western Europe account for ca 50% of world production of these fibers. The consumption of the different types of fibers by the U.S. textile industry in the manufacture of various apparel, household, and industrial products is shown in Table 5. Cotton, nylon, and polyester are the dominant fiber types in the United States; polyester is the most widely used since 1977 (see Polyamides; Polyester fibers).

The cellulosic fibers were the first man-made fibers to become significant commercial materials. They were introduced in the decade following World War I, and they grew in importance until the 1960's when the synthetic polymer fibers began to make their major impact (see Cellulose acetate and triacetate fibers; Rayon). The first synthetic polymer fiber, nylon, introduced in 1940, was the commercial result of pioneering research on condensation polymerization reactions by Carothers and his

Table 5.   Fiber Consumption by the U.S. Textile Industry, 10^6 Metric Tons [a]

Fiber	1971	1976	1980
cotton	1.81	1.54	1.41
wool	0.14	0.04	0.04
rayon	0.45	0.27	0.23
acetate	0.23	0.14	0.14
polyester	0.91	1.50	1.59
nylon	0.77	0.95	1.04
acrylic	0.23	0.27	0.27
polyolefin	0.14	0.23	0.32
glass	0.23	0.32	0.41
Total	*4.91*	*5.26*	*5.45*

[a] Ref. 2.

associates at the DuPont Company (3). Polyester fibers, also based on condensation polymerization reactions, were introduced in the 1950s (4). Addition (vinyl) polymerization reactions, studied extensively in those years, eventually yielded the acrylic and olefin fibers (qv) (5) (see Acrylic and modacrylic fibers).

Man-made fibers are formed by extrusion processes known as melt, dry, or wet spinning (6). In melt spinning, molten polymer is forced under pressure through a metallic dye, known as a spinneret, containing many tiny holes. The emerging fluid jets solidify in a cooling zone to form filaments, which are then wound on containers. The diameter of the filament is determined primarily by the rates of extrusion and take-up. The cross-sectional shape of the filaments can be controlled by the shape of the holes in the spinneret. Melt spinning is the most economical filament extrusion process, but it requires polymers that are thermally stable under melt conditions. Nylon, polyester, and olefin filaments are produced by this method, and glass filaments are produced by an essentially equivalent process. In dry spinning, used for acetate, triacetate, and acrylics, the polymer is dissolved in a suitable solvent, and the solution is forced through a spinneret into a heated zone where the filament solidifies as the solvent evaporates. In wet spinning, the polymer or a chemical derivative of the polymer is dissolved in a suitable solvent, and the solution extruded into a liquid bath where the filament is formed by a combination of precipitation, coagulation, and regeneration mechanisms. Rayon (qv), spandex, and some acrylics are formed by this method.

The spinning or extrusion of filaments is normally followed by the operation known as drawing (6). In this step, the newly formed filaments are irreversibly extended and stabilized by setting or crystallization processes. The combination of spinning and drawing is used to develop a molecular organization characterized by a high degree of orientation and crystallinity. In general, both man-made and natural fibers can be described as having a regular and well-developed molecular organization with considerable structural anisotropy, which is reflected in large differences between axial and transverse physical properties of fibers. In this regard, fibers differ from other polymeric materials such as rubbers and plastics where less well-organized molecular structure predominates.

An important new development in synthetic polymer fiber manufacture is high speed spinning (7–8). In this technology, the spinning (extrusion) and drawing steps are combined into a single operation. The extremely high take-up speed (up to 6000 m/min) induces the high degree of orientation and crystallinity necessary for adequate fiber physical properties. The favorable economics for high speed spinning over the traditional two-step spin-and-draw process suggests that this method will become increasingly dominant, especially for polyester fibers.

With the exception of silk (qv), naturally occurring fibers have finite lengths and generally require several cleaning and purification steps prior to processing into yarns and fabrics (9–10). Man-made fibers, on the other hand, are produced in the form of continuous filaments and can be processed directly into multifilament yarns and subsequently into fabrics. Alternatively, continuous man-made filaments may be cut into discrete lengths to form staple fiber, which can subsequently be processed into yarns and fabrics. Silk is a continuous filament formed by solidification of a protein fluid extruded by the cultivated silkworm, *Bombyx mori* (11). It can be used as a continuous filament or cut for staple fiber. In general, yarns formed from staple fibers (natural, man-made, or blends) are bulkier, softer and have a rougher surface texture than continuous multifilament yarns. In recent years, a number of texturing processes have been developed to impart bulk and softness to continuous multifilament yarns to simulate the characteristics of typical staple-fiber spun yarns (12–13).

The manufacture of textile fabrics from fibers is based on processes that were developed to accommodate the special geometric and physical properties of naturally occurring fibers (9–10). Thus, there is a cotton system of yarn manufacture, a wool system, and a worsted system. Wool and worsted differ in that worsted yarns are manufactured from longer wool fibers that have been drawn parallel by combing operations. Similarly, there are combed-cotton yarns and carded-cotton yarns. In combed-cotton yarns, some of the short fibers have been removed by the combing, and the fibers in the yarn are more parallel, producing a smoother, more lustrous yarn. Carded-cotton yarns are somewhat rougher with fiber ends protruding from the yarn surface. The operations to which fibers are subjected during yarn manufacture are typically opening and cleaning, picking, carding, combing (optional), drawing (referred to as gilling in the case of worsted yarns), roving, spinning, twisting, plying, and winding. Yarns produced by these operations are referred to as spun or staple yarns (13–14). Today, cotton-, woolen-, and worsted-type yarns may be manufactured from staple man-made fibers or from mixtures of natural and man-made fibers. Such blended yarns have found increasing importance since it is possible to combine the desirable properties of man-made and natural fibers.

Yarns used for the production of woven fabrics are classified either as warp yarns or as filling yarns. During weaving, warp yarns are held in the loom under considerable tension whereas filling yarns are inserted and interlaced in various patterns at right angles to the warp. Warp yarns are more highly twisted and less extensible than filling yarns. Yarns used for the manufacture of knit fabrics are, in general, even less twisted and more extensible to produce soft and rather bulky knit fabrics.

After the fabric-formation process, textiles are subjected to either dyeing or printing, and to a variety of finishing operations that may be mechanical or chemical in nature. The specific nature of these dyeing and finishing operations depends largely on the fiber type and on the intended use of the finished fabric.

Dyeing is normally conducted in an aqueous solution of dyestuff (or mixture of dyestuffs) and involves a partition or distribution of the dye from the aqueous phase to the solid fiber phase. Although in most instances dyeing is performed on fabrics, it may also be performed on raw stock (staple fiber) or on yarns. The dye is generally absorbed by the fiber and held by secondary forces, but there are a number of chemically reactive dyes that bond covalently to the fiber polymer. In order to achieve reasonable rates of dyeing, commercial dyeing is normally conducted at 100°C, and in the case of some fibers, particularly polyester, dyeings are conducted under pressure to allow temperatures of approximately 130°C to be used (15). There are many different classes of dyes, including acid, direct, basic, sulfur, vat, disperse, and reactive, each of which has special affinity for certain fiber types. Through the use of dye mixtures and blended yarns and fabrics, it is possible to achieve colorful patterns by dyeing. However, printing, utilizing insoluble pigments instead of dyes, is used more extensively to achieve multicolor patterns and designs. The printing process is much more rapid and requires considerably less energy than dyeing (16) (see also Printing processes).

Mechanical finishing is primarily designed to produce special surface effects. Typical operations in this category are napping, shearing, embossing, and calendering.

Chemical finishing operations may involve either the addition of chemicals or the chemical modification of some or all of the fibers comprising the fabric. It is not always possible to draw a clear distinction between additive chemical finishing and chemical modification. The primary purpose of chemical finishing is to confer special functional properties. Typically, procedures may be designed to impart dimensional stability, wrinkle resistance, wash-and-wear or durable-press characteristics, soiling resistance, water repellency, and flame retardance.

Chemical finishing procedures, other than scouring and possibly bleaching, are performed principally on those fabrics that are composed either entirely or in part of cellulosic fibers (ie, cotton and rayon). These fibers have chemically reactive sites (OH groups) that lend themselves to chemical modification and additive chemical treatments. Of particular importance in this regard are treatments of rayon- and cotton-containing fabrics with difunctional reagents that are capable of chemically cross-linking cellulose chains (17). Such reactions improve fiber resilience, thereby enhancing wrinkle resistance and recovery. The same reactions can be used to impart wash-and-wear and durable-press characteristics. Wool-containing fabrics are also subjected to chemical-finishing treatments (see Wool). The main purpose of these treatments is to control felting shrinkage when wool fabrics are subjected to mechanical agitation at elevated temperatures in the wet state, conditions that occur in normal washing procedures. Felting shrinkage is caused by an inward migration of fibers in a yarn or fabric (18). This, in turn, is caused by a unique scalar structure of wool, and also of other animal fibers, that causes the frictional coefficient to be significantly higher in the against-scale direction than in the with-scale direction (see Felts). The chemical-finishing treatments are designed to minimize this differential friction by inactivating the scale edges. This is achieved by a variety of treatments, including chlorination, oxidation with permanganate solutions, polymer deposition, and combinations of these (18). Wool fabrics can also be chemically after-treated to impart wash-and-wear and durable-press properties. Fabrics composed of synthetic polymer fibers are frequently subjected to heat-setting operations. Because of the thermoplastic character of some such fibers, for example, nylon, polyester, olefin, and triacetate, it is possible to set such fabrics into desired configurations. These heat treatments involve recrystallization mechanisms at the molecular level, and are thus permanent unless the fabrics are exposed to more severe thermal conditions than were used in the heat-setting process.

Although textiles are usually associated with apparel and household products, industrial uses of fibers are of increasing importance. Of particular significance in this regard is the use of fibers to reinforce plastic materials, producing high performance fiber-reinforced composites (see Composite materials; Laminated and reinforced plastics). Such materials are increasingly used in construction and in the automotive, aerospace, and sports equipment industries. The use of textiles in civil engineering applications, such as road construction, erosion control, and heat-exchange devices, is also becoming more widespread. The term geotextiles has been adopted to describe fabrics used in these types of applications.

BIBLIOGRAPHY

1. Textile Products Identification Act.
2. *Textile Organon*, Vol. 51, Textile Economics Bureau, Inc., Roseland, N.J., 1981.
3. W. H. Carothers in H. Mark and G. S. Whitby, eds., *Collected Papers*, Interscience Publishers, New York, 1940.
4. J. R. Whinfield, *Text. Res. J.* **23,** 289 (1953).
5. P. J. Flory, *Principles of Polymer Chemistry*, Cornell University Press, Ithaca, N.Y., 1953.
6. H. F. Mark, S. M. Atlas, and E. Cernia, eds., *Man-Made Fibers Science and Technology*, Vols. 1–3, Interscience Publishers, a division of John Wiley & Sons, Inc., New York, 1967, 1968, and 1969.
7. U.S. Pat. 4,134,882 (Jan. 16, 1979) and U.S. Pat. 4,195,051 (March 25, 1980), H. R. E. Frankfort and B. H. Knox.
8. H. M. Heuvel and R. Huisman, *J. Appl. Polym. Sci.* **22,** 2229 (1978).
9. M. L. Rollins and D. S. Hamby, eds., *The American Cotton Handbook*, John Wiley & Sons, Inc., New York, 1965, 1966.
10. W. von Bergen, ed., *Wool Handbook*, Vols. 1–11, John Wiley & Sons, Inc., New York, 1963, 1968.
11. R. H. Peters, *Textile Chemistry*, Vol. I, Elsevier, New York, 1963, Chapt. 11, pp. 302–314.
12. D. K. Wilson, *Text. Prog.* **9**(3), (1977); **10**(3), (1978); see also S. Backer, *Sci. Am.* **227**(6), 47 (1972).
13. B. C. Goswami, J. G. Martindale, and F. L. Scardino, eds., *Textile Yarns: Technology, Structure, and Applications*, John Wiley & Sons, Inc., New York, 1977.
14. P. R. Lord, *The Economics, Science, and Technology of Yarn Production*, North Carolina State University, Raleigh, 1979.
15. D. M. Nunn, ed., *The Dyeing of Synthetic Polymer and Acetate Fibres*, Dyers Co. Pub., Trust (Society of Dyers and Colourists), Bradford, UK, 1979.
16. L. W. C. Miles, ed., *Textile Printing*, Dyers Co. Pub. Trust (Society of Dyers and Colourists). Bradford, UK, 1981.
17. T. F. Cooke, J. H. Dusenbury, R. H. Kienle, and E. E. Linekin, *Text. Res. J.* **23,** 1015 (1954).
18. K. R. Makinson, *Shrinkproofing of Wool*, Marcel Dekker, Inc., New York, 1979.

LUDWIG REBENFELD
Textile Research Institute

FINISHING

Finishing of textiles includes, in its broadest sense, any process used to improve a knitted, woven, or bonded textile fabric for apparel or other home or industrial use. The principal classes of textile fibers and fabrics include cellulosic, synthetic, protein, glass, and blends. Because of the great differences in the chemical and physical nature of these fibers, finishing processes vary widely among classes and, to a lesser extent, within each class. Mechanical, heat-related, and chemical treatments have been developed over the years since the first attempts to convert fibers into yarns and then into a fabric.

Animal and plant fibers have been used for thousands of years, but the introduction of synthetic fibers into consumer use after World War II caused the most dramatic change in the finishing of natural fibers. Prior to this time it was generally necessary, after washing and drying, to pull or press a garment back into its original shape before wearing. Consumers accepted these disadvantages because there was no real alternative until the introduction of new fibers and their fabrics. When laundered with care, these new fabrics require little or no ironing and retain their shapes because shrinkage and stretching are controlled within narrow limits. These advantages came to be demanded of the garments made from natural fibers. As cotton and rayon were the primary fibers in use, a massive effort was put forth to develop new finishing processes to give cotton and rayon garments the same ease of care and versatility as garments made from synthetic fibers. Similarly, improvements were sought in finishing processes for the synthetics.

Thus, the general topic of textile finishing is very broad, for there are at least ten commercially important textile fibers today, and finishing processes enhance a wide variety of properties including crease retention, ease of care, dimensional stability, abrasion resistance, flame retardance, smolder resistance, odor control, soil release, variable moisture contents, and control of static charges, stretch, water repellency, waterproofing, and resistance to weathering (including sunlight and microorganisms), as well as improving the hand (or feel) and surface appearance of fabrics. Many of these specific topics are presented elsewhere in this Encyclopedia under headings that relate to a specific property, eg, Waterproofing and water repellency or Flame retardants for textiles; to a chemical or a chemical system, eg, Amino resins or Formaldehyde; or to a specific textile fiber, eg, Cotton, Wool, Polyesters, Rayon, etc.

Some of the more important mechanical finishing processes are compressive shrinkage (Sanforized process) and calender finishes, which include schreinering (to produce a silklike appearance), chintz finishing (usually a glazed fabric with bright prints), and embossing. These finishes are semidurable but are made more durable, especially if the base fabric is cotton or rayon, by including a treatment with a conventional N-methylol agent (see Amino resins).

There are many types of mechanical finishing processes employing many types of rollers, but unfortunately there is relatively little discussion of them in the published literature. Trends in fashion have reduced the demand for calendered fabrics, but the processes are still of considerable interest. Permanent effects with fabrics of thermoplastic material are obtainable if the heat and pressure are sufficient to soften the material. Calendering operations have been used to a considerable extent to improve the surface appearance of polyester and nylon fabrics and their blends (see general references).

Chemical Treatments

Generally, chemical treatments are divided into processes involving topical treatments and those involving formation of a chemical bond with the substrate. Most chemical treatments of synthetic fibers are topical because of the general inertness of the substrate. Because the cellulosics and wool are reactive, these fibers may readily form chemical bonds with the finish, as well as act as substrates for nonreactive topical treatment.

Durability of the topical treatments to the various laundering conditions varies widely, but some treatments possess good durability. The properties obtained on synthetics are more permanent when the chemical can penetrate and dissolve in the substrate. Durability of a topical treatment is often improved if the chemicals increase in molecular weight during processing and become entangled with the substrate. Those treatments that involve formation of a chemical bond with the substrate usually produce permanent finishes. However, some of these bonds are broken by hydrolysis during laundering. Some amount of hydrolysis of the finish can be tolerated, and the reactant systems are designed to minimize this reaction. Emphasis in this article has been placed on chemical treatments for the most widely produced commercial products, especially those that involve chemical bond formation with the substrate.

The expression ease of care includes a number of related properties when applied to apparel and related textile goods. The expressions wash-and-wear and durable press generally refer to the performance of a garment after machine washing and line or tumble drying. Since different countries use different washing and drying equipment and conditions, the evaluation for performance (ie, acceptance or rejection by the consumer in a specific country) is varied. What is acceptable in Europe may not be acceptable in the United States, and vice versa. The U.S. market is large, and for the most part, its requirements are of primary concern in this review.

Knitted fabrics generally shed their wrinkles more quickly and completely than woven fabrics regardless of the method of laundering. However, dimensional stability (shrinkage or stretching) of knits is more difficult to control than in woven fabrics. Fortunately, most of the processes that improve the smooth-drying appearance of a fabric also improve its dimensional stability. For example, the cross-linking that occurs during finishing of cellulosic fabric to make it dry smoothly also produces a fabric that resists shrinkage or stretching. Often fabrics are cross-linked with a low level of cross-linking agent to stabilize the fabric. Denim fabrics, for example, are often treated in this manner because a high level of smooth-drying properties is not required. For polyester–cellulosic blends, a heat-setting process is combined with the cross-linking process to achieve maximum performance.

The conventional finishing process for cellulosics involves three steps commonly referred to as pad–dry–cure (Fig. 1). In the first step, the fabrics are immersed in an aqueous finishing formulation and then squeezed to remove excess formulation and to assure even distribution of the reagents in the fabric. The pickup in weight at this point is carefully controlled but usually varies from 60–100% depending upon the process. Interest is growing in new padding technology (1) wherein the increase in weight during padding is even lower than with prior methods. In this new technology,

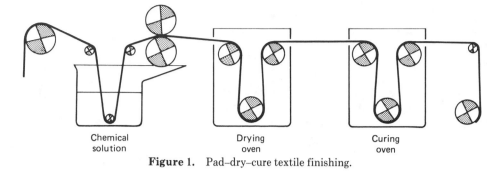

Figure 1.   Pad–dry–cure textile finishing.

which includes use of foam or minimum application techniques, the pad bath is more concentrated since the concentration of solids on the dry fabric must be of the same magnitude regardless of the padding technology used.

A finishing pad bath usually consists of a cross-linking agent, a curing catalyst, and hand modifiers. For conventional durable-press processes of cellulosics and cellulosic blends, using a 60–100 wt % wet pickup, the bath contains ca 5–9 wt % cross-linking agent, enough acidic catalyst to effect a cure (0.4–3.5 wt %, depending upon catalyst strength), and 1–3 wt % emulsified polymer which generally acts as a fiber lubricant or a stiffening agent. A topical treatment for a synthetic fabric generally consists of only 1–3 wt % emulsified polymers.

Except for special applications, the drying process simply involves removal of water from the flat fabric on a continuous range. Often, drying and curing are performed in one continuous process at rates that often exceed 100 m/min. On the other hand, the dried fabrics may be made into garments prior to the curing process. This, of course, requires different curing equipment than that used for a flat, continuous length of fabric.

The padding and drying steps are especially important in the conventional treatment of cellulosics. The fabric must be prepared in such a state that, in general, the reagents readily penetrate the fibers to achieve an even treatment within the fiber. Migration of the reactants to the surface of the fiber during drying and curing is a problem and, in an extreme case, leads to a stiffened, unsatisfactory product. Research continues on this subject (2).

After drying, the padded fabric may be allowed to lose some heat or it may be placed in a roll immediately. Depending upon the formulation used in the pad bath, some reaction with cellulose starts during the drying process and continues if the dry roll is hot. The handling of this hot, dry fabric is important for future processing.

In addition to dimensional stability and smooth-drying appearance, ease of care includes crease retention. In general, adequate creases are obtained by chemical finishing of cellulosics and heat setting of polyester and many other synthetics. Blends of cellulosics and polyester are usually subjected to both processes for optimum performance.

To many in the trade, textile finishing refers primarily to chemical finishing (or resin finishing, as it is called in much of the older literature). The primary chemicals used in the early finishing processes were aminoplasts capable of cross-linking the cellulose molecule but also capable of homopolymerization. Not only were these chemicals capable of improving the appearance and the dimensional stability of the fabric, but they also retained pigments used in printing, provided protection for outdoor exposure, and conveyed other related properties desirable in textiles. A rather long and spirited debate developed over the mechanism involved in developing a smooth-drying cellulosic fabric. By the mid or late 1950s, most textile chemists had decided that cross-linking of cellulose and not polymer formation within or about the cellulose molecule was primarily responsible for the desired smooth-drying properties. However, even today the role of polymer formation with aminoplast resins in promoting smooth-drying properties is not completely discarded.

Fabric Preparation.   A number of processes are performed after the fabric has been woven or knitted; these are not considered finishing steps. Scouring, mercerization, and bleaching are usually referred to as fabric preparation. Chemicals applied prior to dyeing are generally removed to avoid problems in the dyeing procedures. If the fabric is to be colored, the dyeing or printing process is not considered part of the finishing process. On the other hand, fluorescent brightening agents are often applied during a finishing process (see Brighteners, fluorescent; Dyes, application). However, because of energy costs and other considerations, combination of some of these processes with finishing procedures is receiving much attention from both production and research personnel.

Fabric preparation is very important for the quality of the subsequently dyed or printed fabric. A properly prepared fabric is white and free of contaminating chemicals, and it accepts dyes and other chemicals readily. Also, the fabric should not lose appreciable strength during either the preparation or the dyeing steps (3). Some fabric strength is lost during finishing, but most of this loss is due to embrittlement of the fiber because of cross-linking. Care must be exercised during chemical finishing in the selection of a dye–finishing agent combination to avoid an unwanted change in hue during processing or after exposure to sunlight.

Results of Cross-Linking Cellulosic Fabrics.   Essentially any bifunctional compound, except those that cannot penetrate the fiber or that have widely separated reactive groups, will, on reaction with cotton cellulose, produce a fabric with improved smooth-drying appearance.

$$2\,\text{CellOH} + y - X - y \rightleftharpoons \text{CellOXOCell} + 2\,\text{Hy} \tag{1}$$

Fabric finished in this manner is said to possess a memory. For example, a fabric cross-linked in the flat, dry state is expected to return to that state after washing when it is given an opportunity to shed its wrinkles in a tumble dryer. Similarly, a fabric cross-linked in the flat, wet state returns to that state, ie, sheds its wrinkles, when immersed in similar media. This concept was developed using pad–dry–cure and wet-cure processes with formaldehyde as the cross-linking agent (4).

$$2\,\text{CellOH} + \text{CH}_2\text{O}.\text{H}_2\text{O} \overset{\text{H}^+}{\rightleftharpoons} \text{CellOCH}_2\text{OCell} + 2\,\text{H}_2\text{O} \tag{2}$$

Cross-linking a creased or wrinkled fabric likewise causes retention of the crease or wrinkle for the life of the cross-link. Most cross-linking reactions are reversible, and therefore, the properties imparted by cross-linking are reversible by removing the cross-link through hydrolysis.

Some undesirable properties are imparted by cross-linking cellulose. The most objectionable of these are reductions in breaking strength, tearing strength, and abrasion resistance. In addition, the ability of the fabric to retain or regain moisture is generally reduced by cross-linking. Moisture-regain properties can be retained somewhat by use of swelling agents during cross-linking or, occasionally, by selection of the cross-linking agent (5).

Cross-linking of cellulosic fibers also reduces the accessibility of the fiber. In other words, the cross-linking reaction in a heat-cure process takes place after the moisture is removed while the fiber is in a collapsed state. Therefore, dyeing of fabrics is generally performed prior to cross-linking by a heat cure because the dye cannot penetrate the cross-linked fiber (6). As noted below, this property can be used to advantage.

The appreciable loss of strength and abrasion resistance when cellulosics are cross-linked can be minimized, but this is usually accompanied by a loss in smooth-drying performance. One solution to this problem of loss of strength and abrasion resistance in cellulosics is to blend polyester fibers into the yarns that make up the fabric. The cellulosic portion of the fabric still loses strength because of cross-linking, but the effect is masked by the polyester.

Early Cross-Linking Agents. Formaldehyde, urea–formaldehyde, and melamine–formaldehyde were the first cross-linking systems used in commercial finishing or studied extensively in the early days of chemical finishing. They are still being used today, but other reactant systems have replaced them in general use.

Formaldehyde is the simplest and most economical cross-linking agent known. Despite much research (6–8), formaldehyde has had only limited use in practical heat-cure or vapor-phase applications. It produces a durable, inert cross-link that resists hydrolysis and is unaffected by washing, heat, bleaching, sunlight, or other conditions encountered by a finished textile. A 100% cotton fabric finished with formaldehyde, however, has reduced strength and abrasion resistance compared to a fabric finished with an N-methylol agent to a comparable level of durable-press performance. Loss of a great deal of formaldehyde in a conventional pad–dry–cure treatment is another disadvantage, but the primary reason for not using this agent is that it has never been satisfactorily scaled up for continuous industrial production. One of the main failings is the inability to produce a good product consistently. This finish is so stable that errors are essentially impossible to correct.

Urea–formaldehyde or melamine–formaldehyde reactants are resin formers for a large number of products. They have been used as polymers to coat and react with cellulose and as monomeric or oligomeric cross-linking agents. A reactant that coats a fiber generally stiffens it. In the case of a limp fabric, this can be desirable; however, too much resin on the surface will produce a "board." A stiffened, cross-linked fabric usually does not dry smoothly after laundering unless the surface polymer is worked mechanically during washing or in a machine that disturbs the coating. Melamine resins are still used as pigment binders and in binding other additives to cellulosic fabrics. However, during the 1950s both these early resins were rejected as cross-linking agents for cellulosic fabrics primarily because of discoloration and loss of strength upon hypochlorite bleaching. The problem precipitated research into chemical finishing of rayon and cotton aimed at development of new cross-linking systems that would resist the effects of hypochlorite bleach.

Chemistry of N-Methylol Cross-Linking Systems. The urea–formaldehyde reactant system is not dimethylolurea, as depicted in equation 3, but generally a methylolated urea in which the ratio of formaldehyde to urea is 1.3–1.8:1. This ratio provides a reasonably stable pad bath, and the reactant is capable of cross-linking cellulose, polymerization, or both. Catalysis is usually provided by metal salts or selected amine

$$HOCH_2NHCNHCH_2OH + 2\ CellOH \rightleftharpoons CellOCH_2NHCNHCH_2OCell + 2\ H_2O \qquad (3)$$

hydrochlorides.

During bleaching with hypochlorite, some of the N—H groups in the finish are converted to N—Cl. When the bleached fabric is ironed, hydrogen chloride is formed by a series of reactions and the polymer chain of the cellulose substrate is hydrolytically decomposed. The effects can be extreme, ranging from slight discoloration generally accompanied by a slight loss of strength to the fabric falling apart during ironing. Other effects, including removal of part of the finish by the bleaching process, are also encountered (see Bleaching agents).

The chemical mechanisms in chlorine retention and the decomposition of resultant chloroamides have been studied in detail (9–10), and the relation between molecular structure of the agent and its chlorine uptake and release was established. Chloroamides present in finished goods decompose at different rates. Unfortunately, in urea–formaldehyde finishes the chloroamide forms and decomposes readily, producing a heavily damaged fabric (11).

One way of reducing or eliminating chloroamide formation is to completely methylolate the agent and thus limit the number of N—H groups in the finished fabric. The amide–aldehyde reaction is reversible and can approach completion in some cases (eqs. 4 and 5) with excess formaldehyde.

$$\underset{\|}{\overset{O}{C}}NH_2 + HCHO \rightleftharpoons \underset{\|}{\overset{O}{C}}NHCH_2OH \tag{4}$$

$$\underset{\|}{\overset{O}{C}}NHCH_2OH + HCHO \rightleftharpoons \underset{\|}{\overset{O}{C}}N(CH_2OH)_2 \tag{5}$$

Methylolations are generally carried out under alkaline, aqueous conditions because acidic catalysis usually leads to formation of polymers or to methylenebisamides.

$$\underset{\|}{\overset{O}{C}}NHCH_2OH + \underset{\|}{\overset{O}{C}}NH_2 \rightleftharpoons (\underset{\|}{\overset{O}{C}}NH)_2CH_2 + H_2O \tag{6}$$

Methylolations of primary or unsubstituted amides that approach completion (eq. 5) include those of formamide, carbamates, and melamines. Urea, which has four N—H groups, forms only the trimethylol derivative readily.

Addition to secondary or monosubstituted amides is usually difficult to complete (eq. 7) and forms a mixture of products.

$$\underset{\|}{\overset{O}{C}}NHR + CH_2O \rightleftharpoons \underset{\|}{\overset{O}{C}}N\overset{\diagup R}{\diagdown CH_2OH} \tag{7}$$

Thus, N,N'-dimethylol-N,N'-dialkylamides resist formation, and reaction mixtures containing these agents are not effective cross-linking systems for cellulose. On the other hand, cyclic ureas containing secondary NH groups generally methylolate readily, and these agents have been the basis for the improved systems developed to replace urea–formaldehyde.

Glyoxal has been used to produce substituted N-methylol reactants capable of reaction with cellulose. Factors that limit the addition of amides to formaldehyde and glyoxal have been studied for imides, carboxylic acid amides, carbamates, and ureas. In general, complete substitution of all amino hydrogens available is limited by steric effects or by low amide acidity in some cases. Complete substitution of all amido hydrogens becomes more general if cyclization occurs during addition to the aldehyde, or if a cyclic amide with nitrogen in the ring, rather than an N-alkylamide, acts as nucleophile (12).

Formation of a fully methylolated cross-linking agent does not assure that the finish will be immune from hypochlorite bleaching damage. The curing step does not proceed to completion, even for the most effective cross-linking agent, and some N—H groups (formed by loss of formaldehyde) and some N—CH$_2$OH groups are found in the finish. Furthermore, laundering conditions can cause hydrolysis of a link to cellulose to form the easily hydrolyzable N—CH$_2$OH group. Thus, continued washing and bleaching with hypochlorite can alter the original finish and produce one that is susceptible to fabric damage. Most laundry washes are alkaline, but commercial laundering also includes an acid stage. A satisfactory finish must be able to withstand both types of laundering.

$$\overset{O}{\overset{\|}{-\!\!\overset{|}{C}NHCH_2OCell}} + H_2O \rightleftharpoons \overset{O}{\overset{\|}{-\!\!\overset{|}{C}NHCH_2OH}} + CellOH \qquad (8)$$

Reaction Mechanisms for Formation and Hydrolysis of Finishes. It was in the early 1960s that research on these reaction mechanisms, as applied to chemical finishing of cellulosic fabrics, commenced. The formation and hydrolysis (acid) reactions were shown to pass through the same intermediate immonium–carbonium ion (13).

$$\overset{O}{\overset{\|}{-CNHCH_2OH}} + H^+ \rightleftharpoons \overset{O}{\overset{\|}{-CNHCH_2\overset{H^+}{O}H}} \qquad (9)$$

$$\overset{O}{\overset{\|}{-CNHCH_2\overset{H}{\underset{+}{O}}H}} \rightleftharpoons \overset{O}{\overset{\|}{-CNHCH_2^+}} + H_2O \qquad (10)$$

$$\overset{O}{\overset{\|}{-CNHCH_2^+}} \leftrightarrow \overset{O}{\overset{\|}{-C\overset{+}{N}H\!=\!CH_2}} \qquad (11)$$

$$\overset{O}{\overset{\|}{-CNHCH_2^+}} + CellOH \rightleftharpoons \overset{O}{\overset{\|}{-CNHCH_2OCell}} + H^+ \qquad (12)$$

Based on the principle of microscopic reversibility, it was reasoned (14) that a highly reactive agent would form an easily hydrolyzable product. This proved true in practice, and the industry rapidly adopted the phrase that "easy on means easy off" (15).

Hydrolysis resistance of a finish is also important during storage of cured goods and during the final pressing and creasing after cured fabrics have been made into garments. Cotton and rayon fabrics regain moisture on standing; the amount regained depends on the blend level and the finishing process (16–17). Since cured, flat fabrics or finished garments are rarely washed prior to sale to the consumer, the acid curing catalyst remains on the fabric until it is laundered by the consumer. This combination of moisture and acidic conditions has considerable time to act upon the finish and thus, again, the resistance of the finish to hydrolysis is of utmost importance. Errors made in processing are often corrected by removal of the finish by acid hydrolysis or "stripping." Stripping must be accomplished under conditions that do not cause excessive strength loss of the cellulosic substrate.

All these factors attest to the importance of an understanding of the acid-catalyzed hydrolysis reaction for the practical development of a superior commercial product. This knowledge is also of great theoretical importance because it is the best and, at times, the only way to determine the chemical structure of the finish. Equation 1 represents the ideal situation in which X remains unaltered during finishing, garment manufacture, and storage. It has been suggested that the true chemical nature of modified cellulose is not properly represented by a simple cross-link (18–19).

A number of standardized conditions for acidic hydrolysis have been used by industry and researchers. Dilute hydrochloric acid is effective but requires careful control if consistent results are to be obtained. Hydrolysis conditions of 30 min at 80°C in an aq soln containing 1.5% phosphoric acid and 5% urea has been found to give reproducible results with a large number of cured fabrics.

In a recent study (20), more than 50 fabrics were hydrolyzed with this urea-phosphoric acid solution. The fabrics were prepared from mono- or polyfunctional N-methylol agents under similar curing conditions, and the level of nitrogen content on the fabric before and after hydrolysis was determined. It was concluded that the resistance to hydrolysis was determined primarily by chemical factors, ie, electronic effects of substituent groups on the intermediate immonium–carbonium ion (eqs. 11 and 13).

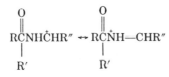

$$\underset{R'}{\overset{O}{\underset{|}{R\overset{\|}{C}NH\overset{+}{C}HR''}}} \leftrightarrow \underset{R'}{\overset{O}{\underset{|}{R\overset{\|}{C}\overset{+}{N}H—CHR''}}} \qquad (13)$$

Agents used included formaldehyde and glyoxal adducts of ureas, melamines, carbamates, formamides, acetamides, hydrazides, pyrrolidone, and similar amides. These and related experiments demonstrated that most N-methylol and similar agents react with cotton with little or no modification of their structure during curing. Loss of formaldehyde (the reverse reaction in eq. 7) from N-methylol groups can be appreciable for some reactant systems (21). Likewise, polymer formation in conjunction with cross-linking occurs in some systems, but this can be controlled (22).

Caution must be used in comparing hydrolysis resistance of cellulose finishes prepared from N-methylol agents. A small change in degree of cure, ie, the overall strength of the curing conditions, can make a large difference in the resistance to hydrolysis of a specific finish (23). A trace of base remaining in the fibers after fabric preparation, a change in catalyst type or amount, or variation in cure time or temperature can result in a finish with vastly different resistance to hydrolysis. The reason for this difference in hydrolysis resistance is presently under investigation. However, it has been shown that the variation occurs in finishes prepared from both monofunctional and polyfunctional agents. These differences, especially for a finish prepared from a monofunctional agent, are counter to results obtained from similar determinations performed in homogeneous systems.

Alkaline conditions found in laundering have generally not been strong enough to cause significant hydrolysis of conventional finishes. In research, cross-links are often designed to resist an acid rinse without taking their stability to alkaline conditions into consideration (24). The alkaline hydrolysis of alkoxymethylureas (25) is initiated by removal of a proton from the nitrogen atom followed by the elimination of the alkoxy group.

$$\underset{\|}{\overset{O}{\text{H}_2\text{N}\overset{-}{\text{C}}\text{NCH}_2\text{OR}}} \rightleftharpoons \underset{\|}{\overset{O}{\text{H}_2\text{N}\text{C}\text{N}=\text{CH}_2}} + \text{RO}^- \tag{14}$$

The mild conditions required for hydrolysis of alkoxymethylureas were considered surprising, and ethers of methylol compounds of carboxylic acid amides were found to be comparatively stable.

The base-catalyzed cross-linking of cotton with dimethylolurea and the synthesis of bis(methoxymethyl)urea from dimethylolurea and methanol using alkaline catalysis (26) support the above conclusions and the suggested mechanism involving a series of reversible steps involving eq. 14. Further support is provided by the susceptibility of similar finishes to alkaline hydrolysis (27) and their tendency to release considerable amounts of formaldehyde after a relatively strong alkaline home wash (28–29).

Leaving-Group Effects. The first leaving-group effect to be noticed in textile finishing was a minor deactivating effect of the methoxy group in comparison to the hydroxyl (30). This finding later provided a way of improving the resistance of highly reactive *N*-methylol agents to hydrolysis (31). The factors or principles involved are illustrated in the following equations:

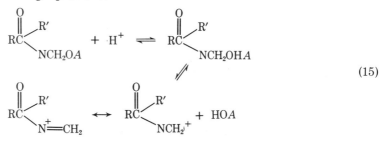

$$\tag{15}$$

Ion stability is determined largely by the electronic and steric effects produced by R and R′. Activation by the leaving group depends upon basicity of the leaving group. The activation by an alkoxy leaving group for formaldehyde-based or glyoxal-based agents (32–34) produces at least a (3–5):1 reaction rate ratio for isopropoxy compared to methoxy. Partially isopropylated agents have been used commercially to produce a better cure at lower cure temperatures or times (23,32). Use of acetoxy derivatives produces even a greater activation (as much as 30:1 when compared to the same methoxylated reagent) (31,33).

Leaving-group effects should help explain the durability to acidic hydrolysis developed by strong curing conditions (35). The anhydroglucose unit in cellulose contains one primary and two secondary hydroxyl groups. On reaction with an *N*-methylol compound, these groups in part determine the resistance to hydrolysis of the bond to cellulose. In other words, —OCell contains groups that can function as either poor or good leaving groups.

Curing Catalysts.   Many acids have been used successfully in finishing formulations to produce a durable-press finish. Selection of the specific catalyst is often determined by secondary factors, such as effluent standards, influence on dyes, formaldehyde release, discoloration of the fabric, effect on chlorine retention, formation of odors, and storage conditions. Many of the references cited include discussions on the selection of a catalyst for the specific reactant system under investigation. The chemical supplier generally specifies a certain catalyst or a selection of catalysts for use in its durable-press formulation. Frequently, the exact content of the formulation is unknown to the finisher.

Curing catalysts have been classified as mineral or organic acids and as latent acids such as ammonium salts, amine salts, and metal salts (36). Metal salts such as zinc nitrate and magnesium chloride have been the most widely used. Magnesium chloride, used by itself, is a very mild catalyst and is not recommended for curing some agents. Triazone and melamine-based reactants, for example, require a strong catalyst because of their buffering action. Mixed catalyst systems, eg, magnesium chloride with glycolic acid or citric acid, are often used to increase production rates by decreasing the curing time or lowering the curing temperature. Acetic acid and stronger acids, ie, those with a dissociation constant $\geq 10^{-6}$ at room temperature, are suitable (37).

The catalytic activity of metal salts is a complex result of several contributing factors, including acidity and oxidation state of the cation, nature of the anion, ionic size and ion mobility, hydrolysis reactions which the salts undergo in aqueous solution, covalent character, and temperature effects (38). Both Lewis-acid and Brønsted-acid mechanisms operate concurrently to different extents with different catalysts.

One of the greatest disappointments for researchers in this field has been the inability to isolate and characterize a complex of an *N*-methylol reactant and a Lewis acid salt. It is assumed (32,39) that coordination of the metal ion with the methylol oxygen (possibly by transfer from the proximate carboxyl oxygen) would be sufficient to initiate the mechanism shown in equations 9–12.

Physical Nature of Chemically Modified Cellulose.   All cross-linking systems applied to 100% cellulosic fabrics under controlled conditions cause similar high losses of strength and abrasion resistance. To date, most finishing research directed at producing an acceptable smooth-drying fabric with improved strength properties involves oligomer and/or polymer formulation with limited cross-linking. Despite numerous independent efforts (40–43), only small advantages have been attained with 100% cellulosics.

Observations with the electron microscope have proven useful in delineating the regions of the cotton cellulose fiber affected by a variety of treatments, including dyeing, mercerization, acid and enzymatic hydrolyses, derivative formation, and resin impregnation (44). Methods to differentiate between impregnation and reaction within the cotton fiber or reaction on the surface of the fiber because of migration to the surface or a failure of the resin to penetrate the fiber have been described (44). Knowledge of this type is of primary importance in understanding the results of experiments designed to distribute reactants to certain portions of the fiber and to determine their mode of reaction, ie, polymerization, reaction with cellulose, or both.

Since cross-linking of cotton reduces strength, abrasion resistance, and other desirable qualities, the relationship between cotton morphology and textile performance continues to be a key research area. Some of the main approaches to improve the physical properties of chemically modified cotton fabrics are examined in ref. 45. The results from conventional pad–dry–cure cross-linking, cross-linking the fiber in the swollen state, simple substitution, and polymer deposition are considered. Attention is given to the pretreatments and the condition of the cotton fiber prior to finishing, eg, ammonia or caustic mercerization and effects of restretching of these pretreated fabrics, the never-dried cotton fiber, the dried fiber conventionally used in commerce, and fibers reswollen with strong alkalies or amines.

Chemically modified cotton fabrics having crease-resistance properties not anticipated on the basis of previous theories (40–41) have been cited (46). These inconsistencies include experimental data obtained from cotton treated with inert solvents, epoxy compounds capable of reacting with fibers preswollen in aqueous base, long-chain monobasic acid chlorides in selected nonaqueous media, and monomers or prepolymers capable of cross-linking or being deposited or grafted. Factors that should be considered in the development of a new theory consistent with all experimental facts include the hydrogen-bonded structure of cotton cellulose, the hydrogen-bonded structure of water, and the interactions between cotton or chemically modified cottons and water.

Further Development of N-Methylol Agents. Urea– and melamine–formaldehyde resins, regardless of how they are prepared, are unstable and difficult to keep in solution in a pad bath unless they are alkylated (usually methylated).

$$>NCH_2OH + CH_3OH \overset{H^+}{\rightleftharpoons} >NCH_2OCH_3 + H_2O \qquad (16)$$

These alkylated resins are often used in combination with other resin systems. Differences between these cross-linking systems are chemical in nature because one commercial N-methylol cross-linking system produces essentially the same set of physical test data for a finished fabric as another.

Melamines and Other Amino-s-Triazines. These finishes are unlike urea–formaldehyde finishes in that hypochlorite does not cause fabric degradation, but it does yellow the fabric (47). The N-chloromelamines are relatively thermally stable, but a dilute solution of sodium bisulfite will remove color from the fabric. Conventional melamine agents are poor from the standpoint of loss of methylol groups, and even a fully methoxymethylated melamine or amino-s-triazine resins will, after processing and laundering, develop sites for color formation.

A large number of N-substituted melamines have been prepared, and only those in which no NH_2 group can occur produce nonyellowing finishes. Examples of these types of finishes have been prepared from the formaldehyde addition products of N,N',N''-tris(2-hydroxyethyl)melamine (48). It is believed that N-substituted melamines of this type have not been used in textile finishing. The research demonstrates that the yellow chromophore results from the attack of chlorine on NH_2 groups in the finish and that the color formed is not directly related to the amount of chlorine retained in the finish.

The commercial melamine derivatives used in finishing since the 1940s have generally been mixtures. A partially methylated di- or trimethylolmelamine is an excellent polymer former and also an effective cross-linking agent. Its shelf life is poor, however, and the solution thickens on standing. On the other hand, hexakis(methoxymethyl)melamine is a waxlike crystalline solid that, unlike products with lower levels of methylolation, resists polymer formation and acts almost completely as a cross-linking agent. Some effective melamine cross-linking systems used for durable-press finishing were highly methylolated (ca 5 mol formaldehyde per melamine molecule), partially methylated, and contained methoxymethylated ureas. Resins of this type produced smooth-drying fabrics with good hand, little or no susceptibility to strength losses due to chlorine bleaches, and only a slight yellowness from chlorine after multiple washes.

Specially prepared nonmethylated dry methylolmelamines have been commercially available, but these have not enjoyed wide usage as finishing agents. In addition to their use with cellulosics for durable-press finishing and for regulating the stiffness or hand of these fabrics, melamine resins have been used to shrinkproof wool, rotproof cellulosics, as a binder for pigments, and in numerous other applications.

Guanamines have been suggested for many of the uses discussed above, but it is unlikely that substantial interest will develop because their chemistry is similar and yellowing from hypochlorite is a problem (see Cyanamides).

Alkyleneureas, Triazones, and Urons. These cyclic ureas were the first commercial N-methylol products offered as cross-linking agents for reducing or eliminating the hypochlorite bleaching problem. They are still available commercially and are used throughout most of the world. The products are similar in structure and, with a slight excess of formaldehyde, the fully methylolated agent is obtained. The systematic names for the 1,3-bis(hydroxymethyl) compounds and their common names are as follows: 2-imidazolidinone or ethyleneurea, tetrahydro-2-(1H)-pyrimidinone or propyleneurea, and tetrahydro-1,3,5-triazin-2(1H)-one or triazone.

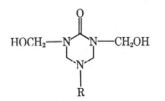

1,3-dimethylolethyleneurea (DMEU)

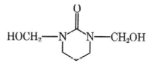

1,3-dimethylolpropyleneurea (DMPU)

1,3-dimethylol-5-alkyltriazone

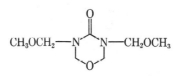

1,3-bis(methoxymethyl)uron

In many cases these cyclic ureas were mixed with urea–formaldehyde or melamine–formaldehyde and used in a number of different formulations. In general, preparation of pure triazone and uron resins is too difficult and too expensive. The main impurity is urea–formaldehyde, but because of the basic nature of the amino nitrogen in the triazone addition, urea–formaldehyde can usually be added to the crude triazone and the formulation can be used to produce finishes that retain strength in hypochlorite bleaching. Similarly, melamine resins can be added to crude uron solutions and the resulting formulations used for commercial finishing. Published work with uron and triazone finishing agents usually describes a method of preparation or states that the uron or unmethylolated triazone was purified before use. Unfortunately, the literature contains much work on formulations of unspecified composition. For this reason a brief review of selected references dealing with the synthesis and chemical finishing with relatively pure materials is given below.

The development of triazone synthesis and use in chemical finishing is described from the synthesis of the first member of its class in 1874 until its development into a generally used finishing agent in 1957 in ref. 49. Methylol derivatives of pure 5-ethyl-, 5-(2-hydroxyethyl)-, and 5,5'-ethylenebistriazinone were compared against four commercial formulations believed to be triazones, a urea–formaldehyde, a melamine–formaldehyde, and a dimethylolethyleneurea formulation. On the positive side, the triazones displayed greater resistance to damage from hypochlorite bleach and to acid hydrolysis.

The longer chain 5-alkyltriazones (from C_8 to C_{18}) are not obtained by the usual methods of synthesis because of formation of a formaldehyde–amine condensate rather than the triazone (50). However, the use of a high boiling solvent and removal of water (by azeotroping with benzene) produces the desired C_8, C_{12}, and C_{18} triazones. None of these triazones penetrates cotton cellulose well, and only the C_8 reactant produced a significant improvement in wrinkle recovery. The C_{18} reactant, on the other hand, appeared to attach itself to the surface of the fiber and to act only as a lubricating agent. These results help provide limits of molecular size and hydrophobic character of a reactant for cellulosic fibers.

The synthesis of triazones by the methods of Burke (51) or Paquin (52) or as noted previously (50) appears to be fairly general. It is a type of Mannich reaction in which the product may form either from the addition of dimethylolurea to the primary amine or from addition of the amine–formaldehyde condensate to urea. Evidently, systems capable of forming stable amine–formaldehyde condensates, eg, substituted hydrazines, resist triazone formation.

Triazones prepared from ethylamine have received the most commercial interest (53). Mixed triazone–urea–formaldehyde resins are said to be cheaper and more effective, and to develop less discoloration on curing at high temperatures than the pure triazone resins. The odor problems noted with triazones could be blamed on either formaldehyde or a volatile amine. Pure triazones are odorless, white, crystalline materials that have been kept for years at room temperature without decomposition or odor development. A solution to the odor problem was thorough washing of the cured fabrics; this procedure has been recommended for all fabrics finished with nitrogenous agents (53).

1,3-Dimethylolhydroxyethyltriazone or 1,3-dimethylol-5-(2-methoxyethyl)-triazone are considered to minimize or eliminate the main failings of triazones. These agents could also be used without an afterwash (54). The pure triazone resins were used alone or with urea–formaldehyde, and no odor other than formaldehyde was noticed.

Possibly the main problem with the triazones and the other agents mentioned in this section is the level of resistance to acid hydrolysis. As is noted below, a relatively low acid stability for a finish correlates with a relatively high level of formaldehyde release. Triazones have not been used generally for many years in the United States, but finishers have not discounted their future use.

The chemistry of urons and their use in textile finishing has been well studied. They are of interest because they represent a potentially excellent cross-linking system based on low-cost starting materials, ie, urea, formaldehyde, and methanol. Both crude reaction mixtures and a fairly pure 1,3-bis(methoxymethyl)uron were developed for commercial use. The unpurified product, expected to contain large quantities of methylated methylolureas, imparts smooth drying properties to cotton fabrics, but the finished fabric is susceptible to strength losses during hypochlorite bleaching and ironing. The pure product, which is readily obtained by vacuum distillation, is an effective cross-linking agent for cotton, and the finished fabric is resistant to chlorine damage even through repeated washes (55).

The uron ring is subject to cleavage during finishing by hydrolysis (eq. 17), but there is strong evidence that this does not occur on finished fabrics (55–57). The

$$(17)$$

normally expected precursors ((1), R = H and (1), R = CH$_2$OH) are difficult to isolate, but both have been purified and characterized (56–57). An alternative method of preparing this compound has been suggested (58). Because of its low cost as a crude product, (1) (R = CH$_2$OCH$_3$) continues to be of commercial interest in the United States but is not widely used.

Of the alkyleneureas, DMEU had wide usage in the United States as a replacement for urea- and melamine–formaldehyde, whereas DMPU was used more by the European industry. These materials are still in limited use today, but they have been replaced for the most part with systems that produce a finish with greater resistance to acid hydrolysis.

The ease of hydrolysis of a DMEU-treated fabric has been used to produce bicolored cotton fabrics (59). This was accomplished by applying a thickened DMEU solution to the pile of the fabric, curing the resin, and dyeing the fabric. The pile resists dyeing because of the presence of the cross-links. The bicolored effect is obtained by removing the DMEU cross-links with a mild hydrolysis and then overdyeing the entire fabric.

The cyclic ureas solved the problems caused by hypochlorite bleaching, but complaints about formaldehyde odors in garment-manufacturing plants and in piece-goods stores and demands for a more permanent finish placed an additional requirement on the textile-finishing industry. This new demand grew in strength during the early 1960s, and a significant change in cross-linking systems occurred. The industry stopped talking about wash-and-wear and shifted to products referred to as "durable press" or "permanent press." These products required a finish with greater resistance to hydrolysis and lower formaldehyde release than those in use at that time.

The selected listing of references on the preparation, physical, and chemical properties and pertinent information relative to uses for ethyleneurea is given in reference 60. Some of the firms or research organizations obtaining patents include American Cyanamid Co., J. Bancroft and Sons, Co., BASF, Bayer, Celanese Corp., CIBA, Cities Service Chemical Corp., Courtaulds, E. I. du Pont de Nemours, Gagliardi

Res. Corp., Heberlein, ICI, Monsanto Chem. Co., Norwich Pharm. Co., Rohm and Haas, Sumitomo Chem. Co., Ltd., Sun Chemical Corp., Tootal Broadhurst and Lee Co., Union Carbide, and USDA.

Most commercially prepared DMEU was not only an excellent cross-linking system, but it was also the first general-purpose nitrogenous agent that was not a mixture of materials. Thus, it constitutes an excellent system for basic research because of its relatively high chemical purity and its pertinence to commerce. Nevertheless, finishing research has labored over the years because of the difficult problem of understanding small chemical changes in a complex physical and chemical substrate. Questions concerning cross-link chain length, reasons for chlorine retention, competitive reactions, etc, have been studied, but the conclusions reached have differed (61–64). DMEU may represent the ideal reactant for finishing research, but it has not yet yielded completely satisfying answers to these questions.

Only the chemical cross-linking of prepared cellulosic fabrics with little or no chemical modification have been considered in this article. Similar treatments have been applied to the chemically modified celluloses, but there are relatively few reports of these results. Pretreatments of cotton cellulose, such as etherification of the hydroxyl groups to methyl, hydroxyethyl, cyanoethyl, and carboxymethyl, cause different responses to cross-linking with DMEU. The effect of replacing cellulosic hydroxyl groups depends on the nature of the substituent and the tendency of the etherified cotton to swell (65). Chemically modified celluloses of these types are of interest for a number of practical uses, but they have not been used commercially as substrates for cross-linking to form ease-of-care fabrics.

Delayed Cure and Permanent Press.   In the early and mid 1960s, many different processes and many cross-linking agents were used, but none became dominant. For a brief period, two-step cross-linking reactions were used commercially, but more widely in Europe than in the United States. In general, a wet cross-linking treatment was followed by a dry cure. Numerous variations have been tried and continue to be researched, but despite a number of seemingly early successes no commercial process of this type made significant impact in the United States (see general references).

As performance levels increased and special processing conditions involving swelling treatments and two-set applications failed to solve the strength–abrasion problem with cotton, the switch to blends of cotton–nylon or cotton–polyester began in earnest. At high performance levels, the usual points of failure in cotton garments are collar tips, the edge of sharp creases, seams, and the points or other portions of cuffs, depending upon how they were fabricated.

With the emphasis on high performance or permanent press, the curing step is often delayed. Fabric is padded as before with a cross-linking formulation, dried, and fabricated into the garment, which is then cured. In a popular variation of this process, fabrication of the garment follows curing, and creases are formed by high pressure, high temperature presses. The presses in this case are curing the cross-linked fabric a second time by breaking cross-links formed in the initial cure of the flat fabric and allowing them to reform in the fabric in the press. If the fabric in the press is creased, then the reformed cross-links keep the fabric in this state. The cellulosic fibers and yarns in the blended fibers suffer great strength loss in the process, but the synthetic fibers, generally polyester, provide the needed strength for a satisfactory garment. There is, of course, a limit to the severity of conditions that are used in the cure or recure of cellulose–polyester fabrics.

By 1965, at least six permanent- or durable-press processes were recognized and the patent situation on processing was confused. Those in the industry recognized an entirely new set of practical problems to be solved involving such detailed items as preparation of uncured fabric for shipping, type of curing catalyst, time and temperature involved in shipping, cutting and sewing procedures, effects on dyes, etc. Detailed information on these practical problems is provided from the viewpoints of chemical suppliers, researchers, a licensor, and cutters in a series of publications (66–67).

The enthusiastic customer acceptance of permanent press not only forced renewed effort to improve the abrasion resistance of cross-linked cotton, but it prompted a shift from the cyclic ureas, ie, alkyleneureas, urons, and triazones, that had solved the hypochlorite bleach problem to a cross-linking system that was also low in formaldehyde release and more hydrolysis resistant. Some interesting situations resulted. High performance permanent-press garments with durable properties do not normally require even touch-up ironing. As a result, the level of resistance to chlorine damage of the permanent-press garment did not have to be as high as the earlier garments that required some touch-up. On the other hand, since finishers were now only drying or partially curing fabrics, the garment maker was handling uncured goods and completing the cure in buildings heretofore used only for cutting, sewing, and touch-up pressing. This shift necessitated drastic reductions in the amount of formaldehyde released from all types of treated fabrics.

Formaldehyde Analysis and Sources of Formaldehyde in Finished Fabrics. In the finishing of textiles, there are three main areas where various types of formaldehyde analyses are of interest: formaldehyde in the air; free formaldehyde in cross-linking reagents and in the pad baths prepared from these reagents; and free and releasable formaldehyde in finished fabrics and garments. Because of recent studies on the toxicity of formaldehyde, the allowable maximum concentration of formaldehyde in the atmosphere of working places is being reconsidered. A comparison of allowable limits in various countries has been published (68). Concentrations significantly greater than one part per million are believed unusual in areas where finishing workers are confined. Great strides have been made in reducing the free formaldehyde content in finishing formulations and this factor, coupled with the use of improved ventilation and cross-linking systems and a genuine concern for reducing exposure to formaldehyde vapors, has reduced the concentrations in air found in finishing areas.

Free formaldehyde is generally considered to consist of formaldehyde, hydrated formaldehyde, and low molecular weight oligomers. Thus, free formaldehyde relates directly to the odor of a fabric or a padding solution or to the amount of formaldehyde transferable from a treated fabric to a wearer, assuming that the wearer is not exposed to conditions that would cause degradation of the finish (28). Since all finishes degrade somewhat, factors such as extractable formaldehyde and releasable formaldehyde must be considered. The real exposure is difficult to measure, a fact reflected in the number of test procedures used throughout the world.

Extraction tests are used in Japan and Europe, and a release test is used in the United States; standard tests have been compared based on the sources of formaldehyde present in a finished fabric (28,68–70). Finished fabric contains both free formaldehyde and formaldehyde released from the following groups: unreacted $>NCH_2OR$, where R = H, CH_3, or $CH(CH_3)_2$, depending upon the type of agent in the pad bath; $>NCH_2OCell$; $>NCH_2N<$; and $>NCH_2OCH_2N<$. With the types of cross-linking agents in current use, the last two groups are expected to be at very low or zero levels.

The analytical methods (28,71) generally used for free formaldehyde are based on a low temperature (0–5°C) titration that involves the reaction of sodium sulfite and formaldehyde.

$$HCHO + Na_2SO_3 + H_2O \rightarrow HOCH_2SO_3Na + NaOH \tag{18}$$

Performed at the specific temperature and pH ranges, the analyses can be carried out in the presence of N-methylol groups. On fabric, the formaldehyde bisulfite compound is decomposed by addition of excess sodium carbonate, and the liberated sulfite is titrated with 0.1 or 0.01 N iodine solution (28). The limits of this method have not been studied extensively, but values of free formaldehyde on laundered fabric are generally <25 ppm and confidence in numbers below this level is not good. On the other hand, commercial fabrics are generally first washed and dried by the consumer, and the free-formaldehyde contents start at ca 50 ppm and may go up to several hundred ppm, depending on finishing and storage conditions.

The determination of free formaldehyde in commercial reagents and pad-bath formulations presents a similar situation in that the usual conditions of analysis allow hydrolysis of the N-methylol groups. Formaldehyde content usually ranges from less than one percent to several percent. This analysis, also performed at 0–5°C, involves the quantitative determination of NaOH formed (eq. 18) with hydrochloric acid using an adaptation of the Morath-Woods method (71).

Because of an incomplete reaction of sulfite with free formaldehyde, these low temperature methods are believed (72) to analyze only part (usually 80–90 wt %) of the free formaldehyde present in solution. Technique becomes very important and there is a possibility of either a low or a high error. The low temperature titrations on fabric are used primarily for research and in trouble-shooting experiments.

Methods using extractions at higher temperatures are used in a number of countries. Unfortunately, the values obtained are referred to as free formaldehyde content even though part of the formaldehyde reported originates in the nitrogenous finish or unreacted cross-linking reagent on the treated fabric. Presumably, the purpose of these methods is to reduce the exposure of the consumer and the worker to formaldehyde as a skin allergen. For the most part, these test methods have originated with industry; however, in Japan a legal regulation effective in 1975 resulted. The mandatory standards involve an extraction at 40°C. An industrial method used in Europe requires extraction at 22°C followed by a determination of the total formaldehyde content of the extract (73). Use of this method identifies finished fabrics with relatively high levels of unreacted cross-linking agents.

A number of researchers have noted the release of formaldehyde by a chemically treated fabric under prolonged hot, humid conditions (74–77). A method, developed in the United States and used extensively for ca 20 yr, measures the formaldehyde release as a vapor from a fabric stored over water in a sealed jar for 20 h at 49°C or 4 h at 65°C. Results from this test have been used by industry in the U.S. to eliminate the less stable finishes. In this respect the method, AATCC Test Method 112 (78), has been very effective and has led to a new set of finishing agents to replace the alkylene-ureas, triazones, and urons.

A urea-glyoxal formaldehyde adduct, 1,3-dimethylol-4,5-dihydroxyethyleneurea (DMDHEU) (79), has been the primary cross-linking agent used in the United States since the combination of formaldehyde release, as measured by the AATCC Test Method 112, and an emphasis on durable properties, ie, hydrolysis resistance, essentially eliminated the other standard agents from general use. Another finish, based on *N,N*-dimethylolcarbamates (80), is the only other durable low-formaldehyde treatment developed industrially, and its use is only about one-fifth that of DMDHEU.

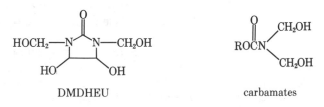

DMDHEU   carbamates

Prior to 1965, it was not unusual for unwashed, finished fabrics to release 3000–5000 ppm formaldehyde when tested by AATCC Test Method 112. Free-formaldehyde contents of these fabrics were not determined, but cotton cellulose is capable of retaining quantities of free formaldehyde in excess of these figures. A reduction in the release to ≤2000 ppm was achieved largely through use of DMDHEU and dimethylolcarbamates. The level of formaldehyde release was later reduced to 1000 ppm by some organizations, and today levels of ca 100 ppm are achievable with additives and modifications of the DMDHEU reactant system. It is believed that very little fabric is finished commercially at the 100 ppm level, and it is likely that considerable yardages of fabric are currently being finished at levels >1000 ppm release.

The changes that helped reduce formaldehyde release included the following factors (not necessarily listed in the order of their implementation):

Attention to certain factors in synthesis, formulation, and processing caused a large drop in release values. Factors such as careful pH control, degree of cure, degree of methylolation, free formaldehyde content, and catalyst selection provide significant reductions, even at high resin usage. Residues left from fabric preparation can be neutralized by corrections in pad-bath formulation.

As attention focused on details, the amount of resin solids on the weight of the fabric came under scrutiny and levels began to drop. Performance may have dropped in some cases, but the common practice of applying as much resin to a 50% cotton–50% polyester fabric as is applied to a 100% cotton fabric is unnecessary (81). The lower level of resin solids produces a similar lower level of formaldehyde release.

Nonformaldehyde additives were found to reduce the losses in performance that accompany changes in resin formulation. A number of polymer systems have been applied with varying degrees of success. Those based on acrylate systems appear to be the most widely used. A drop in performance level of durable-press properties may not be noticed immediately, but the use of significantly lower resin contents and reliance on additives usually result in noticeably poorer appearance after repeated tumble or line drying.

Alkylation of *N*-methylol cross-linking agents lowers the formaldehyde-release values significantly (21,82–84). Generally, DMDHEU is methylated to ca 25–50% of theory, considering four hydroxyl groups available for methylation. An optimized treatment with partially methylated DMDHEU at relatively low solids produces a value of 200–400 ppm releasable formaldehyde from an unwashed fabric. A further reduction to ca 100 ppm may require the use of a number of techniques including a scavenger to tie up free-formaldehyde or free-*N*-methylol groups in the finish (84–88).

It has been suggested (89) that these new finishes cannot be satisfactorily monitored for formaldehyde release by procedures that use the Nash reagent or chromotropic acid because of the low values, ie, $\leq 200–300$ ppm, being reported. For release values in the range of 15–175 ppm, a colorimetric procedure based on the use of 3-methyl-2(3H)-benzothiazolone is recommended. The question is far from settled because the Nash reagent is specific for formaldehyde and considered accurate to 10 ppm (90). Further, the significance of these low values has not been established and as noted previously, the AATCC method for release does not relate to odor of the fabric.

Once the fabric is in the hands of the consumer and has been laundered, a new situation exists for formaldehyde control. Release values do not correlate with the transport of formaldehyde (28), and thus these values do not correlate with the consumer's exposure to formaldehyde. Should formaldehyde transport from fabric to consumer come under control, then a method of testing for free-formaldehyde content and easily released formaldehyde in the fabric would be more appropriate. However, free-formaldehyde values are very low, even in many unwashed commercial fabrics. In normal consumer garments (finished by today's technology) that have been laundered several times, it is questionable whether the transportable formaldehyde content is high enough to be measured by conventional chemical means.

Despite the encouraging results and conclusions discussed above, it is still considered (91) that formaldehyde release values using the AATCC test procedures can be reduced to below the limit of 500 ppm currently imposed on textile finishers by some members of the garment industry in the United States.

Comparisons of results from the analytical methods used in the United States, Japan, and Europe (28,68) on a single fabric demonstrate that the strength of the hydrolysis conditions used in the test determine the numerical value of the results. Therefore, most formaldehyde release should occur in the washer, in the dryer, or on line drying. The effort to develop new test methods may result in the development of new technologies for finishing. Also being considered is the development and use of nonformaldehyde-containing cross-linking agents (90).

DMDHEU and Carbamates.   Unlike triazone and uron cross-linking agents, DMDHEU can be prepared commercially from starting materials in an almost pure state, and no purification steps are necessary. Commercial glyoxal, however, is a mixture of products (92), and its shelf life is limited. Therefore, early commercial DMDHEU agents often varied appreciably in content, but finishing results were generally consistent except for some extreme cases. The yellowing of the fabric that can occur upon curing has been controlled by use of additives (93–94).

Since protonation of either pendant methylols or ring hydroxyls in DMDHEU should lead to the formulation of resonance-stabilized immonium–carbonium ions, it followed that the ring hydroxyls alone could be used to cross-link cellulose (24,95–97). The agent, 1,3-dimethyl-4,5-dihydroxyethyleneurea (DMeDHEU), has been used in Japan since 1974 and has also some limited use in the rest of the world as a non-formaldehyde cross-linking agent for cellulose.

A number of possibilities for finishing with various dihydroxyethyleneureas are shown below.

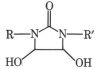

DMDHEU: R = R′ = CH$_2$OH
MDHEU: R = H, R′ = CH$_2$OH
DHEU: R = R′ = H
MeMDHEU: R = CH$_3$, R′ = CH$_2$OH
DMeDHEU: R = R′ = CH$_3$

These agents were studied and compared to DMEU with regard to performance (crease recovery of treated fabric) and durability of the finish to hydrolysis (27). All these agents are capable of producing cross-linked fabrics with improved crease resistance. However, not only does the presence of NH groups in the finish result in susceptibility to fabric damage from chlorine bleaching, but these finishes are also susceptible to alkaline hydrolysis. Finishes prepared from DHEU (no formaldehyde present) are readily removed by normal alkaline home washing conditions; the one (for MDHEU) or two moles (for DMDHEU) of added formaldehyde reduces this susceptibility to alkaline hydrolysis. Addition of one or two N-methyl groups to DHEU lowers the resistance of the finish to acid hydrolysis. Thus, finishes prepared from DMEU and DMeDHEU are readily removed by acidic solutions whereas DMDHEU finishes resist removal under the same conditions. DMEU represents an upper limit of reactivity for useful finishing. In general, agents should fall within the range of reactivity of DMEU and DMDHEU.

A properly cured DMDHEU finish is permanent and is not removed by standard acid stripping solutions. It can, of course, be removed by stronger acid stripping solutions, but these conditions weaken the cellulose. Similarly, an agent appreciably less reactive than DMDHEU requires curing conditions that lower the strength of the fabric because of hydrolyses or degradation of cellulose.

Thus, DMDHEU represents at present the optimum finishing agent. Pendant and ring hydroxyls appear to have equal reactivity toward small alcohols; however, larger alcohols such as isopropyl alcohol and cellulose are sterically hindered in approaching the 4,5- or ring positions, so that only monosubstitution at these positions is accomplished with ease. As a result, DMDHEU is expected to favor reaction with cellulose through its pendant dimethylol groups with active competition occurring from the 4,5-hydroxyls until one group is substituted (98).

As noted in equations 4 and 5, the amide–aldehyde reaction can approach completion in some cases if excess formaldehyde is used. In practice, a finishing bath cannot tolerate the large excess of formaldehyde needed to force completion, and therefore the reactants present in the practical case are the N-methylol and the N,N-dimethylol amide and unreacted formaldehyde. Generally, little or no free amide is present, but it is expected to form by demethylolation during drying or curing. A mixture of this type is expected to form the following cross-links:

CellOCH₂OCell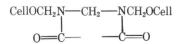

and the following "grafts":

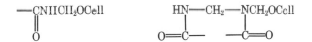

The grafts are sites for N-chloroamide formation and subsequent loss of strength on ironing. Because the level of free formaldehyde in the pad bath is relatively high compared to most cyclic urea reactants, the level of formaldehyde cross-links can also be relatively high, thereby weakening the finished fabric.

The utility of this class of agents was first demonstrated with formamide (99) but only the carbamates (24,80) gained wide commercial usage. Because of the thermal stability of the N-chlorocarbamates, properly applied carbamate finishes, even though they form N-chloro derivatives on bleaching, do not cause fabric damage (100). Methyl, ethyl, isopropyl, isobutyl, hydroxyethyl, and methoxyethyl carbamates are among those that have been used commercially. Mixtures of alkyl carbamates have also been sold in large quantities. Often the manufacturer does not specify the actual carbamate in the formulation. Use of ethyl carbamate is thought to be minimal because of toxicity considerations.

At present, the main advantage of carbamates is price. They are generally a few cents per kilogram less than DMDHEU and are often preferred for some specific uses, eg, sheeting fabrics. The agents may be used in a two-stage curing process (101) or a delayed cure (102). Like other agents, the carbamates can have an adverse effect on the color fastness of dyed fabrics to light, and these effects vary from one carbamate to another. Manufacturers and formulators are the best source of information on lightfastness of specific formulations.

Carbamate finishing agents suffer from the many different types of reaction with cellulose they can produce as noted by the above structures. As a result, the best carbamate crease-recovery and the durable-press ratings are not as high as those obtained from the urea-based agents. In addition, the possibility of formaldehyde cross-links and the subsequent additional loss of strength has, for the most part, limited use of carbamates to cotton–polyester blends.

It is unlikely that carbamate finishing agents will replace DMDHEU-based finishing agents as the principal finishing agent, unless there are unusual quality or price changes in glyoxal manufacturing.

Other N-Methylol Agents.   Cost has been an important factor in the choice of finishing system. Once an economical formulation is optimized and proven in production, a finisher usually resists change unless another factor enters the picture. The use of different dyes or finishing of white fabrics generally causes a minor change in a formulation. However, as noted before, the susceptibility of a finish to chlorine damage, to hydrolysis, or to release formaldehyde are the main factors that have resulted in significant changes in finishing systems. Further changes might be necessitated by results of toxicity testing of the agents themselves or by the introduction of new dyes. If this should happen, there are a number of other N-methylol-type agents that should be considered.

Dimethylolpropyleneurea (DMPU), as noted previously, has been used in other parts of the world but has had only limited use in the United States. Other propyleneureas, including the 5-hydroxy and the 4-methoxy-5,5-dimethyl derivatives, have been developed and provide a wide selection of chemical reactivity comparable to that obtained with ethyleneurea and substituted ethyleneurea reactant systems. A comparison of these systems with respect to reaction mechanisms, rates, structure, and properties is given in reference 103. The same author has provided a comprehensive review of methods of synthesis of cyclic ureas (104).

The characteristics of internally buffered N-methylol reagents and their reactions with cotton have been reviewed (105). Included are methylol derivatives obtained from the Michael addition of ammonia or amino acids to acrylamide, triazones containing an additional tertiary amino group, or carbamates containing a tertiary amino group. These reagents in combination with conventional triazones provide a wide range of cross-linking systems containing basic substituents.

In addition to DMeDHEU, a number of other glyoxal-based cross-linking systems with little or no formaldehyde have been evaluated (106). Work continues on the optimization of linear and cyclic compounds containing $>$N—CH(OH)—CH(OH)—N$<$ groups. The vicinal hydroxyls have been found to possess varying reactivities based on the substituents on the nitrogen atoms and on the size of the nucleophile. These glyoxal-based agents have proven capable of cross-linking cellulose through their vicinal hydroxyls; however, the efficiency of the cross-linking reaction has not been as high as desired, even for the linear N,N'-dihydroxyethylenebisamides which should be free to rotate and relieve the steric hindrance.

Miscellaneous Finishing Processes

Finishing Synthetics and Blends. The changing needs and desires of the consumer are met primarily by the fiber producer and the finisher. The primary natural fibers, cellulosics and wool, possess chemical or physical properties that are amenable to chemical modification and allow the finisher to make substantial changes in the physical and chemical nature of fabrics containing them. On the other hand, the synthetic fiber producer develops a fiber that is designed to require little or no chemical finishing and is chemically inert except to strong or unusual conditions. Modern technology has developed many fiber blends and utilizes many finishing techniques to improve textile products.

The finishing of synthetic fibers is limited primarily to heat setting, soil release, hand or comfort effects, and antistatic agents (qv). The main advantages for synthetic fibers are their greater strength, abrasion resistance, chemical inertness, and thermoplasticity. Some of their disadvantages are caused by their oleophilic character, which results in low moisture content, soiling or staining, buildup of static charges, and difficulty in dyeing. These factors differ for specific synthetic fibers and for specific blends of these fibers with natural or other synthetic fibers (see general references).

Heat setting of synthetic, thermoplastic fibers can occur at almost any stage of processing and is often applied during fabric preparation (107 108). Conditions may include either a dry atmosphere or steam; either procedure improves dimensional stability, pilling, and durable-press performance. Stresses introduced into a thermoplastic fiber during its manufacture are reduced by heat treatments. The combined effects of heat treatments in processing a synthetic fabric are a very important factor in determining the properties of the finished fabric, and the finisher must be aware of the extent of these treatments. Similarly, cellulosic fabrics have been durably set with elevated temperatures and mechanical pressure, but the mechanism is not understood (109).

The strong acceptance of durable press around 1961 led to rapid growth in the use of polyester fibers, particularly in comparison to the growth rates of polyamides and acrylics. In blends with cellulosic fibers, polyester has been most widely used but nylon has also been used as the reinforcing fiber. However, polyester's success with cellulosic blends has tended to mask its shortcomings in other areas, eg, soft yarns, knits, and wool blends. These problems and solutions have been reviewed (110).

Pilling is a problem associated with fabrics made from staple yarns and long associated with wool, but one that has increased with the introduction of the new, stronger synthetic fibers. The unsightly surface effect is generally the result of an abrasive action that forms surface fuzz and entanglement of the fuzz into balls. Some of the pills may be removed during wearing of the garment or during laundering, depending on the abrasion resistance and tenacity of the fibers (111). Electron micrographs show that polyester fiber skin flakes cause the entanglement of fibers. In a cotton–polyester blend most of the fibers in the pill are polyester, and the cotton fibers are only passive participants (112). Heat setting is most important in reducing pilling in fabrics that contain polyester staple, regardless of fabric construction.

Topical treatments, often in conjunction with heat setting, are applied to synthetic fabrics to impart hydrophilic properties (113). One of the first polymers used is a condensation product of dimethyl terephthalate, ethylene glycol, and polyethylene glycol. It worked well with polyester but not with nylon or acrylics. Another hydrophilic polymeric finish is based on poly(acrylic acid). Polyester has also been treated with ionizing radiation from a variety of sources to generate free radicals on the surface and then treated with a selected monomer such as acrylic acid (see Initiators; Radiation curing). Because of costs, it is unlikely that radiation-initiated grafting will find much use in the textile industry. Increased hydrophilicity is also obtained by surface hydrolysis of polyester using sodium hydroxide, but this process is thought to have only limited commercial use at present.

The textile industry's experience with commercial oily-soil release treatments for cotton–polyester fabrics has been described in detail. Also, it was noted that the introduction of liquid ammonia mercerization in fabric preparation should reduce the severity of soil-release problems because it involves less cross-linking and no polyester is required for some fabrics (114).

The hydrophilic treatments discussed above are intended to improve soil-release, antistatic, and comfort properties of fabrics containing synthetic fibers. Other types of topical finishes, including fluorochemical, silicone, acrylate, and hydrocarbon systems, have been applied to textiles to obtain similar properties; to alter the hand of fabric; to improve appearance, strength, and abrasion properties; or to provide water repellancy. In the past, tacky topical finishes have resulted in adsorption of particulate matter which produced a gray or dirty look on freshly laundered fabrics. Some of these additives have been evaluated under several conditions involving conventional, wet, and moist curing (115) (see also Surfactants and detersive systems).

Obtaining Properties of Synthetic Fibers in Cotton. Some of the finishing treatments described above are intended to make polyester more like cotton or other natural fibers. On the other hand, cotton and cotton–polyester blends have been chemically modified (116) to accept disperse dyestuffs and thereby rival polyester and other synthetics as a substrate in transfer printing. Successful treatments include use of glycols with cross-linking agents, use of melamine–formaldehyde resins, and acetylation. Attempts to improve cotton and cotton–polyester resistance to wrinkling using polymers or polymer formers have been extensive and are discussed in some of the general references. Film properties of some of the elastomers used in these attempts have been studied and compared to the results obtained from application of these elastomers to cotton (117). For achieving durable-press performance, no elastomer developed equals the conventional N-methylol cross-linking reactants. These systems are useful because they can be used to replace part of the N-methylol reactants and thereby reduce formaldehyde release in the treated fabric.

Elastic properties needed for stretch textile fabrics are inherent or readily developed in most synthetic, thermoplastic fibers. Considerable production of cotton stretch fabrics for apparel was noted in 1964, and stretch cottons are also used as a backcoating for coated fabrics. The process involves slack mercerization with shrinkage, removal of sodium hydroxide, and setting the yarn crimp by conventional cross-linking. The finishing of cotton stretch fabrics has been reviewed (118), and theoretical aspects including limitations in obtaining stretch properties in cotton have been developed (119–120).

Outdoor Fabrics.   Water repellency as a finishing process is discussed in many of the general references at the end of this article and in ref. 121 (see Waterproofing and water repellency). Some fabrics are sufficiently hydrophobic to be water repellent without addition of a finish. For example, polyalkylene and similar synthetic fabrics repel water readily (see Olefin fibers). Similarly, tightly woven fabrics from synthetic or natural fibers that retain natural or processing oils have significant repellent action. However, most consumer goods are treated with a topical finish to provide adequate and fairly durable properties. The level of treatment depends primarily on the nature and construction of the fabric being treated.

The earliest waterproofing treatments were coatings of a continuous layer impenetrable by water. In later improvements the fabrics were made water repellent, but also permeable to air and moisture, thus enhancing the comfort of the wearer. Aluminum and zirconium salts of fatty acids, silicone polymers, and perfluoro compounds are examples of water repellents that have been applied to both synthetic and natural fibers. The water repellency of a fiber increases with the contact angle of water on the surface of the fiber. Hydrophobic fibers exhibit higher contact angles than cellulosics but, in general, they still require a finish (122).

Outdoor fabrics must withstand the combined effect of mildew and rot, sunlight, airborne acids, alternate wetting and drying, and abrasion. Cotton has been widely used, and reference 123 reviews resistance to weathering of such products as awnings, tarpaulins, and tents as of 1960. The lighter, stronger, microbial-resistant synthetic fabrics have made large inroads into markets held by cotton even though the service life of outdoor cottons has been significantly increased since 1960 (see also Fungicides; Industrial antimicrobial agents).

The microorganisms that grow on cotton textiles cause several undesirable effects. Loss of strength is the most serious effect, but they may also stain the fabric or cause defective dyeing. The problem is one that can be greatly aggravated by climatic conditions. Warm, moist conditions found in the tropics can cause unprotected cotton fabrics in contact with the ground to deteriorate in one to two weeks; the same fabric should last for a considerably longer period of time in the northern countries.

Three different types of agents or processes are used to protect cotton against microorganisms. In general, they poison the microorganism, block its action with a protective layer, or change the fiber chemically so that it cannot be utilized by microorganisms. The protective finish must be nontoxic to human beings, durable, colorless, and inexpensive. In addition, the finish must be stable to light, must not accelerate the photochemical degradation of the cotton fabric, and must not adversely affect the physical properties native to cotton. These are exacting specifications and no agent in present use can satisfy them completely. Concerns of a similar nature apply to both synthetic and cotton outdoor fabrics.

Certain phenols, organometallic compounds, and inorganic salts are toxic to microorganisms. However, the phenols generally accelerate photochemical degradation, thereby weakening the fabric. Screening agents are often successfully used with phenols. Many of the highly effective organometallic compounds (particularly those of mercury) are highly toxic to humans, and improved replacements have been developed (124). Copper salts have been widely used (125), particularly copper 8-quinolinolate, but because it is a colored material, its use is limited.

Many of these inorganic and organometallic compounds can be readily applied to cotton using zirconium compounds such as zirconyl acetate and zirconyl ammonium carbonate (126–127). The antimicrobial agents become water soluble by reaction with one of the zirconium compounds. Cotton fabrics are padded with this aqueous solution and heated to 145°C for three to five minutes. This forms a stable, insoluble product with zirconium that resists leaching. A variety of colors can be obtained.

Other processes used to protect cotton against microorganisms include the blocking mechanism and chemical modification of cellulose. These effects are difficult to separate in some instances. However, the protection obtained by cyanoethylation and acetylation of cotton is obviously the effect of chemical modification and not blocking with a polymer. Resin impregnation by N-methylolamides appears to involve both effects; however, there is some controversy as to which is more important (128–130). In one process an insoluble melamine–formaldehyde resin is deposited inside and outside the fibers in the presence of water. The cure, catalyzed by hydrogen peroxide, must be complete before the fabric is dried. Curing is obtained at room temperature or in steam (relatively faster). Because this treatment does not prevent the growth of mildew, fungicides (qv) must be included in the process. A similar process also utilizes a melamine–formaldehyde adduct with a strongly acidic catalyst and a dry cure at elevated temperatures. Thiourea is added to reduce cross-linking by a competitive reaction. This treatment is said to limit strength losses and to develop a durable finish.

Flame Retardants. Flame-retardant finishes provide protection to cotton during weathering, and there has been a trend to develop flame-retardance standards for outdoor fabrics. A comparision of experimental and commercial finishes has been made (131). This subject has been covered in detail in other publications (132–136) (see Flame retardants for textiles; Flame retardants).

Flame retardance of textiles has been defined in many different ways. The problem to the researcher is to develop a process that meets the need defined by military, political, economical, consumer, and other considerations. Most textiles are organic polymers and regardless of the treatment will ignite and burn under some condition. Thus, criterion is not complete fire resistance but whether a specific test or standard is met. Smoldering, which can be introduced by a cigarette or another ignition source, is an entirely different process from flaming combustion and will not be considered in detail in this review. Resistance to smoldering appears to have little to do with flame retardance, but more with the ability of the textile to "avoid" the heat source. Thermoplastic fibers that melt away from a heat source provide a satisfactory textile for smolder resistance, but in practice these materials may be the fuel source for an eventual conflagration.

Similarly, thermoplastic fibers that melt at relatively low temperatures are considered flame retardant because the melted polymer drips away from the flame. The present Children's Sleepwear Standard allows many untreated polyester and nylon fabrics to be used for this purpose. Therefore, the finishing of textiles for flame retardancy is limited to a few products such as tentage and work clothing in some selected instances.

Tris(2,3-dibromopropyl) phosphate, generally referred to as Tris, is an effective chemical retardant for polyester, acetate, and triacetate fabrics. In the finishing process it is generally padded onto fabric and forced into the fiber using a thermal treatment. Many millions (10^6) of meters of fabrics especially for children's sleepwear were treated with Tris before the Consumer Product Safety Commission banned its use in 1977. A lowering of the standards followed this ban, and the large-scale flame-retardance finishing of children's sleepwear that existed in 1977 essentially ceased. Other retardants similar to Tris have been evaluated for childrens sleepwear, and some have had limited commercial use.

Tris is not an effective retardant in finishes for cellulosic fabrics, but a number of other phosphorus compounds have been used effectively to flameproof cellulosics for use in garments. Currently, the finishes based on tetrakis(hydroxymethyl)phosphonium chloride (THPC) or the sulfate (THPS) appear to be the most widely used. Formulations of THPS or THPC either as is or partially neutralized and in combination with the dimethyl ether of trimethylolmelamine and urea are padded, dried, and cured to produce a durable finish. A number of variations have been successfully applied using other melamine-based reactants or treatments that eliminate the urea or melamine components and optimize the total formaldehyde content of the systems (see Flame retardants).

Neutralization of these phosphonium salts produces a reactive methylol phosphorus system referred to as THPOH. A comparable reactant system can be prepared from tris(hydroxymethyl)phosphine and formaldehyde. THPOH can be used in a number of pad–dry–cure processes with methylolureas or methylolmelamines, but the most widely used process with cotton uses ammonia gas to form a polymer with THPOH. In this process the polymer is deposited within the cotton fiber, little or no reaction occurs with cellulose, and the strength of the fabric is not affected.

As a general rule, lightweight fabrics (70 g/m^2) are more difficult to flameproof than heavier-weight fabrics. The lighter fabric generally requires a higher add-on. If the finish fixes too much resin-type retardant, the hand of the treated fabric often becomes too stiff for use as a garment. Washing or other mechanical handling of the treated fabric may restore a desirable hand. Cellulosic fabrics treated with phosphonium-based reactants require an oxidative wash to stabilize the finish (P(III) is converted to the oxide, P(V)) and to remove odors.

Tentage fabrics are generally treated with an antimony oxide–halocarbon topical finish. Some of the preferred halocarbon systems include emulsifiable, film-forming, chlorocarbons, or appropriate bromocarbons with a binder. Some finishes of this type do not require a cure and are effective on all types of tentage fabrics. Stiffening of the fabric may occur, but the hand is not as important as for garments. A number of states have enacted flame-retardant tent standards, and the Canvas Products Association International has had success in obtaining uniformity through its efforts.

The level of activity of flame-retardance research on textiles is relatively low and appears to be decreasing. The intensive search for improved finishes began in the late 1960s and lasted ca 10 yr. Some inherently flame-resistant fibers, eg, wool, modacrylics, and aramids, have specialized uses, but only limited consumer appeal. The popular cotton–polyester-blend fabrics, especially the 50–50 blend, has proven extremely difficult to retard with retention of other acceptable properties (137).

Finishing of Wool. Wool (qv) competes for markets primarily with the noncellulosic, synthetic fibers where warmth and wrinkle recovery are desired. Its disadvantages are fuzzing, shrinkage, and difficulty in creasing. Chlorination, modification of the disulfide cross-links, and polymer treatments appear to be chemical reactions of choice.

Untreated fabrics of both wool and cellulosic fibers shrink when immersed in water because of a tendency of the fibers to swell and also because of relaxation of stresses developed during earlier mechanical processing. The degree of shrinkage by this mechanism is greater for cellulosics, but wool with its scaly structure also shrinks because of felting. Mechanical action, eg, washing, causes fibers to move about and become entangled. The roughness of the scales reduces or prevents the fibers from returning to their original configuration, which results in permanent shrinkage. The surface appearance is also altered and appears matted (see Felts).

The problem of felting has been solved by altering the scaly structure of wool. For 100 years or more, commercial processes have generally used chlorination or some other form of oxidation for this purpose. Many different processes of this type have been developed to achieve a more uniform surface and to prepare the surface for additional treatments. The chlorination step is often followed by application of a polymer to form a film on the surface of the wool. Various urea–formaldehyde and melamine–formaldehyde resin systems, vinyl and acrylic polymers and copolymers, silicones, and other polymers have been used for this purpose. Numerous procedures have been used with the ultimate objective of obtaining a smoother fiber without stiffening or discoloring it.

Wool fabrics in general resist creasing and are flame retardant. The disulfide cross-links are readily broken with reducing agents such as thioglycolic acid, sodium bisulfite, and formaldehyde. Such processes, similar to those used in permanent waving of hair, are used to set pleats or creases in wool (138) (see Hair treatments).

Recent efforts continue to utilize ozone (qv) to reduce shrinkage (139), and blends of wool, cotton, and polyester have been finished to provide improved flame-retardancy, durable-press, and shrinkage properties (140). Fabrics from the latter study were intended for uniforms and occupational apparel. The development of wool and wool-blend fabrics has been examined from the point of view of the protection offered in a variety of industrial situations (141).

Hand, Comfort, and Bioactivity. The aesthetic appeal of a garment can be the most important factor in the consumer's purchase or rejection of a finished article. A judgement is made on such properties as hand, appearance, levels of protection, and comfort. The industry is aware of these needs, and chemicals or fibers that irritate the skin must be avoided or neutralized.

Factors involving the thermal and moisture-transport aspects of comfort are possibly the simplest to evaluate and are considered in finishing of textiles, especially the synthetics. Stiffness of a fabric is of primary concern and this has been covered in previous discussions. The combination of both objective and subjective aspects makes the subject of comfort of textiles a complex one (142–145).

A variety of chemical products and finished fabrics said to have antibacterial finishes have been or are on the market. Claims range from statements about preventing odors to the prevention of spread of infections (146). Certainly, this would be a highly desirable product, but no finish appears to have gained appreciable acceptance. Two promising new finishes of this type are based on an organosilicon quaternary ammonium chloride compound (147) or on peroxide-containing complexes of zinc acetate (148).

Economic Aspects

The consumption of textiles (149–150) is usually considered on the basis of the general uses, ie, apparel, household, and industrial. In the lower-income countries, apparel uses account for almost all the total fiber consumption, whereas in the United States, apparel uses account for less than half of the total market. Despite a rapid growth in the use of synthetic fibers, cotton remains the world's number one fiber in terms of total production. The percentage of cotton used in the lower-income countries is generally higher than that used in developed countries. However, it is not unusual for finished fabrics and garments of synthetics or blends of fibers to be a main export item for the lower-income countries. In the past, the fluctuating price of raw cotton and temporary shortages of petroleum for manufacture of synthetic fibers have caused concern.

Costs and Volume. Coloring a textile, either by dyeing or printing, is generally more costly than finishing the textile. Dyes and textile chemicals are sold in the United States by many companies (151–152), generally through a smaller firm, which supplies the user with the desired mix and often with technical service. Many of the larger mills have their own chemical supplier, but it is estimated that their share of the market, even with sales to other mills, is <20%. About 300 companies supply textile chemicals. The small supplier, ie, with annual sales <5×10^6, controls ca 80% of the market (152).

The markets for textile chemicals and for finishing agents have been estimated in 1968 and again in 1978 (152–156). Estimates for these markets have been projected for 1980, 1983, and 1988 (153). The current general business slump and increased import of finished textiles will undoubtedly result in a revision of these figures unless the trends are reversed (157).

Total sales were 457×10^6 in 1978 (153), which represents 852,000 metric tons (most products contain >50% water) with an average price of 52.8¢/kg. Backcoating products, which are used primarily on carpets and upholstery fabrics, represented ca 36% of this dollar volume. Other products include nonwoven binders (16%), with water and stain repellents (13%), durable press resins reactants (8%), and softeners (8%). Price per kilogram for specific product types (21 listed) ranges from 26¢ (50% active) for catalysts for durable press to $9.23 (30% active) for soil-release finishes. Many of the leading products and their suppliers are identified (154). A buyer's guide for dyes, pigments, and resin-bonded colors and textile chemical specialties is published yearly by the American Association of Textile Chemists and Colorists (AATCC).

The amount of cross-linking agent used in a pad bath generally adds 2–5 wt % to the cured and washed cotton-containing textile. Efficiency of the reaction is generally ca 85%, but this value can vary. Storage of unwashed goods can result in considerable formaldehyde release and hydrolysis of cross-links, thereby reducing the efficiency of the cure.

Durable, low formaldehyde agents are currently selling at ca 44–88¢/kg for a 45% aqueous solution of tank-truck quantities. The chemical cost of a nonformaldehyde cross-linking system not widely used is ca three times that of a conventional finish. The chemical cost for the catalyst and the auxiliaries for a conventional finish is roughly the same as for the cross-linking agent. However, this cost ratio can vary widely because of special properties desired. For example, some finishers are using relatively high amounts of auxiliaries and lower amounts of cross-linking agent to reduce formaldehyde release in the finished textile.

Elimination of cross-linking agent has been suggested for a specific polyester fiber blended with high wet-modulus rayon or combed cotton in an 80% polyester–20% cellulosic blend (158). Increasing polyester content of blends reduces the need for a chemical finish, but it does not meet the demands for the selection of goods from the stylist or the consumer. A wide range of useful blend levels is available to consumers in the developed countries. The lower limit of polyester in cotton–polyester blends for full-performance treated fabrics has been set at 25% (159).

Another way to eliminate or reduce the use of cross-linking agents has been provided by ammonia mercerization. The first commercial treatment of fabrics in the United States was in 1975 and was based on the efforts of the Sanforized Company. The process is generally considered a pretreatment for chemical finishing, but often the level of smooth-drying properties and dimensional stability obtained is sufficient to eliminate the cross-linking treatment for some fabrics. The practical and theoretical aspects of ammonia mercerization have been the subject of numerous papers (45,160–163).

Analysis of Finishes or Residues

Analytical methods discussed in this review have generally not been concerned with finish or residue analysis. The previously discussed analysis for free formaldehyde on fabric (a residue) is the lone exception, and it is not applicable in all cases because of lack of precision and interference by impurities. The textile chemist uses many analytical methods for fabric analysis at all stages of preparation and finishing. These methods have been gathered in an excellent monograph by the AATCC (164).

Health and Safety Factors

Textiles, especially those used in apparel, are likely to be washed several times during processing, but rarely after entering the finishing plant and before reaching the consumer. It is unlikely that a general health-related problem will show up among consumers because textile workers and garment makers are exposed to much higher levels of obnoxious or potentially toxic materials than the consumer.

The principles of toxicology for textile dyeing and finishing have been reviewed (165). Hundreds of substances are used; the LD_{50}s for common finishing chemicals range from 800 for formaldehyde to 7070 for glyoxal. Many of the polymeric materials used to coat textiles are durably bonded to the fiber and are essentially inert even if removed.

The Consumer Product Safety Commission (CPSC) has been active in a number of textile-related matters involving potential carcinogenicity of a chemical. The most notable of these, ie, the ban on Tris and the transfer of formaldehyde from durable-press and other fabrics, have been discussed previously. The latter subject is currently under review by the CPSC. Some members of the industry concerned with the outcome of the matters have expressed their views (166–167).

Contact textile dermatitis has been traced to dyes, formaldehyde, unreacted monomer from synthetic fibers, and an early flameproofed cotton fabric (168). Skin irritations have also been related to the harshness of the fiber (142–145). A recent review of the literature demonstrates a continuing interest in formaldehyde resins (169), a decrease in the incidence of contact sensitivity to formaldehyde related to government controls in Japan (170), and contact dermatitis on synthetic fibers due to spin finishes (oils, surfactants, and possibly biocides) for polyester (171). Apparently, dermatitis from textiles is generally limited to a small proportion of individuals with apparent high sensitivities or to those (often industrial workers) exposed to chemicals that remove protective oils and fats from their skins. The percentages are low but at times the discomfort is high.

BIBLIOGRAPHY

"Fire-Resistant Textiles" under "Textile Technology" in *ECT* 2nd ed., Supplement Vol., pp. 944–964, by George L. Drake, Jr., U.S. Dept. of Agriculture; "Soil-Release Finishes" under "Textile Technology" in *ECT* 2nd ed., Supplement Vol., pp. 964–973, by Samuel Smith, Minnesota Mining and Manufacturing Company.

1. H. B. Goldstein and H. W. Smith, *Text. Chem. Color.* **12**, 49 (1980).
2. N. R. Bertoniere, W. D. King, and S. P. Rowland, *Tex. Res. J.* **51**, 242 (1981).
3. J. Varghese and D. M. Pasad, *Text. Res. J.* **47**, 802 (1977).
4. W. A. Reeves, R. M. Perkins, and L. H. Chance, *Text. Res. J.* **30**, 179 (1960).
5. R. W. Liggett, H. L. Hoffman, A. C. Tanquary, and S. L. Vail, *Text. Res. J.* **38**, 375 (1968).
6. D. D. Gagliardi and A. C. Nuessle, *Am. Dyest. Rep.* **39**, P12 (1950).
7. S. J. O'Brien and W. J. van Loo, *Text. Res. J.* **31**, 276 (1961).
8. W. A. Reeves, M. O. Day, K. R. McLellan, and T. L. Vigo, *Text. Res. J.* **51**, 481 (1981).
9. L. B. Arnold and co-workers, *Am. Dyest. Rep.* **49**, P843 (1960).
10. H. Petersen in H. Mark, N. Woodling, and S. M. Atlas, eds., *Chemical Aftertreatment of Textiles*, Interscience Publishers, a division of John Wiley & Sons, Inc., New York, 1971, p. 135.
11. S. J. O'Brien, *Am. Dyest. Rep.* **54**, P477 (1965).
12. S. L. Vail and A. G. Pierce, *J. Org. Chem.* **37**, 391 (1972).
13. W. A. Reeves, S. L. Vail, and J. G. Frick, *Text. Res. J.* **32**, 305 (1962).
14. S. L. Vail, *Text. Res. J.* **39**, 774 (1969).
15. H. Petersen, *Text. Res. J.* **41**, 239 (1971).
16. R. A. Gill and R. Steele, *Text. Res. J.* **32**, 338 (1962).
17. R. J. Harper, J. S. Bruno, E. J. Blanchard, and G. A. Gautreaux, *Text. Res. J.* **46**, 82 (1976).
18. R. D. Gilbert and J. H. Rhodes, *J. Polym. Sci. Part C* **30**, 509 (1970).
19. S. P. Rowland, S. M. Stark, V. O. Cirino, and J. S. Mason, *Text. Res. J.* **41**, 57 (1971).
20. S. L. Vail, *Text. Res. J.* **42**, 360 (1972).
21. S. L. Vail and A. G. Pierce, *Text. Res. J.* **43**, 294 (1973).
22. S. L. Vail and G. B. Verburg, *Text. Res. J.* **43**, 67 (1973).
23. S. L. Vail and G. B. Verburg, *Text. Res. J.* **42**, 367 (1972).
24. S. L. Vail, *Chem. Ind. (London)*, 305 (1967).
25. J. Ugelstad and J. de Jonge, *Acta Chem. Scand.* **10**, 1475 (1956).
26. R. L. Arceneaux and J. G. Frick, *Text. Res. J.* **31**, 1075 (1961).
27. S. L. Vail, G. B. Verburg, and A. H. P. Young, *Text. Res. J.* **39**, 86 (1969).
28. S. L. Vail and R. M. Reinhardt, *Text. Chem. Color.* **13**, 131 (1981).
29. R. M. Reinhardt, B. A. K. Andrews, and R. J. Harper, *Text. Res. J.* **51**, 263 (1981).
30. S. L. Vail, F. W. Snowden, and E. R. McCall, *Am. Dyest. Rep.* **56**, 856 (1967).
31. S. L. Vail, R. M. H. Kullman, W. A. Reeves, and R. Barker, *Text. Res. J.* **40**, 355 (1970).
32. H. Petersen, *Textilveredlung* **8**, 412 (1973).
33. S. L. Vail and H. Petersen, *Ind. Eng. Chem. Prod. Res. Dev.* **14**, 50 (1975).
34. H. Petersen, *Textilveredlung* **5**, 437 (1970).
35. S. L. Vail and W. C. Arney, *Text. Res. J.* **41**, 336 (1971).
36. S. Buchholz, *Am. Dyest. Rep.* **56**, 1025 (1967).
37. U.S. Pat. 3,374,107 (Mar. 19, 1968), J. F. Cotton (to West Point Pepperell, Inc.).
38. A. G. Pierce, R. M. Reinhardt, and R. M. H. Kullman, *Text. Res. J.* **46**, 420 (1976).
39. A. G. Pierce, S. L. Vail, and E. A. Boudreaux, *Text. Res. J.* **41**, 1006 (1971).
40. H. Tovey, *Text. Res. J.* **31**, 185 (1961).
41. J. L. Gardon and R. Steele, *Text. Res. J.* **31**, 160 (1961).
42. J. T. Marsh, *Text. Manuf.*, 157 (1967).
43. A. Hebeish, A. T. El-Aref, E. Allam, and Z. El-Hilw, *Angew. Makromol. Chem.* **80**, 177 (1979).
44. V. W. Tripp, A. T. Moore, I. V. deGruy, and M. L. Rollins, *Text. Res. J.* **30**, 140 (1960).

45. S. P. Rowland, M. L. Nelson, C. M. Welch, and J. J. Hebert, *Text. Res. J.* **46,** 194 (1976).
46. R. R. Benerito, *Text. Res. J.* **38,** 279 (1968).
47. W. F. Herbes, S. J. O'Brien, and R. G. Weyker in ref. 10, p. 319.
48. S. L. Vail, J. G. Frick, and J. D. Reid, *Am. Dyest. Rep.* **51,** P622 (1962).
49. J. D. Reid, J. G. Frick, R. M. Reinhardt, and R. L. Arceneaux, *Am. Dyest. Rep.* **48,** P81 (1959).
50. S. L. Vail, J. G. Frick, P. J. Murphy, and J. D. Reid, *Am. Dyest. Rep.* **50,** P200 (1961).
51. W. J. Burke, *J. Am. Chem. Soc.* **69,** 2136 (1947).
52. A. M. Paquin, *J. Org. Chem.* **14,** 189 (1949).
53. R. L. Wayland, *Text. Res. J.* **29,** 170 (1959).
54. S. L. Vail, C. M. Moran, and J. D. Reid, *Am. Dyest. Rep.* **52,** P712 (1963).
55. R. L. Arceneaux and J. D. Reid, *Ind. Eng. Chem. Prod. Res. Dev.* **1,** 181 (1962).
56. M. T. Beachem and co-workers, *J. Org. Chem.* **28,** 1876 (1963).
57. C. D. Egginton and C. P. Vale, *Text. Res. J.* **39,** 140 (1969).
58. A. A. Eisenbraun, D. S. Shriver, and C. R. Walter, *J. Org. Chem.* **29,** 2777 (1964).
59. E. J. Blanchard, R. J. Harper, J. S. Bruno, and G. A. Gautreaux, *Text. Chem. Color.* **8,** 92 (1976).
60. P. K. Shenoy and J. W. Pearce, *Am. Dyest. Rep.* **57,** 352 (1968).
61. J. G. Frick, B. A. Kottes, and J. D. Reid, *Text. Res. J.* **29,** 314 (1959).
62. P. C. Mehta and J. R. Mody, *Text. Res. J.* **31,** 951 (1961).
63. J. D. Reid, L. W. Mazzeno, R. M. Reinhardt, and A. R. Markezich, *Text. Res. J.* **27,** 252 (1957).
64. J. G. Roberts, *J. Text. Inst.* **58,** P418 (1964).
65. R. M. H. Kullman, J. G. Frick, R. M. Reinhardt, and J. D. Reid, *Text. Res. J.* **31,** 877 (1961).
66. H. B. Goldstein and J. M. May, *Am. Dyest. Rep.* **54,** P738 (1965); R. L. Stultz, p. 744; W. L. Beaumont, p. 746; A. S. Cooper, W. A. Reeves, A. M. Walker, M. J. Hoffman, and H. B. Moore, p. 749; R. Nirenberg, p. 755; J. Midholland, p. 757; R. W. Malburg, p. 759.
67. R. M. Reinhardt, *Text. Bull.* **91**(4), 64 (1965).
68. H. Bille and H. Petersen, *Melliand Textilber.* **57,** 155 (1976).
69. R. M. H. Kullman, A. B. Pepperman, and S. L. Vail, *Text. Chem. Color.* **9,** 195 (1978).
70. K. Schliefer and U. Beines, *Melliand Textilber.* **60,** 960 (1979).
71. J. C. Morath and J. T. Woods, *Anal. Chem.* **30,** 1437 (1958).
72. C. M. Moran and S. L. Vail, *Am. Dyest. Rep.* **54,** 185 (1965).
73. G. Lund, *Shirley Inst. Bull.* **48,** 17 (1975).
74. O. C. Bacon, M. F. Parker, and L. F. Horn, *Am. Dyest. Rep.* **46,** 933 (1957).
75. A. C. Nuessle, *Am. Dyest. Rep.* **55,** P646 (1966).
76. J. D. Reid, R. L. Arceneaux, R. M. Reinhardt, and J. H. Harris, *Am. Dyest. Rep.* **49,** 490 (1960).
77. P. X. Riccobono, R. N. Ring, and A. Roth, *Text. Chem. Color.* **8,** 108 (1976).
78. *Method 112–1975*, technical manual, American Association of Textile Chemists and Colorists (AATCC), Research Triangle Park, N.C., 1975.
79. U.S. Pat. 2,764,573 (Sept. 25, 1956), B. V. Reibnitz, A. Woerner, and H. Scheuermann (to Badische Anilin-and Soda-Fabrik Akt.).
80. R. L. Arceneaux, J. G. Frick, J. D. Reid, and G. A. Gautreaux, *Am. Dyest. Rep.* **50,** 849 (1961).
81. R. J. Harper and J. S. Bruno, *Text. Res. J.* **42,** 433 (1972).
82. B. A. K. Andrews and R. J. Harper, *Text. Res. J.* **50,** 177 (1980).
83. B. A. K. Andrews, R. J. Harper, and S. L. Vail, *Text. Res. J.* **50,** 315 (1980).
84. H. Petersen and P. S. Pai, *Text. Res. J.* **51,** 282 (1981).
85. B. A. K. Andrews, R. J. Harper, R. D. Smith, and J. W. Reed, *Text. Chem. Color.* **12,** 287 (1980).
86. A. Hebeish and K. Schleifer, *Text. Res. J.* **46,** 465 (1976).
87. R. S. Perry, C. Tsou, and C. S. Lee, *Text. Chem. Color.* **12,** 311 (1980).
88. J. D. Turner and N. A. Cashen, *Text. Res. J.* **51,** 271 (1981).
89. W. C. Floyd and S. H. Yoon, *Text. Res. J.* **51,** 276 (1981).
90. S. L. Vail and B. A. K. Andrews, *Text. Chem. Color.* **11,** 14 (1979).
91. P. S. Pai and H. Petersen, *Book of Papers*, 1981 AATCC National Technical Conference, Research Triangle Park, N.C., 1981, p. 205.
92. E. B. Whipple, *J. Am. Chem. Soc.* **92,** 7183 (1970).
93. U.S. Pat. 3,576,291 (Apr. 27, 1971), M. R. Cusano and R. D. Featherston (to Proctor Chemical Company, Inc.).
94. U.S. Pat. 3,765,836 (Oct. 16, 1973), R. L. Readshaw and G. H. Lourigan (to Union Carbide Corporation).
95. U.S. Pat. 3,260,565 (July 12, 1966), M. T. Beachem (to American Cyanamid Company).
96. U.S. Pat. 3,112,156 (Nov. 26, 1963), S. L. Vail and P. J. Murphy (to USA as represented by the Secretary of Agriculture).

97. S. L. Vail, P. J. Murphy, J. G. Frick, and J. David Reid, *Am. Dyest. Rep.* **50,** 550 (1961).
98. S. L. Vail and W. C. Arney, *Text. Res. J.* **44,** 400 (1974).
99. S. L. Vail, J. G. Frick, P. J. Murphy, and J. D. Reid, *Am. Dyest. Rep.* **50,** 437 (1961).
100. A. H. Reine, J. D. Reid, and R. M. Reinhardt, *Am. Dyest. Rep.* **55,** 711 (1966).
101. E. I. Pensa, G. C. Tesoro, R. D. Rau, and P. H. Egrie, *Text. Res. J.* **36,** 279 (1966).
102. R. M. H. Kullman, R. M. Reinhardt, and J. D. Reid, *Am. Dyest. Rep.* **53,** 167 (1964).
103. H. Petersen, *Text. Res. J.* **38,** 156 (1968).
104. H. Petersen, *Synthesis,* 143 (1973).
105. S. P. Rowland, C. P. Wade, and W. E. Franklin, *Text. Res. J.* **44,** 869 (1974).
106. S. L. Vail, *Textilveredlung* **14,** 436 (1979).
107. G. R. Turner, *Text. Chem. Color.* **13,** 246 (1981).
108. E. J. Elliott and R. E. Silva, *Text. Chem. Color.* **13,** 257 (1981).
109. C. G. Tewksbury and B. G. Kidda, *Text. Res. J.* **35,** 347 (1965).
110. E. V. Burnthall and J. Lomartire, *Text. Chem. Color.* **2,** 218 (1970).
111. D. Gintis and E. J. Mead, *Text. Res. J.* **29,** 578 (1959).
112. B. C. Goswami, K. E. Duckett, and T. L. Vigo, *Text. Res. J.* **50,** 481 (1980).
113. B. M. Latta in N. R. S. Hollies and R. F. Goldman, eds., *Clothing Comfort,* Ann Arbor Science Publishers, Inc., Ann Arbor, Mich., 1977, p. 33.
114. B. M. Latta and S. B. Sello, *Text. Res. J.* **51,** 579 (1981).
115. H. E. Bille and H. A. Petersen, *Text. Res. J.* **37,** 264 (1967).
116. E. J. Blanchard, J. S. Bruno, and G. A. Gautreaux, *Am. Dyest. Rep.* **65**(7), 26 (1976).
117. H. R. Rawls, E. Klein, and S. L. Vail, *J. Appl. Polym. Sci.* **15,** 341 (1971).
118. S. L. Vail, *Text. Inst. Ind.* **3,** 309 (1965).
119. A. S. Cooper, H. M. Robinson, W. A. Reeves, and W. G. Sloan, *Text. Res. J.* **35,** 452 (1965).
120. R. L. Colbran and T. M. Thompson, *J. Text. Inst.* **58,** 385 (1967).
121. J. L. Moilliet, ed., *Waterproofing and Water-Repellency,* Elsevier Publishing Company, London, 1963.
122. E. I. Valko in H. F. Mark, S. M. Atlas, and E. Cernia, eds., *Man-Made Fibers,* Vol. 3, Interscience Publishers, a division of John Wiley & Sons, Inc., New York, 1968, p. 499.
123. J. W. Howard and F. A. McCord, *Text. Res. J.* **30,** 75 (1960).
124. C. J. Conner and co-workers, *Text. Chem. Color.* **10,** 70 (1978).
125. A. G. Kempton, H. Maisel, and A. M. Kaplan, *Text. Res. J.* **33,** 87 (1963).
126. C. J. Connor, A. S. Copper, W. A. Reeves, and B. A. Trask, *Text. Res. J.* **34,** 347 (1964).
127. C. J. Conner, G. S. Danna, A. S. Cooper, and W. A. Reeves, *Text. Res. J.* **37,** 94 (1967).
128. R. J. Boyle, *Text. Res. J.* **34,** 319 (1964).
129. A. Ruperti, *Am. Dyest. Rep.* **50,** 762 (1961).
130. W. N. Berard, G. A. Gautreaux, and W. A. Reeves, *Text. Res. J.* **29,** 126 (1959); **30,** 70 (1960).
131. D. A. Yeadon and R. J. Harper, *J. Fire Retardant Chem.* **7,** 228 (1980).
132. V. M. Bhatnagar, ed., *Advances in Fire Retardant Textiles,* Technomic, Westport, Conn., 1975.
133. J. W. Lyons, *The Chemistry and Uses of Fire Retardants,* Wiley-Interscience, New York, 1970.
134. W. A. Reeves, G. L. Drake, and R. M. Perkins, *Fire Resistant Textiles Handbook,* Technomic, Westport, Conn., 1974.
135. R. B. LeBlanc, *Text. Ind.* **143**(2), 78 (1979).
136. *Ibid.,* **144**(2), 82 (1980).
137. R. H. Barker and M. J. Drews, *Development of Flame Retardants for Polyester/Cotton Blends,* NBS-GCR-ETIP 76-22, final report to National Bureau of Standards, Washington, D.C., 1976.
138. H. D. Fedtman and B. E. Fleischfresser, *Text. Inst. Ind.* **6,** 334 (1968).
139. W. J. Thorsen, D. L. Sharp, and V. G. Randall, *Text. Res. J.* **49,** 190 (1979).
140. J. V. Beninate, B. J. Trask, and G. L. Drake, *Text. Res. J.* **51,** 217 (1981).
141. P. N. Mehta, *Text. Res. J.* **50,** 185 (1980).
142. D. G. Mehrtens and K. C. McAlister, *Text. Res. J.* **32,** 658 (1962).
143. N. R. S. Hollies, A. G. Custer, C. J. Morin, and M. E. Howard, *Text. Res. J.* **49,** 557 (1979); N. R. S. Hollies and R. F. Goldman, eds., *Clothing Comfort,* Ann Arbor Science Publishers Inc., Ann Arbor, Mich., 1977.
144. L. Fourt and N. R. S. Hollies, *Clothing Comfort and Function,* Marcel Dekker, Inc., New York, 1970.
145. E. T. Renbourn in J. G. Cook, ed., *Physiology and Hygiene of Materials and Clothing,* Merrow Publishing Co., Ltd., Watford, England, 1971.

146. D. D. Gagliardi, *Am. Dyest. Rep.* **51,** 49 (1962).

147. P. A. Walters, E. A. Abbott, and A. J. Isquith, *Appl. Microbiol.* **25,** 253 (1973).

148. G. F. Danna, T. L. Vigo, and C. M. Welch, *Text. Res. J.* **48,** 173 (1978).

149. M. Stuart, *Textiles (Shirley Inst.)* **10,** 41 (1981).

150. *Cotton Counts Its Customers*, National Cotton Council of America, Memphis, Tenn., 1981.

151. *Mod. Text.* **60**(1), 16 (1979).

152. *Chem. Eng. News*, 10 (May 29, 1978).

153. B. J. Obenski, *Text. Ind.* **144**(9), 116 (1980).

154. *Mod. Text.* **60**(4), 27 (1979).

155. S. C. Stinson, *Chem. Eng. News*, 24 (Oct. 4, 1971).

156. *Am. Text. Rep.* **82**(46), 39 (1968); **83**(26), 17 (1969).

157. *Bus. Week*, 66 (Apr. 9, 1979).

158. F. S. Moussalli and C. L. Browne, *Text. Chem. Color.* **3,** 202 (1971).

159. G. F. Ruppenicker and R. M. H. Kullman, *Text. Res. J.* **51,** 590 (1981).

160. T. A. Calamari, S. P. Schriber, A. S. Cooper, and W. A. Reeves, *Text. Chem. Color.* **3,** 234 (1971).

161. S. A. Heap, *Text. Inst. Ind.* **16,** 387 (1978).

162. B. J. Trask, T. A. Calamari, and R. J. Harper, *Text. Chem. Color.* **9**(10), 33 (1977).

163. S. L. Vail, G. B. Verburg, and T. A. Calamari, *Text. Chem. Color.* **7,** 175 (1975).

164. J. W. Weaver, ed., *Analytical Methods for a Textile Laboratory*, 2nd ed., AATCC Monograph No. 3, Research Triangle Park, N.C., 1968.

165. D. A. Bixler, *Book of Papers*, 1979 AATCC National Technical Conference, Research Triangle Park, N.C., 1979, p. 167.

166. O. P. Beckwith, *Text. Ind.* **144**(2), 84 (1980).

167. H. Bille, *Melliand Textilber.* **62,** 811 (1981).

168. I. Martin-Scott, *Br. J. Dermatol.* **78,** 632 (1966).

169. C. Romaguera, F. Grimalt, and M. Lecha, *Contact Dermatitis* **7,** 152 (1981); contains a number of references on the same subject.

170. T. Sugai and S. Yamamoto, *Contact Dermatitis* **6,** 154 (1980).

171. D. Burrows and H. S. Campbell, *Contact Dermatitis* **6,** 362 (1980).

General References

References 10 and 122 are also general references.

J. T. Marsh, *An Introduction to Textile Finishing*, 2nd ed., Chapman and Hall Ltd., London, 1966; early and current methods of chemical and mechanical finishing.

J. T. Marsh, *Crease Resisting Fabrics*, Reinhold Publishing Corporation, New York, 1962; an unusually thorough review.

Textile Finishing, BASF manual, Ludwigshafen, FRG, 1973.

E. Kornreich, *Introduction to Fibres and Fabrics*, 2nd ed., Heywood Books, London, 1962; fabric structures and industrial equipment are described, and trade names of man-made fibers are included.

J. E. Lynn and J. J. Press, *Advances in Textiles Processing*, Vol. 1, Textile Book Publishers, Inc., a division of Interscience Publishers, Inc., New York, 1961; includes chapters on nonwoven fabrics and on bonding and coating fabrics.

H. C. Speel and E. W. K. Schwartz, *Textile Chemicals and Auxiliaries*, Reinhold Publishing Corporation, New York, 1957; industrial processes and products with special reference to surfactants and finishes.

M. E. Carter, *Essential Fiber Chemistry* in L. Rebenfeld, ed., *Fiber Science Series*, Marcel Dekker, Inc., New York, 1971; review of the ten commercially important fibers.

R. J. Harper, *Durable Press Cotton Goods*, in J. G. Cook, ed., *Merrow Monographs*, Merrow Publishing Co., Ltd., Watford, England, 1971.

R. H. Manly, *Durable Press Treatments of Fabrics—Recent Developments*, Chemical Technology Review No. 68, Noyes Data Corporation, Park Ridge, N.J., 1976; detailed description of treatments for cellulosics and for polyester blends and wool based on U.S. patents issued since 1970.

R. Williamson, *Fluorescent Brightening Agents*, Vol. 4 of *Textile Science and Technology*, Elsevier Scientific Publishing Company, New York, 1980.

SIDNEY L. VAIL
Southern Regional Research Center, USDA

TIRE CORDS

A pneumatic tire is a remarkable engineering achievement; many of its properties are derived from the high strength cords it contains. A tire cord gives the tire its size and shape, bruise and fatigue resistance, and load-carrying capacity. It also affects the wear, comfort, and maneuverability of the tire. Yet tire cords comprise only ca 10 wt % of tires.

When R. W. Thompson patented the first pneumatic tire in the UK in 1845, he used textile fabric as the strength member. The term tire cord came into usage in 1893 when tire cord was patented in the UK and the United States.

After World War I, with the ascendancy of the automobile in the United States, square woven cotton tire fabric was replaced by several bias plies of warp fabric made with cotton cords and held together by a few cotton picks and carcass rubber.

In the 1930s, rayon (qv) was introduced for tire application. However, it lacked the adhesion of cotton to the carcass rubber. This problem was solved with the development of the resorcinol–formaldehyde–latex (RFL) adhesive system (1) (see also Phenolic resins).

Although nylon-6,6 had been developed by DuPont in the 1930s, it was not until World War II that it was tried in airplane tires where cost was less important than toughness and light weight. Again adhesion was a problem, but RFL with poly(2-vinylpyridine) latex as part of the latex component improved adhesion (2). Both nylon-6,6 and nylon-6 bias tires had a drawback called flatspotting, which was unacceptable to automobile manufacturers, and rayon-bias tires were preferred during the 1950s and 1960s.

Partially because of the competition between nylon and rayon, properties of tire cords were continually improved (3–4); for example, two-ply rather than four-ply rayon tires were successfully commercialized in 1961. They remained standard original equipment on most passenger cars through 1969.

The flatspot resistance of nylon was improved in a mixture of 70 parts nylon-6 and 30 parts polyester, trade named EF-121 (5), or a melt blend of 70 parts nylon-6,6 and 30 parts poly(hexamethyleneisophthalamide), trade named N-44 (6). However, these yarns were rejected for original-equipment passenger tires because of flatspotting and poor thermal stability during processing (see Polyamides; Polyesters).

Although polyester appeared to be the best replacement for nylon in tires, it proved difficult to obtain good cord-to-rubber adhesion. For mass production, water-based adhesives were needed, and several two-step and one-step dip systems were developed (7). In the early 1960s, polyester passenger tires were introduced. They gave excellent field results, but attempts to extend their use to airplanes or trucks did not succeed because of the poor dynamic performance of polyester at temperatures above those normally found in passenger tires, eg, 82°C, which is the glass temperature T_g of polyester. However, various process modifications have permitted the use of polyester as the carcass reinforcement material for light truck tires and radial truck tires.

The introduction of the radial and belted-bias constructions had a tremendous influence on tire-cord reinforcement. In the radial tire, one or more plies of carcass cords lie radially. In addition, several plies of cords, in the form of a belt, are placed under the tread in such a manner that the belt cords lie in an angle close to the circumferential direction (see Fig. 1). The Michelin Company of France was the first to commercialize this type of tire construction in 1947 (8). Their passenger tires had rayon cords in the carcass with a special three-ply wire belt, and later they introduced all-wire radial truck tires. Radial tires have several advantages over bias tires, eg, up to 100% more tread wear, 7–10% less rolling resistance, excellent bruise resistance in the tread region, greater cornering ability, good control during nibbling (eg, driving on or off the edge of the highway), and generally greater endurance because of lower operating temperatures. Their disadvantages are mainly lower bruise resistance of sidewalls, more difficult retreadability, poorer high speed performance, and poor ride characteristics with soft automobile suspensions. Although rayon belts offer better ride and high speed endurance than steel belts, the latter have longer tread life and better bruise resistance and cornering performance.

The radial constructions became very popular in Europe. The preferred reinforcing materials for passenger tires were rayon in the carcass and rayon or steel in the belts. All steel was preferred in heavy-duty tires.

Although U.S. manufacturers recognized the advantages of the radial construction and BF Goodrich introduced an all-rayon radial passenger tire to the U.S. market, no rapid changeover from bias to radial tires took place because of the tremendous capital investment required. Instead, belted-bias tires, which could be manufactured on existing equipment, were developed.

Belted-bias construction is more efficient than simple bias construction. It allows the use of high modulus cords such as glass to reinforce the crown region of the tire, which increases tread life up to 150% of that of conventional bias tires. The conversion to the belted-bias construction was facilitated by the development of a new high modulus glass tire cord. Owens-Corning introduced glass as a tire cord in a composite impregnated with 15–30% RFL (9), which coated and protected the individual glass filaments. The first commercial glass belted-bias tires had a nylon carcass and a glass belt. Other high modulus materials like rayon, vinal (vinyl alcohol polymer fibers), and steel were also used. Because of the success of polyester as a flatspot-resistant carcass material for bias tires, it became the preferred cord for belted-bias carcasses.

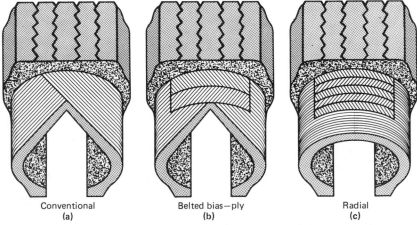

| Conventional | Belted bias—ply | Radial |
| (a) | (b) | (c) |

Figure 1. Schematic diagram of tire cords in commercial tire constructions.

After the 1973 oil embargo, new motor vehicles were equipped with radial tires in order to increase fuel efficiency, and today radial tires dominate the market.

The aramid fiber, Kevlar, developed in the 1970s, was a breakthrough in fiber technology (10) (see Aramid fibers). This aramid has a tensile strength of 1.98 N/tex (22 gf/den), over twice that of nylon, and a tensile modulus approaching that of steel. It is the only organic textile fiber developed for reinforcement. It is used in the belts of a few premium radial passenger tires and appears to offer advantages in the carcass of radial heavy-duty tires. If its high strength could be fully utilized in a tire cord, it should permit lighter tires, save materials, reduce processing energy, and improve tire performance.

New forms of vinal, a textile fiber developed and used in Japan since the 1950s, have tensile strengths and moduli considerably higher than nylon, rayon, and polyester. Although vinal has been described as a super rayon, it is more sensitive than rayon to moisture at high temperatures. Vinal tire cord acts as reinforcement in both radial carcass and radial-belt tires.

Many methods have been suggested for making radial tires by a one-step process in order to use the conventional, bias-building and curing equipment. For radial passenger tires, a slip technique was developed for sliding the green belts and tread over the carcass during the tire lift into the curing press. For large tires, this technique was impractical. However, the invention of stretch cord permitted BF Goodrich to make large radial farm tires with circumferential rayon stretch-cord belts on their conventional tire-building equipment (11).

Economic Aspects

During 1981, ca 3×10^5 metric tons of tire cords (almost one third of which was steel) was consumed in tire manufacture in the United States; world consumption exceeded 4.5×10^5 t.

Cotton was replaced by rayon which, in turn, was replaced by nylon (see Fig. 2). Actually, polyester replaced rayon, whereas use of glass and steel increased as the materials for the belted-bias and radial tires. In the United States, rayon still retains a small share of the market, although a larger market exists in Europe. Both nylon and polyester seem to have reached a plateau at 9×10^4 t/yr. Glass consumption remained steady at 16,000 t/yr; it is expected to grow slowly as it replaces some steel and perhaps more of the rayon in the belts of radial passenger tires. Steel consumption continues to grow in both the carcass and belts of truck and heavy-duty radial tires at the expense of nylon. The auspicious entry of aramid into the tire-cord market has so far been cut short, but prospects are good if the price becomes competitive.

In order to compare the economic value of one fiber to another, the cost of building equivalent tires from each fiber should be compared. However, there is no single property for all fibers that limits the amount needed in any equivalent tire. For example, in comparing rayon with nylon, the room-temperature tenacity indicates that 0.6 kg nylon is equivalent to 1 kg rayon; however, when the two kinds of cords are tested at high speeds and elevated temperatures (110°C), 0.8 kg nylon is equivalent to 1 kg rayon. Therefore, room-temperature tenacity alone cannot be used for accurate cost comparison of these tire cords. Yet, tenacity is one of the main performance criteria, and it is often used to compare cost. The 1982 U.S. tire-yarns price information is given in Table 1.

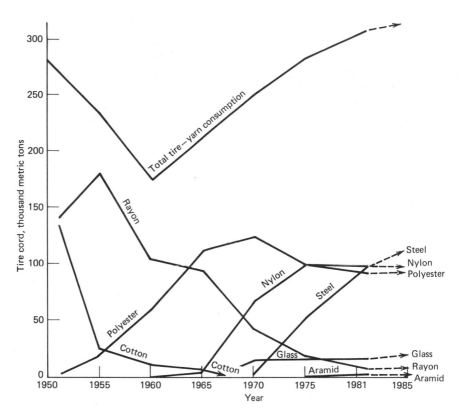

Figure 2.   Tire-yarn consumption in the United States. The arrows indicate the estimated direction on the curves in the next few years (12).

Table 1.   Cost per Tenacity

Type of yarn	Yarn cost, $/kg	Twist and dip cost[a], ¢/kg	Tenacity of cord N/tex[b]	Dipped cost to tenacity ratio, $/(N/tex)[b]
rayon	3.26	26.4	0.36	9.77
nylon-6,6	3.72	33	0.65	6.23
polyester	2.97	44	0.59	5.79
glass (dipped)	2.97		0.56	5.3
steel wire				
passenger tires	2.40		0.30	7.99
truck tires	2.64–4.09		0.30	8.80–13.60
aramid	13.31	44	1.50	9.17

[a] Estd.

[b] To convert N/tex to gf/den, multiply by 11.33.

The cost of the fiber reinforcement in a tire and its performance in a particular design are the principal criteria of choice. Other marketing considerations are availability and the public's perception of a tough and strong material. Since cost per unit tenacity is important, glass is utilized wherever possible. Although the cost–tenacity ratio of low-twist nylon is below that of polyester, development costs have so far prevented commercialization of the former in heavy-duty radial tires, where it has the best chance to succeed. Steel wire offers many advantages in radial tires which explains its growing usage despite its very high cost–tenacity ratio. The slow growth of aramid is due to its high cost and poor strength utilization at the higher twists needed for adequate fatigue resistance.

Processing

Most tire companies produce tire cord from high tenacity, continuous-filament yarn supplied by various fiber manufacturers.

Twisting. Typically, nylon yarn is received on small beams (rolls) with ca 200 ends of 186.7 tex (1680 den) yarn. These yarns are twisted to 394 turns per meter (tpm) in the Z direction (spiraling counterclockwise). Then, two spools of twisted yarn are back-twisted together to 354 tpm in the S direction (spiraling clockwise) to make 186.7/2 (1680/2) nylon tire cord (the number in the denominator indicates the number of plies) (see Fig. 3). If 186.7/3 cord is desired, 3 spools of Z-twisted yarn are back-twisted together in the S direction.

Weaving. Twisting is followed by loosely weaving by 1000–1500 cords (warp) with $\frac{1}{3}$–1 pick (or weft) per centimeter of light cotton, rayon, polyester–cotton, or nylon yarn threads. A type of fabric is obtained in which the cords in one direction, the warp, are far stronger than the threads in the other direction, the weft. In fact, the weft only

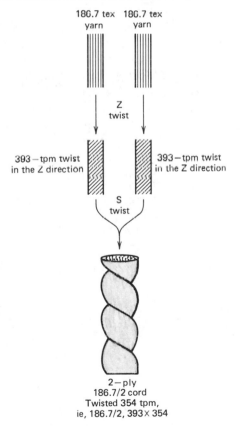

Figure 3. A schematic example of how tire yarn is twisted into tire cord (tpm = turns per meter; 186.7 tex = 1680 den).

serves to give the construction the minimum coherence necessary for the subsequent operations. Rolls of this fabric are then dipped in adhesive, baked, and heat set.

For radial-tire production a stretch pick (Olboplast 25) has been commercialized by Olbo Textilewerke GmbH in the FRG. This material consists of partially oriented, nylon-filament yarn twisted together with cotton fibers for stability during processing. When the so-called green tire plies are lifted into the curing press, enough extension is generated to break the cotton portion of these picks first, and then the pick plus the rubber plies continue smoothly to extend into the mold. The cords separate uniformly and the spread-cord fault is avoided.

Dipping, Heat Setting, and Calendering. A simplified processing train for tire cord is shown in Figure 4. Flexible controls of heat, tension, and speed are required in each processing zone. The fabric, taken off from a roll of 1825 m length, is spliced onto the tail received from the previous roll either with adhesive or with a high speed sewing machine. Continuous operation is maintained during splicing by the fabric stored in the festoon. Then, under controlled tension or controlled length change, the fabric is dipped in an aqueous adhesive RFL dip. After dipping, the fabric passes vacuum suction lines or rotating beater bars to remove the excess dip before entering a drying oven where the water is removed. The dried fabric is baked and heat-set under tension at 175°C for rayon, 230°C for nylon, 245°C for polyester, and 260°C for aramid. After stretching and heat setting, the viscoelastic fabrics (nylon or polyester) are usually relaxed at 175°C under a tension lower than the heat-setting tension. The relaxed fabric is covered on both sides with a precise amount of rubber by a four-roll calender. Tension can also be controlled on viscoelastic fabrics as they are cooled after calendering and before reaching the take-up festoon. At the take-up, a removable liner is inserted to prevent the rubberized fabric from sticking to itself. Usually the dipped, heat-set, and baked fabric is not calendered immediately, but taken up into large rolls, wrapped to protect it from light, ozone, and moisture, and shipped to tire plants for calendering. Before actually calendering the precise amount of rubber onto the fabric, it is dried by reheating to provide better final adhesion.

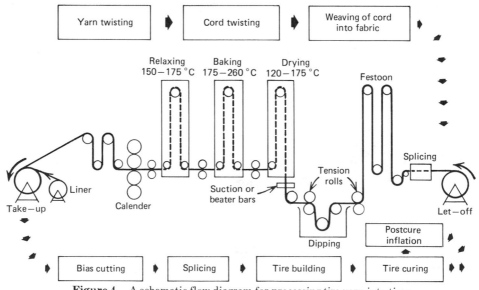

Figure 4. A schematic flow diagram for processing tire yarn into tires.

Typical formulations of aqueous adhesive dips for the organic tire cords are shown in Table 2. For rayon, vinal, and nylon, similar RFL dips are employed in a single-step treatment, whereas polyester and aramid are usually treated with an epoxy-blocked isocyanate dip followed by an RFL dip. Adhesion-activated polyesters reduce the need for the first-step treatment (15), but the strength of adhesion is generally lower than that obtained from the two-step treatment. Several other one-step adhesive systems develop adequate but usually weaker adhesion (16–17). For aramids, a simple first-step dip of epoxy has been recommended by DuPont (18); it avoids the high baking temperatures required to unblock the isocyanates. The serious problem of adhesion degradation caused by exposure of the RFL coating to ozone and humidity has been greatly reduced by the partial substitution of acrylic resins for the RF resin in the basic dip (19). Waxes that tend to bleed to the surface and protect the RFL coating have also been proposed as an RFL modification (20); they are usually contained in glass cords. Further discussion of adhesive dips for tire cords is given in refs. 13, 15, 21–22.

Cutting and Splicing; Tire Building. The calendered fabric is cut to the required angle and width, and the pieces are butt-spliced together automatically into a continuous sheet.

The sheet of parallel cords arranged at the proper angle is cut or torn to the proper length and built into the so-called green tire on a cylindrical drum; the other tire components are included here, eg, beads, chafer, sidewalls, belt plies, cushion gums, undertread, and tread. The green bias or belted-bias tires are then ready for cure or vulcanization.

Table 2. Typical Tire-Cord Adhesive Dips

Dip and composition	Dry, parts	Wet, parts
D-417[a]		
Hylene MP[b] (40% dispersion)	3.60	9.00
NER-010A[c]	1.36	1.36
gum tragacanth (2%)	0.04	2.00
water, soft		87.64
Total	*5.00*	*100.00*
D-5 ammoniated[d]		
resorcinol	11	11
formaldehyde (37%)	6	16.2
sodium hydroxide	0.3	0.3
water, soft		238.5
Total	*17.3*	*266.0*
Final nylon dip		
Gen-Tac (41%)	100	244
resin master	17.3	266
water, soft		60
ammonium hydroxide (28%)		11.3
Total	*117.3*	*581.3*

[a] DuPont. For first-step polyester or aramid (13). Total solids = 5.0%; pH min = 10.3.

[b] Hylene MP is phenol blocked methylenebis(4-phenyl isocyanate) (DuPont).

[c] NER-010A is glycerol polyglycidyl ether, water-soluble polyepoxide (Nagase and Co., Ltd.).

[d] General Tire & Rubber Co. For nylon, rayon, and second step polyester or aramid (14). Total solids = 5.4%; pH min = 10.

Radial tires usually require a second step after the radial carcass has been built onto the cylindrical drum. The carcass is lifted into a toroidal shape before applying the belt plies, cushion gums, undertread, and tread to the green tire. During this process, the carcass cords must separate evenly in order to avoid a blowout and form a uniform tire. This spread-cord fault is minimized in a carcass rubber with high green strength and when stretch picks are used in the fabric.

Curing and Postcure Inflation. The green tires are cured in automatic press molds at 165–180°C and under 1.48–2.86 MPa (200–400 psig). After a minimal curing time of ca 12 min for passenger tires, the tire is automatically ejected and moved to the postcure inflators. Upon release from the press molds, the tire cords are almost completely free to shrink. The tire is reinflated to 0.31–0.51 MPa (30–60 psig), which reloads and stretches the cords. At the same time, cure of the compounded rubber in the tire continues. Control of the postcure inflation process is very important for viscoelastic cords like nylon or polyester because the growth, groove-crack resistance, size, shape, and uniformity of the tire depend on the time, temperature, and load on the cords during postcure inflation. In bias and belted-bias tires, postcure inflation is necessary for polyester tires in order to reduce initial heat generation and increase endurance life. Radial tires, on the other hand, do not need postcure inflation since the stiff belt prevents excessive distortion which may be caused by shrinkage of the viscoelastic cords in the carcass. In fact, the quality of the radial tire sometimes suffers slightly from postcure inflation since it tends to emphasize undulations in the sidewall at splices and uneven cord spacings.

Storage and Shipping. Proper storage and shipping of tire cord is vital to the quality of the fabric and therefore the quality of the tire. The fabric rolls must be packaged to withstand both mechanical and environmental damage. Polyethylene bags, kraft paper, and heavy cardboard are generally employed for packaging. Silica gel reduces the moisture content in the packages and prevents shrinkage of exposed edges of the fabric (so-called tight edges). Tight edges in a roll of greige, dipped, or calendered fabric result in handling difficulties, scrap, and nonuniform tires. For overseas shipment, the fabric rolls are boxed.

Glass and Steel Tire Cords. The processing of glass and steel tire cords is similar to the processing of polymeric tire cords once they are prepared for calendering. However, preparation for adhesion to the rubber compound is different.

Large spools of glass tire cords completely impregnated with a modified RFL are obtained from the supplier. The RFL solids content is >15 wt % in order to thoroughly coat each glass filament. The cord constructions for the belts of bias-belted and radial tires are shown in Table 3.

Steel tire cord is supplied to the tire industry in various constructions. The cords are made from finely drawn high carbon (0.67 wt %) steel and brass plated just before the last draw. The brass coating acts as both a drawing lubricant and an adhesive to the rubber. These wires are twisted together in a closing operation by either cabling or bunching and thus form the appropriate lay length (twist in millimeters per turn) of the strands or cords. Lay lengths of the strands and cords are chosen to hold the wires together and yet obtain the smallest degree of fretting during tire operation. A spiral wrap of a single wire (transfil) is frequently applied to the steel cords in order to increase the cord's compression resistance and bending rigidity. However, this wrap decreases the elasticity of the cord.

For radial passenger tire belts, the 4×0.28 cord has evolved into the preferred

Table 3. Glass[a] Cord Constructions Used in Belted-Bias and Radial Tires

Year introduced	Tire design	Cord construction[b]		Impregnant[c]
1966	belted bias	G 150	10/0	063T
1968	belted bias	G 75	5/0	065T
1970	belted bias	G 15	1/0	075T
1970 (unsuccessful)	radial	G 15	3/0	075T
1976	belted bias	H 15	1/0	086T
1976	radial	H 15	3/0	086T

[a] Type E glass, a high strength glass with a low alkali content and a high alumina content.

[b] G and H designate filament diameters, where G = 8.89 μm dia with 204 filaments per dia, and H = 10.12 μm dia with 158 filaments per dia. The twist of all the above strands is 59 twists per meter. The remaining figures indicate cord length per weight of strand and the number of strands per ply. For example, the first entry describes a cord with G filament dia (8.89 μm dia with 204 filaments), 150 yds/lb (302.4 m/kg) of strand, and 10 strands per ply (10/0).

[c] All impregnants are modifications of RFL with code numbers of Owens-Corning Fiberglass Corporation. Code 086T has a latex with a significantly lower T_g. This improved its low temperature properties so that radial glass belts are operable in cold weather.

construction from the 5×0.25 construction. As shown in Table 4, the 5×0.25 cords tend to form into a pentagonal cross section with the five wires touching each other and sealing off a large hole in the center. The rubber is thus prevented from penetrating into the central hole, but moisture can collect there during cure and reduce the initial adhesion. Furthermore, upon exposure of the tire to high humidity, this hole provides easy access for moisture, which then degrades the adhesive bond. When the tire is accidentally cut, corrosion from salts and water spreads from this hole. The 4×0.28 construction costs less and tends to collapse into a close-packed parallelogram which permits more rubber to flow into the interstices. This flow, in turn, permits more complete surface contact for adhesion and blocks the migration of moisture from adversely affecting adhesion. A more expensive construction that allows better penetration of the rubber is the 2+7×0.22+1×0.15 construction popularized by Firestone. Radial truck tires use larger constructions such as 3×0.20+6×0.35 or 3+9×0.28+1×0.15 in the belt and 3+9+15×0.175+1×0.15 in the carcass.

Steel-wire tire cords have been coated with a thin layer of brass to promote adhesion to the rubber compounds for many years. Yet, experts still speculate as to the exact mechanism of adhesion and adhesion degradation in the presence of moisture.

Such speculation should explain the following phenomena:

1. Maximum adhesion is obtained when free sulfur is present in the initial rubber compound.

2. The maximum initial adhesion is obtained when the rate of cure of the rubber compound is optimized to the rate of adhesion development.

3. Heat aging degrades adhesion.

4. Moisture tends to lower initial adhesion and increase the rate of adhesion degradation.

5. Higher zinc to copper ratios in the brass tend to retard the rate of adhesion development and the rate of adhesion degradation owing to heat aging or moisture.

6. Thin brass coatings tend to resist adhesion degradation better than thick coatings.

Table 4. Nomenclature of Popular Wire Construction[a,b]

construction	5 × 0.25	4 × 0.28	2 + 7 × 0.22 + 1 × 0.15	3 × 0.20 + 6 × 0.35	3 + 9 + 15 × 0.175 + 1 × 0.15	3 × 7 × 0.22 (high elongation, H.E.)
cross section						
lay length, mm/turn	10S	12.5S	6.3S/12.5S/3.5Z	10S/18Z	5S/10S/16Z/3.5S	4S/7S
breaking load, N	620	620	880	1,620	1,730	1,710
tensile modulus, MPa[c]	196,200	198,200	203,000	185,500	193,500	93,000
compression modulus, MPa[c]	47,100	78,100	99,000	93,000	68,000	7,800
principal use	passenger belt	passenger belt	passenger belt	truck belt	truck carcass	off-the-road (OTR) belt

[a] Ref. 23.

[b] ASTM has adopted a wire nomenclature system. The objective of this system is to describe the wire cord construction in the same manner as it is manufactured. The general format is as follows:

$$\text{core } (S \times F \times D) + \text{intermediate part } (S \times F \times D) \ldots + \text{outermost part } (S \times F \times D) + \text{spiral wrap } (S \times F \times D)$$

Where S = number of strands (when S = 1, it is omitted); F = number of filaments; and D = nominal diameter of filaments to the nearest 0.01 mm. The general rules are to work from the center of the cord to the outside. Various parts of layers of the cord are separated by a "+" sign. If the filament diameter is the same for two or more parts (with the exception of the wrap), omit the diameter except for the last part before the change. If the filament diameter of the wrap is different than the filament in the cord, the diameter is given twice.

[c] To convert MPa to psi, multiply by 145.

508

7. Cobalt ions in the rubber compound tend to retain the maximum obtainable level of adhesion for the system more consistantly than without it. Cobalt also improves adhesion to zinc.

8. Copper tends to migrate into the rubber, especially in the presence of moisture.

Two separate articles have been written that try to explain the above adhesion phenomena in light of new data obtained by such modern analytical techniques as esca (24–25).

The principal problem of adhesion degradation, especially in the presence of moisture, has been overcome through improvements in rubber compounds; by avoidance of exposure to moisture of the uncured tire in the tire plants; by development of wire construction that increase rubber penetration; and by development of modified brass coatings, eg, an alloy of 70 wt % Cu–4 wt % Co–26 wt % Zn (25).

Tire Performance Related to Tire Cords

Tire performance characteristics are affected by the physical properties of the tire cords. Although a few characteristics such as burst strength, tire size, and shape are amenable to mathematical analysis based on the physical properties of tire cords (26–30), most relationships are determined empirically and correlated statistically. Tire performance is difficult to define. For example, a tire must safely support a specified load under dynamic conditions with a minimum of power loss, overcome minor obstacles and provide a safe endurance life even beyond its normal life expectancy. And a tire must provide maximum wear, a smooth and quiet ride with good cornering, traction, and skid resistance under various road conditions. A tire must also have a pleasing appearance to complement the vehicle.

Tire performance is therefore measured in a variety of ways and under a variety of conditions in the laboratory, on indoor roadwheels, on outdoor tracks, and on special test fleets. Before commercialization in the United States, a tire has to pass certain qualification tests set up by the tire companies, the Rubber Manufacturers Association (RMA), the DOT, and the original equipment manufacturers, eg, the automobile companies. Competition among tire manufacturers has continually raised the standards of tire performance.

The tire performance most difficult to measure is the endurance life. In order to measure it quantitatively and empirically relate it to the physical properties of the cord, the test tires have to be run to failure in a reasonable length of time. Testing conditions are much more severe than those any tire would normally undergo. Therefore, the accelerated endurance life of a tire usually is measured in combinations of overloads up to 150% of the recommended loads, inflation pressures as low as 75% of the recommended pressure, and speeds >161 km/h. Tests are usually taken on indoor roadwheels where conditions can be most readily controlled. However, prolonged outdoor-track tests are run before an improvement of tire performance is accepted.

Burst Strength. The tensile strength is related to the burst strength by the following relationship (31–32):

$$\text{burst strength} = Nt_uK = \frac{\pi P_B(r_c^2 - r_{max}^2)}{\sin \alpha}$$

where N = total number of cords in a tire; t_u = average ultimate tensile strength of the cords tested at the same rate of elongation as the tire; P_B = burst pressure; r_c = radius from the center of rotation to the crown of tire; r_{max} = radius from the center of rotation to the maximum section width of a tire; α = crown angle between the cord path and the circumferential plane through the crown of a tire; and K = efficiency factor which depends on the distribution of ultimate cord strengths, always <1 in an assembly of cords.

The actual burst strength is always somewhat lower than the calculated strength because the tensile strength and stress distribution among all the cords are never exactly the same. The safety factor used in designing most tires is >10. Thus, a tire normally inflated cold to 274 kPa (25 psig) resists more than 1.82 MPa (250 psig) before burst. When extreme operating temperatures reach 135°C, most organic fibers temporarily lose almost half of their tensile strength, still leaving a safety factor of 5 which is more than adequate to allow for minor weaknesses and normal degradation from oxidation, fatigue, etc.

Bruise Resistance.   Years ago, a tire would blow out when the car was driven over a 15-cm curb at 25–30 km/h. Today, the steel rim may bend, but the tire remains intact even at speeds up to 100 km/h. Thus, bruise resistance has been increased by improved tire cords and tire design and is maintained by quality-control tests.

Bruise resistance is usually tested by measuring the energy required to penetrate a 19-mm cylindrical plunger at a rate of 51 mm/min crosshead speed into the crown of an inflated passenger tire at room temperature. The area under the load–elongation curve is interpreted as the slow penetration bruise or bruise energy of a tire. Since no rigorous mathematical analysis of this test has as yet been made, the correlations with the tensile strength or the tensile product (tensile strength × ultimate percent elongation) of the cord are not exact.

Bruise resistance should be measured at different operating conditions for different materials (33). For example, rupture energy tests at high speeds and elevated temperatures show that rayon resembles nylon at ca 135°C and 6000% elongation per second (34–35). Apparently, rayon dries out, which strengthens it and reduces the normal strength loss expected from increased temperatures. Furthermore, the strengthening effect of high speed straining generally is greater with the high modulus materials. Both of these effects favor rayon over nylon.

A number of high speed bruise tests have been developed which correlate better with field experience than the low speed test (36–38). The DOT has recognized the difference in nylon and rayon tires and specifies that rayon tires with a low speed bruise energy of 186.4 J (137.5 ft·lbf) are equivalent to nylon tires with a low speed bruise energy of 293.8 J (217 ft·lbf). Although high speed tests are more realistic, low speed tests are simpler and faster and are used for the quality control of tires.

Tire Endurance (Separations).   Several different types of separations related to tire cords can occur. Most separations are due to poor adhesion between the cord and the carcass rubber. Initial adhesion between cord and rubber are important, as well as uniformity and adhesion retention. Adhesion uniformity is more important than a high average adhesion since unevenness causes stress concentrations which tend to promote separations.

Degradation of the adhesive system tends to occur because the system is subject to elevated temperatures, moisture, oxygen, and stress–strain cycling. High temperatures weaken all the materials in the system and increase their rate of degradation from oxygen or moisture. Oxygen tends mainly to degrade the rubber portion of the system. Moisture tends to hydrolyze the polyester cords, degrade the brass–rubber bonds of steel wire cords, or swell rayon cords which then tends to increase the stress at the cord–rubber interface and reduce the hydrogen bonds between the adhesive and the cord's surface.

Another separation phenomenon is fatigue failure. At speeds >145 km/h or under loads of ≥20% over those recommended, or inflation pressures of ≥34.5 kPa (5 psi) below those recommended, fatigue failure can occur in the sidewall near the tire-tread shoulder or near the bead where the cord plies end. Both are regions of relatively high stress concentrations and heat generation. When a tire is operated under severe conditions for ≥80 km, it can reach temperatures over 121°C. The combination of heat, oxygen (mainly from the internal air inflation), and stress–strain cycling causes the cord, adhesive, and especially the rubber matrix to degrade and weaken. Continued operation of the tire under such conditions initiates small internal flaws, which then grow in the weakened material and ultimately lead to cord breakage and loss of inflation pressure. Fatigue separations result from interrelated factors that depend on tire-cord construction and design, adhesion, temperature, rubber compound, oxygen concentration, stress concentrations, and severity of the tire's operating conditions (39–41).

High Speed Endurance. During operation, a tire continually stores and releases mechanical energy. Any loss of useful energy develops heat. At very high test speeds (up to 193 km/h), a tire sometimes generates enough heat to cause a tread separation or chunk-out, often starting at a small internal flaw in the hottest and thickest part of the tire, usually the tread shoulder. The heat generated by the tire cords in a bias tire is sufficient to significantly affect the maximum operating speed of the tire. Thus, at high speeds nylon bias tires usually run cooler than rayon bias tires before tread separations occur, whereas rayon radial tires run cooler than rayon bias tires of the same size. The radial tire usually tolerates slightly higher loads than bias tires of the same size but not higher speeds.

Separations are also caused by stress. At high speeds, a tire can go into resonance called a standing wave which violently distorts the tire, causing large stresses and high temperatures to be generated in the regions that distort the most. For passenger bias tires, the onset of resonance generally begins in the crown at speeds >225 km/h. With special designs, onset speed can be increased to >400 km/h. The radial tire is more sensitive to speed. The onset of resonance for a passenger radial tire generally begins in the sidewall at speeds >113 km/h; however, the distortions in the sidewall are mild at first and develop slowly into severe distortions at speeds >137 km/h. Again the onset speed can be raised to much higher values by changing the tire design. For instance, by increasing the resonant frequency of the sidewall (shorter and stiffer sidewall) or reducing the centrifugal force (lower tread or belt weight), the resonant-onset speed can be significantly increased.

Power Loss. In a bias tire, 30–40% of the total power loss, a significant portion of the heat generated, is because of the carcass cords (42). Measurable temperature differences have been found among tires made with different kinds of cords (43–45).

Since power loss is synonymous with heat-generation rate, similar tires made from different cords should show equilibrium operating temperatures corresponding to the heat-generation rates of the cords. However, it is almost impossible to make truly equivalent tires from different cords. A good comparative test demonstrated that under severe conditions, nylon-6-bias truck tires operate 10–25°C cooler than those made of nylon-6,6. These differences apparently correspond mainly to the relative rates of heat generation of the cords.

Tread Wear.   Tread wear partially depends upon load distribution and movement (scuffing) of the tread elements in the tire print. Since the cord modulus contributes to the stiffness and load distribution of the tire print, a higher cord modulus implies less tread wear and longer tread life. Belts improve tread-wear resistance. Ranking the relative wear-rating properties based on the bias tire as 100, the steel-belted radial is ca 200, and the textile-belted radial ca 160 (30,46). Glass-belted radials should rank between the steel and textile belts.

Tire Size and Shape.   The size and shape of an inflated tire depends upon the design of the tire and the modulus of the cord. In addition, some tire growth occurs while the tire is inflated over a long period of time, depending on the creep properties of the cord as affected by the inflation load, tire temperature, and moisture regain (plasticization). Control of moisture regain is especially important for moisture-sensitive rayon and nylon. Because cord processing affects modulus and creep, careful process control must be maintained in order to obtain uniform size and shape.

Tire-Groove Cracking.   When the tread grooves of a tire are subjected to excessive strain, small cracks appear in the rubber at the bottom of the grooves, possibly from ozone attack. Cords with less creep produce less groove cracking. Postcure inflation of nylon tires reduces tire growth and the tendency of the grooves to crack.

Flatspotting and Tire Nonuniformity.   Thermoelastic tire cords such as nylon and polyester tend to shrink readily when they are heated above their T_g. A running tire generates heat, which increases the temperature above the T_g of its cord (usual maximum operating temperature of a passenger tire is 77°C). When the tire is stopped, the cords in the tire-print region are essentially unloaded and free to shrink. The cords in the remainder of the tire are prevented from shrinking by the inflation pressure. When the tire cools below the cord T_g, the cord lengths become frozen, and a flatspot or out-of-roundness remains where the cords shrank. It remains until the tire is reheated to the temperature at which the flatspot was originally introduced. Above that temperature, cord lengths are again equal (47–48).

In a similar manner, a nonuniformity develops in a tire that is postcure inflated hot, and cooled nonuniformly, followed by release of postcure inflation pressure above the T_g of the cord. The hotter cords shrink more than the cooler cords, and the tire is distorted and nonuniform (47). In this simplified analysis, the effect of long-term creep is not considered.

Tire-Cornering Force.   Generally, the higher the cord modulus, the stronger the cornering force developed at any particular steering angle. However, tire design changes, eg, in the cord-crown angle, can often overshadow the modulus effect.

Spring Rate.   Spring rate is a measure of the tire deflection under load. The cord property that most affects spring rate is the modulus.

Noise.   Several types of noise are generated by a running tire. The most irritating noise is a high-pitched whine produced by the interaction of certain tread patterns and tire speeds. A low-pitched rumble or boom generated by irregularities in the road can also be heard. Tire squeal, which develops while accelerating or decelerating the tires, is another type of noise. Different noise levels have been found for tires made from different types of cord. However, the role that tire cord plays in developing tire noise and in transmitting this noise is not well understood and needs further investigation (see also Noise pollution).

Test Methods

Tire cord is tested for quality control and to predict performance. Some tests determine the physical properties, whereas others are intended to simulate processing or use. Properties tested include ultimate tensile strength, ultimate elongation, total work to break, and cord modulus in tension. They are obtained together by recording the load–elongation curve from a tensile-testing machine. The environment is standardized for control purposes but changed for evaluation and design to simulate use conditions. Thus the following testing conditions are varied: temperature, humidity, test speed, and type of loading rate (either constant elongation or constant load).

Yarn-to-Cord Conversion Efficiency. This property is usually measured by relating the ultimate tensile strength of the untwisted yarn to the ultimate tensile strength of the cord (49). Higher cord twist and/or larger diameter of the individual yarns result in lower cord efficiency. This cording loss or loss in efficiency has several causes but is mainly due to the reorientation of the filaments from the axially loaded tensile direction to the twisted angle of the yarn or cord. Higher twist corresponds to higher twist angle (or disorientation) and lower tensile strength of the yarn or cord. Any unevenness in the alignment of these filaments or damage that might have occurred during the twisting operations lowers tensile strength and efficiency. Another cause of tensile-efficiency loss is the inability of the filaments to adjust to the load. Therefore, a lubricating finish increases tensile efficiencies. Because the high modulus yarns are more difficult to adjust to the twisted structure, they tend to have lower twist efficiencies. In addition, it is more difficult to uniformly align the filaments in large twisted structures than those in small ones. Thus, large cords are less efficient.

Length Stability to Moisture. The length of stability of a cord in the presence of moisture is measured by the amount of shrinkage, usually as percent free shrinkage (under minimal load) in hot water. Although swelling and shrinkage in the presence of moisture can aid in the processing when controlled, length stability to moisture promotes and maintains uniformity. Both rayon and nylon are readily plasticized by moisture which tends to release any built-in shrinkage forces. Polyester is not plasticized by moisture and remains stable.

Length Stability to Heat. During cord processing and tire operation, temperatures above the T_g are frequently reached. It is difficult to control length in thermoelastic fibers like nylon and polyester which tend to shrink considerably when heated above their T_g. The amount of shrinkage depends upon the force built into the fibers during their thermal and load history (50). The free shrinkage of a cord is measured by heating the cord under a minimal load and noting its length change. A better evaluation can be obtained by closely simulating cord-processing and tire-operating conditions (47).

Length Stability to Load. Simple creep tests under various loads and temperatures can be related to tire growth, groove cracking, and possible nonuniformity. A low creep rate is desirable.

Second-Order Glass Transition Temperature (T_g). The glass temperature of many polymeric fibers are usually close to operating and processing temperatures. Hence, the cords change from the glassy to the rubbery state as they are heated through their T_g. This affects properties such as dynamic-storage and loss modulus, shrinkage, and even chemical reactivity. There are several methods to measure T_g, and the actual value depends on the measurement. The torsion-pendulum method is useful, but the best method is to test at various temperatures physical changes that are of direct interest to the user. Comparative T_g values have been correlated with tire uniformity and flatspotting (47–48). Table 5 gives the T_g of various tire cords measured by simulating the loads and temperatures that a cord is exposed to during postcure inflation of a tire. Thus, a processed cord is heated to 166°C and loaded to 10 mN/tex (0.9 gf/den). By cooling under load and then releasing this load at various temperatures, the length of the unloaded cord can be measured for each release temperature. A plot of the percent elongation vs release temperature in degrees Celsius develops a curve that can be approximated by straight lines. The slope of these lines is the percent elongation per degree Celsius, or coefficient of retraction (CR). The intersection of these straight lines represents the T_g of the polymer. Additional transitions, T_c, were noted in the cords made from two different polyamides.

Dynamic Properties. Tire cords operate under dynamic conditions. The mode of cycling, construction, internal structure, and temperature affect the rate of heat generation. Although the exact modes of cycling are not known, reasonable stress and strain values can be applied to simulate the cycling modes. Thus, dynamic properties can be measured that indicate how much heat one cord may generate relative to another cord under the same conditions. From the surface temperatures of rotating tires and the material properties and standard heat-transfer equations, the internal tire temperatures and heat generation of the various parts of the tire have been calculated (43,51–52). In bias tires, the heat generation of the lower sidewall carcass correlates with the dynamic properties of the cords in the constant stress-cycling mode. The heat

Table 5. Coefficient of Retraction (CR)[a] and T_g of Tire-Cord Material

Tire-cord material, 186.7[b]/2	CR per °C, % length change		Transition temperatures	
	Value	Range, °C	T_g, °C	T_c, °C[c]
nylon-6	0.029	27–53	54	
	0.050	55–165		
nylon-6,6	0.022	27–47	48	
	0.041	49–165		
nylon-6,6 plus aromatic polymer, 70/30	0.022	27–80	81	120
	0.085	82–119		
	0.033	121–165		
nylon-6,6 plus aromatic copolymer, 70/30	0.022	27–48	49	140
	0.043	49–139		
	0.017	141–165		
nylon-6 plus PET[d], 70/30	0.022	27–59	60	
	0.054	61–165		
nylon-6,6 plus PET[d], 70/30	0.018	27–58	59	
	0.043	60–165		
polyester (PET[d])	0.011	27–84	85	
	0.050	86–165		

[a] After release from a load of 10 mN/tex (0.9 gf/den).
[b] To convert tex to den, divide by 0.1111.
[c] Additional transition.
[d] PET = poly(ethylene terephthalate).

generation of the carcass in the crown and shoulder region of bias tires correlates best with the dynamic properties of the cords in the constant strain-cycling mode. Radial tires perform less work and generate much less heat (44).

Fatigue. During the lifetime of a tire, the cords undergo innumerable stress-and-strain cycles which sometimes lead to degradation to the point of failure. Many tests have been devised in attempts to simulate fatigue failure, but none have been completely successful because of the complexity of true simulation. Fatigue tests are most meaningful when performed under controlled operating conditions. However, a number of laboratory tests have been developed, classified as bare-cord tests and in-rubber tests (53). The bare-cord tests cycle cords between various tensions in controlled atmospheres and temperatures and measure oxidative and mechanical damage (interfilamentary abrasion). They are sensitive to the yarn lubricant, filament placement, cord twist, uniformity, and oxidative resistance at elevated temperatures (54–55).

The in-rubber fatigue tests more closely simulate the fatigue mechanism under use conditions. Such in-depth studies show that the failure mechanism begins with a localized separation of the adhesive system, usually in the degraded rubber next to the cord. This leads to more violent distortions of the cord and its subsequent failure (39–40).

Thermal and Chemical Degradation. Most materials lose strength at elevated temperatures, and some textile materials are degraded by moisture, oxygen, or rubber chemicals. The loss in tensile strength is measured after aging treatments in rubber compounds as well as in other controlled environments. These values are compared with those obtained from tire cords. Minimal strength loss is sought, but compromises are usually made with other properties such as adhesion.

Cord-to-Rubber Adhesion. Adhesion tests are either peel tests or pull-out tests. In the peel test, one layer of cord fabric is peeled from another after they have been cured in rubber. The force required to separate a strip 2.5-cm wide that is pulled at 180° is the peel-adhesion value. It depends not only on the adhesive strength of the system but also on the force required to bend each layer during testing. Since the bending stiffness of each layer is difficult to control or measure separately, more reliance is placed on a subjective appearance rating than the peel force. Pull-out tests measure the force required to pull a cord out of a rubber block. The size of the cord and its embedment length must be standardized for purposes of comparison. Pull-out tests can be very precise and independent of the rubber modulus (56–57). A specially molded pull-out test that measures the energy of adhesion at the surface of the cord has been proposed to ASTM (58). Both the pull-out force and the modulus of the rubber are used to calculate the energy of adhesion. In all adhesive testing, it is important to know where the failure occurs as well as measure the adhesion strength. Adhesion testing should be standardized near the highest normal operating temperature of the tire (100–150°C). With increasing test temperature, most rubber compounds lose strength much more rapidly than either the adhesive or the tire cord. Therefore, at some elevated temperature the rubber compound becomes the weakest link, and the rubber will tear during the test. The limit of good adhesion values is usually obtained at a low temperature when the test breaks cord filaments. Adhesion is improved by strengthening the weakest link in the system.

Compaction and Dip Penetration.   When the adhesive dip penetrates into the filament bundle, it ties the outside filaments together and blocks some of the void space between the filaments. Uniform dip penetration maintains the integrity of each individual yarn bundle making up the cord during flexing. Compaction promotes uniformity and reduces the void spaces, which provide both a reservoir and a conducting tube for air and moisture. The oxygen in the air tends to degrade the cord and surrounding materials especially when hot, whereas the moisture can cause separation problems during cure. In steel cords it degrades the adhesion and increases corrosion.

Compaction and dip penetration are measured by examination of cord cross sections under the microscope. In another method, the time is measured for a standard amount of air to flow through a standard length of cord embedded in a rubber matrix while under a standard air pressure differential (41,59).

Flatspotting and Evaluation of Uniformity.   The flatspot test for tire cord is based on a simulation procedure of what a tire cord "sees" during flatspot formation and elimination (47–48,60–61). A simulation test can be used to predict both the relative tendency for a tire cord to flatspot as well as its tendency to introduce nonuniformity in tires during postcure inflation. The coefficients of retraction or relative length change per degree Celsius to be expected from various cord materials above and below its T_g are shown in Table 5. The higher the CR, the greater is the tendency for nonuniformity and flatspotting. Flatspotting is no problem for radial tires because radials have stiff nonshrinkable belts.

Bending Stiffness.   During tire building the calendered fabric plies are folded around the bead and these ply turnups are held in place by the tackiness of the rubber. Stiff cord is difficult to fold, and it is even more difficult to keep the plies from separating while the green tire is awaiting cure. The stiffness of the processed cord can be measured by the angle of recovery after folding under a standard load.

Tire-Cord Status

Different types of tires require tire cords with different properties. Such tire designs as the radial tire or the belted-bias tire take advantage of these different requirements by placing different types of cords in the belt and in the carcass. The carcass of a bias or belted-bias tire requires a high strength, tough, flexible cord like rayon, nylon, or polyester. For radial tires, high modulus cords like steel or aramid can be used. Successful experiments have been conducted with radial glass carcasses (62). The belt requires a high strength high modulus cord like steel, glass, or high modulus rayon. Airplane tires require lightweight, high strength, tough, flexible cords with low heat generation; nylon seems to fill these requirements best. As airplanes increase in size and speed, new cords will be needed to withstand the high loads and high temperatures. Truck tires, off-the-road tires, high speed tires, and passenger tires all require different cord properties for optimum safe and economical performance. Chemical and fiber producers have increased the number of possible tire cords such as aramids. In turn, new and improved methods of evaluation have been developed.

Nylon and polyester cords are very strong, but rayon is superior to nylon at high speed impact. Polyester resists high speed impact, but loses modulus and strength faster than rayon at 150°C. Nylon and polyester retain tensile strength better than rayon when highly twisted. High modulus nonadjusted fibers, eg, aramid, glass, and steel, lose even more tensile strength than rayon when highly twisted. Polyester is dimensionally stable to moisture, and rayon is dimensionally stable to heat. Polyester creeps the least under load. At 93°C, rayon has a high storage modulus and a low loss tangent; ie, less heat is generated while cycling under constant stress limits. Aramid is even better under stress cycling. Under constant strain limits, nylon generates the least heat. The fatigue life of rayon is good, that of nylon is better, and that of polyester varies from longer than rayon below 93°C to shorter than rayon above 93°C. Under high flex conditions, glass and steel tend to fatigue readily. Aramid needs a relatively high twist for flex-fatigue resistance which lowers its tensile strength. Rayon retains its strength best at 150°C, whereas polyester is least affected by dry air at 150°C. Both nylon and rayon resist hydrolysis and aminolysis at 150°C, whereas polyester is much less resistant. In general, it is easier to process rayon than nylon, polyester, or aramid, except with respect to dimensional stability to moisture, where both rayon and nylon show some shrinkage but polyester and aramid are unaffected. Each tire cord, whether old or new, represents a different combination of properties which have to be optimized. Since the changing demands on tires emphasize different properties, the search for the best tire cord is never-ending.

BIBLIOGRAPHY

"Tire Cords" in *ECT* 2nd ed., Vol. 20, pp. 328–346, by Leonard Skolnik, The B. F. Goodrich Company.

1. U.S. Pat. 2,128,635 (Aug. 30, 1938), W. H. Charch and D. G. Maney (to DuPont).
2. U.S. Pat. 2,561,215 (July 17, 1951), C. J. Mighton (to DuPont).
3. U.S. Pat. 3,282,039 (Nov. 1, 1966), C. J. Geyer and J. B. Curley (to FMC).
4. J. B. Curley, *Rubber Age* **98,** 124 (Apr. 1966).
5. *Chem. Eng. News* **44,** 59 (Apr. 25, 1966).
6. Brit. Pat. 918,637 (Feb. 13, 1963), (to DuPont).
7. G. W. Rye, R. S. Bhakuni, and D. M. Callahan, *Polyester Adhesive Systems*, Apr. 24, 1968, American Chemical Society, Rubber Chemicals Division, Cleveland, Ohio.
8. U.S. Pat. 2,493,614 (Jan. 3, 1950), Bourdon (to Mfg. de Caoutchouc Michelin).
9. U.S. Pat. 3,869,306 (Mar. 4, 1975), A. Marzocchi (to Owens Corning)
10. U.S. Pats. 3,671,542 (June 20, 1972); 3,819,587 (June 25, 1974); and 3,888,965 (June 10, 1975), S. L. Kwolek (to DuPont).
11. U.S. Pat. 3,455,100 (July 15, 1969); U.S. Pat. 3,486,546 (Dec. 30, 1969), J. Sidles, D. P. Skala, and L. Skolnik (to BF Goodrich).
12. *Textile Organon* and *Tire Construction and Reinforcement Forecasts*, Monsanto Chemical Co., St. Louis, Mo., 1971–1981.
13. U.S. Pat. 3,307,966 (Mar. 7, 1967), C. J. Shoaf (to DuPont).
14. *Gen-Tac* (*Vinylpyridine Latex*), Chemical/Plastics Division bulletin, The General Tire and Rubber Company, Akron, Ohio, 1964.
15. A. L. Promislow in *Akron Rubber Group Technical Symposiums*, Akron, Ohio, 1977–1978, p. 98.
16. W. D. Timmons, *Adhes. Age* **10**(10), 27 (1967).
17. Belg. Pat. 688,424 (1967); Fr. Pat. 1,496,951 (1967); and Neth. Pat. 6,614,669 (Apr. 19, 1967), (to Imperial Chemical Industries, Ltd.).
18. Y. Iyen in ref. 15, p. 110.
19. U.S. Pat. 3,968,295 (July 6, 1976), T. S. Solomon (to BF Goodrich).
20. R. E. Hartz and H. T. Adams, *J. Appl. Polym. Sci.* **21,** 525 (1977).
21. R. G. Aitken, R. L. Griffith, J. S. Little, and J. W. McClellan, *Rubber World* **151**(5), 58 (1965).
22. T. Takeyama and J. Matsui, *Rubber Chem. Technol.* **42,** 159 (1969).
23. Bekaert Steel Wire Corp., Pittsburgh, Pa., steel cord catalogue, Mar. 1982.
24. W. J. van Ooij, *Rubber Chem. Technol.* **52,** 605 (1979).
25. G. Haemers, *Rubber World*, 26 (Sept. 1980).

26. R. S. Rivlin, *Arch. Rat. Mech. Anal.* **2,** 447 (1959).
27. W. Hofferberth, *Kautsch. Gummi* **9,** 225 (1956).
28. H. G. Lauterbach and F. W. Ames, *Tex. Res. J.* **29,** 890 (1959).
29. S. K. Clark, *Tex. Res. J.* **33,** 295 (Apr. 1963).
30. J. D. Walters, *Rubber Chem. Technol.* **51,** 565 (1978).
31. C. B. Budd, private communications, BF Goodrich, Brecksville, Ohio, Aug. 9, 1957 and Mar. 30, 1964.
32. B. D. Coleman, *J. Mech. Phys. Solids* **7,** 60 (1958).
33. K. B. O'Neil, M. F. Dague, and J. E. Kimmel, *International Symposium on High Speed Testing*, Boston, Mass., Mar. 7, 1967.
34. J. B. Curley, *Rubber World*, (Sept. 1967).
35. E. W. Lothrop, *Appl. Polym. Symp.* **1,** 111 (1965).
36. F. S. Vukan and T. P. Kuebler, *Determination of Passenger Tire Performance Levels—Tire Strength and Endurance*, Society of Automotive Engineers, Chicago, Ill., May 23, 1969.
37. C. Z. Draves, Jr., T. P. Kuebler, and S. F. Vukan, *ASTM Mat. Res. Stand.* **10**(6), 26 (1970).
38. R. L. Guslitser, *Tr. Nauchno-Issled. Inst. Shinnoi Promsti.* **3,** 154 (1957); transl. by R. J. Moseley, Research Association of British Rubber Manufacturers, translation 817.
39. S. Eccher, *Rubber Chem. Technol.* **40,** 1014 (1967).
40. G. Butterworth, *Chem. Eng. News*, 45 (Sept. 1967).
41. C. Z. Draves, Jr. and L. Skolnik, *Processing Rayon Cord for Optimum Tire Performance*, 2nd Dissolving Pulp Conference TAPPI, New Orleans, La., June 1968.
42. J. M. Collins, W. L. Jackson, and P. S. Oubridge, *Trans. Inst. Rub. Ind.* **40,** T239 (1964).
43. N. M. Trivisonno, *Thermal Analysis of a Rolling Tire*, paper 700474, Society of Automotive Engineers meeting, Detroit, Mich., May 1970.
44. D. J. Schuring, *Rubber Chem. Technol.* **53,** 600 (1980).
45. F. S. Conant, *Rubber Chem. Technol.* **44,** 397 (1971).
46. V. E. Gough, *Rubber Chem. Technol.* **41,** 988 (1968); *Kautsch. Gummi Kunstst.* **20,** 469 (1967).
47. C. Z. Draves, Jr. and L. Skolnik, *Tire Cord Process Simulation and Evaluation*, DKG Rubber Conference, Berlin, FRG, May 1968; *Kautsch. Gummi Kunstst.* **22,** 561, Oct. 1969.
48. P. V. Papero, R. C. Wincklhofer, and H. J. Oswald, *Rubber Chem. Technol.* **38**(4), 999 (1965).
49. J. K. Van Wijngaarden in *Recent Development in Tire Yarns*, American Enka Corp. bulletin 9-6, May 8, 1963.
50. G. Heidemann, G. Stein, and R. Kuhn in B. Von Falkai, ed., *Synthesefasern*, Verlag Chemie, Weinheim, FRG, 1981, pp. 379–423.
51. U.S. Pat. 3,893,331 (July 8, 1975), D. C. Prevorsek, Y. D. Kwon, R. H. Butler, and R. K. Sharma (to Allied Chemical Corporation).
52. Y. D. Kwon, R. K. Sharma, and D. C. Prevorsek, *Tire Reinforcement and Tire Performance*, *Symposium at Montrose*, *Ohio*, *Oct. 23–25, 1978*, American Society for Testing and Materials, Philadelphia, Pa., 1979, pp. 239–262.
53. *ASTM D 885-64*, American Society for Testing and Materials, Philadelphia, Pa., 1967.
54. W. H. Bradshaw, *ASTM Bulletin*, American Society for Testing and Materials, Philadelphia, Pa., Oct. 1945, pp. 13–17.
55. C. B. Budd, *Text. Res. J.* **21,** 174 (Mar. 1951).
56. L. Skolnik, *Rubber Chem. Technol.* **47,** 434 (1974).
57. L. C. Coates and C. Lauer, *Rubber Chem. Technol.* **45,** 16 (1972).
58. G. S. Fielding-Russell, D. I. Livingston, and D. W. Nicholson, *Rubber Chem. Technol.* **53,** 950 (1980).
59. J. B. Curley, *A New Development in Rayon Tire Cord Technology*, Buffalo Rubber Group technical meeting, Mar. 8, 1966.
60. G. W. Rye and J. E. Martin, *Rubber World* **149,** 75 (Oct. 1963).
61. W. E. Claxton, M. J. Forester, J. J. Robertson, and G. R. Thurman, *Text. Res. J.* **35,** 903 (1965).
62. J. A. Gooch, *Akron Rubber Group Technical Symposiums*, Akron, Ohio, 1978–1979, pp. 56–58.

LEONARD SKOLNIK
BF Goodrich

WOOL

Wool, the fibrous covering from the sheep (1), is by far the most important animal fiber used in textile manufacture. World production in 1979–1980 was 1.6×10^6 metric tons (clean) (2); other animal fibers, eg, mohair, alpaca, cashmere, and camel, totaled only 26,000–30,000 t (3) (see also Silk). However, the relative importance of wool as a textile fiber has declined over the decades (Table 1), as synthetic fibers have increasingly been used in textile consumption. Nevertheless, wool is still an important fiber in the middle and upper price ranges of the textile market. It is also an extremely important export for several nations, notably Australia, Argentina, New Zealand, and South Africa, and commands a price premium over most other fibers because of its outstanding natural properties of soft hand (the feel of the fabric), good moisture absorption (and hence comfort), and good drape (the way the fabric hangs) (see Fibers, chemical; Textiles).

Table 2 shows wool production and sheep numbers in the world's principal wool-producing countries.

Table 1. World Production of Cotton, Wool, and Synthetic Fibers, 10^6 t [a]

Year	Raw cotton	Wool (clean)	Synthetic fibers					Total of all fibers
			Synthetic filament	Staple	Cellulosic filament	Staple	Total	
1900	3.162	0.73			0.001		0.001	3.893
1940	6.907	1.134	0.001	0.004	0.542	0.585	1.132	9.173
1950	6.647	1.057	0.054	0.015	0.871	0.737	1.677	9.381
1960	10.113	1.463	0.417	0.285	1.131	1.525	3.358	14.934
1970	11.784	1.602	2.397	2.417	1.391	2.187	8.393	21.779
1980	14.137	1.581	4.731	5.756	1.159	2.085	13.731	29.449

[a] Ref. 4.

Table 2. Wool Production and Sheep Numbers in Principal Wool-Producing Countries, 1979–1980 [a]

	Wool production (greasy basis)		Sheep, 10^6
	1000 t	%	
Australia	713	25.5	136
New Zealand	357	12.8	68
South Africa	103	3.7	24
Argentina	166	5.9	32
Uruguay	70	2.5	17
United States	48	1.7	13
United Kingdom	48	1.7	31
Turkey	57	2.1	46
Eastern Europe and The People's Republic of China	761	27.2	299
other	475	16.9	324
Total	2798[b]	*100*	*990*

[a] Ref. 2.

[b] Clean equivalent is 1.602×10^6 t.

The three main distinctions of wool are fine, medium, and long (Table 3). The outstanding fine wools are produced by merino sheep, an animal first bred in Spain but now prevalent throughout the world. Over 75% of the sheep of the world's largest wool producer, Australia, are merino sheep, which are also bred in large numbers in South Africa, Argentina, and the USSR. Medium wools include types of English origin, eg, southdown, hampshire, dorset, and cheviot, as well as crossbreds, eg, columbia, targhee, corriedale, and polwarth, from interbreeding merino and long wool types. Coarse long wools come from sheep chiefly bred for mutton, eg, lincoln, cotswold, and leicester.

Raw wool from the sheep contains other constituents considered contaminants by wool processors. These can vary in content according to breed, nutrition, environment, and position of the wool on the sheep. The main contaminants are a solvent-soluble fraction called wool grease; protein material; a water-soluble fraction (largely perspiration salts collectively termed suint); dirt; and vegetable matter, eg, burrs and seeds from pastures. Table 4 gives some figures for percentages of fleece constituents in Australian wool, excluding the protein contaminant (5–6). The wools in Table 4 contain little or no vegetable matter, which in certain areas, can be as high as 20% of the raw-wool weight, but it is usually <2%.

Raw-Wool Specification. In buying raw wool, the wool processor is concerned about its quality and the quantity of pure fiber present. The particular properties of concern differ depending on the processing system used and the product. For example, for the main wool market, ie, fine and medium wools for apparel, the single-fiber and staple characteristics that are significant to processing are given in Table 5. The most important are the yield (the percentage content of pure fiber at a standard moisture content); the vegetable-matter content and type; the average fiber diameter; the average length of the fibers; the strength of the staples of fibers and the position of any weak spot along the fibers; the color of the clean wool; and the number (if any) of naturally colored fibers present. For carpet-type wools (long wools) the important properties (8) are yield; fiber diameter; fiber length; color; bulk (the volume occupied by the fibers in a yarn); medullation (see Fiber Morphology); and vegetable-matter content.

Table 3. Main Types of Wool

Type	Breed	Average length, cm	Average diameter, μm[a]	Grade or count[a]
fine	merino	3.7–10	10–30	90s–58s
medium	southdown hampshire dorset cheviot	5–10	20–40	60s–46s
crossbred	corriedale polwarth targhee columbia	7.5–15	20–40	60s–50s
long	lincoln	12.5–35	25–50	50s–36s

[a] Wool fineness is now largely described as the "micron," ie, micrometer. Traditionally, it has been expressed as grade or count which, in the worsted trade, represents the number of hanks, each 560 yd (1512 m) long, of the finest possible yarn that might be spun satisfactorily from 1 lb (0.454 kg) of the wool concerned. The higher the grade or count, the lower is the "micron."

Table 4.   Fleece Composition of Australian Wool, %

Component	Merino	Sheep Crossbred	Skin[a]	Lambs, merino and crossbred
oven-dry wool				
max	66.8	72.2	70.0	68.0
min	29.4	49.3	50.6	54.0
av	48.9	61.0	63.0	60.5
wax				
max	25.4	19.3	20.0	23.5
min	10.0	5.3	9.9	6.4
av	16.1	10.6	15.8	16.0
suint				
max	13.0	13.6	1.4	7.2
min	2.0	4.4	0.1	3.4
av	6.1	8.2	0.6	5.4
dirt				
max	43.8	23.7	21.5	10.0
min	6.3	4.3	6.4	3.8
av	19.6	8.4	11.2	6.4
moisture				
max	12.6	14.2	9.6	14.2
min	8.1	9.5	7.2	9.0
av	9.6	12.0	8.0	11.2

[a] Skin wool is the term applied to fellmongered wool, usually removed from skins by a sweating process involving bacterial action or by action of lime sulfide depilatory.

Table 5.   The Significance of Single-Fiber and Staple Characteristics in Wool Textile Processing[a]

Characteristic	Importance
yield	primary
vegetable matter	primary
fiber diameter	primary
length	primary
strength and position of break	primary
color and colored fibers	primary
fiber diameter variability	secondary
length variability	secondary
cots[b]	secondary
crimp and resistance to compression	secondary
staple tip	minor
age, breed, and category	minor
style, character, and handle	minor

[a] Ref. 7.
[b] Cots are natural tangles of fibers in wool fleece.

Until the late 1960s and early 1970s, the characteristics of fine and medium wools (Table 5) were largely evaluated visually by wool valuers. However, with the development of sampling techniques and equipment capable of rapid and economical measurement of yield, fineness, and vegetable matter (9–10), the sale of wool by sample and objective measurement has become dominant in Australia and South Africa and has grown in use in New Zealand. In sale by sample, cores are drawn from each lot and tested for yield, fineness, and vegetable-matter content in accordance with international standards (11). In addition, a full-length display sample, representative of each lot and obtained by standard procedure (12), is available for buyers to appraise the other characteristics. Sale by sample has reduced costs by reducing the handling of bulk wool in wool-brokers' stores and selling operations, and has also led to more efficient utilization of wool in processing and more accurate alerting of the wool processors' requirements to the woolgrowers.

Techniques for efficiently and economically measuring the other important characteristics, ie, staple (fiber) length (10,13–14) and strength and position of any weakness in the fiber (10,15), are under development. Existing color-measuring equipment can be used to measure the color (whiteness or yellowness) of washed wool, but measuring colored-fiber content remains a problem (10). The overall objective of research underway is to develop a system of "sale by description" for fine and medium wools, whereby the buyer is presented only with measured data on the main characteristics of the raw wool plus an assessment of the less important characteristics by an independent skilled appraiser.

Fiber Growth

The skin of the sheep has two layers, an inner dermis and an epidermis. The fiber grows out of a tubelike structure in the epidermis known as the follicle. The two types of wool follicles are the primary, which develop first in the outerskin of the unborn lamb, and the secondary, which develops later. Figure 1 depicts the structure of the primary follicle. The follicle is formed by growth from the epidermis into the dermis (16). After a short while, the outgrowths of the sebaceous gland, which exudes the wool grease, and the sweat gland develop. There are ca 200×10^6 follicles in the skin of a merino sheep.

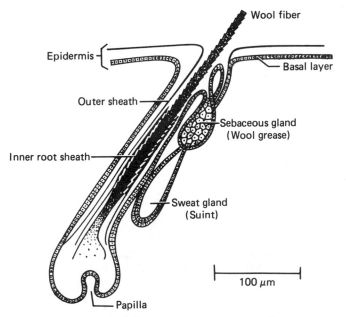

Figure 1.   The primary wool follicle showing production of fiber, wax, and suint.

Fiber Morphology

The two types of cells in fine–medium wool fibers (Fig. 2) are corticle cells, which make up the center of the fiber, and cuticle cells, which form the outer protective cover around the cortex (16–17). The cuticle cells overlap like the tiles on a roof, with the protruding tips of the scales pointing toward the fiber tip (Fig. 3). The orientation of the scales in this way leads to a directional friction effect, ie, a difference in the coefficient of the friction of wool fiber when measured in the "with-scale" and "against-scale" direction. This effect is thought to be the main cause of wool felting (18) (see Shrink-Resistance Treatment). The dimensions of cuticle cells of merino wool are ca $20 \times 30 \times 0.5$ μm (19–20). They are made up of an enzyme-resistant exocuticle and an enzyme-digestible endocuticle (21–22), surrounded by a thin hydrophobic membrane, the epicuticle (19) (see Fig. 3).

Between the cuticle and the cortical cells, and separating the cortical cells themselves, is a cell membrane complex. This complex, ca 25 nm thick, basically "cements" the cells together, and has been called the intercellular cement. Some components of the complex are resistant to enzymes and chemicals, whereas others can easily be extracted with organic solvents or enzymes (23). Mechanically, the cell membrane complex is believed to be the weak link of the fiber (23–24).

The cortex consists of spindle-shaped cortical cells ca 100 μm long and is bilateral in structure with an orthocortex and a paracortex of different chemical reactivity

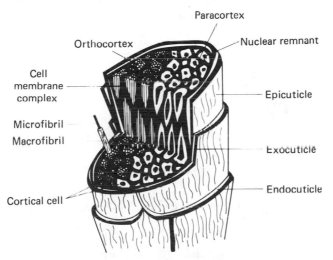

Figure 2.   Diagrammatic representation of the morphological components of a wool fiber.

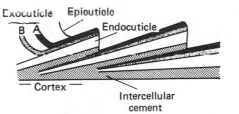

Figure 3.   Diagrammatic representation of the morphological components of the cuticle of a wool fiber.

(25–29). In fine wools such as merino, the paracortex is always located on the inner concave side, and the orthocortex on the outer convex side of the fiber crimp curvature. In contrast, with coarse long fibers such as lincoln, the orthocortex forms a core surrounded by a tubular paracortex.

The cortical cells (Fig. 4) are made up of many highly organized fibrils (30–31). These fibrils in turn are composed of microfibrils of ca 8 nm diameter surrounded by an amorphous protein (32), the matrix. It is believed that these microfibrils are made up of smaller subunits, ie, protofibrils, each of which possibly consists of three α-helical polypeptide chains twisted together. Coarser fibers, such as some carpet wools, sometimes also contain a third component, the medulla, a hollow canal up the center of the fiber.

Physical Properties

That wool is still used as a textile fiber is to a large extent the result of its unique physical properties. In particular, it absorbs large volumes of moisture and hence is comfortable to wear. The absorption of moisture yields heat. The bulk (volume occupied by fibers in a yarn) given to it by its bilateral structure (and the resulting fiber crimp) adds to warmth as a greater volume of air is trapped than in a yarn of the same weight but of other fibers. Wool's unusual elastic properties give it outstanding resistance to and recovery from wrinkles, and outstanding drape. The principal physical properties of the fiber are given in Table 6.

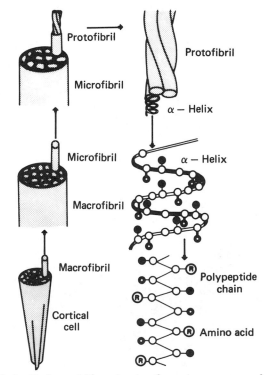

Figure 4. Cortical cell of a wool fiber, showing the various macro- and micro components.

Table 6.   The Principal Physical Properties of Wool Fibers at 25°C[a]

Property	Regain, %							
	0	5	10	15	20	25	30	33
relative humidity, %								
absorption	0	14.5	42	68	85	94	98	100
desorption	0	8	31.5	57.5	79	91.5	98	100
specific gravity	1.304	1.3135	1.3150	1.325	1.304	1.2915	1.2765	1.268
volume swelling, %	0	4.24	9.07	14.25	20.0	26.2	32.8	36.8
length swelling, %	0	0.55	0.93	1.08	1.15	1.17	1.18	1.19
radial swelling, %	0	1.82	4.00	6.32	8.88	11.69	14.57	16.26
heat of complete wetting, kJ/kg wool[b]	100.9	64.5	38.1	20.5	8.8	4.1	1.13	0
heat of complete absorption of liquid water, kJ/kg water[b]	854	624	431	276	142	100	42	33
relative Young's modulus	1.00	0.96	0.87	0.76	0.66	0.56	0.44	0.38
rigidity modulus (torsion), GPa[c]	1.76	1.60	1.26	0.90	0.50	0.28	0.16	0.1
electrical resistivity, 10^6 Ω·cm			4×10^4	800	40	6		
dielectric constant at 10^4 Hz	4.6	5.1	6.2	8.3	12.8			

[a] Ref. 33.

[b] To convert J to cal, divide by 4.184.

[c] To convert GPa to psi, multiply by 145,000.

Moisture Relations.   Wool is the most hygroscopic of common textile fibers (Table 7). Its regain (moisture content) affects its ease of processing, especially in carding, combing, spinning, and weaving; accordingly, the relative humidity of the atmosphere in these mill areas needs to be controlled to optimize processing. The regain depends mainly on the relative humidity, but it is also affected by the state of the fiber, eg, whether it has been damaged chemically or physically in processing or whether it contains acid or alkali.

The relationship between regain and relative humidity is shown in Figure 5. There is a hysteresis effect on drying. The desorption curve shows higher regains at each value of relative humidity. The lower adsorption curve represents the behavior of dry wool

Table 7.   Moisture Regain Values for Wool and Other Fibers at 21°C, %[a]

Fiber	At 65% rh	At 95% rh
Acrilan acrylic	1.5	5
Orlon acrylic	1.5	4
nylon	2.8–5	3.5–8.5
polypropylene	0.01–0.1	0.01–0.1
Dacron polyester	0.4–0.8	
viscose rayon	13	27
acetate	6.5	14
cotton	7	
wool	16	22.5 (90% rh, 25°C)

[a] Ref. 34.

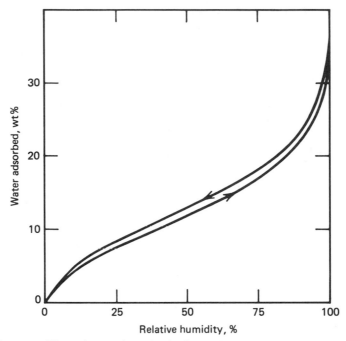

Figure 5.   The moisture adsorption isotherm of 20-μm merino wool at 22°C.

of a given moisture content. In the normal range of relative humidities, there is a difference of ca 2% between the adsorption and desorption isotherm.

As can be seen from Table 6, most physical properties of the fiber are affected
by the regain (see also Table 7). As the moisture content of the fiber increases, there
is a marked radial swelling of the fiber attributable to its oriented structure. At high
regains, fibers are easier to bend and twist and are much better conductors of electricity
and less liable to generate static electricity. Accordingly, prior to spinning, wool tops
should be stored to ensure the highest regain possible.

The regain of wool also has important bearing on commercial transactions where
wool is sold by weight. Standard regains are, therefore, specified for this purpose, and
these are given in Table 8.

Mechanical Properties.   Both in the manufacturing processes of spinning, weaving,
and knitting and in the behavior of the completed textile, the mechanical properties
of wool fibers are the dominant physical factors. The fiber is characterized by high
extensibility and relatively low breaking strength, and it has very unusual elastic
properties particularly when wet. Figure 6 shows the stress–strain curves of wool at
different relative humidities. The stress–strain properties can be considered in terms
of three distinct regions of strain (36). Once the fiber has been straightened (this varies
with the crimp in the fiber under test and for merino fibers can represent a considerable
strain) the stress–strain curve has a nearly linear mechanically stiff region up to a few
percent strain; this region is generally referred to as a Hookean region, although it does
not closely approximate a Hookean spring in its properties (37). The strain of the fiber
then increases rapidly with increase of stress up to strains of ca 25–30%, and this region
is known as the yield region. For extensions beyond the yield region, ie, the postyield
region, the fiber becomes stiffer with further strain.

Table 8.   U.S. and IWTO Standard Regains

| Form | Moisture regain, % | |
	U.S.[a]	IWTO[b]
scoured wool	13.63	17
tops		
oil, combed	15	19
dry, combed	15	18.25
noils[c]		
oil, combed		14
dry, combed		16
scoured and carbonized		17
yarn		
woolen	13	17
worsted, oil spun	13	18.25
worsted, dry spun	15	18.25

[a] Ref. 35.

[b] International Wool Testing Organization.

[c] Noils are shorter fibers separated from longer fibers in combing.

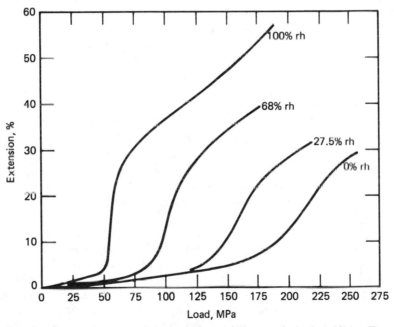

Figure 6.   Load–extension curves for a wool fiber at different relative humidities. To convert MPa to psi, multiply by 145.

There is a hysteresis in the unloading curve as shown by a typical stress–strain curve for a wool fiber in water (Fig. 7). If the fiber is rested overnight in water at room temperature, the cycle can be repeated.

This mechanical behavior of a wool fiber has been explained as being the result of two systems acting in parallel—the microfibrils, which contain the organized α-helical material, and the matrix, the water-sensitive material outside the microfibrils (38).

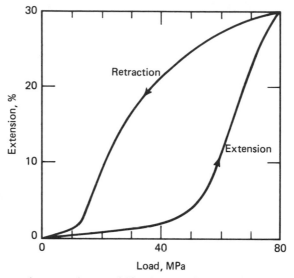

Figure 7.   Load-extension curve for a wool fiber in water. To convert MPa to psi, multiply by 145.

Chemical Structure

Wool belongs to a family of proteins (qv) called the keratins. However, as explained above, morphologically, the fiber is a composite, and each of the components differs in chemical composition. Principally, the components are proteinaceous, but cleaned wool (after successive extraction with light petroleum, ethanol, and water) contains small amounts of lipid material and inorganic ions equivalent to an ash content of ca 0.5–1% after combustion of the fiber.

The simplest picture of the proteinaceous components of the wool fiber is one of polypeptides (qv) composed of α-amino acid residues. Eighteen amino acids (qv) are present in the hydrolysate of the fiber, but the relative amounts of the amino acids (qv) vary from one wool to another. Typical figures for three different samples of merino wool are given in Table 9. The ammonia produced in the hydrolysis of wool is presumed to arise from amide groups and gives a measure of the number of these groups originally present.

The side groups of the amino acids vary markedly in size and chemical nature. The nature of these groups is of course important to the chemical reactions of the fiber. For example, the basic groups (arginine, lysine, and histidine residues) can attract acid (anionic) dyes, and in addition, the side chains of lysine and histidine residues are important sites for the covalent attachment of reactive dyes (see Dyeing).

The main sulfur-containing amino acid present in wool, cystine, plays a very important role, as two of the residues form a disulfide cross-link between neighboring polypeptide chains or between amino acid residues in the same chain:

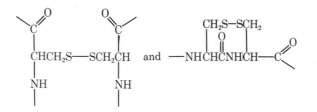

Table 9.   The Amino Acid Content of Merino Wool[a]

Amino acid	Content, mol/g[b]		
	Ref. 40	Ref. 41	Ref. 42
lysine	193	277	269
histidine	58	76	82
arginine	602	613	600
tryptophan	c	c	c
aspartic acid	503	602	560
threonine	547	564	572
serine	860	892	902
glutamic acid	1020	1046	1049
proline	633	561	522
glycine	688	815	757
alanine	417	512	469
$\frac{1}{2}$ cystine	943	1120	922
valine	423	546	486
methionine	37	47	44
isoleucine	234	318	275
leucine	583	721	676
tyrosine	353	380	349
phenylalanine	208	268	257
NH_3	887	855	

[a] Ref. 39.

[b] The results of three separate tests are reported from the indicated references.

[c] Tryptophan is destroyed under the conditions of hydrolysis used for these analyses; values in the range 35–44 mol/g have been obtained by alternative techniques (43–45).

The disulfide cross-links readily rearrange under influence of heat and moisture, facilitating conformational rearrangement of the wool proteins and so leading to relaxation of molecular stress in the fiber. Other cross-links, eg, isopeptide cross-links, are present in the fiber in small amount.

The extractable proteins of the wool fiber are of three different types that differ greatly in structure and properties. The predominant type (58%), SCMK-A (S-carboxymethylkerateine-A), contains a group of low sulfur proteins (sulfur content ca 1.7% (46) compared with the fiber overall at 3.5%). The second fraction (26%) SCMK-B, contains two groups of proteins (47): high sulfur (sulfur content 4–6%) and ultrahigh sulfur (sulfur content 8%) proteins. A third fraction (6%) contains another group of proteins, termed high Gly/Tyr proteins (48) because they are rich in glycine and tyrosine levels (sulfur content 0.5–2%).

The sequencing of amino acids in these protein fractions has been the subject of much research (49). There are at least four different families of proteins in the high sulfur fraction (50), and individual members from three of these families have been sequenced. The low sulfur groups contain at least three families. None of the constituent proteins has been sequenced completely, but partial sequences of helical regions of some of the proteins are now known (50). There are at least six families in the high Gly/Tyr group; one member, containing sixty-one residues, has been sequenced (51).

Wool Processing

The conversion of raw wool into a textile fabric or garment involves a long series of separate processes. There are two main processing systems, ie, worsted and woolen, although an appreciable volume of wool is also processed on the short-staple (cotton) system or on the semiworsted system for carpet use. The main stages in the woolen and worsted systems shown in Figure 8 are described in references 52 and 53, respectively. Nearly all the processing stages involve some aspect of chemistry. For example, although spinning is largely a mechanical operation, the lubrication of the fibers by the spinning oils involves surface chemistry.

Scouring. The first stage in wool processing is removal of the fleece impurities, principally grease, dirt, and suint, by scouring (54). This entails washing the raw wool in aqueous solutions (neutral or slightly alkaline) of nonionic detergent (or, less frequently, soap and soda), followed by rinsing in water. The process is carried out in a series of four or five bowls in which the wool is immersed in the wash liquor and transported through the liquor by mechanical rakes. The liquor flows countercurrent

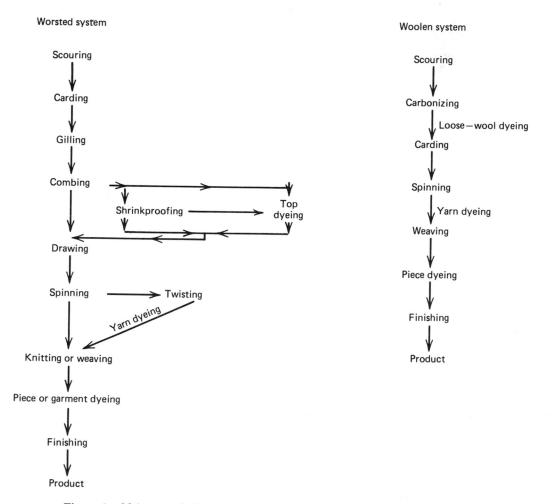

Figure 8.   Main stages in the processing of wool on the worsted and woolen systems.

to the wool, and at the end of each bowl the wool passes through a roller squeeze. Other methods of fiber transport are used to a limited extent, eg, the wool is conveyed under jets of liquor (55) or passed around perforated drums (56), principally to limit the wool movement that leads to fiber felting. Agitation of wool in anhydrous solvents does not cause felting, and several plants (eg, 57–58) have been designed to take advantage of this property, but only a few are in operation commercially (54).

The traditional picture of the contaminants on the wool has been fiber covered with a layer of grease in which suint and dirt (both mineral and organic) are embedded. The principal action of the detergent or soap has been seen as one involving swelling and rolling up of the grease–suint and grease–dirt mixtures and suspension of these in the water (59–60). This simple picture has been questioned by a possible fourth main component, thought to be proteinaceous in nature (61–62), which is present as a layer on the wool fiber and also as small particles dispersed with the dirt and suint in the grease. Under certain scouring conditions (eg, neutral conditions or low ratios of liquor to wool), this material does not swell readily and hinders removal of dirt. As well as affecting the efficiency of scouring if not completely removed, this proteinaceous layer may also influence the color of scoured wool and subsequent processing stages such as carding and shrinkproofing (62).

It is important to remove as much of the wool grease from the waste scour liquors as possible, first, because the product has some commercial value (see Wool Grease), and second, because the pollution load of the discharge is thereby reduced. Wastewaters are therefore normally centrifuged before discharge. Not all of the grease, however, can be removed by centrifuging (normally up to 40% can be removed) because certain components of the grease and dirt form stable emulsions under normal scouring conditions. Because a high concentration of dissolved solids in the liquors reduces the stability of the emulsion, it is possible under certain conditions, to recover ca 85–90% of the grease originally present on the wool (63). This understanding led to the development of the Lo-flo scouring process in which very high concentrations of dissolved solids are produced in the scour liquors by severely restricting water input to special scouring bowls (64). The partial flocculation achieved eases the removal of contaminants by centrifuging. Grease removal is very good, but the removal efficiencies for dirt and suint are less than normal (54).

In most situations, wastewater discharged from a scouring machine still contains large quantities of grease and water-soluble and insoluble material (both organic and inorganic). Disposal or treatment of the wastes to comply with environmental requirements is thus expensive. Probably the cheapest approach to disposal involves biological treatment by irrigation of large land areas (not less than 20 ha per scouring machine is normally required) or maturation in very large shallow lagoons (54).

Physicochemical methods of wastewater treatment, eg, flocculation using inorganic or polymeric flocculants or sulfuric acid, have been investigated, as have physical techniques, eg, membrane processes and solvent extraction, but few processes have passed pilot-scale evaluation. One physical method that has attracted some commercial interest is evaporation; several evaporative plants were installed in Japan in the early 1970s, nearly all followed by incinerators for the sludge produced (65). They are, however, expensive in both capital and operating costs.

Another approach to effluent treatment is to reduce the effluent load and volume by modifying the scouring process. This is the basis of the Lo-flo process mentioned above and also of the WRONZ (Wool Research Organization of New Zealand) rationalized scouring system (66–68).

Carbonizing. Carbonizing is a process used to remove excessive contents of cellulosic impurities, eg, burrs and vegetable matter, from wool. It is carried out on loose wool (69), rags (70), and fabric (71). With loose wool and fabric, the wool is treated with aqueous sulfuric acid and then baked. The cellulosic matter is rapidly destroyed by the hot concentrated acid by which it is converted into friable hydrocellulose, whereas wool is scarcely affected by the treatment. For rags, hydrogen chloride gas is preferred as the mineral acid because it has less effect on the colors. After acid treatment, the carbonized vegetable matter is crushed to facilitate removal. The wool is then normally neutralized in alkali, although some mills omit this stage to facilitate subsequent dyeing with equalizing acid dyes.

Most industrial carbonizing is done on loose wool, and the technology has changed little over the last 40 years. Australian carbonizers now use surface-active agents to reduce fiber damage in the carbonizing process (72–73). The improved results depend on the use of high acid concentrations and rapid throughput (73). Factors important to the role of the surfactants are thought to be the increased efficiency of acid removal in squeezing (74) and the improved spreading of the acid over the fibers in the presence of the surfactant (75).

Attempts have been made to introduce processes and equipment for rapid acidizing, drying–baking, and crushing (76) and for carbonizing wool in sliver (rope) form (77), but neither has been adopted commercially.

Developments in the carbonizing of fabric (78) have been aimed at reducing the volume of acid used (79–80), in order to reduce energy in drying and save neutralization costs, and at the use of generally available textile processing equipment, eg, the pad–mangle, for application of the acid in a pad–dry–bake process (81).

Dyeing. Traditionally, wool is dyed in batch operations in which the textile is boiled in aqueous solutions of dye for periods of at least one hour. Techniques have also been developed for continuously dyeing wool in loose or sliver form (82–83). However, because of the need for large volumes of one color for economic operation, the number of installations of machines using such techniques has been limited.

The chemistry of wool dyeing is complex (84–85). The dyes used industrially are mainly anionic in nature (84). They adsorb on wool as the result of ionic attraction between sulfonic groups ($-SO_3^-$) on the dyes and charged amino ($-NH_3^+$) groups on the fiber and of nonpolar van der Waals forces between the hydrophobic dye anion and hydrophobic parts of the wool adjacent to $-NH_3^+$ groups. Other interactions, eg, hydrogen bonds, are probably also involved. Factors influencing these various interactions include pH and temperature of the dye solution; nature and concentration of electrolytes in the solution; hydration of dye and auxiliaries; effects of contaminants and pretreatments on the fiber surface; affinity of the dye for wool; and mode and depth of penetration of the dye into the fibers (85–86).

Three principal types of acid dyes are used for wool: equalizing acid dyes, milling acid dyes, and supermilling acid dyes.

Equalizing acid dyes have comparatively low affinity for wool, and it is easy to obtain uniform dyeings using them. They combine with the fiber largely because of ionic attraction. They are applied at ca pH 3 to increase the number of positively charged groups acting as sites of attraction. In general, these dyes are not particularly resistant to wet treatments.

Milling acid dyes are of higher molecular weight and possess fewer sulfonic groups than equalizing acid dyes. Because they depend less on ionic attraction for their affinity to wool, a higher pH is used to ensure uniformity. The pH required is usually obtained by adding a weak acid, eg, acetic acid.

Supermilling acid dyes are virtually independent of ionic attraction for good exhaustion and so are applied from an almost neutral dyebath.

Reactive dyes, usually anionic in character, are also important for wool (87–88). Because these dyes contain a group that reacts covalently with the wool fiber, they have outstanding resistance to removal by washing, etc. They also produce very bright hues. Because of these properties, they have significant application in dyeing wools that have been treated to withstand shrinkage in machine washing.

The resistance to wet treatments of some dyes can also be improved by treatment with metal salts, usually those of chromium, to reduce their solubility. These chrome dyes (89) are now supplemented by metal-complex dyes (90) of very high affinity for wool. In these, the metal complex is preformed; it contains one metal atom for each one or two dye molecules, the metal being either chromium or cobalt. The 1:1 complex dyes are applied to wool from strongly acid dyebaths; the 1:2 complexes, which are much more important than the 1:1 complexes, are applied from a very slightly acid bath (pH 5.5–7).

Wool is dyed in loose, sliver, yarn, fabric, or garment form (84), depending on the nature of the product, the equipment available, requirements of the market, and other factors. Traditionally, the process has involved movement of the dye solution through a compressed mass of the textile (eg, yarn package) or transport of the textile (eg, fabric) through the dye solution (see Dyes, application and evaluation). Both approaches entail the use of high volumes of dye solution relative to the volume of the textile being dyed and hence have tended to be expensive both in energy and in water (and consequently effluent disposal). General improvements in design of dyeing equipment to enable a reduction in solution volume have occurred. In package machines, this has involved the use of denser packages and higher rates of solution flow, but sometimes this damages the wool fiber (91). Early versions of jet-dyeing machines for fabric dyeing, originally developed for synthetic fabrics, presented problems in dyeing wool fabrics, but later versions that use gentler actions to achieve fabric movement have proved suitable (92).

Attempts to reduce energy usage, effluent levels, and damage to wool by prolonged boiling (dependent on pH (90)) have generally focused on dyeing below the boil or the use of continuous pad–steam dyeing (87,93). For example, the use of high concentrations of urea in the dye solution allows dyeing of tops or fabric in the cold (94–97). Several other methods have been developed involving the use of special auxiliaries in the dye solution below the boil, eg, ethoxylated nonylphenol surfactants (98) and benzyl and n-butyl alcohols (99–100) (see Dye carriers). CSIRO introduced two machines for pad–steam dyeing, one for loose wool (82), and one for sliver (83). A machine for continuous dyeing of yarns is under development (101). The use of dielectric heating to achieve dye fixation has also been introduced (102) (see Microwave technology).

Printing. The printing of wool is difficult technically, and this is thought to be the reason why less than 2% of total wool production is printed (103–104). The traditional procedure entails pretreatment with chlorine to improve both the color yield and the evenness of prints, followed by neutralization and drying. The design is then printed, usually by screen, and the fabric dried, steamed to fix the dyes, washed to remove print chemicals and unfixed dye, and dried again (see also Printing processes).

For chlorination, dichloroisocyanuric acid or its salts (105) are normally used (see Bleaching agents). Chlorination degrades the fabric, and when preserving the hand (feel) of the fabric is essential, the chlorination step is omitted, with consequent loss of color yield. To avoid the degradation due to chlorination, alternative means to improve color yield have been sought, among them the addition of certain surfactants (106) or fatty acid derivatives (107–108) to the print pastes and pretreatment of the fabric by a shrink-resistance process, eg, the Sirolan BAP process (109).

Incorporation of high concentrations (>300 g/L) of urea in a print paste greatly accelerates the fixation of low molecular weight dyes on wool. This discovery has been used in the development of a cold-print method (110). After being printed, the textile products are left wet and batched with interleaving polyethylene sheets for up to 24 h; they are then washed and dried. No steaming is required, and energy usage in the process is thereby minimized.

Pigment printing has not been widely adopted for wool because of the adverse influence of the encapsulating resins used on the hand of the printed area. Changes in hand can, however, be minimized by careful selection of binders and cross-linking agents (111).

Transfer-print papers marketed for polyester are unsuitable for wool because the sublimable dyes that are used have little affinity for the wool fiber (112). However, processes have been developed that enable this technique to be applied to wool (103). In the Fastran process (113–114), for example, conventional wool dyes are printed on the paper and are heat-transferred to wool (usually knitted garments) that has been padded with a thickened solution. After being printed, the garment is washed to remove thickening agent and unfixed dye. Another process (115–116), developed by the CSIRO in collaboration with the International Wool Secretariat, entails the transfer printing of certain sublimable metal-complexing dyes onto fabric pretreated with a mixture of an anionic surfactant, urea, lactic acid, and chromic chloride, followed by steaming. The process yields prints in a wide range of shades with good resistance to washing and light. The process requires neither prior chlorination nor washing of the fabric after printing and is the simplest method so far devised for printing wool.

Bleaching. Bleaching is carried out to reduce the natural yellowish tinge of normal wool (117–119). Either oxidation (with, principally, hydrogen peroxide, sodium peroxide, or sodium perborate) or reduction (with thiourea dioxide, sulfur dioxide, acidified sodium bisulfite, or sodium dithionite) is used, and sometimes both types of treatment are applied, in which case oxidation must precede reduction (see Bleaching agents).

Peroxide bleaching is the most common approach. The wool is treated at pH 9 (sodium silicate or a phosphate) with 0.6% hydrogen peroxide (0.6% H_2O_2 = 2 cm^3 O_2 liberated per mL of liquor). Treatment begins at 50°C, and the wool is soaked for 3–5 h or overnight. It is then rinsed, treated with dilute acetic acid, and rinsed again. It is important that no metal except stainless steel be present, since metals, particularly copper, act as catalysts for the decomposition of hydrogen peroxide and may cause localized degradation of the wool. Continuous processes utilizing hydrogen peroxide have also been introduced (78).

Whitening. One of the problems with bleached wool is its tendency to yellow on exposure to sunlight, particularly when wet. The mechanisms by which wool is bleached or yellowed by light are complex, and although considerable work (120–121) has been done on the chemical changes that occur on photoyellowing, much remains to be elucidated in this area.

Significant efforts have been made to develop chemical treatments that inhibit light degradation. For example, a sulfonated $2H$-Z-(2-hydroxyphenyl)benztriazole retards phototendering but not photoyellowing (122). Ultraviolet absorbers (123) and antioxidants (qv) (124) do not significantly retard photoyellowing of wool (see Uv stabilizers). One of the most effective antiyellowing procedures is treatment with thiourea–formaldehyde mixtures or precondensates (125).

Insectproofing. The principal insects that attack wool are the case-bearing clothes moth (*Tinea metonella*, *Tinea dubiella*, and *Tinea pellionella*); the common clothes moth (*Tineola bisselliella*); the brown house moth (*Hofmannophila pseudospretalla*); and the variegated carpet beetle (*Anthrenus verbasci*). Wool is protected from attack of the larvae of these insects by application of an insecticide, usually from the dyebath (126) (see Insect-control technology).

The requirements for the insecticide are quite restrictive. For example, as well as being toxic to the insects at low levels of application, it must be stable to the application conditions (high temperature, aqueous solution), resistant to washing and light, and effective for prolonged periods on the textile (>15 yr for wool carpets). As a consequence, few general insecticides are suitable for insectproofing wool. The range commercially available (in 1982) is shown in Table 10 (126), and the structures of their active constituents in Figure 9 (126).

The most commonly used products are Eulan WA new and Eulan U33. Mitin LP and the pyrethroid-based formulations, Perigen and Mothproofing Agent 79, are more recently introduced products. Because of its high cost, Mitin FF has use only where high resistance to wet conditions is required since it is superior to the other products in this respect.

Conditions of application of insectproofing agents are dictated by the needs for satisfactory application of the dyes in the same bath (127–128). Conditions such as the presence of fibers other than wool, particularly nylon, and prolonged boiling at or above pH 5 can adversely affect the uptake of formulations based on polychloro-2-(chloromethylsulphamido)diphenyl ethers, eg, the Eulans, Mitin LP, and Molantin. Similarly, auxiliaries (particularly leveling agents) present to assist dye uptake and distribution may reduce application efficiency and increase residues in the dyebath effluent.

A wide variety of insecticides developed for general use have been investigated for their suitability for insectproofing, eg, organophosphorus, carbamate, and py-

Table 10.   Currently Available Industrial Mothproofing Formulations[a]

Product	Manufacturer	Active constituent[b]
Eulan WA new	Bayer	1
Eulan U33	Bayer	1
Mitin LP	CIBA-GEIGY	1 and 2
Mitin FF high conc	CIBA-GEIGY	3
Mitin N	CIBA-GEIGY	2
Molantin P	Chemapol	1
Perigen	Wellcome	4
Mothproofing agent 79	Shell	4

[a] Ref. 126.
[b] See Figure 9.

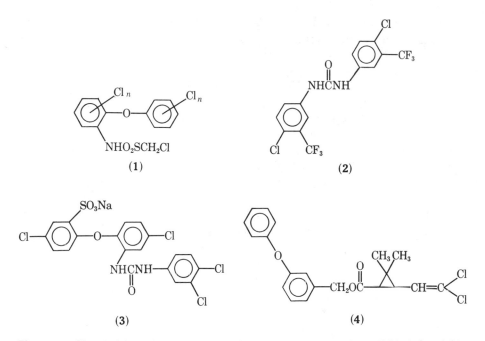

Figure 9. Chemical formulae of active constituents of some currently available industrial insect-proofing agents (see Table 10) (126).

rethroid insecticides (126). Phosvel (129) has shown the best durability among the organophosphorus products. Carbamates, which are generally more active toward the carpet beetle than toward the common clothes moth, proved unsatisfactory because of their poor affinity for wool (130–131). The pyrethroids have proven to be particularly suitable, especially permethrin, fenpropanate, and fenvalerate (132).

Other chemicals investigated have included quaternary ammonium compounds, anionic surfactants, organotin compounds (eg, triphenyltin chloride and acetate), thioureas, and 2-thienylethylene (126). Some of these have shown activity against wool-eating insects, but all have disadvantages that have precluded their commercial use. Similarly, other forms of chemical control by compounds that affect juvenile-hormone activity have been investigated but have not met commercial requirements (126).

One novel approach has involved modification of certain organophosphorus insecticides that show activity against wool-eating pests but suffer from problems of volatility and poor affinity for wool. The modification entails attachment to the insecticide of a group that reacts covalently with the wool fiber (133). Once bonded to the fiber, the group becomes inactive, and losses owing to volatilization, photochemical degradation, and wet treatments are minimal. The insecticide is reactivated when an ester group strategically placed between the fiber-reactive and insecticidal portion of the compound is cleaved in the digestive tract of the wool-eating pests during digestion of the treated wool, as shown diagrammatically in Figure 10. The active organophosphorus insecticide is thus released and rapidly kills the insect.

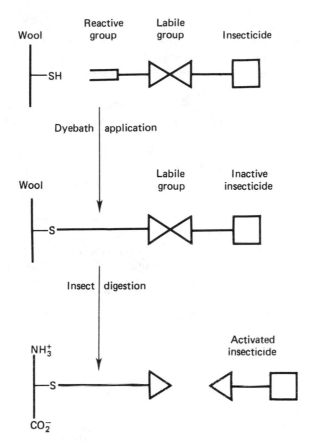

Figure 10.   A diagrammatic representation of the principle of a fiber-reactive insectproofing agent (126).

Setting.   Setting is an important part of the finishing of wool fabrics or garments to impart dimensional stability, shape, and hand appropriate to the product. Twist is set into yarns by steaming, and certain fabrics are set after weaving to prevent undue distortion during subsequent wet-finishing operations, a process known as crabbing. This is normally done by winding the fabric tightly on a roller and treating it in water at 60–100°C. Fabrics are often given a setting treatment to impart a certain amount of dimensional stability. This has traditionally been done by steaming again while the fabric is tightly wound on a roller and is known as blowing or decatizing. More recently, continuous machines, which are often integrated with other mechanical processes, eg, cutting, brushing, shrinking, and conditioning, have been adopted (78). Some of these machines operate at atmospheric pressure, and others at high pressure; some utilize chemicals (often sodium bisulfite or thioglycolate) to assist setting. The effectiveness of the setting treatments involving water or steam depends on fabric tension, the pH of the wool (optimum ca pH 8.5), the moisture content of the wool, the temperature, and the time (134).

The mechanism of setting wool involves bond fission and rebuilding; hydrogen bonds and disulfide and probably other covalent cross-links are involved (135–137). Disulfide rearrangement is promoted by thiols, which accounts for the efficiency in setting of agents that are capable of reducing cystine to cysteine residues, eg, bisulfite and thioglycolate. The rearrangement is catalyzed by the thiol anion (RS⁻):

$$WSSW + RS^- \rightleftharpoons WSSR + WS^-$$

where W is wool. In acid solutions, thiols are largely nonionized, so that wool acquires little set under these conditions. Under alkaline or neutral conditions, thiols exist, at least partly, in their ionized state, and therefore wool sets well (138) (see also Hair preparations).

Commercial methods for setting durable creases in wool garments include (139): wet setting, eg, the Si-Ro-Set (140) and Immacula processes (141); hot-head pressing; dry creasing, eg, the Lintrak process (142); dry setting (143); and autoclave steaming. One of the techniques currently used for all-wool trousers and slacks of high wool content (>70%) blends is spraying the crease area with monoethanolamine sulfite and then steam pressing for approximately one minute (139). Alternatively, the Lintrak process, in which a thin line of silicone adhesive is laid into the inside of preformed trouser creases, may be used (142). For wool-blend slacks containing more than 30% polyester, acceptable crease durability can be obtained by hot-head or electric pressing at 170–180°C for 30 s. For wool and wool-blend skirts, the preferred method (139) is pleat formation followed by autoclave steaming at 110°C for 10 min.

Chemical setting alone will not stabilize wool garments sufficiently to maintain their shape during washing. Unless the garment is rendered shrink-resistant and the set shape stabilized, the garment will undergo felting shrinkage (see Durable-Press Wool).

Shrink-Resistance Treatment.   Two main types of shrinkage can occur when wool fabrics or garments are washed: relaxation and felting.

Relaxation Shrinkage.   Fabrics are often dried under tension because many drying machines are difficult to operate otherwise. The dimensions of most wet wool fabrics are greater than their dry dimensions. Hence, even when a cloth is dried at its exact wet dimensions, the fabric develops strains, which are released when the fabric is wet again and allowed to dry without tension. Such shrinkage is called relaxation shrinkage. Its control requires careful attention to finishing procedures (139).

Felting Shrinkage.   This occurs when individual fibers move during laundering or other wet treatments. Because of the difference in the friction coefficient of a wool fiber with and against the direction of the scale, any movement of fibers always takes place towards the fiber root. How this leads to felting is explained simply in Figure 11. The mechanisms have been discussed in detail (144–145).

Two main types of shrink-resistance treatment are used industrially: degradative treatments and additive treatments. In degradative treatments, the proteins in the scales of the fiber (Fig. 12) are modified by treatment with, for example, chlorine (oxidizing), so that the scales are softened or sometimes removed (Fig. 13). Chlorination, the most common degradative treatment in use industrially, breaks both disulfide cross-links and peptide chains, and it has been postulated that water-soluble, high molecular weight peptides are produced from proteins in the cuticle without disrupting the epicuticle (146). Consequently, upon immersion of the wool in water, the cuticle absorbs water and becomes very soft. The differential friction effect is greatly reduced by the softening of the cuticle so that preferential movement of the fibers and hence felting shrinkage is reduced. Additive treatments entail application of a polymer to the fiber. The polymer masks the scale structure (Fig. 14), bonds adjacent fibers together (Fig. 15), or both.

Other chemical treatments that impart resistance to felting but have not had industrial acceptance include extraction with hot alcohol followed by treatment with

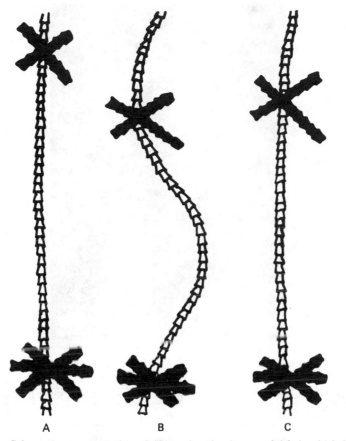

Figure 11.   Schematic representation of fiber migration in a wool fabric which leads to felting shrinkage. A fiber is firmly entangled (A) with other fibers at its lower end but only partly entangled at its upper end. When the fabric is compressed during washing (B), the two ends of the fiber are brought together and it is temporarily bent. The scales are oriented so that the fiber can straighten elastically by sliding through the upper entanglement (C). The fibers in the two entanglements are now drawn closer together permanently because the scales permit fiber movement in one direction only. The fabric has shrunk.

aqueous ammonium thioglycolate (147); treatment with concentrated solutions of sodium hydroxide (148); and treatment with 1,4-benzoquinone or mercury(II) acetate (149) or aryl monoisocyanates (150). Significant research effort has been put into developing polymers and processes suitable for shrink-resistance treatment of wool (145,151–154). Industrially, shrink-resistance polymers are applied to wool mainly while it is in the form of top or of fabric, or garment form. Table 11 lists some of the polymers that have been used.

The effective shrink-resistance polymers on wool tops (slivers of parallel fibers), eg, those used in the chlorine–Hercosett (156–158) and Dylan GRC (159) processes, function by masking the scale structure (Fig. 14). For effective scale masking, these commercial polymers must spread over the surface of the wool fiber to encapsulate it completely. At present, a prerequisite for this is modification of the surface by low level chlorination to increase its surface energy (160). Because chlorination pretreatment requires careful control to avoid fiber damage and to give even treatment, development of polymers and processing conditions that enable encapsulation and

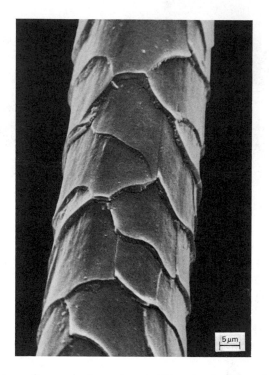

Figure 12.   Electron micrograph of a merino wool fiber showing the scales on the surface.

Table 11.   Some of the Polymers that Have Been Applied to Wool to Impart Shrink Resistance [a]

Trade name	Manufacturer	Backbone	Reactive group
Zeset TP	DuPont	poly(ethylene-*co*-vinyl acetate)	carboxylic acid chloride
Hercosett 57	Hercules	polyaminoamide	azetidinium
Primal K3	Rohm and Haas	polyacrylate	N-methylolamide
DC-109	Dow-Corning	dimethylpolysiloxane	silanol (in conjunction with an aminosilane cross-linker)
Oligan 500	CIBA-GEIGY	poly(propylene oxide)	thiol
Synthappret LKF	Bayer	poly(propylene oxide)	isocyanate
Synthappret BAP	Bayer	poly(propylene oxide)	carbamoyl sulfonate

[a] Ref. 155.

adhesion of the polymer to the fiber without pretreatment is an important research objective. Industrial chlorination has been improved by the introduction of the Kroy process, ie, the continuous chlorination of wool tops and sliver by passage of the material through a U-tube bath containing hypochlorous acid (161). This has been used both as a straight degradative process for shrink-resistance treatment and as a pretreatment before application of the Hercosett 57 (Hercules) resin.

Chlorination–resin processes are also used for shrink-resistance treatment of wool in fabric or garment form. However, for these products, there are also available processes that require no degradative pretreatment. The principal process used for fabric is the Sirolan BAP process (162–163), in which the fabric is padded with a mixture

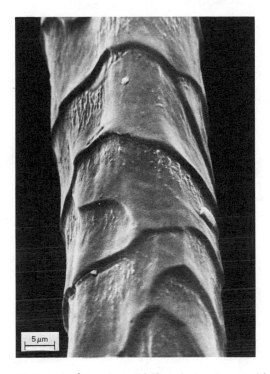

Figure 13. Electron micrograph of a merino wool fiber after treatment with chlorine, showing slight degradation of the scales.

of a reactive water-soluble polyurethane prepolymer and a polyurethane dispersion, with the polymers then cured by heating (see Urethane polymers).

Application of shrink-resistance polymers to garments or full-fashioned garment pieces is often done from organic solvents in modified drycleaning machines. Although several polymers have been developed for this purpose, only those based on silicones have received wide acceptance because of the soft hand they impart to the garment (164).

Three possible mechanisms by which polymers produce shrink-resistance in wool treated at the fabric or garment stage (145,165) are immobilization of fibers by fiber–fiber bonding which (Fig. 15) arises from small resin deposits cementing fibers together, thereby preventing fiber movement; scale masking or encapsulation of the fibers by a thin layer of polymer (Fig. 14); and deposition of large aggregates of polymer on the surface of fibers to prevent the scaly surfaces from interacting (166). The principal mechanism is believed to be fiber–fiber bonding, but probably all of these mechanisms operate to some extent depending on the type of polymer, the level of pretreatment (if any), the method of application, and subsequent processing (154).

Because of the high cost of polymer treatments for garments and the need for drycleaning equipment, research effort has been directed at developing processes in which polymers can be applied from aqueous solution by exhaustion (167–169).

Durable-Press Wool. Durable press is a description of garment performance implying that the garment is machine washable, retains any imparted creases or pleats, and requires minimum ironing after washing and drying. As already mentioned, chemical setting alone does not stabilize wool garments sufficiently to maintain their

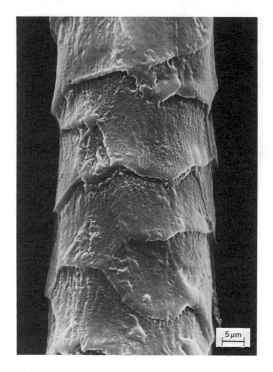

Figure 14. Electron micrograph of a merino wool fiber after treatment by the chlorine-Hercosett process, showing the polymer layer spread over the surface of the fiber.

shape during washing. The garment must also be rendered shrink-resistant. Thus, good durable-press effects on wool can be obtained in garments made from shrink-resistant wool fabric by a suitable setting treatment (139,170). However, careful selection of setting process is required to ensure the stability of the set to machine-wash conditions. Only two processes have, in fact, been applied commercially (139), ie, the CSIRO solvent–resin–steam process (171–172) and the International Wool Secretariat's durable-press process (173).

In the CSIRO process, a reactive polyurethane prepolymer is applied to the made-up garment from perchloroethylene. The garment is re-pressed to remove wrinkles and then hung in a steam oven. After a short warm-up period, saturated steam is introduced for 2 h. The polymer is cured on the garment in the required shape and then is capable of maintaining that shape while the free-hanging garment is set in steam.

The IWS process originally used a thiol-terminated prepolymer as the shrink-resistance agent, but this is no longer available commercially and may be replaced by the polymers used in the Sirolan BAP process (162–163). The polymer is applied to the fabric by padding and is cured. The fabric is then finished normally. After making-up, the garments are set using monoethanolamine sulfite.

Flame-Resistance Treatment.   Wool is naturally a very flame-resistant fiber, and in most applications, no special treatment is needed. However, consumer demand for higher standards of flammability in, for example, aviation furnishings, drapes, and carpets has led to the development of treatments for improving this property. The treatments evolved from the observation that chrome mordanting of wool markedly

Figure 15. Electron micrograph of merino wool fibers in a fabric that has been treated with a typical shrink-resistance polymer, showing fiber–fiber bond formation.

improves its flame resistance (174). However, this approach did not prove practical for all situations because of the discoloration of undyed wool by the chrome mordant. Subsequently, it was shown that titanium or zirconium compounds give similar results (175), but the zirconium compounds did not share the discoloration disadvantage. The process, ie, the IWS Zirpro process, has been used extensively for wool in aircraft interiors to meet FAA requirements and for curtains and carpets in public buildings and similar areas where high flame resistance is required by legislation (176–177) (see Flame retardants in textiles).

Shrink-resistance processes compatible with Zirpro-treated wool have also been developed (178).

Wool Grease

In wool scouring, the contaminants on the wool, mainly grease, dirt, suint, and protein material, are washed off the fiber and remain in the wastewaters either in emulsion or suspension (grease, dirt, protein) or in solution (suint). Centrifugal extraction of the wastewaters produces a grease contaminated with detergent and suint. This product is called wool grease.

Lanolin is wool grease that has been refined to lighten its color and reduce its odor and free fatty acid content. Wool wax is the pure lipid material of the fleece, extractable with the usual fat solvents such as diethyl ether and chloroform. Wool grease is a mixture of compounds that are classed as waxes. However, it does not have the physical characteristics usually displayed by waxes. It is soft and slightly sticky with a greasy appearance. Some centrifugally recovered greases are light buff or ivory in color and practically odorless. Those recovered by other methods contain dark-colored and odorous impurities as well as free fatty acids. Grease recovered by acid-cracking of scour wastewaters can be used in lanolin production, but the centrifugally recovered products are more suited to the alkali refining, bleaching, and deodorizing required in preparing pharmaceutical-grade lanolin. Table 12 gives some physical and chemical data for wool wax.

Chemical Composition. Wool wax is a complex mixture of esters of water-insoluble alcohols (180) and higher fatty acids (181) with a small proportion (ca 0.5%) of hydrocarbons (182). Considerable efforts have been made to identify the various components, but results are complicated by the fact that different workers use wool waxes from different sources and different analytical techniques. Nevertheless, significant progress has been made, and it is possible to give approximate percentages of the various components. The wool-wax acids (Table 13) are predominantly alkanoic, α-hydroxy, and ω-hydroxy acids. Each group contains normal, iso, and anteiso series of various chain length, and nearly all the acids are saturated.

The alcohol fraction is likewise a complex mixture of both aliphatic and cyclic compounds (Table 14). The principal components are cholesterol (34%) and lanosterol and dihydrolanosterol (38%). The aliphatic alcohols account for ca 22% of the unsaponifiable products. Sixty-nine components of aliphatic alcohols had been reported up to 1974 (latest reported work). The hydrocarbons (ca 0.5%) show structural similarity to the wool-wax acids or aliphatic alcohols and contain highly branched alkanes as well as cycloalkanes.

Wool-Grease Recovery. Current systems and research on wool-grease recovery have been reviewed comprehensively (187). The principal recovery process in use involves centrifuging in a cream-separator type of centrifuge modified by the addition of peripheral jets or other mechanical devices to remove dirt. Liquor is usually with-

Table 12.   Some Physical and Chemical Data for Wool Wax [a]

color	yellow to pale brown
density (at 15°C)	0.94–0.97
refractive index (40°C)	1.48
melting point, °C	35–40
free acid content, %	4–10
free alcohol content, %	1–3
iodine value (Wijs)	13–30
saponification value	95–120
molecular weight (Rast method, in phenyl salicylate)	790–880
fatty acids, %	50–55
alcohols, %	50–45
acids	
melting point, °C	40–45
iodine value (Wijs)	10–20
mean molecular weight	330
alcohols	
melting point, °C	55–65
iodine value (Wijs)	40–50
mean molecular weight	370

[a] Ref. 179.

Table 13. Summary of the Average Composition of Wool-Wax Acids[a]

Acids[b]	Chain length[c]	Wool-wax acids, %
normal acids	C_8–C_{38}	10
iso acids	C_8–C_{40}	22
anteiso acids	C_7–C_{41}	28
normal α-hydroxy acids	C_{10}–C_{32}	17
iso α-hydroxy acids	C_{12}–C_{34}	9
anteiso α-hydroxy acids	C_{11}–C_{33}	3
normal ω-hydroxy acids	C_{22}–C_{36}	3
iso ω-hydroxy acids	C_{22}–C_{36}	0.5
anteiso ω-hydroxy acids	C_{23}–C_{35}	1
polyhydroxy acids		4.5[d]
unsaturated acids		2[d]

[a] Ref. 181.

[b] In the iso series, a branching methyl group is attached at the penultimate position, and in the anteiso series, at the antepenultimate position.

[c] Ref. 183.

[d] Tentative data.

Table 14. Summary of the Average Composition of Wool-Wax Alcohols[a]

Alcohol[b]	Chain length[c]	Approximate % of wool-wax alcohols
normal monoalcohols	C_{14}–C_{34}	2
iso monoalcohols	C_{14}–C_{36} ⎱	13
anteiso monoalcohols	C_{17}–C_{35} ⎰	
normal alkan-1,2-diols	C_{12}–C_{25}	1
iso alkan-1,2-diols	C_{14}–C_{30} ⎱	6
anteiso alkan-1,2-diols	C_{15}–C_{29} ⎰	
Total		22
cholesterol		34
lanosterol ⎱		38
dihydrolanosterol ⎰		
Total		72
hydrocarbons		1
autooxidation products		5
undetermined		

[a] Refs. 180, 183–186.

[b] See footnote [b], Table 13.

[c] Ref. 183.

drawn from the second bowl of a five-bowl scouring plant, heated to ca 90°C, and passed into a large settling tank to remove the heaviest dirt particles. The hot emulsion is then passed through a centrifuge designed to remove suspended solids, reheated, and run through the cream-separator type of centrifuge. This separates most of the grease remaining in the liquors. The liquor is cooled to ca 50°C and returned to the scouring plant or held in a reserve tank for later use.

With the introduction of scouring with nonionic detergents instead of soap and soda, grease-recovery rates increased (188); however, the quality of the grease decreased slightly because of increased recovery of the grease from the tip of the fiber (oxidized grease) (189). The recovery of grease by centrifuging has been optimized for normal scouring conditions in the WRONZ Comprehensive Scouring System (190), and the Lo-flo system of scouring (64) yields grease recoveries of ca 85% of the grease originally on the wool, although the quality of the grease is again reduced.

Other techniques aimed at improving grease recovery (and often attempting also to improve the scouring process itself) have included solvent degreasing of the wool (57–58), solvent extraction of the liquor or sludge (191), aeration (192–193), and physical and chemical destabilization (187).

Grease Refining and Fractionation. Lanolin to be used in pharmaceuticals and cosmetics must conform to strict requirements of purity, such as those in the U.S. and British Pharmacopoeias (194–195). These include specifications for the maximum allowable content of free fatty acids, moisture, ash, and free chloride. Lanolin intended for certain dermatological applications may have to meet further specifications in relation to free-alcohol and detergent contents (196–197).

The refining process most commonly used involves treatment with hot aqueous alkali to convert free fatty acids to soaps followed by bleaching, usually with hydrogen peroxide, although sodium chlorite, sodium hypochlorite, and ozone have also been used. Other techniques include distillation, steam stripping, neutralization by alkali, liquid thermal diffusion, and the use of active adsorbents (eg, charcoal and bentonite) and solvent fractionation (187).

Uses of Wool Grease. The uses of wool grease, lanolin, and lanolin derivatives are wide, ranging from pharmaceuticals and cosmetics (qv) to printing inks (qv), rust preventatives, and lubricants (see Lubrication and lubricants).

In pharmaceutical uses, the general inertness of lanolin and its derivatives, together with their ease of emulsification (198), have been important criteria. These properties are also important in cosmetics, but with emphasis also on the ability to absorb large quantities of water or, after suitable modification, to stabilize o/w emulsions.

The anticorrosion properties of lanolin have been utilized over a considerable period, and the product has the status of a temporary corrosion inhibitor (199) (see Corrosion and corrosion inhibitors). There is probably potential for greater use of the poorer qualities of lanolin for long-term storage and protection of machine parts.

Other uses have been reviewed in detail (187).

BIBLIOGRAPHY

"Wool" in *ECT* 1st ed., Vol. 15, pp. 103–134, by John Menkart, Textile Research Institute; "Wool" in *ECT* 2nd ed., Vol. 22, pp. 387–418, by M. Lipson, Division of Textile Industry, CSIRO, Geelong, Victoria, Australia.

1. *Textile Terms and Definitions* 7th ed., Textile Institute, Manchester, England, 1975.
2. *Wool Quarterly* (2), 19 (1981).
3. International Wool Secretariat, London, 1982.
4. *Organon World Cotton Statistics*, Textile Economics Bureau, Roseland, N.J., 1981.
5. M. Lipson and U. A. F. Black, *J. R. Soc. N.S.W.* **78**, 84 (1945).
6. R. B. Sweeten, *J. Text. Inst. Trans.* **40**, T727 (1949).
7. *Report to the Australian Wool Corporation by Specialist Working Group on Sale by Description*, Melbourne, Australia, 1978.
8. G. A. Carnaby and A. J. McKinnon, *Australas. Text.* **2**(2), 11 (1982).
9. *Objective Measurement of Wool in Australia*, Australian Wool Corporation, Melbourne, Australia, 1973.

10. B. H. Mackay, *Proc. Int. Wool Text. Res. Conf.*, (*Pretoria*) **I,** 59 (1980).

11. *Core Test Regulations, Test Methods, IWTO—19-76, 28-75, 3-73, 20-69,* International Wool Textile Organization, London, 1968.

12. *Australian Standard 1363—1976,* Standards Association of Australia, Melbourne, Australia, 1976.

13. A. Baumann, *A Wool Staple Length Meter,* CSIRO Division of Textile Physics, Ryde, New South Wales, Australia, in preparation.

14. J. Grignet and G. Bertoni, *Proc. Int. Wool Text. Res. Conf.* (*Aachen*) **V,** 393 (1975).

15. R. N. Caffin, *J. Text. Inst.* **71,** 65 (1980).

16. R. H. Peters, *The Chemistry of Fibres,* Vol. I of *Textile Science,* Elsevier, Amsterdam, The Netherlands, 1963, p. 263.

17. J. A. Maclaren and B. Milligan, *Wool Science: The Chemical Reactivity of the Wool Fibre,* Science Press, Sydney, Australia, 1981, p. 2.

18. K. R. Makinson, *Wool Shrinkproofing,* Marcel Dekker, Inc., New York, 1979, p. 147.

19. J. H. Bradbury and J. D. Leeder, *Aust. J. Biol. Sci.* **23,** 843 (1970).

20. H. M. Appleyard and C. M. Greville, *Nature* **166,** 1031 (1950).

21. M. S. C. Birbeck and E. H. Mercer, *J. Biophys. Biochem. Cytol.* **3,** 203, 215 (1957).

22. J. H. Bradbury and K. F. Ley, *Aust. J. Biol. Sci.* **25,** 1235 (1972).

23. H. Zahn, *Proc. Int. Wool Text. Conf.* (*Pretoria*) **I**(supplement), (1980).

24. C. A. Anderson, J. D. Leeder, and D. S. Taylor, *Wear* **21,** 115 (1972).

25. H. Horio and T. Kondo, *Text. Res. J.* **23,** 373 (1953).

26. J. H. Dusenbury and J. Menkart, *Proc. Int. Wool Text. Res. Conf.* (*Australia*), F-142 (1956).

27. J. H. Dusenbury, E. H. Mercer, and J. H. Wakelin, *Text. Res. J.* **24,** 890 (1954).

28. G. V. Chapman and J. H. Bradbury, *Arch. Biochem. Biophys.* **127,** 157 (1968).

29. P. Kassenbeck, *Proc. Int. Wool Text. Res. Conf.* (*Paris*) **I,** 367 (1965).

30. C. W. Hock, R. G. Ramsay, and M. Harris, *J. Res. Natl. Bur. Stand.* **27,** 181 (1941).

31. *Ibid.,* p. 234.

32. J. L. Farrant, A. L. G. Rees, and E. H. Mercer, *Nature* **159,** 535 (1947).

33. *Wool Research,* Vol. 2, Wool Industries Research Association, Leeds, UK, 1955.

34. "Man-Made Fiber Chart," *Text. World,* (1961).

35. W. von Bergen, ed., *Wool Handbook,* 3rd ed., Interscience Publishers, a division of John Wiley & Sons, Inc., New York, Vol. 1, 1963, p. 180.

36. J. B. Speakman, *J. Text. Inst. Trans.* **18,** T431 (1927).

37. E. G. Bendit, *J. Macromol. Sci-Phys.* **B17,** 129 (1980).

38. M. Feughelman, *Proc. Int. Wool Text. Res. Conf.* (*Pretoria*) **I,** 35 (1980).

39. Ref. 17, p. 6.

40. D. H. Simmonds, *Aust. J. Biol. Sci.* **8,** 537 (1955).

41. J. J. O'Donnell and E. D. P. Thompson, *Aust. J. Biol. Sci,* **14,** 740 (1962).

42. J. H. Bradbury, G. V. Chapman, and N. L. R. King, *Aust. J. Biol. Sci.* **18,** 353 (1965).

43. B. Milligan, L. A. Holt, and J. B. Caldwell, *Appl. Polym. Symp.* **18,** 113 (1971).

44. M. Cole, J. C. Fletcher, K. L. Gardner, and M. C. Corfield, *Appl. Polym. Symp.* **18,** 147 (1971).

45. L. A. Holt, B. Milligan, and L. J. Wolfram, *Text. Res. J.* **44,** 846 (1974).

46. I. J. O'Donnell and E. F. Woods, *J. Polym. Sci.* **21,** 397 (1956).

47. J. M. Gillespie in A. G. Lyne and B. F. Short, ed., *Biology of the Skin and Hair Growth,* Angus and Robertson, Sydney, Australia, 1965, p. 377.

48. H. Zahn and M. Biela, *Eur. J. Biochem.* **5,** 567 (1968).

49. W. G. Crewther and F. G. Lennox, *J. Proc. R. Soc. N.S.W.* **108,** 95 (1975).

50. W. G. Crewther, *Proc. Int. Wool Text. Res. Conf.* (*Aachen*) **I,** 1 (1975).

51. T. A. A. Dopheide, *Eur. J. Biochem.* **34,** 120 (1973).

52. A. Brearley and J. A. Iredale, *The Woollen Industry: An Outline of the Woollen Industry and Its Processes from Fibre to Fabric,* Wira, Leeds, UK, 1977.

53. A. Brearley and J. A. Iredale, *The Worsted Industry: An Account of the Worsted Industry and Its Processes from Fibre to Fabric,* Wira, Leeds, UK, 1980.

54. G. F. Wood, *Text. Prog.* **12**(1), (1982).

55. *Text. J. Aust.* **44**(6), 34 (1969).

56. *Text. Rec.,* 69 (Jan. 1964).

57. *CSIRO Division of Textile Industry Report No. G.10,* CSIRO Division of Textile Industry, Geelong, Australia, 1960.

58. J. Brach, *Wool Sci. Rev.* (36), 38 (May 1969).

59. A. S. C. Lawrence, *Nature* **183,** 1491 (1959).
60. R. P. Harker, *J. Text. Inst. Trans.* **50,** T189 (1959).
61. C. A. Anderson, *J. Text. Inst.* **73,** 289 (1982).
62. C. A. Anderson, *Australas. Text.* **2**(5), 28 (1982).
63. C. A. Anderson, J. R. Christoe, A. J. C. Pearson, J. J. Warner, and G. F. Wood, *Developments in Scouring and Carbonising 1980, Disc 80,* Australian Wool Corporation, Melbourne, 1980, p. 1.
64. G. F. Wood, A. J. C. Pearson, and J. R. Christoe, *CSIRO Division of Textile Industry Report No. G39,* CSIRO Division of Textile Industry, Geelong, Australia, 1980.
65. *Japanese Wool Textile Industry's Treatment of Effluent from Wool Scouring,* Japan Wool Spinners' Association, Tokyo, Jpn., 1973, p. 53.
66. R. G. Stewart, G. V. Barker, P. E. Chisnall, and J. L. Hoare, *WRONZ Report No. 25,* Wool Research Organization of New Zealand, Inc., Christchurch, New Zealand, 1974.
67. R. G. Stewart, *WRONZ Report No. 27,* Wool Research Organization of New Zealand, Inc., Christchurch, New Zealand, 1974.
68. R. G. Stewart, P. E. Chisnall, J. M. Flynn, and R. G. Jamieson, *WRONZ Communication No. 36,* Wool Research Organization of New Zealand, Inc., Christchurch, New Zealand, 1975.
69. T. A. Pressley and W. G. Crewther, *Wool Sci. Rev.* (30), 16 (1966); (31), 1 (1967).
70. N. C. Gee, *Shoddy and Mungo Manufacture,* Emmott & Co., Bradford, UK, 1950.
71. H. K. Rouette and G. K. Kittan, *Text. Praxis Int.* **36,** 784 (1981).
72. W. G. Crewther, *Proc. Int. Wool Textile Res. Conf. (Australia)* **E,** 408 (1955).
73. W. G. Crewther and T. A. Pressley, *Text. Res. J.* **28,** 67 (1958).
74. A. E. Davis, A. J. Johnson, and L. R. Mizell, *Text. Res. J.* **31,** 825 (1961).
75. M. S. Nosser, M. Chaikin, and A. J. Datyner, *J. Text. Inst.* **62,** 677 (1971).
76. M. S. Nosser and M. Chaikin, *Proc. Int. Wool Text. Res. Conf. (Aachen)* **IV,** 136 (1975).
77. H. J. Katz, D. E. A. Plate, and G. F. Wood, *Wool Sci. Rev.* (45), 28 (1973).
78. M. A. White, *Text. Prog.* **13**(2), (1983).
79. H. K. Rouette and A. Ohm, *Melliand Textilber.* **62,** 583 (1981).
80. H. K. Rouette, A. Gotz, and A. Ohm, *Int. Text. Bull. Dye. Print. Finish.* (3), 211 (1981).
81. M. A. White, *Proc. Conf. Recent Dev. in Wool and Wool Blend Processing,* CSIRO Division of Textile Industry, Geelong, Australia, 1983, p. 20.
82. I. B. Angliss, *CSIRO Division of Textile Industry Report No. G23,* CSIRO Division of Textile Industry, Geelong, Australia, 1973.
83. I. B. Angliss, P. R. Brady, J. Delmenico, and R. J. Hine, *Text. J. Aust.* **43**(4), 17 (1968).
84. C. L. Bird, *The Theory and Practice of Wool Dyeing,* Society of Dyers and Colourists, Bradford, England, 1972.
85. C. H. Giles in C. L. Bird and W. S. Boston, ed., *The Theory of Coloration of Textiles,* Society of Dyers and Colourists, Bradford, England, 1974, p. 78.
86. R. H. Peters, *The Physical Chemistry of Dyeing,* Vol. III of *Textile Chemistry,* Elsevier Publishing, Amsterdam, Oxford, England, and New York, 1975, p. 203.
87. D. M. Lewis, *Rev. Prog. Color.* **8,** 10 (1981).
88. Ref. 84, p. 127.
89. *Ibid.,* p. 106.
90. R. V. Peryman, *Proc. Int. Wool Text. Res. Conf. (Australia)* **E,** 17 (1955).
91. R. L. Holmes-Brown and G. A. Carnaby, *WRONZ Communication No. 67,* Wool Research Organization of New Zealand, Inc., Christchurch, New Zealand, 1980.
92. R. R. D. Holt and F. J. Harrigan, *Melliand Textilber.* **9,** 745 (1979).
93. I. B. Angliss, *Text. Prog.* **12**(4), (1982).
94. J. F. Graham and R. R. D. Holt, *Colourage* **25**(10), 45 (1978).
95. P. Spinaci, A. Panelli, and N. Cicchello, *Int. Dyer* **159,** 349 (1978).
96. J. F. Graham, *Int. Dyer* **150,** 558 (1973).
97. J. F. Graham and I. Seltzer, *Textilveredlung* **9,** 551 (1974).
98. J. R. Hine and J. R. McPhee, *Proc. Int. Wool Tex. Res. Conf.,* Paris III, 183 (1965).
99. L. Peters and C. B. Stevens, *J. Soc. Dyers Colour.* **72,** 100 (1956).
100. W. Beal and G. S. A. Corbishley, *J. Soc. Dyers Colour.* **87,** 329 (1971).
101. I. B. Angliss and R. J. Hine, *Proc. Int. Wool Text. Res. Conf. (Pretoria)* **V,** 629 (1980).
102. *Engineer* **244,** 28 (Apr. 7, 1977).
103. P. R. Brady, *Text. Prog.* **12**(4), (1982).
104. *Wool Statistics 1974–75,* Commonwealth Secretariat, London, 1975, p. 35.

105. K. Reincke, *Text. Praxis Int.* **25,** 419 (1970).
106. Brit. Pat. 1,377,085 (Aug. 19, 1966), (to Hoechst A.G.).
107. Brit. Pat. 1,165,253 (May 1, 1967), (to Sandoz Ltd.).
108. A. I. Matekskiĭ, V. A. Kuz'menkov, and N. I. Volkova, *Tektil. Prom.* **30**(7), 69 (1970).
109. P. R. Brady and P. G. Cookson, *Proc. Int. Wool Text. Res. Conf. (Pretoria)* **V,** 517 (1980).
110. D. M. Lewis and I. Seltzer, *J. Soc. Dyers Colour.* **88,** 327 (1972).
111. P. R. Brady and R. J. Hine, *CSIRO Division of Textile Industry Report No. G35*, CSIRO Division of Textile Industry, Geelong, Australia, 1979.
112. P. R. Brady and P. G. Cookson, *J. Soc. Dyers Colour.* **97,** 159 (1981).
113. G. A. Smith, *Br. Knit. Ind.* **44,** 89 (Dec. 1971).
114. *Ibid.*, **46,** 77 (July 1973).
115. P. R. Brady, P. G. Cookson, K. W. Fincher, and D. M. Lewis, *J. Soc. Dyers Colour.* **96,** 188 (1980).
116. P. R. Brady, P. G. Cookson, K. W. Fincher, and D. M. Lewis, *Proc. Int. Wool Text. Res. Conf. (Pretoria)* **V,** 609 (1980).
117. Ref. 84, p. 54.
118. C. Frommelt, *Bayer Farben Rev.* (27), 46 (1977).
119. *Process Data Sheet CTE-054e*, BASF AG, Leverkusen, FRG, 1978.
120. B. Milligan, *Proc. Int. Wool Text. Res. Conf. (Pretoria)* **V,** 167 (1980).
121. C. H. Nicholls in N. S. Allen, ed., *Developments in Polymer Photochemistry*, Applied Science Publishers, London, 1980, p. 125.
122. P. J. Waters, N. A. Evans, L. A. Holt, and B. Milligan, *Proc. Int. Wool Text. Res. Conf. (Pretoria)* **V,** 195 (1980).
123. I. H. Leaver, P. J. Waters, and N. A. Evans, *J. Polym. Sci. Polym. Chem.* **17,** 1531 (1979).
124. L. A. Holt, B. Milligan, and L. J. Wolfram, *Text. Res. J.* **44,** 846 (1974).
125. B. Milligan and D. J. Tucker, *Text. Res. J.* **34,** 681 (1964).
126. R. J. Mayfield, *Text. Prog* **11**(4), (1982).
127. R. J. Mayfield, *CSIRO Division of Textile Industry, Report No. G40*, CSIRO Division of Textile Industry, Geelong, Australia, 1979.
128. R. J. Mayfield and G. J. O'Loughlin, *CSIRO Division of Textile Industry Report No. G41*, CSIRO Division of Textile Industry, Geelong, Australia, 1980.
129. R. M. Hoskinson, R. J. Mayfield, and I. M. Russell, *J. Text. Inst.* **67,** 19 (1976).
130. R. M. Hoskinson and I. M. Russell, *J. Text. Inst.* **65,** 387 (1974).
131. A. L. Black, R. M. Hoskinson, and I. M. Russell, *J. Text. Inst.* **67,** 68 (1976).
132. R. J. Mayfield and I. M. Russell, *J. Text. Inst.* **70,** 53 (1979).
133. F. W. Jones, R. J. Mayfield, and G. J. O'Loughlin, *Proc. Int. Wool Text. Res. Conf. (Pretoria)* **V,** 431 (1980).
134. C. S. Whewell, *J. Text. Inst. Proc.* **47,** P851 (1956).
135. J. B. Caldwell, S. J. Leach, S. J. Meschers, and B. Milligan, *Text. Res. J.* **34,** 627 (1964).
136. H. D. Weigmann, L. Rebenfeld, and C. Dansizer, *Proc. Int. Wool Text. Res. Conf. (Paris)* **II,** 319 (1965).
137. W. G. Crewther, *J. Soc. Dyers Colour.* **82,** 54 (1966).
138. Ref. 17, p. 294.
139. M. A. White, *CSIRO Division of Textile Industry Report No. G44*, CSIRO Division of Textile Industry, Geelong, Australia, 1982.
140. A. J. Farnworth, *Am. Dyest. Rep.* **49,** 996 (1960).
141. *Bull. Cent. Text. Contr. Rech. Sci. (Roubaix)* **43,** 39 (1959).
142. Technical brochures, IWS Technical Center, Ilkley, UK.
143. J. R. Cook and J. Delmenico, *Proc. Int. Wool Text. Res. Conf. (Paris)* **III,** 419 (1965).
144. K. R. Makinson, *Wool Sci. Rev.* **24,** 34 (1964).
145. Ref. 18, p. 233.
146. K. R. Makinson, *Text. Res. J.* **44,** 856 (1974).
147. A. J. Farnworth, *J. Soc. Dyers Colour.* **77,** 483 (1961).
148. M. Lipson, *J. Text. Inst. Proc.* **38,** P279 (1947).
149. T. Barr and J. B. Speakman, *J. Soc. Dyers Colour.* **60,** 783 (1944).
150. J. E. Moore, *Text. Res. J.* **26,** 936 (1956).
151. *Wool Sci. Rev.* **36,** 2 (1969); **37,** 37 (1969).
152. T. Shaw and J. Lewis, *Text. Prog.* **4**(3), 1 (1972).
153. T. Shaw, *Wool Sci. Rev.* **46,** 44 (1973); **47,** 14 (1973).

154. J. Lewis, *Wool Sci. Rev.* **54,** 2 (1977); **55,** 23 (1978).
155. Ref. 17, p. 236.
156. J. R. McPhee, *Text. J. Aust.* **72,** 60 (1972).
157. P. Smith and J. H. Mills, *Chemtech*, 748 (1973).
158. J. Lewis, *Textilveredlung* **11,** 214 (1976).
159. *Hosiery Trade J.* **79,** 82 (1972).
160. R. H. Earle, R. H. Saunders, and L. R. Kangas, *Appl. Polym. Symp.* **18,** 707 (1971).
161. Brit. Pat. 1,524,392 (Sept. 13, 1978), (to Kroy Unshrinkable Wools Ltd.).
162. K. W. Fincher and M. A. White, *CSIRO Division of Textile Industry Report No. G30*, CSIRO Division of Textile Industry, Geelong, Australia, 1977.
163. M. A. Rushforth, *IWS Technical Information Bulletin No. 12*, International Wool Secretariat, Ilkley, England, 1977.
164. R. Bowrey, C. Brooke, and B. Robinson, *Textilia* **51,** 41 (1975).
165. A. Kershaw and J. Lewis, *Text. Month*, 40 (1976).
166. Ref. 18, p. 268.
167. D. Allanack, M. Rushforth, and T. Shaw, *Proc. Text. Inst. Conf., Edinburgh*, Textile Institute, Manchester, England, 1978.
168. A. Bereck, *Text. Res. J.* **49,** 233 (1979).
169. T. Jellinek, A. De Boos, and M. A. White, *Proc. Int. Wool Text. Res. Conf. (Pretoria)* **V,** 125 (1980).
170. A. G. De Boos and M. A. White, *Proc. Hung. Text. Conf., Budapest*, (1979).
171. M. A. White, *Text. J. Aust.* **48**(6), 20 (1973).
172. M. A. White, *CSIRO Division of Textile Industry Report No. G21*, CSIRO Division of Textile Industry, Geelong, Australia, 1977.
173. T. Shaw, *Wool Sci. Rev.* **47,** 14 (1973).
174. M. J. Koroskys, *Am. Dyest. Rep.* **60**(5), 48 (1971).
175. L. Benisek, *Text. Manuf.* **99,** 36 (Jan.–Feb. 1972); *Melliand Textilber.* **53,** 931 (1972).
176. L. Benisek, *J. Text. Inst.* **65,** 102, 140 (1974).
177. P. Gordon and L. Stephens, *J. Soc. Dyers Colour.* **90,** 239 (1974).
178. L. Benisek, G. Edmondson, and B. Greenwood, *Text. Inst. Ind.* **14,** 344 (1976).
179. E. V. Truter, *Wool Wax*, Cleaver-Hume Press Ltd., London, 1956, p. 32.
180. K. Motiuk, *J. Am. Oil Chem. Soc.* **56,** 651 (1979).
181. *Ibid.*, **56,** 91 (1979).
182. *Ibid.*, **57**(4), 145 (1980).
183. F. Fawaz, C. Miet, and F. Puisieux, *Ann. Pharm. Fr.* **31**(1), 63 (1973).
184. K. E. Murray and R. Schoenfeld, *Aust. J. Chem.* **8,** 424 (1955).
185. D. H. S. Horn, *J. Sci. Food Agric.* **9,** 632 (1958).
186. D. T. Downing, Z. H. Krantz, and K. E. Murray, *Aust. J. Chem.* **13,** 80 (1960).
187. R. G. Stewart and L. F. Story, *WRONZ Tech. Paper No. 4*, Wool Research Organization of New Zealand, Inc., Christchurch, New Zealand, 1980.
188. R. E. Wolfrom, *Am. Dyest. Rep.* **43,** P372 (1954).
189. C. A. Anderson and G. F. Wood, *Nature* **193,** 742 (1962).
190. R. G. Stewart, G. V. Barker, P. E. Chisnall, and J. L. Hoare, *WRONZ Report No. 25*, Wool Research Organization of New Zealand, Inc., Christchurch, New Zealand, 1974.
191. C. Moxhet, *Ind. Text. Belge* **7/8,** 51 (1971); **9,** 83 (1971); *World Text. Abstr.* 8105 (1971).
192. L. F. Evans and W. E. Ewers, *Aust. J. Appl. Sci.* **4,** 552 (1953).
193. USSR Pat. 104,212 (Dec. 1956), A. S. Salin and co-workers.
194. *The United States Pharmacopoeia XX (USP XX–NF XV)*, The United States Pharmacopeial Convention, Inc., Rockville, Md., 1980.
195. *British Pharmacopoeia*, H.M.S.O., London, 1980.
196. E. W. Clark, E. Cronin, and D. S. Wilkinson, *Contact Dermatitis* **3,** 69 (1977).
197. R. Elder, *J. Environ. Path. Toxicol.* **4**(4), 63 (1980).
198. A. Castillo and J. M. Sure, *Ars Pharm.* **9,** 367 (1968).
199. U.S. Pat. 2,473,614 (June 21, 1949), E. Snyder (to American Chemical Paint Co.).

General References

Refs. 17, 18, 35, 52, 53, 54, 84, 126, and 179 are also general references.

P. Alexander and R. F. Hudson in C. Earland, ed., *Wool: Its Chemistry and Physics*, 2nd ed., Chapman & Hall Ltd., London, 1963.

W. J. Onions, *Wool: An Introduction to its Properties, Varieties, Uses and Production*, Ernest Benn Ltd., London, 1962.

R. S. Asquith, ed., *Chemistry of Natural Protein Fibers*, Plenum Press, New York and London, 1977.

R. G. Stewart and L. F. Story, *Wool Grease: A Review of Its Recovery and Utilisation*, Wool Research Organization of N.Z. (Inc.), Christchurch, 1980; comprehensive review of the literature from 1947 to mid-1979.

Papers, 1st International Wool Textile Research Conference, Australia, 1955, CSIRO, Melbourne, Australia.

J. Text. Inst. Trans. **51,** T489 (1960).

Papers, 3rd International Wool Textile Research Conference, Paris, l'Institut Textile de France, Paris, 1965.

L. Rebenfeld, ed., *Proceedings of the 4th International Wool Textile Research Conference*, International Publishers, a division of John Wiley & Sons, Inc., New York, 1971.

H. Ziegler, ed., *Proceedings of the 5th International Wool Textile Research Conference*, Deutsches Wollforschungsinstitut an der Technischen Hochschule, Aachen, 1975.

Proceedings of the 6th International Wool Textile Research Conference, South African Wool and Textile Research Institute (SAWTRI) of the Council for Scientific and Industrial Research (CSIR), Pretoria, S. Africa, 1980.

Wool Science Review, International Wool Secretariat, London; published on an irregular basis.

W. S. Boston
CSIRO

INDEX

A

Abaca, 173
Abrasive papers
    fibers in, 186
Absorbit, 399
Abutilon avecennae gaetn., 182
Abutilon theophrasti, 173, 182
Acala cotton, 103
Acelan, 3
Acele, 84
Acesil, 84
Acetate
    generic name, 120
Acetate fibers, 460
Acetate filament
    price, 134
Acetic acid [*64-19-7*]
    in textile finishing, 475
Acetic anhydride [*108-24-7*]
    cellulose acetate manufacture, 62, 122
    novoloid fiber bleaching, 311
Acetone
    cellulose acetate solvent, 74
    in extrusion of cellulose acetates, 78
    modacrylic fiber solvent, 26
Acetylation
    of cellulose, 75
Acetylation catalysts, 74
Acid dyes
    for wool, 532
Acribel, 2
Acrilan, 1
Acrylamide [*79-06-1*]
    flame-retardant treatment, 199
Acrylamides
    in modacrylics, 127
Acrylates
    in textile finishing, 483
Acrylic
    generic name, 120
Acrylic and modacrylic fibers, 1
Acrylic fibers, 1, 126
    commercial processes, 10
    manufacture, 8
    moisture absorbent, 19
    producers, 2
    properties, 127
Acrylic textile staple
    price, 134
Acrylonitrile [*107-13-1*], 114
    in acrylic fibers, 1
    copolymers, 9
    polymer fibers, 126
Acrylonitrile copolymers, 25
Activated-carbon fibers, 316
Adhesives
    for nonwoven textiles, 294

Adipic acid [*124-04-9*]
    in nylon, 123
Admel, 258
Aerotex UM, 39
Afterglow
    in textile burning, 190
Agave, 173
Agave cantala, 173, 178
Agave fourcroydes, 173, 178
Agave funhana, 178
Agave letonae, 178
Agave lophanfu, 178
Agave sisalana, 173, 177
Agerite, 93
Agilon, 362
Agilon process, 132
Aircraft carriers
    arrestor gear in, 357
Airplane tires
    nylon-6,6 in, 499
Aksa, 3
Alanine [*302-72-2*]
    in silk, 454
    in wool, 529
Albene, 84
Alcantara, 228, 156
Alginate fibers
    chemical, 123
Alkali cellulose
    in rayon mfg, 400
Alkoxymethylureas
    alkaline hydrolysis of, 474
N-Alkylamides
    in textile finishing, 472
Alkyleneureas
    as cross-linking agents, 477
    in textile finishing, 477
Allophanic acid [*625-78-5*]
    esters in spandex fiber, 143
Alloys
    for metal fibers, 232
All-Pro turf, 435
Aloe, 178
Alumina [*1344-28-1*], 440. (See also *Aluminum oxide*.)
    fibers, 131
    refractory fibers, 439, 440
Alumina fibers, 440
Alumina–silica–boria fibers, 440
Alumina–silica–chromia fibers, 440
Alumina–silica fibers, 440
Aluminum acetate, basic [*142-03-0*]
    refractory fibers from, 445
Aluminum borosilicate [*12794-53-5*]
    refractory fibers, 445
Aluminum borosilicate fibers, 445
Aluminum oxide, mixture with silicon dioxide and boron trioxide [*65997-17-3**], 440
Aluminum oxide, mixture with silicon dioxide and chromium trioxide [*65997-17-3**], 440

Aluminum oxychloride [13596-11-7]
    refractory fibers from, 445
Amara, 156
Amcel, 84
Ames mutagenicity test, 207
Aminized cotton, 113
Amino acids
    in wool, 529
4(4′-Aminobenzamido) benzoyl chloride
        hydrochloride [25086-59-3], 52
m-Aminobenzoic acid [99-05-8], 52
p-Aminobenzoic acid [150-13-0], 52
m-Aminobenzoyl chloride hydrochloride, 52
p-Aminobenzoyl chloride hydrochloride [3016-
        64-6], 52
Aminocaproic acid [1319-82-0]
    in nylon, 124
2-Amino-3,5-diiodobenzoic acid [609-86-9]
    stabilizer in nylon, 356
Amino resins
    in textile finishing, 469
Amino-s-triazines
    in textile finishing, 476
Ammonia [7664-41-7]
    cotton treatment, 109
    as mercerizing agent, 494
    in wool, 529
Ammonia cure method
    for flame retardants, 201
Ammonium bromide [12124-97-9]
    textile flame retardant, 192
Ammonium phosphate [7783-28-0]
    cellulose flame retardant, 197
Ammonium sulfamate [7773-06-0]
    textile flame retardant, 192
Ammonium sulfate [7783-20-2]
    as dehydrating agent, 407
    as leatherlike material filler, 223
Ammonium thioglycolate [5421-46-5]
    in shrink-resistance treatments, 538
Ampholytes
    in polyamides, 367
Anilana, 3
Anthrenus verbasci, 535
Antibacterials
    in textile finishes, 492
Antiblaze 19
    flame retardant, 205
Antiblaze 19T
    flame retardant, 205
Antimicrobial finishes
    for cotton, 115
Antimicrobial textile finishes, 489
Antimony oxide [1314-60-9]
    in textile flame retardants, 204
Antimony oxychloride [7791-08-4]
    flame retardant, 192
Antistatic agents
    for nylon, 367

for textiles, 167
Aphid, 104
L-Arabinan [9060-75-7]
    in ramie fibers, 181
Arachne, 300
Aramid
    generic name, 120
Aramid fibers, **32**, 125
    production, 57
    in tire cords, 501
Aramids, 347.   (See also Polyamides, polyamide
        fibers.)
Area-bonding
    spunbondeds, 266
Arenka, 34
Arginine [74-79-3]
    in silk, 454
    in wool, 529
Arnel, 84
Aromatic polyamide fibers, 125
Arrestor ropes, 357
Aryl monoisocyanates
    in shrink-resistance treatments, 539
Asahi Kasei Spandex, 148
Asahi spunbond, 258
Asbestos [12001-29-5]
    fibers from, 438
Asbestos fibers, 460
Asclepias, 183
Ashi, 84
Aspartic acid [56-84-8]
    in silk, 454
    in wool, 529
Astroturf Stadium Surface, 435
Astroturf track surface, 435
A-Tell, 381
Attalea funifera, 173, 179
Avisco, 84
Avlin, 394
Avril, 399
Avril III, 399
Axtar, 258
Azlon
    generic name, 120
Azlon fibers, 460
4,4′-Azodibenzoyl chloride [10252-29-6], 47
Azodicarbonamide [123-77-3]
    blowing agent in coated fabrics, 96
    in spunbondeds mfg, 274
Aztran, 216

B

Bahia piassava, 179
Balsa fiber, 183
Banana plant, 176
Ban-lon yarns, 132

Barex, 228
Basketball courts
    surfaces, 425
Bass fiber, 179
Bast (soft) fibers, 173
BCF nylon, 124
BDP.   See *Bundle-drawing process.*
Bedding
    nonwoven, 302
Bell Flam, 399
Bemberg rayon process, 120
Bem-liese, 258
Benares hemp, 182
Bending
    of fibers, 71
Benzoate fibers, 129
1,4-Benzoquinone [*106-51-4*]
    in shrink-resistance treatments, 539
Benzyl alcohol [*100-51-6*]
    as dye carrier, 533
Beryllium
    fibers, 131
Beslon, 3, 156
Bias tires
    for passenger cars, 499
Bico.   See *Bicomponent fibers*
Bicomponent acrylic fibers, 19
Bicomponent fibers, 137, 371, 152
Bicomponent fibers (bico), 152
Bicomponent modacrylic fibers, 27
Biconstituent fibers, 372
Bidim, 258
Biguanide [*56-03-1*]
    bromobutyrate reaction, 200
Binders
    for nonwoven textiles, 294
    in spunbondeds mfg, 269
Biocidal textile finishes, 489
4,4′ Biphenyldicarbonyl chloride [*2351-37-3*], 47
Birefringence
    of cellulose acetate, 64
Bis(4-aminocyclohexyl)methane [*1761-71-3*]
    in Qiana, 374
Bis(*p*-aminocyclohexyl)methane [*1761-71-3*]
    in nylon, 374
N,N′-Bis(*m* aminophenyl)isophthalamide
    [*2658-07-3*], 55
N,N′-Bis(4-aminophenyl) oxamide [*19532-78-6*],
    46
Bis(aziridinyl)chloromethylphosphine oxide
    [*13846-34-9*]
    flame-retardant treatment, 203
Bischloromethyl ether [*542-88-1*]
    from thpc, 201
Bis(hydroxyethyl) sulfone [*2580-77-0*]
    cotton treatment, 113
1,3-Bis(hydroxymethyl)-5-ethyltetrahydrotriazin-
    2(1*H*)one [*134-97-4*]
    in textile finishing, 478

1,3-Bis(hydroxymethyl)-5-(2-methoxyethyl)triaz-
    inone [*83543-08-2*]
    in textile finishing, 478
Bis(methoxymethyl)urea [*141-07-1*], 474
1,3-Bis(methoxymethyl)uron [*7388-44-5*]
    in textile finishing, 477, 479
Biuret 1108-19-0]
    in spandex fiber, 143
Black fleahopper, 104
Blankets
    refractory, 448
Bleaching
    silk, 455
    of wool, 534
Block copolymers, nylon, 373
Blown fibers
    novolac, 310
Blue C, 394
Boehmeria nivea, 173, 180
Boll weevil, 103
Bollworm, 103
Bombyx mori, 451, 463
Bonafill, 155
Bonding
    nonwovens, 294
Book covers
    nonwoven, 274
Borax [*1330-43-4*]
    textile flame retardant, 192
Boria, 439.   (See also *Boron oxide.*)
    in refractory fibers, 439
Boric acid [*10043-35-3*]
    textile flame retardant, 192
Boron [*7440-42-8*]
    fibers, 131
    nitridization, 446
    in refractory fibers, 110
Boron carbide
    fibers, 131
Boron fibers, 440
Boron nitride [*10043-11-5*]
    fibers, 131
    in refractory fibers, 440
Boron nitride fibers, 440
Boron oxide [*1303-86-2*]
    from boron nitride, 438
Boron trichloride [*10294-34-5*]
    reduction, 446
Boron–tungsten
    fibers, 131
Bowstring hemp, 173, 179
Brass coating
    for steel tire cords, 506
Brazilian pita, 178
Brazilian sisal, 178, 179
Breaking strength
    of fiber, defined, 349
Breaking toughness
    of fiber, defined, 349

Brighteners
  in textile finishing, 469
British Cotton Industry Association, 67
Bromoform [75-25-2]
  flame-retardant treatment, 201
Broom root, 173
Brown hemp, 182
Brown lung disease, 116
Bruise resistance
  of tires, 510
Brushes
  from vegetable fibers, 174
Bulana, 2
Bundle-drawing process
  for metal fibers, 238
Burning
  mechanisms in textiles, 190
Burst strength
  of tires, 509
n-Butanol [71-36-3]
  as dye carrier, 533
Butyl benzoate [136-60-7]
  dye carrier, 63
Butyl rubber, 92
Byssinosis, 116

C

Cabuya plant, 178
Cadillo, 173
Cadmium ethylenediamine hydroxide [14874-24-9]
  cotton solvent, 108
Cadmium salts
  olefin fiber dyes, 322
Cadmium selenide [1306-24-7]
  in spin-dyeing polyamides, 368
Cadmium sulfide [1306-23-6]
  in spin-dyeing polyamides, 368
Cadmium sulfoselenide [12626-36-7]
  in spin-dyeing polyamides, 368
Calamus, 187
Calcium alginate [9005-35-0]
  fibers, 123
Calcium arsenate [7778-44-1]
  boll weevil control, 103
Calcium carbonate [471-34-1]
  in coated fabrics, 95
Calcium oxide [1305-78-8]
  in refractory fibers, 444
Calcium thiocyanate [2092-16-2]
  silk reagent, 454
Calendering
  in rubber coating fabrics, 93
Caliban F/R P-44
  flame retardant, 205
Cal Track, 436

Cambrelle, 155
Cannabin, 180
Cannabis sativa, 173, 180
Cantala, 173
Cantrece, 371, 155
Canvas Products Association International, 491
Capilair, 228
Caprolactam [105-60-2]
  in nylon, 124
Carbamates
  methylolations, 471
  in textile finishing, 486
Carbon [7440-11-5]
  in refractory fibers, 440
Carbon black [1333-86-4]
  uv stabilizer for olefin fibers, 322
Carbon disulfide [75-15-0]
  in rayon manufacture, 121
  in rayon production, 403
  in vinyon manufacture, 128
Carbon fibers, 126, 440
  from novoloids, 314
  properties and uses, 130
Carbonizing
  in wool processing, 532
Carbon monoxide [638-08-0]
  silicon carbide from, 446
Carbon oxysulfide [463-58-1]
  in rayon production, 403
Carding, 291
  polyamides, 364
Caroa, 173
Carolan, 84
Carothers, Wallace H, 118
Carpet
  production from polypropylene, 342
Carpeting
  polyester, 396
Carpets
  fibers for, 334
  flame tests, 5
  indoor-outdoor, 424
  multicomponent fibers in, 157
  nonwoven, 302
  nylon, 363
  tufting process for, 433
Casein [9000-71-9]
  chemical fibers, 123
Cashmilon, 2, 155
Castellation
  in fiber production, 165
Casting
  liquid metal for fibers, 240
Cataban process, 420
Catalysts
  in textile finishing, 475
Cattail fiber, 183
CBEA. See *p-Cyclohexanebis(ethylamine)*.

CBMA. See *p-Cyclohexanebis(methylamine)*.
Ceiba pentranda, 173, 183
Celanese, 84
Celaren, 84
Celcon, 335
Celcorta, 84
Celestra, 258
Cellobiose [528-50-7]
    from cotton, 108
Cellulose [9004-34-6]
    acetylation, 62
    cross-linking of, 485
    flame retardants for, 188
    nonwovens from, 286
    percentage in wood, 73
    rayon, 118
    rayon from, 120
Cellulose acetate [9004-35-7], 62
    fibers of, 119
    manufacture, 122
    moisture regain, 321
    nonwoven fabrics from, 288
    properties, 122, 320
    Schering process, 73
    secondary acetate processes, 73
    solvent process, 73
Cellulose acetate and triacetate fibers, **62**
    acetylation catalysts, 74
    acetyl value, 62
    chemical resistance, 71
    manufacture, 78
    uses, 83
Cellulose acetate and triacetate flakes
    manufacture, 73
Cellulose acetate fibers
    production, 82
Cellulose fabrics
    cross-linking finishes, 469
Cellulose nitrate [9004-70-0]
    fibers, 118
Cellulose triacetate [9012-09-3], 62
    dyeing of, 392
    manufacture, 122
Cellulose xanthate [9032-37-5]
    in bonding nonwovens, 299
    in rayon production, 403
Cellulosic fabrics
    finishing of, 467
Cellulosics, 119
    production volumes, 460
Ceramic fibers, 445
Ceramics
    novoloids fibers in, 314
Cerex, 256
Cerister, 2
Chamaerops humilis, 173
Chardonnet silk, 118
Chemical fibers
    acrylic, 126

alginate, 123
aromatic polyamides, 125
benzoate, 129
carbon, 130
cellulose acetate, 122
glass, 130
inorganic, 131
metal, 131
novoloid, 130
nylon-6,6, 123
PeCe, 128
physical modifications, 131
physical properties of, 120
poly(benzimidazole), 130
polychlal, 129
polyester, 125
poly(hydroxyacetic acid ester), 129
protein, 123
Qiana, 125
Saran, 128
spandex, 129
spinning processes, 119
vinal, 129
vinyl, 128
Vinyon, 128
volume, 110
Chemical stability
    of aramid fibers, 41
Cheviot wool, 520
Chevron 400 track, 436
Chevron 440 track, 436
China grass, 181
China jute, 173, 182
Chintz finishes, 466
Chlorination
    PVC to PeCe fiber, 128
    in shrink-resistance treatments, 538
Chlorine–Hercosett process, 539
Chloroamides
    in textile finishes, 471
1-Chloro-*N*-[5-chloro-2-(4-chlorophenoxy)pheny-
    l] methane- sulfonamide [20132-55-2], 536
1-Chloro-*N*-[5-chloro-2-(2,4-dichlorophenoxy)-
    phenyl]methane- sulfonamide [69588-63-2],
    535
5 Chloro-2-[4-chloro-2-[[[(3,4-dichlorophenyl)-
    amino]carbonyl] amino]phenoxy]
    benzenesulfonic acid [3567-25-7], 535
1-Chloro-*N*-[4,5-dichloro-2-(2,4-dichlorophenoxy)-
    phenyl]methane- sulfonamide [60787-08-8],
    535
N-Chloromelamines
    in textile finishing, 476
Chloromethylphosphonic acid [2565-58-4]
    flame retardant, 197
Chloromethylphosphonic diamide [6326-70-1]
    flame retardant treatment, 203
2-Chloro-*p*-phenylenediamine [615-66-7], 46
Chloroterephthaloyl chloride [13815-87-7], 47

1-Chloro-*N*-[2,3,4,5-tetrachloro-6-(2,2-dichlorop-henoxy)phenyl] methane- sulfonamide [*60787-06-6*], 536
1-Chloro-*N*-[2,3,4-trichloro-6-(2,4-dichlorophen-oxy)phenyl] methane- sulfonamide [*60787-09-9*], 535
1-Chloro-*N*-[3,4,5-trichloro-2-(2,4-dichlorophen-oxy)phenyl] methane- sulfonamide [*60787-10-2*], 536
Cholesterol [*57-88-5*]
  in wool wax, 544
Chrome dyes
  for wool, 533
Chromia [*1308-38-9*], 439. (See also *Chromium oxide.*)
  in refractory fibers, 439
Chromic chloride [*10025-73-7*]
  in fabric printing, 534
Chromotropic acid [*148-25-4*]
  in formaldehyde analysis, 483
Chrysella, 84
Cigarette filters, 85
  cellulose acetate in, 83, 122
Cigarette paper
  flax tow and sunn in, 186
Circo oil, 94
Citric acid [*77-92-9*]
  in textile finishing, 475
Clarino, 215, 216
Claus reaction, 420
Clearspan, 143, 148
Clevyl T, 206
Cloth
  refractory, 449
Clothing
  fibers for, 334
  polyester, 396
Clubturf, 435
CME. See *Crucible melt extraction.*
Coated fabrics, **90**
  as leather substitutes, 213
  processing, 95
Coating processes
  for leather substitutes, 218
Cochlospermum gossypium, 183
Cocos nucifera, 173, 182
Coefficient of retraction (CR)
  in tire cord, 514
Coir, 173
Coker cotton, 103
Colback, 258, 155
Colbond, 258, 155
Columbia wool, 520
Comfort
  of textiles, 492
Composites
  novoloid fibers in, 312
Cone-drop test, 279
Conex, 33

Coning oil, 366
Continuous belt xanthator, 403
Conwed, 258
Copper–cobalt–zinc alloy (35:2:13)
  in tire cords, 509
Copper number
  of cotton, 108
Copper 8-quinolinolate [*10380-28-6*]
  in textile finishes, 490
Copper sulfate [*1344-73-6*]
  in rayon manufacture, 121
Copper sulfide [*1317-40-4*] (1:1)
  recovery in cuprammonium process, 419
Corchorus capsularis, 173, 181
Corchorus olitorius, 181
Cordage
  from vegetable fibers, 174
Cordelan I, 206
Cordelan II, 206
Cordley, 228
Corfam, 215
Corovin PP-S, 256
Corriedale wool, 520
Corrosion inhibitor
  lanolin as, 546
Cotswold wool, 520
Cotton, **99**, 173
  in coated fabrics, 90
  finishing of, 466
  flame retardants for, 190
  moisture regain, 321
  nonwoven fabrics from, 287
  production volumes, 460
  properties as vegetable fiber, 175
  world production, 115
Cotton gins, 104
Courtelle, 155
CPE. See *Polyethylene, chlorinated.*
CR. See *Coefficient of retraction.*
Creep
  defined, 349
Creslan, 3, 155
Crestfil, 156
Crilenka, 3
Crilenka Bico, 155
Crimp
  of multicomponent fibers, 157
Crimping
  polyamides, 364
Crin vegetal, 173
Cristobalite [*14464-46-1*]
  in refractory fibers, 442
Crotalaria juncea, 173, 182
Crowntex, 258
Crucible melt extraction, 244
Crumeron, 2
Crylor, 2
Crysel, 3
CSIRO solvent–resin–steam process, 542

Cumene
  acrylic fiber coagulant, 15
Cuprammonium [9004-34-6] rayon process, 120
Cuprammonium hydroxide [17500-49-1]
  cotton solvent, 108
Cuprammonium rayon, 412
Cupriethylenediamine [15243-01-3]
  cellulose solvent, 197
Cupriethylenediamine hydroxide [18745-03-4]
  cotton solvent, 108
Cut rubber, 138
Cutworm, 104
Cyanoethylation
  of cotton, 113
Cyanuric chloride [108-77-0]
  in bonding nonwovens, 300
p-Cyclohexanebis(ethylamine) [13234-45-2]
  in nylon, 375
p-Cyclohexanebis(methylamine) [2549-93-1]
  in nylon, 375
1,4-Cyclohexanedicarbonyl dichloride [13170-66-
    6], 47
Cystine [485-35-8]
  in silk, 454
  in wool, 529

D

Dabco 33 LV
  amine-type catalyst, 434
Dacron, 394
Daisee, 181
Darcy's Law, 235
DC-109 [63394-02-5], 540
Decabromobiphenyl [13654-09-6]
  flame retardant, 167
Decabromodiphenyl oxide [1163-19-5]
  flame retardant treatment, 205
Decatizing, 537
Deccan hemp, 181
Decitex, 120
Defoliants
  cotton, 104
Dehydrating agents
  for rayon, 407
Delnet, 258
Deltapine cotton, 103
Delustering agents, 291
Delustrants
  for nylons, 368
Delweve, 258
Denier
  definition, 86
Density, fiber
  defined, 348
Desiccants
  as cotton harvest aids, 104

Dexon, 381
DHEU. See 4,5-Dihydroxy-2-imidazolidinone.
Diacetyl, 2,3-butanedione [431-03-8]
  in bonding nonwovens, 295
Diacid chlorides
  diamine reactions, 48
9,9-Dialkyldihydroacridines
  polyamide stabilizers, 356
Dialkyl phosphites
  in cotton flame retardants, 199
Dialkylphosphonopropionamides
  in flame retardants, 199
Diallyl phthalate [131-17-9]
  dye carrier, 63
Diamines
  diacid chloride reactions, 48
Diaminobenzanilides
  polyamides from, 55
3,3'-Diaminobenzidine [91-95-2]
  copolymer fibers, 130
3,5-Diaminobenzoic acid [535-87-5]
  polyamide from, 54
2,4-Diamino-6-diethoxyphosphinyl-1,3,5-triazine
    [4230-55-1]
  flame-retardant treatment, 199
2,4-Diamino-6-(3,3,3-tribromopropyl)-1,3,5-triaz-
    ine [62160-38-7]
  flame retardant for cotton, 199
Diammonium phosphate [7783-28-0]
  textile flame retardant, 192
Diaper covers
  nonwoven, 302
2,3-Dibromopropyl acrylate [19660-16-3]
  in modacrylic fibers, 25
Dicel, 84
4,4'-Dichloro-3,3'-bis(trifluoromethyl)carbanilide
    [370-50-3], 535
3-(2,2-Dichloroethenyl)-2,2-dimethylcyclopropa
    necarboxylic acid, (3-phenoxyphenyl) methyl
    ester [52645-53-1], 536
Dichloroisocyanuric acid [2782-57-2]
  in wool bleaching, 534
Diethyl chlorophosphate [814-49-3]
  flame-retardant treatment, 196
Diethylene glycol [111-46-6]
  polyester by-product, 384
Diethyl phthalate [86-66-2]
  dye carrier, 63
Dihydrolanosterol [79-62-9]
  in wool wax, 544
N,N'-Dihydroxyethylenebisamides
  as cross-linking agents, 487
4,5-Dihydroxy-1-(hydroxymethyl)-2-imidazolidi-
    none [20662-57-1] (DMDHEU)
  in textile finishing, 484
4,5-Dihydroxy-1-(hydroxymethyl)-3-methyl-2-
    imidazolidinone [20662-56-0] (McMDHEU)
  in textile finishing, 484
4,5-Dihydroxy-2-imidazolidinone [3720-97-6]
    (DHEU)
  in textile finishing, 484

Dilauryl thiodiproprionate [123-28-4]
    as olefin fiber antioxidant, 321
Dimethylacetamide [127-19-2], 48
    as polyamide solvent, 54
    polymer solvent, 126
    in rayon solvent-spinning, 412
    solvent for acrylic fiber production, 10
Dimethylacetamide (DMAC)
    modacrylic fiber solvent, 26
3,3'-Dimethylbenzidine [119-93-7], 46
1,3-Dimethyl-4,5-dihydroxyethyleneurea [3923-
    79-3] (DMeDHEU)
    in textile finishing, 484
Dimethylformamide [68-12-2], 48
    modacrylic fiber solvent, 26
    solvent for acrylic fiber production, 10
    triacetate solvent, 78
Dimethylol alkylcarbamate
    cotton treatment, 111
1,3-Dimethylol-5-alkyltriazone
    in textile finishing, 477
N,N-Dimethylolcarbamates
    in textile finishes, 483
1,4-Dimethylolcyclohexane [105-08-8]
    polyester fiber from, 126
N,N'-Dimethylol-N,N'-dialkylamides
    in textile finishing, 471
Dimethyloldihydroxyethyleneurea [1854-26-8]
    cotton treatment, 111
1,3-Dimethylol-4,5-dihydroxyethyleneurea
    [1854-26-8] (dmdheu)
    in textile finishing, 482
Dimethylolethyleneurea [136-84-5]
    cotton treatment, 111
1,3-Dimethylolethyleneurea [136-84-5] (DMEU)
    in textile finishing, 477
Dimethylolmelamine [5001-80-9]
    flame-retardant treatment, 203
Dimethylolpropyleneurea [3270-74-4]
    cotton treatment, 111
    in textile finishing, 486
1,3-Dimethylolpropyleneurea [3270-74-4]
    (DMPU)
    in textile finishing, 477
Dimethylolurea [25155-29-7]
    cotton treatment, 111
    cross-linking of cotton by, 474
    in textile finishing, 471
Dimethyl phosphite [868-85-9]
    cellulose flame retardant, 196
    flame-retardant treatment, 203
Dimethylpolysiloxane [9016-00-6]
    in shrink-resistance treatments, 540
Dimethyl sulfoxide [67-68-5], 48
    modacrylic fiber solvent, 26
    solvent for acrylic fiber production, 11
    triacetate solvent, 78
Dimethyl terephthalate [120-61-6] (DMT)
    polyester fibers, 125

polyester fibers from, 381
Dioctyl phthalate [117-84-0]
    in coated fabrics, 95
Diolen, 155
Diolen Biko, 155
Diolen ultra, 155
Diphenyl isophthalate [744-45-6]
    copolymer fibers, 130
Dipryl, 258
Disco, 212
Disodium hydrogen phosphate [13708-85-5], 201
Disperse dyes
    for acetate fibers, 63
    for polyesters, 390
Dispersol Fast Scarlet B 150 [2872-52-8]
    polyester dye, 391
Disposable nonwovens
    nonwoven, 302
Disposable textiles
    nonwovens, 279
Distearyl thiodiproprionate [693-36-7]
    as olefin fiber antioxidant, 321
Disulfide cross-links
    in wool, 528
Dithiocarbonic acid [75-15-0]
    in rayon mfg, 403
Divinyl sulfone [77-77-0]
    cotton treatment, 113
    precursors, 113
DMDHEU. See 1,3-Dimethylol-4,5-
    dihydroxyethyleneurea.
DMeDHEU. See 1,3-Dimethyl-4,5-
    dihydroxyethyleneurea.
DMEU. See 1,3-Dimethylolethyleneurea.
DMPU. See 1,3-Dimethylolpropyleneurea.
DMT. See Dimethyl terephthalate.
Docan process, 267
Dodecanedioic acid [143-07-7]
    in qiana, 374
Dolan, 2
Dorlastan, 147, 148
Dorset wool, 520
Dowtherm, 358
Dralon, 2, 155
Dralon MA, 23
Drawing
    polyamides, 360
Drawtwister, 360
Drying
    of acrylic fibers, 18
Dry spinning
    of acrylic fiber, 12
    of spandex, 144
Dunlop, 435
Duon, 258
Duplex bonding
    spunbondeds, 271
Durable press
    polyester, 392

Durable-press textiles, 467
Durable-press wool, 541
Durette, 39, 206
Dyeability
    of aramid fiber, 41
Dye carriers
    for polyester dyeing, 390
Dyeing
    acetate fibers, 63
    of acrylic fibers, 18
    polyamides, 368
    silk, 456
    textile, 464
    wool, 532
Dyeing of acrylic fibers, 20
Dyes
    for olefin fibers, 322
    for polyesters, 390
Dylan GRC process, 539
Dynac, 258
Dynel, 22

E

EA fiber, 155
Easy-care textiles, 467
Eftrelin, 374
Elastic fibers
    from polyolefins, 338
Elasticity
    defined, 349
Elastic limit
    of fiber, defined, 349
Elastomers
    in coated fabrics, 92
    fibers of, 136
Electrical properties
    of aramid fibers, 39
Electrical resistivity
    cellulose acetate, 67
Electrodischarge machining, 365
Electron-beam milling, 365
Elmendorf test, 279
Elongation recovery
    acetate fibers, 71
Elura wigs, 23
Emulsion spinning, 130
Encron, 394
Enkamat, 258
Enkaswing, 148
Enkatron, 155
Ensolite, 437
Epitropic fiber, 155
Epoxides
    cotton treatment, 112
Erifon process
    flame retardancy, 192

Erosion control
    nonwovens in, 281
Escaine, 156
ES fiber, 155
Eslon, 84
Espa, 148
Esparto, 186
Estron, 84
Etherification
    of cotton, 113
Ethyl carbamate [51-79-6]
    in textile finishing, 486
5,5'-Ethylenebis[tetrahydro-1,3-bis(hydroxy-
        methyl)-1,3,5-triazin-2(1H)-one] [4832-94-4]
    in textile finishing, 478
Ethylene carbonate [96-49-1]
    solvent for acrylic fiber production, 11
4,4'-Ethylenedianiline [14755-35-2], 46
Ethylene glycol [107-21-1]
    polyester fibers, 125
    polyester fibers from, 384
Ethyleneimine [151-56-4]
    cotton treatment, 113
Ethylenethiourea
    in rubber-coated fabrics, 94
Ethyleneurea [120-93-4]
    in textile finishes, 479
Ethylene-vinyl acetate copolymer [24937-78-8]
    in multicomponent fibers, 156
Ethyl γ-tribromobutyrate [62160-36-5]
    biguanide reaction, 200
Eulan U33, 535
Eulan WA, 535
Euroacril, 2, 155
Evolution/Evolution II, 258
Exlan, 3, 155
Exrolod, 220
Extar FR, 206
Extruded latex, 139
Extrusion
    acetate fibers by, 78
    fabrics, 274
    olefin fibers, 328
Extrusion processes
    for fibers, 462

F

Fabric bases
    nonwoven, 302
Fabrics
    coatings for, 90
    urethane-coated, 214
Fabrics, bonded, 285. (See also Nonwoven
        textile fabrics.)
Fabrics, engineered, 285. (See also Nonwoven
        textile fabrics.)

Fabrics, formed, 285. (See also *Nonwoven textile fabrics*.)
Fade-Ometer, 67
Felting, 538
    of wool, 492
Felts
    refractory, 448
Fenilon, 33
Ferrox process, 420
Fertilizers
    for cotton, 103
Fiber 700, 399
Fiber B, 33
Fiber blends
    flame retardants, 205
Fiberization
    of molten refractories, 443
Fiber morphology
    wool, 523
Fibers, 459, 152. (See also *Yarns*.)
    acetate and triacetate, 62
    antistatic, 160
    bicomponent, 137, 152
    bilaminar, 152
    chemical modifications, 133
    composite, 152
    conjugated, 152
    consumption volume, 462
    glass, 152
    manufacture, 129
    matrix–filament, 161
    novoloid, 305
    phenol–formaldehyde, 305
    production volumes, 460
    resistivity, 67
    sea–island, 153
    structure of, 462
Fibers, bicomponent, 371
Fibers, chemical, **118**, 460. (See also *Textiles, survey*.)
Fibers, elastomeric, **136**
Fibers, metal, 231. (See also *Metal fibers*.)
Fibers, multicomponent, **152**
Fibers, polyamide, 347
Fibers, polyester, 381
Fiber structure
    wool, 523
Fibers, vegetable,
    172, 183
    preparation, 175
    production, 183
Fibertex 200, 256
Fibrillation
    of polyolefin films, 336
Fibroin [9007-76-5]
    in silk, 452
Fibrous insulation, 440
Filaments, refractory. See *Refractory fibers*.

Fillers
    novoloid fibers, 314
    refractory, 448
    refractory fibers as, 448
Film
    olefin fibers from, 336
Films
    fibrillated, 336
Filter fabrics
    fibers for, 334
Filters
    nonwoven, 253
Filtration media
    nonwoven, 302
Fina, 3
Finel, 3
Finishing
    textiles, **466**
Fire resistance
    vertical flame test, 193
Fire-resistant textiles. See *Flame retardants for textiles*.
Fisisa, 3
Flags
    nonwoven, 274
Flame-resistance treatment
    for wool, 542
Flame-resistant fibers, 34
    phosphorus compounds for, 410
Flame-resistant materials
    novoloid fabrics, 312
Flame retardance
    of modacrylic fibers, 25
Flame retardants
    textile finishes, 490
    for textiles, 491
    water soluble, 192
Flame retardants for textiles, **188**
Flame retardant textiles
    of acetate and triacetate fibers, 84
Flamgard, 399
Flammability
    of acrylic carpets, 5
    manufacture, 8
Flatspots
    in tires, 512
Flax, 173, 179, 355
Flexural rigidity
    acetate fibers, 71
Flooring radiant panel test, 431
Fluoroesters
    polyamide-fiber lubricant, 367
Foam bonding
    nonwovens, 298
Foamed plastics
    as leather substitutes, 214
Foam spinning, 274
Football field
    surface, 425

Footware
   leather substitutes for, 214
Formaldehyde [50-00-0]
   cotton treatment, 111
   flame-retardant treatment, 203
   in novoloid fiber curing, 310
   in novoloid fiber mfg, 309
   in textile finishes, 481
   in textile finishing, 470
   use in flame retardants, 199
   as wool additive, 535
Formaldehyde bisulfite [75-92-3]
   in textile finishes, 482
Formamide [75-12-7]
   methylolations of, 471
Fortrel II, 381
French cylinder test, 278
Fujibo Spandex, 148
Fumaroyl chloride [627-63-4], 47
Furcraea gigantea, 173, 178
Furnace linings
   refractory fibers in, 449
Fyre fix
   flame retardant, 203
Fyrol 76
   flame retardant, 203
Fyrol FR2
   mutagenicity, 207

G

Garnett cards, 292
Gaskets
   fibers in, 186
Geotextiles, 281, 465
Ginning
   cotton, 104
Glass
   fibers, 131
   generic name, 121
   in tire cords, 501
Glass fibers
   bicomponent, 152
   in coated fabrics, 91
   manufacture, 131
   refractory, 438
Glass, oxide [65997-17-3*], 440
Glass-transition temperature
   of tire cord materials, 513
Glassy carbon
   novoloid based, 317
Glospan, 143, 148
l-Glucosan [498-07-7]
   from cellulose, 191
Glucose [50-99-7]
   from cotton, 108
Glutamic acid [6899-05-4]
   in silk, 454

   in wool, 529
Glycerol
   acrylic fiber coagulant, 15
   in spandex fiber, 143
Glycine [56-40-6]
   in silk, 454
   in wool, 529
Glycolic acid [79-14-1]
   in textile finishing, 475
Glyoxal [107-22-2]
   in textile finishing, 472
Gossypium arboreum, 100
Gossypium barbadense, 100
Gossypium herbaceum, 100
Gossypium hirsutum, 100
Gossyplum, 173
Graboxan, 228
Grand Turf 7000, 435
Graphite [7782-42-5]
   in refractory fibers, 438
Graphite fibers, 438
Gräs, 435
Grass Sport 500, 435
Grass surface
   in playing fields, 428
Greige good, 92
Guanamines
   in textile finishes, 477
Guanidine [113-00-8]
   flame-retardant treatment, 189, 203
Guanylurea [141-83-3]
   flame-retardant treatment, 189
Gymnasium floor
   surfaces, 427

H

Hampshire wool, 520
Harvester
   for cotton, 104
Hawsers
   fibers for, 186
Heat-resistant fibers, 34
Helanca process, 362
Hemicellulose [9032-32-6]
   percentage in wood, 73
Hemp, 173, 180
Henequen, 173
Hercosett 57 [25212-19-5], 540
Herculon, 155
Heterocyclic units
   in polyamides, 56
Heterofilaments, 152
Heterofils, 152. (See also Heterofilaments.)
Heteroyarns, 153
Hevea brasiliensis, 139

Hexakis(methoxymethyl)melamine [*3089-11-0*]
    in textile finishing, 476
Hexamethylenediamine [*124-09-4*]
    in nylon, 375
Hexamethylenetetramine [*100-97-0*]
    in novoloid fiber curing, 310
Hexamethylphosphoric triamide [*680-31-0*]
        (HPT), 48
    toxicity, 58
Hexanetriol
    acrylic fiber coagulant, 15
HFG fiber, 399
Hibiscus *altissima*, 181
Hibiscus *sabdariffa*, 181
Hibiscus cannabinus, 173, 181
Hibiscus sabdarifa, 173
Hilake, 156
Histidine [*71-00-1*]
    in silk, 454
    in wool, 529
HMD. See *Hexamethylenediamine.*
Hoch Modul 333, 399
Hofmannophila pseudospretalla, 535
Hole–above–hole technique
    in fiber production, 165
Hollow-fiber membranes
    cellulose acetate fibers, 87
Hollow fibers
    rayon, 410
Hosiery. (See also *Stockings*.)
    nylon, 352
    nylons, 371
Hottenroth test, 405
HP rayons. See *High performance rayons.*
HPT. See *Hexamethylphosphoric triamide.*
HT-4 polymer, 39
HWM rayon. See *High wet modulus rayons.*
Hycar, 92
Hydrazide–polyamide fibers, 43
Hydrazine
    the urethane chain extender, 144
Hydrochlorides
    of amides in polymer solvents, 49
Hydrogen peroxide [*7722-84-1*], 490
    silk bleaching, 455
    in wool bleaching, 534
Hydrolysis
    of textile finishes, 472
Hydroxybenzimidazoles
    stabilizers in nylon, 356
2-Hydroxybenzothiazole [*934-34-9*]
    stabilizer in nylon, 356
2(2-Hydroxy-3',5'-di-*tert*-butylphenyl)-5-chloro-
        benzotriazole [*3864-99-1*]
    polypropylene light stabilization, 322
p-(Hydroxyethoxy)benzoic acid [*33070-39-2*]
    polymer, 129
5-(2-Hydroxyethyl)-1,3-bis(hydroxymethyl)-tetr-
        ahydro-1,3,5-triazin-2(1*H*)-one [*1852-21-7*]
    in textile finishing, 478

Hydroxyethyl carbamate [*5395-01-7*]
    in textile finishing, 486
Hydroxyethyltriazone [*1852-21-7*]
    in textile finishing, 478
N-Hydroxymethylbromoacetamide [*71990-02-8*]
    in flame retardants, 199
N-Hydroxymethylchloroacetamide [*2832-19-1*]
    in flame retardants, 199
N-Hydroxymethyliodoacetamide [*71990-03-9*]
    in flame retardants, 199
2-Hydroxy-4-*n*-octyloxybenzophenone [*1843-05-
        6*]
    polypropylene light stabilization, 322
5-Hydroxypropyleneurea [*13348-19-1*]
    in textile finishing, 486
8-Hydroxyquinoline [*148-24-3*]
    stabilizer in nylon, 357
Hypalon, 92
Hypochlorite bleaches
    effect on textile finishes, 470
Hypochlorous acid [*7790-92-3*]
    in shrink-resistant treatment, 540

I

Immacula process, 538
Impact tester
    for artificial surfaces, 428
Indian hemp, 182
Initiators
    for acrylic polymers, 9
Insecticides
    for cotton, 103
Insectproofing
    of wool, 535
Instant Turf, 435
Instron tensile tester, 69
Insulation
    refractory fibers for, 440
Insulation, fibrous, 440
Interfacial polycondensation, 48
Interlock rubber floor systems, 437
Ionizing radiation
    effect on aramid fibers, 41
Ironing
    of acetate fibers, 65
Iron oxide [*1309-37-1*]
    olefin fiber dye, 323
Isobutyl carbamate [*543-28-2*]
    in textile finishing, 486
Isoleucine [*73-32-5*]
    in silk, 454
    in wool, 529
Isophthalic acid [*121-91-5*]
    in aramid fibers, 125
    in spunbondeds, 259
Isophthaloyl chloride [*99-63-8*], 33
    polyamide from, 55

Isopropyl carbamate [1746-77-6]
   in textile finishing, 486
Istle, 173
Itaconic acid [97-65-4]
   in modacrylic fibers, 26
IWS durable-press process, 542
IWS Zirpro process, 543

J

Jaumave istle, 178
Jekrilon, 2
Jute, 173, 181

K

Kabipor, 228
Kanebo acryl, 3
Kanebo nylon 22, 156
Kanecaron, 23
Kanekalon, 23
Kanekaron, 206
Kaolin [1318-74-7]
   refractory fibers, 439
Kapok, 173, 183
Kelvar
   in tire cords, 501
Kenaf, 173, 181
Keratins [9008-18-8]
   wool, 528
Kevlar, 125, 206
Kevlar-29, 34
Kevlar-49, 34
Kimcloth, 258
Knitted fabrics, 467
Knitting process
   for artificial surfaces, 433
Knot tenacity, 348
Kodel, 126
Kodel II, 126, 381
Ko-Kneader unit
   for xanthation, 403
Koplon, 399
Krasil, 84
Kridee, 258
Kroy process, 539
Kumbi, 183
Kynol, 130, 206, 305
Kyrel, 258

L

Lactic acid [50-21-5]
   in fabric printing, 534
Lammus, 155
Lanalbene, 84
Lancer, 435
Land reclamation
   nonwoven uses in, 274
Langley units, 66
Lankart cotton, 103
Lanolin [8036-49-5], 543
Lanosterol [79-63-0]
   in wool wax, 544
Latex
   extruded thread, 139
Latex binders
   for nonwovens, 296
Latexes
   nonwovens from, 286
Laundering
   effect on fire retardancy, 189
Lauvest, 155
Leacril, 2, 3
Leacril BC, 156
Leaf (hard) fibers, 174
Leafhopper, 104
Leafminer, 104
Leather
   substitutes, 212
Leatherlike materials, 212.   (See also Leather.)
Leather, patent, 226
Leavil, 206
Leaving-group effects
   in textile finishes, 474
Lecithin [8002-43-5]
   in PVC coated fabrics, 95
Leicester wool, 520
Leucine [61-90-5]
   in silk, 454
   in wool, 529
Life preservers
   fibers for, 186
Lignin [9005-53-2]
   percentage in wood, 73
Lillionette 506, 256
Limiting oxygen index
   novoloid fiber, 306
Limiting oxygen index (LOI), 39
Lincoln wool, 520
Linters
   cotton, 101
Lintrak process, 538
Linum usitatissimum, 173, 179
Lithium chloride [7447-41-8]
   in rayon solvent-spinning, 412
Lo-flo process, 531

Lo-flo scouring system, 546
Loft
  of nonwovens, 284
Logwood [8005-33-2]
  silk dye, 456
LOI. See *Limiting oxygen index.*
Lonzona, 84
Loop tenacity, 349
Loteyarn, 84
LRC-100
  flame retardant, 205
304L stainless steel, 235
Lubricants
  for acetate fibers, 81
  for nylon processing, 367
Lufnen, 206
Lutrabond, 258
Lutradur H7210, 256
Lutrasil, 258
Lycra, 148, 335
Lygus bug, 104
Lynel, 148
Lysine [56-87-1]
  in silk, 454
  in wool, 529

M

Magnesium chloride [7786-30-3]
  in textile finishing, 475
Magnesium oxide [1309-48-4]
  in refractory fibers, 444
Makrolan, 3
Malimo, 300
Malipol, 300
Malivlies, 300
Maliwatt, 300
Malon, 3
Manila hemp, 172
  nonwoven fabrics from, 287
Manila maguey, 178
Man-made fibers. See *Fibers, chemical.*
Maps
  nonwoven, 274
Marijuana, 180
Marix, 258
Marubeni/Toray GS-2, 435
Mateflex, 436
Mats
  from vegetable fibers, 174
Mattresses
  fibers for, 186
Maurer unit
  for xanthation, 403
Mauritius, 173
Mauritius hemp, 178
Maxisorb, 399

MDA. See *4,4'-Methylenedianiline.*
MDHEU. See *4,5-Dihydroxy-1-
  (hydroxymethyl)-2-imidazolidinone.*
Melamine
  in aramid fibers, 39
Melamine–formaldehyde resin [9003-08-1]
  in textile finishing, 470
Melamines
  methylolations, 471
Melana, 3
Melt-blown textiles, 273
Melt extrusion
  of acrylic fibers, 20
Melt fiberization, 444
Melt-spinning, 118
  elastomers fibers, 142
  metals for fibers, 240
MEMDHEU. See *4,5-Dihydroxy-1-
  (hydroxymethyl)-3-methyl-2-imidazolidino-
  ne.*
2-Mercaptobenzomethylthiazole [149-30-4]
  stabilizer in nylon, 356
Mercerization, 109
  by ammonia, 494
  nonwovens, 300
Mercury(II) acetate [1600-27-7]
  in shrink-resistance treatments, 539
Merino wool, 520
Mesta, 181
Mesyl cellulose [34377-95-0]
  bromination, 196
Metal-complex dyes
  for wool, 533
Metal fibers, **231**
  properties, 231
Metalian, 156
Metalian antistatic fiber, 160
Metallic coatings, **252**
Metallic fibers, 131
Methacrylate
  cotton treatment, 111
Methanesulfonyl cellulose [37377-95-0]
  bromination, 196
Methionine [63-68-3]
  in silk, 454
  in wool, 529
4-Methoxy-5,5-dimethylpropyleneurea [13747-
  12-1]
  in textile finishing, 486
Methoxyethyl carbamate [1616-88-2]
  in textile finishing, 486
Methyl acetate [79-20-9]
  triacetate solvent, 78
Methyl acrylate [96-33-3]
  in acrylic fibers, 8, 127
3-Methyl-2(3H)-benzothiazolone [2786-62-1]
  in formaldehyde analysis, 484
Methyl carbamate [598-55-0]
  in textile finishing, 486
Methylenebisamides
  in textile finishing, 471

Methylenebis(4-phenyl isocyanate) [101-68-8]
in spandex, 144
Methylene chloride [75-09-2]
cellulose triacetate manufacture, 122
triacetate solvent, 73, 78
4,4'-Methylenedianiline [101-77-9], 46
in nylon, 375
Methyl methacrylate [80-62-6]
in acrylic fibers, 8
N-Methylmorpholine N-oxide [7529-22-8]
in rayon mfg, 400
for solvent-spun rayon, 411
Methylolacrylamide [924-42-5]
flame-retardant treatment, 203
N-Methylolacrylamide [924-42-5]
in bonding nonwovens, 295
Methylolamides
cotton treatment, 111
Methylolated melamine [1017-56-7]
cotton treatment, 111
Methylolation
in textile finishing, 471
Methylolmelamine [937-35-9]
flame-retardant treatment, 200
Methylolmelamines, 477
Methylphosphoric acid [993-13-5]
flame-retardant treatment, 198
Methylphosphoric diamide [4759-30-2], 203
N-Methylpyrrolidinone [872-50-4]
as polyamide solvent, 54
in rayon mfg, 400
in rayon solvent-spinning, 412
Methyl salicylate [119-36-8]
dye carrier, 63
Methyl zimate, 93
Mexican henequen, 178
Michael addition, 114
Micronaire reading, 106
Micro-Quartz
refractory fibers, 438
Milkweed floss, 183
Mill fever, 116
Mirafi, 258
Mite, 104
Mitin FF, 535
Mitin LP, 535
Mitrelle, 457
Mobilon, 142, 148
Modacrylic
generic name, 121
Modacrylic fibers, 22, 126
producers, 23
Modal fibers, 409
Modulus
acetate fibers, 70
textile fibers, 68
Moisture absorption
by acetate and triacetate fibers, 64
Molantin, 535

Mondo Sport Surfaces, 436
Monoethanolamine sulfite [13427-63-9]
for setting wool, 538
Monophenyl phosphate [701-64-4]
cellulose flame retardant, 197
Monvelle, 137, 372, 156
Morus alba, 452
Morus nigra, 452
Mothproofing, 535
Mothproofing agent 79, 535
Moths
silk from, 451
Muhlenbergia macroura, 173, 179
Mullen test, 279
Mullite [1302-98-8]
refractory fibers, 439
Musa sapientum, 176
Musa textilis, 173, 176
Mutagenicity
flame retardants, 207
Mylar
in patent leather mfg, 226

N

N-44
in tires, 499
Napara, 156
Nash reagent
formaldehyde analysis, 483
Natural rubber
in elastomeric fibers, 136
Needle punching
nonwovens, 300
Needling
spunbondeds, 266
Neochrome, 3
Neoglaziovia variegata, 173, 179
Neopentylene glycol [126-30-7]
polyamide-fiber lubricant, 367
Neoprene [31727-55-6]
as leather substitute, 212
Netlon, 258
Netlon process, 275
Net 909 P520, 256
Net 909 process, 274
Nets
fibers for, 334
spunbonded, 275
Nettle
ramie source, 180
Newcell process, 400, 411
Newsprint
kenaf in, 186
New Zealand flax, 179
New Zealand hemp, 179
Nextel
refractory fibers, 439

Ni-bis[*O*-monoethyl(3,5-di-*tert*-butyl-4-hydroxy-
     benzyl)]phosphonate [*30947-30-9*]
  polypropylene light stabilization, 322
Nitron, 3
Nomex, 32, 55, 125
Nomex 111, 206
Nonwoven fabrics
  coating, 91
Nonwovens
  fibers for, 334
Nonwoven textile fabrics
  spunbonded, **252**
  staple fibers, **284**
Nonwoven textile fabrics, spunbonded, 252. (See
     also *Textiles, nonwoven*.)
Nonylphenol [*25154-52-3*]
  surfactants, 533
Novaceta, 84
Novaweb AB-17, 256
Novolac resin, 305
Novoloid
  generic name, 121
Novoloid fibers, 130, **305**
  acetylated, 311
Nylon [*32131-17-2*], 118, 347, 462. (See also
     Polyamides, polyamide fibers.)
  BCF, 124
  in coated fabrics, 90
  crimped for hosiery, 166
  flame retardants for, 204
  generic name, 121
  generic term, 32
  in leatherlike materials, 223
  nonwoven fabrics from, 287
  in nonwovens, 256
  properties, 124
  spinning, 123
Nylon-4 [*24938-56-2*], 376
Nylon-6 [*25038-54-4*], 124, 347
  in elastomeric fibers, 137
  manufacture, 124
  in multicomponent fibers, 155
  in nonwovens, 256
  nonwovens from, 288
  properties, 331
  in recreational surfaces, 426
  in tire cords, 499
Nylon-7 [*25035-01-2*], 347
Nylon-8 [*25035-02-3*], 347
Nylon-9 [*25035-03-4*], 347
Nylon-11 [*25035-04-5*], 347
Nylon-12 [*24937-16-4*], 357
Nylon-17 [*79392-50-0*], 347
Nylon-6,6 [*9011-55-6*], 123, 347
  moisture regain, 321
  in multicomponent fibers, 155
  in nonwovens, 255
  nonwovens from, 288
  properties, 320, 331

  in recreational surfaces, 426
  thermal transitions of fiber, 321
  in tire cords, 499
Nylon filament
  price, 134
Nylon-6,6–hexamethylene isophthalamide
     copolymer [*26353-66-2*] (80:20)
  in bicomponent fibers, 372
Nylon-6–hexamethylene terephthalamide
     copolymer [*25776-72-1*]
  in bicomponent fibers, 372
Nylon-6,6–nylon-6 copolymer [*24993-04-2*] (80:
     20)
  in bicomponent fibers, 372
Nylon-6,6–nylon-6,10 copolymer [*26572-08-7*]
     (50:50)
  in bicomponent fibers, 372
Nylon–polyester blend
  in multicomponent fibers, 158
Nylon staple, 364
Nytril
  generic name, 121

O

Oakes process, 220
Ochroma pyramdale, 183
Olboplast 25, 504
Olefin
  generic name, 121
Olefin fibers, **318**, 462. (See also *Polyolefin
     fibers*.)
Oleic acid
  rayon fiber finish, 409
Oligan 500 [*51059-99-5*], 540
Opelon, 148
Organosol
  in leatherlike material, 217
Orlon, 1, 2, 23, 355, 155
Outdoor fabrics
  flame-retardant treatment, 193
Oxydianiline [*27133-88-6*], 46
Oxygen index
  acetate, 196
  acrylic, 196
  aramid, 196
  cotton, 196
  fabrics, 196
  flame-retardant cotton, 196
  flame retardant textiles, 189
  nylon, 196
  polyester, 196
  polypropylene, 196
  rayon, 196
  wool, 196
Ozone [*10028-15-6*]
  in textile finishing, 492

P

Packaging
  fibers for, 181
PACM. See *Bis(p-aminocyclohexyl)methane.*
Pad-bake process, 111
Padding
  in textile finishing, 467
Palanil navy blue RE [*12235-96-0*]
  polyester dye, 391
Palma istle, 178
PAN. See *Polyacrylonitrile; Acrylonitrile polymers.*
Paper
  of Nomex, 39
  from novoloid fibers, 308
  vegetable fibers in, 186
Paper products
  from vegetable fibers, 174
Pa–Quel antistatic fiber, 156
Paracril, 92
Par weight
  of silk, 455
Passenger tires
  polyester in, 499
Patora, 228
Paymastir cotton, 103
PDME. See *Pendant-drop metal extraction.*
PeCe fiber, 128
Pectin [*9000-69-5*]
  in cotton, 106
Peel adhesion value
  of tire cords, 515
Peel test
  rubber adhesion, 515
Pendant-drop metal extraction, 246
Pentaerythritol [*115-77-5*]
  polyamide fiber lubricant, 367
Pentamethylphosphorotriamide [*10159-46-3*]
  flame-retardant treatment, 203
Perchloric acid [*7601-90-3*]
  acetylation catalyst, 74
Perchloroethylene [*127-18-4*], 72
Perigen, 535
Perluran, 347
Permanent press
  polyester, 392
Permanent-press textiles, 479
Permeability
  darcy unit, 235
PET. See *Poly(ethylene terephthalate).*
Petex, 258
Petromat, 258
Pewlon, 2
PFR fiber, 399, 410
PFR rayon, 206
Phenol [*108-95-2*]
  novoloid fibers from, 309

Phenol–formaldehyde resin, 305. (See also *Novolac resin.*)
Phenol–formaldehyde resin fibers, 130
Phenolic fibers, 130
Phenolic resins
  in novoloid fibers, 305
Phenylalanine [*63-91-2*]
  in silk, 454
  in wool, 529
N,N'-m-Phenylene-bis(*m*-aminobenzamide) [*2362-23-4*], 55
m-Phenylenediamine [*108-45-2*], 33, 46
  in aramid fiber, 33
  in aramid fibers, 125
p-Phenylenediamine [*106-50-3*], 33
  in aramid fibers, 33
Phenylone, 33
Phormium, 173
Phormium tenax, 173, 179
Phosphamic acid [*2817-45-0*]
  flame retardant treatment, 191
Phosphates
  polyamide-fiber lubricant, 367
Phosphites
  in synthesis aramides, 48
Phosphonitrilic acid [*28883-29-6*]
  in flame-resistant fibers, 410
Phosphoric acid [*7664-38-2*]
  cellulose flame retardant, 197
  flame-retardant treatment, 189
  in hydrolysis of fabric finishes, 473
Phosphorus [*7723-14-0*]
  in aramid fibers, 43
  in textile flame retardants, 204
Phosphorus oxychloride [*10025-87-3*]
  cellulose flame retardant, 197
Phosphorus pentoxide [*1314-56-3*]
  cellulose flame retardant, 197
Phosphorus trichloride [*7719-12-2*]
  cellulose flame retardant, 197
Phosphorylation
  in polyamide formation, 51
Phosvel
  mothproofing agent, 535
Photooxidation
  of olefin fibers, 321
Photoyellowing
  of wool, 535
Phthalocyanine Blue [*147-14-8*]
  in spin-dyeing polyamides, 368
Phthalocyanine Green [*1328-53-6*]
  in spin-dyeing polyamides, 368
Piassava, 173
Pigments
  in acetate fiber, light stability, 67
Pilling, 487
Pillows
  fibers for, 186
Pink bollworm, 103

Pipe-in-pipe technique
  for fiber production, 164
Pita, 178
Piteira, 178
Plastisols
  in leather substitute mfg, 217
Playfield, 435
Playing courts
  artificial surfaces, 425
Pneumoconiosis
  from silicon carbide whiskers, 448
Point-bonding
  spunbondeds, 266
Polcorfam, 228
Poligras, 435
Polishing
  of fabrics, 27
Polwarth wool, 520
Polyacrylates
  in bonding nonwovens, 295
Poly(acrylic acid) [9003-01-4]
  in fiber finishing, 488
Polyacrylonitrile [25014-41-9] (PAN), 1
  in bonding nonwovens, 295
  moisture regain, 321
Poly(allyl phosphonitrilate)
  flame-retardant treatment, 201
Polyamide–hydrazides, 43
Polyamides
  polyamide fibers, **347**
  in propylene fibers, 338
  synthesis, 52
Poly(m-benzamide) [25735-77-7] (PMB), 52
Poly(p-benzamide) [24991-08-0] (PPB), 33
Poly(p-benzamide) fiber, 43
Poly(benzidine terephthalamide) [25667-70-3], 54
Poly(benzimidazole) [26985-65-9]
  fibers, 130
Poly(butadiene-co-acrylonitrile) [9003-18-3]
  in suedelike materials, 226
Poly(1-butene) [9003-28-5]
  fibers, 318
Poly(1-butene) isotactic [25036-29-7]
  thermal transitions of fiber, 321
Poly(butylene terephthalate) [26062-94-2], 381
  in self-crimping filaments, 166
Polycaproamide. See Nylon-6.
Polycaprolactone [25038-54-4]
  in contraceptive device, 169
  in elastomers fibers, 142
Polycarbonate [25971-63-5]
  in multicomponent fibers, 160
Poly[(5-carboxy-m-phenylene)isophthalamide]
  [31808-02-3], 39
Polychlal
  fibers, 129
Poly(1,4-cyclohexylenedimethylene
  terephthalate) [24936-69-4], 381

Poly(dimethylterephthalate-co-ethylene glycol)
  [9003-71-8]
  in textile finishing, 488
Polyester, 462
  in coated fabrics, 90
  generic name, 121
  nonwoven fabrics from, 287
  in nonwovens, 256
  POY, 132
Polyester EF-121 [25038-59-9]
  in tires, 499
Polyester fibers, **381**
  flame retardants for, 204
  manufacture, 125
Polyester filament
  price, 134
Polyester staple
  price, 134
Polyester Type G.H, 206
Poly(ethyl acrylate) [9003-32-1]
  in bonding nonwovens, 295
Polyethylene [9002-88-4]
  fibers, 318
  in nonwovens, 256
Polyethylene, chlorinated [64754-90-1]
  composite with novoloid fiber, 314
Polyethylene (high density) [9002-88-4]
  thermal transitions of fiber, 321
Poly(ethylene-co-naphthalene-2,6-dicarboxylic
  acid), 167
Poly(ethylene 4-oxybenzoate) [25248-22-0], 381
Poly(ethylene terephthalate) [25038-59-9]
  for artificial surfaces, 432
  in bonding nonwovens, 295
  fiber, 381
  fibers, 291
  Kodel, 126
  moisture regain, 321
  nonwovens from, 288
  properties, 331
  spunbondeds, 255
  thermal transitions of fiber, 321
Poly(ethylene terephthalate) polyester
  in matrix–filament fibers, 153
Poly[ethylene-co-vinyl acetate] [24937-78-8]
  in bonding nonwovens, 295
  in shrink-resistance treatments, 540
Poly[(ethylimino 1,4-cyclohexanediyl(ethylimino)
  1,12-dioxo-1,12-dodecanediyl)] [52277-94-8],
  375
Polyfelt, 258
Polyglycerols
  polyamide-fiber lubricant, 367
Poly(glycolic acid) [26124-68-5; 26009-03-0], 129
Polyglycolide [26124-68-5], 381. See
  Poly(glycolic acid).
Poly(hexamethyleneisophthalamide) [25722-07-0]
  in tires, 499

Polyhydroxyacetic acid.   See *Poly(glycolic acid)*.
Poly(hydroxyacetic acid ester)
   fibers, 129
Poly(iminocarbonyl-1,4-phenylene
      carbonylimino-1,6 hexanediyl) [24938-70-3],
      375
Poly[imino-1,4-cyclohexanediyl methylene-1,4-
      cyclohexanediylimino (1,12-dioxo-1,12-
      dodecanediyl)] [25035-12-5], 374
Poly[iminomethylene-1,4-phenylenemethylene
      imino (1,10-dioxo-1,10-decanediyl)] [31711-
      07-6], 375
Poly(imino-1,4-phenyleneiminocarbonyl-1,4-phe-
      nylenecarbonyl) [24938-64-5]
   in tire cords, 501
Polymers
   in shrink-resistance treatments, 539
Polymers, thermoplastic
   in nonwovens, 255
Poly(3-methyl-1-butene) [25085-05-6]
   fibers, 318
Poly(3-methyl-1-butene) (isotactic) [26703-14-0]
   thermal transitions of fiber, 321
trans-Poly[(methylimino)-1,4-cyclohexanediyl
      (methylimino) (1,10-dioxo-1,10-dodecane-
      diyl)] [53830-66-3], 375
trans-Poly[(methylimino)-1,4-cyclohexanedyl
      (methylimino)(1,12-dioxo-1,12-dodecane-
      diyl)] [53830-66-3], 375
Poly(4-methyl-1-pentene) [25068-26-2]
   fibers, 318
Poly(4-methyl-1-pentene) (isotactic) [24979-98-4]
   thermal transitions of fiber, 321
Poly(methyl sorbate) [25987-01-3]
   in olefin fibers, 341
Polynosic fibers, 399
Polyolefin fibers, 318.   (See also *Olefin fibers*.)
   properties, 127
Poly(*m*-phenyleneisophthalamide) [24938-60-1]
      (MPD-I), 32
Poly(*p*-phenyleneterephthalamide) [24938-64-5]
      (PPD-T), 33
Poly(phenyl ether)s
   polyamide fiber lubricant, 367
Polypivalolactone [24937-51-7], 381
Polypropylene [9003-07-0]
   in leatherlike materials, 223
   in multicomponent fibers, 160
   in nonwoven fabrics, 91
   in nonwovens, 256
   nonwovens from, 288
   in recreational surfaces, 426
Polypropylene fibers, 127
Polypropylene (isotactic) [25085-53-4]
   fibers, 318
   properties, 331

   thermal transitions of fiber, 321
Poly(propylene oxide) [25322-69-4]
   in shrink-resistance treatments, 540
Polypropylene textile staple
   price, 134
Poly(sodium acrylate) [9003-04-7]
   in nonwovens, 289
Polystyrene [9003-53-6]
   in matrix–filament fibers, 154
   in propylene fibers, 337
Poly(styrene-*co*-butanediene carboxylates)
   in bonding nonwovens, 295
Polystyrene (isotactic) [9003-53-6]
   properties, 331
Polytetrafluorothylene [9002-84-0], 130
Poly(tetramethylene ether) glycol [25190-06-1]
   in spandex, 144
Poly(tetramethylene-*co*-naphthalene-2,6-dicarbo-
      xylic acid), 167
Poly(2,2,6,6-tetramethyl-4-piperidylamino-1,3,5-
      triazine) [77136-90-4] (CR 141 and 144)
   polypropylene light stabilization, 322
Poly-Turf, 435
Polyurethane [27416-86-0]
   in elastomeric fibers, 137
   fibers, 129
   for recreational surfaces, 425
   in shrink-resistant treatment, 541
   use in fabric coating, 96
Polyurethanes
   in coated fabrics, 94
Poly(vinyl acetate) [9003-20-7]
   in bonding nonwovens, 295
   in material backing, 432
   nonwoven binder, 287
   in prep of poly(vinyl alcohol), 129
Poly(vinyl acetate-*co*-acrylic acid) [24980-58-3]
   in bonding nonwovens, 295
Poly(vinyl alcohol) [9002-89-5]
   fibers, 129
   nonwovens from, 288
Poly(vinyl bromide) [25951-54-6]
   flame-retardant treatment, 205
Poly(vinyl chloride) [9002-86-2]
   in bonding nonwovens, 295
   in coated fabrics, 91, 94
   consumption in coating fabrics, 96
   fibers, 128
   in leatherlike materials, 217
   in material backing, 432
Poly(vinylidene chloride) [9002-85-1]
   in material backing, 432
Poly(vinylpyridine) [9003-47-8]
   dyeing of, 323
Poly(2-vinylpyridine) [25014-15-7]
   in tire cords, 499

Polyweb, 258
Poromerics, 94, 214
    sueded, 226
Port a Pit, 437
Porvair, 215
Potassium iodide–iodine complex [12298-68-9]
    stain for poly(vinyl acetate) binder in
        nonwoven fabric, 287
Potassium oxide [1236-45-7]
    in refractory fibers, 444
Precursor process
    for refractory fibers, 445
Prima, 399, 409
Primal K3 [9077-54-7], 540
Primary acetate. See Cellulose acetate.
Print-bonding, 270
    nonwovens, 297
Printing
    of wool, 533
Printing papers
    esparto in, 186
Proban
    flame retardant, 201
Progesterone [57-83-0]
    in contraceptive device, 169
Proline [147-85-3]
    in silk, 454
    in wool, 529
Pro-Pit, 437
Protein fibers
    chemical, 123
Pull-out test
    rubber adhesion, 515
Punking
    of novoloid fibers, 307
PXD. See p-Xylylenediamine.
Pyrethroids
    mothproofing agents, 535
Pyroset CP
    fire retardant, 197
Pyrovatex CP
    flame retardant, 199
Pyrrolidone [616-45-5]
    nylon 4 preparation, 135

Q

Qiana, 374
    manufacture, 125
Quest, 437
Quindo Magenta [16043-40-6, 980-26-7]
    in spin-dyeing polyamides, 368

R

Radial tires
    for passenger cars, 499
Radiant panel test, 5
Radiation
    effects on cotton, 114
Radiation curing
    of textiles, 487
Raffia, 187
Ramie, 173, 180
Raphia raffia, 187
Raschel fabrics, 148
Rattan, 187
Rayon [61788-77-0], 399
    in coated fabrics, 91
    finishing of, 466
    generic name, 121
    in leatherlike materials, 223
    manufacture of, 120
    moisture regain, 321
    nonwoven fabrics from, 287
    properties, 320
    spunbonded, 276
    in tire cords, 499
Rayon fibers, 460
Rayon staple, regular
    price, 134
Rayon tires, 499
Reaction spinning
    elastomeric fibers, 143
Reactive dyes
    for wool, 532
Recaflex, 436
Recreational surfaces, 424
Redolen, 2
Reemay, 258, 396
Refractive index
    acetate fibers, 64
Refractory cloth, 449
Refractory fibers, 438
Refractory filaments. See Refractory fibers.
Refractory yarns, 449
Refrasil
    refractory fiber, 438
Reinforcing fibers
    refractory, 449
Rekortan, 436
Relaxation
    of acrylic fibers, 18
    of fibers, 352
Resilience
    of fibers, 70
Resilite, 437
Resistivity
    various fibers, 67
Resole [9003-35-4]
    composite with novoloid fiber, 314

Resorcinol–formaldehyde–latex (RFL) adhesive system
  for tires, 499
Retting
  flax, 180
RFL. See *Resorcinol–formaldehyde–latex adhesive system.*
Rhodia, 84
RMA. See *Rubber Manufacturers Association.*
Ropes
  fibers for, 334
Roselle, 173
Rotavapor flask, 445
Rotoformer, 294
RoyalDek, 436
Royalene, 92
Rubaturf track, 436
Rubber
  adhesion of in tires, 507
  fiber, 137
  thread, 137
Rubber in coated fabrics, 92
Rubber Manufacturers Association (RMA), 509
Rubber thread, 150
Running shoes, 427
Running tracks
  surfaces, 427

S

S–28, 156
Saffil
  refractory fibers, 439
Salt-index test
  for viscose coagulation, 405
Samuela carnerosana, 178
Sandoflam 5060
  in textile flame retardants, 204
Sanforized process, 466
Sanitary napkins
  nonwoven, 302
Saniv, 23
Sansevieria, 173, 179
Saponification
  of acetate and triacetate fibers, 72
Saran
  generic name, 121
Saran fibers, 128
  manufacture and properties, 128
Savina, 155
Sayelle, 155
Schering process, 74
Schotten-Baumann reaction, 48
Schreinering, 466
Schusspol, 300
Schweitzer's reagent, 413
Scouring
  wool, 530

Scroop
  of silk, 456
Scutching
  flax, 180
Sea–island fibers, 152
Secondary acetate. See *Cellulose acetate.*
Seed-hair fibers, 173
SEF, 23, 206
Septum
  for fiber production, 164
Sericin [60650-88-6]
  in silk, 452
Sericulture. See *Silk.*
Serine [56-45-1]
  in silk, 454
  in wool, 529
Setting, 537
Sharnet, 258
Sheep
  wool from, 520
Ship cables
  fibers for, 186
Shock absorption
  in playing fields, 428
Shoe manufacturing
  costs, 229
Shrink-resistance treatment
  for wool, 538
Sideria, 155
Silheim, 258
Silica [7631-86-9], 440. (See also *Silicon dioxide.*)
  in refractory fibers, 440
Silica–alumina [37287-16-4], 440
Silica fibers, 440
Silicate esters
  polyamide-fiber lubricant, 367
Silicon carbide [409-21-2]
  fibers, 131
  in refractory fibers, 440
Silicon carbide fibers, 440
Silicones
  polyamide-fiber lubricant, 367
Silicon monoxide [10097-28-6]
  refractory fibers from, 446
Silicon nitride [12033-89-5]
  fibers, 131
  in refractory fibers, 440
Silicosis
  from refractory fibers, 447
Silk, **451**, 463
  production volumes, 460
Silk-cotton tree, 183
Silk gum, 452
Silk-moth caterpillar, 451
Silk waste, 452
Silkworm, 451
Silk yarns, 452
Silpalon, 3

Sini process, 402
Siracril, 2
Sirolan BAP process, 534, 540
Si-Ro-Set process, 538
Sisal, 173
Sneakers, 229
Societé Rhodiaceta solution process, 77
Sodium alginate [9005-38-3], 123
Sodium bisulfite [7631-90-5]
    in fabric bleaching, 476
    setting wool, 537
    in wool bleaching, 534
Sodium chloride [1647-14-5]
    as leatherlike material filler, 222
Sodium dicellulose phosphate [72017-65-3]
    flame retardant, 197
Sodium dihydrogen orthophosphate [7558-80-7]
    in flame retardant treatment, 201
Sodium dithionite [7775-14-6]
    in wool bleaching, 534
Sodium hexametaphosphate [14550-21-1]
    cellulose flame retardant, 197
Sodium hydrosulfite [7681-38-1]
    silk bleaching, 455
Sodium p-methallyloxybenzene sulfonate [1208-
    67-9]
    in modacrylic fibers, 26
Sodium methallyl sulfonate [1561-92-8]
    in acrylic fibers, 8
    in modacrylic fibers, 25
Sodium oxide [1313-59-3]
    in refractory fibers, 444
Sodium perborate [7632-04-4]
    in wool bleaching, 534
Sodium peroxide [1313-60-6]
    in wool bleaching, 534
Sodium phosphate dodecahydrate [10101-89-0]
    textile flame retardant, 192
Sodium silicate [1344-09-8]
    in wool bleaching, 534
Sodium styrene sulfonate [27457-28-9]
    in acrylic fibers, 8
    in modacrylic fibers, 25
Sodium sulfate [7681-38-1]
    as dehydrating agent, 407
    in rayon manufacture, 121
Sodium sulfite [7757-83-7]
    formaldehyde test, 482
Sodium thioglycolate [367-51-1]
    setting wool, 537
Sodium trithiocarbonate [534-18-9]
    in rayon mfg, 403
Sodium tungstate dihydrate [10213-10-2]
    textile flame retardant, 192
Sodospun, 258
Sofrina, 156
Soil-burial tests
    on fibers, 72
Soil-release properties
    of textile finishes, 487

Sol process
    for refractory fibers, 445
Solubility parameters, 411
Solvents
    for modacrylic fibers, 26
Solvent-spun rayon, 400, 410
Southdown wool, 520
Spandelle, 143
Spandex, 136, 462. (See also Fibers,
    elastomeric.)
    generic name, 121
    production, 141
Spandex fibers, 129
    composition, 137
Spanzelle, 143, 148
Specific gravity
    acetate fibers, 64
Spiders
    silk from, 451
Spinnerettes
    for multicomponent fiber production, 161
Spinning
    acetate fibers, 78
    of aramid fibers, 57
    high speed, 463
    of silk, 455
Sportan resilient surfacing, 436
Sport-Tred, 436
Spunbondeds, 252. (See also Nonwoven textiles,
    spunbonded.)
Spunize process, 362
Spunlace textiles, 301
Staple
    nylon, 353
    polyester, 387
Staple fibers, 285. (See also Nonwoven textile
    fabrics.)
Staple yarns
    silk, 452
Staudinger, Herman, 118
Stearamidomethylpyridinium chloride [4261-72-
    7]
    as water repellant, 113
Stearic acid [57-11-4]
    in rubber-coated fabrics, 94
Steel
    fibers, 131
    melt spinning, 243
    in tire cords, 501
Steel wool, 240
Steiner tunnel test, 5
Stipa tenacissima, 186
Stockings
    nylon, 352
Stoneville cotton, 103
Stress
    SI unit for, 4
Stress–strain behavior
    of nylons, 350

Stretch modulus, 349
Stuffing wool, 448
Styrene
    cellulose grafting, 114
Styrene–butadiene rubber (SBR) [9003-55-8]
    leather substitute, 229
Sualen, 258
Suede
    synthetic, 225
Suede 21, 156
Suedes
    synthetic, 167
Suint, 520
Sulfur [7704-34-9]
    rubber reaction in tire cords, 507
Sulfur dioxide [7446-09-5]
    in wool bleaching, 534
Sulfuric acid [7664-93-9]
    in wool processing, 532
Sunn, 173
Sunn hemp, 182
Supac, 258
Supersuede, 212
SuperTurf, 435
Supreme Court, 437
Surfaces, 424. (See also Recreational surfaces.)
Surfactants
    in wool dyeing, 533
Surgical packs and gowns
    nonwoven, 302
Suture
    polyglycolide fibers, 129
SVM fiber, 43
Swiss Flex, 436
Syntex, 258
Synthappret BAP [69670-71-9], 540
Synthappret LKF [57596-47-1], 540
Synthetic fibers, (See also Fibers, chemical.)
    finishing of, 487
    nonwoven fabrics, 134
    nylon-6, 124
    polyamide, 123
    polyolefins, 127
    Teflon, 130
Synthetic polymers
    production volumes, 460

T

TA. See Terephthalic acid.
Taber and Schiefer abrasion test, 430
Tablecloths
    nonwoven, 273
Tack spinning
    textiles, 276
Tafnel, 258

Tairylon, 2
Tampons
    nonwoven, 302
Tape
    fibers for, 334
Tapilon, 156
Tapyrus, 258
Targhee wool, 520
Tartan resilient surfacing, 436
Tartan Turf, 435
Taslan, 362
Taslan yarns, 132
Taylilan, 2
TCF, 258
Tea bag paper
    abaca fibers in, 186
Tea bags
    nonwoven, 289
Teflon fibers
    properties and manufacture, 130
Teklan, 23
Telloy, 93
Tenacity
    acetate fibers, 68
    of fiber, defined, 348
Tenera, 228
Tennis courts
    surfaces, 425
Tensile modulus, 349
Tensile properties
    of aramid fibers, 34
    of nylons, defined, 348
Tensile strength
    of fiber, defined, 349
Tercryl, 2
Terephthalic acid [100-21-0] (TPA)
    polyester fibers, 125
    polyester fibers from, 381
Terephthaloyl chloride [100-20-9], 33, 47
Tergal X–403, 156
Terlon, 43
Terram, 155
Terram 1000, 256
Tetoron 88, 156
Tetrakis(hydroxymethyl)phosphonium chloride
        [124-64-1]
    in aramid fibers, 39
    flame retardant, 200
    textile flame retardant, 491
Tetrakis (hydroxymethyl)phosphonium
        hydroxide [512-82-3]
    flame-retardant treatment, 202
Tetrakis (hydroxymethyl)phosphonium sulfate
        [55566-30-8]
    flame retardant, 201
    textile flame retardant, 491
Tetramethylolacetylenediurea [5395-50-6]
    cotton treatment, 111
Tetramethylurea [632-22-4] (TMU), 48

Tex
definition, 4
Texizol, 258
Textile fabrics
flame retardants, 188
Textile finishing, 466
Textiles, 459
coatings for, 90
finishing, **466**
flammability tests, 193
flash spun, 273
melt blown, 273
survey, **459**
tack spun, 276
from vegetable fibers, 174
Textiles, nonwoven, 252. (See also *Nonwoven textile fabrics*.)
Texturing
yarns, 132
Texturing/bulking methods for fibers, 132
T_g. See *Glass-transition temperature*.
Thermal bonding
spunbondeds, 266
Thermosol dyeing, 390
Thermosol process, 205
2-Thienylethylene [*1918-82-7*]
for insectproofing wool, 536
Thinsulate, 258
Thiocarbonic acid [*10016-32-7*]
in rayon mfg, 403
Thiokol, 92
Thiourea [*62-56-6*]
in flame-retardant formulation, 201
flame-retardant treatment, 207
as wool additive, 535
Thiourea dioxide [*1758-73-2*]
in wool bleaching, 534
THP. See *Tris(hydroxymethyl)phosphine*.
THPC. See *Tetrakis(hydroxymethyl)-phosphonium chloride*.
THPOH. See *Tetrakis(hydroxymethyl)-phhosphonium hydroxide*.
THPS. See *Tetrakis(hydroxymethyl)-phosphonium sulfate*.
Threonine [*72-19-5*]
in silk, 454
in wool, 529
Thrip, 104
Tientsin jute, 182
Tile interlocking
for tennis court surfaces, 434
Tiles
refractory, 439
Tinea dubiella, 535
Tinea metonella, 535
Tinea pellionella, 535
Tineola bisselliella, 535
Tin oxide [*21651-19-4*]
flame retardant, 167

Tire cord
aramids in, 125
nylons, 357
polyester, 126
rayon, 122
Tire cords, **499**
from multicomponent fibers, 160
Tire performance
testing of, 509
Tires, 499. (See also *Tire cords*.)
Tire yarn, 348
Titanium [*7440-32-6*]
in flame-resistance treatment, 543
Titanium acetate chloride
flame retardant, 192
Titanium dioxide [*13463-67-7*]
acetate fiber pigment, 79
for acctate fibers, 63
in acrylic fibers, 11
in chemical fibers, 133
delusterant for spandex fiber, 143
as delustering agent, 291
elastomeric fibers, 143
olefin fiber dye, 322
as polyester delusterant, 381
Titanium sulfate [*13825-74-6*]
flame-retardant treatment, 197
Titanox FR process
flame retardancy, 192
TMM. See *Trimethylolmelamine*.
p-Toluene sulfonyl cellulose [*9048-42-4*]
bromination, 196
Toraylina, 156
Toraylon, 3
Toraylon unfla, 23
Tossa, 181
Tosyl cellulose [*9048-42-4*]
bromination, 196
Totes, 212
Toughness
of fiber, defined, 349
Towels
nonwoven, 302
Town flower, 2
Toxicity
of aramid fibers, 58
formaldehyde, 494
formaldehyde from textile finishes, 481
Fyrol 76, 207
Fyrol FR2, 207
glyoxal, 494
of hexamethylphosphoric triamide, 58
textile finishes, 494
Tris, 207
Toyobo Spun-Bond, 258
Tracklite, 436
Track surface
artificial, 424
Transfer-print papers
for wool, 534

Tree wool, 99
Trevira, 355, 394
Trevira 271, 133, 206
Trevira 30/150, 256
Trevira-Spunbond, 259
Treviron, 206
Triacetate.   See *Cellulose acetate and triacetate fibers.*
    preparation, 76
Triacetate fibers, 460
Tri-A-Faser, 84
Triallyl phosphate [1623-19-4]
    flame-retardant treatment, 200
Triana, 2
Triazones
    as cross-linking agents, 477
    synthesis of, 478
    in textile finishing, 477
Tricel, 84
Trichloroethylene
    effect on triacetate fibers, 72
Tricot fabric, 396
Tricot knitting, 83
Triethanolamine [102-71-6]
    flame-retardant treatment, 200
Triethanolamine phosphate [10017-56-8]
    flame retardant treatment, 191
Trilan, 84
Trilbene, 84
Trimethylolglycouril [496-46-8]
    flame-retardant treatment, 205
Trimethylolmelamine [1017-56-7]
    flame-retardant treatment, 199, 202
Trimethylolmelamine dimethyl ether [1852-22-8]
    textile flame retardant, 491
Trimethylolpropane [77-99-6]
    in spandex fiber, 143
Trimethyl phosphite [121-45-9]
    flame-retardant treatment, 199
Trimethylphosphoramide [6326-72-3]
    flame-retardant treatment, 207
Triphenyl phosphite [101-02-0], 52
    in polyamide formation, 52
Triphenyltin acetate [900-95-8]
    for insectproofing wool, 536
Triphenyltin chloride [639-58-7]
    for insectproofing wool, 536
Tris, 133
    flame retardant, 491
Tris(1-aziridinyl) phosphine oxide [545-55-1]
    flame-retardant treatment, 201
Tris(2,3-dibromopropyl) phosphate [126-72-7]
    flame retardant treatment, 204
    in textile finishing, 491
N,N′,N″-Tris(2-hydroxyethyl)melamine [4403-07-01]
    in textile finishes, 476

Tris(hydroxymethyl) phosphine [2767-80-8]
    flame-retardant treatment, 202
Tris(hydroxymethyl)phosphine oxide [1067-12-5], 202
Trithiocarbonic acid [594-08-1]
    in rayon mfg, 403
Truck tires
    polyester in, 499
Tryptophan [73-22-3]
    in silk, 454
    in wool, 529
Tuex, 93
Tufting process
    for grass-like surfaces, 433
Tula istle, 178
Turf
    artificial, 425
    artificial, manufacturers, 437
Tussur.   See *Silk.*
Twines
    fibers for, 334
Typar 3351, 256
Typha, 183
Tyrosine [60-18-4]
    in silk, 454
    in wool, 529
Tyvek, 274
Tyvek 1073-B, 256

U

Ultraspan, 2
Ultrasuede, 227, 156
Ultraviolet
    effect on aramide fibers, 41
Unisel, 259
Upholstery
    fibers for, 334
    leather substitutes for, 213
    of nonwoven fabrics, 92
    polypropylene usage, 342
Urea [57-13-6]
    cellulose flame retardant, 197
    flame-retardant treatment, 189
    hydrolysis of fabric finishes with, 473
    methylolations of, 472
    in textile finishing, 471
    in wool dyeing, 533
    in wool printing, 534
Urea–formaldehyde resin [9011-05-6]
    in textile finishing, 470
Ureas
    cyclic, 476
Urena lobata, 173, 182
Urethane polymers
    in elastomeric fibers, 136
    as leather substitutes, 212, 220

Urons
    as cross-linking agents, 477
    in textile finishing, 476
Uv stabilizers
    for olefin fibers, 322

V

Vacuum-forming process
    of refractory fibers, 448
Valine [72-18-4]
    in silk, 454
    in wool, 529
Vantel, 228
Vegetable fibers, 172
Velicren, 2, 156
Velion, 84
Verel, 22, 206
Viledon-M, 259
Viloft, 399, 410
Vinal
    generic name, 121
Vinal fiber
    manufacture and properties, 129
Vinal fibers
    in tire cords, 501
Vincel, 399
Vinyl acetate [108-05-4], 129
    in acrylic fibers, 8
    PVC copolymer in fibers, 128
Vinyl bromide [593-60-2]
    in modacrylic fibers, 25
Vinyl chloride [75-01-4]
    in modacrylic fibers, 1, 26
    in modacrylics, 127
    PeCe fiber from, 128
Vinyl chloride–vinyl acetate copolymer [9003-22-
        9]
    as coated fabric slip, 96
Vinyl fabrics
    expanded, 214
    as leather substitutes, 214
Vinyl fibers, 128
Vinylidene bromide [593-92-0]
    in modacrylic fibers, 25
Vinylidene chloride [75-35-4]
    copolymer with vinyl chloride, 128
    fibers from, 128
    in modacrylic fibers, 1, 26
    in modacrylics, 127
Vinyl plastisol
    in leatherlike materials, 217
Vinyl polymers
    as leather substitutes, 212
Vinylpyridine [1337-81-1]
    dyeing of in fibers, 323
Vinylpyridines
    in acrylic fibers, 8

Vinyon [9003-22-9], 128
    generic name, 121
    manufacture, 128
    nonwovens from, 288
Vinyon HH, 289
Viscoelastic behavior
    of fibers, 70
Viscose rayon, 400
    nonwoven fabrics from, 288
Viscose rayon process, 120
Viscous flow
    darcy units, 235
Vistanex, 92
Viton, 92
Vivelle, 259
Vnivlon, 43
Voltex, 300
Vonnel, 3
Vonnel H-704, 23
Vonnel V–57, 156
Vyrene, 143

W

Wash-and-wear textiles, 467
Wastewater treatment
    wool processing, 531
Waterproofing
    textiles, 489
Water repellency
    of textiles, 489
Wax
    in cotton, 107
    wool, 543
Weaving
    of artificial surfaces, 433
Wet look, 212
Wetrelin, 374
Wet spinning
    of acrylic fiber, 12
    elastomeric fibers, 146
Whiskers, 440.   (See also Refractory fibers.)
Whiskers, silicon carbide
    silicosis from, 448
White fly, 104
White jute, 181
Wiedemann-Franz relation, 235
Wipes
    nonwoven, 302
Wire
    melt spinning, 243
Witten process, 384
Wolpryla, 2
Wood pulp
    cellulose from, for acetate fibers, 73
    nonwoven fabrics from, 287

Wool, **519**
    finishing of, 492
    grease, 543
    moisture regain, 321
    physical and chemical properties, 524
    production volumes, 460
Wool grease, 520
    recovery and refining, 544
    uses, 546
Work of rupture
    fibers, 70
Work-to-break
    of fiber, defined, 349
Wrinkle-recovery angle
    of fabrics, 199
WRONZ scouring system, 531, 546

X

Xanthation
    in rayon mfg, 403
Xylan [9014-63-5]
    in ramie fibers, 181
p-Xylene [106-42-3]
    oxidation of, 384
p-Xylylenediamine [539-48-0]
    in nylon, 375

Y

Yarn lubricants, 367
Yarns, 119, 152. (See also *Fibers*.)
    cotton type, 463
    from olefin fibers, 336
    polyester, 396
    refractory, 449
    silk, 452
    wool type, 463
    worsted type, 463
Yarn texturing, 361
Yield point
    of fiber, defined, 349
Young's modulus, 68
Yttrium oxide [1314-36-9]
    phase stabilizer, 445

Z

Zefkrome, 3
Zefran, 3
Zeset TP [25267-86-1], 540
Ziegler catalysts
    for polypropylene fiber, 318

Zinc acetate [557-34-6]
    in antibacterial finishes, 493
Zinc nitrate [7779-88-6]
    in textile finishing, 475
Zinc sulfate [7733-02-0]
    in rayon manufacture, 121
Zircar
    refractory fibers, 439
Zirconia [1314-23-4], 440. (See also *Zirconium oxide*.)
    fibers, 131
    in refractory fibers, 440
Zirconia fibers, 440
Zirconium [7440-67-7]
    in flame-resistance treatment, 543
Zirconium acetate [14311-93-4]
    refractory fibers from, 445
Zirconium oxychloride [12769-92-5]
    refractory fibers from, 445
Zirconyl acetate [5153-24-2]
    antimicrobial peroxide complexes, 115
    in textile finishes, 490
Zirconyl ammonium carbonate [32535-84-5]
    in textile finishes, 490
Zirconyl chloride [13520-92-8]
    in fiber production, 445